Modern Cable Television Technology

Video, Voice, and Data Communications

WALTER CICIORA
Consultant

JAMES FARMER
Antec Corporation

DAVID LARGE
Media Connections Group

Morgan Kaufmann Publishers, Inc.
San Francisco, California

Senior Editor Jennifer Mann
Director of Production and Manufacturing Yonie Overton
Production Editor Cheri Palmer
Editorial Assistant Karyn Johnson
Cover Design Ross Carron Design
Cover Photograph XXXXXXXX
Text Design Rebecca Evans & Associates
Copyeditor Robert Fiske
Proofreader Jennifer McClain
Composition and Illustrations Technologies 'n Typography
Indexer Steve Rath
Printer Courier Corporation

Designations used by companies to distinguish their products are often claimed as trademarks or registered trademarks. In all instances where Morgan Kaufmann Publishers, Inc. is aware of a claim, the product names appear in initial capital or all capital letters. Readers, however, should contact the appropriate companies for more complete information regarding trademarks and registration.

Morgan Kaufmann Publishers, Inc.
Editorial and Sales Office
340 Pine Street, Sixth Floor
San Francisco, CA 94104-3205
USA
Telephone 415 / 392-2665
Facsimile 415 / 982-2665
Email *mkp@mkp.com*
WWW *http://www.mkp.com*
Order toll free 800 / 745-7323

Library of Congress Cataloging-in-Publication Data

Ciciora, Walter S.
 Modern cable television technology : video, voice, and data
communications / Walt Ciciora, James Farmer, David Large.
 p. cm.
 Includes bibliographical references and index.
 ISBN 1-55860-416-2
 1. Cable television. I. Farmer, James, (date) II. Large,
David, (date) III. Title.
TK6675.C53 1999
621.388'57—dc21
 98-35328
 CIP

To Our Families

Foreword

Archer S. Taylor
The Strategis Group

Modern Cable Television Technology seems destined to become the standard engineering reference work in the rapidly developing field of cable television. This is the first and only publication that covers, in such a thorough and authentically expert manner, all aspects of the architecture, engineering design, evaluation, and operation of modern hybrid fiber-coax (HFC) cable TV networks, from the headend to the customer interface. This is no academic exercise. Through the pages of this unique work, the reader is provided access to the outstanding professional expertise of the authors, developed through their long and practical careers at the cutting edge of cable TV research, design, and system operation.

Readers with a basic understanding of electrical and electronic circuits and phenomena should find in the book a wealth of practical information. Readers with more advanced background will find theory and mathematical support readily traceable to the basic principles of physics. An extensive bibliography is provided for those who wish to investigate further. For readers whose primary experience is in telephony or computer science, the detailed descriptions of cable TV technology indicate the extent to which these technologies are parallel and interdependent. Readers already well versed in cable TV technology are assured of adding to their knowledge and understanding.

The modern cable television system is a telecommunications facility, utilizing specialized termination hardware and software at the headend and customer interfaces or even at intermediate points throughout the distribution network. This book's treatment of the classical technology for the distribution of video, voice, and data on coaxial cable and optical fibers is unique and comprehensive. For the first time, the specialized characteristics of the transmitted signals and termination facilities required for telecommunications service are described as they relate to HFC distribution networks. For example, the critical issues of two-way network reliability and availability are discussed in depth, especially with respect to return, or "upstream," transmission. A tutorial section describes the algorithms specified by the Moving Picture Experts Group (MPEG) for source coding and the high spectral efficiency modulation methods of channel coding for digital television (DTV). The data-over-cable service interface specification (DOCSIS) protocol for high-speed cable modems is described, as well as spectrum sharing and other data transmission issues.

Legend has it that voice communication by telephone originated in 1876 when Alexander Graham Bell spilled acid on his clothes and inadvertently transmitted, over a primitive telephone circuit in the laboratory, his now famous call for help: "Mr. Watson. Come here. I want you!" At about the same time, in 1874, Paul Nipkow patented the renowned "Nipkow disc" from which television eventually evolved. Telephones were already commonplace in 1939, when RCA's David Sarnoff introduced modern electronic television to the public at the New York World's Fair. Until 1952, however, wartime and postwar policy considerations blocked widespread deployment of over-the-air television broadcasting. Impatient with being denied the much-touted wonders of television, enterprising entrepreneurs extended the coverage range of the few existing TV stations with broadband, wired distribution networks called community antenna TV (CATV), or simply cable TV.

Then, in 1977, Steve Jobs and Steve Wozniak introduced Apple II, the first commercially successful personal computer based on microprocessor technology and software programming. As the power and availability of PCs boomed, functional distinctions between the converging telephone, cable TV, and computer technologies have become increasingly diffuse. The Internet is basically a computer network, which is converging with both telephony and cable TV through software-based Internet Protocol (IP) telephony without circuit switching. Information is being transmitted with high-speed cable TV modems at much higher rates than is apparently possible over existing copper telephone circuits. Computer software is already driving the microprocessor-controlled cable TV customer interface, enabling users to transact business, make purchases, and interact with visual images. Helping customers navigate comfortably through the crowded sea of entertainment and information services on cable without foundering is a task for computer science. The program capacity of cable TV bandwidth has been expanded manyfold with digitally compressed video programs, generated by computer technology. All that remains for the computer display to become a TV set is the adoption of interoperability standards, which are still locked in vigorous contention.

Modern Cable Television Technology is the most thorough presentation of cable TV engineering fundamentals yet published in a single volume. But beyond that, it provides important information regarding the convergence of cable TV, telephone, and computer technologies. Its arrival is timely and welcome. Although innovation in concepts and protocols may outdate, perhaps quickly, a few of the practices and details described in the book, the underlying technology and theory will remain fundamental and sound.

Contents

Part 1 Once Over Lightly

Part 3 Headends

Part 5 System Architecture

Preface

Overview

The major characteristic of the technologies covered in this book is *accelerated change*. The authors have been privileged to observe and, in some small part, to contribute to the growth of these technologies. Cable, commercial television, the transistor, and other technologies we use in our work did not exist when we were born. We grew up with them. We have had to engage in "life-long learning" just to keep up with these technologies. This is both the curse and the blessing of our chosen field of endeavor. We hope that you already share our enthusiasm for cable technology. If you have not yet caught this contagion, we would like this book to be a carrier.

Intended Audience

The target audience includes several constituencies. Most important are our colleagues in the industry, particularly the members of the Society of Cable Telecommunications Engineers (SCTE). It is our intention that this book be the most useful study tool for those preparing for the SCTE's certification examinations. The cable industry is made up of those who build cable systems, operate them, connect to them, and supply the equipment used in them. All are part of the intended audience. The consumer electronics industry produces products that come to life when connected to our cables. A better understanding of cable technology can facilitate equipment and system designs that will help our subscribers and their customers derive more enjoyment out of purchases of hardware and services. The technologists of the computer industry are on a path to convergence with cable. Just as we need to learn more about their technologies, so they will benefit by understanding broadband interconnects with computer systems. We hope to reach others working in the various embodiments of the telecommunications arts and sciences. Included are those who may see us at one time or another as competitors. We recognize that in this rapidly changing world, competitors must cooperate too. Often competitive situations converge into collaborative efforts. This book will be of value to professionals who are not in a traditional cable environment. A number of telephone companies have seen the advantages of us-

ing hybrid fiber-coax (HFC) systems for full service networks delivering a package of communication services to customers. Additionally, industrial, commercial, campus, and other networks benefit from the cable architecture. The contents of this book will be of interest to these efforts. Lastly, we hope that the general technical fraternity might find satisfaction for a curiosity in the technology of this wonderful delivery mechanism.

Although we have not dealt more thoroughly with metric measurements, the PAL and Secam formats, and non-U.S. regulations, the technology differences are relatively minor, and most of the material should be of interest and value to non–North American audiences.

Fitting the material to the audience is a challenge. The audience comes with different experiences and knowledge. Some of you will be experienced engineers in related fields who are unfamiliar with the details of cable technology while intimately familiar with television. Others will be familiar with the various methods of transmission of signals but will have never dealt with television. Some will be experienced in radio frequency techniques but not the signals conveyed on cable. Still others will be experts in manufacturing and wonder if there is a market for their products in the cable business.

Though some of you will be engineers with graduate degrees, others will be self-taught. Even those who have previously studied these topics may benefit from some refreshing. Another aspect of the problem of fit is the scope of interest. Some may find the history of cable fascinating, but others will have no interest in it at all. Some would like to know details, but others just want the results.

The task is to write so that the majority derives benefit. There must be sufficient information for nearly all to have their interests served. But there also will be information that will be of little or no interest to individual readers. This means minimizing the gaps for those interested in details and avoiding the extraneous material for those who are coldly results oriented. Since these goals are almost contradictory, this is a difficult challenge. A little cooperation from you is requested. Boldly skip over what seems to be beyond your interest. Dig deeper into the references listed for more detail. Seek us out at the next convention and ask about what is missing.

Approach

The level of this book is aimed at the competent technologist who understands the basics and is looking for a brief refresher of things not recently considered before tackling new territory. Mathematics is held to a minimum and is used where it helps understanding rather than to "prove" a theory. It is not our purpose to develop concepts from first principles or to be scientifically rigorous. Rather, we hope to build a practical and useful understanding of the technology. Our intention is to create a text that will have lasting value in an era of rapid technological change. This requires an emphasis on the fundamentals that are the foundations upon which the advanced technology is built.

The approach used in ordering the flow of material is to start with the signals. The characteristics of the signals determine the nature of the infrastructure. Next, we consider the headend as the place where signals are gathered and generated and prepared for transmission. The components of the distribution system that carry the signals and the architecture that assembles these components are then studied. Finally, the customer interface issues governing how signals are converted into sensory stimuli are covered.

Content

The content of this book is a foundation grounded in television's analog basics; carrying through to the digital technologies that are the present excitement; and fully covering the fiber optic, coaxial, satellite, and radio frequency transmission paths all the way to the interface to the consumer's television, VCR, Web appliance, and personal computer.

The book is organized into six major parts:

Part 1: Once Over Lightly

Part 2: The Signals

Part 3: Headends

Part 4: Broadband Distribution Systems

Part 5: System Architecture

Part 6: Customer Interface Issues

The parts and even the chapters stand alone reasonably well. Dependencies on material from other chapters are clearly indicated. If you have specific needs or a generalized knowledge of cable, you are invited to dive directly into the chapters of interest. If you are beginning a study of cable television technology, it is well to approach the processes iteratively. First a light treatment. Then more detail from a sequential reading of the chapters. Even so, skipping a chapter of little interest is acceptable.

Part 1, Once Over Lightly, consists of Chapter 1, Introduction to Cable Television, which gives a casual review of cable technology business in the United States. If you are not familiar with the cable industry, you will find Chapter 1 a quick once-over that will put the following chapters in perspective. The first chapter evolves from an updated and revised version of the CableLabs publication "Cable Television in the United States, An Overview."

Part 2, The Signals, covers in four chapters the technical details of the signals that convey the programming and other services to customers. Chapter 2, Analog Television, covers the signals of the service that is the original motivation for cable television. The analog television system has served well for more than 50 years and will continue to do so for a decade or more. The section on Modern Analog Television covers the major compatible additions to the television signal:

color, stereo sound, and ancillary signals. Chapter 3 Digitally Compressed Television, covers the topic of most current excitement. Redundancy in the analog signal is explored along with ways of removing it. This facilitates high-definition television (HDTV) in the same 6 MHz that used to carry just one ordinary analog signal. Alternatively, multiple programs can be carried in the same spectrum. Chapter 4, Cable Data Transport, covers the basics of packaging data for transport on a system where multiple subscribers are connected to the same datastream. Different techniques are used for modulating signals in the downstream and upstream directions. Chapter 5, Cable Telephony, describes the implementation of a telephony service on a broadband cable system using the fundamentals taught in the previous chapter. The basics of traffic engineering are covered. The issues of reliability and availability are especially important to telephony.

Part 3, Headends, covers the facility that launches the signals onto the broadband hybrid networks of fiber-optic and coaxial cables. Part 3 consists of three chapters covering the methods by which signals enter headends, how those signals are processed in headends, and how signals leave headends. Chapter 6, Signal Reception, covers the acquisition of signals. Chapter 7, Headend Signal Processing, explains the details of how signals received at the headend are prepared and assembled into a spectral package. Chapter 8, Headend Operation, discusses the way in which the signals are launched onto the cable system. Both coaxial and optical systems are currently used. The management of these resources is described, including the preservation of quality.

Part 4, Broadband Distribution Systems, explains the components of cable's infrastructure in six chapters. Modern technology offers many alternatives to building this infrastructure. Chapter 9, Coaxial RF Technology, describes the original workhorse technology that gave the industry its name. The details of coaxial technology are fully explored in this pivotal chapter. Chapter 10, Coaxial Distribution System Design, delineates the basics of system structure and the all-important cable drop connection. Chapter 11, Linear Fiber-Optic Signal Transportation, describes one of the most important technical innovations in cable's most recent decade and a half. The fiber-optic world thought that only digital signals could be conveyed on glass fibers. The important breakthrough that opened cable's horizons was the realization that lasers and light receivers could be operated in a linear manner so that cable's traditional analog modulation could continue to be used. Chapter 12, Other Supertrunking Alternatives, discusses ways of covering large distances while preserving signal quality. Microwave techniques are important since they can be operated in linear mode, making the reception site simple and cost effective. Supertrunks are coaxial systems operated to maximize linearity over long distances and many amplifiers. Chapter 13, End-to-End Performance, shows the requirements and the calculations necessary for predicting performance. Real-world results are given. Chapter 14, Upstream Issues, covers unique factors affecting communications back to the headend. There are many difficult challenges in making the return path practical and cost effective.

Part 5, System Architecture, tells how the components go together to form a system. Chapter 15, Service-Related Architecture Requirements, shows how the nature of the services delivered to subscribers determines the specifications on the cable delivery system. Chapter 16, Architectural Elements and Examples provides samples of structures for various purposes and budgets. Chapter 17, Network Reliability and Availability, covers the means of computing the expected performance of a cable system under various situations.

Part 6, Customer Interface Issues, deals in three chapters with the final link to the subscribers' eyes and ears. This final link consists of an optional element provided by the cable operator and the subscribers' owned equipment. Chapter 18, Program Denial Technologies, explores the methods of conditional access that support the economics of premium services. Chapter 19, Consumer Electronic Interface, describes the problems consumers have when they attempt to use the features and functions of consumer electronics products with cable service. Chapter 20, Equipment Compatibility, covers the necessary technical specifications for equipment to be "cable ready" from the perspective of avoiding signal impairments.

Several appendices provide useful data that supplements more than one chapter and is made available in convenient form for ready reference.

Background

We, the authors of this book, have written hundreds of technical papers and trade magazine articles, delivered countless presentations, and presided over dozens of technical conference sessions during combined activities approaching a century in the cable industry. During that time, we have received a great deal of feedback, both positive and negative. That feedback has shaped the way we write and how we present the material. All three of us have done some teaching and lecturing both in the cable industry and outside it, which also has helped the process of preparing material. Each of us has different areas of specialization and experience. The three areas of expertise interlock and complement one another. We have enjoyed learning from one another and broadening our own knowledge while writing the book. You will perhaps recognize some style differences among the chapters. If you have worked in the industry, read its trade journals, and attended its conferences, you will probably be able to tell who wrote what just from the differences in style. Every chapter, however, has had the benefit of multiple reviewers whose comments have helped bring a dose of clarity where needed.

This book started about a decade ago as a monograph at American Television and Communications. It was intended as a brief introduction to cable for potential foreign suppliers of equipment. It was extensively revised multiple times to become a CableLabs monograph. Morgan Kaufmann Publishers asked if the monograph could be expanded into a book. Wisely, the original author, Walt Ciciora insisted on help, so a second author, Dave Large, was brought on. The two authors are consultants who thought the book would be a productive way to fill

spare time between clients. Fortunately for the consultants, but unfortunately for the book, the consulting business boomed. The publisher then strongly suggested adding a third co-author. Jim Farmer was invited. Today, almost nothing is left from the original monograph. The idea of simply expanding the original monograph quickly fell to efforts to improve the flow, increase the coverage, and create an improved product.

Acknowledgments

First, we thank our "book widows," Jeanette Ciciora, Kathy Farmer, and Sally Large, who have seen less of us than we would wish. Although all have persevered through many years of marriage, they have had to share their spouses with days of travel, clients, labwork, paperwork, and recently, "that damn book." Perhaps the best we can offer besides our thanks is a promise of no more books (at least until next time). And, of course, our apologies go to our children and grandchildren, who have seen less of us than might otherwise be the case.

Special thanks to Jennifer Mann of Morgan Kaufmann Publishers. She is our editor but, more important, our very good friend. Though she has been patient with our distractions and the demands of our clients for our time, she has also advised, coordinated, and cajoled us and reminded us of our commitments. When the load of work seemed to be way beyond anything anticipated, Jennifer's contagious enthusiasm for the project would provide the needed energy boost. Karyn Johnson, officially our editorial assistant but unofficially "wonder woman," attended to the many details and made sure we all had our morning email every day. Our production editor, Cheri Palmer, helped put the pieces together to produce a product we can be proud of.

Our reviewers included our wives, our friends, and those enlisted by Jennifer Mann as proposal reviewers: Alex Best, Cox Cable; Rex Bullinger, Hewlett Packard; Keith Cambron, Pacific Bell; Jim Chiddix, Time Warner Cable; Brian Colbry, Bell Labs; Joseph Glaab, General Instrument Corporation; Andrew Glasspool, Farncombe Technologies UK; Paul Shumate, Bellcore; and Archer Taylor, The Strategis Group.

Chapter reviewers did the heavy lifting: Tim Brophy, General Instruments Corporation; Bob Dickinson, Dovetail Surveys, Inc.; Keith Fleming, Electronic System Products/Antec; Joseph Glaab, General Instrument Corporation; David Grubb, General Instrument Corporation; Nick Hamilton-Piercy, Rogers Cablesystems Inc.; Dan Harris, PhD, Corning Glass; Ted Hartson, Scottsdale Television Labs; Jud Hoffman, Panasonic; Steve Hughey, Antec; Doug Jones, U.S. West; John Kenny, Antec; Keith Kreager, Antec; Bill Lagoni, Thomson Consumer Electronics; Michael Maslaney, Arris Interactive; M.C. "Mitch" Matteau, Aris Interactive; Marv Nelson, SCTE; Jim Raiser, Scientific-Atlanta, Inc.; Gerald Robinson, Scientific-Atlanta, Inc.; Blair Schodowski, Electronic System Products/Antec; Oleh Sniezko, TCI Cable; Merrill Weiss, Merrill Weiss Group; and Norman Woods, Cable AML, Inc. Special acknowledgments go to Andrew Large for the treatment of user-perceived reliability, Nick Hamilton-Piercy for the contribution

of recent failure analysis data, and Ted Hartson for additional help on the fine points of the history of television technology. For help with difficult details, we are grateful to Vic Williams, Scientific-Atlanta, Inc.; Dave Womble, Scientific-Atlanta, Inc.; Scott Ramsdell, Antec UK; John Vartanian, Viewers' Choice; Mitch Matteau, Arris Interactive; Mark Bugajski, Arris Interactive; and Keith Kreager, Antec.

Archer Taylor has been a most remarkable reviewer. Not only have his efforts and generosity with his time been extraordinary, but the breath of his experience and knowledge of the cable industry has been an irreplaceable asset. He provided a number of interesting historical notes. Archer is to be credited for a suggested notation for the addition of cascaded noise and distortion in distribution systems. Our heartfelt thanks, Archer!

Our thanks to all the contributors. If we have failed to thank someone for his or her assistance (or misspelled a name in spite of strenuous efforts to avoid that failure), we apologize. If we have misinterpreted your suggestions or stubbornly refused to take your good advice, or taken someone else's advice instead of your suggestions, we take full responsibility for the result.

Lastly, we wish to thank those in the industry we have been privileged to know, our colleagues, and those who have gone before us and left us a rich legacy of tools with which to further build. We have enjoyed the adventure and look forward to years of further challenges.

Acknowledgments

Chapter 1 figures are based on Walt Ciciora, *Cable Television in the United States.* May 1995. Cable Television Laboratories, Inc., Louisville, CO. © 1995 Cable Television Laboratories, Inc.

Figure 3.9 is based on Charles Poynton, *A Technical Introduction to Digital Video.* Page 4, 1996. New York: John Wiley & Sons. © 1996 John Wiley & Sons.

Figures 3.11(a), 3.11(b), 3.12, 3.15, 3.16, 3.19, 3.22, and 3.23 are based on S. Merrill Weiss, *Issues in Advanced Television Technology.* Pages 33, 73, 74, 90, and 141, 1996. London: Focal Press. © 1996 Focal Press.

Figures 3.11(c), 3.13, and 3.14(a) are based on Keith Jack, *Video Demystified.* Pages 517, 518, and 519, 1996. Eagle Rock, VA: HighText Publications. © 1996 LLH Technology Publishing.

Figures 3.21, 3.24, 3.25, 3.26, and 3.29 are based on *Guide to the Use of the ATSC Digital Television Standard,* Doc. A/54. Pages 24, 68, 86, 97, and 146, October 1995. Advanced Television Systems Committee, Washington, D.C. © 1995 ATSC Organization.

Figure 3.28 is based on John Watkinson, *An Introduction to Digital Video.* Page 218, 1994. London: Focal Press. © 1994 Focal Press.

Figure 9.12 is based on data furnished by Howard Carnes of Antec Corporation.

Figure 11.6 is based on Ronald C. Cotton, *Lightwave Transmission Applications.* Page 108, September 15, 1993. Cable Television Laboratories Inc., Louisville, CO. © 1993 Cable Television Laboratories, Inc.

Figures 11.9 and 11.10 are based on Dogan A. Atlas, Fiber Induced Distortion and Phase Noise to Intensity Noise Conversion in Externally Modulated CATV Systems, *1996 National Cable Television Association Technical Papers.* Page 291 and 292, April, 1996. National Cable Television Association, Washington, D.C. © 1996 National Cable Television Association.

Figure 11.12 is based on Dan Harris, Primer on Baseband Digital Transmission on Single-Mode. Page 94, August 1996. *Communications Technology,* Potomac, MD. © 1996 Phillips Publishing International, Inc.

Tables 12.1 and 12.2 are adopted from material furnished by Cable AML, Inc., September 5, 1997. Cable AML, Inc., Torrance, CA.

Table 12.3 and the supporting text and equations are based on material furnished by Dr. Francisco Bernues of CableAML, Inc. September 5, 1997. CableAML, Inc., Torrance, CA.

Figures 12.4, 12.5 and 12.6 are reprinted with permission from the International Telecommunication Union. ITU-R Recommendation ITU-R PN.837-1, *Characteristics of Precipitation for Propagation Modeling.* Figures 1, 2 and 3. © 1996 International Telecommunication Union. Telecommunications Union, General Secretariat, Sales and Marketing Service, Place des Nations, CH-1211, Geneva 20, Switzerland.

Appendix A is based on EIA/ANSI-542, *Cable Television Channel Identification Plan.* Electrical Industries Association. Available from Global Engineering Documents, 800-854-7179. Used with permission of Global Engineering Documents.

Appendix B is based on EIA/TIA-250C, *Electrical Performance for Television Transmission Systems.* Electrical Industries Association. Available from Global Engineering Documents, 800-854-7179. Used with permission of Global Engineering Documents. Also based on Margaret Craig, *Television Measurements: NTSC Systems.* Pages 76–78. Tektronix, Beaverton, OR.

Endsheets are based on *NCTA Recommended Practices for Measurements on Cable Television Systems.* Section 5, August, 1985 and October, 1993. NCTA Engineering Committee, Standards Subcommittee. National Cable Television Association, Washington, D.C. © 1993 National Cable Television Association.

Part One
Once Over Lightly

Chapter 1
Introduction to Cable Television

1.1 Introduction

Currently, cable television service is enjoyed by nearly 65 million U.S. households. This is a market penetration of nearly 67%. Cable service is available to 96.7% of U.S. television households.[1]

Cable television is an important part of the way in which the citizens of the United States are informed and entertained. It is a means of providing large numbers of television channels to more than two-thirds of the U.S. population in a cost-effective way.

Because cable television has been so successful and has enjoyed such vigorous growth and acceptance, it has spawned competitors, including prerecorded media (tape and disk sales and rentals), direct broadcast satellite (DBS), as well as the interest of the telephone industry. Cable's high visibility has also attracted the attention of regulators and legislators. Important public policy issues are involved. Understanding what cable television is, how it works, and its economics will help decision makers in these arenas. This understanding will also aid technologists in determining which technologies are appropriate for cable television applications and which are not.

Television has long been the preferred source of information for the general public, and cable television offers viewers more choice than ever before. In fact, if cable had to be described with one word, that word would be *choice*. Choice is made possible by cost-effective bandwidth to deliver programming to almost 70% of the households having televisions in the United States. Since cable is available to more than 90% of television households, nearly everyone has the opportunity to subscribe to cable. In the United States, more people have televisions than telephones, and most of those people have cable.

There are more than 160 programs on the satellites that feed cable headends. Cable systems must select some of the programs they will offer to their subscribers from all those available since no analog cable system has sufficient capacity to carry them all. In addition, cable systems carry the local broadcasts and some programs locally originated or designated for local citizen access. In some cable systems, the local schools, library, or government has channels to deliver programs of special interest. In this way, the needs of a democratic society are served while citizens are well informed and entertained.

The broadband cable infrastructure can deliver an appealing analog service while adding hundreds of digital channels. This hybrid service is important for the transition to digital television since an abrupt changeover from analog to digital television would be chaotic. Most citizens will continue to enjoy analog television for many years to come. Even those who purchase a digital television receiver for their principal viewing room will have analog receivers in the rest of the house.

The same broadband cable can carry high-speed data service to subscribers, making Internet-based and other information services fast and convenient. The cable infrastructure is inherently two way, requiring only modest additional in-

vestment and maintenance to implement. When activated, this capability facilitates interactive cable services and other nontraditional cable services such as telephony and Internet telephony.

The success of cable has stimulated other broadband competitors, all using the programming services made possible by cable's investments and commitments. The cable industry has financed and developed these program services, tested their marketplace demand, sorted out those wanted by subscribers from those not wanted, and demonstrated that subscribers have a voracious appetite for programming. Cable's competitors merely have to supply an alternative infrastructure and attempt to attract customers to a proven business model. The major risks have already been taken.

In the following chapters, you will gain a solid understanding of the technologies required to deliver broadband services to the home. You will see how the pieces fit together to make up a complete system for the delivery of information and entertainment choices to consumers. If yours is a related business, you will better understand how it fits with the cable industry. If you are already knowledgeable in some aspects of the broadband networks used by cable, this book will fill in the gaps.

1.2 Historical Perspective

The cable television system was not originally intended to be a general-purpose communications mechanism. Only recently has cable's role expanded to include more general-purpose communications services in some cable systems. But even in those cases, the new services are built, as much as possible, on cable's older foundations. Cable's primary and often sole purpose is the transportation of entertainment television signals to subscribers. In those applications, it needs only to be a one-way transmission path from a central location, called a *headend*, to each subscriber's home, delivering essentially the same signals to each subscriber. The signals are intended for use with the consumer electronics equipment that subscribers already own. This equipment is built to operate on the current U.S. television technical standard called NTSC, after the organization that created it in 1941, the National Television Systems Committee. This black-and-white television standard was modified in 1953 to provide compatible color information to color television receivers and again in 1984 to add compatible stereo sound.

The original purpose for cable television was to deliver broadcast signals in areas where they were not received in an acceptable manner with an antenna. These systems were called community antenna television (CATV). In 1948, Ed Parson of Astoria, Oregon, built the first CATV system consisting of twin-lead transmission wire strung from housetop to housetop. In 1950, Bob Tarlton built a system in Lansford, Pennsylvania, using coaxial cable on utility poles under a franchise from the city. Martin Malarkey opened his cable system in Pottsville,

Pennsylvania, almost simultaneously (within a week of Tarlton). Malarkey used Radio Corporation of America (RCA) Antennaplex equipment that was originally developed for the in-house distribution of the National Broadcasting Company's (NBC's) signals in the RCA building in New York. Tarlton used Jerrold Corporation's antenna booster amplifier. Malarkey is a pioneer who organized and developed the National Community Television Association (NCTA), which later was renamed the National Cable Television Association.[2]

In most of the original cable systems, viewers found that off-air signals either were not available or were very weak because of the terrain or the distance from television transmitters. In some areas, such as New York City, multiple signal reflections and shadows cast by buildings made reception difficult. In both of these environments, a hard-wired method of delivery of signals to subscribers was welcomed. The first operators of these systems were entrepreneurial broadcasters or retail TV receiver dealers who sought to expand the market for the sale of their products by also providing the signals the products required. By the late 1960s, nearly all the areas of the United States that could benefit from a community antenna had been served.

In the mid-1970s, an embryonic technology breathed new life into cable television. This technology was the satellite delivery of signals to cable systems. The pioneering movie channel, Home Box Office (HBO), showed the way. HBO's dramatic relay of the Ali–Frazier heavyweight prizefight in October 1975 from Manila to cable systems in Mississippi and Florida effectively triggered the development of the satellite network industry. Although satellites and earth stations were very expensive investments, these programming pioneers understood that the costs could be spread over many cable operators who, in turn, serve many subscribers.

Three categories of signals came into existence: (1) "superstations," local stations that are distributed nationally over satellite and became mininetworks (the Turner Broadcasting System of Atlanta, Georgia, pioneered the concept); (2) specialized channels for news, sports, weather, education, shopping, and so on; and (3) movie channels such as HBO that sparked new excitement in the business. Cable television became much more than just a community antenna for areas with poor reception. Cable television became a means of receiving programming otherwise unavailable from broadcasters.

At present, subscribers are offered a variety of video services. The foundation service all subscribers are required to take is called *basic*. Off-air channels, some distant channels, and some satellite-delivered programs are included. The satellite programs include the superstations and some of the specialty channels. Pay television constitutes premium channels, usually with movies and some special events, which are offered as optional channels for an extra monthly fee. Some cable systems offer pay-per-view (PPV) programming, which is marketed on a program-by-program basis. Recent movies and special sports events are the mainstay of PPV programming. Impulse pay per view (IPPV) allows the subscriber to order the program spontaneously, even after it has begun. The ordering mech-

anism usually involves an automated telephone link or, occasionally, two-way cable.

Ways of providing conditional access to allow for a limited selection of service packages at differing price points are often included in the cable system. Simple filters remove the unsubscribed channels in some systems, whereas elaborate video and audio scrambling mechanisms involving encryption are used in other cable systems.

A wide variety of other services have been repeatedly offered to cable subscribers: videotext, teletext, other forms of electronic publishing and information age services, home security, and digital audio programming. Subscribers to date have shown a remarkable lack of interest in these services. It is interesting to speculate on the reasons for this indifference, but the negative experience has been without exception. Although subscribers have been very enthusiastic about video, particularly entertainment-oriented video, other services have left them cold. This has caused cable operators to concentrate their cable system design efforts on efficient, cost-effective video delivery.

Since cable television systems must utilize the public right-of-way to install their cables, they, like power, telephone, and gas companies, must obtain a franchise from the local governmental authorities. This is a nonexclusive franchise. However, experience with multiple cable systems has shown that the economics of the business generally supports only one system per community.

Cable television's most fundamental concept is spectrum reuse. Historically, the over-the-air spectrum has been assigned to many other purposes. Much of it is therefore not available for the carriage of television. The result is an inadequate supply of spectrum to serve viewers' needs. Cable can reuse spectrum that is sealed in its aluminum tubes.

1.3 Spectrum Reuse

Compared with nearly any other communications need, video is a bandwidth hog. Whereas telephone-quality voice needs only 4 kHz of spectrum and high-fidelity sound takes 20 kHz or so (40 kHz for stereo), the current baseband video standard consumes 4.2 MHz. When modulated, the television signal takes up 6 MHz of spectrum. High-definition television (HDTV) requires about 30 MHz for each of the red, green, and blue signals that make up a color picture. Extensive bandwidth compression techniques will reduce the amount of spectrum required by HDTV to more acceptable amounts. These signals must then be modulated onto carriers to deliver multiple signals to the consumer's equipment.

For NTSC, each television channel is assigned 6 MHz because of the need to share limited spectrum with other services. With double sideband modulation, the 6-MHz straitjacket permits only about 2.5 MHz for the demodulated video, after allowing room for the sound carrier. In 1941, vestigial sideband amplitude modulation (VSB-AM) was invented. This modulation approach enabled

television engineers to achieve higher resolution within the 6-MHz limit. Compared with double sideband amplitude modulation, VSB-AM transmits one complete sideband and only a vestige of the other. At the time the standard was created, the design requirements of practical filters determined the amount of sideband included. The consumer's receiver selects the channel to be watched by tuning a 6-MHz portion of the assigned spectrum.

In the terrestrial broadcast environment, channels must be carefully assigned to prevent interference with each other. The result of this process is that most of the terrestrial broadcast television spectrum is vacant. Better television antennas and better television circuits would allow more of the spectrum to be utilized. However, with more than 250 million receivers and more than 150 million VCRs in consumers' hands, the changeover process to upgraded systems would be costly and require 20 or so years. Consumers continue to purchase about 25 million color receivers and about 15 million VCRs per year. The reliability of the products is phenomenal. Their "half-life" is about 10 to 12 years.

The rest of the terrestrial spectrum that is not assigned to broadcast has other important uses. These include aircraft navigation and communications, emergency communications, and commercial and military applications. The terrestrial spectrum is too limited to supply the video needs of the U.S. viewer.

Cable television is made possible by the technology of coaxial cable. Rigid coaxial cable has a solid aluminum outer tube and a center conductor of copper-clad aluminum. Flexible coaxial cable's outer conductor is a combination of metal foil and braided wire, with a copper-clad, steel center conductor. The characteristic impedance of the coaxial cable used in cable television practice is 75 ohms. The well-known principles of transmission-line theory apply fully to cable television technology.

The most important characteristic of coaxial cable is its ability to contain a separate frequency spectrum. The properties of that separate spectrum allow it to behave like an over-the-air spectrum. This means that a television receiver connected to a cable signal will behave as it does when connected to an antenna. A television set owner can become a cable subscriber without an additional expenditure on consumer electronics equipment. The subscriber can also cancel the subscription and not be left with useless hardware. This ease of entry and exit from an optional video service is a fundamental part of cable's appeal to subscribers.

Since the cable spectrum is tightly sealed inside an aluminum environment (the coax cable), a properly installed and maintained cable system can use frequencies assigned for other purposes in the over-the-air environment. This usage takes place without causing interference to these other applications or without having them cause interference to the cable service. New spectrum is "created" inside the cable by the "reuse" of spectrum. In some cable systems, dual cables bring two of these sealed spectra into the subscriber's home, with each cable containing different signals.

The principal drawback of coaxial cable is its relatively high loss. Coaxial cable signal loss is a function of its diameter, dielectric construction, temperature,

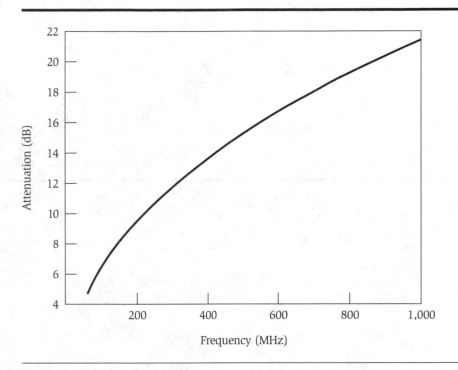

Figure 1.1 Coax cable attenuation versus frequency.

and operating frequency. A ballpark figure is 1 dB of loss per 100 feet. Half-inch diameter aluminum cable has around 1 dB of attenuation per 100 feet at 181 MHz; at 1-inch diameter, the attenuation drops to 0.59 dB per 100 feet. The attenuation of cable varies with the square root of the frequency. Thus, the attenuation at 216 MHz (within TV channel 13) is twice that of 54 MHz (within TV channel 2) since the frequency is four times as great. If channel 2 is attenuated 10 dB in 1,000 feet, channel 13 will be attenuated 20 dB. Figure 1.1 demonstrates this relationship for 1,000 feet of 5/8-inch aluminum cable.

1.4 Cable Network Design

Since cable television originally was not a general-purpose communications mechanism, but rather a specialized system for transmitting numerous television channels in a sealed spectrum, the topology or layout of the network was customized for maximum efficiency. The topology that has evolved over the years is called tree-and-branch architecture.[3]

There are five major parts to a cable system: (1) the headend, (2) the trunk cable, (3) the distribution (or feeder) cable in the neighborhood, (4) the drop

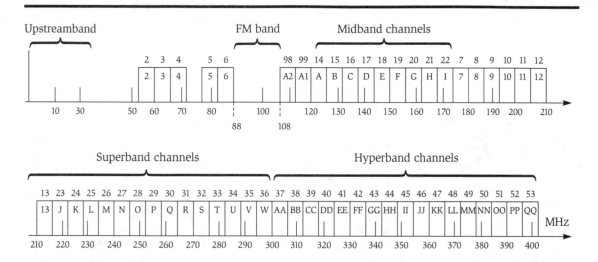

Figure 1.2 Cable frequency plan. Numbers above the rectangles are the EIA-542 designations. Historical designations are inside the rectangles. Current cable plant bandwidth can exceed 1 GHz.

cable to the home and in-house wiring, and (5) the terminal equipment (set top terminals and consumer electronics hardware).

Flexible coaxial cable is used to bring the signal to the terminal equipment in the home. In the simplest cases, the terminal equipment is the television set or VCR. If the TV or VCR does not tune all the channels of interest because it is not "cable compatible," a converter is placed between the cable and the TV or VCR tuner.

Broadcast channels 2 through 13 are not in a continuous band. Other radio services occupy the gaps. Cable can reuse these frequencies because its spectrum is self-contained within the coaxial environment. The cable converter has a high-quality broadband tuner and output circuitry that puts the desired cable channel on a low-band channel not occupied in the local off-air spectrum. Typically, this is channel 2 or 3 and occasionally 4 or 5. The TV or VCR is tuned to this channel and behaves as a monitor. Premium programming is typically "scrambled" so that it is not clearly visible on an ordinary television receiver. The picture appears torn up, and the pieces jump around on the screen in the usual display. The subscriber wishing access to premium programming pays an additional fee and is supplied with a *descrambler*, which is a component of the set top terminal. The subscriber who has purchased a product labeled "cable ready" may expect to avoid the set top terminal. The requirement to use a set top terminal with a descrambler for premium services can be a sore point.

Figure 1.2 shows the cable frequency plan. Although Figure 1.2 goes up to only 400 MHz, cable systems have been built to slightly in excess of 1 GHz with more than 150 analog channels. Cable systems with bandwidths of 750 MHz are not uncommon, and 550 MHz is typical for upgrades and new construction. The home is connected to the cable system by the flexible drop cable, typically 150

Figure 1.3 Terminal equipment and cable drop.

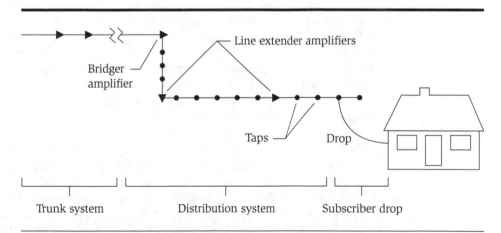

Figure 1.4 Distribution plant.

feet long (see Figure 1.3). The distribution cable in the neighborhood runs past the homes of subscribers. This cable is tapped so that flexible drop cable can be connected to it and routed to the residence. The distribution cable interfaces with the trunk cable through an amplifier called a bridger amplifier, which increases the signal level for delivery to multiple homes. One or two specialized amplifiers called line extenders are included in each distribution cable. Approximately half of the system's cable footage is in the distribution portion of the plant, and half is in the flexible drops to the home (see Figure 1.4).

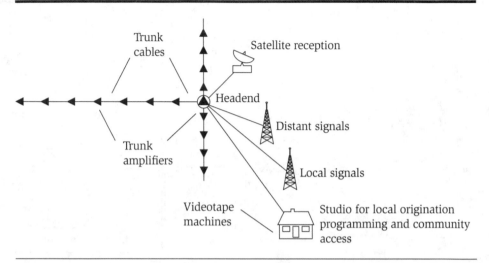

Figure 1.5 Cable system headend.

The trunk part of the cable system transports the signals to the neighborhood. Its primary goal is to cover distance while preserving the quality of the signal in a cost-effective manner. Broadband amplifiers are required about every 2,000 feet depending on the bandwidth of the system. The maximum number of amplifiers that can be placed in a run or cascade is limited by the buildup of noise and distortion. Twenty or 30 amplifiers may be cascaded in relatively high-bandwidth applications. Older cable systems with fewer channels may have as many as 50 or 60 amplifiers in cascade. Approximately 12% of a cable system's footage is in the trunk part of the system.

The headend is the origination point for signals in the cable system. It has parabolic or other appropriately shaped antennas for receiving satellite-delivered program signals, high-gain directional antennas for receiving distant TV broadcast signals, directional antennas for receiving local signals, machines for playback of taped programming and commercial insertion, and studios for local origination and community access programming.

Local origination is programming over which the cable operator has editorial control. It can range from occasional coverage of local events to a collection of programming almost indistinguishable from that of an independent broadcaster. Often, mobile coverage of events is provided with microwave links back to the headend or backfeed of the signal up the cable system to the headend. Local origination also includes alphanumeric text as well as studio productions and videotape. In many systems, text may be the only form of local origination.

Community access is programming access for community groups mandated by the franchise. The cable system typically cannot exercise editorial control over quality or content of community access programming (see Figure 1.5). When the whole picture is assembled, the tree shape of the topology is evident. The trunk and its branches become visible (see Figure 1.6).

Figure 1.6 Tree-and-branch cable topology.

The ultimate goal of the cable system is to deliver pictures of adequate quality at an acceptable price while satisfying stockholders, investors, and holders of the debt generated by the huge capital expenses of building the cable system plant.

1.5 Signal Quality

Delivering signal quality to subscribers is a difficult balancing act. It would be a simple matter to deliver very high quality video if cost were not a consideration. Experience teaches that subscriber satisfaction is a complex function of a variety of factors led by program quality and variety, reliability of service, video and sound quality, and the amount of the subscriber's cable bill.

The principal picture impairments can be divided into two categories: coherent and noncoherent. Coherent impairments result in a recognizable interfering pattern or picture. They tend to be more objectionable than noncoherent impairments of equal strength.

The principal noncoherent picture impairment is noise. Random noise behavior is a well-understood part of general communications theory. The familiar Boltzmann relationship, noise figure concepts, and so forth apply fully to cable television technology. Random noise is the consequence of the statistical nature of the movement of electric charges in conductors. This creates a signal of its own. This noise is inescapable. If the intended analog signal is ever allowed to become weak enough to be comparable to the noise signal, the signal will be polluted by the noise, yielding a snowy pattern in pictures and a seashore-sounding background to audio.

Noise levels are expressed in cable system practice as ratios of the visual carrier to the noise in a television channel. This measure is called the carrier-to-noise

Table 1.1 Signal-quality target values.

Parameter	Symbol	Value
Carrier to noise (C/N)	C/N	46 dB
Carrier to composite second order	C/CSO	53 dB
Carrier to composite triple beat	C/CTB	53 dB
Signal level at TV		0 to +3 dBmV

ratio (C/N) and is given in decibels (dB). The target value for C/N is 45 to 46 dB. Subjective data was compiled in 1991 in a project sponsored by the cable industry's research and development consortium, Cable Television Laboratories, or CableLabs. The work, supervised by Dr. Bronwen Jones, indicates "perceptible but not annoying" C/N at about 47 to 50 dB. The "slightly annoying" category was within about 1 dB of 41 dB.[4]

Coherent interference includes ingress of video signals into the cable system, reflections of the signal from transmission-line impedance discontinuities, cross modulation of video, and cross modulation of the carriers in the video signal. This latter phenomenon gives rise to patterns on the screen called beats. These patterns often look like moving diagonal and horizontal bars or herringbones.

The evaluation of signal quality takes place on two planes, objective and subjective. In the objective arena, measurements of electrical parameters are used. These measurements are repeatable. Standardized, automated test equipment has been developed and accepted by the video industries. Table 1.1 lists the parameters usually considered important and the values of good, current practice. They are described in the remaining text.

The ultimate performance evaluation involves the subjective reaction of viewers. One example of the difficulties experienced is the fact that different frequencies of noise have differing levels of irritation. High-frequency noise tends to become invisible, whereas low-frequency noise creates large moving blobs that are highly objectionable. Subjective reactions to these phenomena are influenced by such factors as the age, gender, health, and attitude of the viewer. The nature of the video program, the characteristics of the viewing equipment, and the viewing conditions also affect the result.

Signal processing in the TV receiver changes the impact of signal impairments. Noise in the band of frequencies used to transmit color information is demodulated and converted into lower-frequency, more objectionable noise. Noise in the synchronization part of the TV signal can cause the picture to break up entirely, resulting in much greater impairment than the same strength noise confined to other portions of the signal.

Subjective assessment of viewer reaction to picture impairments has been made using various methods. In 1959, the Television Allocations Study Organization (TASO) studied the amount of noise, interference, and distortion viewers will tolerate in a TV picture.[5] The results were expressed in a six-point scale with

grades named excellent, fine, passable, marginal, inferior, and unusable. This is very old data, and the scale is no longer accepted by psychometric experts because it mixes impairment and picture-quality assessments.[6,7] Additionally, the lack of a midscale category causes difficulties in the statistical analysis. However, the broadcast industry still uses this scale even though it is considered obsolete by psychometric practitioners. The most widely accepted standard for subjective measurements of impairment is now a five-point scale that was also used in the tests sponsored by CableLabs. However, the most widely published data on subjective impact of video impairments is based on the Bell Telephone Laboratory seven-point scale, which has not been used in any other study.

It is important to realize that the demand for signal quality is a function of time. Five to 10 years ago, consumer electronics products were not capable of displaying the full resolution of the NTSC signal. Gradually, these products improved until high-end models were capable of more performance than the NTSC signal could deliver. The Super VHS and Hi-8 videotape systems have greater resolution than broadcast NTSC. As time progresses, the level of performance of consumer electronics will continue to increase. As advanced television (ATV) and HDTV are introduced, still more demands will be made on cable system performance. The trend to larger screen sizes also makes video impairments more evident.

1.6 Cable System Trade-Offs

The experienced cable system designer has learned how to balance noise, nonlinear distortions, and cost to find a near optimal balance.

Signals in cable systems are measured in decibels relative to 1 mV across 75 ohms. This measure is called *dBmV*. This unfortunate name is the source of a great deal of confusion. Since the name of the measure ends in V, it is natural to assume it refers to a voltage. It does not. The measure dBmV is a *power* ratio with reference to 13.33 nanowatts of power, defined as 1 millivolt (rms) across 75 ohms.

Applying the well-known Boltzmann noise equation to 75-ohm cable systems yields an open-circuit voltage of 2.2 microvolts in 4 MHz at room temperature. When terminated in a matched load, the result is 1.1 microvolts. Expressed in dBmV, the minimum room temperature noise power in a perfect cable system is −59.17 dBmV.

Starting at the home, the objective is to deliver at least 0 dBmV to the terminal on the television receiver, but no more than would produce overload in the consumer's tuner. Lower numbers produce snowy pictures, and higher numbers overload the television receiver's tuner, resulting in cross modulation of the channels. Unfortunately, the overload point of television receivers is not well defined. As a result, the FCC rules merely state that it is to be avoided.

If a converter or descrambler is used, its noise figure must be taken into account. There are two reasons for staying toward the low side of the signal range: cost and the minimization of interference in the event of a signal leak caused by a

faulty connector, damaged piece of cable, or defect in the television receiver. Low signal levels may cause poor pictures when excessive signal splitting without prior amplification is used to serve too many receivers. Working our way back up the plant, we need a signal level of 10 to 15 dBmV at the highest frequency in the passband at the tap to compensate for losses in the drop cable.

The design objectives of the distribution part of the cable system involve an adequate level of power not only to support the attenuation characteristics of the cable but also to allow energy to be diverted to subscribers' premises. Energy diverted to the subscriber is lost from the distribution cable. This loss is called *flat loss* because it is independent of frequency. Loss in the cable itself is a square-root function of frequency and is therefore contrasted to flat loss. Because of flat losses, relatively high power levels are required in the distribution part of the plant, typically, 48 dBmV at the input to the distribution plant. Note that 0 dBm (dB with respect to 1 mW) is equivalent to 48.75 dBmV. These levels force the amplifiers in the distribution part of the plant to reach into regions of their transfer characteristics that are slightly nonlinear. There is no sharply defined "nonlinear" region; it is rather a matter of degree. As a result, only a small number of amplifiers, called line extenders, can be cascaded in the distribution part of the plant. These amplifiers are spaced 300 to 900 feet apart depending on the number of taps required by the density of homes.

Because the distribution part of the plant (also called the *feeder*) is operated at higher power levels, nonlinear effects become important. The television signal has three principal carriers: the visual carrier, the aural carrier, and the color subcarrier. The visual carrier is by far the strongest, with the color carrier the weakest. These concentrations of energy in the frequency domain give rise to a wide range of "beats" when passed through nonlinearities. To minimize these effects, the aural carrier is attenuated about 15 dB below the visual carrier.

When cable systems carried only the 12 VHF channels, second-order distortions created spectrum products that fell out of the frequency band of interest. As channels were added to fill the spectrum, second-order effects were minimized through the use of balanced, push–pull output circuits in amplifiers. The third-order component of the transfer characteristic dominates in many of these designs. The total effect of all the carriers beating against each other gives rise to an interference called composite triple beat (CTB). CTB should be measured with the system fully loaded with all channels carried. The *NCTA Recommended Practices* (second edition revised October 1993) provides the following formula for the number of beats at the Mth channel with a total of N carriers:

$$\text{beats} = \frac{N^2}{4} + \frac{1}{2} \cdot (N - M) \cdot (M - 1) \qquad (1.1)$$

For a 35-channel system ($N = 35$), the center channel ($M = 18$) will experience 450 beats.

Third-order distortions increase about 6 dB for each doubling of the number of amplifiers in cascade. A 1-dB reduction in amplifier output level will generally improve CTB by 2 dB. If these products build to visible levels, diagonal lines will be seen moving through the picture. When these components fall in the part of the spectrum that conveys color information, spurious rainbows appear. The dominant nonlinear effect in the feeders is the triple beat between multiple visual carriers, since the visual carriers are the points of concentration of most of the energy in the television signal.

If we assign a design level of noise and nonlinear distortion at the subscriber's television receiver that is below the threshold of visibility, we can conceive of a budget of noise and distortion to be "spent" in the various parts of the system design. The distribution part of the system has relatively high power and has used up most of the budget for nonlinear distortions. On the other hand, little of the noise budget has been consumed. It can be allocated to the trunk part of the system, which brings the signals into the neighborhood.

The design objective of the trunk part of the cable system is to move the signal over substantial distances with minimal degradation. Because distances are significant, fiber and/or lower-loss cables are used. One-inch and 3/4-inch cable is common in the trunk, whereas 1/2-inch cable is found in the distribution. Signal levels in the trunk at an amplifier's output are 30 to 32 dBmV depending on the equipment used.

It has been determined through analysis and confirmed through experience that optimum noise performance is obtained when the signal is not allowed to be attenuated more than about 20 to 22 dB before being amplified again in all-coaxial systems. Amplifiers are said to be "spaced" by 20 dB. The actual distance in feet is a function of maximum frequency carried and the cable's attenuation characteristic. High-bandwidth cable systems have their amplifiers fewer feet apart than older systems with fewer channels. But with the advent of hybrid fiber-coax (HFC) systems described later, which yield very short amplifier cascades and high channel capacity, the trend is toward spacing as high as 30 to 40 dB. Since attenuation varies with frequency, the spectrum in coaxial cable develops a slope. This is compensated with equalization networks in the amplifier housings.

The attenuation of the cable is a function of temperature and aging of components. Modern amplifiers use a pilot signal to regulate automatic gain control (AGC) circuits. A second pilot signal at a substantially different frequency than the first allows the slope of the attenuation characteristic to be monitored and compensation to be introduced with automatic slope control (ASC) circuits. Thus, long cascades of amplifiers can, once properly set up, maintain their performance over practical ranges of temperature and component aging.

Since the signal is not repeatedly tapped off in the trunk part of the system, high power levels are not required to feed splitting losses. As a result, signal levels are lower than in the distribution portion of the plant. Typical levels are about 30 dBmV. For the most part, the amplifiers of the trunk are operated within their linear regions. The principal challenge of trunk design is keeping noise under

control. Each doubling of the number of amplifiers in the cascade results in a 3-dB decrease in the C/N at the end of the cascade and a 6-dB increase in the amount of CTB.

If the noise at the end of the cascade is unacceptable, the choices are to employ lower noise amplifiers, shorter cascades, or a different technology such as microwave links or fiber-optic links. Fiber-optic links are the technology of choice for these applications.

1.7 Hybrid Fiber-Coax

The hybrid fiber-coax (HFC) architecture is an optimized combination of fiber in the trunk and coaxial cable in the feeder and drop. HFC has made it possible to cost-effectively increase bandwidth, signal quality, and reliability, while reducing maintenance costs and retaining a craft-friendly (that is, easy-to-work-with) plant. It also reduces the technical challenges of providing two-way cable service.

1.7.1 The Hybrid Fiber-Coax Architecture

The HFC approach breaks the cable system into a multitude of much smaller cable systems with amplifier cascades limited to four to six amplifiers. Each of these small cable systems is fed with a fiber link to the headend (see Figure 1.7).

Surprisingly, the trunk portion of the cable plant covers only about 12% of the total footage. The feeder portion of the cable system supports taps for subscribers. It is usually constructed to no more than a mile and a half in length. This limitation comes from the fact that energy is tapped off to feed homes. Consequently, the power levels must be relatively high. These higher power levels reach into the slightly nonlinear regions of the amplifiers. As a result, only a few amplifiers can be used before the distortion begins to degrade the picture quality. Approximately 38% of the total footage in a cable system is in the feeder portion of the plant. The drop is the flexible cable that goes to the home. It is usually constructed to no more than 400 feet in length; typically, it is 150 to 200 feet long. Approximately half of the total footage of cable in a system is in the drop and the flexible wiring in the home.

The feeder portion of the cable plant is a hotbed of activity. Every day, new subscribers are added and others are removed. Approximately 20% of Americans move to a new residence every year. When they move, the drop is usually disconnected. When they move to their new residence, that drop must be reconnected. It is important that the technology used in the feeder portion of the trunk supports this constant activity. It must be craft friendly. The trunk portion of the cable plant is relatively stable. Few changes are made.

The concepts behind HFC are straightforward. The bandwidth of coaxial cable has no sharp cutoff. It is the cascade of amplifiers that limits the bandwidth. Twenty to 40 amplifiers in cascade not only reduces bandwidth but also consti-

Headend

Fiber node area
Coaxial cable
Fiber optics

Figure 1.7 Cable systems with hybrid fiber-coax.

tutes a severe reliability hazard. Laying low-loss fiber over the trunk portion of the plant eliminates or reduces the number of trunk amplifiers. Wider bandwidth is thus facilitated. Two-way cable operation becomes practical because of two benefits of HFC. First, the cable system is broken up into a large number of smaller cable systems, each isolated from the others by its own fiber link to the headend. If ingress should cause interference in one of these small cable systems, that interference will not impair the performance of the other portions of the cable plant. Second, the fiber itself is not subject to ingress of interfering signals.

Today, it is common practice to install passive components (taps, splitters, amplifier housings) that will pass 1 GHz. Actual 1-GHz amplifiers are still too costly. However, some systems are being built "GHz ready" so that the addition of GHz amplifier modules will upgrade the system with little or no waste of installed plant. GHz capability is a future possibility, but 750-MHz upgrades are done today, and 550 MHz is commonplace.

The lasers that drive the fibers are expensive, costing many thousands of dollars. The receivers that convert the optical energy into signals that TVs and VCRs can use are also modestly expensive. To be practical, these components must serve hundreds of subscribers each so that the costs can be shared. This is accomplished by fibers that feed small coaxial systems, which in turn serve a few hundred to a couple of thousand subscribers.

A major attraction of HFC is that its implementation cost is low. Fiber is "overlashed" onto the existing trunk plant. The in-place cable is broken into

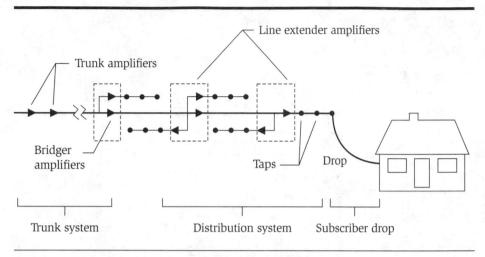

Figure 1.8 "Superdistribution" or express feeder plant.

segments and used for the small-scale cable systems. Some of the amplifiers are reversed in direction. Nothing is wasted.

1.7.2 Distribution Plant Improvements

If we now turn our attention to the distribution plant, we find that we must run higher signal levels to support the tapping of energy to serve drops to customers. The higher signal levels mean that we begin to reach into nonlinear areas of the amplifier's operating characteristic. Nonlinear distortion builds up. In addition, the taps are not perfectly impedance matched to the cable. Consequently, the signal is reflected between the taps, resulting in a smearing of the picture. This phenomenon is called *microreflection* for two reasons: the strength of the reflection is low, and the time delay of the reflection is short.

Superdistribution is an answer to these difficulties. In one form of its implementation, line extender amplifiers are structured to have up to three hybrid amplifier chips. One feeds the next line extender amplifier, one feeds half of the taps back to the previous line extender, and the third feeds half of the taps to the next line extender. The existing tapped feeder cable is cut in half between line extenders. New, untapped cable is overlashed to connect line extenders. The consequences are that the signal level between amplifiers is lower since that cable is not tapped. The signal level on the tapped runs is lower because they are shorter. Nonlinearities are reduced, and they do not build up as in the previous structure. Also, signal leakage may be less of a problem because of the lower signal levels. In addition, the number of taps in series in any cable is drastically reduced, thereby reducing the amount of microreflections experienced by any one subscriber. This technique effectively cures the ills of the distribution portion of the plant (see Figure 1.8). With fewer amplifiers in series, the constraints on their de-

sign and operation are reduced. Higher bandwidths become practical. Costs are reduced.

Costs are just one part of the business equation. Revenues are another important factor. The collection of revenues is possible only if there are not easy ways of getting the goods or services for free. The next section considers techniques for restricting access to just those who are willing to pay for the service.

1.8 Signal Security Systems

Signal theft is a "victimless crime" that has no physical transfer of property. Stealing the signal does not directly prevent its use to other viewers. Because of these factors, people who would never even think of shoplifting seem to have few problems with their consciences when it comes to helping themselves to cable service without paying. Unfortunately, signal theft is a serious issue. The members of the creative community that provides the programming viewers enjoy are entitled to compensation. The motivation to excel is destroyed when the monetary rewards are not available. The programmer, network operator, and cable operator as well as their investors, employees, and holders of debt all deserve compensation. Theft of service means that those who are honest will have to cover the costs avoided by those who steal. Signal security systems are the means of prevention of signal theft.

Means of securing services from unauthorized viewership of individual channels range from simple filtering schemes to remote-controlled converter/descramblers. The filtering method is a commonly used means of signal security and is the least expensive.

1.8.1 Trapping Systems

There are two types of filtering or trapping schemes: *positive trapping* and *negative trapping*. In the positive trapping method, an interfering jamming carrier is inserted into the video channel at the headend. If the customer subscribes to the secured service, a positive trap is installed at the customer's house to remove the interfering carrier. The positive trapping scheme is the least expensive means of securing a channel in systems where less than half of the basic customers subscribe to the premium service.

A drawback to positive trap technology is its defeatability by customers who obtain their own filters through theft, illegal purchase, or construction. Another drawback is the loss of resolution in the secured channel's video caused by the filter's effect in the center of the video passband. Preemphasis is added at the headend to correct for the filter's response, but loss of picture content in the 2–3 MHz region of the baseband video signal remains. New positive trap schemes take advantage of sharper surface acoustic wave (SAW) or crystal filter technology. This approach allows the interfering carriers to be positioned a few kilohertz away from the secured channel's visual carrier in contrast to 2.25–2.75 MHz

away with the conventional approach. Illegal construction of these filters is much more difficult. The new technology has the promise of achieving greater signal security and a higher-quality picture for the authorized customer, but at a higher price.

Negative trapping removes signals from the cable drop to the customer's home. The trap is needed for customers who do not subscribe. This is the least expensive means of securing a channel in systems where more than half of the basic customers subscribe to the premium service. The negative trap is ideal. There is no picture degradation of the secured channel because the trap is not in the line for customers who take the service. A drawback occurs for customers who do not subscribe to the secured service but want to view adjacent channels. These customers may find a slightly degraded picture on the adjacent channels because of the filter trapping out more than just the secured channel. This problem becomes more significant at higher frequencies, due to the higher Q (efficiency) required of the filter circuitry. Negative traps have a problem from a security standpoint. The customer merely has to remove the negative trap from the line to steal an unauthorized service. Maintaining signal security in negative trap systems depends upon ensuring that the traps remain in the drop lines.

1.8.2 Scrambling and Addressability

There are two classes of scrambling technologies: (1) RF synchronization suppression systems and (2) baseband scrambling systems.

The concept of addressability should be considered separately from the scrambling method. Nonaddressable converter/descramblers are programmed via internal jumpers or a semiconductor memory chip called a programmable read-only memory (PROM) to decode the authorized channels. These terminals' authorizations must be physically changed by the cable operator. Addressable converters are controlled by a computer-generated data signal originating at the headend and placed either in the vertical blanking interval (VBI) or in a separate RF carrier. This signal remotely configures the viewing capabilities of the converter. Impulse-pay-per-view (IPPV) technology is supported by addressable converter/descrambler systems.

1.8.3 RF Synchronization Suppression Systems

Converter-based scrambling systems that perform encoding and decoding of a secured channel in an RF format comprise the commonly used scrambling technology. There are two basic RF scrambling formats. The more common is known as *gated* or *pulsed synchronization suppression*. With this method, the horizontal synchronizing pulses (and with some manufacturers, the vertical synchronization pulses) are suppressed by 6 dB and/or 10 dB. This is done in the channel's video modulator at the IF frequency. The descrambling process in the converter/descrambler occurs at its channel output frequency. This is accomplished by restoring the RF carrier level to its original point during the horizontal synchroni-

zation period. Variations of this method pseudorandomly change the depth of suppression from 6 to 10 dB or only randomly perform suppression.

The older, and now less popular, format is known as *sine-wave synchronization suppression.* In this format, the scrambling effect is achieved by modulating the visual carrier signal with a sine wave. This causes the synchronization to be suppressed, as well as changing the characteristic content of the basic video information. This encoding process is performed at the IF frequency in the channel's video modulator at the headend. The decoding process in the converter/descrambler is accomplished by modulating the secured channel with a sine wave of the same amplitude and frequency as in the headend, but whose phase is reversed.

A phase-modulated RF scrambling technique based on precision matching of SAW filters constructed on the same substrate has been introduced. This low-cost system has extended operators' interest in RF scrambling techniques for use within addressable plants.

1.8.4 Baseband Scrambling Systems

Baseband converter/descrambler technology provides a more secure scrambling technology for delivering video services. The common encoding format is a combination of random or pseudorandom synchronization suppression and/or video inversion. More complex and effective schemes have been invented but have not found broad acceptance because of cost. Because the encoding and decoding are performed at baseband, these converter/descramblers are complex and expensive.

Maintenance of the system's video quality is an ongoing issue. The encoders are modified video processing amplifiers. They provide controls to uniquely adjust different facets of the video signal. The potential for setup error in the encoder, in addition to the tight tolerances that must be maintained in the decoders, has presented challenges to the cable operator.

A recent extension to baseband video scrambling involves the addition of digitized and encrypted audio to the video signal's vertical and horizontal synchronization pulse intervals. Though there are issues of stereo compatibility, providers of these products claim that the combination of video scrambling with audio encryption provides a significantly higher degree of security.

1.8.5 Off-Premises Systems

The off-premises approach is compatible with recent industry trends to become more consumer electronics friendly, and it removes security-sensitive electronics from the customer's home. This method controls the signals at the pole rather than at a decoder in the home. This increases consumer electronics compatibility since authorized signals are present in a descrambled format on the customer's drop. Customers with cable-compatible equipment can connect directly to the

cable drop without the need for converter/descramblers. This allows the use of all VCR and TV features.

Signal security and control in the off-premises devices take different forms. Although there were several attempts to take modified addressable converter/descramblers and enclose them on the pole, this approach was not successful. The system was costly and less consumer friendly than having the converter/descrambler in the home because it delivered only a single channel at a time to the customer's equipment.

Interdiction technology involves a scheme similar to that of positive trap technology. In this format, the pay television channels to be secured are transported through the cable plant without being scrambled. The security is generated on the pole at the subscriber module by adding interference carriers to the unauthorized channels. An electronic switch is incorporated, allowing all signals to be turned off. In addition to being consumer electronics friendly, this method of security does not degrade the picture quality on an authorized channel.

The whole purpose of a cable system is to deliver signals to the customer's premises for enjoyment on equipment owned by subscribers in the privacy of their homes. The next section considers the issues involved with connecting to consumer-owned equipment.

1.9 The Signal and the Customer's Equipment

The technology that makes inexpensive TVs, VCRs, and other consumer electronics possible has created a variety of features and functions that are confusing to the subscriber. Connecting this equipment together and to the cable system, and then using the most desirable of the features, can be challenging to most customers. Too often, the consumer does things that prevent the full utilization of equipment features or limit access to services.

1.9.1 Signal Splitting at the Customer's Premises

The common devices at the cable drop to the customer's home are grounding safety devices called ground blocks and a two-way signal splitter that sometimes has a built-in grounding terminal.

Some systems use ground blocks or two-way splitters that incorporate a high-pass filter. These filters are used in two-way plant to minimize RF ingress into the cable plant in the 5–40 MHz reverse (upstream) spectrum from noise sources in the home. These noise sources include appliances with motors and consumer electronics products with spurious emissions in these frequencies. These filters have a crossover frequency low enough not to cause group delay in the downstream video channels.

Splitters or ground blocks should have little effect on picture quality provided there is adequate signal to handle the splitter's loss. The signal strength may be

below specifications because of an excessively long drop or more activated cable outlets in the house than the cable design anticipated. To compensate, some systems use an ac-powered drop amplifier. These amplifiers can create problems—a reduced C/N or increased distortions.

Consumer electronics switching devices, designed to allow convenient control and routing of signals between customer products and cable systems' converters/descramblers, have built-in amplification stages to overcome the losses associated with the internal splitters. These amplifiers add distortions or noise. When cable systems were designed, consumer electronics switching devices were not taken into account because they did not exist.

Signal splitting in VCRs can be a problem. To compensate for recording S/N deficiencies, low-grade VCRs sometimes split the signal unequally between the bypass route and the VCR tuner. This gives the VCR a stronger signal than the TV receiver to improve VCR performance. In addition, this strategy reduces the quality of the signal routed to the TV. When it is compared with VCR playback, the disparity in performance is reduced.

1.9.2 Consumer Electronics Compatibility

In the early 1980s, a Joint Engineering Committee (JEC) was formed under the sponsorship of the Electronic Industries Association (EIA) and the NCTA. The purpose of the JEC was to bring together technical representatives from the cable and the consumer electronics industries to attempt to find ways to make cable service and consumer electronics products more compatible. It was soon recognized that the cable subscriber and the consumer electronics customer were one and the same. In order to better serve that customer, cooperation between the two industries was necessary.

The two industries had for decades evolved slowly, but without coordination. As technology began to accelerate, a serious divergence became apparent, and it was the consumer who ultimately suffered. The high-minded goal of improving the situation for the consumer ran into a few practical problems.

Problems on the consumer side stem from critical cost considerations. Additional expense is hard to recover by raising prices because of the extreme competitiveness of that industry. That competitiveness is a consequence of the ability to produce larger quantities of very reliable, high-quality products than the marketplace can easily absorb. This overcapacity makes prices almost impossible to sustain, much less raise. The nature of technology is such that it is impossible to contain it. Any technical advance quickly spreads throughout the industry, making differences between most brands difficult for the consumer to distinguish.

Problems on the cable side come from the need for signal security and the desire to experiment with new services. A prerequisite to offering a subscription service is the ability to deny that service to those who choose not to pay. Without payment, programming cannot be purchased, and those who create that

programming cannot be compensated. So signal security is critical to the delivery of subscription services. Technology makes possible experimentation in new services. Because these services are offered in set top terminals owned by the cable operator, the consumer is protected from financial loss if service fails.

Early on, it was clear that three things were needed: (1) a standardized channel plan versus frequency, (2) tuner specifications, and (3) a decoder interface. With these three elements, much of the difficulty could be resolved.

A major feature of cable television is its self-contained spectrum. Because this spectrum is isolated from the spectrum in the environment, the usual restrictions placed on broadcast television do not apply. The entire spectrum can be filled with television signals. Early cable practice utilized a variety of methods for labeling these channels. This diversity made it difficult to use a common labeling scheme for consumer electronics products with extended tuning range. A frequency plan was necessary to allow common nomenclature.

These cable signals can be very well controlled in their relative strength. This is in sharp contrast to broadcast signals. To prevent broadcast co-channel interference, channels are spaced. The VHF and UHF bands have gaps where others have claim on the spectrum. It is common for the broadcast tuner to attempt to tune a distant weak signal centered between nearly adjacent strong local signals. Thus the demands on tuners designed for broadcast and for cable are very different. It is difficult to make a tuner that performs well in both situations.

One serious difference in the requirements of tuners designed for cable and those designed for broadcast is the issue of direct pickup. A broadcast tuner needs minimal electromagnetic shielding to protect it from outside signals. The spectrum being tuned is occupied only by the broadcast signal. A cable tuner has another problem. Because the propagation speed through cable is about two-thirds that of free space, the broadcast signal arrives at the TV first. A few microseconds later, the cable signal arrives. If the tuner directly responds to the over-the-air signal, the two signals will be mixed, and interference will result. The solution to this problem is covered by a now-expired U.S. patent (3,333,198 issued July 25, 1967), the Mandell patent. The cable signal is selected by a tuner specifically designed for the purpose. It is well shielded and has the ability to discriminate against immediately adjacent channels. One constraint is relaxed. Signal levels across the tuned band differ only slightly in level. This tuner's output is converted to a single channel chosen to be one not used locally. Therefore, there is no off-air signal to be directly picked up by the TV and mixed with that delivered by the cable system. The problem of direct pickup is solved only because the output of the terminal is a single channel on a frequency not used locally.

The use of the set top terminal solves one problem but makes the TV's tuner useless. Those features that depend on the tuner are also made inoperable. In particular, the remote control that came with the TV or VCR is used only to turn those devices on and off. The set top's remote control changes channels. In the case of the VCR, the timer that changes channels is crippled because it cannot change the channels of the terminal. Taping different channels consecutively becomes impossible.

In order for a "cable-ready" TV or VCR to provide acceptable performance when connected to cable, it has to have a minimum amount of electromagnetic shielding to make it immune to direct pickup interference. Also, it must not radiate spurious signals up the cable. These signals can come from the local oscillator or from the harmonics of all the digital signals that provide on-screen displays and drive other features. These signals can cause interference to other receivers in the same residence or in adjacent residences if they are allowed to enter the cable. All these issues have been covered by a cable RF specification.

Cable's requirement for signal security caused the need for set top terminals to move beyond mere frequency conversion. Set top terminals took on the function of descrambling signals in the early 1980s. If a "cable-ready" TV or VCR were designed with adequate tuning range, adequate electromagnetic shielding against direct pickup, and the ability to handle a full spectrum of adjacent signals without overload, it might still require a set top terminal. This would duplicate the circuitry in the TV, provide slight additional signal degradation, and interfere with certain features built into TVs and VCRs. The solution to this problem is the decoder interface connector. This is a 26-pin connector on the back of TVs and VCRs that brings video and audio signals out of the consumer electronics product for external processing. The descrambled signals are returned to the consumer electronics product for display or recording. The main advantage of the decoder interface is that it makes scrambling nearly transparent to the user.

In 1992, Congress passed the Cable Act, which has far-reaching implications for cable system operation and economics. An amendment attached to the Cable Act requires the FCC to create rules with the intent of improving the compatibility between consumer electronics products and cable operations. These rules regulate the conditions under which scrambling can be employed and provide a technical definition for "cable-ready" products. That definition included the channel plan created by the JEC, specifications on tuner performance, and the inclusion of a decoder interface connector. Products that do not comply with the Commission's technical specifications cannot be sold as "cable ready" or "cable compatible," or under any other term that might mislead consumers.

As complex and difficult as the decoder interface has been, the digital television future presents even more interesting challenges. Digital television (DTV) takes advantage of the fact that we can now afford to put millions of digital transistors into consumer products to process signals. Additional millions of transistors can be included in memory elements. This tremendous processing power offers a wide variety of options for digital video compression and new interactive and multimedia services. The biggest unknowns center on the consumer. Which services will find acceptance? How much tolerance exists for the digital artifacts that this new signal processing brings?

The 1996 Telecommunications Act requires the FCC to establish rules for the availability of set top terminals at retail. As we move to subscriber ownership of the electronics that implement services, new issues become important. In the old paradigm, the service provider answered questions and supplied training in the use of the equipment. The equipment was repaired at no additional cost. If the

service failed to enjoy sustaining support, the cable operator absorbed the equipment losses. In the new world of subscriber ownership, the retailer and manufacturer of the equipment must stand ready to answer questions on usage and provide training and repair. The consumers' expectations must be adjusted to accept situations in which the equipment becomes no longer useful because the service failed or was replaced with an improved version. Cable operators will no longer be absorbing the losses since the equipment is no longer theirs.

An additional complication arises. If subscriber-owned set top terminals introduce interfering signals into the cable system, they will have to be disconnected. If the subscriber insists on using them, that subscriber will have to be disconnected. There will be charges for service calls to solve these problems. Consumers will be angry when the hardware they purchase cannot be used on the cable system because of interference it may cause.

As we read the trade press, we see a massive interest in set top terminals for interactive television, multimedia, the National Information Infrastructure (NII), Telco-provided video, and so on. These set top terminals will make it possible for consumers with older TVs and VCRs to try out new services and new service providers. They will allow purchasers of new, inexpensive TVs and VCRs to also participate in these new services.

The history of cable television is the quest for "more": more programming, more channels, more options, more spectrum. Digital video compression is the next quantum step in this pursuit.

1.10 Digital Video Compression

Digital video compression (DVC) has given massive capacity to the cable spectrum. DVC gained its first momentum in the pursuit of HDTV. The analog bandwidth of signals from an HDTV camera consists of something like 30 MHz each red, blue, and green. Nearly 100 MHz of analog signals would require more than a gigabit per second for straight digital transmission. The FCC, which sets federal regulations on spectrum usage, allows only 6 MHz! As a result, there is a need to remove a lot of the redundancy in the picture. The same technology can be used to squeeze multiple standard definition signals into the same 6 MHz. When DVC is used to put multiple current quality signals into 6 MHz, the result is called standard definition television (SDTV).

DVC was first used by consumers in the Direct Broadcast Satellite (DBS) environment. Both DirecTV and PrimeStar are using a version of the Moving Picture Experts Group (MPEG) standard. DVC compresses signals by removing redundancy in each still picture, by removing redundancy between the series of still pictures that make up a motion sequence, and by deleting those aspects of the image that have minimal visibility to the human visual system.

The reasons for the lower data rate with movies stems from two main factors. First, movies have a 24-frames-per-second rate, whereas video has a 30-frames-per-second rate. Although motion reproduction in video is better than in movies,

there is a 20% increase in data required for the same image quality. Second, movies can be processed iteratively. The compressor parameters can be adjusted on a scene-by-scene basis. Much more processing can be brought to bear since the computations do not have to be done "in real time."

The other factor that determines the packing density of movies in the 6-MHz television channel is the modulation method. The two main competitors are quadrature amplitude modulation (QAM) and VSB modulation. These come in various data speed capacities. The two most interesting for cable applications are 256-QAM or 16-VSB. Both are *double data rate* systems in that they can deliver data payloads of about double the data rate required to deliver one HDTV signal in 6 MHz. The Advanced Television Test Center (ATTC) completed tests of 16-VSB on cable in Charlotte, North Carolina. The 16-VSB system tested delivered 38.5 Mb/s in all locations. Some older cable systems were included in the tests. The longest amplifier cascade was 48! Fiber links and microwave amplitude modulated links (AMLs) were also included. 256-QAM delivers similar results. Thus it is possible to transmit 12 DVC movies in 6 MHz at 3 Mb/s each.

Current thinking is to allocate the lower end of the downstream cable spectrum to analog signals. This covers the 50 to 450 MHz (or 550 MHz) frequencies. A basic unscrambled tier of broadcast signals would be at the low end of the spectrum. At higher channel numbers, trapped pay services or analog scrambled services would be supplied. Above 450 or 550 MHz, digital signals would be provided to the band edge located at 750 MHz or 1 GHz. Thus, from 200 MHz (33 6-MHz channels) to 550 MHz (91 6-MHz channels) will be available. With 12 movies per 6 MHz, 396 to 1,092 movies may be offered simultaneously. Of course, digitized video signals (rather than movies) may also be offered by reducing the total number of programs.

The fundamental differences between science and technology are the economic factors and the emphasis on practical application. The main thrust of our concerns is technology, but the technology does not exist in a vacuum. An understanding of the basic economics of the cable business is essential to making appropriate technical decisions.

1.11 Cable Economics

It is important to understand the dimensions of the cable industry and some of its economic parameters in order to place the technology in context. The cable industry is a massive business that affects the lives of many citizens, yet the entire cable business is about the size of the smallest Regional Bell Operating Company (RBOC). In 1997, the U.S. cable industry operated almost 11,517 cable headends. These range in size from a small system with only a few hundred customers to the largest system, which serves more than a million subscribers. Figures from 1997 show that the cable industry employed 120,530 people and had annual revenues of about $30 billion from subscriber services and another $8 billion from advertising.[8]

The economics of cable television is governed by a number of factors, including the cost of construction, the cost of programming, the willingness of subscribers to pay for the service, and the operating costs of the system.

Cable system construction costs depend greatly on location. The geography, local laws, and local cost of labor are important factors. The two main forms of cable plant construction are aerial and buried. Aerial construction is mounted on utility poles or, less commonly, from house to house. The cost of aerial construction is roughly $30,000 per mile. Buried construction is at least double the cost of aerial. Since most systems have a blend of aerial and buried plant, a blended cost figure ranges from $40,000 to $50,000 per mile depending on the ratio of the two styles. Clearly, home density is a major factor in the total system's cost. The number of homes per mile varies depending on whether the system is urban, rural, or country. A very attractive density would be about 150 homes per mile. A good density is 100 homes per mile, whereas average density is about 70. Fifty homes per mile is about the minimum that's reasonable, and 35 is uninteresting. Average cable penetration is about 68% and is slowly climbing. With an average of 70 homes per mile and an average of 68% penetration, there are nearly 50 subscribers per mile. At that rate, the cable plant would cost more than $1,200 per subscriber for buried, half that for aerial, and somewhere in between for a blended configuration. Two-way cable adds about another $2,000 per mile to the construction costs.

Since cable headend costs are spread over a large number of subscribers, each subscriber's share of the headend for modest-size systems is less than $50. This means that the set top terminal is the second most expensive capital investment for the cable operator. Analog set top terminals cost from $120 to $170 depending on features. Digital set top terminals are introduced at costs between $300 and $450, again depending on features. The prices of digital set top terminals are expected to decrease substantially.

Figure 1.9 shows an approximate distribution of cable construction costs. The labor to install each component is included with the cost of that component. Hardware includes the mechanism for supporting the cable. In aerial construction, a strong, heavy wire called strand is mounted on the poles. A thinner wire, called lashing, is spirally wrapped around the cable and strand to support the cable. Strand also supports amplifiers and passive components. The passive devices split the signal and tap into the cable for distribution to the home. Electronics includes amplifiers and microwave links.

The market value of cable systems recently sold varies from $1,800 to a high of almost $2,500 per subscriber. These values are determined by both the current cash flow and the potential for increased penetration and new revenue streams. Cash flow ranges from about $150 to $210 per subscriber per year. The cable business is a capital-intensive business, which is currently experiencing dramatic increases in programming costs.

According to statistics gathered in 1997 by NCTA, subscribers paid about $26 per month for basic service and $8.29 per month for each optional pay service

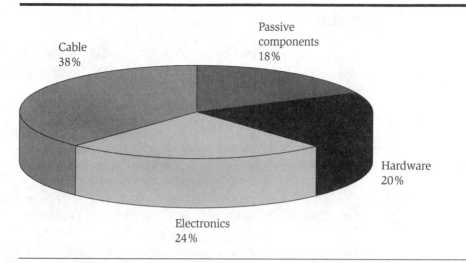

Figure 1.9 Distribution of cable plant costs.

such as HBO. Although not every basic subscriber takes an optional pay service, some take more than one. Average per month revenue is about $30.

1.12 Summary

In this chapter, we have presented a brief overview of cable television practice in the United States. Starting with a few comments on cable's history, we have reviewed the structure of the cable industry and have described the fundamental concept of spectrum reuse and the factors that make it possible. An overview of the design of cable systems builds a picture that helps visualize how the pieces fit together. Also, we discussed signal-quality goals and trade-offs. We introduced the hybrid fiber-coax architecture, which was the major force of change in the industry over the last decade. Signal security and the interface to the consumer's equipment bring the signal into the home and present it to those who subscribe while denying it to those who do not. Digital video compression is the major force of change in the industry's next decade. It facilitates the delivery of truly massive increases in programming. We finish the chapter with the reality factor of the engineer's world: economics. Attention to the economic aspects keeps the service affordable for the general public while rewarding the industry's investors and employees. The real world demands attention to economics.

With this brief overview, we have introduced the general concepts of cable television in the United States. The rest of the book will fill in the details.

It's time to roll up our sleeves and dive in!

Endnotes

1. *Cable Television Developments,* fall 1997, NCTA, Washington, D.C.

2. Private correspondence from Archer Taylor, The Strategis Group, Inc., Washington, D.C.

3. Walter S. Ciciora, *Cable Television in the United States, An Overview.* CableLabs, Louisville, CO, 1995.

4. Bronwen Lindsay Jones and James A. Turner, *Subjective Assessment of Cable Impairments on Television Picture Quality.* CableLabs, Louisville, CO, June 1992.

5. C. E. Dean, Measurements of the Effects of Interference in Television Reception. *Proceedings of the IRE,* Institute of Radio Engineers, New York, June 1960.

6. Method for the Subjective Assessment of the Quality of Television Pictures, Recommendation 500-4, *Recommendations and Reports of the CCIR,* Vol. XI, Part 1. CCIR, Dubrovnik, Croatia, 1986.

7. B. L. Jones and L. E. Marks, Picture Quality Assessment: A Comparison of Ratio and Ordinal Scales. *SMPTE J,* Vol. 94, No. 12, December 1985.

8. *Cable Television Developments,* fall 1997, NCTA, Washington, D.C.

Part Two
The Signals

Chapter 2
Analog Television

2.1 Introduction

One of the most successful technologies of our time is based upon the standard created by the National Television Systems Committee (NTSC). Like the U.S. Constitution, it seems to have anticipated much of its future and adaptively followed the evolution of its world. Starting with vacuum tubes and progressing to integrated circuits, analog television has survived and prospered. Though much of the current excitement centers on digital television, it is reasonable to assume that analog television will be with us for many more years. Analog television receivers will be found in the homes of the less well to do for many years while they wait for the prices of digital receivers to come down to affordable levels. Even for the well off, the digital receiver will serve the principal viewing room while analog receivers will be found throughout the rest of the residence. A major advantage of cable technology is sufficient bandwidth to serve both technologies well. Without cable's bandwidth, the transition from analog to digital would be either harsh and burdensome or simply impossible.

2.2 Baseband Monochrome Video and Monaural Audio

The technology of analog television is the foundation of the cable industry. An understanding of analog television is key to fully appreciating the rest of the subject matter of this book. Understanding the television signal is a basic requirement for designing systems to deliver it to subscribers.

2.2.1 Television Briefly

The Federal Communications Commission (FCC) adopted standards for commercial television broadcasting in the United States in 1941.[1] This television technology is an analog system in which the picture is transmitted in a vestigial sideband modulation format on the visual carrier, and the sound component is transmitted as frequency modulation on a separate sound carrier. In 1954, the National Television Systems Committee "compatibly" extended the NTSC system to include color information by increasing the utilization of the 6-MHz spectrum occupied by the television channel. The extension to color was compatible because it allowed continued use of essentially all the previously used receivers with little degradation of their performance. This compatibility was not perfect. Some receivers had noticeable degradation, whereas others performed quite acceptably. In 1976, the FCC reserved line 21 for the closed captioning system, and in 1983 a teletext service was authorized. In 1984, the NTSC system was again compatibly extended to include stereo sound. Most recently, in June 1996, the FCC authorized additional ancillary data services in analog television. These technologies will be considered in this chapter.

The NTSC television system is termed *analog* because the signals representing the picture and sound information can take on *any* value between the minimum and maximum values. There are an *infinite* number of possibilities for these

values. In the analog system, the strength of the transmitted signal is inversely related to the picture information, with black portions of the picture having the most transmitted power and white portions of the picture having the least transmitted power. This negative modulation of the video information was found to reduce the impact of Gaussian white noise on the displayed image. Periodic pulses are included at powers greater than those used to represent black areas in the picture. These pulses provide the timing information required to synchronize the transmitter and the receiver, allowing the picture to be painted correctly on the screen.

Cable television utilizes the same NTSC standard, permitting most consumers to directly connect their television receivers to the cable system and enjoy access to at least some of the signals without the requirement of any additional equipment. This means consumers can try cable service with no risk that newly acquired equipment will be useless if they discontinue the service. It also means that the basic service remains available to them even if they move to another location. Premium services may require additional equipment such as set top terminals.

Understanding television technology is aided by some appreciation of its history.

2.2.2 Scanning

A picture is made up of many points of light called picture elements or *pixels*. The information for each of these points of light must be derived from the object or scene to be televised, organized in a logical manner, conveyed to the receive site, and then displayed as a pleasing image. Since the number of points of light needed to make up a picture is huge, it is not practical to convey them each in their own transmission channel. A method of simplifying the process is required. Scanning is that simplifying process.[2]

The process of scanning was first used on a large scale to transmit still images for use in newspapers (see Figure 2.1). A photograph is placed on a rotating drum and a light-sensitive device is focused on it. As the drum rotates, the light-sensitive device follows a path across the image. With each revolution of the drum, the light-sensitive device is advanced down the length of the rotating drum. This creates a helical scan that follows the path of an imaginary coil spring across the photograph on the scanning drum. The electrical output of the light-sensitive device is proportional to the reflectance of the image, bright areas yielding a larger electrical signal than dark areas. At the receive location, the process is reversed. A light-sensitive film is attached to a rotating drum, and a light source scans the same helical path across the film. High electrical input results in a bright light and a full exposure or blackening of the film. Conversely, low electrical input yields a lower exposure or a more transparent area on the film. In this way, a negative image is produced. Standard photographic techniques yield a comprehensible image. It is important that the rotating drums at the transmit site and the receive site be synchronized. They must rotate at the same speed, and the start of scanning must occur simultaneously in both locations. Likewise, the rate

Figure 2.1 Scanning of photograph on rotating drum for transmission to receiver with light source and photosensitive film.

of advance of the helical path down the drum must be the same in both locations. If this synchronization does not occur, then geometric distortions are introduced in the reproduced image. Modest distortions involve stretching or compressing of the reproduced image as if it were printed on a rubber sheet and then the rubber sheet distorted. More destructive distortions are possible if the speeds are very different in the two locations.

Although scanning works well when the receive site includes a photosensitive film that can accept and retain the image, the problem becomes a bit more complex when the receiver is the human eye. The human eye exhibits a phenomenon termed *persistence of vision* that allows it to continue "seeing" a light stimulation for a brief time after the light is removed. The application of the scanning principle to direct human vision requires that the scanning process be executed repetitively and fast enough for the eye's persistence of vision to fuse all the scan lines into an image.

In 1883, the German engineer Paul Nipkow invented a scanning method consisting of a rotating mechanical disk with appropriately located holes arranged in a spiral (see Figure 2.2). As the disk rotates, only one hole is located on the image to be transmitted (see Figure 2.3). As one of the holes completes a left-to-right scan, another hole commences scanning from the left slightly below the path of the previous hole. When the bottom hole completes its scan, the scanning disk will have made a full revolution, and the top of the image is again scanned. Nipkow was granted a German patent (patent number 30105 applied for January 6, 1884) on this invention but did not actually build the system. In 1923, John Logie Baird of the United Kingdom built a crude system based on the Nipkow scanning disk. He used two disks, one at the sender location and the other a few feet away at the receiver location. The sender disk had a selenium light-sensitive cell behind it, and the receiver disk had a light source driven in proportion to the signal derived from the selenium cell. Once the rotating disks were synchronized, the holes in each disk would be at the same relative location on the image in both

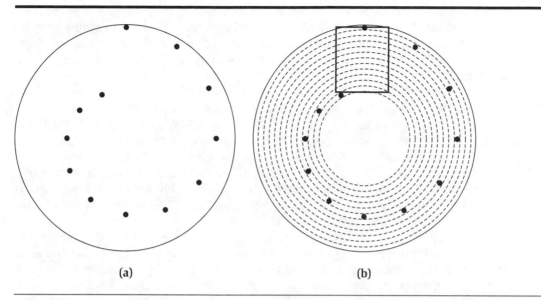

Figure 2.2 Nipkow disk operation. (a) Nipkow disk with a spiral of apertures. (b) Nipkow disk's scanning trajectory through the image plane.

Figure 2.3 Only one hole at a time is exposed to the image on the sending side and the light source on the receiving side.

the sender and the receiver. Then, instant by instant, the light sensed by the selenium cell at the sender would cause the lamp at the receiver to emit a proportional amount of light. The scanning repeated with each rotation of the disk. The eye's persistence of vision fuses the scanning disk's light output into a crude image.

"Mechanical television" was clumsy and produced poor pictures, but the principle of scanning was confirmed with these experiments. Television did not become practical until electronic display tubes and pickup tubes were introduced. In 1923, Vladimir Zworykin, a Russian immigrant who became a U.S. citizen, applied for a patent for a completely electronic television system. The interest in television was intense at that time, and many inventors were active in the field. Space does not allow a discussion of the numerous contributions and the interesting patent fights between inventors. This fascinating story is described elsewhere in detail and is recommended reading.[3]

In 1929, Zworykin applied for a patent that made the monochrome (black-and-white) picture tube practical (see Figure 2.4). An *electron gun* launches electrons past a control grid that regulates the strength of the electron beam. The beam then passes through focusing elements that concentrate the electron beam into a small spot on a glass faceplate that is covered with a fluorescent coating on its inside surface. When the electron beam strikes the fluorescent coating, light is emitted in proportion to the strength of the electron beam. Magnetic coils on the neck of the tube deflect the electron beam vertically and horizontally to implement scanning. The scanned rectangle of light on the face of the picture tube is called a *raster*. The picture tube is also called a *kinescope* because it was meant to display moving or kinetic images. Yet another name for the kinescope is the cathode ray tube (CRT). The part of the electron gun that emits the electrons is called a *cathode*. Early in the development of electron tubes, the flow of electrons emitted from the cathode was termed a *cathode ray* by pioneering experimenters. Another early pioneer, Philo T. Farnsworth, patented a device he called an Image Dissector in 1927. The Image Dissector embodied fundamental principles leading the way to practical television cameras.

In October 1931, Zworykin applied for a patent on a practical electronic television camera tube that he called the *iconoscope* (see Figure 2.5). As with the kinescope, an electron gun forms an electron beam which is scanned across a surface by magnetic deflection coils. In the case of the iconoscope, the surface is coated with a light-sensitive material. Light focused on the sensitive material causes it to develop an electrical charge in proportion to the strength of the light. When the electron beam strikes the light-sensitive surface, it interacts with the surface in proportion to the accumulated charge. A signal plate on the back side of the light-sensitive surface produces an electrical signal proportional to the light that fell on the surface. Again, many other inventors were working on similar devices, and more patent struggles occurred.

After a scan line is completed, it is necessary to move the electron beam back to the left side of the screen.[4] Ideally, this would be done instantaneously since no picture information is transmitted during this backward motion, termed *retrace*. Both the scanning drum of Figure 2.1 and the scanning disk of Figure 2.2 did

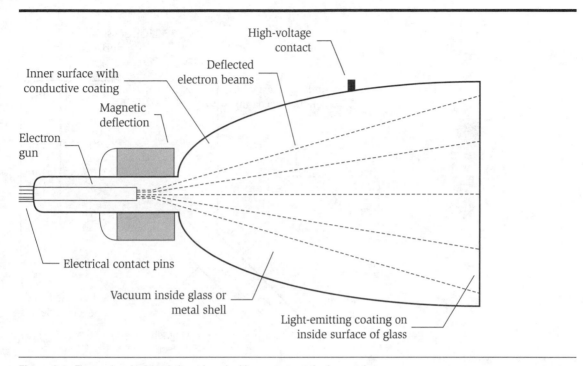

Figure 2.4 Electronic television is based on the kinescope or cathode ray tube.

not require retrace. However, electronic deflection is accomplished with magnetic fields, and magnetic fields store energy. Deflection circuits require time to change the magnitude and direction of the energy stored in the magnetic deflection fields. As a result, it is not possible to instantaneously move the electron beam from the right side of the screen to the left or from the bottom to the top. During the time the electron beam is in retrace, the electron beam must be turned off (a process called blanking) to prevent unintended stimulation of the phosphor screen and the resulting interfering light. The period during which the electron beam is turned off in horizontal retrace is called the horizontal blanking interval (HBI). When the electron beam reaches the bottom of the screen, it must be returned to the top of the screen to continue the process of making pictures. Just as in the horizontal case, the electron beam must be blanked to prevent disturbing light patterns on the screen. This period is called the vertical blanking interval (VBI). The VBI is much longer than the HBI. The combination of the two blanking intervals constitutes around 25% of the total time. This time cannot be used to convey analog pictures. The HBI and the VBI are the price paid for using cost-effective magnetic deflection. It is interesting to note that modern liquid crystal display (LCD) panels do not require retrace time.

Thomas Edison is credited with developing practical *moving pictures*. Moving pictures depend upon a rapid display of a succession of still pictures. In modern motion picture practice, 24 pictures are displayed per second. Earlier motion

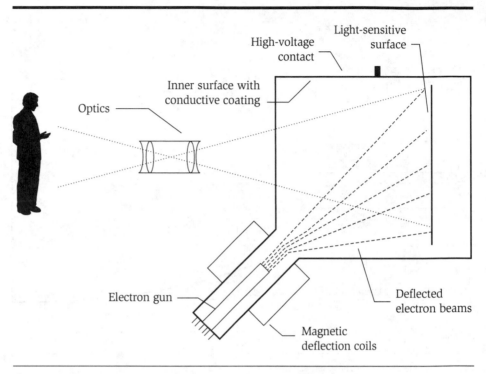

Figure 2.5 Electronic television is also based on Zworykin's iconoscope.

pictures and home movies used fewer pictures per second and consequently displayed an annoying flicker. Flicker was so characteristic of the movie experience that movies became known as "flicks." The human response to flicker is very complex and is related to the color, size, and brightness of the flickering image. The eye's ability to detect flicker is a nonlinear function of brightness. A small increase in the brightness of a flickering image results in a much higher degree of noticeable flicker. Early movie theaters were dark because the movie image must be dim to avoid objectionable flicker at 24 frames per second. To further reduce this flicker, each movie frame is projected twice, yielding a flicker rate of 48 frames per second. This has the positive effect of reducing flicker and the negative effect of reducing brightness because of the time the image is not projected between its two displays. Television developers (in the development of monochrome television) settled on 60 images per second in the United States and 50 images per second in Europe. Because of the technique of interlace described later, 60 (or 50 in Europe) half-images are displayed per second. That is, 30 complete pictures are presented per second in the United States, and 25 are presented in Europe. These choices were originally tied to the power line frequency causing any unfiltered power line "hum" that got into the picture to be stationary on the screen. (It was determined that slowly moving patterns of interference are much more objectionable than stationary patterns.) The eye's tolerance to flicker is

strongly related to the flicker rate. The minimum flicker frequency that is not objectionable is called the *flicker fusion frequency*. A flickering light source at frequencies above that value appears as continuous illumination. For example, fluorescent tubes operated at 60 Hz appear to be continuously "on" because they are operated above the flicker fusion frequency for the levels of brightness in normal situations. Because they extinguish at each zero crossing of voltage, they provide two pulses of light per power cycle. In the European systems with 50 pictures per second, the brightness level at which television flicker is not objectionable is only about 30% of the brightness of television systems with 60 pictures per second. Many viewers accustomed to 60-pictures-per-second television find that the European systems flicker excessively at the brightness levels normally used. However, the human perception system is remarkably adaptive. American visitors to Europe often report that though they first found TV pictures to have objectionable flicker, they become accustomed to it over time.

Many of the specifications for the NTSC television system are based on the desire to minimize the cost of receivers. A great many consumers were able to afford receivers even when vacuum tubes were the only means of signal processing. When we compare today's designs using integrated circuits having millions of transistors per "chip" with the designs constrained to only a dozen or so vacuum tubes, we can appreciate the cleverness of the early television engineers.

Modern circuit design has relaxed many of the earlier constraints. For example, it is now practical to keep the power line hum completely out of video circuits. Consequently, the need to relate image display rate to power line frequency is no longer significant. As further evidence of this, in Japan, 50-Hz power is used in the Osaka region, and 60-Hz power is used in the Tokyo area. The same television receivers and television system serves both markets.

The amount of information required to produce pictures of acceptable horizontal and vertical resolution at a 60-pictures-per-second repetition rate is huge. We will discuss video bandwidth in more detail later, but a few comments are important here. At current resolutions but with a repetition rate of 60 complete pictures per second (that is, without the interlace technique described later), twice the information would be included, and a baseband bandwidth of 8.4 MHz would be required for the video alone. As we shall see, the modulated bandwidth using simple double sideband modulation would be double the baseband bandwidth, or 16.8 MHz—almost three times the current broadcast bandwidth of 6 MHz. The consequence of these choices would be a serious reduction in the number of available television channels. Only about one-third of the current quantity would be possible. Clearly, ways of reducing the bandwidth had to be found. To put it another way, signal compression techniques were required that saved bandwidth but complemented the human visual system in a manner that did not impair picture quality.

One of the earliest signal compression techniques involved modifying the scanning system. Experience with motion pictures confirmed that 60 pictures per second is much more than is needed for acceptable rendition of motion. The movies use 24 complete images per second (each flashed on the screen twice).

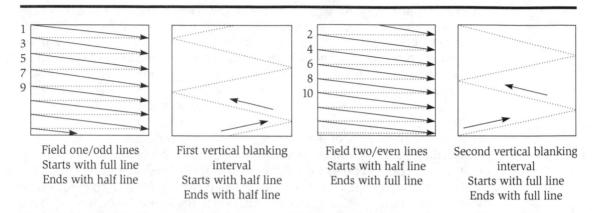

Field one/odd lines	First vertical blanking	Field two/even lines	Second vertical blanking
Starts with full line	interval	Starts with half line	interval
Ends with half line	Starts with half line	Ends with full line	Starts with full line
	Ends with half line		Ends with full line

Figure 2.6 Interlaced scanning to reduce flicker while conserving bandwidth.

Although not perfect, this provides commercially acceptable motion fidelity and yields very few complaints. The reason for 60 pictures per second in television is to create pictures bright enough to watch in daylight with the window drapes open. Since the picture tube technology could produce bright images, the television system itself should not be limiting. The solution to this problem is called *interlaced scanning*. With interlaced scanning, half of the scan lines are displayed in one "field" and the other half of the scan lines in another "field." Two fields make up one "frame." The frame is the complete picture and is transmitted at a rate of 30 images per second. This is 25% more than the 24 frames per second used in movies. So the motion rendition of video is better than movies. Yet the flicker frequency has been doubled to 60 fields per second. The eye fuses the lines from the two fields, and a bright, essentially flicker-free, image results that requires half the bandwidth of a system without interlace. Instead of 8.4 MHz of baseband bandwidth, the interlaced scan system takes up just 4.2 MHz of baseband bandwidth.

Interlaced scanning is illustrated in Figure 2.6. There are a total of 525 scan lines in the raster, yielding 262.5 lines in each field. The odd fields start with a whole line and end with a half line, whereas the even fields start with a half line and end with a whole line.

Even though interlace scanning works well with most "natural" video (that is, that which comes from a camera), interlace scanning in the display can produce inferior results with "synthetic video." Synthetic video is electronically generated and avoids some of the averaging phenomena of practical cameras. For example, the electron beam in a television camera has a distribution of charge of roughly a Gaussian shape. As a result, it accepts and averages light from adjacent scan lines. This averaging causes a smoothing of the image. Synthetic video, especially when producing text or graphics, sharply defines the images of each line. This results in a brightly illuminated edge that is presented at 30 images per second, well below the flicker fusion frequency. This is unacceptable in computer

displays that are bright and utilized at very close viewing range. As a consequence, nearly all computer displays do not use interlaced scanning. They use a simple top-to-bottom "progressive scan" technique. Since computer displays are driven by locally generated video, no transmission path is involved. The higher bandwidth requirement of a progressive scan raster is therefore not a problem.

2.2.3 Synchronization

As previously described, there are two fields in each frame of video.[5] Frames have 525 scan lines and repeat 30 times a second. When 1/30 of a second is divided into 525 equal parts, this results in 63.5 microseconds being allocated per horizontal scan line. The reciprocal of this yields the monochrome horizontal line frequency, which is 15,750 Hz. Since fields occur at the rate of about 60 per second, a field lasts 16.67 milliseconds. Every 63.5 microseconds, a horizontal synchronization pulse is required. Every 16.67 milliseconds, a vertical synchronization pulse is required. The horizontal synchronization pulses coordinate the side-to-side placement of images on the screen, whereas the vertical synchronization pulses coordinate the up-and-down placement of the picture.

Note that though the original television scan rate was 60 fields per second, this was slightly reduced to 59.94 when color was introduced into the standard. As a consequence, the horizontal scanning frequency became 15,734.264 ±0.44 Hz in color television. A more in-depth discussion of this will be found in the description of color television (see Section 2.4.1).

The designers of the NTSC system decided that the best way to separate synchronization (sync) signals from the composite signal was to discriminate on the basis of signal amplitude. They allocated the voltage from 12.5% of maximum to 75% of maximum to carry video signals. About 25% of the signal voltage is devoted to the sync pulses. The signal is blanked about 16.5% of the horizontal line time and 7% of the vertical field time. Thus the transmission of sync information takes up about 25% of the voltage of the signal and about 23.5% of the time. Synchronization is critical to the display of a useful picture, and adequate signal resources have to be assigned to ensure the task is accomplished.

Figure 2.7(a) displays an example of a signal that might come from a television camera. At 75% of the maximum voltage, a blanking pulse is inserted, timed to occur during the retrace time. This is shown in Figure 2.7(b). The blanking pulse is nominally 10.9 microseconds or 16.5% of the horizontal time of 63.6 microseconds. Active video occurs in this time difference of nominally 52.6 microseconds. The blanking pulse is at the video signal's black level, keeping the picture tube from emitting light during the blanking interval. The sync pulse is installed on top of the blanking pulse and is at a level referred to as "blacker than black" (see Figure 2.7(c)). The sync pulse is nominally 4.7 microseconds long.

Because video does not reach the blanking voltage levels, circuits can be designed that trigger between the blanking level and the top of the sync pulse and yield a stream of two level pulses containing only the sync information. In this way, sync pulses can be separated from video information. Figure 2.8 shows the

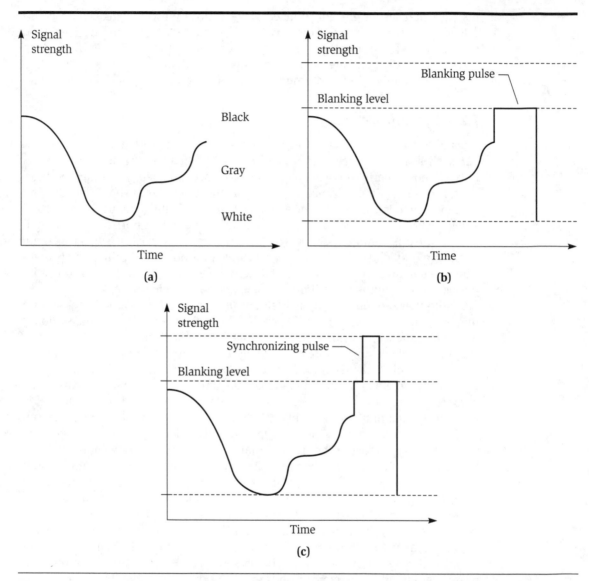

Figure 2.7 Construction of a video signal. (a) Video signal from television camera. (b) Blanking pulse added to avoid diagonal retrace lines. (c) Horizontal synchronization pulse added to facilitate left-to-right timing.

relative levels of black, blanking, and sync pulses in the modulated envelope of the signal.

Figure 2.9 shows the horizontal sync pulses and video of Figure 2.8 plus additional sync pulses for the vertical blanking interval. Each field's VBI is 21 lines long, leaving 483 lines for active video. The first nine lines of the VBI are devoted to pulses associated with the vertical synchronization process. The fourth, fifth,

Figure 2.8 Composite monochrome television signal levels (not to scale).

Figure 2.9 Vertical and horizontal synchronizing pulses of field one and field two. Note locations of half lines (not to scale).

and sixth lines in each field contain the vertical sync pulse. This pulse is serrated, creating pulse edges properly located for the horizontal synchronization to be maintained during the vertical retrace. The vertical sync pulses are called *broad pulses* in comparison to the narrow horizontal sync pulses.

Simple resistor–capacitor filter circuits can be used to separate the vertical pulses from the horizontal pulses.[6] In the lower portion of Figure 2.10(a), the circuit consisting of a series resistor R1 connected to a capacitor C1 to ground will accept charge and bleed it off to ground via the input impedance of the following

Figure 2.10 Receiver synchronization. (a) Sync separation and deflection ramp generation. (b) Vertical synchronization
pulse generation.

circuits. The narrow horizontal pulses will not build up a significant voltage be-
cause it bleeds off during the relatively long time between horizontal sync pulses.
However, the broad pulses will build up significant charge. This is shown in
more detail in Figure 2.10(b). Those who have studied such circuits will recog-
nize this configuration as a simple low-pass filter, also called an *integrator*. A
properly adjusted level-sensitive circuit following the resistor–capacitor network
will provide a vertical timing pulse. This pulse is used to synchronize a vertical
deflection oscillator, which in turn, drives the vertical deflection magnetic coils
on the neck of the picture tube. Six "equalizing pulses" are inserted before and af-
ter the broad pulses to equalize the charge buildup on the capacitor, preserving
the precision of the vertical timing. This is necessary because one field ends with
a half line and starts with a whole line, whereas the other field has the opposite
configuration. Without this precaution, the timing between the two fields will be
upset, and the rasters they produce will not be evenly interlaced. The resulting
distortion is called *pairing* because adjacent lines from both fields pair up and

blur each other's detail while leaving dark lines between them. Although modern digital sync separator circuits do not need the equalizing pulses, the original receivers using vacuum tubes had to rely on tailoring the signal to allow simple circuits.

In the upper portion of Figure 2.10(a), a series capacitor C2 with a resistor R2 to ground will convert the edges of the pulses into positive and negative going spikes that can be detected by appropriate level-sensing circuits. Note that the serrations in the broad pulses allow a continuous flow of uninterrupted pulses at the correct timing for horizontal synchronization. When these pulses occur halfway through a horizontal line, the horizontal circuits ignore them. They are effective only when they are near the correct time. Those who have studied such circuits will recognize this configuration as a simple high-pass filter, also called a *differentiator.*

With the allocation of significant signal resources (about 25% of the signal time and about 25% of the voltage range), rugged synchronization of the video signal has been achieved.

2.2.4 Spectrum of Scanned Images

It is well understood that any repetitive signal can be decomposed into a sum of sine and cosine waves of appropriate amplitude, frequency, and phase.[7] There are at least three ways we know this: (1) mathematical proof, (2) geometric construction, and (3) measurement with spectrum analyzers. The most rigorous method is the mathematical. Equations (2.1), (2.2), and (2.3) display the familiar Fourier series equations.

$$f(t) = a_0 + \sum_{n=1}^{\infty} (a_n \cos n\omega_0 t + b_n \sin n\omega_0 t) \quad (t_0 < t < t_0 + T) \qquad (2.1)$$

Equation (2.1) is the trigonometric Fourier series representation of $f(t)$ over an interval $(t_0, t_0 + T)$. The various constants a_n and b_n are given by

$$a_n = \frac{\int_{t_0}^{(t_0+T)} f(t) \cos n\omega_0 t \, dt}{\int_{t_0}^{(t_0+T)} \cos^2 n\omega_0 t \, dt} \qquad (2.2)$$

and

$$b_n = \frac{\int_{t_0}^{(t_0+T)} f(t) \sin n\omega_0 t \, dt}{\int_{t_0}^{(t_0+T)} \sin^2 n\omega_0 t \, dt} \qquad (2.3)$$

We could derive these equations and apply them to scanned images, but that would take up the rest of this book and would duplicate what is already available in many other places. Alternatively, we could use readily available math computer programs to perform this kind of analysis.

The second way in which we know that any repetitive signal can be decomposed into a sum of sine and cosine waves of appropriate amplitude, frequency, and phase is by geometric construction. Although this is not a formal proof, it is a plausibility argument that, when coupled with the knowledge that our mathematical colleagues have rigorously proved, should be sufficient for most engineering purposes. Figure 2.11(a) is a graphical construction showing how a square wave can be assembled out of a series of sine waves of odd harmonics, the amplitude of each being proportional to the reciprocal of the harmonic number. The first frame of Figure 2.11(a) shows the fundamental sine wave that is of the same frequency as the square wave. The second frame shows the third harmonic of the fundamental at one-third the amplitude and its contribution to the final result. The next three frames of Figure 2.11(a) show the fifth, seventh, and ninth harmonics at one-fifth, one-seventh, and one-ninth the amplitude of the fundamental and their contributions to the final result. It can be appreciated by inspection of this graphical exercise that adding more terms to the series will continuously improve the representation of the square wave. This kind of graphical analysis used to be done tediously by hand but now can be done to arbitrary precision with computer graphics programs. A personal computer spreadsheet program has powerful graphics capability that can be used in these exercises. It has been used to construct Figure 2.11(a). We invite you to try this.

Finally, the modern spectrum analyzer allows us to easily study any repetitive waveform. A spectrum analyzer can be considered to be a very narrow band filter that is tunable across a range of frequencies. The strength of the output of the filter is measured and plotted against frequency. Ideally, the filter would pass just one frequency. Then we would have a measure of the frequency content of the waveform across the spectrum. In practice, the spectrum analyzer's filter is not infinitely narrow. A band of frequencies are averaged and then graphed on the output device.

Collections of sine and cosine waves are graphed as vertical-pointing arrows whose length is proportional to the strength (amplitude) of the sine and cosine waves and whose horizontal position is proportional to the frequency of the wave. These graphs are called *spectrum plots*. Figure 2.11(b) shows the spectrum of the bottom graph of the square wave development in Figure 2.11(a). This is similar to what would be observed on the screen of a spectrum analyzer.

A paper published by Mertz and Gray in the *Bell System Technical Journal* in July 1934 titled "Theory of Scanning and Its Relation to the Transmitted Signal in Telephotography and Television" announced a very important result. Their analysis showed that the periodic scanning of an image yields a spectrum with energy clustered around the horizontal scanning frequency with large gaps in between. This pioneering work was done long before the availability of modern computers. Donald Fink did follow-up work and included the results in his 1957 *Television*

Figure 2.11 Square wave composition in time and frequency. (a) Frequency components of a square wave, time-domain plotted with a personal computer spreadsheet program. (b) Frequency components of a square wave.

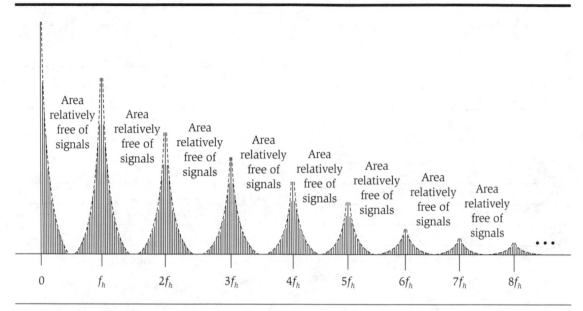

Figure 2.12 Video spectral energy clusters around harmonics of horizontal scan frequency, f_h, all the way up to the upper band edge.

Engineering Handbook.[8] His work clearly showed that the picture's energy is clustered around harmonics of the horizontal scanning frequency. Figure 2.12 shows these principles in a more schematic form that better illustrates the point.

2.2.5 Television Audio

Sound transmission in television signals is limited to 15 kHz because of the presence of strong spurious signals related to the monochrome horizontal scanning frequency of 15,750 Hz. For many years, audio in television was underappreciated. Only in the last decade has it received the attention it deserves as a fundamental part of the viewing experience. Perception experiments with human subjects at the Media Laboratory of the Massachusetts Institute of Technology have demonstrated that subjects shown two identical video presentations with differing accompanying sound qualities believed that the video with the better sound also had better video quality.

2.3 Modulation

Modulation is the process by which an information-bearing signal is impressed upon a carrier signal to facilitate the carriage of multiple signals and to allow practical and efficient transmission.

Baseband frequencies are those from 0 Hz to the maximum frequency content of the signals that normally come out of a television camera or a microphone. When these signals are conveyed by wire, only one can be carried since a second signal, occupying the same frequencies, would interfere. The most important application for modulation in cable communications is to allow multiple baseband signals to occupy different frequency bands in the cable. The modulation process is applied to each of several baseband signals, putting each baseband signal in a different band of higher frequencies. The separately modulated signals are then added together to provide one composite signal for transmission. This process is called frequency division multiplexing (FDM) because it uses multiple frequencies to divide up the signal space and carry multiple signals simultaneously. FDM gives rise to the familiar concept of "channels." Each channel is a band of frequencies associated with one video and audio signal.

There are several ways of approaching the subject of amplitude modulation. A phenomenological approach describes the physical processes and follows the historical path by which amplitude modulation was invented and deployed. A mathematical approach starts from first principles and develops relatively simple equations describing double sideband suppressed carrier modulation, synchronous detection, envelope detection, and single sideband modulation. This approach builds a firm foundation upon which to consider the addition of compatible color to the monochrome television system. Additionally, Chapter 7 explains the modulation process in terms of vector diagrams called phasors. It is best to become comfortable with all three methods of describing amplitude modulation since all three are commonly used.

2.3.1 A Phenomenological Approach to Amplitude Modulation

A television camera produces signals that are proportional to the strength of the light in the image and includes synchronization pulses. Figure 2.13(a) shows the waveform of such a signal. This is an information-bearing baseband signal because the frequencies of this signal extend from zero to the maximum produced by the television camera. The zero frequency component represents the average brightness of the image. The highest frequency conveys the finest detail of the picture. The white areas of the picture are near zero volts, whereas the black areas are of higher voltage. The highest voltages are reserved for the synchronization pulses.

A natural method of conveying the television camera signals would be to modulate the strength of a carrier wave in proportion to the amplitude of the original baseband signal. Figure 2.13(b) shows such a carrier wave. In Figure 2.13(c), the strength of the carrier wave of Figure 2.13(b) is modulated in proportion to the camera signal of Figure 2.13(a).[9] The outline of the maximum excursions of the carrier in Figure 2.13(c) is termed the *envelope* and is seen to be a replica of the original baseband information-bearing signal coming from the television camera and depicted in Figure 2.13(a). This process can be repeated for multiple baseband signals that are modulated onto multiple carriers of different frequencies. All the modulated carriers are added together for transmission.

(a)

(b)

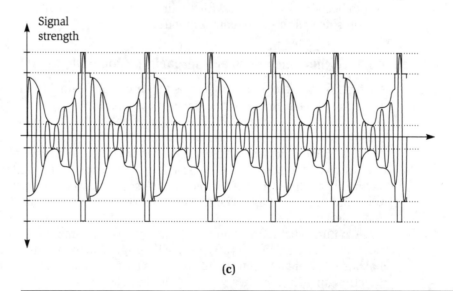

(c)

Figure 2.13 Television modulation. (a) Composite monochrome television signal (not to scale). (b) Carrier wave (shown at much lower frequency for convenience of illustration). (c) Carrier wave (shown at much lower frequency for convenience of illustration) modulated with video signal.

Figure 2.14 Envelope detection with a diode and low-pass filter.

At the receive site, the goal is to separate the signal from one television camera and use it in one television set. This can be done first with a tuned circuit that passes only the desired modulated carrier (actual receivers implement a more sophisticated version of this method based on the superheterodyne receiver principle). The next task is to recover the baseband signal. A simple diode will pass only the positive going portion of the modulated signal, as shown in Figure 2.14. The envelope of the signal is now clearly evident. The waveform after the diode differs from the original baseband signal in that it has half sine waves of the carrier. These are high-frequency components that can be simply filtered out using a low-pass filter (LPF), as shown at the right of the figure. Unfortunately, the phenomenological approach can teach little more.

2.3.2 A Mathematical Approach to Amplitude Modulation

The mathematical approach to amplitude modulation is the most instructive, and leads to the most complete understanding, of the details. Modulation can be explained using the properties of multiplied trigonometric functions. The most important of these relations, called *trigonometric identities,* are proved in courses on trigonometry. The following trigonometric identities are the starting point of our discussion:

$$\cos(A) \cdot \cos(B) = \frac{1}{2}\cos(A - B) + \frac{1}{2}\cos(A + B) \tag{2.4}$$

$$\sin(A) \cdot \sin(B) = \frac{1}{2}\cos(A - B) - \frac{1}{2}\cos(A + B) \tag{2.5}$$

$$\sin(A) \cdot \cos(B) = \frac{1}{2}\sin(A - B) + \frac{1}{2}\sin(A + B) \tag{2.6}$$

If you wish a plausibility argument for these relations rather than the formal proof found in trigonometry, you can substitute in numbers using a hand calculator or, more conveniently, use a spreadsheet program on a personal computer.

2.3.3 Double Sideband Modulation

To put the trigonometric functions to work in the study of amplitude modulation, they must be converted to functions of time. Let $A = 2\pi f_1 t$ and $B = 2\pi f_2 t$ in Equations (2.4), (2.5), and (2.6), where f_1 and f_2 are frequencies in Hz. We need the 2π in front of the f symbol because the argument of a sine or cosine function must be an angle where the angle of the sine or cosine function is specified in mathematical radians. There are 2π radians (or 360°) in a complete cycle of a sine wave or cosine wave. When the time variable, t, increases an amount equal to the period of the frequency of the cosine wave, the periodic function will have completed one cycle. Making these substitutions in Equations (2.4) through (2.6) yields

$$\cos(2\pi f_1 t) \cdot \cos(2\pi f_2 t) = \frac{1}{2}\cos(2\pi f_1 - 2\pi f_2)t + \frac{1}{2}\cos(2\pi f_1 + 2\pi f_2)t$$

$$\cos(2\pi f_1 t) \cdot \cos(2\pi f_2 t) = \frac{1}{2}\cos 2\pi(f_1 - f_2)t + \frac{1}{2}\cos 2\pi(f_1 + f_2)t \tag{2.7}$$

and similarly

$$\sin(2\pi f_1 t) \cdot \sin(2\pi f_2 t) = \frac{1}{2}\cos 2\pi(f_1 - f_2)t - \frac{1}{2}\cos 2\pi(f_1 + f_1)t \tag{2.8}$$

$$\sin(2\pi f_1 t) \cdot \cos(2\pi f_2 t) = \frac{1}{2}\sin 2\pi(f_1 - f_2)t + \frac{1}{2}\sin 2\pi(f_1 + f_2)t \tag{2.9}$$

These very basic equations tell us that if (1) two cosines or (2) two sines or (3) one sine and one cosine are multiplied together, the product will contain two new cosines (or sines), one of which will be at the sum of the original frequencies and the other at the difference of the original frequencies. If f_1 is much greater than f_2, then we have the spectrum of Figure 2.15(a) for the two original frequencies, and from Equation (2.7) the spectrum of Figure 2.15(b) for the multiplication of the two cosines at differing frequencies. In this example, f_2 is the baseband information-bearing signal, and f_1 is the *carrier frequency.* It is called the carrier because its function is to carry the information through the frequency spectrum. Figure 2.15(b) shows why this is termed double sideband modulation. Two mirror images, $(f_1 + f_2)$ and $(f_1 - f_2)$, of the information signal appear on opposite sides of the carrier frequency, f_1. This may more completely be termed *double*

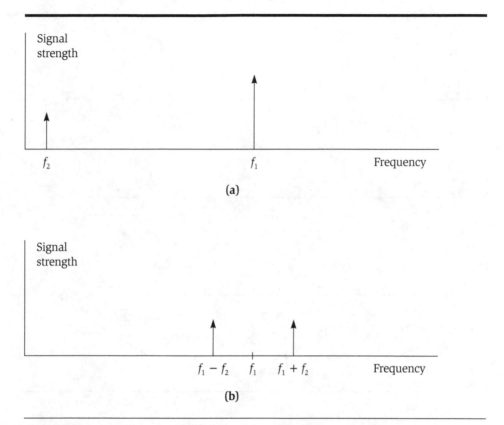

Figure 2.15 Modulation spectra. (a) Carrier and baseband frequency spectral graphs. (b) Double sideband suppressed carrier modulation of cosine wave f_2 onto carrier f_1.

sideband suppressed carrier modulation since the carrier frequency f_1 is not present in the output spectrum of Figure 2.15(b).

We can see now that if we had several situations like Figure 2.15(a) with all the baseband information-bearing signals in the same frequency band but all the carrier signals at substantially different frequencies, we could create a spectrum containing all the baseband signals on different carrier frequencies. Figures 2.16(a) and (b) demonstrate FDM of the information-bearing signals f_2, f_4, and f_6 onto the carrier signals f_1, f_3, and f_5. Three simultaneous channels of communication have been created in this example where baseband techniques would have permitted only one without mutual interference.

2.3.4 Synchronous Detection

The process of extracting the baseband information-bearing signal from the modulated double sideband signal can be explored using simple trigonometric identities.

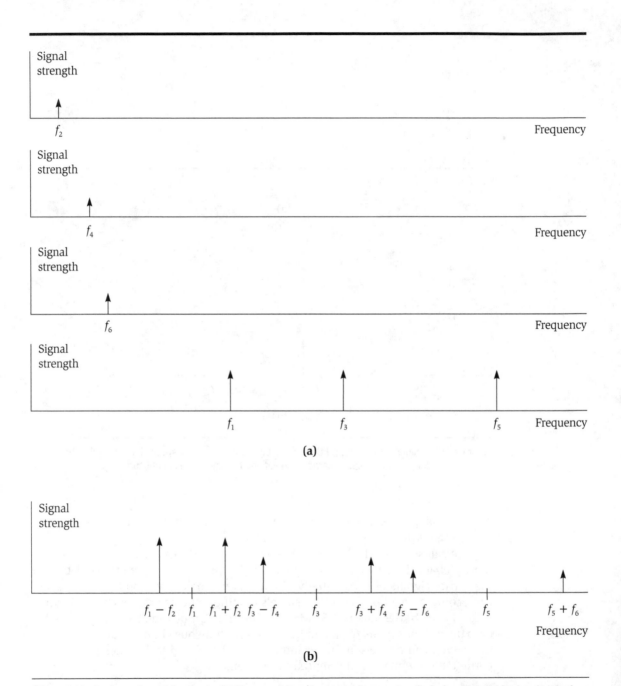

Figure 2.16 Carriage of multiple signals. (a) Spectra of three baseband frequencies, f_2, f_4, f_6, and three carrier signals, f_1, f_2, and f_5. (b) Double sideband suppressed carrier modulation of three cosine wave signals onto three carriers, demonstrating frequency division multiplexing.

It is simpler to work with the basic trigonometric equations with letter angles first and then substitute in the $2\pi ft$ terms later. If we multiply Equation (2.4) by $\cos(C)$ and carry out the terms and again apply the trigonometric identity of Equation (2.4), we get

$$\cos(A)\cdot\cos(B)\cdot\cos(C) = \frac{1}{2}\cos(A-B)\cdot\cos(C) + \frac{1}{2}\cos(A+B)\cdot\cos(C)$$

$$= \frac{1}{2}\left[\frac{1}{2}\cos(A-B-C) + \frac{1}{2}\cos(A-B+C)\right.$$

$$\left. + \frac{1}{2}\cos(A+B-C) + \frac{1}{2}\cos(A+B+C)\right] \qquad (2.10)$$

Now let $C = A$.

$$\cos(A)\cdot\cos(B)\cdot\cos(C) = \cos^2(A)\cdot\cos(B)$$

$$= \frac{1}{2}\left[\frac{1}{2}\cos(A-B-A) + \frac{1}{2}\cos(A-B+A)\right.$$

$$\left. + \frac{1}{2}\cos(A+B-A) + \frac{1}{2}\cos(A+B+A)\right]$$

$$= \frac{1}{4}[\cos(-B) + \cos(2A-B) + \cos(B) + \cos(2A+B)] \quad (2.11)$$

But the cosine function is symmetric about zero, so

$$\cos(-B) = \cos(B)$$

$$= \frac{1}{4}[\cos(B) + \cos(2A-B) + \cos(B) + \cos(2A+B)]$$

$$= \frac{1}{4}[2\cos(B) + \cos(2A-B) + \cos(2A+B)] \qquad (2.12)$$

$$\cos^2(A)\cdot\cos(B) = \left[\frac{1}{2}\cos(B) + \frac{1}{4}(\cos(2A-B) + \cos(2A+B))\right] \qquad (2.13)$$

Now let $C = A = 2\pi f_1 t$, and $B = 2\pi f_2 t$. Then we have

$$= \cos(2\pi f_1 t)\cdot[\cos(2\pi f_1 t)\cdot\cos(2\pi f_2 t)] = \cos^2(2\pi f_1 t)\cdot\cos(2\pi f_2 t)$$

$$= \left[\frac{1}{2}\cos(2\pi f_2 t) + \frac{1}{4}(\cos(2\{2\pi f_1\} - 2\pi f_2)t + \cos(2\{2\pi f_1\} + 2\pi f_2)t\right]$$

$$= \left[\frac{1}{2}\cos(2\pi f_2 t) + \frac{1}{4}(\cos 2\pi(2f_1 - f_2) + \cos 2\pi(2f_1 + f_2)t)\right] \qquad (2.14)$$

(a)

(b)

(c)

Figure 2.17 Signal detection. (a) Synchronous detector output before filtering. (b) Low-pass filtering to recover baseband signal after synchronous detection. (c) Recovered baseband signal.

This tells us that we have the original information-bearing baseband frequency back as f_2 and a double sideband pair at twice f_1. This is illustrated in Figure 2.17(a). If we now introduce a low-pass filter, as in Figure 2.17(b), only the low-frequency f_2 term will pass, and the high-frequency terms centered on twice f_1 will be blocked, as seen in Figure 2.17(c).

Recall that one of the important objectives of modulation is to move multiple sets of baseband information to different frequency bands, allowing them to be

added together to form one composite signal for transmission. The detection process described here has selected one of the frequency bands centered on f_1 and extracted its baseband information, f_2.

This process is termed *synchronous detection* because the multiplying signal represented by $\cos(C)$ is exactly the same as the original carrier frequency represented as $\cos(A)$. That is, $\cos(C)$ and $\cos(A)$ are in synchronism.

It can be seen that if, instead of just the simple situation of Figure 2.17(a), we had the situation of Figure 2.16, the double sidebands around f_3 and f_5 would be moved to other parts of the spectrum, but none of the resulting components of these signals would fall in the LPF passband. Only the desired signal at f_1 would be selected.

Now with the understanding of these fundamentals, we can consider the more complex and realistic case of Figure 2.18. Figure 2.18(a) shows the situation of a broadcast transmitter fed with information-bearing baseband signal $f(t)$ and a carrier $\cos \omega_c t$. The substitution $\omega_c t = 2\pi f_c$ is made to simplify the notation. The symbol ω_c is called the *radian frequency* because it absorbs the 2π that we have been obliged to laboriously carry along with the frequency f_c. Figure 2.18(b) shows the voltage versus time graph of $f(t)$, and Figure 2.18(c) shows the spectrum that would result from either applying the Fourier series Equations (2.1), (2.2), and (2.3) or simply passing the signal of Figure 2.18(b) through a spectrum analyzer. Figure 2.18(d) shows the voltage waveform resulting from the multiplication of the information-bearing baseband signal $f(t)$ times the carrier signal $\cos \omega_c t$. Note that the high-frequency carrier is contained within an envelope (shown by the dotted line in Figure 2.18(d)) that bears some resemblance to the original information-bearing signal in Figure 2.18(b). Closer inspection reveals that the envelope is the mathematical "absolute value" of $f(t)$ and the carrier changes phase at the zero crossings of the envelope. Figure 2.18(e) is the spectrum of the modulated double sideband suppressed carrier signal emitted by the transmitter.

Figure 2.18(f) illustrates a receiver using a synchronous detector that multiplies the received signal by a cosine wave of the same frequency as was used at the transmitter. The multiplier is followed by a low-pass filter. The result is shown in Figure 2.18(g), where the original information-bearing baseband signal is restored.

A mathematical complication is shown in Figure 2.19. The mathematical fiction of "negative frequencies" is introduced. This simply makes the math work out well. In real life, there are no "negative frequencies." (In spite of this, a spectrum analyzer will usually allow the demonstration of negative frequencies.)

Synchronous detectors are implemented in two main ways. The first uses a high-performance phase locked loop to acquire a cosine wave that is identical in frequency and phase to the cosine wave used to generate the signal at the point of origination. The second method is called *quasi-synchronous* since it takes the received signal, strips off any amplitude variations, and amplifies it to very high levels. There are practical advantages and disadvantages to both methods.

A further benefit of modulation is apparent to circuit designers. Tuned circuit design becomes more challenging when the signals processed occupy more than

Figure 2.18 Transmission and reception processes. (a) Double sideband suppressed carrier transmitter. (b) Baseband signal waveform. (c) Baseband signal spectrum. (d) Modulated signal waveform. (e) Modulated signal spectrum. (f) Double sideband suppressed carrier receiver. (g) Double sideband suppressed carrier demodulation.

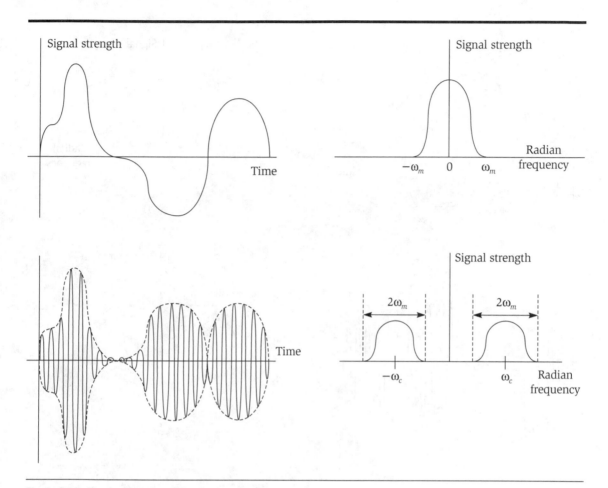

Figure 2.19 The concept of negative frequencies.

one octave, that is, more than one doubling of frequency content. Baseband signals can occupy multiple octaves. Maintaining a constant transmission characteristic over multiple octaves is impractical. When signals are modulated, the resulting energy usually occupies much less than one octave.

Another major benefit of modulation is apparent in the broadcast situation. An efficient antenna's size is related to the wavelength of the signal. The higher the frequency, the shorter the wavelength. This yields a more practical-size antenna.

2.3.5 Envelope Detection

The synchronous detector is moderately complex but relatively inexpensive when implemented with semiconductor-integrated circuits. Implementation with

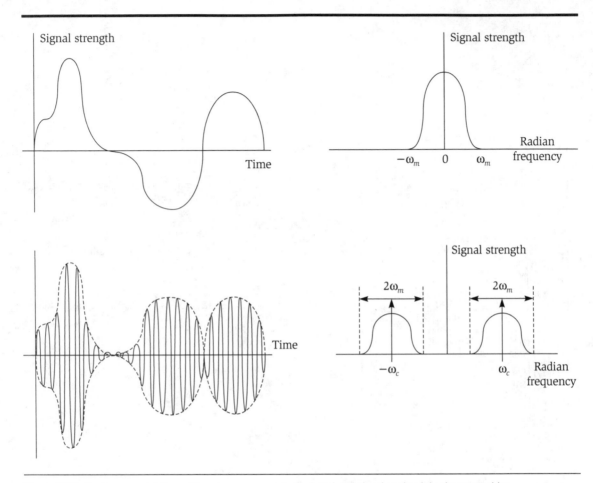

Figure 2.20 Envelope modulation adds a dc component to the baseband signal so that it is always positive.

discrete transistors is much more difficult and expensive. The vacuum tube circuits in use at the time of the creation of NTSC made cost-effective implementation out of the question. A less expensive implementation of the detection function was mandatory if consumers were to afford receivers.

If a dc component is added to the baseband information-bearing signal, preventing it from ever going to negative voltages, the resulting envelope of the modulated waveform is exactly the wave shape of the information-bearing baseband signal. This is shown in Figure 2.20. The information-bearing baseband signal voltage function is shown at the upper left-hand part of Figure 2.20, and its spectrum at the upper right-hand portion of the figure. Modulation intended for envelope detection is accomplished with the addition of a dc component A to the modulating signal, as shown in the lower left-hand part of Figure 2.20. The conse-

Figure 2.21 Envelope detection.

quence of this is the presence of the carrier frequency in strength in the output spectrum shown in the lower right-hand portion of the figure.

A simple receiver for such a signal is shown in Figure 2.21. The tuned circuit that separates the desired channel from the complete spectrum is shown in the figure. The diode removes the bottom (negative going) part of the modulated waveform. The LPF blocks the high-frequency signal and smoothes it out. The capacitor, C, removes the dc component introduced at the transmitter. The received output signal is identical to the input signal seen in the upper left-hand portion of Figure 2.20. Note that a synchronous detector could have been used here as well. It would function exactly as described in the previous section. However, the whole point of adding the dc component to the baseband signal is to allow a cheaper, simpler detector in the receiver. The penalty in the transmitted signal is increased power in a residual carrier that carries no information.

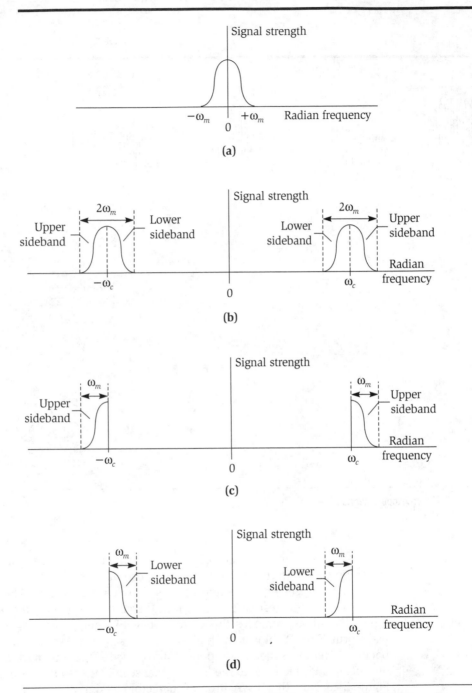

Figure 2.22 Single sideband. (a) Baseband spectrum. (b) Double sideband modulated spectrum. (c) Single sideband modulated spectrum, retaining the upper sidebands. (d) Single sideband modulated spectrum, retaining the lower sideband. (e) Single sideband modulated spectrum at the receiver. (f) Synchronous cosine wave for detection at the receiver. (g) Single sideband demodulation at the receiver.

Figure 2.22 (*continued*)

2.3.6 Single Sideband Modulation

The double sideband modulation method with residual carrier allows the inexpensive diode envelope detector, but at the cost of substantial power for the carrier signal. Closer observation reveals that there is still more "wasted" power. Since the upper and lower sidebands in the modulated signal are mirror images of each other, even in a double sideband signal with suppressed carrier, twice the necessary power is used. Perhaps worse, twice the bandwidth is consumed than is really necessary to convey the information. Figure 2.22 shows the solution. In Figures 2.22(c) and (d), just one of the two mirror image sidebands is used. The

other is filtered out. Both sidebands are equivalent in their information-carrying capacity, so the choice of which one to keep is arbitrary. The bandwidth and the power consumption of the single sideband signal are both cut in half compared to double sideband suppressed carrier modulation.

If the single sideband suppressed carrier signal of Figure 2.22(e) is multiplied by the exact frequency of the original carrier, as shown in Figure 2.22(f), then the spectrum of Figure 2.22(g) results. An LPF then can easily separate out the original information-bearing baseband signal. This is just the familiar synchronous detection process already studied with respect to double sideband signals.

The creation of the appropriate filters to generate single sideband suppressed carrier signals is very difficult and expensive. This is the main drawback of this technique. There are other methods for generating single sideband suppressed carrier signals, but they are not inexpensive.

2.3.7 Vestigial Sideband Modulation

The vestigial sideband method is a compromise between single sideband and double sideband.[10] It employs practical filters to substantially reduce the bandwidth needed for the transmission of visual information. Yet the signal may still be received with an envelope detector yielding only very modest and acceptable distortions. Figure 2.23(a) shows the result of simple double sideband modulation of the video information. The video sidebands extend 4.5 MHz above and below the video carrier, occupying a total of almost 9 MHz of spectrum. Alternatively, if the 6-MHz bandwidth was imposed, less than 3.0 MHz would be available for video information. The result would be significantly reduced resolution.

A filter at the transmitter is used to roll off the lower sideband, as shown in Figure 2.23(b). The resulting video signal is double sideband from 0 Hz to 750 kHz and single sideband from 1.25 to 4.2 MHz, with a rolloff characteristic up to 4.5 MHz where the audio carrier is added in. Between 750 kHz and 1.25 MHz, the spectrum transitions from double sideband to single sideband. Only a "vestige" of the lower sideband remains. The result is a 6-MHz occupied bandwidth instead of the 9 MHz, which would have been the case had only simple double sideband modulation been used.

The video carrier is the reason the vestigial filter is placed at 750 kHz instead of at 0 MHz. A practical filter has an influence on the phase shift of frequencies near its cutoff frequencies. Moving the filter much closer to the carrier would result in a disrupted carrier phase and the possibility that the filter would begin to absorb some of the carrier power. This would be detrimental for the carrier's characteristics and would complicate the filter design still further. The carrier is critical to facilitating the use of simple envelope detectors in the receiver at a time when vacuum tubes were the active elements of circuit design. If the television system was being designed in the age of microcircuits, no powerful carrier would be necessary, and the spectrum would be much closer to single sideband with very little vestige of the removed sideband left, saving still more of the precious spectrum-limited resource.

Figure 2.23 Vestigial sideband television. (a) Double sideband television spectrum alternatives as might have been without the VSB innovation. Either much more bandwidth for the same video resolution or much less resolution in the same bandwidth (normalized to 0 Hz at the visual carrier). (b) Vestigial sideband television spectrum (normalized to 0 Hz at visual carrier) saves 2.95 MHz for the same resolution.

The receiver must deal with this modified spectrum. Figure 2.24(a) again shows the impact of converting the double sideband spectrum to a vestigial sideband version. In Figure 2.24(b), we see what would happen if nothing were done to compensate for the modified spectrum of Figure 2.24(a). The double sideband portion of the spectrum contains twice the power of the single sideband portion because there are two sidebands. The receiver characteristic must compensate for this by having a characteristic that is antisymmetric about the video carrier frequency, as shown in Figure 2.24(c). This antisymmetric filter is called a *Nyquist filter* after the engineer who studied its properties. The output of this receiver filter will be flat at all video frequencies.

The output of the filter characteristic of Figure 2.24(c) must be converted to a baseband spectrum for application to the display device. In the early days of television, synchronous detector methods could have been utilized, but they involve much more electronics and hence were prohibitively expensive. As mentioned earlier, when envelope detection is used, some distortion results from the fact that the envelope of the modulated waveform is not an exact replica of the baseband information. This is shown in Chapter 7. It is, however, a reasonable approximation. The system design is such that this distortion is fully acceptable to viewers. It should be noted that though synchronous detection avoids some of the problems due to vestigial sideband modulation, the vestigial sideband filters used in television transmitters introduce phase distortion, which is not corrected by synchronous detection.

A wide variety of approaches have been employed to maximize viewer enjoyment of the resulting images. They cause a result somewhere between what would be obtained with pure envelope detection and pure synchronous detection. Current practice involves techniques that yield results more like that obtainable from synchronous detectors.

2.3.8 Video Bandwidth

When the NTSC system was being designed, a tremendous number of human visual experiments were undertaken. Estimates were made on the likely course of technology. It is a tribute to the work done that the NTSC system has lasted for more than 50 years and is likely to be in use for at least another decade or two—at least on cable.

NTSC television is meant to be a "five times picture height" system. That is, the compromises made between video quality and economics were such that they would not be noticed if the picture were viewed from a distance of at least five times the picture height. Consideration of this and of the bandwidth available, as well as of the ways of occupying that bandwidth in a manner that would allow affordable receivers, led to the choice of 525 scan lines. As described earlier, 21 lines in each field are allocated to the vertical retrace. This leaves 483 active lines. If a synthetic video pattern is generated by electronically making alternate lines black and white, and if the whole raster is placed in the viewing area of the screen, then 483 picture elements can be resolved in the vertical direction. In the

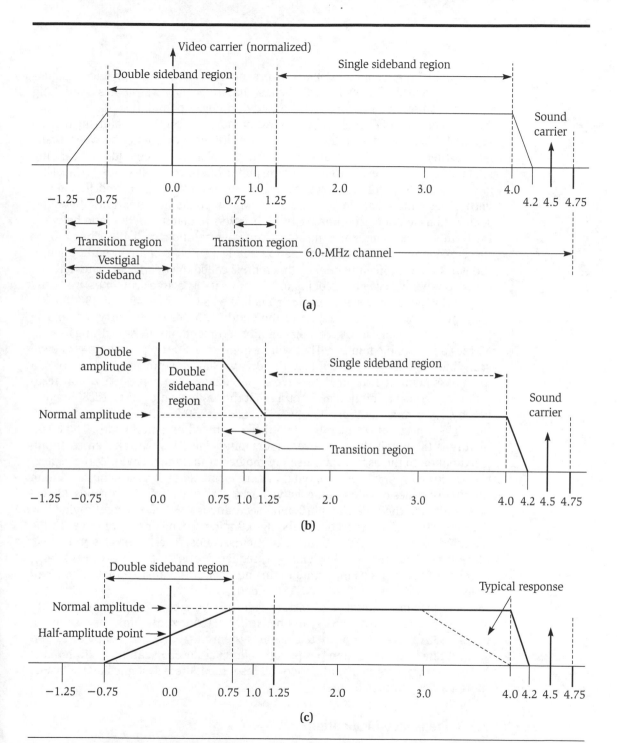

Figure 2.24 VSB reception. (a) Vestigial sideband spectrum normalized to 0.0 Hz as seen at RF (at IF, spectrum is reversed). (b) Output of a "flat response" television receiver as seen at RF (at IF, spectrum is reversed). (c) Idealized and typical TV receiver response curves as seen at RF (at IF, spectrum is reversed).

"real world," a large number of factors prevent the realization of that level of vertical resolution. Just a few include the fact that the scanning system is a sampling system, and raster lines do not fall exactly on object boundaries. Rather than a clean edge, some of the light falls into two or more adjacent lines, blurring the image. Additionally, there is raster instability. Minor movements between scans add further blurring. The scanning spot in the pickup device and the scanning spot in the display device are not precise points but have a bell-shaped (Gaussian curve) distribution of effectiveness. All these complex factors yield less actual vertical resolution than the number of scan lines would indicate. A number of researchers have done experiments with observers to determine the ratio between the number of scan lines and the actual vertical resolution available. The ratio for these experimenters varied from 0.5 to 0.85. R. D. Kell is credited with much of the work on this phenomenon, so this ratio is called the *Kell factor*. The Kell factor is commonly accepted as being 0.7. That is, if 483 active scan lines are transmitted, the effective vertical resolution is taken as 0.7 times 483, or 338 lines.

There is no single derivation of the bandwidth needed to carry video. The best that can be done is to give a plausibility argument for why something like 4.2 MHz is a reasonable number. That will be done here. Starting out with 483 active scan lines and applying a Kell factor of 0.7, we have a vertical resolution of 338 lines. Assuming it is desired to have approximately the same horizontal resolution as vertical resolution and noting that the "aspect ratio" of an NTSC picture is 3 by 4, we must multiply 338 by 4/3, yielding 450 horizontal lines of resolution. The aspect ratio is the ratio of vertical height to horizontal width. The NTSC picture is three units high for every four units wide. This was chosen as an approximation to the aspect ratio used in movies at that time. Dividing the aspect-ratio-adjusted number of horizontal picture elements by 2 yields the maximum number of cycles of video in the active horizontal video time. This yields 225 cycles. Although the time of a horizontal scan line is 1/30 of a second divided by 525 lines, or 63.5 microseconds, about 10.9 microseconds are given over to the blanking interval, leaving about 52.6 microseconds. The 225 cycles must take place in 52.6 microseconds, yielding a frequency of 4.28 MHz. Given the large number of assumptions in getting to this number, it is plausible that 4.2 MHz of video bandwidth is required by NTSC video.

The maximum number of picture elements possible in a video frame constructed electronically is 483 scan lines times 450 horizontal lines of resolution, or 217,350 pixels. European television systems that use the 50-pictures-per-second rate have more time per picture, allowing for 625 scan lines. To retain a balance between vertical and horizontal resolution, these systems have allocated more bandwidth per channel.

2.3.9 Frequency Modulation

The previous discussion of modulation considered only modifying the strength of a carrier signal to impress an information-bearing baseband signal upon it.[11] A co-

sine wave has three parameters that can be varied in a manner to convey information. Equation (2.15) shows them:

$$s_0(t) = A \cdot \cos(2\pi f t + \theta) \tag{2.15}$$

Here A is the amplitude, f is the frequency, and θ is the phase angle of the cosine function. The symbol $s_0(t)$ is the signal as a function of time.

In the work so far, only the amplitude A carried information and was a function of time, $A(t)$, as shown in Equation (2.16).

$$s_1(t) = A(t) \cdot \cos(2\pi f t + \theta) \tag{2.16}$$

If, instead, the phase angle carries the information, we get Equation (2.17).

$$s_2(t) = A \cdot \cos(2\pi f t + \theta(t)) \tag{2.17}$$

As the last alternative, the frequency can be impressed with the information, yielding Equation (2.18).

$$s_3(t) = A \cdot \cos(2\pi[f_1 + g(t)]t + \theta) \tag{2.18}$$

The information-bearing signal is $g(t)$, which is a deviation from the unmodulated frequency f_1. Without loss of generality, the constant phase angle θ can be dropped, yielding

$$s_3(t) = A \cdot \cos(2\pi[f_1 + g(t)]t) \tag{2.19}$$

If the baseband information-bearing signal is a simple cosine wave of frequency f_2, the resulting expression is

$$s_4(t) = A \cdot \cos(2\pi[f_1 + \cos 2\pi f_2 t]t) \tag{2.20}$$

The mathematics of cosines within cosines rapidly becomes very complex, involving advanced mathematical functions and concepts. There are excellent treatments of these details for anyone who is interested. The principle result of these analyses for our purposes is that the bandwidth of the frequency modulated signal can exceed that of an amplitude modulated signal. The negative consequence of this is that more spectrum is consumed. The positive consequence is that the signal is more robust and can survive noise and other distortions better than pure amplitude modulation. Frequency modulation belongs to a class of signals called *spread spectrum* because it spreads the information signal over more bandwidth than it occupies in its baseband version. This technique trades off spectrum inefficiency for signal robustness.

Figure 2.25 Frequency modulation via direct variation of an oscillator's frequency.

The phenomenological concept motivating frequency modulation is the notion that noise, particularly static, was believed to be only an amplitude effect. If the amplitude of the signal did not carry information, but rather the frequency of the signal was impressed with the baseband signal, the impact of noise added to the signal would be negligible. This simplistic model is mostly correct, and the main benefits of FM include much less susceptibility to noise.

The creators of NTSC considered it to be a video addition to radio. The fundamental signal was the audio. The video merely augmented the audio. The intention was that even at the outer limits of reception, the audio signal would survive even if the video failed.[12] Therefore the audio signal was made extremely robust by the spread spectrum technique of frequency modulation. In cable practice, we are able to take advantage of this by reducing the strength of the audio signal. This reduces the overall signal loading of the cable system.

A wide variety of ways of generating frequency modulation have been devised. Here is just one simple example. In Figure 2.25, the information-bearing signal $g(t)$ is applied across a back-biased diode called a "varactor diode." The back-biased diode does not conduct and acts instead like a small capacitor. The value of the capacitor is related to the sum of the signal $g(t)$ and the bias voltage. When the varactor diode is placed in parallel with a fixed capacitor and coil, a variable frequency-tuning element is created. The oscillator, which depends on these elements to determine its frequency, then changes frequency in step with the input information-bearing signal $g(t)$.

The detection of frequency modulation has been accomplished with many different implementations of three fundamental approaches. The first approach converts frequency modulation to amplitude modulation and then uses the techniques already developed for amplitude modulation detection. The second technique uses the mathematical properties of the frequency modulated waveform and a linear multiplying device. The third technique involves time measurement of the distance between the zero crossings of the frequency modulated wave.

A simple example of frequency modulation detection is offered to start the discussion. Figure 2.26(a) shows that the frequency modulated signal $s(t)$ can be viewed as a "walking" spectral element. The movement of the spectral element is in proportion to the information-bearing signal $g(t)$. If this signal is put through a

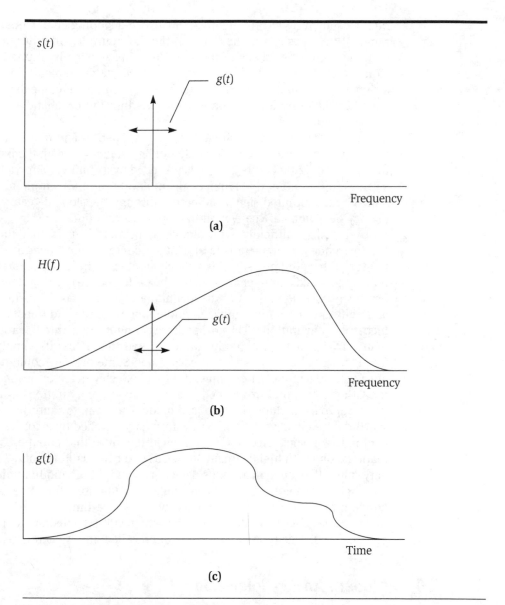

Figure 2.26 FM reception. (a) Frequency modulation as a "walking" spectral line. (b) One possible
filter characteristic for an FM slope. (c) Demodulated output of a slope detector.

filter as shown by $H(f)$ in Figure 2.26(b) with a sloping characteristic across the range of frequencies spanned by $s(t)$, the filter output will differ in strength depending on the frequency of the $s(t)$. The filter is usually a tuned circuit. This simple technique is called *slope detection*. It converts frequency modulation into a signal with amplitude modulation as well. Then an envelope detector can convert the amplitude modulation components into the baseband signal shown in Figure 2.26(c).

If the received FM signal were directly applied to Figure 2.26(b), the amplitude effects of static and noise would not be avoided. To gain the benefits of FM, the circuit realizing Figure 2.26(b) is preceded by a "limiter" circuit. The limiter is a saturating amplifier that removes all amplitude variations from the received signal. When the limited signal is then put through the slope detector, only the frequency deviations are converted to amplitude variations.

A practical difficulty with simple slope detection is that the amplitude-versus-frequency characteristic of a tuned circuit is not linear. If the frequency range of the modulated signal is severely limited to the approximately linear portion of the tuned circuit characteristic, the performance of the system is very poor. Practical implementations of the technique that converts frequency modulation to amplitude modulation utilize multiple tuned circuits and transformers in symmetrical configurations. The goal of these configurations is to linearize the performance by balancing the nonlinearities of two tuned circuits' amplitude-versus-frequency characteristics against each other. Some of the techniques are named after their inventors, for example, the Foster–Seeley detector or the Travis detector. Other techniques are named after the circuit configuration. Examples of the latter approach include the balanced diode discriminator and the ratio detector.

If the frequency modulated waveform is multiplied by a cosine wave, the resulting trigonometric equations will yield one term that contains original information along with higher-frequency terms that can be filtered out. This process is very much like the synchronous detection of amplitude modulation.

Digital techniques have been applied to measure the time between zero crossing of the frequency modulated wave. These time differences are proportional to the baseband information, and the resulting detector can be very linear and accurate. This type of detector is often called a *pulse width demodulator*.

2.4 Modern Analog Television

Modern analog television differs from the original system in only three respects. Modern analog television has the addition of color, stereo sound, and ancillary signals. All this has been accomplished in a compatible manner without making earlier products obsolete. If the very first commercial television receiver were operational and powered today, it would still receive and display television programming.

"Compatible" is an ambiguous term. Older products continue to function but at reduced performance because of the addition of color, stereo, and ancillary signals. These compromises stand the cost–benefit test and have provided the

viewing public with a net positive benefit. Few would argue against these trade-offs. But reductions in performance have been forced upon the older products.

2.4.1 Color Television

Color is the most significant addition to NTSC.[13,14] A large number of trade-offs were necessary in the original design of NTSC to make it practical and affordable, but even more trade-offs were required to add color. The color system carefully matches its design to the human visual system to minimize the need for additional information transmission capacity.

The most important discovery that facilitates the addition of color to NTSC is that the human visual system responds to color in two distinct ways: (1) as a response to the frequency (or wavelength) of the light, and (2) as a response to combinations of primary colors. If we pass sunlight through a glass prism, we decompose the light into individual wavelengths that we see as separate colors. If this were the only way in which the eye perceives color, a color television system or even color printing and photography would be all but impossible. Fortunately, the eye also is able to combine primary colors to perceive nearly all the possible colors.

The three primary colors for light-emitting devices are red, green, and blue.[15] The science of colorimetry is fascinating and complex and well beyond the scope of this brief chapter. Only a few basics can be covered here. In the color receiver, the picture tube is actually three picture tubes in one.[16] There are three electron guns and three sets of colored phosphor elements on the screen. A *shadow mask* in the tube is carefully aligned to allow only the electrons from a particular gun to strike the phosphors of its assigned color. The shadow mask is also called an *aperture mask* because of its array of holes, or apertures (see Figure 2.27). The three electron guns are tilted at an angle with respect to the axis of the picture tube. The three electron beams pass through the same holes in the shadow mask, which is located very close to the screen. Because they aim from different angles in the neck of the tube, the beams land on adjacent but different spots. The spots have phosphors that emit red, green, and blue light. The spots are small and close together, and the eye fuses them into a continuous color from normal viewing distances. The reader is invited to take a magnifying glass and look closely at the screen of a color receiver. Clusters of dots or clusters of small bars of the three different colors will be visible when the screen is illuminated.

If a combination of three complete images in red, green, and blue were required to make a color picture, the required transmission bandwidth would be tripled. This would reduce the number of possible channels to one-third of the current numbers. Even the improvement color brings might not be worth that drastic a loss of choice. Fortunately, the eye is quite pleased with a "colored" television system and does not require a true "color" system. If we consider the Sunday color comics, we realize that we have a detailed black-and-white image with some color thrown on it. The eye and the imagination do the rest. In colored television, the same approach is taken. The detail is in the luminance (brightness) signal. The color signal contains much lower resolution and fills in the

Figure 2.27 Color picture tube structure with colored phosphor dots, aperture (or shadow) mask, and cluster of three electron guns.

monochrome outlines. That this is the case can be seen when colored text is in the picture. For example, a yellow letter T on a blue background will have a yellow top bar because the image is large enough horizontally to be represented by the frequencies involved. However, the vertical component of the T will usually not be yellow. To represent it in yellow would require frequency components beyond what is available.

To make the color signal compatible with older monochrome receivers, it is necessary to combine the signals in a manner that yields the luminance signal plus two narrower bandwidth "color difference" signals. Since the eye is more sensitive to detail in certain colors than in others, it is possible to reduce the bandwidth of some of the colors more aggressively. The color television system manipulates the color signals using a set of linear matrix equations. This separates the signal into a component that contains flesh tones and near flesh tones, and an-

Figure 2.28 Luminance spectral energy clusters around harmonics of horizontal scan frequency f_h all the way up to the upper band edge. Chroma spectra energy components cluster around odd multiples of half horizontal scan frequency and are interleaved with luminance.

other component that has quite different colors. The principal color signal is termed the I signal and has a transmitted bandwidth of around 1.5 MHz. The secondary color signal is termed the Q signal and has a transmitted bandwidth of around 0.5 MHz. In a color receiver, linear networks can then separate the signals into the three primary colors, red, green, and blue. Simultaneously, a monochrome receiver will respond only to the luminance signal and ignore the color signals.

Although the NTSC system uses bandwidths of 1.5 MHz in the I channel and 0.5 MHz in the Q channel, practical color receivers actually implement a bandwidth of only about 0.5 MHz (or less) in each channel. This compares with several megahertz for the luminance signal. The main reason for the reduced bandwidth is the problem of "chroma noise." When the higher-frequency chroma signals are brought to baseband, the noise added at those frequencies is also brought down in frequency. This causes the interference resulting from the noise to increase in size on the screen. Reducing the chroma bandwidth minimizes this unpleasant effect, but at a reduction in color resolution.

With 1.5 MHz of I signal and 0.5 MHz of Q signal, we still need to find a place to put 2.0 MHz of additional information. The principle of compatibility precludes simply adding more bandwidth. This is where the work of Mertz on the spectrum of scanned images provides an answer. Figure 2.12 shows that the signal energy is clustered around multiples of the horizontal scan frequency (15,750 Hz) and that there are significant gaps in the spectrum. The spectrum appears as a comb. Since this is a consequence of scanning and the color information likewise is a product of scanning, it too has spectrum gaps. The trick then is to arrange the two spectra, causing their respective energy clusters to fall into each other's spectrum gaps just as two combs can be meshed together. Figure 2.28 shows that if the color signal is modulated onto a carrier that is an odd harmonic of half of the line frequency, the color signal spectrum will interleave with the

monochrome signal spectrum. The color signal occupies frequencies that are nearly empty in a monochrome signal.

This technique for hiding the chroma signal in the luma can be visualized in the time domain as well (see Figure 2.29). A cosine wave that is an "odd harmonic of half of the line frequency" will start with the opposite phase on adjacent lines. In the basic case, the first harmonic of half the line frequency will make only half of a cosine wave in the time of one horizontal scan. It will begin the next line with the opposite phase. So where the signal is brightest on one line, it will be dimmest on the adjacent line and vice versa. These opposite brightnesses on adjacent lines will tend to average out and reduce visibility of the color carrier. The same is true of higher odd harmonics (even harmonics would cause the signal to have the same phase on adjacent lines, and the bright and dark spots would reinforce each other rather than tend to cancel). At the color subcarrier frequency, the bright and dark spots on the screen are very small and are positioned on adjacent lines relative to each other so as to average out the brightness they cause.

Since there are two color signals, the I and the Q signals, both must be accommodated.[17] The two color difference signals are modulated in quadrature to each other onto the color subcarrier. *Quadrature modulation* means that two carriers are used, one phase shifted from the other. The color signal can be expressed as

$$c(t) = I(t) \cdot \cos 2\pi f_c t + Q(t) \cdot \sin 2\pi f_c t \qquad (2.21)$$

The mathematical sine function is 90° phase shifted from the mathematical cosine function. These functions are said to be "in quadrature" with each other because 90° defines a quadrant (quarter) of a circle. The required detection process is called *quadrature detection*. In the receiver, quadrature detection cleanly separates the two signals. This separation is based on the simple trigonometric identities of Equations (2.7) through (2.9), which are repeated here:

$$\cos(2\pi f_1 t \cdot \cos(2\pi f_2 t) = \frac{1}{2}\cos 2\pi(f_1 - f_2)t + \frac{1}{2}\cos 2\pi(f_1 + f_2)t \qquad (2.7)$$

$$\sin(2\pi f_1 t) \cdot \sin(2\pi f_2 t) = \frac{1}{2}\cos 2\pi(f_1 - f_2)t - \frac{1}{2}\cos 2\pi(f_1 + f_2)t \qquad (2.8)$$

$$\sin(2\pi f_1 t) \cdot \cos(2\pi f_2 t) = \frac{1}{2}\sin 2\pi(f_1 - f_2)t + \frac{1}{2}\sin 2\pi(f_1 + f_2)t \qquad (2.9)$$

Let $f_1 = f_2 = f_c$, which is the carrier frequency for the color information. Then from Equation (2.7),

$$\cos(2\pi f_c t) \cdot \cos(2\pi f_c t) = \frac{1}{2}\cos 2\pi(f_c - f_c)t + \frac{1}{2}\cos 2\pi(f_c + f_c)t \qquad (2.22)$$

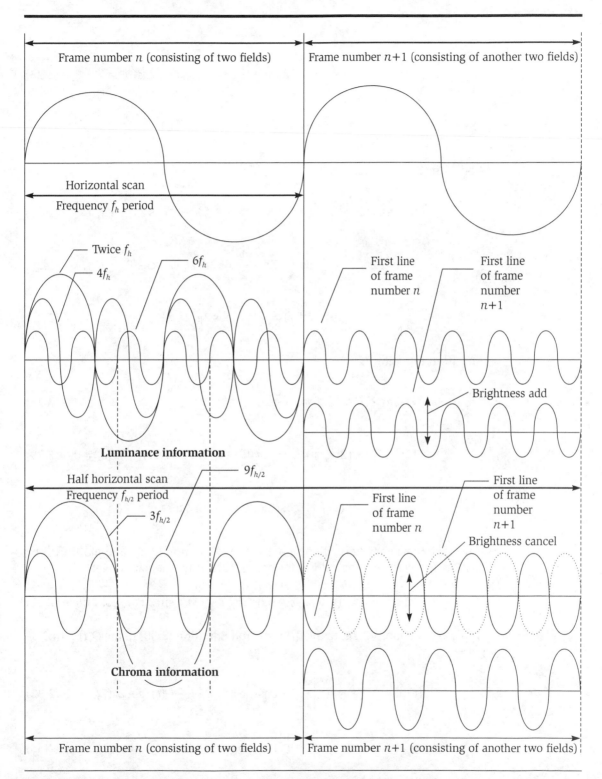

Frame number n (consisting of two fields) Frame number $n+1$ (consisting of another two fields)

Horizontal scan
Frequency f_h period

Twice f_h

$6f_h$

$4f_h$

First line
of frame
number n

First line
of frame
number
$n+1$

Brightness add

Luminance information

Half horizontal scan
Frequency $f_{h/2}$ period

$9f_{h/2}$

$3f_{h/2}$

First line
of frame
number n

First line
of frame
number
$n+1$

Brightness cancel

Chroma information

Frame number n (consisting of two fields) Frame number $n+1$ (consisting of another two fields)

Figure 2.29 Time domain view of the frequency interleaving principle used to hide color information.

$$\cos^2(2\pi f_c t) = \frac{1}{2}\cos 2\pi(0)t + \frac{1}{2}\cos 2\pi(2f_c t) \quad \cos(0) = 1 \qquad (2.23)$$

$$\cos^2(2\pi f_c t) = \frac{1}{2}[1 + \cos 2\pi(2f_c)t] \qquad (2.24)$$

And from Equation (2.8),

$$\sin(2\pi f_c t) \cdot \sin(2\pi f_c t) = \frac{1}{2}\cos 2\pi(f_c - f_c)t - \frac{1}{2}\cos 2\pi(f_c + f_c)t \qquad (2.25)$$

$$\sin^2(2\pi f_c t) = \frac{1}{2}\cos 2\pi(0)t - \frac{1}{2}\cos 2\pi(2f_c t) \quad \cos(0) = 1 \qquad (2.26)$$

$$\sin^2(2\pi f_c t) = \frac{1}{2}[1 - \cos 2\pi(2f_c)t] \qquad (2.27)$$

And from Equation (2.9),

$$\sin(2\pi f_c t) \cdot \cos(2\pi f_c t) = \frac{1}{2}\sin 2\pi(f_c - f_c)t + \frac{1}{2}\sin 2\pi(f_c + f_c)t \qquad (2.28)$$

$$\sin(2\pi f_c t) \cdot \cos(2\pi f_c t) = \frac{1}{2}\sin 2\pi(0)t + \frac{1}{2}\sin 2\pi(2f_c t) \quad \sin(0) = 0 \qquad (2.29)$$

$$\sin(2\pi f_c t) \cdot \cos(2\pi f_c t) = \frac{1}{2}\sin 2\pi(2f_c t) \qquad (2.30)$$

If we take the color information signal, $c(t)$, of Equation (2.21) and multiply it by $\cos(2\pi f_c)$, a cosine at the color information carrier frequency, we get

$$c(t) \cdot \cos 2\pi f_c t = I(t) \cdot \cos 2\pi f_c \cdot \cos 2\pi f_c t + Q(t) \cdot \sin 2\pi f_c t \cdot \cos 2\pi f_c t \qquad (2.31)$$

Applying Equation (2.24) to the $I(t)$ term and Equation (2.30) to the $Q(t)$ term, we get

$$c(t) \cdot \cos 2\pi f_c t = I(t)\left[\frac{1}{2}(1 + \cos 2\pi(2f_c t))\right] + \frac{1}{2}Q(t) \cdot \sin 2\pi(2f_c t) \qquad (2.32)$$

$$c(t) \cdot \cos 2\pi f_c t = \frac{1}{2}I(t) + \frac{1}{2}[I(t) \cdot \cos 2\pi(2f_c t) + Q(t) \cdot \sin \pi(2f_c t)] \qquad (2.33)$$

The first term on the right-hand side of Equation (2.33) is the $I(t)$ signal at baseband frequencies. It has been synchronously demodulated. The bracketed terms are modulated onto carriers at twice the chroma carrier frequency, f_c, and can be easily removed with a low-pass filter. The $I(t)$ signal has been completely separated from the $Q(t)$ signal through quadrature detection.

If we take the color information signal, $c(t)$, of Equation (2.21) and multiply it by $\sin(2\pi f_c)$, a sine at the color information carrier frequency, we get

$$c(t) \cdot \sin 2\pi f_c t = I(t) \cdot \sin 2\pi f_c t \cdot \cos 2\pi f_c t + Q(t) \cdot \sin 2\pi f_c t \cdot \sin 2\pi f_c t \quad (2.34)$$

Applying Equation (2.30) to the $I(t)$ term and Equation (2.27) to the $Q(t)$ term, we get

$$c(t) \cdot \sin 2\pi f_c t = \frac{1}{2} I(t) \cdot \sin 2\pi(2f_c t) + Q(t)\left[\frac{1}{2}(1 - \cos 2\pi(2f_c t))\right] \quad (2.35)$$

$$c(t) \cdot \sin 2\pi f_c t = \frac{1}{2} Q(t) + \frac{1}{2}[I(t) \cdot \sin 2\pi(2f_c t) - Q(t) \cdot \cos 2\pi(2f_c t)] \quad (2.36)$$

The first term on the right-hand side of Equation (2.36) is the $Q(t)$ signal at baseband frequencies. It has been synchronously demodulated. The bracketed terms are modulated onto carriers at twice the chroma carrier frequency, f_c, and can be easily removed with a low-pass filter. The $Q(t)$ signal has been completely separated from the $I(t)$ signal through quadrature detection.

The I signal is the *in-phase* color signal and is amplitude modulated onto a color subcarrier that is located at 3.579545 MHz, the 455th odd multiple of half the line-scanning frequency. The Q signal is the *quadrature* color signal. It is amplitude modulated on a carrier that is derived by a 90° phase shifting of the 3.579545-MHz carrier used for the I signal (see Figure 2.30). Note that the Q signal is a double sideband signal, whereas the I signal has a 0.5-MHz double sideband component and an additional 1.0 MHz of single sideband modulation. However, as already mentioned, practical consumer receivers limit bandwidth in both channels to around 0.5 MHz.

It will be recalled from the study of geometry that there are two common forms of coordinate systems, rectangular and polar. In the rectangular system, a location is defined by the distance in the horizontal direction and the distance in the vertical direction. In the polar system, a location is defined by a distance along an angle. This is like giving driving directions of 3 miles east and 4 miles north. If the traveler has an airplane, he can simply be told to fly 5 miles on a heading of 30 degrees. In both cases, the traveler will arrive at the same place. Both systems are equivalent. The diagram of Figure 2.31 applies these principles to the quadrature color equation (2.21). The color signal $c(t)$ can be given either as the sum of the $I(t)$ signal and the $Q(t)$ signal or as the resulting magnitude $C(t)$,

Figure 2.30 Location and bandwidths of the luminance and chroma signals.

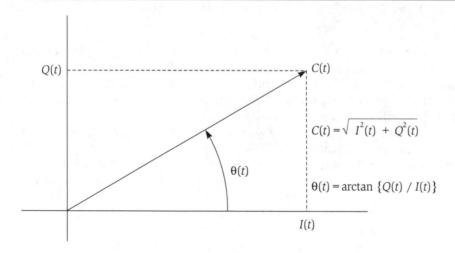

Figure 2.31 Rectangular versus polar representation.

which is the Pythagorean solution of the square root of the sum of the squares of $I(t)$ and $Q(t)$ at an angle given by the inverse tangent trigonometric function of the ratio of $Q(t)$ to $I(t)$. In this "polar" way of interpreting the color signal, the color or hue is defined by the angle, and the intensity of the color (termed its satu-ration) is described by the length $C(t)$. Figure 2.32 shows the location of various colors in a polar graph.

The changes made to accommodate color required the horizontal and verti-cal scan rates to be slightly modified. This is because the difference between the color subcarrier frequency and the aural subcarrier frequency is a frequency well

	C	θ
(B−Y)	1	0°
Q	1	33°
Magenta	82	61°
(R−Y)	1	90°
Red	88	103°
Yellow	62	167°
Burst	1	180°
Green	82	241°
Cyan	88	283°
−I	1	303°
Blue	62	347°

Figure 2.32 Polar representation of television colors.

within the video bandwidth. Any nonlinearity in that signal path will produce a beat product that is in the video passband and can produce an annoying pattern on the screen. This difficulty has been termed the 920-beat problem because the difference between the frequencies selected for the chroma subcarrier and the sound subcarrier is 920 kHz. Two techniques are used to minimize this problem. The existing 4.5-MHz trap strongly attenuates the aural carrier frequency before the video path. However, it was thought that this measure may not be sufficient in existing monochrome receivers that were designed and produced before the color standard was established. The remaining 920-kHz beat will be made less visible if it too is frequency interleaved (just like the chroma signals) with the luminance signal. This requires that the 920-kHz beat be made an odd multiple of half scan line frequency. To accomplish this (since the chroma subcarrier is already an odd multiple of half scan line frequency), the sound carrier must be any multiple of the horizontal scan frequency (that is, an *even* multiple of half

scan line frequency). The 286th multiple of the monochrome horizontal scan frequency (15.75 kHz) is 4.5045 MHz. Using these numbers would require changing the aural subcarrier by 4.5 kHz. Some of the then existing monochrome television receiver's audio circuits might have failed to perform with this much of a shift in frequency, jeopardizing the compatibility mandate. Also, the effectiveness of the aural frequency trap will be reduced. The alternative is to change the horizontal frequency so that it is the 286th submultiple of the original 4.5-MHz aural subcarrier. This yields a horizontal frequency of 15,734.26 Hz. Since there are 525 scan lines in a frame and 2 fields in a frame, the field rate will be the horizontal rate divided by 525 and multiplied by 2, yielding 59.94 Hz. The chroma subcarrier is the 455 multiple of half of this line rate, or 3.579545 MHz.

Synchronous detectors in the television receiver are used to completely separate the two chroma signals. Then, using matrix equations (2.37) through (2.39), the luminance signal E_Y, the in-phase color signal E_I, and the quadrature signal E_Q are matrixed to yield the blue signal E_B, the red signal E_R, and the green signal E_G that drive the tricolor picture tube.

$$E_B = E_Y + 1.70E_Q - 1.10E_I \tag{2.37}$$

$$E_R = E_Y + 0.62E_Q - 0.96E_I \tag{2.38}$$

$$E_G = E_Y - 0.65E_Q - 0.28E_I \tag{2.39}$$

Synchronization of the color oscillator used in the synchronous detection process is facilitated by the addition of a burst of unmodulated color subcarrier on each horizontal blanking pulse after the horizontal sync pulse (see Figure 2.33).

These techniques allow color television to be compatible with monochrome television. This compatibility is not complete because there are compromises. Monochrome receivers built before the introduction of color had video bandwidths of up to 4.2 MHz. This allowed for very sharp black-and-white pictures. When color signals were introduced, these receivers suffered from "dot crawl." The color signal was not adequately rejected by the older receivers and appeared as a moving pattern of annoying dots. This problem was overcome in later monochrome receivers by introducing a notch in the frequency response to eliminate much of the color signal. The consequence was a loss of resolution and sharpness. Alternatively, the video bandwidth of monochrome receivers was rolled off, causing the color signals to be attenuated. This, too, reduced sharpness. These compromises allowed the two types of receivers, color and monochrome, to continue in production and receive the same signals. But this came at the cost of reduced performance in new monochrome receivers and degraded performance in monochrome receivers manufactured before the introduction of color.

This compatibility was critical to the rational introduction of color television into a market already populated with monochrome television receivers. Consumers who purchased color receivers derived more benefits from the same signals that consumers with monochrome receivers continued to enjoy. Those who

Figure 2.33 Color burst relationship to sync pulse.

could not afford a color receiver could buy a new monochrome receiver and still have access to television. No one was disenfranchised by the technological advance to color.

2.4.2 Stereo Sound

The next improvement to NTSC came in the area of sound.[18] Television sound is frequency modulated on a separate carrier that is a fixed 4.5 MHz above the visual carrier. When stereo sound was added to the television system, the requirement of compatibility was again enforced. This avoided the chaos that would

have resulted from obsoleting the existing sound system. Just as with compatible color, there were compromises to monaural receivers when stereo sound was added. But the net benefit to consumers was considered positive. The marketplace gave its approval to both compatible color and compatible stereo sound.

Stereo sound is implemented by first creating a spectrum that includes the sum of the left and right sound channels at baseband. As in FM radio broadcasting and monophonic television practice, the sum signal is preemphasized using a 75-microsecond preemphasis curve. (See the discussion in Chapter 7 under modulators.) The difference between the left and right channels is a double sideband suppressed carrier modulated onto a carrier at twice the horizontal scan frequency ($2 \times 15{,}734 = 31{,}468$ Hz). In order to improve the signal-to-noise ratio, this difference signal is companded; that is, the signal is compressed in amplitude at the transmission point and expanded at the receiver. The compression is frequency selective and is based on an algorithm originally proposed by the dBx Corporation, whose name it bears. This is in contrast to the FM radio stereo system, which is not companded. The FM system also employs the same deviation of the sound carrier, by both the sum and difference carriers. Both changes to the television system were brought about by the need for improved signal-to-noise ratio. A limited bandwidth (10 kHz) monaural second audio program (SAP) channel is frequency modulated onto a carrier at five times the horizontal scan frequency. The SAP channel is intended for second language or other such purposes. A very narrow bandwidth (3.4 kHz) "professional channel" is frequency modulated onto a carrier at six and one-half times the horizontal scan frequency. It is used for television plant intercommunications. This entire complex spectrum is then frequency modulated onto the 4.5-MHz carrier (see Figure 2.34). The relationship between the video and aural carriers is tightly controlled since nearly all television receivers depend on this relationship. The visual carrier is used as the local oscillator to bring the aural spectrum down to baseband. This technique is called the *intercarrier-sound method* of TV receiver design. Since the final aural modulation process is that of frequency modulation, the TV receiver uses a "limiter" circuit to strip off any amplitude modulation that may have become impressed upon the aural signal. Typical sources of this amplitude modulation are noise, particularly impulse noise, and cross modulation in nonlinear elements in the signal path. The TV receiver's aural system then becomes relatively insensitive to amplitude modulation on the aural carrier.

A variation of the previously described aural recovery technique is the *direct-sound method*. It has also been called the split-sound method. In this approach, the aural component is extracted from the composite IF signal prior to video detection. This signal is subjected to limiting to remove incidental amplitude modulation components and presented to an FM detector.

When direct-sound techniques are applied, the aural signal is not dependent upon any properties of the visual carrier, as it is in the intercarrier-sound approach. Though the aural carrier is free from the properties of the visual carrier, it becomes dependent upon the accumulated instability of processing oscillators (heterodyning) along the signal chain. These sources of instability include the

Figure 2.34 Stereo audio baseband spectrum.

transmitter, cable heterodyne processors, and local oscillators of intervening frequency translation devices, such as cable subscriber terminal devices and the tuner of the TV itself.

There are trade-offs in making a choice between intercarrier-sound and direct-sound designs. Some visual modulators suffer from nontrivial amounts of incidental phase modulation of the visual carrier. The intercarrier principle causes these visual carrier instabilities to be transferred to the aural carrier, and this can introduce sound distortions. A direct-sound design would not be subject to this problem. Currently, it is likely that more is gained from the intercarrier design's ability to discriminate against phase instabilities in the multiple local oscillators that heterodyne the signal than is lost in the transfer of transmitter phase instabilities to the aural carrier. That is why essentially all consumer TVs and VCRs use the intercarrier principle.

2.4.3 Ancillary Signals

It was determined that more uses can be made of the electron blanking time. One important application is to use the VBI to carry analog test signals. The analog test signals can be used to measure the transmission characteristics from the signal source to intermediate points along its path to the final point of use. Another use for the VBI is to carry analog signals that represent digital data. The data signals can be of two or more levels that are resolved into data bits by appropriate circuits. Since the "digital" signals are of just a few discrete levels, the data detection circuits can discriminate against significant amounts of noise, distortion, and interference. This makes these data signals more robust than the analog video signal itself for most forms of interference.

The first U.S. attempt to use the VBI for ancillary data purposes was in 1970 when the National Bureau of Standards (NBS—now the National Institute of Standards and Technology, NIST) proposed to use it for the distribution of precise time information nationwide. The American Broadcasting Company (ABC) television network was a partner in that effort. Even though this initiative did

not result in a service, ABC recommended a captioning service for the hearing impaired.

The First National Conference on Television for the Hearing Impaired met in Nashville, Tennessee, in 1971. This was followed in early 1972 with a demonstration by the NBS and ABC at Gallaudet University, a school for the deaf in Washington, D.C. In 1973, the engineering department of the Public Broadcasting System (PBS) initiated development funded by the Department of Health, Education and Welfare (HEW). As a result of this work, the FCC reserved line 21 of field one of the television signal for the transmission of closed captions in the United States in 1976. In 1979, the National Captioning Institute (NCI) was founded to caption programming and to further the cause of captioning. In the early 1980s, Sears Roebuck stores carried a captioning decoder in set top adapter configuration selling for about $250. In 1989, NCI contracted for ITT Semiconductor Corporation to develop a cost-effective caption decoder microchip for use in television receivers. In 1990, Congress passed the Television Decoder Circuitry Act, mandating that after July 1, 1993, new television receivers of 13-inch diagonal display measure or greater must include caption decoding circuits. Approximately 20 million television receivers per year are covered by this requirement. In 1992, NCI worked with the FCC and the Electronic Industries Association (EIA) to develop captioning technical standards. The 1996 Telecommunications Act requires the FCC to institute rules requiring closed captioning on video programming but allowing exemptions for programming that would suffer an "undue burden."

The closed captioning (CC) system is called "closed" because it is turned on or off depending on who is using the television receiver. Those without hearing impairments and who understand the spoken words need not be disturbed by text on their screens. The CC system supplies data to appropriate digital and analog circuits that place carefully timed text on the television screen, making it possible for the hearing impaired to read a description of the conversation taking place and have indications of other relevant sounds. Similarly, those who cannot understand the spoken words may have text in their language, allowing them to follow the program. The CC system uses very low-speed data in order to minimize the impact of transmission path problems such as reflections and interfering signals. The data rate for the CC system is 503.5 kb/s of binary (two-level) data. This allows only two 8-bit characters to be transmitted per VBI line. If only field one is used, there will be 30 of these lines per second. This yields 480 bps or 3,600 characters per minute. If the average word is 5 characters long with a space, then 600 words can be conveyed per minute. The rest of the line is occupied with both a burst of seven sine wave cycles of 503.5-kHz clock run-in and a unique "start bits" pattern placed at the beginning of the line. These signals synchronize the detector circuitry. Since only line 21 is protected for captioning by FCC rule, the rate of transmission is slow, but perfectly adequate for the purpose. The on-screen display consists of a maximum of 15 rows of 32 characters each. The captions usually appear only on rows 1 through 4 and rows 12 through 15. The middle rows are usually transparent to show the action. A text mode provides scrolling text. Further details can be found as part of the EIA standard number EIA-608. The CC system was designed at a time when electronic circuits for the correction of trans-

mission deficiencies were very expensive. Since the target market for CC systems is the hearing impaired, it was desired to make the equipment as affordable as possible. An extremely conservative data rate facilitates inexpensive circuits by minimizing the technical challenge.

The closed captioning signal carries four components. There are two captioning channels and two text channels. The first captioning channel is synchronized to the video programming, causing the words to carefully match the video. The second captioning channel is not synchronized.

The EIA filed a petition with the FCC to expand the captioning standard EIA-608, to allow use of line 21 of field two. This adds two more captioning channels and two more text channels. A fifth channel has been added to carry extended data services (EDS). EDS will carry a wide variety of additional information, including precise time information to set clocks in consumer products. Finally, the original intent of the NBS will be realized.

The channel's name and call letters are included along with current program information such as title, length, rating, elapsed time, types of audio services and captioning services, and intended aspect ratio. Also included is the data for the "V-chip" (violent programming advisory), which is intended to control children's access to programming that parents deem objectionable. Public service announcements such as weather and emergency advisories will be transmitted. Cable system channel layout information will be provided, causing the channel number indicator to use the more familiar channel identification number rather than the number associated with the frequency utilized. This facility will bring the same "channel mapping" benefits subscribers have enjoyed in their cable set top terminals to consumer electronics products.

A subsequent VBI data transmission system, called teletext,[19] was invented to provide ancillary services to television users. The teletext system can display up to 24 rows of 40 characters on the television screen (but a specification of 20 rows was selected for use in the United States). Teletext quickly evolved into a transmission system for more complex data, including the downloading of software to computers. At the time teletext was introduced, electronic components were still relatively expensive, which severely limited the affordable capability of the teletext systems.

Teletext is a more aggressive form of data transmission than CC. It has been successful in Europe but has failed to enjoy commercialization in the United States. Teletext originated in Great Britain, with experimental transmission commencing in 1972. The British Broadcasting Corporation (BBC) branded their service Ceefax, and the Independent Broadcast Authority (IBA) called their service Oracle. France developed a packet-based teletext system called Antiope, based on a transmission system called Didon. Later Canada developed another system called Telidon, which featured higher-resolution graphics. The Japanese system, called Captain, featured "photographic coding" to accommodate the Chinese Kanji characters and the Japanese Kana set.

There are a number of reasons for the difficulties in the United States. Principal among these was the failure to find a strategy that made money. Without this, the system could not be supported. Additional difficulties included the high cost

of memory at the time of implementation. Even though a teletext page requires only about a kilobyte of storage, that small amount of memory was expensive then. Further problems centered on the quality of the graphics. The less expensive World System Teletext (WST), based on the British approach, had crude "Lego-style" graphics in its basic form. The other contender, the North America Presentation Layer Protocol System (NAPLPS), used a higher-resolution graphics system that painted itself on the screen painfully slowly, trying the patience of the user. Still another complication was the FCC's 1983 decision to allow two standards and hope that the marketplace would resolve the confusion. One of the systems is WST, and the other is the NAPLPS evolution of Antiope, Telidon, and efforts by AT&T. A final problem was reliability of data reception. In a test in the Bay Area of San Francisco, only about 25% of installations of the NAPLPS system were trouble free. The remainder suffered from various degrees of multipath impairment. The more robust WST system was not tested in that environment.

Both U.S. teletext systems have a data rate of 5.727272 Mbps, which is 364 times the horizontal rate and 8/5 of the color subcarrier. The data signal has a nonreturn to zero (NRZ) binary format. The WST data line consists of eight cycles of clock run-in (16 bits) followed by a unique 8-bit "framing code" followed by 16 bits of control codes and a payload of 32 8-bit display words. The displayed row has 40 characters, 8 more than can be conveyed in one VBI line of data. This is accommodated by transmitting 4 rows of 32 characters on 4 VBI lines and a supplementary VBI line with the 4 groups of 8 characters needed to finish the previous 4 display rows. This process is called *gearing* because of the ratios between the display row and the transmitted VBI data. Since the page format is 40 characters by 20 rows with an additional "header row," 21 field lines are required plus 5 field lines of supplementary 8-character groups for a total of 26 field lines per full page of WST teletext. The payload of 296 bits per line allocated means that if one VBI line in each field is allocated, a data rate of $296 \times 2 \times 30 = 17,760$ bps is obtained. Ten lines of VBI are possible (line 21 is reserved for captions, and the first nine lines form the vertical synchronization pulses), yielding a maximum of 177.6 kb/s for full VBI utilization.

The WST system maps the data location in the VBI line to memory locations and to screen locations, and always puts data in the same memory place. This allows for a very simple error-protection scheme. Since the instructions in the header are Hamming-code protected, a measure of the quality of the received signal is obtained. If the signal is of low quality, it is not loaded into memory. Only good-quality data is stored. As a result, good data can be accumulated from repetitions of the page until a good page of data is built up. It is also possible to use a "voting" approach to obtain very robust transmission.

The fundamental difference between the WST and the evolving set of Antiope, Telidon, and NAPLPS systems is that the latter systems all used a packet structure. They have been characterized as asynchronous because there is no mapping between the transmission scheme and memory and screen locations.

PBS has developed a packetized data delivery system based on teletext called the PBS National Datacast Network. The standard teletext data rate of 5.72 Mbps

is used, yielding 9,600 baud per VBI line allocated per field. The Datacast network distributes the same signal nationally. The goal is to generate revenue to help support the PBS network. There are a wide variety of commercial applications for this signal. Currently, the StarSight Electronic Program Guide (EPG) signal is distributed via PBS.

A system by the WavePhore Corporation utilizes a teletext-like system in lines 10 through 20 in each field for a data speed of up to 150 kb/s. WavePhore added substantial error detection and protection bits to its structure to protect against multipath and other transmission problems.

Underutilized portions of the NTSC spectrum can be employed to "hide" data. In many cases, the process of hiding the data is incomplete and results in artifacts under certain conditions. In other cases, the preparation of the NTSC signal to better hide the data reduces video quality. So the challenge is to both hide the data and not impair video quality while retaining signal robustness and the potential for an economic implementation.

The National Data Broadcasting Committee (NDBC) was formed in 1993 to establish a single standard for data transmission in video. It issued requests for proposals (RFPs) and narrowed down the selection process to contenders from the WavePhore Corporation and the Digideck Corporation. Laboratory tests were conducted by the Advanced Television Test Center (ATTC) in Alexandria, Virginia, in December 1994. In April 1995, the NDBC selected Digideck for field testing. In June, WavePhore convinced the committee to retest their system after making improvements based on the results of the lab tests.

Meanwhile, the FCC issued a Notice of Proposed Rulemaking (NPRM) in April 1995. On June 28, 1996, the FCC approved digital data transmission in the video portion of broadcast television transmission in its Report & Order (R&O), "Digital Data Transmission Within the Video Portion of Television Broadcast Station Transmissions," in MM docket No. 95-42. This R&O amends FCC rules to allow ancillary data within the video portion of the NTSC signal in four formats. Two of the formats, by Yes! Entertainment Corporation and A. C. Nielsen Co., place low data rate signals in the overscan region of the picture. The other two systems, Digideck and WavePhore, embed the digital signal into the video signal. Both Digideck and WavePhore participated in the voluntary standards committee titled the National Data Broadcasting Committee (NDBC) sponsored by the National Association of Broadcasters (NAB) and the Consumer Electronics Manufacturers Association (CEMA). NDBC has conducted field tests of these systems in Washington, D.C., on WETA, channel 26, and WJLA, channel 7.

The WavePhore system begins by reducing video luminance and chrominance bandwidths. The "luminance" is reduced from its theoretical value of 4.2 MHz to a lower value of 3.9 MHz, and the upper sideband of the color signal is reduced by about 300 kHz. It is then possible to insert a data signal in this region at a carrier frequency of approximately 4.197 MHz above the video carrier and a strength approximately 20 dB above the noise floor of the video system. The data is synchronous with the video carrier and thus with the horizontal line frequency. As an odd multiple of one-quarter of the horizontal scan frequency, it interleaves between the luminance and chrominance bundles of spectral energy.

Data is not sent during the vertical and horizontal blanking intervals. Thirty bits of data are sent per video line. There are 240 available lines per field (not counting the VBI, during which the signal is blanked). This yields a raw data rate of 431.6 kb/s. After error-correction coding, the raw data rate is reduced to approximately the T1 rate divided by 4 or 384 kb/s. This is one-quarter the telephone T1 data rate, so WavePhore calls their system TVT1/4.

WavePhore shuffles the data before applying biphase modulation and filtering out the lower sideband. Shuffling the data reduces its visibility in the video. An adaptive equalizer is used in the receiver. A major advantage of the WavePhore approach is that once inserted into the video, it can be conveyed through the video path without giving it further attention. The WavePhore VBI system and the WavePhore subvideo system can be combined to provide more than 500 kb/s.

There is some degradation of pictures using this system. However, it appears that the regulating body, the FCC, is willing to let the broadcaster determine what his or her individual marketplace values are and to respond to that decision.

The Digideck system adds a differential quadrature phase shift key (DQPSK) signal carrying about 500 kb/s placed 1 MHz below the video carrier. In this regard, it is similar to the European NICAM system for adding digital audio to analog television broadcasts (see Chapter 7). This places the new carrier in the vestigial side band (VSB) region of the signal. To accommodate this, the lower VSB slope is increased. Rather than starting at the traditional 750 kHz below picture carrier, in the Digideck system, it starts at 500 kHz and drops more rapidly. The carrier is about 36 dB below peak power and has a raw capacity of 700 kb/s. Forward error correction and other overhead burdens reduce the data capacity to around 500 kb/s. Digideck calls the new carrier the D-Channel. The data signal is clocked synchronously to the television signal for ease of recovery and for better hiding in the video. The Digideck receiver also depends on an adaptive equalizer. A consequence of the D-Channel is that it must be inserted at the transmitter site and be brought there by an alternative path. Like the WavePhore system, Digideck introduces some artifacts. A marketplace approach will allow the broadcaster to determine acceptability.

The Yes! Entertainment Corporation's system introduces a pulse in the video between 9.1 and 10.36 microseconds after the start of the horizontal synchronization pulse. The data rate is very low, about 14 kb/s. Its application is to deliver audio to a talking toy teddy bear! A. C. Nielsen uses line 22 of one field of the video for transmitting a program source identification. This ID is used to measure the viewing population for statistical purposes. A fifth system by En Technology was denied permission at the time of the R&O. This system allowed data to extend from the VBI into all areas of the picture with the image being constrained to a variable-size box surrounded by the "snow" caused by the data. This system was judged too intrusive.

EnCamera Sciences Corporation has presented papers on a system that compatibly embeds up to 4.5 Mb/s in an analog 6-MHz NTSC television signal. Up to 3.0 Mb/s is modulated using multilevel signaling onto a carrier that is in quadra-

ture with the visual carrier. This approach would normally result in visible arti-facts in the analog video. However, techniques have been developed that avoid these problems. An additional 1.5 Mb/s is amplitude modulated onto the aural carrier using multilevel signaling and specialized filters.

Because of the advent of low-cost, high-performance, single-chip tuners, it is possible to multiplex datastreams onto more than one cable channel simul-taneously. As a consequence, these techniques can be combined to imple-ment high-capacity datastreams over more than one cable channel. Statistical multiplexing can be employed to benefit in these applications.

Since 1.5 Mb/s can provide acceptable video, especially when the source ma-terial is film, a 4.5-Mb/s capacity datastream can supply up to three simultaneous movies while preserving the original analog channel's use. Thus even a modest cable system with 30 analog channels could add up to 90 digital channels without a cable system rebuild and without degrading the analog service already in place. This approach has been called the compatible digital upgrade. More or less data capacity can be assigned to the video channels to match the video quality to the application. This can be done on a dynamic basis as needs change. The capacity can also be partially or completely allocated to data carriage for personal comput-ers or Web appliances. The applications that were anticipated with teletext but that failed in the United States because of low data rates, poor graphics, and high memory costs can now be realized.

2.5 Signal Quality and Customer Expectations

The human response to television pictures is a complex phenomenon. This makes establishing performance specifications difficult. Yet these specifications are important for setting design goals, purchasing objectives, contractual obliga-tions, and regulations.

Psychophysical testing assesses subjective responses of a large number of subjects to television pictures and assigns figures of merit versus objectively mea-sured parameters such as random noise, phase noise, microreflections, composite triple beat, and chroma/luma group delay.

2.5.1 CableLabs Updates Early Testing

Early testing was done by the Television Allocations Study Organization (TASO) in 1958.[20] Further testing was done in 1983 by the network labs of Columbia Broadcasting System (CBS), the National Aeronautics and Space Administration (NASA), and the Canadian Research Council (CRC).[21] Most recently, in 1991, work sponsored by CableLabs added to the base of knowledge concerning signal quality and customer expectations.[22]

The most significant conclusion is that customer expectations are rising. The TASO study proclaimed that in 1958 consumers found a weighted signal-to-noise ratio of about 30 dB to be "somewhat objectionable." In 1983, subjects' expecta-

tions were for a much better picture. About 40-dB signal-to-noise ratio was considered to be "somewhat objectionable." The CableLabs study found 45 to 46 dB to be considered "slightly annoying." This is a very substantial increase in demand for picture quality.

The CableLabs tests were conducted under carefully controlled circumstances and used both expert and nonexpert viewers. The expert viewers were used primarily to set the endpoints of the scales for the different parameters. The selected visual material was presented on a custom-made analog video disc and on D2 tape.

Well-established procedures defined by the International Consultative Committee for Radio (CCIR) were employed.[23,24] The subjective scale ranged from "imperceptible," "perceptible but not annoying," "slightly annoying," "annoying," and "very annoying." The double stimulus pair comparison method used allows subjects to compare pictures to reference images. This eliminates the need to remember what images looked like at other points in time.

Subject material was selected for sensitivity to the subjective impairments under study. Sources for the material included the Society of Motion Picture and Television Engineers (SMPTE), the European Broadcast Union (EBU), and Kodak Corporation. All these entities have long histories in the analysis of images.

Five impairments were studied in the CableLabs test: (1) random noise, (2) phase noise, (3) chroma/luma delay, (4) microreflections, and (5) composite triple beat (CTB). These impairments are briefly described here (see Table 2.1). Random noise is the result of thermal motion of electrical charges and yields "snow" in the picture. Some degree of electrical noise is inescapable. However, cascaded amplifiers and other equipment cause noise to accumulate and rise above acceptable limits. Phase noise is the visual manifestation of instability of oscillators that are used in heterodyning devices to shift frequencies. Several of these systems are in the signal path from point of origination to the subscriber's screen. Cable systems include several of these processing steps. Phase noise usually appears as near horizontal streaking of the picture. Chroma/luma delay results in the colored part of the image misregistering from the luminance part of the picture. The color is smeared and located in the wrong place, giving rise to the a comparison with comic book drawings. Microreflections cause delayed versions of the signal to add destructively or constructively. In many cases, the picture is blurred. In a few circumstances, the subjective sharpness of the picture may appear to improve even though the objective measurements may lead to a different conclusion. Quantitative data was not taken, so it does not appear in Table 2.1. Finally, composite triple beat (CTB) is the creation of new, undesirable frequency components due to the signal passing through nonlinearities. The nonlinearities cause new frequencies to be created that interfere with the quality of the picture.

The results indicate that it takes 45 dB of signal-to-noise ratio to be "slightly annoying." A search was conducted for the "perceptible but not annoying" level. This was found to be between 48 and 51 dB. The limit of "imperceptible" came at 53 dB. Phase noise had to be suppressed by 85 dB in order to be judged as "slightly annoying," and by 90 dB to be "perceptible but not annoying." CTB

Table 2.1 Suppression of impairments for various subjective reactions.

Impairment	Imperceptible	Perceptible but Not Annoying	Slightly Annoying	Annoying
Random noise	53 dB	48 to 51 dB	45 dB	40 dB
Phase noise		90 dB	85 dB	
CTB		50 dB	40 dB	30 dB
Chroma/luma delay			300 ns	400 ns

came in at 40 dB for "slightly annoying" and at 30 dB for "annoying," whereas "perceptible but not annoying" was around 50 dB. Microreflections below 20 dB were not of consequence to most viewers. Below 10 dB, some pictures looked sharper, whereas others looked softer. Chroma/luma delay yielded objectionable results when the delay exceeded 400 nanoseconds.

The most important conclusion is that subscribers have become more critical and are likely to become even more so as they see more images delivered with DTV.

2.5.2 The W Curve and DPU

In 1986, CBS Laboratories and the NAB tested subjective reaction of viewers to co-channel interference.[25] The desired signal was at broadcast channel 23 (visual carrier at 525.5 MHz) and consisted of three images: a gray flat field, a still image (of a boat), and moving video. The undesired signal was a series of sine waves at intervals of 500 kHz, starting at the frequency of the lower band edge of the channel. The ratio of the strength of the interfering sine wave to the strength of the visual carrier was adjusted until expert viewers agreed that the interference was "just perceptible." At the lower band edge, to be "just perceptible," the strength of the interfering sine wave needed to be attenuated so that it was only some 15 dB below the visual carrier strength. As the frequency of the interfering carrier was increased, more attenuation was required to bring it down to levels that were "just perceptible." Maximum attenuation, some 68 dB, below visual carrier strength was required at a frequency 250 kHz above the visual carrier. Less attenuation, about 55 dB, was required at the middle of the channel. Again more attenuation, in the low 60s, at the chroma carrier was required. And then, as the upper band edge was approached, less attenuation was required. These measurements yielded a W-shaped curve, as shown in Figure 2.35. The video content of the desired signal had some influence on the amount of attenuation required in the middle of the channel. Somewhat surprisingly, the flat field hid the interference better in the low to midfrequencies, whereas the moving picture hid the interference better around the color subcarrier. Most surprising was the finding that the difference between the "just perceptible" level and unacceptable levels was just 2 to 3 dB.

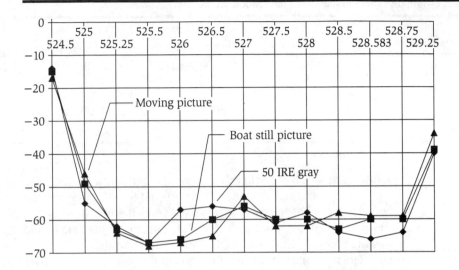

Figure 2.35 W curve: Level of "just perceptible" interfering carriers for three images.

Although co-channel interference is not generally a problem for cable television, direct pickup (DPU) interference is a major headache. In the case of a strong broadcast channel that is carried on the same (or close) frequency on cable, the DPU may result in a ghost or a smearing of the image since the propagation time through the cable is longer than through the air. In the case of other users of the spectrum, interfering patterns may appear on the screen. One manifestation of this is in the cable channel 19 area where pager transmitters are authorized to operate. DPU can occur in the cable plant or in equipment connected to the cable drop. Those DPU signals can propagate between products connected to the same drop and cause picture degradation. If the tuner of a TV or VCR is not adequately shielded, the DPU will cause ghosts and other interference. An occasional problem arises when subscribers connect FM radios to the cable. These products generally have almost no shielding on their tuners and may conduct DPU signals back into the cable system to interfere with reception by other devices. The W curve gives an indication of sensitivity to DPU at various frequencies.

In 1993, the 1986 CBS tests were repeated with the emphasis on their application to DPU. Since the TASO observations were found to no longer be applicable, it was reasonable to wonder whether consumers had become more sensitive to co-channel and DPU problems. The tests were done on three cable channels (6, 12, and 78), at four frequencies relative to visual carrier (+0.25, +0.75, +1.75, and +2.55 MHz), at two desired signal strengths (0 and +10 dBmV), and at two carrier-to-noise ratios (43 and 50 dB). Only the gray flat field was used for the desired signal. The 1993 tests generally confirmed the results of the 1986 CBS experiments. The W shape of the curve remained. Across the three channels, four

frequencies, two desired signal levels, and two carrier-to-noise ratios, the average value of the sine wave strength attenuation relative to the carrier strength to achieve "just perceptible" impairment levels was 56 dB. Since these tests involve human subjects and a wide range of variables, they are difficult to repeat and obtain closely similar results.

2.6 Other Television Systems

The NTSC color system was historically the first to be adopted in the world. Because of the close economic relationships between the North American countries and Japan, all these countries adopted the same color television system, allowing products to be made and sold in these countries interchangeably with only minor modifications due to differing channel frequencies.

Other countries adopted color television later.[26] As a result, they were able to take advantage of the experience of the NTSC countries and make modifications that better suited their priorities. In some cases, these priorities included the desire to protect domestic manufacturing by choosing standards that were different. Controlling the licensing of the intellectual property (patents) erected market-protecting barriers.

In the NTSC system, there are two distinct types of crosstalk between the luma and the chroma signals. Linear crosstalk is a consequence of luma signals (under some circumstances) having spectral components that lie in the chroma regions. The classic example of this is the spurious rainbow pattern that appears on a checkered or tweed jacket. The pattern produces frequencies that stimulate the chroma circuits. Conversely, the 3.58-MHz chroma signals are not completely eliminated from the luma circuits and appear as small moving dots, especially on sharp edges in highly saturated colored areas. Slight nonlinearities cause the luminance and chrominance signals to modulate each other. In particular, the chroma phase and amplitude become a function of the signal's brightness. These phenomena are known as *differential phase* and *differential gain,* and cause infidelities in color reproduction. As circuit technology improved, these problems have become less severe. But in the first years, using vacuum tube circuits, great care and expense was required to keep these problems manageable.

The primary objective of European developments in color television was to minimize the impact of differential phase and differential gain and of other undisciplined phase shifts in the signal path. The principle behind the approach is to shift the phase of one of the chroma signals by 90° on adjacent lines. This approach gives rise to the technique's name: phase alternating at line (PAL) rate. In addition, the chroma reference burst also shifts 90° at the end of each line. The beneficial impact of this approach is that hue shifts on adjacent lines are in opposite directions. Two approaches can be used to take advantage of this. In the less expensive and simpler approach, it is left to the observer's eye to average the hues. This is quite effective for the small phase errors of modern equipment.

However, larger phase errors give rise to horizontal lines sometimes called Hanover bars or Venetian blinds. The more expensive alternative is to use delay lines to electrically average the signals before display. While avoiding the artifacts mentioned, vertical chroma resolution is halved and additional expense results.

PAL also requires a relationship between the chroma subcarrier frequency and the line scan rate. However, because the phase of the chroma burst changes 90° at the end of every line, the relationship must be that the subcarrier frequency is an odd multiple of the sum of one-quarter of the scan line frequency plus one-half the frame frequency. This is in contrast with NTSC's simpler relationship of an odd multiple of one-half of the horizontal scan frequency.

The PAL strategy of alternating phase on adjacent lines and the need to create a relationship with an odd multiple of one-quarter the horizontal scan frequency means that the chroma spectral lines are not centered in the luma spectral gaps. In the time domain, the averaging of brightnesses caused by the chroma signal is even less effective. The saving grace is that the chroma subcarrier frequency is almost a megahertz higher than in NTSC, and therefore the chroma dots are much smaller and less visible.

European television uses the UHF frequencies and thus has more bandwidth available for channelization. Whereas the North American and Japanese video baseband bandwidth is 4.2 MHz, the British have 5.5 MHz and most of Europe has 5.0 MHz. The French and some of the former Soviet Block countries use 6.0 MHz. Additionally, most of the European countries utilize a 50-Hz power line frequency and thus have chosen a vertical monochrome frame rate of 50 frames per second. The consequence of the wider bandwidth and lower frame rate is higher vertical and horizontal resolution. These systems have 625 scan lines per frame compared to the NTSC's 525 scan lines. The horizontal scan line rate is half of 50 times 625, or 15,625 Hz. However, the lower frame rate gives rise to flickering in bright areas of the image and to reduced temporal resolution. The PAL chroma subcarrier is at approximately 4.43 MHz (compared with NTSC's 3.58 MHz). A further benefit of the wider bandwidth is that both color signals in PAL have the same bandwidth, 1.3 MHz. The NTSC strategy of two different bandwidths for matrixed colors matching the human response to flesh tones is avoided. Simple color difference signals are used involving red minus luma and blue minus luma.

The French SECAM system is more complex and uses frequency modulation for the two color difference signals. These two signals are alternated on adjacent lines, yielding half vertical color resolution.

2.7 Summary

The fundamentals of analog television are covered in a manner intended to build an understanding around which to better appreciate cable television technology.

We presented the basic concepts behind the structure of the signal and its modulation. This allows the requirements on the signal processing implemented in cable to be developed in later chapters. Bandwidth compression is important in analog television. The interlaced scanning system and vestigial sideband modulation are both important bandwidth compression techniques. When color is added to the signal, the careful matching of the signal to the human color visual response saves bandwidth.

We surveyed the consequences of the technical compromises forced by the need to have affordable receivers in the early days of television. We explored the important concept of compatibility, demonstrating that complete compatibility is not necessary. Some modest degree of degradation of the performance of earlier products is acceptable in order to advance the performance of new products. A balance is struck.

The story of NTSC is replete with technical compromise. The goals have always been to match image and sound quality to the human who is expected to pay for the hardware and the services. A good compromise and a good bargain have been attained.

Endnotes

1. *Television Standards and Practice: Selected Papers from the Proceedings of the National Television System Committee and Its Panels.* Donald G. Fink, ed. New York: McGraw-Hill, 1943.

2. David E. Fisher and Marshall Jon Fisher, *Tube, The Invention of Television.* The Sloan Technology Series, Washington, D.C.: Counterpoint Publishers, 1996.

3. Ibid.

4. Bernard Grob, *Basic Television,* 3rd ed. New York: McGraw-Hill, 1964.

5. K. Blair Benson and Jerry Whitaker, *Television Engineering Handbook,* revised edition. New York: McGraw-Hill, 1992.

6. Bernard Grob, op. cit.

7. Mischa Schwartz, *Information Transmission, Modulation, and Noise,* 2nd ed. New York: McGraw-Hill, 1970.

8. K. Blair Benson and Jerry Whitaker, op. cit.

9. B. P. Lathi, *Communication Systems.* New York: John Wiley & Sons, 1968.

10. Andrew F. Inglis and Arch C. Luther, *Video Engineering,* 2nd ed. New York: McGraw-Hill, 1996.

11. Mischa Schwartz, op. cit.

12. Donald G. Fink, *Principles of Television Engineering.* New York: McGraw-Hill, 1940.

13. *Proceedings of the IRE, Color Television Issue.* Volume 39, Number 10, October 1951. The Institute of Radio Engineers, New York.

14. *Proceedings of the IRE, Second Color Television Issue.* Volume 42, Number 1, January 1954. The Institute of Radio Engineers, New York.

15. *Color TV Training Manual,* 3rd ed. Indianapolis, IN: Howard W. Sams, 1970.

16. A. M. Morrell, H. B. Law, E. G. Ramberg, and E. W. Herold, *Color Television Picture Tubes.* New York: Academic Press, 1974.

17. Keith Jack, *Video Demystified.* San Diego, CA: HighText Publications, 1996.

18. Theodore S. Rzeszewski, *Television Technology Today.* Piscataway, NJ: IEEE Press, 1985.

19. Special Issue on Consumer Text Display Systems, *IEEE Transactions on Consumer Electronics,* Vol. CE-25, No. 3, July 1979.

20. C. E. Dean, Measurements of the Effects of Interference in Television Reception. *Proceedings of the IRE,* Institute of Radio Engineers, New York, June 1960.

21. G. Chouinard, W. A. Whyte Jr., A. A. Goldberg, and B. L. Jones, Experimental Results Supporting the Determination of Service Quality Objectives for DBS Systems. *IEEE Journal on Selected Areas in Communications,* Vol. SAC-3, January 1985.

22. B. L. Jones and J. A. Turner, Subjective Assessment of Cable Impairments on Television Picture Quality, *1991 NCTA Technical Papers,* p. 92. NCTA, Washington, D.C.

23. Method for the Subjective Assessment of the Quality of Television Pictures, Recommendation 500-4, *Recommendations and Reports of the CCIR,* Vol. XI, Part 1. CCIR, Dubrovnik, Croatia, 1986.

24. B. L. Jones and L. E. Marks, Picture Quality Assessment: A Comparison of Ratio and Ordinal Scales. *SMPTE J,* Vol. 94, No. 12, December 1985.

25. G. Chouinard et al., op. cit.

26. Geoffrey H. Hutson, *Colour Television Theory, PAL-System Principles and Receiver Circuitry.* London: McGraw-Hill, 1971.

Other References of Interest

Donald Christiansen, *Electronics Engineers' Handbook,* 4th ed. New York: McGraw-Hill, 1996.

Margaret Craig, *Television Measurements, NTSC Systems.* Beaverton, OR: Tektronix, 1990.

Jeffrey L. Thomas, *Cable Television Proof-of-Performance.* Englewood Cliffs, NJ: Prentice Hall, 1995.

Ted Rzeszewski, *Color Television.* Piscataway, NJ: IEEE Press, 1983.

Joel Brinkley, *Defining Vision, The Battle for the Future of Television*. New York: Harcourt Brace, 1997.

J. Allnatt, *Transmitted Picture Assessment*. New York: John Wiley & Sons, 1983.

B. L. Jones, Subjective Assessment of Cable Impairments on Television Picture Quality, *1992 NCTA Technical Papers*, pp. 1–25. NCTA, Washington, D.C.

NCTA Technical Papers, yearly, NCTA, Science and Technology Department, Washington, D.C.

IEEE Transactions on Consumer Electronics, quarterly, Institute of Electrical and Electronic Engineers, Piscataway, NJ.

Chapter 3
Digitally Compressed Television

3.1 Introduction

Digitally compressed video is a subject of broad scope. The full story fills volumes and cannot be completely covered here. Our objective is to examine the fundamentals that will facilitate an understanding of the relationship of the topic with the wider purposes of this book. There are a wide variety of implementations of digital video compression, but all that have achieved commercial success are based on the same principles. Different set top terminals are needed for the two or three major DBS systems, the cable systems, and the off-air broadcast system. The details of implementation differ sufficiently to make the systems incompatible. There is not enough space to go into all those details here. The principles are fascinating and are all that is needed for the present investigation. Anyone interested can occupy many months pursuing the details from other sources.

3.2 Digitization of Video

Before considering the digitization of video, we must first review the differences between analog and digital signals and some of the reasons for using digital techniques. These are general principles that apply to any analog signal, not just video analog signals.

3.2.1 Digital Signals

The human sensory systems for images and sounds are analog. Images and sounds start as analog phenomena. To be enjoyed by humans, the phenomena must be converted to signals, and the signals must eventually be displayed as analog stimuli for eyes and ears to enjoy. Unfortunately, as analog signals are transmitted over long distances, they encounter noise, distortion, and interfering signals that degrade the quality of the images and sounds, and eventually make them first unpleasant and then unusable. If the analog signals are converted to digital signals, an arbitrarily small degree of noise is introduced in the conversion process, but all subsequent degradation of the signal can be avoided using practical and well-understood techniques.

The advantages of baseband (unmodulated) digital signals include the ability to completely regenerate the signal, thereby preventing the accumulation of noise and distortion, and the ability to apply computational techniques for the purposes of error detection and correction and redundancy reduction. Redundancy reduction is important because it saves data storage space and transmission time and bandwidth.

To convert an analog signal into a digital signal, it is first sampled in time. This can be accomplished by multiplying the analog waveform by a train of evenly spaced narrow pulses that alternate between zero volts and a constant, convenient voltage (for example, 1 volt) (see Figure 3.1). The information science theorist Harry Nyquist of the Bell Telephone Laboratories proved that if a signal is

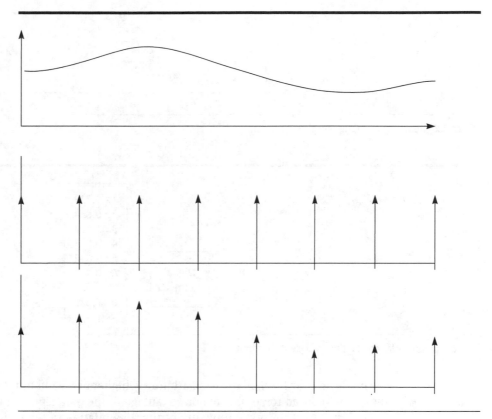

Figure 3.1 Time sampling an analog waveform.

sampled at least as frequently as twice the maximum signal frequency it contains, it can be perfectly recovered with no loss of information.[1,2] This principle is called the Nyquist theorem. Sampled signals are still analog signals because they can take on any value. They are just time quantized.

Nyquist proved that if the sampled signal is low-pass filtered, the original signal is recovered with just a delay caused by the propagation time through the circuits (see Figure 3.2). If each time sample's strength is then measured (for example, with a digital voltmeter) and the resulting measurement represented by a number of limited precision, the sampled analog signal will be converted into a sequence of data (see Figure 3.3). Limited precision numbers have a fixed number of decimal places. The uncertainty in precision of the number is determined by the value of its last decimal place. The information to be transmitted is no longer the original analog signal or its time-sampled version (which can take on any value), but rather another signal that conveys the limited precision numbers describing the strength of the original signal samples. The representation of the signal by a limited precision number introduces an error that can be considered to be a type of noise called quantization noise. The amount of quantization noise can

Figure 3.2 Recovery of a time-sampled waveform.

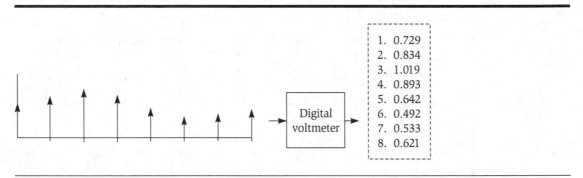

Figure 3.3 Quantization of a time-sampled waveform.

be made arbitrarily small by using arbitrarily higher precision numbers, but it can never be reduced to zero. A major advantage of the data signal approach is that techniques exist for preventing any further degradation of the signal.

The limited precision numbers used to represent the sampled analog signal can be of a variety of forms. Most individuals are accustomed to using a number system based on the value 10. That is, numbers commonly used in human transactions utilize the 10 symbols called *digits:* 0, 1, 2, 3, 4, 5, 6, 7, 8, and 9. This is called a base 10 system. A further characteristic of the base 10 system is that the value of a number is determined by the symbol used and its place in the string of numbers. The rightmost place has only the symbol value. The next place to the left has the symbol value times 10. The next place to the left has the symbol value times 100, and so on. Thus the number 543 has the value 3 plus 4 times 10 plus 5 times 100 (see Figure 3.4). A similar number system can be created using just two symbols, 0 and 1. Such a system is called *binary* because it has only two fundamental symbols. The symbols are called *binary digits,* or *bits* for short. In such a system, higher values are represented by the values assigned to the locations of the symbols in the number. Again, the rightmost location has the symbol value times 1. The location to the left has the symbol value multiplied times 2, or double the place value of the location to its right. The next location to the left has the symbol value multiplied times 4, or double the place value of the location to its right. The next location to the left has the symbol value multiplied times 8, or double the place value of the location to its right. The process continues with

$$
\begin{array}{rcl}
5 \times 100 & = & 500 \\
4 \times 10 & = & 40 \\
3 \times 1 & = & \underline{3} \\
& & 543
\end{array}
$$

Figure 3.4 Base 10 system.

$$
\begin{array}{rcl}
1 \times 4 & = & 4 \\
0 \times 2 & = & 0 \\
1 \times 1 & = & \underline{1} \\
& & 5
\end{array}
$$

Figure 3.5 Base 2 system.

each new place having twice the value of the last. So the binary number 101 is valued, starting from the right-hand side, as 1 times 1 plus 0 times 2 plus 1 times 4. The total is 5 (see Figure 3.5). This same process can be used to determine the value of any binary number.[3]

Advantages of binary numbers include that they have two states that can be represented by simple and inexpensive circuits, and that the impact of electrical noise and distortion can be minimized or even eliminated by proper design. A circuit element, such as a transistor, that processes an analog signal must faithfully reproduce all values of the signal while adding a minimum of distortion and noise. If many such circuit elements process an analog signal, their individual contributions of noise and distortion accumulate, causing substantial signal degradation. On the other hand, a binary circuit can have two well-defined states, on and off, which are easily distinguishable. The on state may represent the binary number 1, whereas the off state can represent the binary number 0. (The opposite choice is equally valid.) The important point is that if the circuit element is mostly off, but not completely off, it will not be confused with the on state. Similarly, if the circuit element is mostly on, but not completely on, it will not be confused with the off state. Thus imperfect performance of the circuit can still faithfully represent the binary values. Only when the on state approaches half of the assigned value or when the off state is almost halfway to the on condition can confusion result. If this degree of deficient performance is avoided, the two states can be discriminated and the signal perfectly resolved (see Figure 3.6). If, as the signal is transmitted, it suffers some noise and distortion degradation, it can still be perfectly recovered as long as the two states—the on state representing a binary 1 and the off state representing a binary 0—can be reliably discriminated. Eventually, sufficient noise and distortion will accumulate so that the two states become confused. If the system is designed so that the signal is regenerated prior to this destructive level of degradation, a fresh binary signal can be substituted for the degraded signal, and all the damage caused by noise and distortion can be completely undone. This process can be repeated an arbitrary number of times allowing error-free communications over arbitrarily long distances. This is something that cannot be accomplished with analog signals.

A further advantage of digital signals is that the circuit design art has progressed to where the transistors that process digital signals are very small and very inexpensive. Gordon Moore, one of the founders of the Intel Corporation,

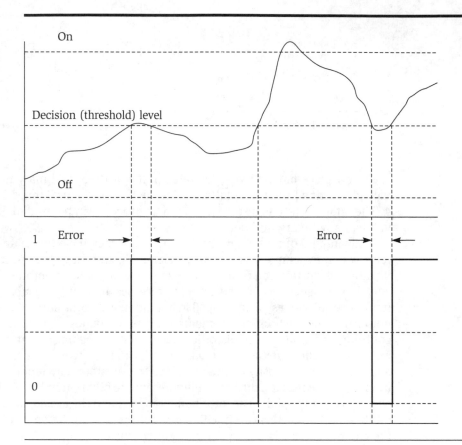

Figure 3.6 Impact of excessive noise.

observed that approximately every 12 to 18 months, the number of digital transistors that can be afforded for a given price doubles. Alternatively, the cost of a given number of digital transistors approximately halves in that period. This process has been continuing for decades and appears likely to continue for some time to come. As an example of this phenomenon, the first personal computers introduced in the early 1980s used an Intel-brand integrated circuit (IC) that had 30,000 digital transistors. The Pentium computer ICs of the mid-1990s have more than 5 million digital transistors. Tens of millions of digital transistors can be expected in consumer products at affordable prices by the end of the millennium. The same experience has not been enjoyed by analog circuits because they must faithfully process the infinite range of values of analog signals. That severe constraint has prevented analog circuits from progressing as fast or as far in complexity and cost reduction.

An additional advantage of digital signals is that they can be mathematically manipulated in very complex ways that lead to methods of determining if trans-

mission errors have occurred and even of correcting some of those errors. Note that there are only two possible types of errors: a binary 1 symbol may be damaged and converted into a binary 0 symbol, or a binary 0 symbol may be damaged and converted into a binary 1 symbol. There are no other alternatives in a binary system. As an example, a common method of identifying an error condition is called parity detection. Binary symbols are grouped into clusters of seven, and an eighth symbol is appended depending on whether the previous seven have an even or an odd number of 1 symbols. If the appended symbol is such that an even number of 1 symbols is always present in each valid group of eight symbols, the method is called *even parity*. A single transmission error will convert a 1 to a 0 or a 0 to a 1, and the result will be an odd number of 1 symbols. The system can detect this error. If two errors occur, a much less probable event, the system will be fooled and think that no error has occurred because there will still be an even number of 1s. However, if three errors occur, the damage will again be detected. The ability to detect certain error conditions has come at the price of an appended symbol that takes up transmission time and that requires additional circuits to process the signal at both the transmission and receiving ends. More complex schemes, called data detection and correction algorithms, can detect multiple errors and even compute the original correct signal. These more complex methods increase the amount of additional, nondata symbols and are said to have increased *overhead*. Also, additional processing complexity is required at both the sending and receiving ends of the transmission path.

Still another advantage of digital signals is that they are amenable to methods of compression that reduce the redundancy in the information. This allows more information to be transmitted per unit time in the same bandwidth or reduces the bandwidth required to transmit a given amount of data. A further benefit of compressed signals is that they require less memory for storage. Just one example of data compression is the technique of run length coding.[4] If a data signal contains a "run" of the same symbol, a coded message can indicate the length of the run with fewer symbols than simply repetitively transmitting the basic symbols themselves. For example, if there are 30 "0" symbols, fewer than 30 symbols are required to code that fact. Another example is the use of special symbol tables defined for the specific set of information to be transmitted. For example, if the information set to be transmitted consists of words from the English language, a special symbol table is used. Information groups with a high frequency of occurrence are assigned short digital codes, and information groups with a low frequency of occurrence are assigned the remaining longer codes. The Morse code is an example of this technique. The letter *e* is the most frequently used letter in the English language. It is assigned the shortest Morse code, the dot. Numbers and punctuation occur much less frequently and so are relegated to the longer series of dots and dashes. All the other characters of the alphabet and the numeric and punctuation symbols are assigned codes of lengths that are inversely proportional to their probability of occurrence in the English language. Huffman developed a method of creating such compression codes, called Huffman codes, when the probability of the message symbols is known. If another language is to be

transmitted, another symbol table is used. If digitized video is to be transmitted, still another symbol table is applied.[5]

Bits are grouped into *words*. The number of bits in a word is a power of 2—2, 4, 8, 16, 32, 64, and so on. A group of eight bits is called a byte. When data is stored, the customary unit of storage is the byte. Larger units of storage are labeled with Greek prefixes and are the powers of 2 nearest to 1,000, 1 million, and 1 billion: kilobyte for $2^{10} = 1,024$ bytes, megabyte for $2^{20} = 1,048,576$ bytes, and gigabyte for $2^{30} = 1,073,741,824$ bytes. The standard published by the Institute of Electrical and Electronic Engineers (IEEE) calls for the following abbreviations: kilobyte, kB; megabyte, MB; and gigabyte, GB. Note carefully which letters are capitalized and which are not. The use of powers of 2 is based on the need to access the locations in memory. This is done with binary signals, and they operate in powers of 2. When data is transmitted, the unit of measure is the bit per second rather than the byte per second. In transmission, it is customary to use the Greek prefixes in their normal manner. Thus, using the IEEE standard notation, a kilobit per second (kb/s) is just 1,000 bits per second of data transmission; similarly, for the megabit per second (Mb/s) and the gigabit per second (Gb/s).

When binary signals are to be transmitted at radio (or television) frequencies, they must be modulated onto a carrier for the same reasons that analog signals are modulated onto carriers. The simplest method is to amplitude modulate the logic levels with two different strengths onto the carrier. At the data receiver, the goal is to recover the data from the baseband signal after amplitude demodulation. If, as is usually the case, the noise and distortion are modest, the most important task of the data demodulator is to convert the analog representation of data into clean logic levels of voltage specified by the circuit designer. This is done with a *slicer* and a sampling circuit driven by a synchronized clock. The slicer is a circuit that compares the input signal strength with a predetermined voltage level called a *threshold*. If the input signal is above the threshold, it is assigned one of the two logic levels. If it is below the threshold, it is assigned the other logic level. The output of the slicer is once again a clean signal, free of noise and distortion. However, it is not yet data since there is ambiguity as to where the data pulses begin and end. A commonly used data format is the nonreturn to zero (NRZ) system. In this method, a string of several logic 1s would be represented by a constant high-signal level, whereas a string of several logic 0s would be represented by a constant low-signal level. Any ambiguity between where one bit ends and the next one begins is resolved by circuits that sample the logic levels at precisely the correct times. This yields data pulses that are suitable for further digital logic processing or for inputting into a microprocessor.

Since digital signals are essentially numbers, numerical techniques can be applied. Nearly all the branches of mathematics have been employed for the processing of digital data. Of special interest is Boolean logic. This branch of mathematics allows statements of logic to be analyzed in terms of logic operators such as EQUALS, GREATER THAN, LESS THAN, AND, OR, NOT. These kinds of statements are the foundations of controllers and programmable machines that operate in a specified sequence. The computational nature of digital signals makes it

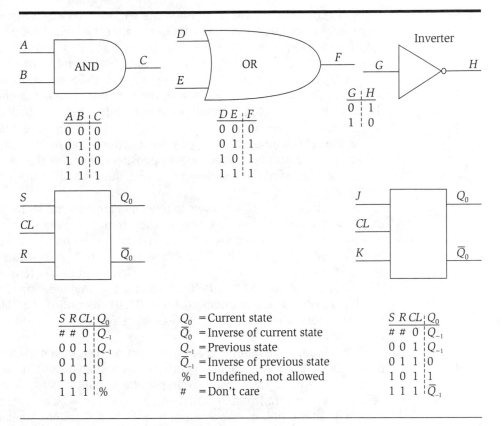

Figure 3.7 Basic logic elements.

possible to implement a great deal of processing in software on more general-purpose processors. The degree of processing can be very complex. In addition, much of the processing can be assisted with dedicated digital circuits. The analysis and design of logic circuits is an extensive study in itself. Several excellent texts exist. However, many logic circuits can be understood with just a few fundamentals. Figure 3.7 shows the fundamental logic circuits as functional blocks along with their *truth tables*. Truth tables are simply exhaustive (that is, all possibilities are included) lists of possible inputs to circuits and the corresponding outputs. The two flip-flop circuits at the bottom of the figure are basic memory elements that operate in digital systems that are timed with *clock pulses*. The term *flip-flop* comes from the fact that these circuits have two defined output states, and the clock signal (designated *CL* in the figures) causes the circuits either to remain in the previous state or to change to the opposite state, depending upon the inputs to the other terminals. The need for the clock signal is to give the memory elements in a complex system time to settle into their new states before executing the next step in the system's operation. Without the clock signal, chaotic operation could result under some circumstances. The flip-flops have two outputs, Q

and its logical inverse. The *RS* flip-flop is a *set, reset* device. When $S = 1$ and $R = 0$, the device goes to its set state with $Q = 1$ after the next clock pulse, independent of what its previous state was. When $S = 0$ and $R = 1$, the device goes to its reset state with $Q = 0$ after the next clock pulse, independent of what its previous state was. The input condition where both $R = 1$ and $S = 1$ is undefined and not allowed. When both $R = 0$ and $S = 0$, the previous state is remembered. When the clock is low, $C = 0$, the circuit is passive. Only when the clock is high, $C = 1$, does the circuit respond to its other inputs. The *JK* flip-flop differs in that the condition where both inputs are at logic 1 has a defined operation. With those inputs, the state of the output changes to its opposite value after the next clock pulse. The name *JK flip-flop* has no known meaning, and its origination is lost in the history of logic design.

Although the use of two levels to represent logic 1s and logic 0s is almost the universal method of designing digital logic circuits, it is not necessarily the only way in which this can be done. If circuit elements were found that had other numbers of very distinguishable and stable states, entire logic systems could be created around them. Consider an electronic element that has four natural states. With four states, two bits could be represented at any one time. This is because two bits have four possible combinations: 00, 01, 10, and 11. Similarly, if an electronic element with eight natural states were available, it could represent three bits at any one time.

Even though the multilevel approach has not yet proved useful in the design of commercially available logic circuits, it is extremely useful in the transmission of data. When only two levels are transmitted, one bit per unit of time is conveyed. The signal transmitted in that unit of time is called a symbol since it carries information. Symbols can be of multiple levels and therefore can carry more than one bit per unit of time. It can be shown that the number of symbols that can be transmitted in a given baseband (unmodulated) bandwidth is approximately twice the maximum frequency carried in that bandwidth. So a 4-MHz baseband bandwidth can carry around eight megasymbols per second when each symbol represents one bit. Such a binary system can tolerate noise levels almost equal to half the excursion between the strength of the signal representing logic 1 and the strength of the signal representing logic 0. In the case of systems where data is embedded in television signals, the main signal is analog television. The noise level at the receiving end of the transmission path must be constrained to modest levels in order to have acceptable television. The noise cannot begin to approach half the excursion between the strength of the signal representing logic 1 and the strength of the signal representing logic 0 without ruining the analog video. Because the noise strength is low, more logic levels can be accommodated.

In the data receiver, after demodulation, three *level slicers* equally spaced between four levels would support transmission of four levels of signal (see Figure 3.8). Four levels of signal would convey two bits of data simultaneously at each symbol time. Similarly, seven level slicers equally spaced between eight signal levels would support transmission of three simultaneous bits of data. This technique greatly speeds up data transmission while requiring only a modest increase

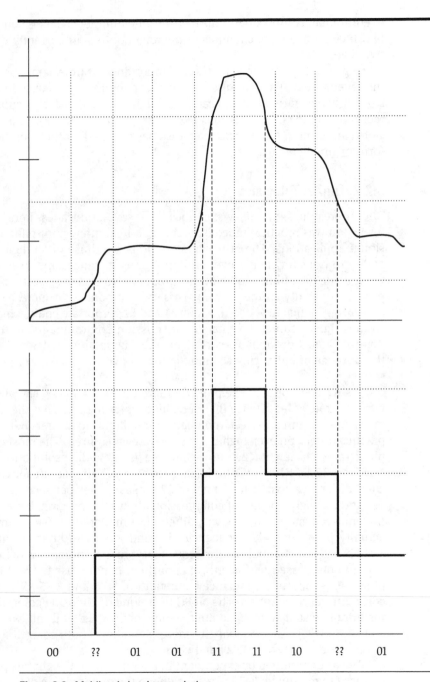

Figure 3.8 Multilevel signal transmission.

in equipment complexity. At the receiver, the multilevel data is converted back to two levels since logic circuits and microcomputers all currently deal with only two levels.

Figure 3.8 also demonstrates the importance of time sampling the output of the level slicers. The output signal may occupy multiple levels in any one time slot. Only one of those values can be valid. It will be the level occupied at the correct moment in time. Situations like those found in the second and eighth time slots can result in errors. They can arise because of distortions of the input waveform or timing errors.

3.2.2 Digital Television

The interest in digital television goes back several decades. Long before digital television was practical for consumer use, it held interest for military and professional applications. The principal attractions are those of all digital signals: the ability to transmit signals over arbitrarily long distances and to store the signals without degradation; the ability to process signals for security and other purposes; the ability to remove unnecessary redundancy to increase transmission and storage efficiency; and the ability to add appropriate coding and redundancy so that signals can survive harsh transmission environments. In the late 1960s, Zenith Radio Corporation worked on the digitization of television signals for the purpose of enhancing scrambling for an over-the-air subscription television service.[6]

As we mentioned, to be properly reproduced, an analog signal must be sampled at a rate at least twice that of its highest frequency. That theoretically minimum rate requires perfect (nonrealizable) filters at the receiving end. To use practical filters, an even higher sampling rate is required. The most common sampling rates chosen for use in digital television with resolution comparable to NTSC are 13.5 MHz for the luminance signal and half that rate for each of the two color difference signals for a total of 27 megasamples per second.

Interestingly, the use of compression allows the program bandwidth and the transmitted bandwidth to be very different. A modern television camera for NTSC actually produces video frequencies beyond what can be transmitted in the 6-MHz broadcast channel. Those frequencies are just thrown away in analog practice. In compressed television, some of them may be used. The 13.5-MHz sampling rate supports a luminance bandwidth of 5.75 and 2.75 MHz in each of the color difference channels. The NTSC technique of choosing color axes that favor the human visual system and limiting one to 0.5 MHz while allowing the other to go to 1.5 MHz is not necessary. Compression allows both color difference signals to have the same 2.75-MHz bandwidth.

It was determined by experiment that a minimum of eight bits per sample are required to minimize visible artifacts due to the quantization of the data. In some applications, ten bits per sample are used. In digital television systems using 27 megasamples per second and eight bits per sample, uncompressed transmission requires 215 Mb/s for the video data. Of course, audio must be digitized as well.

Using our bandwidth rule of thumb, something over 100 MHz of bandwidth would be required. This is not a problem internal to systems that process a single channel or for transmission of a few channels over fiber-optic links. But this kind of bandwidth consumption is not acceptable for broadcast or for multichannel environments where the number of channels is important to subscriber appeal. There is a clear need for methods of reducing the bit rate to acceptable levels.

The big push for consumer digital television came as a consequence of the search for a practical high-definition television (HDTV) system. HDTV was defined as having twice the horizontal resolution, twice the vertical resolution, a wide picture with a ratio of 16 units of width for every 9 units of height, no visible artifacts at reasonable viewing distances, and compact disc quality sound. Before any processing, the analog signal coming from an HDTV camera could consist of 30 MHz of red information, 30 MHz of green information, and 30 MHz of blue information. Almost 100 MHz of analog information is involved. If this is converted to digital form, it is first sampled at more than twice the highest frequency. Then each sample is represented by a byte of data. More than a gigabit per second of data transmission is required. It can be appreciated why all the early HDTV proposals were analog! Methods of reducing this bandwidth appetite had to be found. The work on HDTV began in Japan in 1970 using analog transmission technology coupled with significant digital processing at both the point of origination and at the receiver. The Japanese proposal was called Multiple Sub-Nyquist Sampling Encoding (MUSE). It applied many of the techniques used to minimize NTSC bandwidth. The goal was to match the HDTV system to the human visual response since there is no need to transmit what the eye does not see.

When the FCC decided to pursue HDTV in the United States, it laid down a number of challenges. The first challenge was for a "compatible" HDTV system that would not obsolete current NTSC receivers. The second major challenge was that the original NTSC and the new HDTV signals were to be limited to no more than two 6-MHz slots in the television spectrum. A further restriction required the new signals to avoid undue harm to existing NTSC transmissions. After an extended search for a compatible method of creating HDTV, it became clear that all methods proposed used the original NTSC signal plus in-band and out-of-band "helper signals." All these resources were required to create the compatible signal, and two 6-MHz bands were consumed for each HDTV signal. This approach meant that NTSC would always have to be supported. It also meant that 12 MHz was committed for each HDTV signal. If there were ever to be a transition away from broadcast NTSC, this approach would have to be abandoned.

Zenith Electronics Corporation first broke ranks with the analog proponents by proposing a hybrid system that continued the transmission of the high frequencies of the image in analog form but converted the lower frequencies to a digitized form. This hybrid approach seemed to use the best of both worlds. It recognized that most of the energy in an NTSC signal is in its low frequencies that include the synchronization pulses. In NTSC, the sync pulses have the highest energy because they are separated from the video signal by amplitude discrimination. By digitizing the low frequencies, the majority of their power consumption

was eliminated. Yet the burden on the digital circuits was relaxed because only relatively low frequencies were processed. The high frequencies remained analog and contributed little to the power requirements. The lower data rate digital signals might also be less susceptible to multipath. The remaining problem was that this approach was no longer compatible with existing NTSC receivers. This problem was solved by allowing compatibility to include simulcasting. That is, both the hybrid signal and the NTSC signal would carry the same programming, but at two different resolutions and at two different frequencies. This would preserve the utility of older receivers. And since no successful system that put both NTSC and HDTV into the same 6 MHz had been proposed, two 6-MHz channels would still be required, as in all the other proposed systems.

This approach had a number of major advantages. The low-power HDTV signal was tailored to cause minimum co-channel interference in adjacent locations using the same frequencies. Also, since the NTSC signal was not needed as a component of the HDTV signal, it could eventually be abandoned. The NTSC channel could then be reallocated to other purposes. Even before that happened, the requirement for simulcasting could be relaxed based on policy rather than technological constraints. By this step-by-step process, compatibility was abandoned for the first time in broadcast television. (The noncompatible CBS color system, though temporarily the official system, did not achieve commercial success before it was replaced with the RCA/NTSC compatible color system.)

Shortly thereafter, General Instrument Corporation proposed an all-digital solution. Quickly, all serious proponents with the exception of the Japanese MUSE system converted to all-digital designs. The committee, charged with selecting a winner, found that it could not come to a conclusion. The technical issues were too complex, and the political issues were overwhelming. At the time a decision was to be made, all the proposed systems made unacceptable pictures. The result was an elaborate ruse to score all the systems as acceptable under the condition that a "grand alliance" be formed that allowed the proponents themselves to hammer out a single system. Thus the political battles could go on behind closed doors under the guise of selecting the "best of the best" for a single proposal to the FCC.

The television and computer industries have jointly created standardized toolkits for processing images and maximizing the efficiency of transmission and storage of the resulting digital realizations. Two important systems for accomplishing this are known as the Moving Picture Experts Group (MPEG) standards. The MPEG standards consist of collections of techniques that can be selected depending on the nature of an application. This progress in the area of digital TV bandwidth compression resulted in a national standard being selected by the FCC in December 1996. Included is a modulation scheme that conveys around 20 Mb/s in 6 MHz. If the original raw HDTV signal needed a gigabit per second of data, the resulting compression ratio is 50:1. Roughly 2% of the original information is really needed. Ninety-eight percent of the original information is redundant and can be discarded. Using this standard, a single HDTV program can now be transmitted within the analog broadcast TV channel assignment of 6 MHz rather than

the tens of megahertz once thought necessary. In the case of cable's well-behaved spectrum, double the data transmission rate is possible; two HDTV signals can be carried in 6 MHz.

The same technology that makes HDTV possible in 6 MHz allows multiple standard definition digital signals to be stuffed into 6 MHz. Movies have several advantages over video in this regard. Movies have 24 frames per second versus video's 30. This alone is a 20% reduction in data requirements. Movies have the further significant advantage that they can be processed iteratively. That is, the movie can be run through a processor several times, with adjustment of the processor made to minimize the creation of artifacts on a scene-by-scene basis. Very good results have been obtained with movies at data rates of just 3.0 Mb/s. Quite acceptable results have been seen at 1.5 Mb/s. When compared to the video obtained from a commercially recorded VHS cassette, the digital results have some advantages. Since the HDTV transmission rate is around 20 Mb/s (in 6 MHz), six 3.0-Mb/s movies can be carried in the same spectrum. At 1.5 Mb/s, double that number, 12, is possible. Since cable has a more controlled spectrum, it can approximately further double these numbers, leading to perhaps 24 movies in 6 MHz. This becomes even more practical in systems that use statistical multiplexing.

It is expected that the transmission of standard analog NTSC will occur for many years before a transition to digital television is completed. The British experience with vertically polarized, 441-line monochrome television is instructive. It took some 50 years to abandon that system even with a very small number of receivers. A minority of opinion holds that the U.S. transition may never be completed because of the vast installed base of analog receivers (currently more than 250 million color receivers and more than 150 million VCRs) and the political problems associated with obsoleting consumers' "investment" in products. Low-income viewers and the economically challenged would be especially hard hit. It is difficult simultaneously to be a supporter of "free television" and to require the poor to purchase new, expensive television sets. Of course, the new receivers will rapidly come down in price, but the analog receivers will also continue to decline in price. The shutdown of analog television will not be easy.

3.3 Digital Compression

The history of television technology chronicles a constant struggle to get better pictures in the same or lower bandwidth. The digitization of video makes the need more critical because digitization first vastly increases bandwidth. Fortunately, digital signals facilitate the processing needed for compression.

3.3.1 Video Redundancies

The NTSC television system was created at a time when electronics were expensive and signal processing difficult and limited. The number of programs were

few, and they were relatively expensive to produce. An audience was lacking. A great many compromises were made to allow cost-effective receivers. Without affordable receivers, there would be no point in having efficient use of the spectrum. The priority was rapid penetration of the marketplace and the generation of demand for programming and receivers.

The situation is now quite the opposite. The scarce resource is spectrum. Electronics and electronic processing are affordable and on a predictable path of further cost reduction. The goal now is to create systems that efficiently utilize the spectrum and that will last for several decades without becoming painfully obsolete. The first step is to study the redundancies in television images that can be eliminated to lighten the spectrum burden.

There are a number of significant redundancies in television images. In most frames of video, there are areas of high detail and areas of relatively low detail. NTSC treats them equally, with a constant transmission of information. We can enjoy greater efficiency if we allocate data transmission resources to those areas that need them and do not squander them in areas of low detail.

There are several areas of redundancy to consider: horizontal, vertical, temporal, and spatial (see Figure 3.9). If the image brightness doesn't change in the horizontal direction, there is no need to transmit it again. Since the scanning is done sequentially, this is a relatively easy processing step. The vertical direction is a bit more complex because the information is transmitted at different times corresponding to the raster scan lines. Sufficient data storage is necessary to be able to process this information over nonadjacent segments of time. Interlaced scanning complicates this process further. Changes that take place over longer periods require even more complex processing and more extensive memory. Spatial redundancy involves patterns that repeat themselves or that move to new locations in the picture. Dealing with spatial redundancy is perhaps the most complicated challenge of all.

Better results are obtained if the redundancy removal is matched to the human visual system. NTSC took advantage of the human visual system's differing sensitivities to various colors to reduce the information transmission. The bandwidth of the color signal is a fraction of the bandwidth of the luminance (brightness) signal. Since sensitivity is maximum at some colors, most of the signal resources are concentrated on assuring faithful reproduction of those colors. Other colors can be distorted without most viewers taking note. Since digital television can do much more extensive processing, it can take advantage of other aspects of the human visual system. For example, eye sensitivity to detail is reduced on objects in motion (unless the eye tracks the motion). Motion in the horizontal direction is more noticeable to humans than motion in the vertical or diagonal directions. Perhaps this is because most predators move horizontally. Those of our ancestors who had high visual sensitivity to horizontal movement ate well. Those who had poor sensitivity were eaten and did not reproduce. Vertical sensitivity to movement appears to have been less important. These phenomena help us to allocate resources as we seek to compress images.

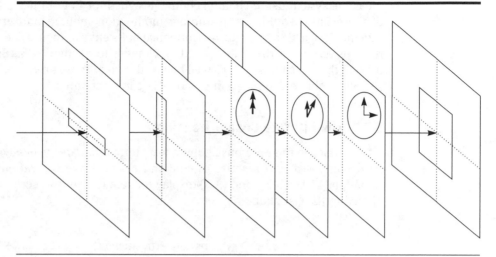

Figure 3.9 Various kinds of redundancies.

3.3.2 Source Coding and Channel Coding

There are two distinct aspects to the digitization of video and audio. The first is called source coding. As the name implies, source coding is the conversion of the source information into digital form and its processing to meet certain objectives. The objectives include the amount of resolution, the tolerance for artifacts, the maximum data rate, and the cost constraints. All these objectives dictate the kind of processing that makes sense for any particular application. As we'll soon see, the variety of objectives means that there is not just one way of coding the source information. There is a whole toolkit that has interrelated, but not necessarily fully compatible, approaches.

Channel coding tailors the signal for the medium used. The characteristics of satellite, broadcast, cable, microwave, twisted pair, fiber optics, video disc, magnetic tape, and so forth are quite different. Some media are essentially noise free. Other media have little random noise, but extensive bursty noise. Still other media are plagued with multipath and co-channel interference, whereas some media exist with essentially none of these difficulties. These varieties of characteristics yield different requirements for modulation and error protection, which are the fundamental techniques of channel coding.

A signal may experience multiple channel codings and recodings. For example, the signal may start out on videotape at a satellite uplink using the type of channel coding appropriate for magnetic media. It might then be conveyed via satellite to a broadcast affiliate using channel coding appropriate for a satellite link. It could then be broadcast over the air using channel coding for that environment. Finally, it might be picked up by a cable system headend and put through

the cable system using channel coding designed for fiber optics and cable. Each of these links would use a different modulation method and error-protection scheme designed to optimize performance in its environment. The source coding remains the same; the channel coding changes to fit the media. Since the spectrum in the cable plant is very well behaved, two or more broadcast source coded signals might be combined in one 6-MHz cable "channel."

3.3.3 The Discrete Cosine Transform

As discussed in the previous chapter, waveforms that are cyclically repetitive can be decomposed into a series of sine and cosine waves of varying amplitudes. Equations (3.1), (3.2), and (3.3) display the familiar Fourier series equations discussed in the last chapter.[7]

$$f(t) = a_0 + \sum_{n=1}^{\infty} (a_n \cos n\omega_0 t + b_n \sin n\omega_0 t) \quad (t_0 < t < t_0 + T) \tag{3.1}$$

Equation (3.1) is the trigonometric Fourier series representation of time waveform $f(t)$ over a time interval $(t_0, t_0 + T)$. The various constants a_n and b_n are given by

$$a_n = \frac{\int_{t_0}^{(t_0+T)} f(t) \cos n\omega_0 t \, dt}{\int_{t_0}^{(t_0+T)} \cos^2 n\omega_0 t \, dt} \tag{3.2}$$

and

$$b_n = \frac{\int_{t_0}^{(t_0+T)} f(t) \sin n\omega_0 t \, dt}{\int_{t_0}^{(t_0+T)} \sin^2 n\omega_0 t \, dt} \tag{3.3}$$

If a waveform is not periodic, such as shown in Figure 3.10(a), it is possible to define a periodic version, as shown in Figure 3.10(b).

The Fourier series in Equations (3.1), (3.2), and (3.3) will be valid over the interval from time t_0 to time $t_0 + T$, but incorrect elsewhere. This is not a problem if we simply reapply the technique of forming periodic extensions for each piece of the waveform. We can then process the signal one piece at a time and load the results in memory.

If a waveform has symmetry about the zero time point, then only cosines are needed to represent it, and all the b_n terms become zero. These cosines occur at multiples—or harmonics—of the fundamental repeat frequency of the waveform.

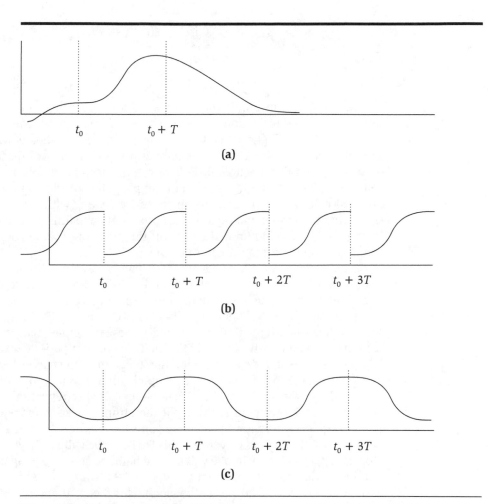

Figure 3.10 Periodic construction. (a) Nonperiodic waveform. (b) Periodic extension. (c) Symmetric periodic extension.

Figure 3.10(c) shows that the periodic extension technique can be applied in a manner that makes the signal symmetric so that its b_n terms will become zero. Again, the Fourier series is valid over only the interval from time t_0 to time $t_0 + T$, but incorrect elsewhere.

The amplitude values of these cosine waves constitute the cosine Fourier series representation of the waveform. The Fourier representation simplifies to just Equations (3.4) and (3.5).

$$f(t) = a_0 + \sum_{n=1}^{\infty}(a_n \cos n\omega_0 t) \quad (t_0 < t < t_0 + T) \qquad (3.4)$$

$$a_n = \frac{\int_{t_0}^{(t_0+T)} f(t)\cos n\omega_0 t \, dt}{\int_{t_0}^{(t_0+T)} \cos^2 n\omega_0 t \, dt} \tag{3.5}$$

We are going to sample the video waveform according to the Nyquist theorem (as discussed in Section 3.2.1). Samples will be taken at more than twice the video waveform's highest frequency. Consequently, the video waveform will no longer be continuous. It will exist only at a finite number of discrete points in time. Therefore the integrals of Equation (3.5) become simple summations because there are only a finite number of sample points to add up rather than a continuous set of values to integrate. Since we are dealing with a bandwidth-limited signal, the summation of Equation (3.4) does not go to infinity, just to the highest frequency in the band-limited signal. In this way, the cosine Fourier series becomes a discrete cosine transform (DCT).

A picture is a two-dimensional object. The same principles that applied in the horizontal direction of the graph of Figure 3.10(a) can be applied in the vertical direction of a picture as well. But instead of the time variable in Figure 3.10, a distance variable is used. This is equally valid. (Of course, distance along a raster-scanned image has a relationship to time!) These principles also apply to both directions simultaneously. To apply the DCT, we need a periodic extension of the image. This is best done by breaking the image up into smaller blocks. The size block chosen in most digital television applications is 8 picture elements (pixels) wide and 8 pixels high. This block can be imagined to repeat in a checkerboard pattern so that it is a symmetrical replication from checkerboard square to checkerboard square. This allows us to utilize only cosines and avoid the sines entirely. But just as with the time waveform, this DCT will be valid only within the chosen block. Each block of the image will have to handled separately. Again, bandwidth limiting of the video means that the image details are limited in the number of variations they can make within the block. This means that the DCT summations have a fixed maximum number and don't have to go to infinity.

The video samples in two directions are transformed into a form of spectrum by the two-dimensional DCT of Equations (3.6), (3.7), and (3.8).[8]

$$F(u,v) = \frac{1}{4}C(u)C(v)\sum_{x=0}^{7}\sum_{y=0}^{7}f(x,y)\cos\left[\frac{(2x+1)u\pi}{16}\right]\cos\left[\frac{(2y+1)v\pi}{16}\right] \tag{3.6}$$

where

x and y = pixel indices within an 8-by-8 video block

That is, x and y are the row and column numbers of the block that contains the measurements of the picture's brightness at each location in the block. Similarly, u and v are DCT coefficient indices within an 8-by-8 spectrum block. That is, u

and v are the row and column numbers of the block that contains the measurements of the picture's spectrum at each location in the block. Here

$$C(u) = \frac{1}{\sqrt{2}} \text{ for } u = 0 \quad \text{and } C(u) = 1 \quad \text{for all other values of } u \qquad (3.7)$$

$$C(v) = \frac{1}{\sqrt{2}} \text{ for } v = 0 \quad \text{and } C(v) = 1 \quad \text{for all other values of } v \qquad (3.8)$$

To summarize, when we take a block 8 pixels wide and 8 pixels high from a bandwidth-limited video signal, there are 64 pixels. Each pixel is sampled in time. Its amplitude is measured, and the measurement is rounded to the nearest quantizing level. This quantizes the amplitude to a resolution of 8 bits (256 levels). We then have an array of 64 numbers in an 8-by-8 square (see Figure 3.11(a)).

If we apply Equations (3.6), (3.7), and (3.8) to this array, we will get another array of 64 numbers in 8 rows and 8 columns (see Figure 3.11(b)). This array represents the frequency components of the original amplitude array. The cells in this array contain numbers called the frequency coefficients. The number in the upper left-hand corner is related to the dc component, that is, the average amplitude across the video square. The cells across the top row are the values of the horizontal frequencies in the block. The cells down the left side of the block are the vertical frequencies in the block. The other blocks represent frequencies in other directions (see Figure 3.11(c)).

The Fourier series works in both directions. If we have an amplitude waveform, we can convert it into a frequency series. If we have a frequency series, we can convert it back to the waveform. Similarly, there is an inverse discrete cosine transform (IDCT), which will convert the frequency coefficients array back into the amplitude array. The equations for the IDCT are in (3.9), (3.10), and (3.11).

$$f(x,y) = \frac{1}{4} \sum_{u=0}^{7} \sum_{v=0}^{7} C(u)C(v)F(u,v) \cos\left[\frac{(2x+1)u\pi}{16}\right] \cos\left[\frac{(2y+1)v\pi}{16}\right] \qquad (3.9)$$

$$C(x) = \frac{1}{\sqrt{2}} \text{ for } x = 0 \quad \text{and } C(x) = 1 \quad \text{for all other values of } x \qquad (3.10)$$

$$C(y) = \frac{1}{\sqrt{2}} \text{ for } y = 0 \quad \text{and } C(y) = 1 \quad \text{for all other values of } y \qquad (3.11)$$

Applying these equations to the array of Figure 3.11(b) yields the array of Figure 3.12, which is recognized as the original amplitude array. This is a lossless conversion (to the limits of computational accuracy).

As described in the previous chapter, color television is based on the discovery that the human eye responds to three colors and mixes them to produce

139	144	149	153	155	155	155	155
144	151	153	156	159	156	156	156
150	155	160	163	158	156	156	156
159	161	162	160	160	159	159	159
159	160	161	162	162	155	155	155
161	161	161	161	160	157	157	157
162	162	161	163	162	157	157	157
162	162	161	161	163	158	158	158

(a)

235.6	−1	−12.1	−5.2	2.1	−1.7	−2.7	1.3
−22.6	−17.5	−6.2	−3.2	−2.9	−0.1	0.4	−1.2
−10.9	−9.3	−1.6	1.5	0.2	−0.9	−0.6	−0.1
−7.1	−1.9	0.2	1.5	0.9	−0.1	0	0.3
−0.6	−0.8	1.5	1.6	−0.1	−0.7	0.6	1.3
1.8	−0.2	1.6	−0.3	−0.8	1.5	1	−1
−1.3	−0.4	−0.3	−1.5	−0.5	1.7	1.1	−0.8
−2.6	1.6	−3.8	−1.8	1.9	1.2	−0.6	−0.4

(b)

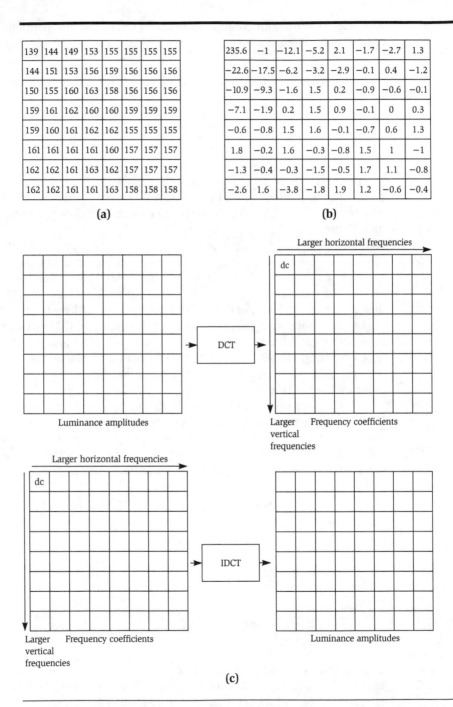

(c)

Figure 3.11 DCT 8-by-8 blocks. (a) Luminance values. (b) Frequency coefficients after DCT. (c) DCT and IDCT transformations.

139	144	149	153	155	155	155	155
144	151	153	156	159	156	156	156
150	155	160	163	158	156	156	156
159	161	162	160	160	159	159	159
159	160	161	162	162	155	155	155
161	161	161	161	160	157	157	157
162	162	161	163	162	157	157	157
162	162	161	161	163	158	158	158

Figure 3.12 Luminance values after IDCT.

essentially all other colors. Color television requires three signals. The three signals start at the camera and end at the picture tube as red, green, and blue. In between, they can be transformed into other formats that are more convenient for processing. In digital television, the color information is sent as two components. The first component is the difference between the luminance value and blue in that part of the picture. The second component is the difference between the luminance value and red in that part of the picture. Since the human eye is less sensitive to detail in color information than in luminance (brightness), the color information can be sent with less detail. This is accomplished in digital television by taking a 16-pixel by 16-line block (called a macroblock) and sampling every other pixel for its two color difference values.

The order of transmission of this information is shown in Figure 3.13. First, four blocks of luminance information are sent. Then one block of each of the two color difference signals is sent. The color difference signals come from sampling every other location in the luminance array of four blocks. All six of these blocks are separately converted by the DCT circuits and separately transmitted to the receiver. The six arrays are combined in the receiver to yield the R, G, and B signals the picture tube needs. This process is repeated for each 16-by-16 macroblock in the frame of video at a rate of about 30 frames per second.

3.3.4 Compression

Let's review what has been accomplished so far: the video signal has been time sampled and quantized in a manner that gives the values of the luminance and color difference components at specific points in the picture. Those luminance points have been organized into blocks of 8 pixels wide by 8 pixels high. The

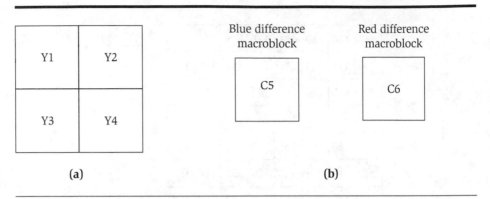

Figure 3.13 Order of transmission. (a) Luminance blocks. (b) Color difference macroblocks.

same has been done for the color difference signals, except every other pixel has been skipped to form macroblocks that span 16-by-16 luminance pixels. Then the values of these pixels have been converted to frequency values using the DCT in a manner that allows the original values to be again computed using the IDCT. This digitization and conversion by the DCT has not, by itself, done anything to reduce the redundancy of the image. More processing is needed to eliminate redundancy.[9]

The first step in the process is to note that there is a much wider range of values in the frequency domain cells than in the original video domain cells. Many of the frequency numbers are very small. In order to efficiently transmit the information in the frequency cells, we must reduce the amount of information necessary to convey it. It is already sampled and digitized, so the next step is to quantize the values in the frequency cells.

The transform coefficients represent different frequencies in brightness and color. Studies have shown that the human visual system has differing sensitivities to this information depending on the amount of motion, the amount of detail, and the color of the image. There is a lot more information in this data than the human visual system requires for a satisfactory image. As a consequence, the values in the cells can be approximated by a smaller set of values. The actual cell value is compared with the smaller set, and the nearest set value is chosen to represent the actual value. This process is another quantization mechanism. The cells are quantized differently according to the results of the research on human perception. A *quantizer matrix* is assigned to the transform coefficient matrix that prescribes the coarseness of quantization for each cell. Predetermined quantizer matrices are built into the system. Additionally, specific quantizer matrices can be downloaded as part of the signal for special cases.

Coarser quantization requires fewer bits, and the video matrix consisting of 64 cells with 8 bits in each row is transformed into 64 cells with fewer total bits. A most important result is that many of the high-frequency blocks contain co-

efficients that are so small that they become zero when quantized to a coarse scale. Thus the coefficient matrix is loaded with zeros. In addition, a quantizer coefficient is applied to the entire quantizer matrix to adjust its coarseness relative to other blocks in the picture. In this way, some areas of the picture are assigned more bits than other areas in relation to their importance to the image.

The order of transmission of the values in the coefficient matrix is very important. A process called zigzag scanning is employed. Figure 3.14(a) illustrates this. In Figure 3.14(b), the frequency coefficients from Figure 3.11(b) are plotted as a graph in the order scanned in Figure 3.14(a). (In Figure 3.14(c), the large dc term is not plotted so that the relative sizes of the remaining terms can be better seen.) The frequency coefficients quickly become small in value as the scanning progresses through the block. Most can be rounded to zeros, yielding a "run of zeros." Rather than transmit all the zeros, a pair of numbers is transmitted. The first number of the pair tells the number of zeros. The second number tells the value of the cell that terminates the run of zeros. When a condition is reached where all the remaining cells have zeros, an end of block code is sent. Already, the number of bits to be transmitted has been significantly reduced.[10]

Some of the image's blocks will have little detail and lots of zeros in their coefficient matrices and thus will use few bits. Other blocks will have a great deal of detail and relatively few zeros in their coefficient matrices and will require many more bits. In this manner, bits are allocated to the parts of the picture that need them and are not squandered on the parts of the picture that do not. The quantizer matrix helps balance the flow of bits so that a frame's bits are transmitted in time to decode the picture. In the case of prerecorded material, it is possible for human intervention to guide the allocation of bits so that the important parts of the picture are adequately encoded and there is little squandering of bits on unimportant parts of the picture that might just be "busy" anyway.

Another technique is employed to reduce the amount of data to be transmitted. The run length information occurs at a predetermined set of probabilities. Because of this, it is possible to employ the methods variously known as Huffman codes or entropy codes. These methods are similar to the way in which the Morse code is constructed. That is, the most frequently occurring patterns are represented with the shortest codes, and the less frequently occurring patterns are assigned longer codes. This further reduces the number of bits needed to transmit the information.

3.3.5 Intraframe and Interframe Coding

The discussion to this point has described the transmission of a single picture. This process is called *intraframe coding* because it is completely contained within just one frame of video. Most of the benefits of compression come from reducing the redundancy *between* frames of video. This process is called *interframe coding*.

Video consists of approximately 30 frames per second. Movies have 24 frames per second. In both cases, most of the time, successive frames are very

0	1	5	6	14	15	27	28
2	4	7	13	16	26	29	42
3	8	12	17	25	30	41	43
9	11	18	24	31	40	44	53
10	19	23	32	39	45	52	54
20	22	33	38	46	51	55	60
21	34	37	47	50	56	59	61
35	36	48	49	57	58	62	63

Luminance amplitudes

Zigzag scanning

(a)

(b)

(c)

Figure 3.14 DCT spectral block transmission order. (a) Zigzag scanning of the amplitude blocks. (b) Frequency coefficients scanned in zigzag order including dc component. (c) Frequency coefficients scanned in zigzag order without dc component.

Figure 3.15 Interframe encoder.

similar. If we take the limiting case of a movie or video of a still image, say, a landscape with no activity, there are no differences between frames. Sending one frame tells all. Succeeding frames simply can be repeated from the receiver's memory. The successive coefficient matrices would be identical.

Figure 3.15 shows a way in which the waste of sending the same frame over and over can be avoided. Figure 3.15 is the processor at the signal origination site. At the moment the equipment is turned on, all memories are empty. The source makes the first frame available, and the DCT processor creates the frequency coefficient matrices for the luminance and color blocks of the picture, and sends them to the IDCT decoder at the receiver site and also to a local IDCT decoder. The local decoder produces the same image that will exist at the receiver. This image is then compared to the contents of the previous picture memory. Since there was no previous picture, the previous picture memory is empty, and there is nothing to add or subtract. The entire decoded picture is loaded into the picture memory. When the second frame comes in from the source, the previous picture is subtracted from it. Since they are identical, there is nothing coming out of the subtractor, nothing to process in the DCT unit. The coefficient matrix is all zero. The local decoder puts out nothing. The previous picture memory adds its output to the zero output of the local decoder, and the result is recycled into the picture memory. The process repeats itself with no further data being transmitted to the receiver.

Figure 3.16 displays the receiver. Initially, its memories are empty. When the first frame's coefficient matrix arrives, it is decoded in the IDCT unit. This is presented for display and also for storage in the picture memory. At the time of the next frame, a zero coefficient matrix arrives. Nothing is added to the previous picture memory, and its contents are presented for display and are reloaded into the picture memory.

Figure 3.16 Interframe decoder.

Watching a still picture on television is not very interesting. Now let's as-
sume that the picture contains something that moves. Recall our landscape image
shown in Figure 3.17(a). Now consider an airplane entering the picture (see Fig-
ure 3.17(b)). Back at the origination site, Figure 3.15, the previous picture will be
subtracted from the new picture. All that remains is the airplane (see Figure
3.17(c)). The airplane is encoded in the DCT, and then a new coefficient matrix is
transmitted to the receive site and the local decoder. Note that this coefficient ma-
trix will have very little data. It needs only to convey the airplane. Everything else
is unchanged. In the processor at the origination site, Figure 3.15, the coefficient
matrix of the airplane is decoded to create an image of the airplane as now exists
at the receive site. This image is added to the previous image memory output, and
the new total picture of the old landscape and the new airplane is loaded into the
picture memory. At the receive site, Figure 3.16, the coefficient matrix of the air-
plane is received and added to the previous landscape stored in the previous pic-
ture memory. It is presented for display and stored in the picture memory.

Next, the airplane progresses across the sky. It is in a new location in the next
frame (see Figure 3.18(a)). When this image enters the subtractor of Figure 3.15,
two things must change. The image of the airplane's previous position must be
deleted and replaced with the image of the sky that has been uncovered, and
an image of the airplane's new position must be introduced. These two image
changes (see Figure 3.18(b)) are what comes out of the subtractor, are encoded in
the DCT, and are transmitted to the receive site and the local decoder. The image
of a patch of sky at the airplane's previous location and the image of the airplane
at its new position are added to the previous picture memory output. These then
are stored in the picture memory. At the receiver, Figure 3.16, the coefficient ma-
trix is converted in the IDCT processor into the patch of sky and the new plane lo-
cation. These are added to the contents of the previous memory location, the
result presented for display and stored in the picture memory.

Two kinds of coefficient matrices have been transmitted in our example. The
first coefficient matrix was a complete picture. This is an intraframe (or an initial-
ization frame). It has been termed an *I frame*. The I frame was followed by a
coefficient matrix that contained only the changes between the I frame and the
succeeding frame. Since a complete second frame was not transmitted, the addi-

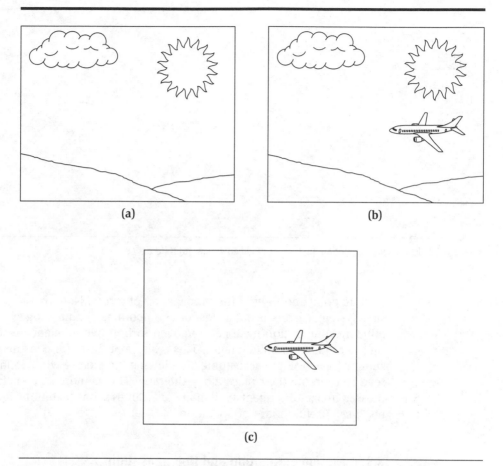

Figure 3.17 Differential transmission. (a) Original image. (b) Modified image. (c) Difference.

tional information is a predictor of what changed. This simplified description includes the fundamental principles involved. In Figures 3.15 and 3.16, the block labeled "Previous picture memory" actually includes additional processing to form a prediction of the next frame. The same prediction process is used in both the origination equipment and the receiving equipment. The frame after the I frame contains the information needed by the predictor to modify the contents of the picture memory for display. That frame is called a *P frame.*

In a perfect world, we'd need just one I frame and all the rest could be P frames. There are a couple of reasons why this is insufficient. First, if the viewer missed the I frame while "surfing" channels, he or she would not get a correct image. In our example, the viewer might just get the airplane and not the landscape! A second problem is that there are imperfections and distortions that accumulate. It is important to reset the process every so often to limit the propagation of errors. So I frames are sent several times a second to allow channel surfers to have a

Figure 3.18 More complex motion. (a) Further modified image. (b) Differences.

starting point and to limit the propagation of errors. Note that in a digital television system, surfing is not as fast or as smooth as in an analog system. Since redundant information has been removed to improve efficiency of transmission, time is required to build images. This is also a problem for consumer VCRs. Consumers have become accustomed to viewing the picture while in fast forward or reverse to locate their place along the tape. This is more challenging with compressed images. Frequent availability of I frames is one technique for accomplishing these useful modes of operation.

3.3.6 Motion Estimation and Compensation

Reviewing what was covered so far, the amount of information to be transmitted has been substantially reduced by

1. Sampling the analog video according to Nyquist
2. Digitizing the analog video samples to 8-bit (256-level) amplitude resolution
3. Dividing the image into 8-pixel by 8-line blocks of luminance information
4. Dividing the color information into 8-by-8 blocks that cover 16 luminance pixels and 16 luminance rows to take advantage of the eye's lower sensitivity to color
5. Transforming these blocks into spectral information with the discrete cosine transform
6. Quantizing the spectrum data in a manner that matches the human visual system and allocates bits to important areas of the image while avoiding squandering them on unimportant areas
7. Zigzag scanning of the coefficient matrix

8. Run length coding the coefficient series

9. Entropy (Huffman) coding the run length data

10. Using an end of block code when all the remaining coefficients have been quantized to zero

11. Encoding the differences between frames rather than separately coding each frame

Additional techniques can be employed to significantly reduce the redundancy still further. If we return to our landscape and airplane example of Figure 3.17, we note that the image of the airplane is the same in all appearances. It is redundant to have to retransmit the airplane. Once should be sufficient! It should be adequate to simply convey an instruction to move the airplane to the new position. Of course, it is not that simple.

Motion estimation is based on the assumption that much of the motion in images consists of a displacement of an element of the image. In its simplest form, motion estimation consists of taking each of the 8-by-8 blocks and comparing it with all possible 8-by-8 blocks formed by making horizontal and/or vertical displacements in increments of one pixel. Some motion estimation algorithms work in half-pixel increments. When the best fit is found, instructions are conveyed to the receiver to work with the contents of the blocks and, when done, relocate them to their new positions. In our example, the encoder motion estimator would determine that the airplane has simply been moved horizontally. Instructions to the receiver would tell it to take the contents of a number of blocks in memory and move them to other locations that are some number of pixels displaced horizontally and vertically.

The motion compensator may find that the match is not perfect. In that case, it finds the errors and transmits only them. In our example, if the airplane made a slight rotation, simply moving it horizontally and vertically would not be enough. The difference between its previous pixels and its new pixels would be encoded. In most cases, this would be just a small fraction of the total pixels. Most would be unchanged. Motion estimation allows the receiver to compensate for moving parts of the picture that do not change much. The resulting instructions to the receiver include taking blocks from memory, modifying them slightly, and then relocating them. When they are relocated, parts will overlay other blocks, and parts will be covered by other blocks. That information must be conveyed as well. Holes in the image will be created. They must be filled. The various parts of the image may move in different directions by different amounts. The motion estimation process can be very sophisticated to aggressively reduce the amount of information that needs to be transmitted (see Figure 3.19). Areas in the preceding frame of Figure 3.19 are found that best match the areas in the current frame. These are transmitted along with instructions for their new position and the differences needed to update them. This minimizes the amount of information that must be conveyed to the receiver.

Comparing Figure 3.15 with Figure 3.16, we see that the system's complexity is asymmetrical. It takes more processing to encode the signal than to decode it.

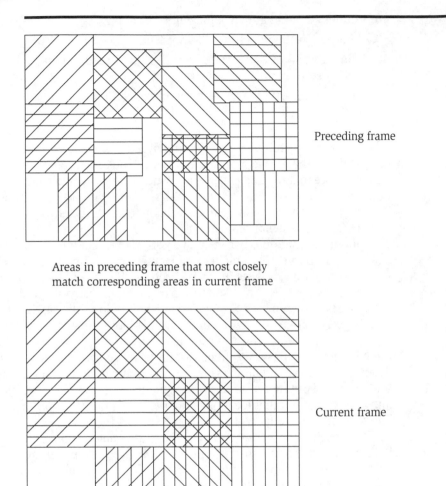

Preceding frame

Areas in preceding frame that most closely
match corresponding areas in current frame

Current frame

Figure 3.19 Motion compensation.

This is intentional since there are many more decoding sites than encoding sites.
The goal is to make receivers inexpensive. The most complex processing is in the
motion estimation. Since this is done at the origination point, viewers will enjoy
improved results with their existing decoders as encoder motion compensation
technology advances.

There is at least one area where low-cost encoding is important: camcorders.
Consumer camcorders will have to encode as well as decode. The trade-off there
is likely to be less aggressive compression to reduce the encoding electronics, at
least initially. A consequence of this will be greater tape consumption. An ad-
vantage of less aggressive compression is the facilitation of viewing the pic-

ture while it is in fast forward and reverse modes. This is considered an important feature.

3.3.7 B Frames and Groups of Pictures

Consider the airplane and landscape example as shown in Figure 3.20. Five locations of the airplane occur. When the airplane moves from one position to another position, it uncovers a piece of sky. If that piece of sky is identical to adjacent sky, motion estimation and compensation will fill it in. If it is different, the difference will have to be transmitted. Note that the piece of sky uncovered when the airplane moves from its first position to its second position is present in subsequent images. If we had one of the subsequent images, for example, Figure 3.20(c), already in memory, we could take the piece of sky needed from a "future" image. We could "predict backward" in time. The ability to predict both forward and backward in time is possible if we introduce a slight delay in the transmission and add memory to the receiver for more than one frame of video. The frames that allow bidirectional prediction are called *B frames.* Although they add cost to the decoder, they increase transmission efficiency. The trade-off in cost and complexity is governed by the cost of the memory. Since the cost of memory has come down substantially, B frames are currently considered a worthwhile investment since bandwidth is the ultimate scarce resource.

We have defined three kinds of frames: I frames, P frames or predictor frames, and B frames or bidirectional frames. They are clustered into groups of pictures (GOPs) (see Figure 3.21). In Figure 3.21(a), the frames are presented in the order in which they are displayed. In order to minimize receiver complexity, they need to be transmitted in the order shown in Figure 3.21(b). In a sophisticated encoder, the flow of the GOPs is interrupted at a scene change. A new I frame is transmitted. I frames require much more data than other frames. So the flow of data is very uneven. Buffers of sufficient size are required to even out the flow.

3.3.8 Digital Artifacts

Since each block of the image is treated separately, it is easy to appreciate that there may be differences in decoded video at the boundaries of the blocks. Techniques exist for smoothing over these boundaries. A digitally compressed image with inadequate data will occasionally display the boundaries of the blocks. This is called *blocking noise.*[11]

Recall that the 8-by-8 DCT and IDCT losslessly convert between the video and frequency space representations to the limits of computational precision. But once the coefficients are quantized, the lossless nature of the computations is abandoned. Some distortion is introduced. If the distortion is minor, it might not be noticeable or, if noticeable, it might not be objectionable. It may just blend in and get lost. However, under some circumstances, sharp edges will yield

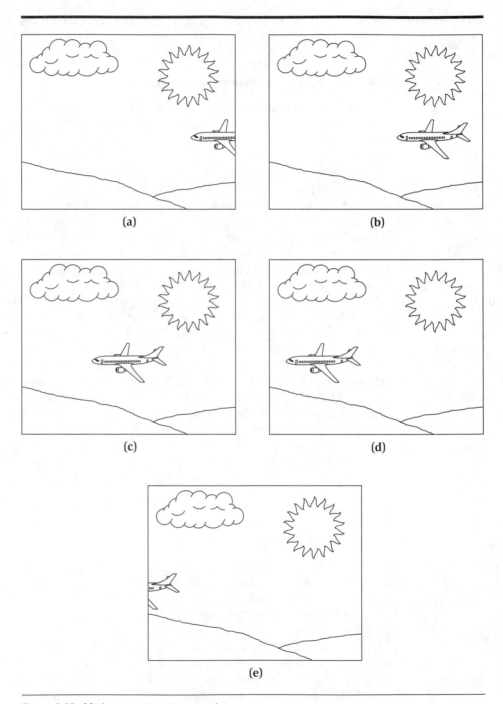

Figure 3.20 Motion compensation example.

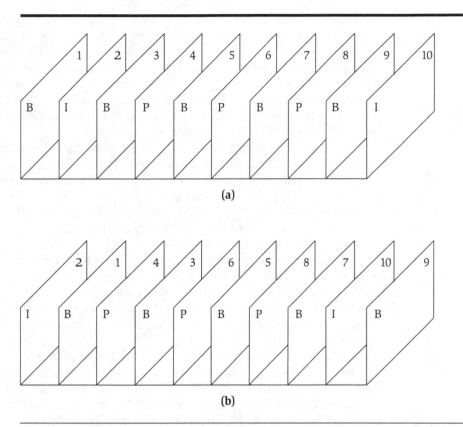

Figure 3.21 Order of transmission and display of I, P, and B frames. (a) Frame type and display order. (b) Frame type and transmission order.

attenuated ripples. This is called Gibbs phenomenon after the mathematician who studied it.

If a sharp point exists in the image, the Gibbs phenomenon will exist in two dimensions. The resulting image distortion has been called *mosquito noise* or *the attack of the killer bees*. This artifact appears as a cluster of mosquitoes or killer bees surrounding the sharp point in the image. If a sharp edge exists in the image, ripples may precede and follow it.

3.3.9 Digital Audio

The human auditory system is considerably less tolerant of imperfections than the human visual system. Compression of quality audio is much more difficult than the compression of video. Compression ratios of only 4:1 or 8:1 are achieved in audio, whereas video enjoys compression ratios of 30:1 or 50:1. Fortunately, the audio bandwidths are small, and the resulting data rates are not too burdensome.[12]

Two major competitive methods of audio compression are proposed for use with video compression: the European Musicam system and the Dolby AC-3 method. Much of Europe and the U.S. DSS satellite system have chosen Musicam, whereas the U.S. cable industry and the U.S. broadcast standards use AC-3. The digital video disc (DVD) technology uses Musicam for 50-Hz scanning systems (mostly European) and AC-3 for 60-Hz systems (mostly U.S. and Japanese). The fundamentals of the two systems are the same, but the details vary. Because of space limitations, we will concentrate on Dolby AC-3.

The Dolby AC-3 system is said to have 5.1 channels of audio: front left, front right, center, surround left, surround right, and low-frequency enhancement (LFE). The bandwidth of the channels is around 20 kHz with the exception of the LFE channel, which is limited to 120 Hz. This gives rise to the ".1" of the 5.1 channels. Since low-frequency sounds are not localized, only one speaker is needed for the low frequencies, and it can carry the low frequencies of all the channels. Additionally, the LFE speaker can be placed anywhere in the room.

Audio is compressed by first converting it to digital form with a sampling rate of 48 kHz. The audio is then segmented into *frames* of samples 1,536 samples long. The analog audio is quantized to at least 16-bit resolution. The frame is divided into six 512 sample *blocks*. Since the human auditory system is so sensitive to imperfections, half of the samples in each block overlap half of the samples from the adjacent blocks. That is, there is a 25% overlap with the preceding block and a 25% overlap with the following block. The sensitivity of the human auditory system to discontinuities requires doubling the number of transmitted samples! The samples are *windowed;* that is, the set of samples is multiplied by a weighting function that tapers the amplitude of the samples at the beginning and at the end of the sample set. The receiver blends the ending tail from one set with the beginning tail from the next set to avoid sudden changes or clicks or pops.

The audio samples are transformed into the frequency domain as a set of spectral data points using a modified version of the discrete cosine transform called the time division aliasing cancellation (TDAC) transform. Its special properties allow recovery from the need to have block overlap that would otherwise generate twice as many frequency coefficients. The result is that the 512 samples of an audio block yield only half as many, 256, frequency coefficients. These frequency coefficients divide the audio spectrum into 256 subbands of less than 100 Hz each. Up to this point, the processing of the digital samples is fully reversible, and only digitization has been accomplished. Compression has not yet been achieved.

As with video, compression is achieved by the quantization of the frequency components of the signal. Audio compression relies on the psychoacoustic phenomenon of audio masking. Audio masking renders lower-level sounds inaudible when there are louder sounds in the same band of frequencies. The phenomenon is quite complex and very nonlinear. It is more effective when the quieter sound to be suppressed is higher in frequency than the louder sound. The sound to be suppressed is the noise resulting from the quantization of the frequency compo-

nents. The psychoacoustic model dictates the degree of coarseness that can be accepted in the quantization of the audio spectrum data. The frequency coefficients are transmitted in the mathematical format known as *scientific notation.* A basic number is multiplied by an exponent. The basic number is called the mantissa. The same exponent generally applies to several adjacent blocks of samples, and data transmission can be reduced by taking advantage of this. Only changes are conveyed. Extensive techniques exist for reducing the amount of data that must be transmitted. The optimum bit rate for 5.1-channel Dolby AC-3 is 384 kb/s although other rates are possible, yielding varying qualities of the resulting decoded audio signals.

3.3.10 JPEG, MPEG-1, and MPEG-2

The Joint Photographic Experts Group (JPEG) created a standard set of tools to be used for the compression of still pictures based on the DCT. This highly successful approach was soon adopted by some to convey movies and video since these are just a succession of still images. Although redundancy in each image is substantially reduced by JPEG, no advantage is taken of the redundancy found between successive pictures.

MPEG created a standard based on the discrete cosine transform using 8-by-8 blocks intended to be used with computers and CD-ROMs. It introduced differential encoding and motion estimation and compensation, which reduced the redundancy between frames. I frames, P frames, and even B frames are used in some extensions to the system. It is intended for progressively scanned images only. No system layer has yet been defined, and Musicam 1 is the audio compression scheme. Conditional access is not provided. As consideration turned to the problems of NTSC and PAL television, the first MPEG system became known as MPEG-1.[13]

Those anxious to apply MPEG to video made nonstandard adaptations and advancements and called their approaches MPEG+ or MPEG-1+. These were proprietary adaptations of the original MPEG.

For video applications, a new version was devised and called MPEG-2. MPEG-2 is specifically designed for moving video and supports interlaced rasters. A systems layer is provided with packetized transport. Two audio systems are commonly used with MPEG-2 video, an advanced Musicam 5.1 and Dolby AC-3, which is also a 5.1 system. Audio is time stamped to the video so that synchronism between the two can be maintained.

MPEG-2 is a toolkit comprising a collection of image-processing components. This means that a specific application uses appropriate portions of the standard. Figure 3.22 shows the way in which the common configurations of components are listed. The "level" describes the amount of video resolution. The "profile" describes the principal features of that collection of components. The general-purpose profile is called the main profile. The simple profile drops the B frames and simplifies the decoder by reducing the memory and processing required. The

		Profile					
Level	Simple	Main	S/N	Spatial		High	
High		1,920				1,920	960
		1,152				1,152	576
		60				60	30
High-1,440		1,440		1,440	720	1,440	720
		1,152		1,152	576	1,152	576
		60		60	30	60	30
Main	720	720	720			720	352
	576	576	576			576	288
	30	30	30			30	30
Low		352	352				
		288	288				
		30	30				

Notes: 1. Matrix intersections with values shown are the only ones defined.

2. Values given in cells represent, from top to bottom, maximum samples per line, maximum lines per frame, and maximum frames per second.

3. Matrix intersections having two cells with values are defined as having two spatial resolution layers—an "enhancement" layer, shown on the left, and a "lower" layer, shown on the right.

4. The "S/N" and "spatial" profiles allow for scaled coding in which the scaling occurs in the dimension indicated by the profile name.

Figure 3.22 Levels and profiles.

trade-off is that the bit rate generally goes up. The high profile is for HDTV. The S/N profile is intended to be scalable for different video–signal-to-noise situation (through control of quantization), whereas the spatial profile scales resolution.

3.3.11 Other Approaches to Compression

There are a variety of other approaches to compression besides the DCT. DCT has enjoyed the most attention because it is based on the Fourier transform, which has been studied more than any other transform. The Fourier transform has other broad applicability to the study of spectra and signal analysis. It has received extensive funding in research on radar, sonar, and deep-space signal analysis. Efficient computational methods have been developed to apply it. Custom-designed integrated circuits have been constructed to implement very fast conversion of waveforms to spectra and vice versa. The DCT is a natural starting point for video compression.

Other approaches will continue to be studied and may at some point close the gap that currently exists between DCT-based and other compression techniques. Efficient software and hardware may develop for these other approaches, and eventually the DCT may be displaced.

Probably the three most talked about alternatives are wavelets, fractals, and vector quantization. Wavelet transforms appear to be more "scalable" than other techniques. That is, they have the ability to serve a low-resolution display and then, with little waste of what has been transmitted, add more and more detail. A low-resolution display is able to use simpler circuits to extract what is needed, whereas a higher-resolution display can support the cost of more complex circuits and extract more information from the signal. A special class of filters is used to implement wavelets. As more research is done, more cost-effective implementations will become available.

Vector quantization involves approximating the image out of a prestored table of image elements (sometimes called *stamps*). This "library" is used to build up the image. Choosing the elements to place in the library is critical to the success of vector quantization. Simpler images place lower demands on the library. More complex images are more challenging.

Briefly, fractals are equations that describe images. They may be the most efficient compression method representing the most complex images with the least data. The challenge with fractals has been in finding the equations.

It is reasonable to expect that there will be further advances in video and audio compression. It is therefore important to avoid building regulatory or other mechanisms that will stand in the way of implementing advancing technology. The recent history of the personal computer industry has shown us that the marketplace will tolerate some degree of difficulty with interoperability and compatibility if the technical advances are substantial. This comes at a heavy toll in consumer frustration and sometimes anger. However, the old model of rigid compatibility may fall to the rapid advance of technology. In the end, the marketplace decides.

3.4 Packetized Multiprogram Datastream Formation

A packet of data is a collection of bits intended for a specific purpose and having a common destination. The packet of data always begins with a header and sometimes ends with an optional trailer. The header includes a synchronization word that establishes the timing of the packet. It defines where to begin assigning bits to words. Sometimes a packet will include *clock run-in* bytes for the synchronization of the receiver system clock. Depending on the nature of the system, this may be essential or unnecessary. In a system where data is embedded in another signal, such as closed captioning or teletext, the clock run-in is critical. In systems that are all data, it is superfluous.

3.4.1 Packetization and the PES

The major strength of a packetized data system is that packets for different purposes or for different destinations can be intermixed. The different purposes include different applications as well as different recipients. The ability to send packets to different recipients connected to the same path gives rise to the

concept of a *virtual circuit*. A *physical circuit* is a dedicated physical path between a transmitter and a receiver. When multiple receivers are connected to the same path and the nature of the signal in the path determines which can receive the message and which cannot, the circuit becomes a virtual circuit. The virtual circuit approach affords a way of sharing the same physical structure for multiple users and purposes while maintaining privacy and control in an economic manner. This is one of the strengths of packetized data technology.

Packetization of signals requires that there must be a method of separating them out for their specific purposes and intended users. The header contains details on how this is to be done. It contains information about the destination of the data, whether there is more data in another packet, and details on how to reconstruct the complete message from its pieces when they are spread between multiple packets. Some packet systems include data to define the application of the packet and even the source of the packet.[14]

There are packetization systems that have fixed length packets and other systems that have variable length packets. The trailer on a packet is optional. In fixed length systems, the receiver can easily determine the end of the packet by simply counting pulses. In variable length packet systems, the header might define the length of the packet, in which case an "end of transmission" message is not required.

The *payload* is the actual information to be conveyed, stripped of any headers or trailers. The nonpayload data is "overhead." The amount of overhead required depends on the goals of the system and the quality of the transmission path.

Forward error correction (FEC) is a process by which additional data is added that is derived from the payload by an assigned algorithm. It allows the receiver to determine if certain classes of errors have occurred in transmission and, in some cases, allows other classes of errors to be corrected. FEC contributes to both overhead and equipment complexity.

MPEG-2 defines a *systems layer* that gives structure to the packetization method. The fundamental packet in the MPEG-2 system is the packetized elementary stream (PES). PESs contain data from one source only. There are video PESs, audio PESs, and data PESs. The PES is the fundamental building block of two other important datastreams: the program stream and the transport stream. Program streams consist of variable length packets grouping together the data from the PESs necessary to convey a program. They are intended for error-free environments and do not require error protection and detection. Transport streams are designed for the particular channels through which they are transmitted. They will be different for satellite, broadcast, cable, twisted pair, and so on. Figure 3.23 displays the relationship between these kinds of packetized streams.

Figure 3.24 shows the structure of the PES packet as used in the ATSC digital television standard for the United States. Space does not allow (and attention span probably won't tolerate) a complete description of the packet structure. A few comments will illustrate the fundamentals. The start code prefix sets the timing for the packet. The packet is of variable length with the maximum size for

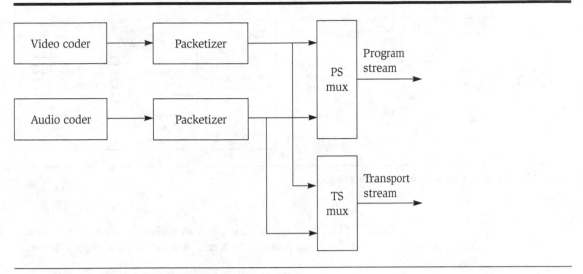

Figure 3.23 Coding, packetization, and multiplexing.

3 bytes	1 byte	2 bytes	2 bits	14 bits	1 byte	Variable length	
Packet start code prefix	Stream ID	PES packet length	10	PES header flags	PES header length	PES header fields	PES packet data block

Figure 3.24 Packet structure.

nonvideo applications set at 2^{16} bytes. Video packets have no maximum length. A great deal of information is provided in the header. Since this type of packet is intended for error-free environments, there is no error detection and protection.

3.4.2 Time Division Multiplexing and the Transport Stream Packet

A common goal of communications system design is to provide cost-effective communications. Central to this goal is the need to share expensive resources. This is done by multiplexing. In the previous chapter, we saw how modulation, tuners, and filters can be used to share the spectrum either over the air or in a hard-wired medium like cable. That technique is known as *frequency division multiplexing* because it divides up the spectrum resource into groups of frequencies. *Space division multiplexing* is a term that has been used to describe switched physical circuits. The space between an originator and some recipients is used to convey appropriate messages to the desired individuals. By taking different routes between the originator and the many recipients, the space between them is divided into various paths. The local phone system uses this method. *Time*

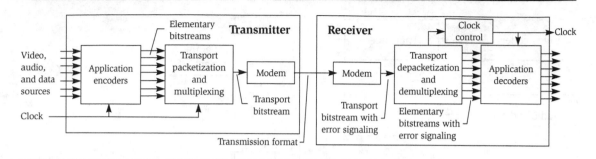

Figure 3.25 Transport packetization and multiplexing.

division multiplexing (TDM) allocates time slots in a channel to multiple users. This can be according to a fixed pattern, or it can be adaptively changed depending on the demand for capacity.

Figure 3.25 displays how PESs are time division multiplexed and further packetized to form a transport stream. The various PESs are multiplexed together into different time slots. The video data is sent in a different packet than the audio data. This facilitates the routing of the data to the different circuits and subsystems in the receiver that process and decode those datastreams into the output signals that drive the presentation devices. Video and audio data must be intermixed in a manner that ensures that sufficient audio and video data arrive for their respective processors so that the sound and picture are reproduced in synchronism. This is time division multiplexing within the transport stream for a particular program. This figure is generic, and the concept applies for all transmission paths. The transport stream can be a combination of a variety of PESs. For example, the same video PES can be associated with several audio PESs in multiple languages. Likewise, the same audio PES can be associated with several video PESs, providing multiple views of the same athletic event. This packetization is necessary if there is to be local insertion of programming and commercials. This, of course, is a part of the current business structure in analog systems and is a critical part of the economics of the business. Supporting structures for conditional access for pay programming are included, but the standards do not specify the details.

Figure 3.26 shows an MPEG-2 transport stream header. Space does not permit a complete description of all the bits, but a few are worthy of note. The sync byte identifies the start of the packet and sets its timing. The program identification (PID) portion provides the definition of the purpose of the packet. If the data capacity of the channel is sufficient to carry more than one program stream, the streams can be multiplexed into a combination of channels. The PIDs make it possible to demultiplex the transport stream at the receiver and select the program of interest and ignore the rest of the data. Alternatively, one program can be processed for display while another one (or more) is recorded for later use. The

A Sync byte value = 47h
B 1 bit—transport error indicator
C 1 bit—PES packet start indicator
D 1 bit—transport priority
E 13-bit packet ID (PID)
F 2 bits—transport scrambling control
G 2 bits—adaptation field control
H 4 bits—continuity counter
I Adaptation field or packet payload

Figure 3.26 Packet structure.

programs recorded do not need to be further processed. That can take place at playback. In this way, multiple programs can be time division multiplexed together.

The signal structure can be extremely complex. Individual PESs are time division multiplexed together to form a program stream that can include multiple audio streams, multiple video streams, ancillary signals such as captioning or teletext, and so forth. Multiple program streams can be time division multiplexed into a transport stream carrying up to 38 Mb/s. This complex structure is then modulated onto carrier signals forming a frequency division multiplex of several hundred MHz. In a cable system, space division multiplexing can be employed to route different programming to different neighborhoods; for example, Spanish programming to the predominantly Spanish neighborhoods, Chinese programming to the Chinese neighborhood, extensive business programming to the downtown area, and so on.

3.4.3 Statistical Multiplexing

The goal of digital video compression is to reduce wasteful redundancy in the programming signal. Since programs have different amounts of redundancy at any given time and since the amount of redundancy varies over time in a given program, there is a capacity management challenge. If there is never to be a shortage of channel capacity, then there must almost always be an underutilization of the channel so that peak demands can be met. This is not a practical approach. An alternative approach is to provide a buffer memory in the receiver that smoothes the flow of data to the decoders. This introduces a latency to

accommodate buffer filling and emptying, and it also requires a standardization on the minimum buffer size. Although the buffer memory allows some flexibility, the problem of managing the flow of data is not completely eliminated. A level of tolerable buffer overflow and a receiver strategy for dealing with it must be established.

When the capacity of the channel compared with the average data rate of the programming allows multiple programs to be carried, an opportunity results. When there are multiple programs, the likelihood is that there will be a distribution of complexities among the channels that is amenable to statistical analysis. This is a reapplication of a phenomenon applied by long-distance phone companies decades ago. It was determined that much of conversation involves pauses when nothing is being said. If a large number of conversations are combined, the dead time of one conversation could be used to carry other conversations. If a half dozen conversations are combined, there is sufficient statistical dead time for the addition of one or two more conversations. Efficiency is dramatically increased. The scarce resource of transmission capacity is better utilized. The larger the number of combined conversations, the better the statistics and the better the results. This process is called *statistical multiplexing*. It applies equally well to video signals that have varying amounts of relatively dead time. The dead time in video is the time when little changes from frame to frame, and the need for transmission capacity is minimal.

Buffer memories for each of the multiplexed channels facilitate the process of statistical multiplexing but require sophisticated algorithms to manage their utilization. As a buffer memory reaches its capacity, provisions must be made for it to be relieved. The better the algorithm, the more effective the result. The management of the buffer memories can be done at the origination point. As better algorithms become available, better results can be obtained using the same old receiving hardware.

Another tool can be employed in multichannel systems. The distribution of programming between channels can be dynamically reassigned when the statistics in one channel start to deteriorate. The receiver is automatically controlled to go to a different frequency and multiplex time slot. The viewer continues to use the same program identification means and has no idea that the system has been reconfigured. This dynamic channel reassignment process further improves the ability of the system to carry multiple programs.

3.4.4 Challenges

Compressed digital video in a multiplexed environment provides challenges to the normal operational practices. Insertion of local commercials, newsbreaks, and so on are common analog practice and are important to the economic welfare of the business. When several program streams are multiplexed together, the details of this process become very complex.

Local emergency alerts are currently provided by a crawling text overlay along the bottom of the screen. Sometimes they are accompanied by audio tones. In other applications, text and graphics are overlaid on the video or inlaid into the

video. Though this is common analog practice and relatively simple to implement, it is hard to imagine doing this while the signal is in digital form. The process of compression and the DCT make it all but impossible to understand what to do with the bits in the spectrum representation in order to insert text and graphics into the signal so that they appear on the screen when the compression and IDCT process is implemented in the receiver. The alternative is to bring the signal down to analog form for text and graphics insertion. This would then require recompression. The compression hardware is very expensive.

Cable operators are packagers of programming. This is part of their editorial function in the electronic publishing model. When multiple multiplexed transport streams arrive at the cable headend, repackaging the programming is a challenge especially if statistical multiplexing has been employed. Cable operators would like to select some programs that are appropriate for their audiences and not carry others that may have little appeal. This should be done in a way that maximizes the utilization of capacity. Small cable operators will be economically challenged by this task.

The carriage of HDTV signals presents some interesting challenges. As will be discussed later, the broadcast standard involves one type of modulation, 8-VSB, whereas most cable systems are settling on 64-QAM. Also, it has been demonstrated that, in general, cable systems have twice the data capacity per channel as broadcasters. The broadcast standard includes 16-VSB, which is intended for cable use and is capable of two HDTV programs. 256-QAM has been demonstrated on cable. It too can carry two HDTV signals. How should cable operators carry HDTV? Should they just put the broadcaster's signal on directly in its 8-VSB format? Should they convert it to 64-QAM? Should they combine two HDTVs into one 16-VSB or 256-QAM signal? If there are an odd number of HDTV programs, should they fill in the remaining capacity in the channel with standard definition television?

If the broadcast must-carry rules that apply to analog transmission are found to apply to digital broadcasts, additional challenges will be experienced. If the must-carry rules apply to some of the standard definition signals but not others, how can they be demultiplexed? Can other signals be multiplexed in their place? Must they be carried in one modulation scheme or another? These are some of the still unanswered difficult questions.

When programming from different sources is combined, the differences in implementation details must be accommodated. Currently, there are multiple implementations of the DCT compression technique and of the transport layer. Conditional access methods are different. Combining programs from multiple sources while maintaining signal quality is a challenge. There are noticeably more artifacts when signals are compressed, uncompressed, and recompressed.

3.5 Modulation

Modulation serves the same purpose for digital signals as it does for analog signals. Although digital signals can conveniently utilize time division multiplexing,

frequency division multiplexing remains an important part of the plan. There are several reasons for this. Practically, digital circuits are limited in their speed. They cannot support a TDM structure for the entire GHz spectrum of cable. Breaking the spectrum into 6-MHz pieces relaxes the speed requirement on the digital circuits. TDM can be utilized within each of the 6-MHz channels.

Additionally, FDM facilitates the transition from analog television to digital television. No cable system is known to have plans to convert all signals to the digital format. Those cable systems that are adding digital are doing it for those customers who want more. They are adding it to the current collection of analog channels. In many cases, the analog scrambling is left in place. The investment in analog set top terminals is too great to abandon. By separating the spectrum into two areas, one for analog channels and one for digital channels (including TDM), a rational transition to digital technology is accomplished.

Unfortunately, the modulation methods chosen for use on cable (QAM) and for over-the-air transmission are different. The ATSC standard selected a vestigial sideband modulation (VSB) method with a reference carrier. This was done because the broadcast environment must contend with co-channel interference between multiple digital television signals, between digital television and NTSC, and between NTSC and digital television signals. The VSB method has been specifically designed for this environment. The reference carrier aids the television tuner in acquiring the signal when the channel is changed.

Two VSB structures were created. The over-the-air 8-VSB structure has a payload of almost 20 Mb/s and can carry one HDTV signal with relative robustness. Since cable has a more disciplined spectrum with fewer challenges, a 16-VSB version was created that has less error immunity and double the data capacity. It can carry two HDTV signals. Cable has chosen to use 64-QAM modulation and, in some circumstances, 256-QAM. Chapter 4 contains more detail on these modulation techniques.

3.6 Error Correction

One of the most important features of digital signals is the ability to judiciously add redundancies that allow error detection and correction. The understanding of error detection and correction begins with the realization that an error is the conversion of a 1 to a 0 or the conversion of a 0 to a 1. The fact that the error is limited to just those two cases is very powerful. Earlier in this chapter, we discussed the simplest of error-detection schemes, parity checking. If parity is taken into two dimensions, substantially more powerful results can be obtained.

Figure 3.27(a) shows seven binary words of seven bits each. In Figure 3.27(b), spaces are introduced to help identify the clusters of seven bits. In Figure 3.27(c), the seven binary words of seven bits each are arranged in a two-dimensional array. An even parity bit is attached to each row. Recall that even parity means that the appended bit will be a 1 if an odd number of 1s appear in the word and a 0 if an even number of 1s appear. Parity allows us to determine if an odd number of

0110100111000100110011100110011110110000011010101

(a)

0110100 1110001 0011001 1100110 0111101 1000001 1010101

(b)

Word	Parity		Word	Parity
0 1 1 0 1 0 0	1		0 1 1 0 1 0 0	1
1 1 1 0 0 0 1	0		1 1 1 0 0 0 1	0
0 0 1 1 0 0 1	1		0 0 1 1 0 0 1	1
1 1 0 0 1 1 0	0		1 1 0 1 1 1 0	0
0 1 1 1 1 0 1	1		0 1 1 1 1 0 1	1
1 0 0 0 0 0 1	0		1 0 0 0 0 0 1	0
1 0 1 0 1 0 1	0		1 0 1 0 1 0 1	0
0 0 1 0 0 1 1	1		0 0 1 0 0 1 1	1

(c)	(d)

0110100111000100011001111001100011101110000010101010101000100111

(e)

0110100111000100011001111 0 111000111101110000010101010101000100111

(f)

Figure 3.27 Block code example.

errors has occurred in any of the words. But it does not allow us to determine where the error occurred.

Now let us form parity bits in the vertical direction as well. This will give us an eight-bit byte at the bottom of the array. We now have 8 columns and 8 rows for 64 bits. This array can be strung out into a serial stream, as shown in Figure 3.27(e).

If an error occurs as shown in Figure 3.27(f), it will convert a 0 to a 1, or it will convert a 1 to a 0. In this example, a 0 has been converted into a 1. If at the receive site, we rearrange the received bits into a block as shown in Figure 3.27(d) and apply the parity-checking principle to the rows and columns, we will find a violation in row 4 and column 4. This tells us we have an error and it tells us where!

Since an error can only be an inversion of the logic, we can perfectly correct the error. A little experimentation with this example will reveal that if we have two or more errors that do not occupy the same row or column, we can locate them and correct them. This would work for up to seven errors as long as they were not in the same row or column. If, however, two errors occurred in the same row (or the same column), we could determine that we have two errors, but we could not locate the row (column) in which they occurred. Knowing that two errors exist is valuable information.

In this example, a penalty is paid in complexity. Thirty percent more information is sent. Also the hardware at the sending and receiving end is more complex. This example is termed a *block code* since the data is partitioned into blocks for processing. More complex codes exist that are beyond the scope of this text to explain. They involve advanced mathematical concepts and are very effective.

Three error-combating strategies, including two FEC codes, are employed in the U.S. digital television standard. A most important error-combating strategy is called interleaving. Figure 3.28 shows how this works. The goal is to take adjacent bits and spread them out in time so that a noise burst does not destroy adjacent bits. When the interleaving is undone in the receiver, the group of destroyed bits is distributed over a length of time so that damaged bits are not adjacent. Then techniques like the block code example can detect and correct the errors. Interleaving converts bursty noise into something more like random noise.

A variety of advanced forms of FEC have been developed. A description of their principles is beyond the scope of this text and the available space in this volume. For those interested, there are excellent references listed at the end of the chapter. Among the two most important FEC techniques are Reed–Solomon codes and Trellis codes.

The version of Reed–Solomon FEC code selected for the U.S. digital standard introduces redundant bits in the ratio of 20 for every 187 data bits. The Reed–Solomon FEC is applied to both the broadcast and cable modes. It uses a mathematical algorithm to compute the presence of a class of errors and their locations.

The broadcast environment is more difficult and requires more protection. The Trellis code is added to the 8-VSB system. It introduces a great deal of redundancy. One bit in every three is consumed. The result, however, is a very robust transmission system. Finally, the broadcast mode includes a comb filter that can be switched in when interference from NTSC is detected. The comb filter has notches near the picture, color, and sound carriers of the interfering NTSC signal. This is not necessary in the cable environment.

Figure 3.29 displays the results of these efforts. Of note is the symbol rate of 10.76 million symbols per second. The raw data rates in the broadcast mode are three times that, and in the cable mode are four times that. The payload rates are 19.28 and 38.57 Mb/s at C/N thresholds of 14.9 and 28.3 dB, respectively. The payload rates are in a 2:1 ratio. The C/N thresholds are well below what is necessary for acceptable NTSC television.

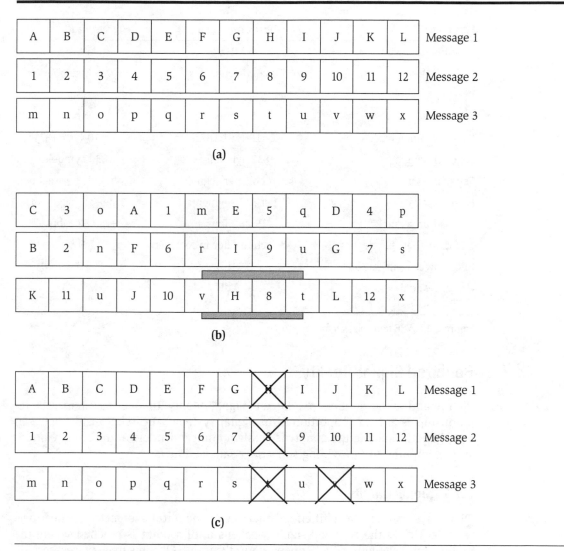

Figure 3.28 Interleaving to protect against burst errors. (a) Order of interleaving. (b) Interleaved messages suffering a burst of noise. (c) Deinterleaved messages suffering nonadjacent losses.

Parameter	Terrestrial Mode	High Data Rate Mode
Channel bandwidth	6 MHz	6 MHz
Symbol rate	10.76 Msymbols/s	10.76 Msymbols/s
Bits per symbol	3	4
Trellis FEC	2/3 rate	None
Reed–Solomon FEC	T=10 (207,187)	T=10 (207,187)
Segment length	832 symbols	832 symbols
Segment sync	4 symbols per segment	4 symbols per segment
Frame sync	1 per 313 segments	1 per 313 segments
Payload data rate	19.28 Mbps	38.57 Mbps
NTSC co-channel rejection	NTSC rejection filter in receiver	N/A
Pilot power contribution	0.3 dB	0.3 dB
C/N threshold	14.9 dB	28.3 dB

Figure 3.29 VSB transmission modes.

3.7 Required Signal Quality

Picture and sound quality are related to signal quality. In analog systems, the relationship is almost proportional. In digital systems, the relationship between signal quality and picture quality is highly nonlinear. Understanding this relationship is important to providing subscriber satisfaction.

3.7.1 Signal Quality

Digital systems have a "cliff effect" that allows them to have perfect or near perfect pictures as the received signal weakens until a point is reached where the picture fails altogether. It happens very abruptly. All viewers have essentially the same quality picture and sound, no matter where they are in the service area. When the signal strength falls below a threshold, the picture essentially disappears.

Noise in an analog picture appears as snow and begins to mask the image and get in the way of the enjoyment of the picture. If the programming is compelling, viewers will put up with this and continue watching even very poor pictures. If the programming is marginal, viewers will switch to something else. Viewers of digital images won't get the chance to make that decision. The picture will be as noise free at the home as it was in the studio. When the noise becomes excessive, the picture will be useless. It will fall off the cliff.

Multipath distortions (ghosts) that degrade an analog picture either destroy the digital picture by pushing it over the threshold or have no impact at all. The digital broadcast system's built-in multipath canceler system brings reception to areas not previously served.

Field tests in North Carolina have indicated that the broadcast digital system is so robust that anywhere an analog picture was even marginally acceptable, the digital picture was perfect. In many cases where the analog picture was unwatchable, the digital picture came in just fine.

Cable spectrum is much better behaved than the over-the-air broadcast media. Either the ghost canceler is not needed or its design parameters are much more relaxed compared with the broadcast case. If the analog signal is minimally acceptable, the digital signal will provide essentially perfect pictures (to the extent that the compression quality is acceptable).

3.7.2 Picture Quality

Picture quality is a function of the residual digital artifacts. These are determined primarily by the aggressiveness of the compression, the detail and motion in the picture, and the quality of the encoder. There is little data on the interaction of these parameters.

Though a number of expert viewers can easily spot digital artifacts in currently available digital video, there is little to no complaint from consumers. This may be just because there are still relatively few consumers, and they are enthusiastic about their new digital services. It may be that once they gain experience, they will be able to detect and object to digital artifacts. Or it may be that the design well matches the tastes of consumers and they will be well pleased with it.

Movies encoded using MPEG-2 at main level and main profile with 3.0 Mb/s appear in general to be very acceptable to very good. Consumers who have found consumer grade, VHS videotape satisfactory will be pleased with the result. At 1.5 Mb/s, the frequency of artifacts begins to be noticeable. A lot depends on the quality of the encoder and whether human interaction has been applied to adjust compression parameters with skill. Above 3.0 or 5.0 Mb/s, results are excellent with very few exceptions. Video results are more variable because video has fewer imposed constraints.

3.8 OpenCable

CableLabs, the cable industry's research and development consortium, has created a number of purchasing specifications on behalf of the industry. These are not standards since CableLabs is not an accredited standards-setting organization. Rather, these specifications define the functionality and external parameters of products used in cable but leave the internal details to the manufacturers. For example, CableLabs will not specify a single microprocessor or operating system for these applications. But to ensure interoperability, CableLabs establishes a

certification program to verify that the products comply with the specification and will perform satisfactorily.

OpenCable is one such specification. The goal for OpenCable is to create a multivendor interoperable environment of digital set top terminals. OpenCable is intended to facilitate a range of new interactive services and functions that are more like a digital set top computer than just a digital set top terminal. An important objective is competitive supply of all the elements of the system, including the advanced set top terminal, its operating system, core applications and plug-ins, conditional access system, platform management system, and billing, provisioning, and subscriber management systems.

The procedure began with a request for information (RFI) sent to leading computer and consumer electronics companies. Twenty-three companies responded. From these submissions, CableLabs solicits vendors to author specific elements of the suite of specifications. An intellectual property (IP) pool is created with cross licensing of all IP required for the specification. Implementation IP not inherently required by the specification is left to vendors to license in their normal fashion.

Key OpenCable interfaces include the formats for digital cable television, consumer privacy systems, copyright protection system, interfaces for high-speed connection to the Internet, and interfaces required to author interactive applications.

A "Harmony Agreement" for digital standards has been achieved. Five open layer standards, including MPEG-2 video, the Dolby digital audio AC-3 system, MPEG-2 transport multiplex, ATSC service information tables, and the International Telecommunications Union ITU-J.83 Annex B modulation system, are included in the agreement. The modulation system calls for 64-QAM carrying 27 Mb/s and 256-QAM carrying 40 Mb/s with concatenated Trellis code plus enhancements such as variable interleaving depth for low latency in delay-sensitive applications. All these agreements have been endorsed by CableLabs and most have been adopted as digital standards by the Society of Cable Telecommunications Engineers.

OpenCable will use an implementation of the federally published core data encryption standard (DES). By utilizing common core encryption, operators will not have to duplicate the transmission of programming and related content in order to implement dual conditional access. Although operators would still need to maintain separate headend control systems to control each supplier's set top terminals, the conditional access data can be carried in a single unified bitstream.

CableLabs will implement a vendor's certification process, whereby specific products can be tested for compliance with the OpenCable specifications. The certification process will allow these devices to be made available to consumers through retail outlets.

CableLabs recommends that its members utilize the specifications for their procurements. However, each member will make its own procurement decisions. This is not a joint purchasing agreement.

3.9 The Digital Video Standard

The Society of Cable Telecommunications Engineers (SCTE) is an accredited American National Standards Institute (ANSI) organization that utilizes the due process methods required by ANSI to set standards for the cable industry. This is the mechanism by which digital video for use on cable is standardized. These SCTE standards are an integral part of OpenCable.

This digital video transmission standard for cable television describes the framing structure, channel coding, and channel modulation for a cable television realization of a digital multiservice distribution system. The standard applies to signals that can be distributed via satellite directly to a cable system, as is the usual cable practice. The specification covers both 64-QAM and 256-QAM.

The modulation, interleaving, and coding is tailored to the characteristics of cable system practice in North America. The modulation is selected by the cable system to be either 64-QAM or 256-QAM. The forward error-correction approach utilizes a concatenated coding approach that produces low error rates at moderate complexity and overhead. The system goal is one error event per 15 minutes. The input data format is MPEG-2 transport.

Two modes are supported: Mode 1 for use with 64-QAM with a symbol rate of 5.057 Mb/s, and Mode 2 for use with 256-QAM with a symbol rate of 5.361 Mb/s. Higher-order QAM will be considered in the future as warranted by confidence.

The cable channel, including optical fiber, is a bandwidth-limited linear channel, with a balanced combination of white noise, interference, and multipath distortion. The QAM technique used, together with adaptive equalization and concatenated coding, is well suited to this application and channel.

The transport layer for MPEG-2 data comprises fixed length packets having 188 bytes. One byte is for synchronization purposes followed by three bytes of header data containing service identification and scrambling and control information. This is followed by 184 bytes of MPEG-2 or auxiliary data. The data-stream is transmitted serially with the MSB first. The first packet byte is a sync byte having a fixed value of 47 HEX. Interleaving is used to enable the correction of burst noise errors.

3.10 Summary

We have discussed the fundamentals of digital technology as applied to video and as they apply to cable television. We presented the discrete cosine transform (DCT) and its inverse transform. The DCT is applied to still images, and differential techniques are utilized to reduce redundancy between frames. We introduced error-detection and correction techniques and their applications.

The application of digital techniques to cable dramatically increases the number of programs and the quality of the delivered signals while facilitating new services. At the same time, cable's unique bandwidth allows existing

television receivers and VCRs to continue to receive analog signals and provide entertainment and information choices to subscribers who cannot afford digital hardware or who wish to initially limit their investment in digital approaches to the principal viewing room of their residence.

Endnotes

1. H. Nyquist, Certain Factors Affecting Telegraph Speed. *Bell System Technical Journal*, 3(1924):324–346.

2. H. Nyquist, Certain Topics in Telegraph Transmission. *AIEE Transactions*, 47(1928):617–644.

3. F. J. Hill and G. R. Peterson, *Introduction to Switching Theory & Logical Design*, 2nd ed. New York: John Wiley & Sons, 1968.

4. K. Sayood, *Introduction to Data Compression*. San Francisco: Morgan Kaufmann, 1996.

5. D. A. Huffman, A Method for the Construction of Minimum Redundancy Codes, *Proceedings of the IRE*, Vol. 40, Sept. 1952, pp. 1098–1101. Institute of Radio Engineers, New York.

6. Joel Brinkley, *Defining Vision, The Battle for the Future of Television*. New York: Harcourt Brace, 1997.

7. Mischa Schwartz, *Information Transmission, Modulation, and Noise*, 2nd ed. New York: McGraw-Hill, 1970.

8. S. Merrill Weiss, *Issues in Advanced Television Technology*. Boston: Focal Press, 1996.

9. Charles Poynton, *A Technical Introduction to Digital Video*. New York: John Wiley & Sons, 1996.

10. John Watkinson, *An Introduction to Digital Video*. Boston: Focal Press, 1994.

11. A. N. Netravali and B. G. Haskell, *Digital Pictures, Representation and Compression*. New York: Plenum Publishing, 1988.

12. J. Watkinson, *Compression in Video & Audio*. Boston: Focal Press, 1995.

13. J. Watkinson, *An Introduction to Digital Video*. Boston: Focal Press, 1994.

14. *Guide to the Use of the ATSC Digital Television Standard*, Doc A/54, Advanced Television Systems Committee, *http://www.atsc.org*, October 1995.

Other References of Interest

Blair Benson and Jerry Whitaker, *Television Engineering Handbook*. New York: McGraw-Hill, 1992.

Keith Jack, *Video Demystified*. San Diego, CA: HighText Publications, 1996.

Stanford Goldman, *Frequency Analysis, Modulation and Noise.* San Diego, CA: McGraw-Hill, 1948.

B. P. Lathi, *Communication Systems.* New York: John Wiley & Sons, 1968.

Bill Waggener, *Pulse Code Modulation Techniques.* New York: Van Nostrand Reinhold, 1995.

Alan V. Oppenheim and R. W. Schafer, *Digital Signal Processing.* Englewood Cliffs, NJ: Prentice Hall, 1975.

R. H. Stafford, *Digital Television, Bandwidth Reduction and Communication Aspects.* New York: John Wiley & Sons, 1980.

R. C. Gonzalez and P. Wintz, *Digital Image Processing,* 2nd ed. Reading, MA: Addison-Wesley, 1987.

W. W. Hodge, *Interactive Television.* New York: McGraw-Hill, 1995.

D. Minoli, *Video Dialtone Technology.* New York: McGraw-Hill, 1995.

Theodore S. Rzeszewski, *Digital Video, Concepts and Applications Across Industries.* Piscataway, NJ: IEEE Press, 1985.

IEEE Transactions on Consumer Electronics, quarterly, Institute of Electrical and Electronic Engineers, Piscataway, NJ.

NCTA Technical Papers, yearly, NCTA, Science and Technology Department, Washington, D.C.

Chapter 4

Cable Data Transport

4.1 Introduction

Perhaps the most discussed new service in cable television today is data transport. The primary emphases are cable telephony and Internet access services via cable, as well as digital video transmission. Factors making cable an attractive medium include economy, speed, and connectionless service.

This chapter does not purport to be a treatise in data communications. Rather, it offers concepts that are particularly germane to the practice of data transmission on HFC networks. It is intended primarily to be an introduction to data communications for cable practitioners who have not previously been exposed to the subject. The examples of modulation methods and communications protocols are restricted to those that appear to be most germane in cable data communications for the next few years.

4.2 Modulation Methods

The object of data transmission is to get a stream of bits from here to there. *Bit* stands for *binary digit*, and is a simple voltage that can take on only one of two levels. We arbitrarily call these levels 1 and 0. Often, but not always, they are represented in electronic circuits by the voltages +5 and 0 volts (newer systems use 3.3 volts or so). A combination of many of these bits make up *digital data.*

It is possible to transmit bits of data over an RF carrier using the same variables as are used in analog transmission. One can vary, or "modulate" the amplitude, frequency, or phase of the carrier, with the information to be transmitted. Frequency modulation is often used for very simple data communications links, due to the ability to construct low-cost FM receivers capable of very good performance. However, frequency modulation does not provide good "spectral efficiency."

4.2.1 Spectral Efficiency

One figure of merit used to compare different modulation formats is *spectral efficiency,* expressed in bits per hertz: the number of bits per second of data, divided by the RF bandwidth required to transmit it. A modulation type with high spectral efficiency can transmit more bits per second of data in the same number of hertz of RF bandwidth. On the other hand, it can be expected to cost more and be less robust (more susceptible to noise and transmission distortion). For example, we will show that the theoretical spectral efficiency of BPSK modulation is one bit per hertz. That is, when using BPSK modulation, the theoretical bandwidth required to transmit the signal is equal to the bit rate: a 1-Mb/s (1 megabit per second) data signal will require 1 MHz of RF bandwidth. On the other hand, if the same signal were transmitted using 256-QAM, the theoretical bandwidth required is only 0.125 MHz because the spectral efficiency is eight bits per hertz $(1/8 = 0.125)$. However, all else being equal, 256-QAM requires a carrier-to-noise

ratio (C/N) more than 13 dB better, and the cost of the hardware is considerably higher. In all cases, the actual occupied bandwidth is wider than the theoretical because of practical considerations.

4.2.2 Frequency Shift Keying (FSK)

The simplest commonly used modulation method is frequency modulation, called *frequency shift keying* (FSK). Information is transmitted by shifting an oscillator between two frequencies. The receiver consists of a conventional frequency demodulator. It is followed by a shaping section, which restores the pulses to their original shape.

FSK systems tend to be popular in cable applications for relatively low-speed data transmission where receiver cost is paramount. Applications include downstream data transmission to addressable set top terminals, upstream transmission from RF impulse-pay-per-view (IPPV) converters, and status monitoring. Spectral efficiency is very poor, but this is relatively unimportant in low-speed applications. The transmission is robust (not easily damaged by noise or distortion), and required hardware is quite low in cost. Single-chip data receivers have been available for many years. Because of its popularity, the technique is described in some detail.

Figure 4.1 illustrates a simple FSK communications system. Data, in the form of a series of 1s and 0s, is applied to a low-pass filter whose function is to minimize the number of sidebands produced by the modulation process. The filtered waveform is applied to a frequency modulated oscillator, whose frequency deviation is proportional to the voltage applied at the modulating input. A parameter of an FSK system is the peak-to-peak deviation, or the difference between the two frequencies of the oscillator, when 1 and 0 are applied.* The wider the deviation, the more immune the system is to noise, but the more spectrum the signal occupies.

After transmission to the receiving end, the signal is usually passed through a bandpass filter (BPF), then applied to a discriminator, which is very similar to one used for analog signal recovery. Low-pass filtering (LPF) is usually used after the discriminator to reduce noise picked up during transmission. Capacitive coupling is shown after the filter to make sure the signal can be "sliced" at its average value. From the capacitor, the signal is supplied to a data slicer, a comparator whose output is 1 when the input voltage is above ground and 0 when it is below ground (or vice versa). The slicing process restores the wave shape of the original signal.

* In the analog world, deviation is normally measured on a peak basis—the difference between either extreme frequency and the center frequency. Peak-to-peak deviation is twice the peak deviation. In digital transmission, the carrier never rests at the center frequency, so peak deviation tends to have less meaning. We must be precise, however, in distinguishing between peak and peak-to-peak deviation.

Figure 4.1 A simple FSK system.

Most transmission systems require that the dc value of the data waveform be removed for transmission. This means that, in some manner, the waveform averaged over some reasonable interval must have a value of 0 volts. This might be violated if the datastream contains a long string of 1s or 0s. If the signal doesn't have the same quantity of 1 and 0 states when averaged over some short time interval, C1 will take on a charge, and the slicing will not occur in the center of the waveform. If the signal has more 1s than 0s or vice versa, it is said to have a dc component, which is undesirable. Later we show several ways to ensure that a signal doesn't have a dc component.

Filtering in FSK Transmission Systems

At least three stages of filtering are used in the transmission system. A low-pass filter at the transmitter limits the spectrum of the baseband signal (before modulation). In many other modulation methods, the filter may be implemented either at the baseband or at RF, after modulation. However, in FSK the nonlinear nature of frequency modulation precludes doing so. The receiver contains a bandpass filter at RF (or IF), which is used to limit the energy applied to the discriminator. After demodulation, a baseband low-pass filter is used to remove as much noise as possible.

In low-cost FSK systems, the filtering philosophy often recognizes that either the transmitter or receiver is in the subscriber's home, and this end must be built at minimum cost. Placing most of the filtering at the headend helps keep the cost of the home equipment down. This somewhat compromises the noise performance and spectral efficiency in favor of low cost. For example, a data communications system that communicates from the headend to addressable set top terminals will employ significant filtering at the transmitter. At the receiver, the IF filter frequently employs a low-cost ceramic filter made for FM receiver use

Figure 4.2 Manchester encoded data.

(180–200 kHz bandwidth), and the deviation is set to make the occupied spectrum match the width of the filter.

Manchester Encoding

We now describe one method for adding a clock to the data signal. It is independent of the modulation method but is introduced at this point because it is often used with FSK modulation. In data transmission, it is necessary not only to recover the transmitted data but also to recover a clock signal, which is used to define when a data state should be sampled. This is necessary in order to allow the receiver to "know" when a new bit arrives. Clock recovery, or making this determination, can be done in many ways, but Manchester encoding is often used in low-cost systems because of its simplicity. The data is said to be "self-clocking." The penalty is that the spectral efficiency of the transmission is halved compared with what it would be otherwise.

Figure 4.2 illustrates one form of Manchester encoding, which is applied to the data before it enters the modulator. At the top of the figure is a string of data represented in NRZ (nonreturn to zero) format, in which the data bits appear one after another, with nothing in between. A separate clock signal is typically used to transfer NRZ data from one place to another. However, in a communications link, a separate clock signal path does not exist. Therefore, in some manner, the clock must be embedded with the data. Manchester encoding provides a simple, though bandwidth inefficient, way to do this.

The center portion of Figure 4.2 shows the bit cell boundaries: the time when the signal transitions from one bit to the next. At the bottom of the figure is shown the Manchester encoded data. In the middle of each bit cell, Manchester encoding causes the signal to transition from one level to the other. This regularly spaced set of transitions forms a clock for the receiver. The first half of each cell contains the value of the bit, and the second half contains the complement (opposite state) of the bit. The opposite polarity is used in some systems. A variation is differential Manchester encoding: at the beginning of each cell containing a 0, the encoded signal makes another transition; at the beginning of each cell containing a 1, the transition is omitted.

Note that Manchester encoding is *not* a modulation method. Rather, it is a baseband coding method that is useful in low-cost systems to provide clock recovery and removal of a dc component of the data.

4.2.3 Baseband Spectrum of a Data Signal

The spectrum of a data signal may be understood with reference to the waveform of Figure 4.2 and the low-pass filtering of Figure 4.1. Figure 4.1 shows that it is possible (and usually desirable) to low-pass filter the data waveform so that when a string of alternating 1s and 0s (or a string of like bits in Manchester encoded data) is encountered, the waveform looks substantially like a sine wave. No useful data is lost in the filtering process, and the occupied bandwidth of the signal is thus minimized. Practical filters are somewhat wider than this, but the principle remains.

When a string of 1s and 0s is encountered in an NRZ signal and optimally filtered, the resulting waveform is substantially a sine wave at a frequency of one-half the bit rate. To understand this, recall that a complete cycle of a sine wave occupies the time from one positive going zero crossing to the next positive going zero crossing. In order to produce such a waveform, two bits, 1 followed by 0, are required.

In general, a data signal can contain alternating 1s and 0s. This pattern defines the highest baseband frequency, equal to one-half the bit rate, assuming perfect filtering. However, that data can also contain arbitrarily long strings of 1s or 0s, which represent a frequency approaching 0. Because of this, the spectrum of the baseband (unmodulated) data signal extends from one-half the bit rate, down to very close to 0. Some systems limit the number of consecutive 1s or 0s transmitted (another way to remove a dc component) because many clock recovery techniques cannot tolerate an infinitely long string with no transitions. Also, if the frequency gets too low, slicing thresholds can be thrown off.

In summary, the spectrum of a digital signal extends from very low to at least one-half the bit rate (really the symbol rate, a concept to be introduced later in the chapter). Practical systems have frequencies that extend above this due to real-world filtering constraints. The excess bandwidth may range from a few percent to much higher depending on the application.

4.2.4 Introduction to Bit Error Rate

An important figure of merit for any data communications system is the *bit error rate* (BER) of the system. This expresses the proportion of the bits transmitted, which are received incorrectly. If, on the average, two bits are in error for every million (10^6) bits transmitted, then the bit error rate is

$$\text{BER} = \frac{\text{number of bits received in error}}{\text{total number of bits transmitted}} = \frac{2}{10^6} = 2 \times 10^{-6}$$

The BER is a function of the type of modulation, the characteristics of the modulator and demodulator, and the quality of the transmission channel. A common way to compare different modulation systems is to plot BER versus signal-to-noise ratio (data communications practitioners tend to not differentiate between "signal" and "carrier" to noise). This is a convenient way to compare the capabilities of different modulation techniques and to predict performance in the presence of noise. This topic is covered in Section 4.2.13. Sometimes BER is plotted versus signal-to-noise ratio in the presence of certain other impairments, such as group delay distortion, interference, and phase noise.

Obtaining an equation relating carrier-to-noise ratio to BER for an FSK system is difficult because of the nonlinear nature of the frequency modulation process. The literature contains only a few attempts, and contradictions exist. Within the parameters commonly used in cable television work (very wide deviation compared with the bit rate), the BER can be quite good at very low C/N. The penalty is that very significant bandwidth is consumed, compared with that required for other modulation methods operating at the same bit rate.

4.2.5 ASK and PSK

Less commonly used, but sometimes encountered, forms of digital modulation include amplitude shift keying (ASK), in which the amplitude of the carrier is changed between two states to represent 1 and 0. Phase shift keying (PSK) varies the phase of the carrier between two states to represent 1 and 0. If the phase shift between the two states is 180°, the modulation is called BPSK.

4.2.6 Biphase Shift Keying (BPSK)

Biphase shift keying (BPSK) is the simplest of a class of modulation that is used extensively in cable television work. It is somewhat more complex to detect than is FSK though producing a BPSK signal is simple. In a BPSK system, the phase is shifted 180° between transmission of a 0 and a 1.

Figure 4.3 illustrates a simple BPSK transmission system. An NRZ datastream is low-pass filtered (LPF1) to remove higher-order harmonics. The filtered signal is passed to a balanced amplitude modulator (M1), whose RF input is supplied from a local oscillator.

The figure illustrates the output of the modulator, which consists of a carrier envelope that drops to zero amplitude in the transitions between 0 and 1. The modulated carrier has one phase relation for a 0 and the opposite phase when a 1 is being transmitted. Note the phase of the carrier with respect to the timing tick marks shown below the RF envelope. The reduction in amplitude to zero between states is a direct result of the filtering action of LPF1. Were it not for this filtering, the transition between 1 and 0 would not require the amplitude of the carrier to drop, but the occupied bandwidth would theoretically be infinite.

At the receiver, the signal is bandpass filtered in BPF1. From the filter, the signal is supplied to mixer M2, the demodulator. The second input to M2 is from

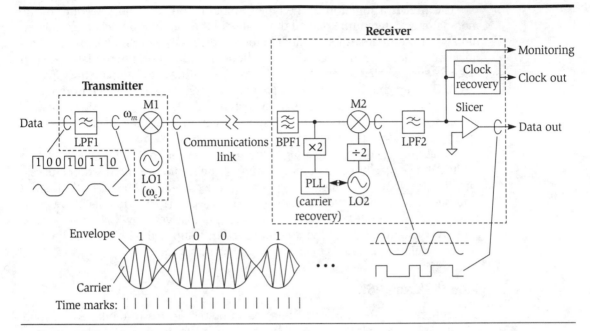

Figure 4.3 BPSK transmission system.

local oscillator LO2. This oscillator must be phase locked to the incoming RF carrier. A number of different carrier recovery circuits can be used. The circuit shown utilizes the fact that the second harmonic of the carrier contains no phase shift (when the frequency is doubled, a 180° phase shift becomes a 360° phase shift, or no phase shift at all). The incoming frequency is doubled by the times two multiplier, ×2, then used to lock LO2. The output of LO2 is divided by 2 and supplied as the second input to M2.

Basic Math of BPSK

You might be tempted to look at the waveform of Figure 4.3 and conclude that BPSK is a phase modulation technique because the carrier reverses phase between transmission of 1s and 0s. Indeed, the name of the modulation leads to that conclusion. The correct way to visualize BPSK is as double sideband suppressed carrier modulation. This type of modulation was described for analog signals in Chapter 2.

To understand the modulation process, assume that the data signal is an alternating series of 1s and 0s, and has been filtered to a sine wave, following the logic of earlier sections of this chapter. Then the modulating signal can be expressed as

$$e_m = M \sin \omega_m t \tag{4.1}$$

where

$_m$ = the modulating frequency

$\omega_m = 2\pi f_m$

As implied by Figure 4.3, the modulation process involves multiplying this modulating frequency by the carrier frequency, represented as

$$e_c = A \sin\omega_c t \tag{4.2}$$

The resulting modulated signal is expressed as follows, expanded using common trigonometric relations:

$$
\begin{aligned}
e_r &= e_c \cdot e_m \\
&= AM \sin\omega_c t \sin\omega_m t \\
&= \frac{AM}{2} \cos(\omega_c - \omega_m)t - \cos(\omega_c + \omega_m)t
\end{aligned}
\tag{4.3}
$$

The center row shows the phase shift with modulation: when the sinusoid representing the data modulation ($\sin \omega_m t$) is positive, then the carrier has one phase. When it is negative, the carrier phase is reversed, as illustrated in Figure 4.3. Equation (2.9) expresses the same thing, with frequency expressed in hertz rather than radians ($\omega = 2\pi f$), and f_1 and f_2 used in place of ω_c and ω_m.

The bottom expression shows that the frequency components of the modulated signal exist at the carrier frequency ω_c minus the modulating frequency ω_m, and at the sum frequency. Thus, the bandwidth occupied by the signal is twice the modulating frequency, as is the case with conventional double sideband amplitude modulation. A component at the carrier frequency does not exist. The demodulation process follows the mathematics of synchronous detection in Chapter 2.

4.2.7 Some Fundamental Principles of Data Communications

Having introduced the concept of BPSK modulation, a fundamental digital modulation method, we use it to introduce a number of concepts that are useful in understanding many modulation formats. The principles presented in this section are applicable to many types of digital modulation.

Filtering Methodology

This section introduces the filtering methodology that is followed in many data communications applications and for many types of modulation (except FSK, which we covered earlier). The filters at the transmitter and receiver are effectively in cascade in the signal path, and together it is desirable that they form a filter that is just wide enough to convert the highest modulating frequency

into a sine wave. (As a practical matter, the filter will be somewhat wider than this.) A particularly useful class of filters is called *raised cosine* filters by the data communications industry and *sine squared* filters by the television engineering community.

The equivalence is seen in the trigonometric relationship

$$2\sin^2 \alpha = 1 - \cos 2\alpha \qquad (4.4)$$

The left side of the equation shows the origin of the television usage of the term *sine squared,* whereas the right side shows the basis of the data communications term *raised cosine.* The 1 on the right "raises" the cosine term, which by itself, ranges from -1 to $+1$. In television work, a sine squared filter is used to produce the well-known sine squared pulse and bar, which are useful in analyzing video channel performance. In data communications, the cascaded performance of the low-pass filters at the transmitter and receiver, possibly in addition to the response of the IF filters, form a raised cosine filter, which ensures a fast transition halfway between the states, consistent with minimal spectrum occupancy.

It is desirable to heavily filter at the transmit end in order to reduce the spectrum usage, and heavy filtering is desirable at the receive end to minimize noise added during transmission. However, the total filtering applied must conform to the raised cosine shape. In many communications systems, the designer places one-half of the filter function at each end of the communication channel. When this is done, the filter response at each end is the square root of the complete filtering function, so the filter used on each end is often described as a *root raised cosine* filter. Note that the filters at the transmit and receive ends must be matched to each other and collectively are often referred to as *matched filters.*

Required Bandwidth of the Signal

The bandwidth requirements of BPSK can be understood by considering the worst case of an alternating string of 1s and 0s. This will produce a square wave, one cycle of which comprises an alternate 1 and 0, or two bits. Thus, the frequency of the square wave, f_m, is one-half the bit rate. Figure 4.4 illustrates the spectrum of the data signal at baseband. The square wave of alternating 1s and 0s illustrates the highest frequency of the data signal. A square wave consists of all odd harmonics of the fundamental frequency, as shown by the vertical lines in the spectrum plot of the figure. In real data systems, the data is not an alternating series of 1s and 0s. Rather, it is all possible combinations of strings of 1s and 0s. The resulting spectrum will fill in the spectrum from the maximum frequency (f_m) down to close to 0 frequency (and all harmonics thereof).

However, this total spectrum is not needed to convey the information contained in the datastream. If all frequencies above f_m are removed by filtering, then an alternating series of 1s and 0s will produce a sine wave of frequency f_m. Transitions that occurred at a slower rate (that is, from a series of like states) would follow a sinusoidal transition path, but would be flattened on top.

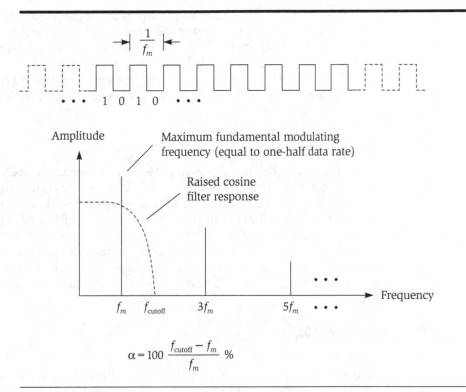

Figure 4.4 Baseband spectrum of BPSK signal.

The *theoretical* (not practical) spectral efficiency of a BPSK signal is developed as follows. The alternating 1–0 sequence defines the highest frequency of the data baseband signal. If it is filtered by removing all harmonics, leaving a sine wave, then the frequency of the sinusoid is one-half the data rate (one cycle comprises a set of 1 and the following 0). The bandwidth is twice the modulating frequency (the sum less the difference of the carrier and modulating frequency), which in turn is equal to one-half the bit rate. Thus, the spectral efficiency is two times one-half the data rate, or one times the bit rate, one bit per hertz.

A filter that completely cuts off just above f_m may be ideal, but is hardly practical. A practical filter has the raised cosine shape and is somewhat wider than f_m. The sharper the filter cutoff (that is, the less bandwidth required between the filter passband and the stopband), the less bandwidth is required. However, the sharper filter is more difficult to build, and more group delay is likely, depending on the realization. The filter width is described by the "alpha" of the filter, where alpha is defined as the excess of occupied bandwidth compared with the minimum required bandwidth:

$$\alpha = 100 \frac{f_{cutoff} - f_m}{f_m} \% \tag{4.5}$$

Recall that the common filtering practice is to put the square root of the raised cosine filter at the transmitter, and the same at the receiver. For example, as shown later, the filter specification for the transmitter of a DOCSIS-compliant downstream modulator operating in the 64-QAM mode is an 18% root raised cosine filter, meaning that the filter is 18% wider than the minimum required bandwidth.[1] This may also be interpreted that the occupied bandwidth is 18% wider than the theoretical minimum.

Clock Recovery

Although self-clocking formats, such as Manchester encoding (see Section 4.2.2), could be used with BPSK, this is normally not done. The higher number of transitions associated with Manchester encoding would double the required bandwidth of the signal. Rather, an NRZ datastream is transmitted. Clock recovery consists of synchronizing a clock to the transitions in the received data. Where no transition occurs, the clock "freewheels" through the place where the transition should occur. If a long transmission of 0s or 1s occurs, it is possible to lose clock synchronization, so *scrambling* is often used to force a state change every so often.

Data Scrambling

As a method of dc removal, scrambling is often used. In order to effect proper operation of a number of circuits in the receiver, including clock recovery, dc removal is necessary. Scrambling used in the data transmission sense is not the same as scrambling used in the traditional cable television sense. In traditional cable television, the word *scrambling* refers to the process of rendering a video and/or audio signal unintelligible to a normal receiver. A special circuit must be used to restore the signal to its intelligible form.

In data transmission, scrambling is the process of combining a datastream with a pseudorandom bit sequence (PRBS), with the intent to remove any possible dc component of the datastream. An equivalent way of expressing the function of the scrambler is to say that it equalizes the power density over the occupied bandwidth. It can be shown that, if any data signal is exclusive ORed with a pseudorandom signal having no dc component, the resultant signal will not have a dc component. The signal can be recovered at the receiver by exclusive ORing the signal with the same pseudorandom signal.

The exclusive OR operation is a logical operation on two input bits of data. If the two bits are the same, the output of the exclusive OR operation is a logical 0. If the two bits are different, the output is a logical 1. One of the input bits is from the datastream and the other from the PRBS.

Differential Encoding

The phase locked loop used for carrier recovery in Figure 4.3 creates a problem, in that a phase error ambiguity exists: if LO2 divided by 2 is in phase with the car-

rier applied to M1, the recovered data will have one polarity. If it is out of phase, the recovered data will have the opposite polarity. Since the loop is closed at twice the frequency of the carrier, there does not exist a simple way to determine the correct phase. One way would be to detect a code word buried in the data and correct for a phase error.

A method frequently used is *differential encoding.* In differential BPSK, a 0 is represented by no change in phase from the previous state. A 1 is represented as a 180° phase change. The detector is arranged to compare the phase of the previous bit with the phase of the present bit. If the phase doesn't change, a 0 is delivered as the received bit. If the phase is changed, a 1 is delivered (or vice versa). Differential encoding is used with a number of different modulation methods.

Evaluation of the Quality of a Digitally Modulated Signal

The recovered waveform at the "monitoring" point shown in Figure 4.3 should be the same as the waveform applied to modulator M1. However, distortion in the transmission path will change the wave shape, as suggested in Figure 4.5 (the dotted line is the original wave shape, and the solid line is the received wave shape). Noise will put "ripple" on the waveform as seen on an oscilloscope. If enough noise is added, then the state of the signal may be misinterpreted, producing a bit error. In addition, group delay can distort the wave shape, as can echoes and amplitude tilt in the channel.

A useful oscilloscope display is called an *eye diagram,* as shown on the oscilloscope. It is produced by supplying the received waveform (before the slicer) to the vertical input of the oscilloscope. The horizontal time base is set to display more than one bit cell and is triggered by the recovered clock. (For simplicity, the diagram shows only one bit cell width, though typically some additional time would be displayed.) The eye diagram is so named for its resemblance to an eye. When a bit is undergoing a transition from 0 to 1 at the trigger time, the waveform will be positive going on the left. When it is in transition from 1 to 0, it is negative going. It may also be the same as the last bit, in which case it forms a flat line at the top or bottom of the display. Similarly, the display toward the right is formed by transitions to the next bit after the one displayed.

The slicer interprets the signal to be either a 1 or a 0. The sample time should be in the center of the bit cell, when the eye is as "open" as possible. A figure of merit is the "eye opening," or the percentage of the peak (fully opened) eye, that is seen as open in the eye diagram.

One of the practical error sources, which is often not considered in computing BER versus noise, is errors in the timing of the sample: if the timing is in error owing to problems in the clock recovery circuit (perhaps because of noise, interference, or waveform distortion), then it is more probable that the bit will be misinterpreted.

Eye closure can be caused by a number of transmission problems. The effect of noise is illustrated in Figure 4.5. Besides noise, errors in the filter shape can

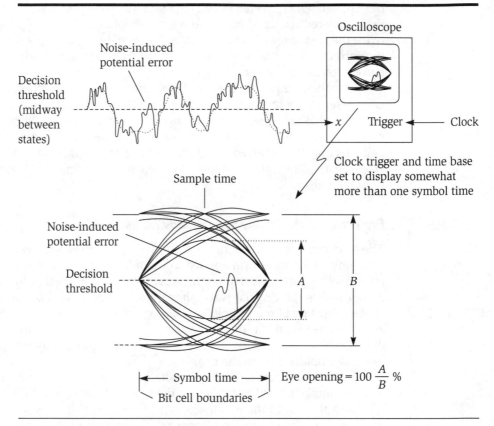

Figure 4.5 BPSK received waveform and eye diagram.

cause pulse distortion, which will tend to close the eye. If the transmission path exhibits excessive group delay, or if an echo exists, then one bit may not completely "go away" before the next bit starts. This condition is called *intersymbol interference* (ISI). The word *symbol* is used rather than *bit* for reasons to be introduced later.

Illustrated in Figure 4.5 is a potential bit error, in which a noise spike carries the waveform so far from its correct position that it actually crosses the decision threshold of the slicer. If the clock causes the slicer output to be read at the time of the noise spike, then a bit error would occur.

Peak-to-Average Ratio

Though we have argued that the BPSK signal is properly viewed as an amplitude modulated signal, the two signal states have the same amplitude. Were it not for the filtering of LPF1 (Figure 4.3), the signal would not have any amplitude char-

acteristics. However, since the signal is filtered, the transition between states takes some time, and during that time, the carrier amplitude passes through zero, again as shown in Figure 4.3. Because of this, the peak signal level is slightly higher than the average level. Digital signals in cable television work are usually measured based on the average level, yet the peak level is sometimes important.

4.2.8 Quadrature Phase Shift Keying (QPSK)

BPSK modulation provides very good immunity to noise, is simple, and is low in cost. The problem is that it is inefficient in its use of the spectrum. In order to make better use of the spectrum, it is necessary to take some small steps back toward (but not reaching) the analog domain, in that possible states are represented by different analog levels and/or phases. This is done by transmitting more than one bit at a time. The next modulation formats to be discussed are expansions of BPSK, which improve on the spectral efficiency by transmitting more than one bit at a time. The penalties are an increase in complexity and reduced immunity to noise and other transmission impairments.

Figure 4.6 illustrates a basic QPSK modulator and demodulator. Data into the transmitter is split into two channels. For reasons that will soon be apparent, the two are called the I (in-phase) and Q (quadrature) channels. Two bits are transmitted simultaneously, one in each channel. Spectral efficiency is improved because the spectrum required to transmit the two bits is no wider than the spectrum required to transmit one bit using BPSK.

After the data is split into the I and Q channels, each channel is modulated onto a carrier, just as in the case of BPSK. The same carrier frequency is used for each channel, but the phase of the two are 90° apart, or *in quadrature*. It can be shown that the two can be transmitted through a common communications channel, ideally without interfering with each other. (In the real world, several mechanisms can cause the two channels to interfere, or crosstalk, but they can be controlled.)

After the two channels are modulated, the modulated carriers are combined for transmission to the receiver. At the receiver, a phase locked loop (PLL) locks a local oscillator, LO2, to the incoming carrier. The signal from LO2 is split into in-phase and quadrature components. The in-phase component is sent to demodulator M3, and the quadrature component to the quadrature demodulator, M4. An encoder combines the two bits from the I and Q channels into the resulting datastream. A clock recovery circuit reconstructs the clock signal from the incoming data.

Bit Rate and Baud Rate

Perhaps the most common mistake made in communications terminology today is to confuse the rate at which *data* is transmitted with the rate at which *symbols*

Figure 4.6 Basic QPSK modulator and demodulator.

are transmitted. The rate at which bits are transmitted (the *bit rate*) is measured in *bits per second.* However, in QPSK and most other modulation methods, more than one bit is transmitted at a time. The bits that are transmitted simultaneously are grouped into a *symbol.* The rate at which symbols are transmitted is the *baud rate.*

In an older telephone modem that employed QPSK, 2,400 bits were transmitted each second. However, since two bits were transmitted simultaneously in a symbol, the *symbol rate* was only 1,200 baud. (Note that the plural of *baud* is *baud,* without an *s.*) As we will show, 256-QAM modulation groups eight bits into a symbol, so the baud rate is only 1/8 of the bit rate. The occupied bandwidth of the signal is a function of the baud, or symbol, rate.

The Constellation Diagram

It is useful to display the eye diagram of a QPSK signal independently for the I and Q channels. Another, particularly useful, representation of QPSK and higher-order signals is the constellation diagram, shown at the bottom of Figure 4.6. It is formed by routing the demodulated I signal, before slicing, to the *x* axis of an oscilloscope, and the Q signal to the *y* axis. Each of the four possible positions of the symbol is formed from the I- and Q-channel data: the position along the *x* axis being determined by the I data, and that along the *y* axis, by the Q data. The peak amplitude of the signal is proportional to the length of the vector from the origin to any one of the four states, which are equidistant from the origin. (Note that, in the absence of any transmission impairments, the four points in the constellation are small points. They are made bigger in the illustration to make them easier to see.)

The *y* axis forms the decision threshold for the I channel since it is halfway between the I-channel states. Similarly, the decision threshold for the Q channel is defined by the *x* axis. Computation of bit error rate (BER) is based on determining the proportion of time that any one "spot" in the constellation could be forced to cross a threshold into another state's place in the constellation. When this happens, at least one, and sometimes more, bits are received in error. Obviously, the closer together the states in the constellation (that is, the "denser" the modulation), the less noise or other disturbance can be tolerated before a state can be misinterpreted as that of a neighbor.

The peak amplitude of the signal is equal to the length of a vector from the origin to the farthest point in the constellation. In the case of QPSK, all four points are equidistant from the origin, so any of them can be used as the peak amplitude. Note that the reading of peak amplitude is a voltage, not power, reading on the constellation diagram.

Though the QPSK signal has four states of the same amplitude, between states the amplitude of the carrier drops because of filtering at the modulator. This filtering is represented by LPF1 and LPF2 of Figure 4.6. Also see Figure 4.3 for an illustration of the resultant modulated waveform. Since QPSK is nothing more than two BPSK signals combined 90° out of phase, the same waveform of Figure 4.3 would generally apply to a QPSK signal: between symbols, the signal amplitude passes through 0 if the state of both bits in the symbol changes. This usually doesn't cause much concern in cable television work. However, in the case of systems having high AM/PM (pronounced "AM to PM") conversion, where a change in amplitude of the signal can cause a change in its phase, the amplitude of the signal dropping to 0 can be a problem. A satellite communications link is such a system, where the power amplifier on the satellite is driven slightly into saturation. To reduce the effect, satellite communications often use offset QPSK, in which the transition times of the data in the I and Q channels are offset by one-half of a bit cell. Since the two channels don't change at the same time, the amplitude tends to be more constant.

AM/PM conversion is a phenomenon in which changes in the amplitude of a signal result in a change to the phase of the signal. In data communications, it can distort the phase reference of a receiver, making accurate recovery of the data less likely. It can be caused by certain nonflatness of the channel and by capacitance modulation in an amplifier that is operating out of its linear range.

The Effects of Interference on the Constellation Diagram

The constellation diagram is useful for understanding the effect of interference on the signal. In Figure 4.7, quadrant I illustrates the effect of a CW interfering signal in the passband of the receiver. An interfering carrier creates a beat note, which forces each point in the constellation to spread out in a circle around the desired point. The radius of the circle is proportional to the relative amplitude of the interfering carrier. Unlike the situation that exists with analog video signals, the impact of the interfering carrier tends to be independent of its location within the passband, except for some effect caused by nonflatness of the receive filter.

Quadrant II illustrates the effect of phase noise on the signal. Phase noise is introduced when a local oscillator has significant phase noise. It can also be shown to occur when the transmission channel does not have flat response. The noise is transferred to the modulated signal. It causes the constellation to rotate about the origin.

Random noise (quadrant III) causes each point in the constellation to "blur," somewhat as with the CW interference, except that the spreading of the point tends to be more nearly uniform within the circle and the circle does not have a well-defined radius. Again, recall that the spreading of points in the constellation becomes a problem when the spread points cross a threshold to another state. With QPSK, that threshold is defined by the axes of the constellation diagram. With more dense modulation, the thresholds are closer together, so less interference will cause the threshold to be crossed. As a practical matter, an error may occur when a point is forced close to a threshold boundary: real demodulators don't work quite as well as the theory predicts.

Finally, reflections, which cause ghosting in analog transmission, cause a replication of the constellation around each point in the constellation, as illustrated in quadrant IV. The constellation is rotated with respect to the "direct path" constellation, due to phase shift between the direct and reflected RF carriers. Usually, systems that employ denser modulation formats include adaptive equalizers, which can compensate for echoes, if the echo doesn't change too quickly. QPSK systems don't usually include such equalizers because the high noise immunity of QPSK doesn't justify the cost.

If the peak amplitude of the signal is 1, as shown in quadrant IV, the distance from the nominal location of any of the four states to the nearest state boundary is 0.707. This result is easily obtained from simple geometry, where the amplitude vector, of length 1, is the hypotenuse of a right triangle, and the distance to the nearest decision threshold is one of the two other sides of the triangle. For consis-

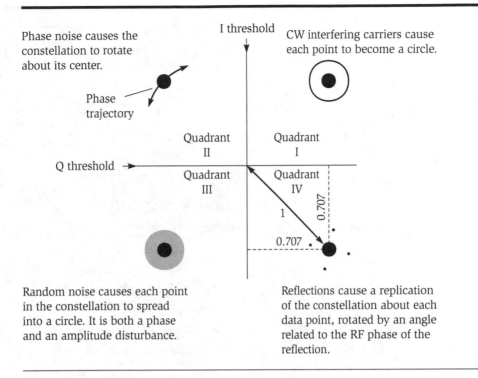

Phase noise causes the constellation to rotate about its center.

Phase trajectory

I threshold

CW interfering carriers cause each point to become a circle.

Quadrant II

Quadrant I

Q threshold

Quadrant III

Quadrant IV

1

0.707

0.707

Random noise causes each point in the constellation to spread into a circle. It is both a phase and an amplitude disturbance.

Reflections cause a replication of the constellation about each data point, rotated by an angle related to the RF phase of the reflection.

Figure 4.7 The effect of interference on the constellation diagram.

tency, we compare different constellations by normalizing them to the same peak amplitude of 1. However, bear in mind that it is more common to measure the amplitude of a digital signal by using the average amplitude, which is a concept we will explore.

4.2.9 Higher-Order QAM Modulation

16-QAM Modulation

The next denser modulation format commonly found in cable television work is 16-QAM (16-state quadrature amplitude modulation). It has been proposed for return services that require higher bandwidth efficiency than that offered by QPSK. It is similar to QPSK, except that each axis is divided into four levels so that each channel (I and Q) can carry the information of two bits. Each of two phases carries two bits at a time, so there are four bits per symbol: the baud rate is 1/4 of the bit rate.

Figure 4.8 illustrates a basic 16-QAM modulator and demodulator. Note the similarities between this and the QPSK modulator of Figure 4.6. (Clock recovery

Figure 4.8 16-QAM modulator and demodulator.

is done as illustrated in Figure 4.6.) Each channel of the 16-QAM modulator can take on not only the two phases of QPSK but also the intermediate amplitude values. Data is split into two channels, the I and Q channels. Instead of only one bit being routed to each channel at a time, two bits are routed to each. They are added as shown in the table beside the Q-channel modulator: for example, if the two bits are 11 (pronounced "one-one," since we are working in binary numbers), then the output of the adder is +3, and the Q-channel modulator puts out an amplitude of 3, with a positive phase (calling this the "positive" phase is arbitrary). If the data set is 10 (one-oh), then the adder supplies an analog signal of amplitude +1 to the modulator. Similarly, an 01 data set yields an analog level of −1, which causes the modulator to supply a 1 level signal with the opposite phase. An 00 data set will cause the adder to supply −3 to the modulator, producing a higher-amplitude negative phase signal.

The same thing takes place in the I channel. Each channel can thus take on any of four states. With the I channel in any of four states, and the Q channel independently in any of four states, a total of 16 states is possible. This is the number of states required to represent 4 bits ($2^4 = 16$). The modulation technique, 16-QAM, derives its name from the 16 possible states of the modulated signal.

The receiver contains in-phase and quadrature demodulators, as does the QPSK receiver. However, the output of each demodulator must be further processed to recover two bits. This is done for the I channel, using the three data slicers shown. The data slicers have thresholds set by the four voltage divider resistors, at levels of T1, T2, and T3. These levels are set halfway between the received states in the I channel. The thresholds are shown in the constellation diagram at the bottom of Figure 4.8. By reading the outputs of the three comparators, A1 through A3, the encoder can deduce the transmitted state. The constellation diagram is formed on the oscilloscope exactly as it was for QPSK, by supplying the I-detector output to the x input, and the Q-detector output to the y input.

Binary Numbering

We showed groups of two bits whose value is designated as 00, 01, 10, or 11. The correct way to speak this is oh-oh, oh-one, and so on. These are not to be confused with the base 10 numbers that look the same. If confusion between base 2 (binary) and base 10 (our familiar numbering system) is possible, then the number will be followed by a subscript 2 or 10. The following are equivalent numbers in base 2 and base 10: $010_2 = 2_{10}$ and $1101_2 = 13_{10}$.

For downstream transmission, the industry employs modulation formats that are similar to 16-QAM except that they have more states. Put another way, they transmit more bits per symbol, so the constellation is denser. The distance between states is smaller, so the susceptibility to noise and transmission path distortion is greater. On the other hand, the spectral efficiency is greater, meaning that more bits can be transmitted in the same RF bandwidth.

64- and 256-QAM

Figure 4.9 illustrates the constellation diagrams for 4-QAM (that is, QPSK), 16-QAM, 64-QAM, and 256-QAM. The diagrams are drawn to the same peak scale, so it is easy to see the higher density of the higher-order modulation formats. Also shown for comparison are several characteristics of the different modulation formats. Most of the characteristics are shown measured in decibels with respect to a QPSK signal.

For each modulation type, the theoretical spectral efficiency in bits per hertz is shown first. The spectral efficiency is a figure of merit, defining how many bits per second can be transmitted per hertz of RF bandwidth. The actual spectral efficiency will be less than this, due to the practical need for excess filter bandwidth. Next is shown the approximate power–peak-to-average ratio. This is computed by averaging the power of each state, ignoring the effects of filtering.

Figure 4.9 Constellation diagrams, 4-, 16-, 64-, and 256-QAM.

Section 4.2.11 shows how this is done. A more precise peak-to-average ratio must include the effects of filtering, and so cannot be stated in general.

Below the peak-to-average ratio is shown the relative distance between a state and the closest state boundary, expressed in decibels. The relative distance between a state and the boundary is defined in the QPSK constellation, as is the peak amplitude. It is the decibel measurement of how much closer to each other the states are in the denser modulation formats. The reference is a QPSK constellation, and the distance is normalized to the peak, not average, amplitude.

The relative amplitude entry is derived from the BER versus signal-to-noise ratio of the equation graphed in Figure 4.13. It is an indication of the increase in signal-to-noise ratio required to maintain a 10^{-6} bit error rate. Recall that the signal level to which we refer in this case is the average, not the peak level.

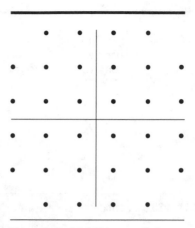

Figure 4.10 32-QAM constellation.

32-QAM Modulation

A similar modulation format to those shown earlier is 32-QAM, which has 32 states and a theoretical spectral efficiency of 5 bits per hertz. The constellation is shown in Figure 4.10. It is distinguished from the QAM formats shown above in that the four corner states are not used (this would yield 36 states). A total of 32 states represents 5 bits ($2^5 = 32$). With 16-, 64-, and 256-QAM, each axis, I and Q, can represent a certain number of bits, since the number of states along each axis is a power of 2. On the other hand, 32-QAM does not have this correspondence between bits and states. The five bits making up a symbol can, however, be encoded into one of the 32 possible states.

4.2.10 8- and 16-VSB Modulation

The FCC has designated 8-VSB modulation as the standard for off-air digital television in the United States. It is very similar in characteristics to 64-QAM, and 16-VSB is similar to 256-QAM. In VSB modulation formats, only one phase of the carrier is used: there is no quadrature component. All bits making up the symbol modulate the same phase of the carrier. That carrier (which is almost completely suppressed) is placed very near the band edge rather than in the center, and most of one sideband is removed by filtering.

Both 8-VSB and 64-QAM have essentially the same spectral efficiency. 64-QAM carries data on two quadrature phases of the carrier but must place the carrier in the center of the passband. Where carriers in quadrature are employed, it is necessary to transmit both sidebands to allow for separation of the two channels. 8-VSB uses only one phase of the carrier, so separation of components is not an issue. Compensating for the lack of the quadrature phase, 8-VSB is able to transmit twice the symbol rate in the same bandwidth since the second sideband is not transmitted.

Figure 4.11 VSB modulator.

Figure 4.11 illustrates the principle of VSB digital modulation. A set of bits is converted to a parallel word. The number of bits taken at a time is three if 8-VSB ($2^3 = 8$) is being transmitted (proposed for over-the-air use), or four if 16-VSB is being transmitted. These bits are added, and the result is converted to analog and supplied to a balanced modulator. The modulator output is filtered in a vestigial sideband (VSB) filter. Ideally, the VSB filter wouldn't pass any of the lower sideband, but it would pass the carrier. As a practical matter, some of the lower sideband must be transmitted, as in the case of analog transmission. Advances in filter technology since the NTSC television system was developed have made a much smaller vestigial sideband possible.

A small amount of carrier is inserted in the VSB signal to facilitate carrier recovery. With QAM modulation formats, the carrier may be recovered by taking advantage of the symmetrical sideband nature of the signal. It may be shown that where symmetrical sidebands exist, a suppressed carrier can be recovered by doubling the frequency of the received signal, which yields a second harmonic of the carrier. This is not the case with VSB, so a small carrier component must be transmitted to allow recovery.

The susceptibility to transmission impairments of a VSB signal and the corresponding QAM signal are similar. VSB has a claimed advantage in the presence of phase noise. As shown in Figure 4.7, phase noise causes the constellation to rotate around the origin. With VSB transmission, the distance to the state boundary on the phase trajectory is longer than it is with QAM.

4.2.11 Peak-to-Average Ratio

Higher-order modulation techniques exhibit a higher peak level than the average level read on a power meter. Figure 4.12 illustrates one reason for this, based on a 16-QAM signal. To the left is the same constellation diagram depicted in Figure 4.8. The constellation is scaled such that the amplitude of any one of the four corner states is 1. This represents the peak level of the signal. The chart to the right computes the average signal level based on equal occupancy of any of the 16 pos-

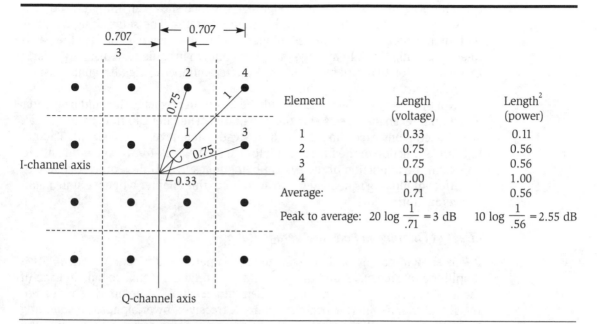

The table within the figure:

Element	Length (voltage)	Length2 (power)
1	0.33	0.11
2	0.75	0.56
3	0.75	0.56
4	1.00	1.00
Average:	0.71	0.56

Peak to average: $20 \log \dfrac{1}{.71} = 3$ dB $10 \log \dfrac{1}{.56} = 2.55$ dB

Figure 4.12 Peak and average level of a 16-QAM signal.

sible states. Since all four quadrants behave the same way, it is necessary to average over only one quadrant.

The voltage amplitude of each state is determined by simple geometry, assuming that the highest amplitude state has an amplitude of 1. It follows then that the coordinates of that state are (0.707, 0.707), from the geometry of a right triangle. The state closest to the center has a distance on each axis of one-third of this, so its total distance is one-third that of the farthest state. States 2 and 3 each have one coordinate of 0.707, and one coordinate of one-third of that. Application of the Pythagorean theorem shows that the magnitude is

$$\sqrt{0.707^2 + (0.707/3)^2} = 0.75.$$

When looked at on a voltage basis, the voltage average is the average length of the four vectors that define the states in the quadrant. The power average is obtained by squaring the voltage vector for each state because power is equal to the square of the voltage, divided by the resistance:

$$P = \frac{E^2}{R}$$

The resistance would cancel out in the computation, so it is omitted. The right column in Figure 4.12 computes the power–peak-to-average ratio, based on squaring the voltage vectors.

The peak-to-average ratio is not completely characterized by the state distribution as implied, however. It is also a function of the filtering at the

transmitter. This was illustrated in Figure 4.3, which shows an amplitude dip in a BPSK signal when the transmitted bit changes states. The same thing happens with higher-order modulation techniques, and that somewhat changes the peak-to-average ratio. Peak-to-average ratio is of concern in cable systems, particularly in the operation of return plant, where a limited number of digital signals exist on return spectrum handled by the return laser.

Not only can the average amplitude be less than predicted, it could be greater if certain filter shapes are used at the transmitter. The filter can ring, causing peak signal excursions exceeding the highest steady-state level of the signal. The details are a function of the filtering employed, so the peak-to-average ratio will, to an extent, be a function of the system being measured. Unless otherwise noted, the effect of filtering is not taken into account in this chapter when we state peak-to-average ratio.

Effect of Filtering on Peak-to-Average Ratio

It was shown that the peak-to-average ratio is partially a function of the relative amplitude of each state, and the percentage of time the signal spends in each of those states. Another factor, not yet taken into account, is the filtering applied to the signal. As the filtering becomes more aggressive (lower alpha corresponding to smaller excess RF bandwidth), the filter tends to ring more, meaning that the signal will be of even greater amplitude than predicted, for a small percentage of the time.

Ito[2] studied this theoretically for a filter having an alpha of 11.52%. He computed the percentage of time the power of a 64-QAM signal exceeded its average by various amounts. The curve follows the percentage computed based on state occupancy, until the percentage of time is less than the percentage of time the signal spends in the highest state. He found that the signal exceeded the average by 6 dB or more for about 0.1% of the time, and was 9 dB above the average for about 0.00001% of the time (6.25% of the time the 64-QAM signal is in the highest amplitude state, which is 3.68 dB higher than the average power).

Multiple carrier modulation such as OFDM (described in Section 4.3.3) has a much higher peak-to-average ratio. It exceeds the average by 12 dB for 0.00001% of the time. This is because multiple carrier modulation signals tend to have the amplitude distribution of random noise, which has a finite probability of being very high for a very short time. The subject of peak-to-average ratio is revisited in Chapter 14.

4.2.12 Measuring the Level of a Digitally Modulated Signal

Analog television signals are measured at the peak amplitude of the signal, which corresponds to the sync tips. This level is constant with modulation. The formal definition of the signal level is "the RMS value of the carrier amplitude during sync tips."

Digital signals are sometimes measured based on the peak level, but it is standard in cable practice to measure the signal based on average level. Measurement

of the average level is likely more accurate since it can be measured by supplying the signal to a thermocouple type RF power meter, and making a reading, usually in decibels with respect to 1 milliwatt (dBm). Such meters are simple and quite accurate. A thermocouple meter forms the basis for the signal level definition commonly used in cable television work.

The formal interpretation of "signal level" as applied to digital signals is as follows:

> The definition of the level of a digital signal shall be the power level as measured by a power meter which uses a thermocouple as a transducer. That is, the measurement shall be the average power in the signal, integrated over the actual occupied bandwidth of that signal. The signal level should be presented in decibels with respect to one millivolt RMS in a 75 ohm system. Thus the measurement reported is the RMS value of the sinusoidal voltage that would produce the same heating in a 75-ohm resistor, as does the actual signal.[3]

This definition simply means that the reported signal level is the level of a CW signal that would produce the same heating in a resistor as does the actual signal. The signal level is converted from dBm (commonly reported by power meters) to the equivalent value in decibels with respect to 1 millivolt in a 75-ohm system (dBmV), the common unit of measure of signal strength in cable television practice. The relationship between signal level measured in dBm and measured in dBmV is the same as in the analog domain:

$$\text{signal level (dBmV)} = \text{signal level (dBm)} + 48.75$$

4.2.13 Bit Error Rate Versus Carrier-to-Noise Ratio

It is appropriate to discuss the bit error rate versus the carrier-to-noise ratio of the signal. Usually in data communications, though, the independent variable is not considered to be carrier-to-noise ratio, but is expressed in terms of the energy per bit divided by the noise per hertz of bandwidth. We explain this concept and then discuss the requirements for various modulation formats.

The Concept of Eb/No

For a given modulation format, as you increase the data rate, the required transmission bandwidth will increase proportionally. If a 1-Mb/s datastream requires 0.5 MHz of transmission bandwidth, then a 2-Mb/s datastream will require 1 MHz of bandwidth. The bandwidth required has doubled, doubling the noise that of necessity must be admitted by the filtering, but the data rate has also doubled.

In order to compare the performance of different modulation formats, which may be used to transmit different bit rates, we use the concept of energy per bit divided by noise power per hertz, Eb/No (pronounced "E B over N zero"). The parameter removes data rate and noise bandwidth from the expression. It serves the same purpose as a carrier-to-noise ratio in analog video transmission, but it is not the same as carrier-to-noise ratio. You can think of this parameter as the

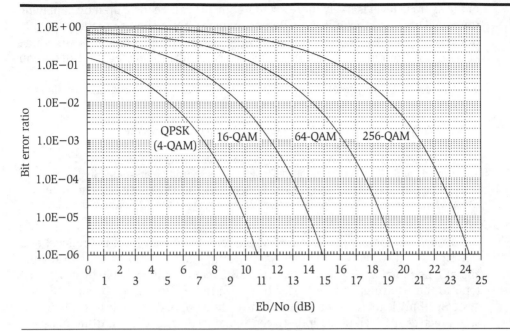

Figure 4.13 Bit error ratio versus Eb/No for several levels of QAM.

average carrier-to-noise ratio per bit. In a given case, you can convert to carrier-to-noise ratio by multiplying Eb/No by the bit rate, and then dividing by the noise bandwidth. In logarithms, this is done as follows:

$$\frac{C}{N} = \frac{Eb}{No} + 10 \log(\text{bit rate in b/s}) - 10 \log(\text{noise bandwidth in Hz})$$

The use of Eb/No permits you to compare the performance of different modulation systems without having to correct for different bit rates between the systems.

BER Versus Eb/No

The bit error rate of various levels of QAM modulation can be approximated by the following equation:[4]

$$P_M = 2\left(1 - \frac{1}{\sqrt{M}}\right)erfc\left(\sqrt{\frac{3k\gamma_b}{2(M-1)}}\right)\left[1 - \frac{1}{2}\left(1 - \frac{1}{\sqrt{M}}\right)erfc\left(\sqrt{\frac{3k\gamma_b}{2(M-1)}}\right)\right]$$

where

M = number of states per symbol (for example, 64 for 64-QAM)

γ_b = average C/N ratio per bit (*Eb/No*)

The error function, *erfc,* is a mathematical expression of the probability that a normal random variable (such as random noise) will exceed a certain level.

Graphing this expression for a number of levels of QAM yields the chart of Figure 4.13. This graph is normalized for a carrier-to-noise ratio expressed as Eb/No, the energy per bit divided by the noise per hertz of bandwidth. This is a common expression for carrier-to-noise ratio, which is independent of bit rate and bandwidth, as described in the last section.

Note that digital communications engineers often refer to this as *signal*-to-noise ratio rather than *carrier*-to-noise ratio. Cable television engineers often reserve the term *signal*-to-noise ratio for the baseband measurement, and use *carrier*-to-noise ratio for the RF measurement. Digital communications engineers don't always make this distinction.

4.3 Forms of Spectrum Sharing

This section discusses several forms of spectrum sharing, in which multiple users share the same spectrum.[5] The basic reason for spectrum sharing is better efficiency of a scarce resource (spectrum). Some of the forms shown also claim advantages in the efficacy of data transmission. Note that these forms do *not* represent new modulation techniques—they generally use some level of QAM modulation—but they do divide the spectrum differently. Nonetheless, some writers have used the term *modulation* to describe them. You will note similarities between some of the techniques; differences may sometimes be subtle.

4.3.1 Time Division Multiple Access (TDMA)

One mature method of sharing spectrum is time division multiple access (TDMA). It is being used extensively in early two-way data systems on cable because of its relative simplicity and maturity. It is quite useful for return path transmission, in which a number of return transmitters need to communicate to the headend.

Figure 4.14 illustrates time division multiple access. The reference system is a number of transmitters in one node, each of which must communicate back to the headend. In TDMA systems, each transmitter transmits on the same frequency, but at different times. After each transmitter has its turn, the sequence begins again. One set of transmissions from each transmitter is called a *frame.* Ideally, the level of each is adjusted such that a consistent receive level is measured back at the headend. TDMA systems are covered in more detail in Chapter 5. TDMA is used in several early entry cable telephony systems and is specified as the return sharing method in the DOCSIS purchase standard, covered in this chapter.

Some guard time must be allowed between individual transmissions to allow for errors in the transmission time of each individual transmitter. The receiver must also be able to lock to each transmission when it starts. Of course, since each transmitter is not transmitting full time, the transmission rate must be fast

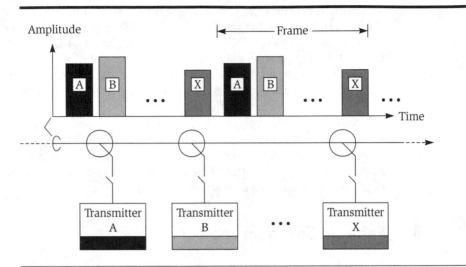

Figure 4.14 Time division multiple access.

enough to allow for each transmitter to transmit its complete set of bits during its turn. The bandwidth required is increased accordingly. A synchronizing signal must be sent from the headend, which defines the time slots.

It is not necessary that each time slot (A, B, and X) be of equal length. Several protocols, including DOCSIS, allow the length to be varied depending on the amount of data to be transmitted. This is explained in Section 4.5.4.

For downstream signal transmission, where only one transmitter is involved, a similar system can be used. The only difference is that it is not necessary to turn off one transmitter and turn on another. This transmission method is called *time division multiplexing* (TDM). It is covered in more detail in Chapter 5.

4.3.2 Frequency Division Multiplexing (FDM)

Another established method of sharing spectrum is frequency division multiplexing (FDM), as illustrated in Figure 4.15. FDM systems use a separate narrow subchannel for each transmission, dedicating the channel to that transmitter. Compared with TDMA systems, FDM requires multiple receivers at the headend though arguably digital signal-processing (DSP) techniques significantly ease that burden. (DSP uses a specialized microprocessor to implement signal-handling functions digitally. Modulation and demodulation functions, for example, can be described mathematically. Traditional circuitry approximates the appropriate mathematical functions. DSP techniques directly implement the appropriate mathematics.)

A guard band is required between channels to allow for practical filtering limitations. This causes some loss of efficiency though arguably TDMA systems also exhibit some loss of efficiency due to the need for guard times between transmissions. Any suitable modulation method may be used by each transmitter. It is

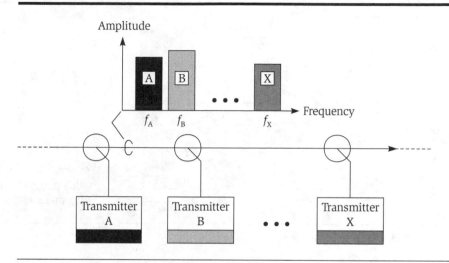

Figure 4.15 Frequency division multiplexing.

possible to monitor the error rate on each and to vary the modulation format to suit the quality of the channel.

4.3.3 Orthogonal Frequency Division Multiplexing (OFDM)

Figure 4.16 illustrates the idea of OFDM transmission. The incoming data is broken into a number of individual datastreams, each of which is then QAM modulated onto its own carrier. The carriers are spaced just far enough apart to prevent undue adjacent channel interference. Each carrier has modulation sidebands around it, as illustrated by the dotted lines in the figure (separated for clarity). The carriers may be moved closer together than you would normally expect based on the data rate if the timing of bits on adjacent carriers is carefully controlled. The adjacent signals are then said to be "orthogonal."

Any convenient modulation may be used on each of the carriers. QPSK is implied in the figure, but other techniques may be used. To generalize the type of modulation, communications engineers often use the term *m-ary* to refer to a modulation of any type. Note that OFDM is not itself a modulation technique though you will often see it discussed as if it were. However, it is a method of frequency sharing, analogous to TDM and FDM.

By dividing the spectrum as shown, you replaces a single band of high-speed data with a large number of bands of lower-speed data. Since demodulation is done over a narrower band, the effects of impulse noise and group delay are reduced, and the effective carrier-to-noise ratio is improved in each channel by the narrowing of the passband.

It is possible to monitor the error rate on each carrier individually. If one channel develops an unacceptable error rate, its density of modulation can be reduced (for example, go from 16-QAM to QPSK modulation). This will necessitate

Many carriers, each modulated with a portion of the data, using m-ary QPSK modulation

Figure 4.16 Principle of OFDM transmission.

reducing the bit rate on the carrier, meaning that either some bits will have to be reassigned to another channel or the overall bit rate will have to be reduced.

For return path services, it is possible to allow bits on different subchannels to originate from different transmitters. In this case, the technique has been called orthogonal frequency division multiple access (OFDMA). It is also possible to utilize each subchannel in a time division multiple access form, where each subchannel is shared by many transmitters, one at a time. This has been called variable constellation multitone (VCMT).

A closely allied sharing method is synchronous discrete multitone (S-DMT). Adjacent subchannels are synchronized such that spectrum overlap is minimized, allowing closer spacing of carriers. This effectively negates the disadvantage of having to allow guard bands between channels. It does require careful synchronization between transmitters, however. Adjacent channels so synchronized are said to be *orthogonal*. Yet another similar method is known as discrete wavelet multitone (DWMT). The method again assigns small subchannels to carry data in long symbols (a low data rate in each subchannel). You might observe that some of these methods are different names for the same basic technique.

4.3.4 Code Division Multiple Access (CDMA)

Code division multiple access (CDMA) is a form of spread spectrum communications that is being used for, among other applications, personal communications systems (PCSs), a newer form of cellular telephone service using particularly small cells. (A second PCS system uses TDMA.) Some vendors use it in cable telephony return path transmission. The method is also known as SCDMA, or synchronous CDMA.

Figure 4.17 illustrates the principle of CDMA transmission. The basic modulation is often QAM of some level. The data to be modulated on each axis is exclu-

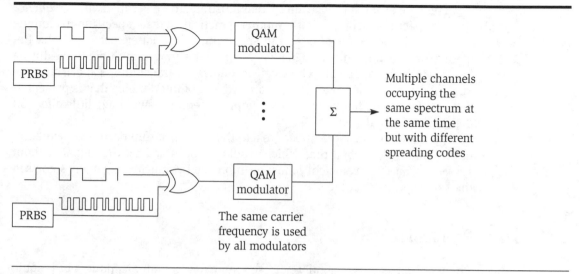

Figure 4.17 Principle of CDMA transmission.

sive ORed with a higher-speed pseudorandom bit sequence (PRBS), also known as a *spreading code* or *spreading sequence*. The signal is then modulated, normally using m-ary QAM. The effect is to spread out the carrier over a wider bandwidth than necessary for carriage of the information.

At the receiver, the demodulated signal is again exclusive ORed with the same PRBS. The result is a collapse of the transmission bandwidth back to that required to convey the information signal without the PRBS. Recovery of the original signal can then proceed. In the process, any interference that is added to the signal during transmission is spread, most of it out of the receive channel. In addition, if an impairment, such as a suck-out, affects part of the channel, the signal quality is minimally degraded because most of the energy lies outside the affected portion of the passband. (For the benefit of data communications engineers, the term *suck-out* is used in the cable television industry to mean a narrow frequency-selective reduction in the response of a transmission path.)

Figure 4.17 illustrates that many intentional transmissions may occupy the same frequency band simultaneously without interfering with each other, so long as each uses a different spreading sequence. All signals may be received simultaneously, with each having its own unique *despreading* sequence applied to separate it from the other signals. As each signal is separated with its despreading sequence, power in the other signals is spread out as if it were background noise. All spreading codes used must have the mutual (that is, shared) property of *orthogonality*. That is, they must be noninterfering with each other. A practical limitation on the number of signals that can be transmitted in the same bandwidth is the number of orthogonal spreading codes available. In turn, the number of spreading codes depends on the number of bits in the spreading sequence: the longer the sequence, the more orthogonal codes exist. Spreading code design is a complex subject.

The type of spread spectrum system illustrated here is called *direct sequence* spread spectrum. There exists another practical spread spectrum technique known as *frequency hopping* spread spectrum. In the latter technique, the frequency on which the digital signal is transmitted is moved multiple times during the transmission of one bit. The receiver follows the transmitter's spectrum hopping. You can see intuitively that a narrow band of interference that is present at one or a few of the frequencies in the hopping sequence will have little effect on the received signal.

With any spread spectrum approach, the total spectrum occupied by the signal is much wider than that which would be occupied by the signal without spreading. Spectrum efficiency must be improved by transmitting many signals in the same spectrum at the same time.

4.4 Protocol Issues

The word *protocol* is generally not used in analog communications, but it could be applied. When two pieces of equipment communicate with each other, there must be an established set of rules for that communication. In digital communications, the set of rules is called a *protocol*. For example, if we were to apply the word to the analog world, we might say that the protocol for FM radio broadcasting in monaural would be as follows:

Deviation: 75 KHz peak

Preemphasis: 75 μs

Frequency: channels spaced 200 kHz, beginning at 88.1 MHz through 107.9 MHz (North America)

This would define reasonably well a protocol to which monaural FM receivers could be designed, with assurance that a receiver built by one company would be compatible with a transmitter built by someone else. Data communications protocols are more complex, but are intended for the same purpose: to make sure that a talker will be understood by the listener.

4.4.1 The Concept of Protocols

Data protocols can be introduced using the analogy of a letter. A letter is written by one person and intended to be read by another. In the process of sending the letter from the writer to the reader, it is handled by a number of different people and packaged in different ways with other letters. Those carrying the letter have no interest in, and no need to know, the content of the letter, which is "hidden" in an envelope. Similarly, a digital message from one entity to another is "packaged" with other messages and transported by a transport system that has no knowledge of the contents of the message.

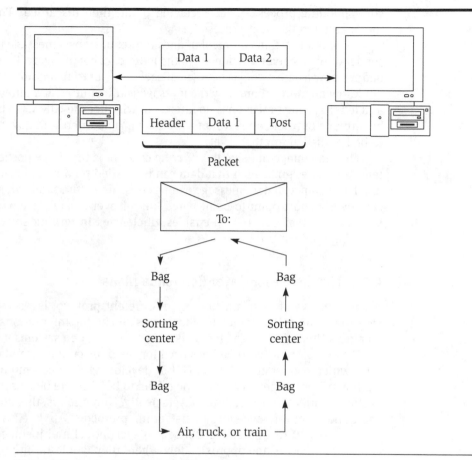

Figure 4.18 Concept of a data protocol.

A letter, as shown in Figure 4.18, is sent by the sender, who addresses it to the intended recipient. It is placed in a bag with other letters, for transport to a sorting center, then rebagged with other letters going to the same general destination. The bag is placed on a conveyance (truck, train, or airplane), which carries it close to its destination, where the process is reversed, before the letter is ultimately delivered to the intended recipient, who removes it from the envelope and reads it. In data transmission, the "envelope" is replaced, typically, with a *header* and *postheader,* which define that a message to be transported exists between them. Subsequent handling steps don't pay attention to what is enclosed, nor would they know what to do with the contents if they did look.

Just as a letter can be placed in an envelope, the envelope placed in a bag, and the bag placed in a train, different layers of the protocol may place their own header and postheader on the message. Those headers are interpreted by the

corresponding process at the receiving point, then discarded. The message is then passed to the next process.

A communications protocol defines the set of steps a message takes from the sender to the receiver. Analogous to the letter concept of Figure 4.18, the protocol defines how a data signal is handled at each stage of transmission. One layer does not know anything about how the message is transmitted at a different layer. Just as a letter-sorting process has no interest in whether the letter is to be transported by airplane or truck, one layer of a protocol has no interest in how the message is to be handled at another level.

One advantage of using the layering concept is that large portions of the system can be designed such that data can be carried on a number of different media. For instance, a data message can be constructed the same whether it is to be sent over a short point-to-point wired link or over a satellite link that is shared with many other users. This enables efficiencies in writing software and constructing hardware.

4.4.2 The OSI Seven-Layer Reference Model

A common way to explain the role of different protocol levels is the so-called seven-layer model promulgated a few years ago by the International Telecommunications Union (ITU). The model is known as the open systems interconnection (OSI) model.[6,7,8] ISO is sometimes used for the International Standards Organization, which cooperates with the ITU in developing data communications specifications. Some experts consider the model to be obsolete because many modern protocols don't fit into the seven layers neatly. Nonetheless, the model provides a useful frame of reference for understanding protocols.

Figure 4.19 illustrates the OSI seven-layer model and includes examples of common protocols and approximately where they would fit into the model. The model involves transmitting information from, for example, the left column (sender) to the right column (recipient). At the top of each column ("stack") is the application, such as a Web browser or an email system. The bottom three layers are usually of the greatest interest to engineers working on data transmission. These three layers define the transmission network, whereas the higher layers define the end user system.

Physical Layer

The lowest layer is the *physical* layer, which defines the functions required to get data from one place to another. Besides specifying such things as the modulation format (to which the first portion of this chapter was dedicated), the physical layer takes care of such functions as synchronization. *Synchronization* refers to the process of ensuring that data recovery at the receiver is timed properly with the data being transmitted. The actual transmission system, such as an HFC network, is not considered part of any layer of the protocol, but obviously the protocol must have knowledge of the function below (and above) itself.

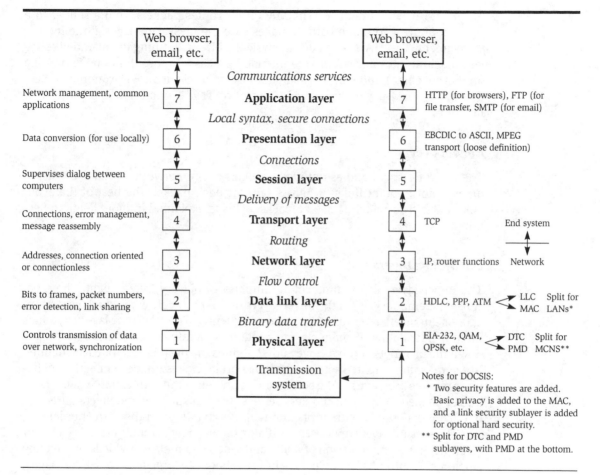

Figure 4.19 The OSI seven-layer model.

Data Link Layer

The *data link layer* handles such tasks as gathering up sets of bits for transmission as packets and making sure that the packets get from one end to the other in the correct order. Packets can actually arrive in the incorrect order in some cases, so the data link layer must correct this before passing the data up to the next layer. In addition, recognizing that physical layer transmission sometimes introduces errors, the data link layer handles error detection (and sometimes correction).

When the OSI model was introduced, data communications were point to point. With time, data communications have become much more multipoint to multipoint, so the data link layer was divided to recognize multipoint needs. With the split of the data link layer, the traditional functions of the layer went to the upper portion, called the *logical link layer*. A new sublayer, *media access*

control (MAC), was defined. The MAC layer routes packets from a sender to a receiver along a common path. It makes sure the message arrives at the intended recipient. The MAC layer adds a physical address, defining the intended recipient machine (not necessarily the intended process if more than one is running on that machine) and controls shared access to a resource. For example, in Section 4.3.1, TDMA was introduced. Control of TDMA operation is a MAC layer function.

Network Layer

The network layer handles routing of packets over a network that may consist of many point-to-point links. It figures out, for each packet, the best of the available routes over which to send a packet. This is explained in more detail in Section 4.4.4.

End System Layers

The upper layers of the protocol are important, but are treated lightly here because the emphasis is on the lower layers of the protocol, with which communications engineers tend to interact more frequently. The transport layer combines multiple low data rate channels into a higher data rate channel, and then breaks down the channels at the receive end. It defines the destination for each channel. In a connection-oriented service, the transport layer sets up the connection, delivers error-free messages in the correct order, and ends the connection when it is no longer needed. In a connectionless service, the transport layer delivers messages one by one, possibly with errors and with no guarantee of the correct order.

The session layer supervises the dialog between two computers. It is used, for example, in the transmission of audio and video channels in a videoconferencing session. Synchronization points are inserted between large numbers of messages, such as the continuous messages of video and audio files. Synchronization points are packets that divide the sequence of messages into groups.

The presentation layer converts between different representations of data. For example, some computers represent words by placing the most significant bit first, whereas other computers do the opposite. The presentation layer adapts the data to be used on the host machine if necessary. MPEG video encoding may be loosely thought of as a presentation layer function. The MPEG transport layer probably would be better thought of as some combination of the transport layer and some functions from lower layers. (MPEG does not really conform to the seven-layer model, nor does the Internet.)

The application layer interfaces with end application programs that require communications functions. The application programs, such as Web browsers, don't *reside* in the application layer themselves: they are above all the communications services. The application layer provides file transfer, terminal emulation, remote log-in, directory service, and remote job execution. These services are used by the end application.

4.4.3 MAC Layer Issues of Interest in Cable Modem Applications

Of particular interest to users of cable modems is the MAC sublayer of the data link layer. It falls to the MAC layer to provide an efficient protocol to allow operation of modems in the HFC environment. This physical environment is different from that of many traditional data networks. A significant issue is that each upstream "talker" cannot "hear" other upstream talkers unless the upstream transmission is repeated in the downstream direction. Also, unlike local area networks, the propagation delay on the HFC plant is significant and, if not taken into account, can create inefficiencies in data communications.

The following discussion of the use of time slots presupposes a time division multiple access (TDMA) protocol in the upstream direction. This is the approach taken by the DOCSIS modem purchase specification and other early entry cable modem systems. Other protocols, such as frequency and code division multiplexing, are possible and may become popular at some time in the future.

Reservation of Time Slots

In TDMA systems, some method must exist to allow for determining which of many transmitters gets to transmit in a given time slot. One such method is generally known as *reservation*. In a reservation protocol, each talker who wants to talk requests, and is then assigned, one of the available time slots. When that time slot arrives, he or she talks, then is quiet while others talk. The management and assignment of the time slots comes from some central authority, which in the case of a cable modem system would be the headend controller. A reserved time slot allows a modem to transmit a known amount of information in a known period. However, if a modem does not wish to use its time slot, the time slot is not used by anyone else until the controller decides that the time slot is available and assigns it to someone else. This can result in inefficiencies when there are a lot of modems, each with little to say. On the other hand, when a modem has much to say, reservation can be more efficient than the alternatives.

Contention for Time Slots

An alternative is to have the modems contend for time slots: when a modem needs a time slot, it simply speaks up and requests the time slot. In the case of cable modems, where one modem cannot "hear" any other modems, the talking modem will have to wait until the headend can acknowledge that it received the transmission. If another modem tried to talk at the same time—even if only part of the transmissions overlapped—then neither would be heard, so neither would be acknowledged. The transmissions are said to have *collided*. When that happens, both modems wait for an arbitrary time, and then try to transmit again. If they again collide, or if one collides with any other modem, they must again back off for a random time and try yet again. With some protocols, such as Ethernet, the talkers can hear each other, so they know when a collision occurs.

When talkers have relatively little to say, but need to say it quickly when they do have something, contention works fairly well. In cable systems, though, contention is particularly difficult because one talker cannot hear other talkers unless the upstream transmission is repeated in the downstream direction. This is done in some early cable modem systems, which are adaptations of Ethernet. However, the propagation delay of a round-trip on cable plant can be long compared with the length of a message. Consequently, by the time two modems know that they collided, a lot of time has been wasted. In addition, the need for upstream data to be repeated in the downstream direction is an inefficient use of the spectrum resource.

Combination Protocols

Contention protocols are sometimes called *aloha* protocols. Simple aloha protocols do not feature time slots, but allow any modems that need to talk the ability to do so at any time. If the protocol follows the discipline of time slots into which any talker's transmission falls, it is sometimes called a *slotted aloha* protocol. If some time slots are available for reservation, whereas others are available for contention, the protocol might be called a *reserved aloha* protocol.

The difficulties with contention systems on cable might lead one to prefer some sort of reservation system. It is certainly possible to reserve for each modem on a particular channel a permanent time slot for upstream transmission. However, such a permanent assignment is a problem when each modem is in use at most for a few hours each day, and has little to say when it is used. The rest of the time, the time slot cannot be occupied by other users. Some sort of method should be used to assign time slots on an as-needed basis so that if a modem was not using its time slot that slot could be used by another modem.

If a modem is assigned a time slot only when it needs one, there must be some way provided to allow a modem to request a time slot. Since the headend can have no a priori knowledge of when a particular modem will need a time slot, it is necessary to allow a modem to request one. This request implies that there should be some mechanism of contention in a cable modem protocol. The mechanism can be one where one time slot in a TDMA system is allocated to all modems not presently talking, in which they can request a time slot. They must do this in a contention mode. A modem needing a time slot waits for the contention time slot, then makes a request to the headend. If it is the only modem doing so, its request will be acknowledged in the return datastream, and a time slot will be assigned if available. If a second modem has also transmitted a request at the same time, then neither will be acknowledged, and they must both back off, waiting for a random time before trying again in the contention time slot.

4.4.4 Connection-Oriented and Connectionless Service

A communication link may be differentiated on the basis of whether it is a connection-oriented service or a connectionless service. Connection-oriented services are better known to the public, as represented by the public switched

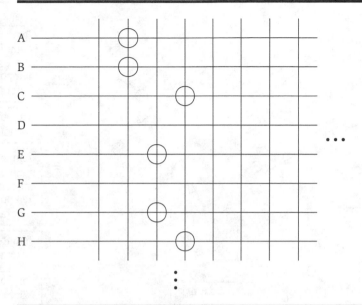

Figure 4.20 Representation of a connection-oriented service.

telephone network (PSTN). Connectionless service is a less familiar concept, which has advantages to the user and to the carrier. Both are described here.

Connection-Oriented Service

Figure 4.20 represents a connection-oriented service, such as telephone service. It shows a section of a switch, which couples one user (users A through H are shown) with any other user. A switch is often thought of as a matrix with users connected to horizontal lines, and vertical lines used to interconnect users on demand. The circles represent connections between the horizontal and vertical lines. As shown, A is connected to B, C to H, and E to G. D and F are not in communication with any users.

A connection must be set up whenever communication is desired between any two points, then "torn down" when the communication is completed. Because time is required to set up (and tear down) a communications channel, the user must wait until a channel is set up before he or she can use it. Resources are dedicated to the communications channel whether or not communication is taking place at the moment. For example, during a voice telephone call, a channel is dedicated to that call even when one party has left the phone to get some information.

Connectionless Service

By contrast, a connectionless system passes information when information is available, and the rest of the time, the resources are available for other users.

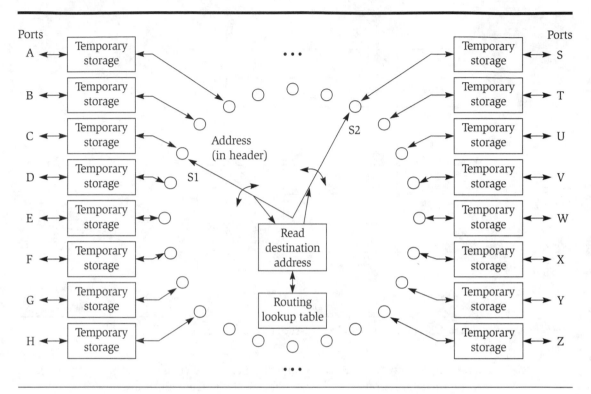

Figure 4.21 Router, representing a connectionless system.

Figure 4.21 illustrates the concept of routing, a key component of a connection-less circuit. The concept is used in the Internet backbone. The figure shows communication ports A through H and S through Z, each connected to a temporary storage facility, which accepts a message from the user and stores it for a short time until it can be routed to the next destination. Each interface can be thought of as either a user or another point in a multipoint route between two endpoints (users). All communications routes are two way.

A rotary switch has two commutators, S1 and S2, each of which can independently connect to any interface. Conceptually, S1 connects in turn to each port. If there is data there to be routed further, that data will contain a header that defines the destination. The destination header is read, and a lookup table is consulted. The lookup table tells the router which of the ports moves the signal most expeditiously toward its destination. Commutator S2 connects to this port, moving the signal toward the next point on the journey to its destination. (Real routers don't use switches to make connections, but this is a convenient mechanism with which to illustrate the function.)

If A is talking to B (as in Figure 4.20), then S1 will connect at some point, to A, the destination address B will be read, and the lookup table will be consulted.

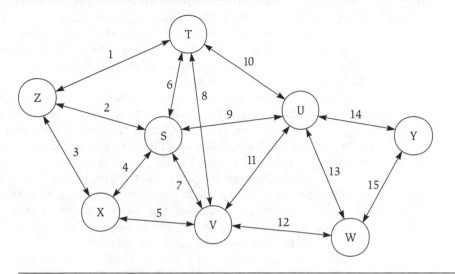

Figure 4.22 Multiple point routing.

In this case, the lookup table would say to connect to B directly since the destination port is connected to the router.

Figure 4.22 shows a larger view of a network. Each of the circled letters represents a router as in Figure 4.21, each in a different location. Each router is connected to a number of adjacent routers. The routing table in each must know of every ultimate destination and must know how to efficiently route a packet toward that destination. In several cases, multiple routes are possible, so if one is not available another can be used. For example, to send a packet from Z to X, the most efficient route is 3, which connects the two directly. If route 3 is not available, then the packet could be sent to S on route 2, then to X on route 4. Route 3 might not be available because it has failed, been taken out of service, or does not have the capacity to handle all the traffic being fed to it at the moment. The routing table in each router must be aware of the best route and alternatives, and must be aware of the present condition of each route.[9]

Differences Between Connection-Oriented and Connectionless Service

In summary, connection-oriented services, an example being the public switched telephone network, dedicate a communications path to two users for the duration of a communications session. The session may last less than a second, or it may last for days. When communications are desired, a circuit must be set up and is then dedicated to that communicating pair for the duration. At the end of the session, the circuit is torn down, releasing any common resources for other users. The device at the heart of a connection-oriented service is called a switch.

In contrast, connectionless circuits are established on an ad hoc basis when they are needed for an individual packet of information. As soon as the packet passes, the resources are available for other users. Each packet of information is required to have a destination address attached to it. The destination address is read by the router, which then consults a routing table to learn the best route on which to send the packet at that moment.

To the user of a connection-oriented service, it is necessary to establish the circuit (for example, by placing a call), wait for the connection to be established, then use it. The delay, once the circuit is established, is short and predictable. A user of a connectionless service is not aware of the establishment of a path for each packet, that function being handled automatically in protocol. So far as the user is concerned, the communications path is always available to him or her. On the other hand, because of route diversity, adjacent packets may travel on different paths, and since some routes may be faster than others, it is not guaranteed that packets will arrive on a regular basis or even in the right order. It is the responsibility of the protocol to rearrange packets into the correct order and to present them to the using function correctly. (With reference to the OSI model of Figure 4.19, this function would be handled by the network layer.)

Connectionless services use system resources more efficiently since no resources are tied up between users who have established a communications path that they are not actively using at the moment. On the other hand, the user of a connectionless service "thinks" a communications channel is available at all times even though he or she may experience delays. Of course, many of the delays are small on human terms, and packets received out of order are properly ordered without user awareness.

Connectionless Service Applied to Cable Modems

The concept of routing was used to introduce the function of a connectionless service. However, when looking at a cable modem service, another point of view of connectionless service may be preferable. RF return cable modem service is connectionless but does not involve routing in the traditional sense. (Of course, when a user enters the Internet from a cable system, he or she will encounter routing functions.)

The connectionless nature of cable modem service relates to the protocol used. Each modem, when it has something to send, contends for a portion of the upstream resources. When something comes downstream for a particular modem, it is accommodated in the downstream data multiplex as soon as possible. Because details of the signal routing are handled by the protocol, the user is not aware of the steps taken to complete communications with him or her. The concept is explored further later, when the DOCSIS protocol is introduced.

The Internet backbone is based on connectionless concepts. The end link experienced by most home users today, though, is connection oriented. When a user wants to do something on the Internet, he or she must place a call to an Internet service provider (ISP), and once the call is established, the user can do what he or she needs to do. At the end of the session, the user hangs up. Con-

trast this to the experience with a two-way cable modem, which offers a connectionless experience. With a two-way cable modem (not using telephone return), the user seems to be connected at all times. When the user is not communicating with the Internet, he or she is not tying up resources. When the user does have something to communicate, the modem system negotiates a time slot for his or her use, sends the message, then relinquishes the time slot to another user. This is done transparently to the user.

So-called push technologies work poorly in a connection-oriented environment, but they work well in a connectionless one. For example, suppose a home user subscribes to a service that automatically downloads certain news stories to him or her as they become available. With a telephone modem (connection oriented), the download cannot occur when the information becomes available. Instead, it must be held until the user calls and requests it. In a connectionless environment, though, the information is simply "pushed" through to the user's computer as soon as it is available. The information is delivered faster and without action by the user.

4.5 The DOCSIS Protocol for Cable Modems

Early cable modem systems used proprietary protocols, many derived from Ethernet. A group of cable television operators and Cable Television Laboratories (CableLabs) collaborated in the development of a purchase specification to be used by participants and other interested parties, as a procurement standard for cable modems. The collaboration operates under the name MCNS Holdings, L.P., an acronym for Multimedia Cable Network System. The set of specifications is referred to as DOCSIS, for the formal name of the specification set, data-over-cable service interface specification.[10]

DOCSIS is an important advance in cable modem standards since it is the first opportunity to allow modems, or any other cable equipment, to be owned by the consumer. Standardization is necessary to ensure that any modem bought by any consumer can connect to the network. Further, the subscriber can assume that if he or she moves to another cable system, the same modem will work in the new system. The standard allows multiple vendors to make competing products. With competition comes innovation and lower costs.

The IEEE is working on a separate standard for cable modems, 802.14. It has been agreed that DOCSIS and 802.14 will share a common physical layer, but will be different at the MAC layer. However, the two standards will exist in the same downstream channel. In addition, 802.14 will study upstream physical layer alternatives such as CDMA and DMT.

4.5.1 DOCSIS Reference Model

Figure 4.23 illustrates the headend-to-home portion of the DOCSIS reference architecture.[11] The architecture is fairly straightforward. In the downstream direction, the data signals, bubble A, are coupled with other downstream signals and

Figure 4.23 DOCSIS reference architecture, headend to home.

supplied to the downstream HFC network. (Chapter 8 of this book covers head-end RF architectures in more detail.) At the cable modem, the signal is supplied to diplex filter D1, which separates the downstream signal (above 54 MHz) from the upstream signal (below 40 MHz, though some regard the upstream band to extend to 42 MHz).

The downstream signal is supplied to the receiver in the cable modem, whose output is the downstream data.[12] In turn, that data is supplied to interface logic, which converts the signal to one of several defined interface protocols. One such interface is the 10baseT protocol. This protocol is a particular physical layer implementation of Ethernet, which operates at a data rate of 10 Mb/s using four-conductor CAT-5 cable and RJ-45 connectors (the interconnecting cable looks like a telephone cable with bigger connectors). It has been selected as a standard interface because of its suitable speed and the availability of 10baseT interfaces in many computers. For computers lacking built-in 10baseT interfaces, suitable add-in cards are available.

Two other interfaces are defined for DOCSIS-compliant modems. One is the universal serial bus (USB).[13] The other is the peripheral component interface (PCI).[14] USB is gaining acceptance as an interface between computers and a number of external devices. It is a 12-Mb/s bus that allows many external devices to be daisy-chained to the PC. The PCI is used with internal modems and other hardware. It has a maximum transfer rate of 132 MB/s and is backward compatible with the older ISA bus.

A cable modem transmitter handles return signals bound for the headend. The cable modem acts as an Ethernet bridge at the MAC layer. This means that it

passes traffic from and to the user's computer, but does not pass the remainder of the traffic on the cable. Thus, a user doesn't see traffic bound to other users on the node.

A removable security module is a portion of the security built into the system. Security is considered important because traffic bound for all modems in a particular node (or even group of nodes) is delivered to all modems attached to the node(s). This contrasts with a telephone modem (in which the modem itself contains no security at all), where only traffic bound for a particular user enters his or her home. For this reason, considerable effort has been expended to ensure the integrity of the system, ensuring users that their data will not be compromised.

An optional telephone return interface is shown. Internet traffic to individual subscribers is usually asymmetrical (much more data *to* the subscriber than *from* the subscriber). Thus, some early entry systems use telephone return, with cable-delivered data to the subscriber. This allows modem service on cable systems that have not upgraded to two-way capability.

Figure 4.24 illustrates the headend portion of the reference architecture. Bubbles A and B correspond to the same bubbles in the previous figure. The *cable modem termination system* (CMTS) includes the demodulator and modulator, which interface to the cable system. It also includes the interface to the network. Several options are defined for interfacing with the Internet backbone, either locally or (probably more commonly) by reaching a remote Internet backbone point of presence via a wide area network. The cable modem termination system also interfaces with a security controller and an optional (but recommended) operational support system.

4.5.2 Downstream RF Characteristics

Table 4.1 summarizes key parameters of the downstream RF signal. The relationship between symbol rate and bit rate is described earlier in this chapter, as is the significance of the filter shape. With both modulation formats, 64-QAM and 256-QAM, the signal is contained within one 6-MHz channel. The channel edges are specified to comply with the standard, IRC or HRC band plans shown in ANSI/EIA-542 (see Chapter 8). The DOCSIS document release existing at the time of this writing refers to the band plan by its original interim designation, IS-6, but this has been replaced with ANSI/EIA-542. ANSI/EIA-542 specifies picture, color, and sound carrier frequencies, which obviously are not germane to the data signals, but the channel boundaries are to be honored.

Additional Downstream Physical Layer Features

The downstream physical layer is divided into two sublayers (see Figure 4.19). The parameters described so far reside in the physical media dependent (PMD) sublayer. On top of the PMD sublayer resides the downstream transmission convergence (DTC) sublayer, which conforms to MPEG-2 (ITU-T Recommendation H.222.0). The DTC sublayer allows for different types of data, such as digital

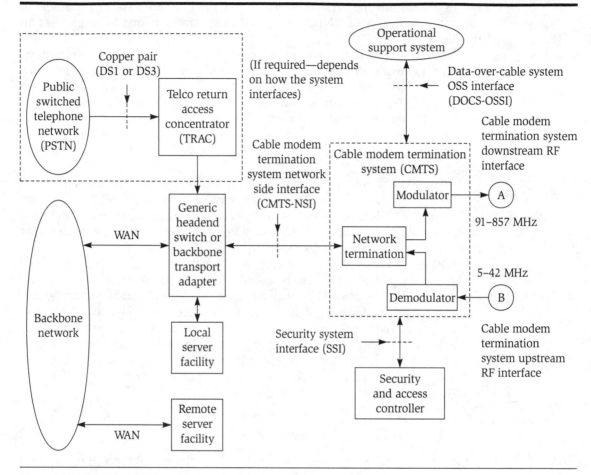

Figure 4.24 DOCSIS reference architecture, headend.

video streams, to be delivered to modems. Virtually any data service can be packaged within MPEG packets as desired. The DOCSIS documentation doesn't define these services.

The PMD sublayer, the lowest layer, specifies, in addition to the parameters already discussed, concatenation of Reed–Solomon block code and Trellis code for error detection and correction. Variable depth interleaving supports both latency-sensitive and latency-insensitive data. Chapter 3 describes error correction.

4.5.3 Upstream RF Characteristics

Table 4.2 summarizes some of the upstream RF parameters. Recognizing that the upstream spectrum is more difficult to use than is the downstream spectrum, the modulation types specified are not as aggressive: they are more forgiving of noise

Table 4.1 DOCSIS downstream signaling parameters.

Parameter	Value
Frequency	91–857 MHz
Modulation type	64- or 256-QAM
64-QAM parameters	
Symbol rate	5.056941 Msym/s
Corresponding bit rate	30.342 Mb/s
Filter	18% root raised cosine
256-QAM parameters	
Symbol rate	5.360537 Msym/s
Corresponding bit rate	42.884 Mb/s
Filter	12% root raised cosine

Table 4.2 DOCSIS upstream signaling parameters.

Parameter	Value
Frequency	5–42 MHz
Modulation type	QPSK or 16-QAM
QPSK parameters	
Symbol rates	160, 320, 640, 1,280, 2,560 ksym/s
Corresponding bit rates	320, 640, 1,280, 2,560, 5,120 kb/s
Filter	25% root raised cosine
Corresponding occupied bandwidth	200, 400, 800, 1,600, 3,200 kHz
16-QAM parameters	
Symbol rates	160, 320, 640, 1,280, 2,560 ksym/s
Corresponding bit rates	640, 1,280, 2,560, 5,120, 10,240 kb/s
Filter	25% root raised cosine
Corresponding occupied bandwidth	200, 400, 800, 1,600, 3,200 kHz

by about 9 dB (based on average power) than are the corresponding downstream modulation formats. Also, the filter parameters are looser—25% excess bandwidth compared with 12% or 18% in the downstream direction—resulting in somewhat easier demodulation. (Some experts question the viability of 16-QAM as a return signal modulation format, especially at higher data rates, due to deficiencies in the return path.)

Note that the data rates shown are the physical layer data rates. They contain overhead in addition to payload data (the data desired by the user), so the user will experience data rates that are somewhat lower than those shown. The ratio of payload data to gross data rate depends on the amount of error correction being used. In turn, the amount of error correction depends on the quality of the

transmission channel at that moment. Also, because of limitations in the Internet backbone, it is possible to experience much slower data rates than what we would be limited to by the cable modem system.

Limitations in the current Internet backbone have resulted in some service providers adding caching. A caching server periodically goes out to popular sites on its own and copies Web pages to itself. When a subscriber requests those pages, the caching server supplies them. Thus, when a subscriber retrieves those sites, the request need go only as far as the nearest caching server, which is typically operated by the service provider. It is not necessary to retrieve data over the Internet when these sites are accessed, so the subscriber experiences much faster download. The drawback to using a caching server, besides cost, is that a Web site may be updated, but the subscriber doesn't get the update because the caching server has not gotten the update.

The modems must be frequency agile (capable of operating on any frequency in the prescribed bands) in both the upstream and downstream directions. Downstream agility is important in order to allow the operator to move services around so as to maximize efficiency in system operation. Upstream agility is especially important, given the nature of upstream interference, where the modem may have to be moved in frequency to avoid interference that develops during operation.

However, the preferred method of avoiding interference is not frequency agility. Most systems will want to use other countermeasures first, and change frequency only as a last resort. Other countermeasures available for upstream interference mitigation include dropping modulation from 16-QAM to QPSK. The system can also reduce the symbol rate or increase the error-correction depth. Any of these countermeasures reduces the effective bit rate in the return path, but doesn't require frequency changes. Frequency changing, though often effective, does tie up frequencies not needed for actual data transmission.

Additional Upstream Physical Layer Features

The physical layer in the upstream direction includes other physical media dependent (PMD) features. The access method is time division multiple access (TDMA), in which a number of modems use the same channel at different times. The standard supports both fixed and variable length messages to improve the efficiency of the return path. Modems may be assigned a time slot, or they may contend for an open time slot. Programmable Reed–Solomon block coding allows trade-offs in error correction and overhead. The standard provides means for controlling the frequency of the modem, the time at which it transmits, and the power level with which it transmits. (Chapter 5 contains important information related to upstream level control.)

4.5.4 Data Link Layer—MAC

The lowest sublayer of the data link layer is the media access control (MAC) sublayer. The MAC protocol includes the following services:

- Bandwidth allocation, controlled by the cable modem termination system (CMTS)
- A stream of minislots in the upstream direction
- Dynamic mix of contention- and reservation-based upstream transmit opportunities (see Section 4.4.3.)
- Extensions provided for future support of ATM (asynchronous transfer mode) or other data PDUs (packet data units)
- Class-of-service support
- Extensions for security as well as virtual LANs at the data link layer
- Support for a wide range of data rates

Bandwidth allocation refers to the process of determining the data rates and frequencies to use in communications. This is done by the modem headend system, the CMTS, based on programmed parameters and existing conditions. (Note that the CMTS could be placed at a hub or other location—it does not have to be located at a headend.)

In order to achieve improved efficiency, and acknowledging that many upstream messages are very short, the protocol permits upstream time slots to be subdivided, allowing access by a larger number of modems. Also, data may be sent in packets of variable length rather than require the same length in all packets regardless of the length of the message contained.

Basic privacy is provided by an extension of the MAC layer; that is, DOCSIS has added security features that were not a part of MAC layers in other standards. These features are known as the baseline privacy interface (BPI). The security added at this layer is intended to offer the user privacy of data being carried on the plant, but falls short of providing complete security. All DOCSIS cable modems must provide BPI. It offers single DES encryption and is believed to meet most of the needs of data traffic. Additional security is optionally provided, as described in the next section.

4.5.5 Security Sublayer (SSI)

Besides the basic privacy feature of the MAC, a link layer security sublayer is added by DOCSIS to provide hard security in the cable modem system. The sublayer is called the security system interface (SSI). Cable modems are free to implement the SSI or not. It is believed that BPI will provide adequate security for all but very high-value assets.

In addition, security can be added at an application if so desired. For example, even though there is encryption provided by the modem system's SSI, a program that transfers the user's instructions to pay bills may want to add security of its own. This can be done, but the cable modem system does not need to know about it.

Security in the cable modem environment is multifaceted: it is necessary to protect the cable system from unauthorized use of the services by parties who are not paying for them, and it is also necessary to protect users from having their

data compromised by someone seeking to steal data. This is to be accomplished in an environment in which a user is allowed to purchase his or her own modem at retail, and to take it when moving from one cable system to another.

Based on the industry's experience in protecting premium television programming from access by pirates, the industry insists on a renewable and replaceable security system. If the security is compromised, it should be possible to change security systems simply, without compromising investment in the system. For instance, a smart card that subscribers plug into their modems could be sent to them. The smart card would replace some portion of the security system built into the modem. If that system was defeated, a new smart card could be sent to subscribers, at a small fraction of the cost of replacing all modems.

The algorithms employed are the RSA public key encryption system and the DES symmetric key encryption system. A *key*, as used in this context, is a digital word that is used to encrypt data such that the data cannot be read without the reader knowing the key. A multiple-key system uses more than one key to improve security. The cable modem security is based in a device called a subscriber security module (SSM). The module may be a physical module through which data is routed, but more commonly, it will be a logical function contained within a larger functional block.

When the SSM is manufactured, it is provided with a public/private key (PPK) pair. The private key is stored in a protected area of memory within the SSM, where it cannot be read later. The manufacturer also stores in the SSM the public key, date of manufacture, serial number, and manufacturer's identification. This becomes the certificate of authenticity, and can be created only by that manufacturer, using his or her protected private key. It doesn't have to be stored in a protected area of memory.

When a subscriber gets the modem, he or she must contact the cable operator's customer service representative (CSR) to request service. This contact may be made by telephone with a real CSR, or it may be made through communications with a service-provisioning server. The customer provides the identity of the modem manufacturer and the MAC address of the modem, along with billing and credit information. The CSR then authorizes service and notifies the access control system (ACS, one of the functions of the security system interface block of Figure 4.24) of the entitlements of this modem. The entitlements are the list of services authorized for that modem.

An automatic process then validates the modem. The SSM, in the subscriber's modem, and the ACS, in the headend, conduct an activation process whereby a security association is established. The modem sends a message requesting service. The ACS receives the modem's manufacturer certificate of authenticity from the modem. If the certificate is valid, then the ACS provides the SSM with its security profile and certificate chain. The certificate chain is used by the SSM to validate messages received from the ACS.

The SSM must still be provided keys and associated entitlement information, which enable it to access the services to which the subscriber is entitled. It sends an entitlement request to the ACS, which in turn sends the modem a full complement of its entitlements. These include both the unicast service key, used to de-

crypt data bound for that one modem only, and a set of multicast service keys. The multicast keys are used to decrypt contents intended for multiple subscribers. These keys in turn are used to generate frame keys that decrypt individual frames of information transmitted to the modem.

4.5.6 Data Link Layer—LLC and Higher Layers

Above the link layer security sublayer, the DOCSIS standard merges with other standards and so is not specified completely in DOCSIS documentation. The logical link control sublayer (class 1 only) is provided in accordance with ISO/IEC 10039. The network layer is the Internet Protocol (IP), currently version 4, with an anticipated migration to version 6 as it becomes available. Services residing above the network layer use IP as a bearer, and the services are transparent to the cable modem system.

In addition, several network management and operation capabilities are specified, including the following:

- SNMP (simple network management protocol) for network management
- TFTP (trivial file transfer protocol), a file transfer protocol for downloading software and configuration information
- DHCP (dynamic host configuration protocol), a framework for passing configuration information to hosts on a TCP/IP network
- A security management protocol as defined by DOCSIS

4.6 Error Detection and Correction

An important concept in data transmission is error correction. It is possible, and usually likely, that certain bits will be received in error. Techniques have been developed to allow detection and, in many cases, correction, of errors. This is done by transmitting extra bits that can be shown to permit recovery from certain errors. The subject is dealt with in Chapter 3.

4.7 Summary

This chapter has attempted to introduce the key points related to data communications on hybrid fiber-coax networks. It is not intended to constitute a course in data communications, and many important topics in data communications were omitted. Those that were included were selected because of their particular relevance to data communications as it is being practiced on cable television systems today. We can expect many changes in techniques in the next few years as the art matures.

The next chapter will deal with one particular system that transmits data on HFC plant, the cable telephone system. Many of the concepts presented in this chapter will be useful in the next.

Endnotes

1. MCNS Holdings, L.P., *Data-over-Cable Interface Specifications, Radio Frequency Interface Specification*, SP-RFII01-970326, March 26, 1997, p. 37. CableLabs, Louisville, CO.

2. Dr. Yasuhiro Ito, Performance of OFDM Transmission System—Peak-to-Average Power Ratio and Clipping Effect. *SPECS Technology*, June/July 1995, CableLabs, Louisville, CO.

3. National Cable Television Association, NCTA Recommended Practices: Upstream Transport Issues. This document is a recent addition to the *NCTA Recommended Practices* manual. Though the citation is oriented to return path issues, the definition is generally applicable.

4. John G. Proakis, *Digital Communications*. New York: McGraw-Hill, 1983, p. 187.

5. Portions of this information are supplied by M. Bugajski of Arris Interactive, Duluth, GA, in a summary prepared for internal use.

6. Larry Peterson et al., *Computer Networks*. San Francisco: Morgan Kaufmann, 1996, p. 35ff.

7. Jim Stratigos, ISO/OSI Model Introduction and Ethernet Overview. Preconference tutorial presented at the 1997 Conference on Emerging Technologies, Nashville, TN, by the Society of Cable Telecommunications Engineers, Exton, PA.

8. Jean Walrand and Pravin Varaiya, *High-Performance Communication Networks*. San Francisco: Morgan Kaufmann, 1996, p. 92ff.

9. An interesting historical perspective of the Internet, which explains the routing concept somewhat incidentally, may be found in *Where Wizards Stay Up Late*, by Katie Hafner and Matthew Lyon, New York: Simon & Schuster, 1996.

10. *http://www.cablemodem.com*

11. MCNS Holdings, L.P., *Data-over-Cable Interface Specifications, Radio Frequency Interface Specification*, SP-RFI-I02-971008, released October 8, 1997. CableLabs, Louisville, CO.

12. MCNS Holdings, L.P., *Cable Modem to CPE Interface*, SP-CMCI-I02-980317, released March 17, 1998. CableLabs, Louisville, CO.

13. *http://www.usb.org*.

14. *http://www.pcisig.com*.

Chapter 5
Cable Telephony

5.1 Introduction

This chapter covers the principles of telephony over a hybrid fiber-coax (HFC) network. Principles of telephone operation are discussed for the benefit of cable TV engineers not familiar with the telephone network. The discussion of telephone technology concentrates only on those principles that aid in understanding the operation of telephony on an HFC network. The chapter is not intended to be a dissertation on the operation of the telephone system.

The elements of a modern telephone system, generally from the switch to the subscriber, are outlined in order to aid in understanding the telephone-side interface of an HFC telephony system. This is followed by a brief introduction to the digital hierarchies used in telephone interfaces, which will help in understanding the headend interfaces. A sample HFC telephony system is then described. It is not intended to represent any of the telephone systems on the market but is intended to outline some of the issues of importance to understanding the operation of any HFC telephony system.

5.2 Modern Telephone System Architecture

Today's telephone systems may be broken into several elements. A convenient way to look at the telephone system is to start with the switch, the device that connects telephones. We could look at the side of the switch that interfaces to other switches, or we could look at the side that interfaces to telephones. We spend most of this chapter looking at the telephone-side interfaces since this is where changes are made to accommodate cable telephony. We do, however, mention the other side for the benefit of cable engineers who are not familiar with the telephone system.

5.2.1 The Local Exchange

Modern telephone systems in North America tend to be constructed in a relatively "flat" architecture, as illustrated in Figure 5.1.[1] Previously, a different architecture with a hierarchy of five different types of switches was used, and some of the terminology is left over from those architectures. In Figure 5.1, each telephone is connected to an *end office*, also called a *central office, local exchange,* or *class-5 local exchange.* Local exchanges are identified by the first three digits of the seven-digit local telephone number. If two telephones connected to the same end office talk to each other, then the signal goes no farther than that end office. End offices connect to other end offices and to *tandem offices,* or *intraLATA toll offices.* Tandem offices are the location at which operator services are normally provided ("provisioned" in telephone terminology). These tandem offices connect calls to other end offices and also to long-distance carriers, called *interLATA carriers.* Note the distinction between intraLATA and interLATA.

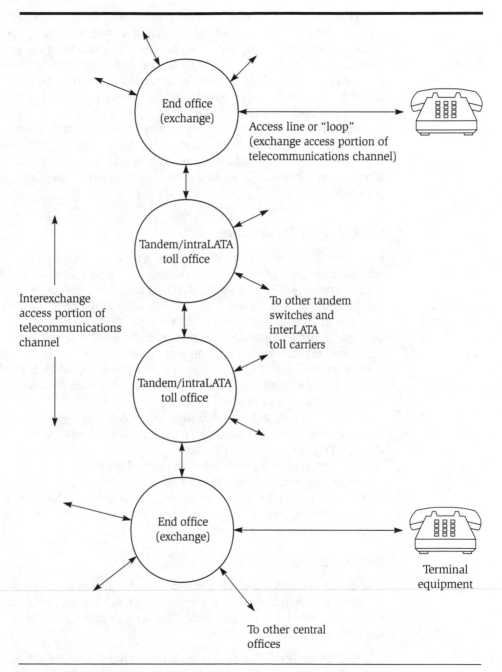

Figure 5.1 Modern telephone system architecture.

The term *LATA* stands for *local access and transport area.* It is the area in which a local telephone company transports long-distance calls itself without passing the call to a long-distance company. (As this is being written, the rules are changing to allow a local telephone company to offer long-distance services.) If a call is between telephones in the same LATA, then the local telephone operating company carries the call on its own facilities (intraLATA) rather than handing it to a long-distance carrier (interLATA).

In North America, switches communicate with each other using a protocol known as signaling system number 7 (SS7). This is an out-of-band signaling system used to provide basic routing information, call setup, and other call termination functions. It operates on a data network independent of all voice channels.

5.2.2 Switch Architecture

Figure 5.2 illustrates the architecture of a modern digital switch. The set of peripheral interfaces shown at the bottom of the figure convert external signals to digital form if they don't already exist that way. They also perform some low-level call processing. Signals are passed to the *switching matrix,* which performs the function of connecting each call as appropriate. Within the peripheral interface set, *line controllers* interface with individual telephone instruments where they are routed into the exchange. These are ordinarily analog inputs, which must be converted to digital. *DLC interfaces* connect with digital loop carrier systems, a concentration system explained in Section 5.2.3. *Trunk controllers* interface with other end offices and tandem switches.

Voice channels are handled digitally in modern switches. The basic digitization method converts each voice channel in each direction to a 64-kb/s datastream. This is explained in Section 5.3.1.

The input/output controller system provides access to the switch for maintenance, billing, routine operations and administration, and loading of software. The central processor is usually duplicated to ensure reliability. It controls call-processing activities, such as assigning time slots, and administering features, such as call forwarding.

Call Processing

It is useful to understand what information originates in the switch, and in what form, since this aids in later understanding what functions can be provided in a particular cable telephony system. The switch provides a number of services to the call, as follows:

> **Call Detection.** The switch determines that the telephone receiver has been lifted ("goes off-hook"). This is normally done by detecting current on the line. The current is set up by a voltage supplied from the switch. When the telephone goes off-hook, a resistance is placed across the phone line. The switch's central processor confirms that the phone is a valid subscriber,

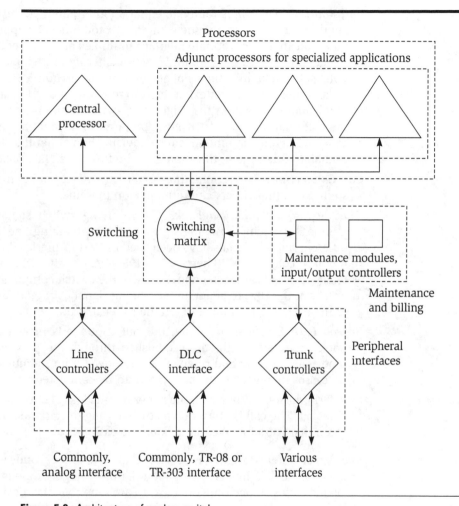

Figure 5.2 Architecture of modern switch.

and marks the line as busy (as opposed to idle). It creates a *call register* to receive and store dialed number data, and a *device register* to handle additional messages to and from the line.

Dial Tone. When a switch detects an off-hook condition, a dial tone generator in the switch provides dial tone to the line.

Digit Collection. The dialed digits are collected, based on either the tones or dial pulses received from the telephone instrument.

Digit Translation. The dialed digits are accumulated until enough digits have been collected to identify where the call is to be routed. If invalid information is detected, the call is directed to a special tone or recorded announcement.

Call Routing. The call is routed to either another phone in the exchange, or another exchange, or a tandem/interLATA toll office, as appropriate. Two time slots in the appropriate interface multiplex are reserved, one for each direction of the call. (A *multiplex* is the sharing of a single communications path—usually a time slot as described in Section 5.3.3—by multiple calls. As used in this context, it refers to reservation of an appropriate communications path from the originating switch to the next point to which the call must be routed.) If no appropriate time slots are available, the caller gets a tone or announcement advising him or her that a circuit is not available. Often the notification is in the form of a "fast busy" signal.

Call Connection. The terminating circuit (the *called line*) is marked "busy" so that no other callers can terminate on that line.

Audible Ringing and Ringback. The receiving switch signals the called number by placing a ringing signal on the terminating line. A number of different ringing signals have been used in North America, but at this time the most common is 20 hertz, 60 to 108 volts rms. If a DLC or similar system (including cable telephony) is used, the switch commands the line interface at the DLC to generate the ringing signal. A *ringback* signal is sent to the initiating telephone to tell the caller that the phone is ringing.

Speech Path Established. When the called phone is picked up (goes offhook), the voice paths are established through the previously reserved time slots. Ringing and ringback are removed from the lines. The central processor then causes the connection to be completed.

Call Termination. When a phone hangs up, the loss of current on that line is detected. The call is "torn down" by releasing the circuits, time slots, and so on that were involved in making the call.

Because of system design, it is possible that up to four seconds will elapse between the time a called line is seized and the time a ringing tone is applied to it. This explains why you occasionally pick up a phone to make a call and are surprised to find someone calling in even though the phone didn't ring.

5.2.3 Digital Loop Carrier

Traditionally, telephone instruments have been connected directly to the switch in the local exchange. However, this requires a lot of copper wire to run between homes and the exchange. A more efficient way of interconnecting telephones and switches has been developed in the form of *digital loop carrier* (DLC) systems. Bulk transport is used between the exchange and a point in the field closer to the users. From that point, individual copper pairs serve each subscriber.

Figure 5.3 may be compared with Figure 5.1, which showed a telephone connected directly to the switch at the exchange. A DLC system adds a *concentrator* in the field, which multiplexes a number of subscriber lines onto a single output line going to the switch. The concentrator interfaces with the DLC interface at the

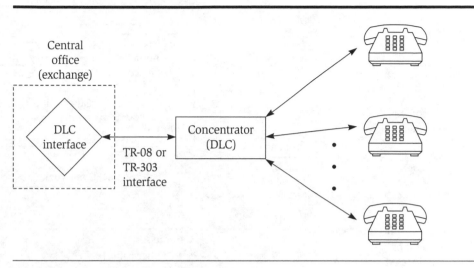

Figure 5.3 Integration of digital loop carrier into system.

lower center of Figure 5.2. The DLC converts the analog telephone signals to digital, as do the line controllers in the switch. Rather than being located in the switch, the DLC is located in the field, close to the subscriber.

Two interfaces are commonly used for DLC systems in North America; other interface specifications are used elsewhere. The more established interface is TR-08,* which is a DS1 interface (see Section 5.3.4) consisting of 24 time slots, each of which can handle one telephone call. The newer interface specification is TR-303, a wire line and fiber-optic (SONET) interface. It can support up to 2,048 lines and many telephone features.

Modification of the DLC Concept to Accommodate Cable Telephony

Figure 5.4 illustrates the way in which a DLC architecture is modified for inclusion of digital telephony over cable television systems. Later sections will cover more information relating to cable telephony.

Compare Figure 5.4 with Figure 5.3. The concentrator of Figure 5.3 is located in the field and collects calls from a number of lines, concentrating them onto one cable for interface with the switch. The host digital terminal (HDT) of Figure 5.4, in the headend, serves the same function and "looks" to the switch as if it were a concentrator. In turn, the HDT interfaces with the telephones connected to the hybrid fiber-coax (HFC) network.

An interface is required at each home subscribing to cable telephony services. This home interface supplies analog signals to the telephones in the home,

* TR stands for *technical reference*, a Bellcore-created technical document with proposed generic requirements for products, interfaces, technologies, or services.

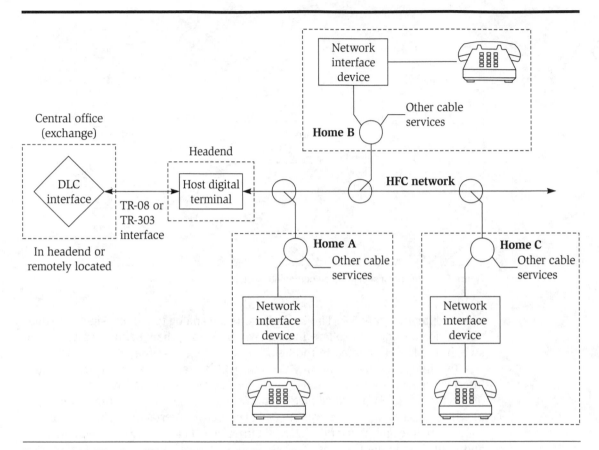

Figure 5.4 Modification of DLC architecture for cable telephony.

just as would a switch or a DLC system. It is known by many names, but here we use the generic term *network interface device* (NID) to mean the equipment at the home that interfaces between the HFC network and the subscriber's telephone instruments. Other names by which the interface is known include network interface unit (NIU), customer access unit, and wallbox.

NID is also sometimes used to mean a plastic or metal box, usually mounted on the side of a house, that contains protection equipment and the subscriber interface port to the telephone system. The term is used in the cable industry at times to describe a similar facility used to house the ground block, maybe a splitter, and maybe an amplifier. It is possible, but not mandatory, to place the telephony interface equipment in a NID according to this definition. In this chapter, we use the term to mean the equipment used to interface between the subscriber telephone equipment and the cable plant.

In general, you would expect to split the cable in front of the NID in order to provide other cable services. The splitter may be provided as part of the NID, or it may be supplied externally. It is possible, and some say desirable, to place a high-pass filter in the "other cable services" side of the splitter output to prevent leakage of spurious signals into the cable plant. This topic is covered in Chapter 14.

5.2.4 Backup Powering

The telephone industry tends to power from a −48-volt dc battery that is float charged from commercial power so that the batteries take over when commercial power fails. In addition, remote access vehicles are available that have on-board batteries to ensure the survivability of power if commercial power is lost for an extended time. Central offices and major field installations have generator backup and battery backup sufficient to hold the office during generator start-up.

5.2.5 POTS Through CLASS

A number of types of telephone service are available today. For residential service, we can talk about the services that are universally offered to all subscribers, as well as suites of upgraded services. The classical services, or plain old telephone services (POTS), include rotary and tone dialing, voice connection, operator services, access to long distance and international long distance, and in most locations in North America, 911 emergency service. (In most of North America, it is possible to summon police, fire, ambulance, or other emergency services by dialing 911. An emergency operator is presented with a screen showing the location from which the call is placed and pertinent information relating to that location.) We would expect any cable-based telephony system to provide POTS.

Advanced features allow more services and new revenue streams for telephone system operators. A second set of services, custom calling services (CCS), also called custom messaging services (CMS), work with any touch-tone telephone. The services include call waiting, call forwarding, three-way calling, speed dialing, added directory number (often called teen service) with distinctive ringing, multiline custom calling, and extension-bridged services. Some or all of these features may or may not be supported by cable-based telephony systems.

Custom local area signaling services (CLASS) are a set of advanced services provided by modern digital switches and the expanded intercommunication between them. These services include automatic callback, automatic recall, calling number delivery (also known as caller ID), customer-originated trace, distinctive ringing, selective call forwarding, and selective call rejection. Table 5.1 summarizes CLASS features.

Many CLASS services depend on delivery of the number of the calling party. The number is delivered to the called phone before it is answered. This

Table 5.1 CLASS features.

Display Features	
Calling number	Transmitted during ringing
Calling name	Local or SS7 database

Security Features	
Customer-originated trace	Recipient of harassing call dials code and makes caller's number available to law enforcement
Calling number delivery blocking	Allows caller to prevent his or her telephone number from being available to called party

Call Screening Features	
Selective call acceptance	Only calls from certain telephone numbers can ring phone
Selective call rejection	Certain numbers cannot ring phone
Selective call forwarding	Forwards calls only from a list of numbers
Anonymous caller rejection	Rejects calls from anyone who has blocked number delivery
Distinctive ringing/call waiting	Distinctive signals from designated phone numbers

Convenience Features	
Automatic callback	Lets caller know when busy called line is free
Automatic recall	Allows called party to quickly return missed call

necessitates transmission of the calling number between applications of the ringing voltage.

Besides POTS, custom calling, and CLASS, other services are often provided by local telephone companies. Network-based voice mail takes the place of answering machines, with the added advantage of being able to answer incoming calls when the phone is in use. Voice messaging, voice menus, fax messaging, and voice forms are offered, primarily to provide small businesses with big-business types of features.

Integrated services digital network (ISDN) services allow digital transmission on the normal copper pair serving businesses and residential users. So far, ISDN has had most of its impact on businesses. It allows simultaneous voice and data transmission on the same copper pair. *Basic rate ISDN* operates at 144 kb/s and allows two 64-kb/s voice channels (b, or bearer channels) and one 16-kb/s data channel (d channel). *Primary rate ISDN* operates at nearly 1.5 Mb/s and offers 23 b channels and 1 d channel.

5.2.6 The Subscriber Loop

Here we present more information on the operation of the subscriber loop, the portion of the telephone system that connects to the telephone instrument.[2] This is of interest to the HFC telephony technologist because it describes the interface between the NID and the telephone instruments in the home.

Residential telephone service is provided using a single twisted pair line, which carries signaling as well as voice in both directions. The two wires are referred to as *tip* and *ring*—from the Model 310 phone jacks used to make circuit connections in the days when the telephone system was operated manually. The 310 jack is also known as a *phone jack* and is used for many audio applications. If a third connection is added for stereo, that connection is called the *sleeve*.

Battery

A station battery voltage of 24 to 48 volts dc is supplied from the central office (the term *battery* is used regardless of the source of the voltage). The *ring* wire is negative with respect to the *tip,* but modern telephone instruments are insensitive to the polarity of the applied voltage. The battery voltage is used to power the telephone instrument and to provide current that indicates an off-hook condition. Current is limited to prevent wire heating and personal injury, and to protect the central office against shorts on a line. In the case of a cable telephony system, the network interface device provides this voltage.

Ringing

When the telephone rings, a sinusoidal alternating voltage is placed on the line, sometimes superimposed on the 48-volt battery voltage. The voltage used varies from about 60 to 108 volts rms, and would likely be toward the low-voltage end of the range for cable telephone interfaces. The higher voltages are not needed because the telephone does not have to be rung from a long distance, so the voltage drop in the twisted pair is low.

A number of different frequencies have been used for the ringing voltage. For party-line systems, in which multiple homes share one telephone line, one method used to ring the telephone of only the called party is to assign each home a different ring frequency, then to place a filter in each drop, which would pass ringing voltage of only the frequency intended for that home. In North America, party lines are rarely used anymore, so a multitude of ring frequencies is not as important. A common ring frequency today is 20 Hz. For HFC telephony systems, the ring signal is generated at the NID rather than being transmitted from the headend.

Seizing the Line

When a receiver is lifted off the hook (goes off-hook), a resistance of around 200 ohms is placed between the tip and ring. That resistance and the battery voltage

set up a current, which is detected as an off-hook condition by the switch. In the case of a cable telephony system, the off-hook condition is detected by the network interface device at the home. The NID then signals the switch of the off-hook condition.

This system, in which the resistance is applied between the tip and ring, is called a *loop start* system and is by far the most common method of seizing a line in residential telephone systems. Another method, used primarily in certain private branch exchanges (PBX) is *ground start,* in which the resistance is placed from either the tip or ring to ground. We are not likely to encounter such a system in cable telephony work.

Dialing

Two methods are used to communicate the called number to the switch. The older method uses a rotary dial. Every time a digit in the telephone number is dialed, the dial is pulled to a stop and released. As it returns to its rest position, it breaks then remakes the circuit a number of times equal to the number being dialed. Recall that as soon as the telephone goes off-hook, a load is placed on the loop, that is, the circuit is "made." The circuit is "broken" by momentarily removing the load, returning the circuit to its on-hook condition. The nominal duty cycle of one break/make is 0.1 second. The circuit is broken for 60 ms and made for 40 ms. A period of 200 ms minimum represents the spacing between digits.

The method of "dialing" used more commonly today is the use of dual-tone multifrequency (DTMF) signaling. These are the well-known touch tones. For each digit, 0 through 9, two sinusoidal tones are generated simultaneously. One tone is from the so-called low group (representing rows on the telephone keypad), and the other is from the high group (representing columns). A total of 16 states can be represented by the tones, of which 12 are normally available on consumer telephone instruments. The last four tones, reserved for special functions, are not used for customer premises equipment.

At the switch, tones are detected using a tone decoder, which supplies digit information to the call register. As the number is dialed, it is interpreted by the switch based on the information placed in the call register. In an HFC telephony system, the tones will be passed through the NID and to the switch.

Two tones are used to communicate each number. Though it is possible to communicate the same information using single tones (16 frequencies would be needed), use of dual tones improves reliability. A single-tone detector could be more easily confused into thinking that a touch tone was being transmitted when it was not. It is much less likely that a transmitted signal would simultaneously include power at the two required frequencies.

Call Progress Signaling Tones

As the call progresses, a number of tones are used to effect signaling and to indicate the status of the call. Figure 5.5 illustrates the call progress tones commonly

Figure 5.5 Call progress and DTMF tones.

used in North America, and also includes the DTMF tones. Table 5.2 describes each call progress tone, as well as a few other tones of interest.

Sidetone and Echo

Though the telephone transmission system is a two-wire system with signals going both ways on the same wire pair, within a telephone instrument, the signals are separated, using a hybrid transformer or an electronic equivalent. The hybrid by itself can provide excellent isolation between the transmit and receive directions. In fact, the isolation can be too good: a talker who doesn't hear his or her voice at the correct level in the receiver will tend to shout. For this reason, it is common in designing telephone instruments to intentionally couple some energy from the transmitter (mouthpiece) to the receiver (earpiece). This is called *sidetone.*

Sidetone is desirable, but it is similar to a phenomenon that is undesirable: echo. Echo occurs when a signal travels from the sending telephone to the receiving telephone, then is reflected to the sending telephone. The caller hears his or her own voice delayed. In long circuits, it is common to use echo suppression to reduce the intensity of the echo. Echo suppression may be very simple, such as increased attenuation in the direction not speaking, or more complex signal-processing circuits can be used. In the latter case, the term *echo cancellation* is often used.

Table 5.2 Call progress tones.

Signal	Tones (North American plan)	Description	Alternative Plans
Dial tone	350 and 440 Hz	Placed online when switch detects off-hook condition	600 and 120 Hz, or 600 and 133 Hz
Busy signal	480 and 620 Hz interrupted 60 times per minute (60 IPM)	Signifies calling line is in use	Dial tone frequencies interrupted 60 IPM
All trunks busy (ATB)	Same as busy except 120 IPM	Switch is unable to complete the call because of unavailability of resources	
Ringback	440 and 480 Hz, 2 sec. on, 4 sec. off	Informs caller that called phone is ringing	420 and 40 Hz
High tone	480 Hz	Confirmation (e.g., phone programming option has been completed successfully)	500 Hz
Bong	697 and 1,477 Hz followed by exponentially decaying dial tone	Prompts for entry of calling card number	
Off-hook	1,400, 2,060, 2,450, and 2,600 Hz	Tells subscriber that phone is off-hook	

In shorter circuits, echo suppression is not used. In order to prevent these circuits from creating objectionable echo, Bellcore has established a standard for delay in the circuit.[3] The Bellcore standard for round-trip delay of a voice signal is given by

$$RTD_{max} = 0.32 + 0.0168M$$

where

RTD = round-trip delay (in milliseconds)

M = one-way route distance (in miles)

This requirement may be broken into two parts: the 0.32 ms allows processing delay at each end of the circuit. This limited processing delay is the reason some manufacturers of cable telephony equipment are reluctant to use error cor-

rection in the cable transmission: error correction could put them outside the delay requirement because of the way that practical error-correction systems work. Error correction is applied over a certain number of bits. The bits must be held at the receiver until all are received; further, they must be examined for errors, and any errors must be corrected. Then those bits can be released for the next processing step. Moreover, most practical error-correction schemes depend on the order of the bits being changed so that a burst of error will not affect a packet beyond the point where errors could be corrected. This rearrangement of the order of the bits adds more processing delay.

The second element in the equation is the propagation delay in the medium, be it fiber or coax. The round-trip delay permitted by the equation is 16.8 μs per route mile. Since a delayed signal traverses the path twice, one can allow propagation delay of 8.4 μs per mile of plant. This is barely adequate for modern fiber-optic cables, and indeed may be slightly below the propagation delay in some cables today. Of course, propagation in coaxial cable is faster, so the coax portion will exhibit less delay.

Line Balance

Telephone signals are normally transmitted on twisted pair lines using balanced transmission. This is contrasted to the unbalanced transmission of most RF signals. In order to control noise pickup on telephone lines, it is important to control the balance between the two wires. We can show that if a transmission path is truly balanced, it neither radiates nor picks up power from the outside world. This follows from Ampere's law, which states that the magnetic field surrounding a conductor is proportional to the *net* current in that conductor. If we observe from such a location that two conductors having equal but opposite currents appear to be in the same place, then no magnetic field exists due to the current. Real balanced paths are, of course, a compromise in that it is not possible to totally balance two wires of a transmission path. However, the better balanced a path is, the less it will radiate or pick up signals.

In order to improve the balance of a circuit, it is useful to twist the two wires making up the circuit. This forces them to be in proximity to each other, which limits any magnetic field radiation (or pickup). Because the wires are often twisted, telephone engineers often refer to a wire pair connecting to a telephone as a twisted pair. The same issues apply to the handling of audio signals in headends, and is covered in more detail in Chapter 7.

5.3 The Telephone Network Digital Hierarchy

We now discuss the telephone network on the transmission side of the switch. This is the side of the switch that connects to other switches, not to telephone sets. The telephone network has been extensively digitized since the 1960s, based on a standardized hierarchy. Digital transmission was introduced in the 1960s,

followed by digital switching in the 1970s. The starting point is the digitization of a single telephone call. The digital resources required for a call are identified, then the multiplexing of a number of such calls is considered.

5.3.1 Sampling of the Voice Signal

The first step in digitizing a voice signal is to sample it. The samples are taken in order to give an analog-to-digital converter a fixed level to sample. In order to promote good-quality voice transmission, it is known that you must transmit, as a minimum, frequency components from about 300 to 3,000 Hz. A somewhat wider range adds naturalness to the voice. It was shown by Harry Nyquist of Bell Laboratories that the minimum number of samples per second that could be taken and still allow the signal to be reproduced was equal to twice the highest-frequency component of the signal.[4] It is essential to filter out any components exceeding one-half the sampling rate. If a higher-frequency component is sampled, then the spectrum of the sampled signal will fold back on itself, resulting in "aliasing" and a very distorted recovered signal.

The signal is sampled at some instant of time, and the sampled level is held in the capacitor, shown in Figure 5.6, for the time required to digitize the voltage level. Then a new sample is taken, and the process begins again. At the top, the signal is sampled, digitized, and sent through the "system" (in this case, the telephone system), then converted back to analog. The sampling process may be accurately modeled as an amplitude modulation process, in which the "carrier" is a square wave that goes between the value 0 (switch open) and 1 (switch closed). The rate at which the signal is sampled in the telephone system is 8,000 samples per second (8 ks/s), resulting in a Nyquist bandwidth, the bandwidth that can be reproduced, of 4 kHz. The practical upper frequency of the voice signal is lower than 4 kHz so as to allow for filter rolloff.

The spectrum generated by the sampling process is shown at the bottom of Figure 5.6. The baseband spectrum is present even after sampling. In addition, the first harmonic of the "carrier," at 8 kHz, is shown, surrounded by the normal modulation sidebands, which are due to the audio. The second harmonic, at 16 kHz, is also present with sidebands, though not shown here, and so on. Note that the spectra of the baseband signal and the harmonics do not overlap.

5.3.2 Analog-to-Digital Conversion

After sampling, the voice signal is converted to a digital representation. It has been found that a suitable minimum signal-to-noise ratio for voice transmission is approximately 30 dB. (The definition of signal-to-noise ratio is somewhat different from that used in video systems.) We can analyze the signal-to-noise ratio of a digitization process by considering the effect of an error in the least significant bit (LSB) of the digitized signal. It may be shown that a system for digitizing analog signals exhibits an encoding error ("quantizing error") of one least significant bit. This is taken as the quantizing noise of the system. Quantizing

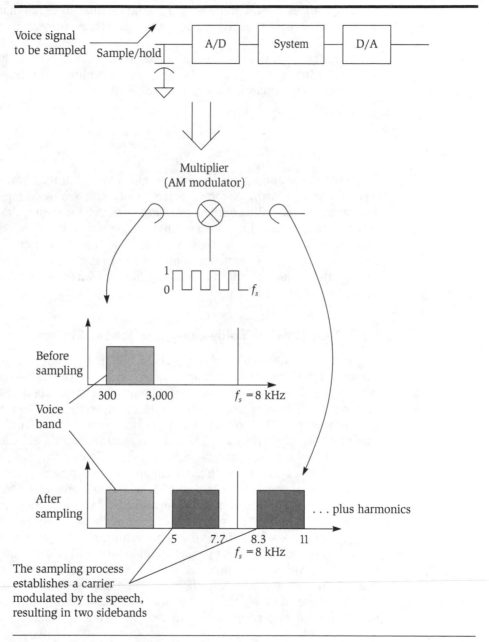

Figure 5.6 The sampling process.

noise acts somewhat the same as the random thermal noise familiar to cable engineers in that it masks the true value of the signal by creating an uncertainty in the value.

Telephone signals are digitized to eight bits. Since $2^8 = 256$, this process represents the analog signal as 1 of 256 levels. One minimum increment is 1 of these 256 levels, so the quantizing signal-to-noise ratio is

$$\frac{S}{N} = 20 \log \frac{1}{256} = -48 \, \text{dB}$$

This provides an acceptable safety margin from the minimum deemed necessary. In order to further improve voice quality, the analog-to-digital conversion process utilizes so-called *mu-law encoding*, in which the digital states don't represent constant voltage increments of the analog waveform. Low-level signals are encoded with less distance between states than are high-amplitude signals. By doing this, we can gain apparent signal-to-noise ratio with fewer bits. Various encoding laws (values of mu) are used in different countries. Outside North America, the A-law is used for encoding.

5.3.3 Time Division Multiplexing in Telephony Systems

As we stated, a voice channel is sampled 8,000 times a second, and each sample is encoded to 8 bits. The data rate required to transmit the signal is thus $8 \times 8,000$ = 64 kb/s (kilobits per second). This data rate is a basic rate of the time division multiplexing hierarchy used in digital telephony systems. The rate is called the *DS0* (pronounced "D S zero"), where DS stands for digital signal. Thus, when a telephone engineer speaks of a DS0, he or she means a 64-kb/s data signal.

A number of signals may be multiplexed to allow carriage of many signals on a single transmission path. In the standard North American hierarchy, 24 DS0s are multiplexed together to form a DS1, which operates at 1.544 Mb/s. The 24 DS0s require 1.536 Mb/s (megabits per second), whereas the remaining 8 kbps are overhead. This DS1 datastream is often called a T1 datastream, though technically T1 applies only when the transport medium is a twisted pair of wires.

European telephone companies tend to use a different multiplexing strategy in which 30 calls are multiplexed in a 2.048-Mb/s datastream, known as an *E1* stream. The data rate of 2.048 Mb/s is fast enough to allow two more voice channels, but those channels are reserved for signaling (network control). The E1 data rate is sometimes used in North America internal to a system.

5.3.4 Asynchronous Digital Hierarchy

Figure 5.7 illustrates the hierarchy used in North America for digital multiplexing of asynchronous digital signals.[5] The basic DS0 datastream may represent one data source, such as a telephone call, in which case it is called DS0-A.[6] It may also

Figure 5.7 North American asynchronous digital hierarchy.

comprise several lower data rate channels, in which case it is called DS0-B. (The terms *DS0-A* and *DS0-B* are used only for digital data service protocols, not for voice work. For voice work, we usually refer to simply DS0.) Twenty-four such signals are multiplexed, along with 8 kb/s of synchronization information, to form a DS1 signal. This 1.544-kb/s signal is very commonly encountered, but is slowly yielding to higher-rate asynchronous signals, and even to higher-rate synchronous signals. Four DS1s may be multiplexed to form a DS2. Only rarely will you encounter a DS2: it is usually just an intermediate multiplexing level in a system that ultimately multiplexes to an even higher level. Seven DS2s may be multiplexed to form a DS3, which can carry 672 simultaneous telephone conversations, or other digital data at the same rate. For example, television networks sometimes use a DS3 to transport a single digital video signal that has been compressed using algorithms most often based on linear predictive coding.

Notice that we have defined only a signal format. We have said nothing about the medium on which the signal is carried. The medium may be twisted pair (at least up through DS1, which is known as T-1 when carried on two twisted pairs—

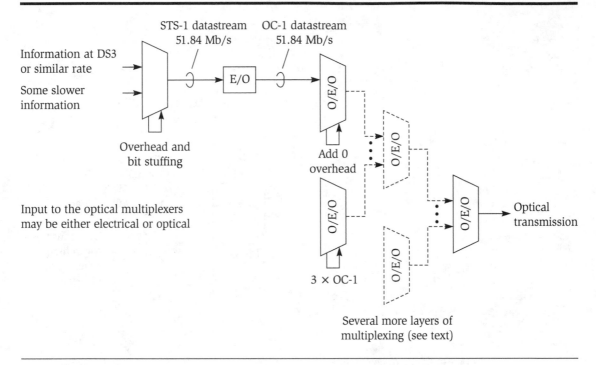

Figure 5.8 Synchronous digital hierarchy.

one in each direction), coaxial cable, microwave radio, optical fiber, or any other suitable medium.

The TR-08 interface to a telephone switch is based on DS1, several of which may be interfaced from one DLC multiplexer. In Europe and other places, a similar interface is known as the channel associated signaling (CAS) interface.

5.3.5 North American Synchronous Digital Hierarchy

A newer format used for optical signaling is the synchronous digital hierarchy (SDH). The word *synchronous* in this context refers to transmission of signals that have a constant bit rate and are derived from a common clock. It may be that the signals are initially developed from different clocks (though this is unlikely), but then they would have to be synchronized before being multiplexed. It builds on the asynchronous hierarchy shown earlier but is extended to much higher speeds. Extra bits are included to allow the data to "float" in the time slot. This means that if the data arrives after the allocated time slot begins, it can be accommodated if it is not too late, in which case it would wait for the following allocated time slot.

Figure 5.8 illustrates the synchronous digital hierarchy. Although there are variations in the multiplexing options, it is convenient to think of the starting

point as a DS3 signal at 44.736 Mb/s. This signal is up-converted to a 51.84-Mb/s datastream by adding some low-rate data and/or "stuffing" extra bits around the payload (the DS3), to add up to the desired number of bits per second. This datastream, 51.84 Mb/s, is known as STS-1 data, where STS stands for "synchronous transport signal." When converted to optical form, the signal is known as an OC-1 signal, where OC stands for "optical carrier." As opposed to cable television practice, the baseband STS-1 signal on/off modulates the laser without use of an RF carrier.

Signals can be multiplexed further. If the signals to be multiplexed are in optical form, the multiplexer must convert them to electrical form to multiplex them, then convert back to optical form. The next data rate is three times the OC-1 data rate, 155.52 Mb/s, and is known as an OC-3. One OC-3 is capable of carrying up to 2,016 telephone calls or an equivalent amount of other data. It is possible to use even higher data rates. Commercial equipment is available at this writing to support OC-192 (9.953 Gb/s) transmission.

The TR-303 switch interface mentioned earlier is capable of operating with either synchronous or asynchronous interfaces. In Europe and some other places, the equivalent interface is known as the V5.2 interface. Unfortunately, V5.2 and TR-303 are incompatible.

5.4 Elements of a Cable Telephony System

As this is being written, cable telephony is very new, so standard practices have not yet arisen. Thus, the descriptions given here necessarily reflect a limited representation of systems in use. The examples are chosen to be representative of practical systems, but variations exist.

Figure 5.4 illustrates the overall architecture of a cable telephone system. The interface hardware in the headend is generically known as a host digital terminal (HDT). It consists of modulators and demodulators on the cable interface side, and either a TR-08 or TR-303 interface on the telephone side (outside North America, V5.2 interfaces are becoming common with newer switches, and CAS interfaces are common on older switches). Between the cable and switch sides of the HDT are a host of functions required to format signals in both directions, control the HFC link, translate protocols, and support a host of telephone services. The intervening equipment also provides operational support (a common term in the telephone industry for a superset of what the cable industry calls *status monitoring*).

5.4.1 The Home Interface

A transition is necessary between the signals on the HFC network and those required to work with standard analog telephone instruments. This equipment may be located in the home, but is often mounted on the side of the house, where it forms an interface between the cable plant and all cable services located in the

Figure 5.9 Telephone interface at home.

home. The box on the side of the home is called a network interface device (NID). Other names have also been given to the box.

Figure 5.9 illustrates the basic functions of an HFC telephone NID. Also shown is a portion of the HFC network, from the optical node, through a bidirectional RF amplifier, the tap at the subscriber's home, and the NID. Cable signals are routed through the NID, where a directional coupler extracts some energy for the telephony system. The rest of the power in the signal continues to the home to provide other cable services. In this path, the manufacturer may choose to provide a service disconnect switch, an amplifier, and/or a filter to remove upstream energy coming out of the home.

From the directional coupler, the signal is routed to a diplex filter, which separates the downstream frequencies above 54 MHz, from the upstream frequencies, usually below 40 MHz. The high-frequency output from the diplexer is supplied to a receiver, which receives the downstream telephony signal. The automatic gain control shown is conventional and corrects for any variations in signal level as a result of system operating point.

The low-frequency side of the diplexer is used as an input for the upstream signal that comes from the NID transmitter. The attenuator, AT4, controls the output level from the transmitter. As explained in Chapter 14, a long loop ALC sys-

tem is used, in which the level received at the headend is compared with the expected level, and attenuator AT4 is commanded to adjust the transmitter output level until the level received at the headend equals that expected.

The modem logic is responsible for tuning the transmitter and receiver to the correct frequencies, using time slots (or equivalent) assigned from the HDT, adjusting transmit level according to instructions received from the headend, format conversion, and control of the NID. The interface circuitry provides standard analog signals to the telephone instruments in the home. It provides 24- or 48-volt dc battery voltage, ringing voltage, and dial tone, as well as handling conversion of voice signals between the analog and digital domains.

One or two lines are typically supported in a NID. It is possible in some systems to replace the analog telephone function for one line with a 64-kb/s data channel. A different subscriber interface is required, as is a different interface at the switch.

5.4.2 Powering the NID

A very important consideration involves powering the NID. At the time of this writing, four powering philosophies are being considered by the industry:

1. Powering from the plant using the center conductor of the drop cable.
2. Powering from the plant using a twisted pair bonded to the drop cable.
3. Powering from the home ac supply, with a backup battery.
4. Powering from the home without a battery, and foregoing lifeline service.

A principle embedded in the first three powering philosophies is that telephone service should be maintained in the event of commercial power failure, as is done by the incumbent phone company. This requires some sort of battery power and possibly generator power also. The fourth philosophy implies that backup power is too hard and too expensive to supply, so the cable company will not attempt to provide lifeline service and will concentrate on serving the second-line market.

Early generations of NIDs drew 5–10 watts of power. The industry would like to see power drain reduced to 1 or 2 watts. As the power drain decreases, network powering of some sort becomes more palatable. Typically, a NID will draw less power at idle, more while in use, and the greatest power when ringing.

NID Powering Using the Center Conductor

This is a popular method of system powering, where permitted by local code. It can use the existing drop if that drop is in good condition. If the drop is not in good condition, it should be replaced when adding two-way services anyway. Power is delivered over the center conductor of the drop cable from a special power-passing tap (sometimes called a telephony tap). Chapter 9 discusses power-passing taps.

Of concern with using the center conductor to pass power to the NID is that if arcing occurs, due to a faulty center conductor contact, the arc will transfer a very

significant amount of power to the upstream plant, likely causing interference with all users of the reverse spectrum. On the other hand, an arc often causes healing of a bad contact and thus can be self-extinguishing.

NID Powering Using a Twisted Pair

A similar power-passing tap places power on a separate pair of terminals, ground and hot, for each drop. A special drop cable is used that has a pair of wires molded into the same outer jacket that covers the coaxial cable. This cable is often called *Siamese cable.* The tap used with Siamese cable has connectors that are used to fasten the wire pair of the Siamese cable.

The advantage of the twisted pair (which may not actually be twisted) over center conductor powering is that you do not have to tolerate the signal loss associated with connecting RF blocking inductors to the tap ports. The problem with using this configuration is that a special drop cable must be run. In all-new installations this is not a problem, but in retrofit situations, it may not be economical or desirable.

A further concern with Siamese power-passing taps is the possibility of ingress due to signal pickup on the wire pair. The signals picked up can be introduced into the coaxial cable. Adequate filtering of the voltage connections usually is sufficient to prevent problems.

Home Powering with Battery Backup

This method of powering relieves the cable operator of the burden of powering the NID, but requires the use of a battery in each home. This raises issues of long-term maintenance: who is responsible for maintaining the battery and ultimately changing it? What happens if it is the subscriber's responsibility, but he or she doesn't do it and loses an emergency call as a result? In addition, the installer will have to go inside the home and locate a power outlet from which he or she knows the power supply will never be removed. In some cases, this will necessitate installing a new power outlet, a job for a licensed electrician in most locations.

Closely related to the issue of battery maintenance and change-out is the need to know the condition of the battery. Though average battery life is moderately well known, given a battery type and conditions of use, some batteries will inevitably fail earlier, either from unusual environmental conditions or from random failures. By monitoring the voltage of individual battery cells, it is possible to deduce the condition of the battery with acceptable accuracy. Each cell is monitored, and the results will probably be telemetered to a central location.

5.4.3 Headend Telephony Equipment

Figure 5.10 illustrates a headend configuration to support an HFC-based telephony system. Variations abound. The downstream HFC plant receives signals

Figure 5.10 Headend elements of telephony system.

from a transmitter, typically part of a bank of transmitters and receivers (modems) dedicated to the telephony system. Typically, the downstream signal will be located toward the higher end of the downstream spectrum, perhaps between 350 and 750 or 870 MHz.

A cable operator may own and operate his or her own switch, or may lease access to the incumbent telephone operator's switch. The transmitter and receiver interface to a telephone switch using one of the interface protocols (TR-08 or TR-303 in North America) described in this chapter. The switch may be co-located with the HFC telephony equipment, or the two may be interconnected with a suitable link, such as a DS1 (TR-08) or SONET (TR-303) interconnection.

Monitoring the Return Spectrum

It is useful to monitor the return spectrum independently of the signals being received by the headend receivers. Return spectrum tends to consist of frequencies that are useful and frequencies that are less useful because of interfering signals. Whatever the source of interference, it is usually the case that some frequencies are not usable at certain times of the day. (Upstream issues are treated in Chapter 14.)

Because of the return spectrum that cannot be used at certain times, many cable telephony systems employ frequency hopping. If a frequency on which the return path is operating becomes corrupted to the point of affecting transmissions, the system will switch to a new frequency. If the system does not have the ability

to monitor unused spectra, then when it is time to change frequencies, the hop would be made blind. It is possible that the system would hop to a new frequency that was as bad as or worse than the one it left. One way to prevent this is to use a separate spectrum-monitoring receiver, as shown in Figure 5.10.

Typically, the spectrum-monitoring receiver monitors a number of nodes sequentially, allowing the modem bank to have a continuously updated database of quiet frequencies, which are candidates for hopping should the need arise. A threshold is established based on received level at the monitoring receiver. When power is detected above that threshold, the frequency is marked as bad. It is taken out of the database of candidate hop frequencies.

The receiver may serve other needs as well. As shown later in the discussion of the marshaling process for TDMA-based return systems, a time offset must be determined for each NID, based on measurement of the round-trip propagation delay between the headend and that NID. One way to obtain the measurement is to have the NID transmit in response to a command from the headend in a designated time slot on the operational upstream channel. This presents a problem, though, because the delay in fiber-optic cable can easily cause the transmission to be received out of the designated time slot, causing interference with another NID. The monitoring receiver can be used to allow an out-of-channel transmission. From this data, the approximate in-channel time delay is established, ensuring that when the NID is moved to the operational channel, it will not transmit out of its assigned time slot.

Finally, the monitoring receiver can be used to monitor the return paths for any problems that would require maintenance attention. With appropriate software, a flag could be generated if an unusual pattern or level of interference developed. Maintenance personnel would be dispatched to uncover the cause.

OAM&P

OAM&P is a term used in the telephone industry for operations, administration, maintenance, and provisioning. It is a generic term for software suites that allow centralized administering of a telephone network. It is a superset of what the cable industry knows as status monitoring. In the telephone industry, cable's "status monitoring" would be a part of the "maintenance" part of OAM&P. This interface allows for determining the status of the telephone system, downloading software upgrades for any part of the system, and configuring the hardware to match services ordered by subscribers.

5.4.4 Channel Sharing

For the sake of efficiency, it is often desirable to share one communications channel between a number of users. This may be done in many ways, which we describe in more detail in Chapter 4. The review here is oriented toward a time-sharing protocol, which is common and related to that used in baseband data

communications, as outlined in Section 5.3. In the downstream direction, time division multiplexing (TDM) is used. Each of the multiple (typically 24 or 30) lines associated with the data carrier is assigned one time slot. Data is transmitted in time slot 1 to line 1, then to line 2 in time slot 2, and so on.

In the return direction, a related concept is used, except that the multiplexing must be done by turning off one transmitter, then turning on another. This is called time division multiple access (TDMA).

In typical telephony practice, the data in either direction is divided into *packets,* or data sets that are transmitted at one time. Each packet will contain several samples (frequently eight) of a single phone call. The transmission of an entire set of packets from each allocated user is called a *frame* (analogous to the television usage, where a frame is one complete picture). Commonly, frames are repeated every 1 millisecond, a frame rate of 1,000 per second. In a frame, there will be 24 packets (if the system accommodates 24 lines), each packet containing 8 samples of 8 bits each, plus overhead.

5.4.5 A Sample HFC Telephony Marshaling Protocol

This section deals, in a fairly general manner, with some of the issues that are addressed when a new NID is added to the system. The exact details of what must be done, and the order in which the steps are taken, are a function of the particular system. This section is intended only to illustrate what must be done. The frame of reference is a TDM/TDMA system, in which the downstream data consists of data supporting, typically, 24 to 30 lines simultaneously.

Marshaling and Provisioning

Acquisition of a new NID presents some interesting problems. When a new NID is added in the field, it must first be *marshaled.* This process includes being recognized by the headend system and being assigned a frequency and level at which to transmit. After marshaling, the NID is able to communicate with the headend, but it is not necessarily enabled to handle telephone calls. It must be *provisioned* by receiving software and/or setup parameters that configure it for the assigned task set.

Frequency Assignment

The first step in marshaling is to communicate the frequencies on which the NID is to receive and transmit. The NID searches its entire receive band until it finds a downstream frequency on which it recognizes data. One of the pieces of information the NID will be listening for is the upstream frequency on which it is to transmit. The initial transmit frequency may be the one to which the NID will eventually be assigned, or it may be a special marshaling frequency, which we explain later.

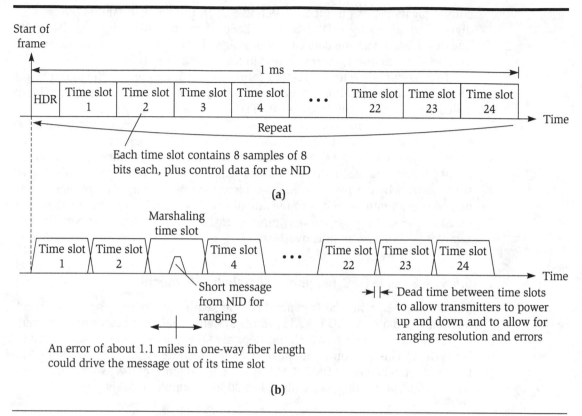

Figure 5.11 TDM and TDMA transmission. (a) Downstream TDM signal stream. (b) Upstream TDMA signal stream with time offset measurement.

TDM/TDMA Summary

Although not all systems will use TDM/TDMA for transmission on the cable, many early entry systems are using it. This section summarizes and illustrates some of the concepts described previously.

Figure 5.11 illustrates several significant aspects of a typical TDM/TDMA signal. The downstream TDM signal is illustrated in Figure 5.11(a). In this example, it consists of 24 time slots, each of which can accommodate one telephone call in the downstream direction. The frame starts with a header, which marks the start of the frame. The frames are 1 ms long, for a total of 1,000 frames per second. Since each telephone call is sampled 8,000 times each second, each time slot must contain a packet of 8 samples, each 8 bits in length. In addition, the time slot must include overhead for operating the NID. Instructions to ring the phone, supply dial tone, adjust upstream level, reconfigure (reprovision) itself, accept a new software download, and other instructions must be sent to the NID. Because of this, a protocol that supports 24 lines will likely operate at a higher data rate

than 1.544 Mb/s, the DS1 rate described earlier, which supports 24 lines. The next standard rate available is the E1 rate of 2.048 Mb/s.

Note that in a TDM system, it is not necessary to allow time between individual packets (time slots). The header defines where each bit occurs between it and the following header. It is necessary only to count the number of bits that have been received since the header to know which bit is now being received.

Contrast this to the TDMA datastream used in the upstream direction (see Figure 5.11(b)). Packets from individual return transmitters "parade" back to the headend. The same packet structure as in the downstream data may be used, but it is necessary to allow some small time between the packets. This time allows one transmitter, which has sent its packet, to turn off, and the next transmitter to turn on and stabilize before it begins transmitting its packet. Because of the need for dead time, it is necessary either to transmit at a higher instantaneous bit rate during a packet or to not transmit some bits in the upstream direction.

Shown in the TDMA signal is a marshaling time slot, which may be reserved, at least at times, for marshaling a new NID. Some options for marshaling the NID were discussed earlier. In one possibility, a new NID transmits in the time slot reserved for marshaling. The problem is that the range (signal transport delay) from the NID to the headend is not known initially. If the assumed range is in error by as little as 1 mile or so of fiber, it is possible that the response from the marshaling NID will fall outside the allocated time slot, interfering with another telephone conversation.

Time Slot Assignment

The switch interface, modem logic, and modem bank controller of Figure 5.10 work to implement a number of functions required by the HFC system. Time slots (in a time division multiplexed system) are assigned to each telephone call. In early systems, this was often done on a permanent or semipermanent basis. Later systems include concentration, in which time slots are assigned dynamically as needed.

NID Identification

It may be that when a NID is issued from the warehouse, it is identified to the modem bank, which begins transmitting its address in a downstream datastream. This downstream frequency may be a separate acquisition frequency, or the new addresses may be included in an operational datastream. Alternatively, a "successive approximation" technique may be used. In successive approximation (so named from the operation of some analog-to-digital converters), a hailing signal is transmitted that says, for example, "all unmarshaled NIDs whose address begins with a 1 reply now." If no replies are received, the headend knows that no unmarshaled NIDs have addresses beginning with 1. It can then ask for NIDs with addresses beginning with 0. A variation of this protocol doesn't initiate the marshaling process unless it is determined that an unmarshaled NID exists on the node. This could be determined by reserving a time slot in which any

unmarshaled NID will transmit. The headend listens for any reply during the time slot. When it hears a NID reply, it begins a successive approximation search to determine the address of that NID. In the unlikely event that more than one NID is unmarshaled and replies, the successive approximation search will find all of them eventually.

NID Transmitter Level

The HFC protocol must manage the level of the transmitter in the NID. As shown in Chapter 14, level control is crucial to proper operation of the cable plant. The most critical need for tight level control is at the input to the upstream laser. However, since it is located in the field, the best the system can do is to observe the return signal level in the headend and command the NID transmitter to increase or decrease its level until the level in the headend is correct. In order to know what is "correct," the engineer must know the gain between the upstream transmitter input and the point in the headend at which gain is measured.

The headend system must continuously monitor the upstream level it receives from each transmitter and must signal the individual NID transmitters when they need to adjust their level up or down. Level resolution of less than 1 dB is recommended. The resolution with which the NID can be set should be somewhat finer than the dead zone of the headend level tolerance. This will prevent continuous "hunting" for the correct level. The headend system must accurately measure the level of each return signal. Some early systems have been reported to measure bit error rate (BER) and to command the NID transmitters to increase level when the BER drops. This is not correct because if the BER is compromised from interference that has developed on the frequency, all this algorithm will do is to overdrive the return laser transmitter. If the NID cannot communicate with acceptable BER and with nominal signal level, it must be moved to a new frequency.

Implicit in any acquisition protocol is the need to get the headend to recognize the presence of the new NID. The NID must transmit to make this happen, but it has no a priori knowledge of the level at which it must transmit. It cannot risk transmitting at too high a level because doing so will cause distortion, and probably clipping, of the upstream laser. Other services using the return path will experience interruption. The rational protocol would be one in which the NID transmits at its lowest level and waits for a reply from the headend. Hearing none, it increases its level somewhat and transmits again. The process continues until the NID gets an acknowledgment from the headend. At that time, the headend can command the NID to the correct level.

Timing Offset

After the NID is recognized by the headend and its transmit level is established, it is assigned a time slot on a TDMA upstream frequency. The time slot may be permanently assigned, or it may be assigned only for the duration of one phone call, then reassigned to another call. Not only must the time slot be assigned, but a time offset value must be assigned, based on the distance from the headend to the

NID. The round-trip signal delay from the headend to the NID and back must be measured, and the NID must be given a time offset, which tells it how much to advance its clock with respect to the headend clock so that when it transmits up-stream, its signal will be received at the headend at the proper time.

The speed of light (or radio waves) in free space is 186,000 miles per second. In coaxial cable, it is typically 80% to 90% of this, depending on construction of the cable. Fiber-optic cable is denser, so the velocity of propagation is around 60% of that in free space, or around 111,600 miles per second. If 10 miles of fiber are used to reach a node, then the signals will experience a round-trip delay equivalent to 20 miles of fiber. The delay is given by

$$delay = \frac{round\text{-}trip\ distance}{velocity\ of\ propagation} = \frac{20}{111,600} = 179\ \mu s.$$

At a data rate of 2.048 Mb/s, 1 bit is transmitted in 0.45 μs, so the delay time is equivalent to about 367 bits. Obviously, the propagation delay must be com-pensated. This compensation can be done by measuring the time delay of a sam-ple signal, then computing an offset time, which is transmitted to the NID. The NID synchronizes its transmission to a "time = 0" marker transmitted from the headend. It counts the number of clock cycles from this marker to determine its transmit time. The number of cycles is offset based on the measured round-trip delay from the headend.

In order to make the time delay measurement, the NID must transmit a mes-sage at a time known to the headend modem logic. The modem logic measures the delay in the time the return signal was received and compares that with the time it expected the reception. The difference in the two is the time advance that must be transmitted to the NID. This measurement may be done on the operating channel, where other NIDs are already operating. The problem is that the initial transmission from the NID must not interfere with communications from another NID transmitting in an adjacent time slot. In order to overcome this problem, the system may initiate a very short transmission in the center of the time slot being used for marshaling. This will limit the fiber distance over which marshaling may be accomplished with no a priori knowledge of that distance to about 1.1 miles. (This is based on a TDMA system having 24 time slots, 1,000 frames per second, and a timing error of one-half of one time slot.)

The distance may be increased by telling the modem logic how much fiber is in the nodal path so that it does have a priori knowledge of the distance. Then only the much shorter delay in the coax portion need be considered. Another way is to use out-of-band marshaling to achieve crude distance measurement, which is then refined if necessary after the NID is moved to its operating frequency.

5.4.6 Implication of Diverse Routing

In order to improve the availability of cable telephony systems, some operators are installing redundant node receivers and transmitters, and connecting them

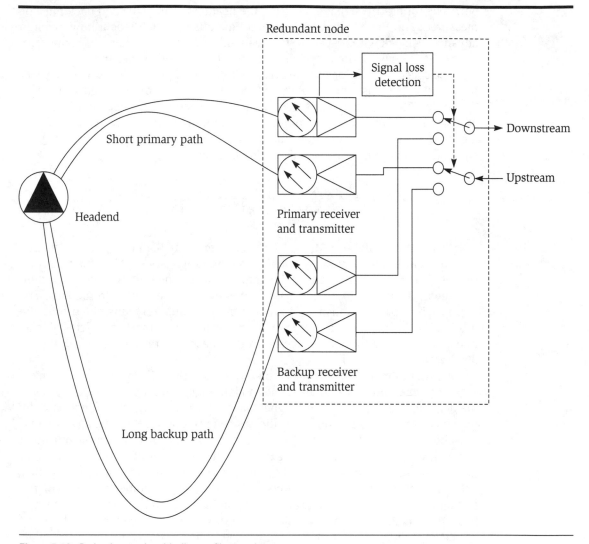

Figure 5.12 Redundant node with diverse fiber routing.

to the headend using fiber-optic cable having diverse routing. Figure 5.12 illustrates use of a redundant node, which has two optical receivers and two optical transmitters. (Chapter 16 includes more information about plant architectures.) Normally, the selection switches are in the positions shown, coupling the primary receiver and transmitter to the headend. If the received signal is lost, this fact is detected, and the switches thrown to the opposite position, to couple to the backup equipment. The loss of signal may be based on either loss of light or loss of RF signals. Normally, it is assumed that if downstream light or RF is lost, the reason is a cut in the fiber cable, so the reverse path is also affected. For this rea-

son, most redundant systems cause the upstream switch to follow the downstream switch.

Typically, a telephone system will tolerate loss of the link for a minimum of 50 ms before the call is impaired, so it is important to detect the failure and effect switching in less time than this. The backup path is almost certainly a different length from that of the primary path. This means that the marshaling of all NIDs will be wrong when the backup path is selected. The problem is only a translation in time if all the NIDs are similarly affected by the change in propagation delay. However, it is possible to split a headend telephone modem, with some of its NIDs in one node and some in another. If this has been done, then complete remarshaling will be necessary.

One proposed solution is to build two marshaling tables into the modem logic and to notify it when the backup route is in use. This means that, at some time before the backup route is needed, it will have to be activated and a second marshaling table built. If changes are made in the backup routing, then the process will have to be repeated. If the same modem serves more than one node, then backup tables must be implemented for each node served, and the modem logic will have to be notified of which node has transferred to its backup path.

5.4.7 Alternative Transmission Methods

Most of the preceding discussion was based on TDM/TDMA transmission protocols. These are currently in widespread use, but other methods have been introduced to the marketplace as well. Alternatives include SCPC, CDMA, and PFDM. They and more are covered in Chapter 4.

5.5 Network Engineering: Quality of Service

The art of network engineering deals with providing adequate equipment such that subscribers don't experience undue *call blocking*. A call is said to be *blocked* if it cannot be made because of lack of network resources. Provision of adequate equipment must be weighed against the need to control equipment costs by not providing more equipment than needed. A fundamental concept of network engineering is the busy hour, the hour in the average day at which traffic peaks. This usually occurs in late morning. If the network was engineered to carry the peak traffic on the busiest day of the year (high day busy hour, HDBH, usually Mother's Day), then at all other times the network would carry traffic with no call blocking. However, the cost of the network would be prohibitive. Most of the equipment would not be used most of the time. Balancing the need for low call blocking and low equipment cost is a domain of network engineers.

The grade of service, or probability that a call will be blocked, assigns a percentage of time (0% to 10%) that calls are allowed to be blocked when the network is congested. A key variable is the amount of time a line is being used. Planners use, as a basic unit of measure, the centum call second (CCS). One CCS

is 100 seconds of line use per hour. For example, a 3-CCS line is in use, on the average, for 300 seconds in that hour. A related term is the *erlang*, the maximum physical capacity of a line or trunk. One erlang is equal to 36 CCS, or 3,600 call seconds (or one call lasting for an hour). A 3-CCS line is thus utilized to $3/36 = 0.0833$ erlang.

A common assumption of utilization of a residential line is 3 CCS (equivalent to usage for five minutes an hour). Business lines are assumed to be 6 CCS, and an ISDN line 12 CCS. With the emerging popularity of second residential lines, often used for Internet access, the usage of residential lines may be increasing. Network engineers estimate usage of a line, and combine this with Poisson distribution tables, or similar tools, for predicting when a call is likely to be initiated and how long it will last. This information is used to predict the quality of service or the relative infrequency that a call will be blocked. Adequate equipment is provided to obtain the desired quality of service.

5.5.1 Call Concentration and Blockage

In order to achieve optimum performance in terms of cost and service level, it is important to *concentrate* calls. That is, since not every telephone is used all the time, the time slots on the HFC plant and the time slots in the interface with the switch may be shared between users. Design of concentration is part of the job of the network engineer. The complete job of network engineering is beyond the scope of this chapter, but information should be available from the chosen telephony vendor. Only the basic principles are presented here.

In HFC telephony, concentration may be applied independently to two different parts of the network. The interface between the HFC telephony system and the switch (if it is a TR-303 interface) can be concentrated. This path is typically shared by a number of headend modems, so the universe over which concentration can be achieved is large.

The other point of concentration is in the RF plant. It is not necessary to assign a permanent ("nailed up") time slot to each line, but rather, the time slot each line occupies can be assigned on a call basis. By allowing all lines to have access to the entire universe of time slots, significant *RF concentration* may be achieved. When a line needs a time slot (that is, is involved in a phone call), any time slot not previously in use can be assigned to it. Lines not currently active in making a phone call are not tying up a time slot.

Figure 5.13 illustrates call concentration. The concentration group illustrated is too small to be of practical significance, but it illustrates the concept. Each call is represented as a rectangle, with time on the horizontal axis. The first number in each call identifies the telephone: telephones 1 through 5 are involved. The number after the dash is the sequential call number from that telephone. Telephone 1 made two calls, telephone 2 made one, and so on.

In Figure 5.13(a), each telephone is assigned its own time slot (or spreading code, or frequency, or copper pair, or whatever resource is to be shared). Concentration is not practiced in this case. Since five time slots are dedicated, the total

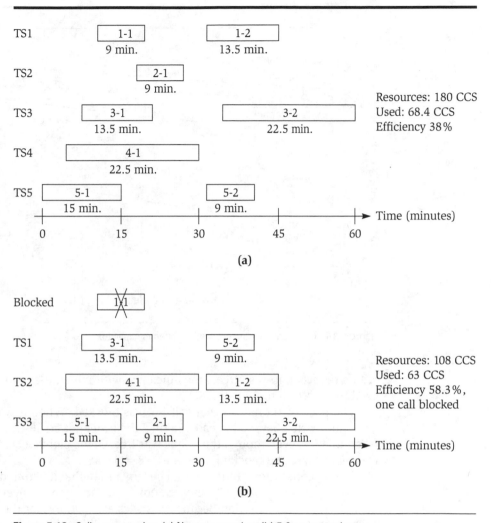

Figure 5.13 Call concentration. (a) No concentration. (b) 5:3 concentration.

capacity is 5 times 36, or 180 CCS, the resource dedicated to handling calls from these five phones. The total call time occupied is 68.4 CCS, so the efficiency is 68.4/180 = 38%.

Figure 5.13(b) illustrates what happens if only three time slots are dedicated to these five telephones. Note that call 3-1 was assigned to time slot 1, the only slot open when that call was initiated. When call 3-2 was initiated, it was assigned to time slot 3. Call 5-1 was assigned to time slot 3, but when call 5-2, involving the same phone, was placed, it was assigned to time slot 1. In this example, since only three lines are dedicated to serving these same five phones, the total resources tied up were only 108 CCS, a savings of 40% in resources. Because

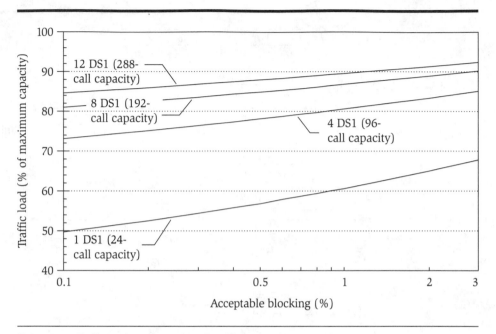

Figure 5.14 Traffic load versus blocking for differing capacities.

of this reduction in resources, the efficiency with which the resources were utilized increased from 38% to 58.3%.

The penalty, though, is that call 1-1 was blocked. When it was initiated, all three lines were busy, so instead of a ring, the caller got an all-trunks-busy signal (see Table 5.2) even though the phone was free. Of the eight calls considered, one was blocked, so the blocking rate was 1/8 = 12.5%.

Real concentration groups should be larger than that illustrated in Figure 5.13 in order to provide the best concentration. The larger the concentration group, the better the concentration because statistics are working for the network engineer: the more lines available, the greater the likelihood that one of them will be available when needed. Figure 5.14 illustrates this principle.[7] This curve plots the percent blockage accepted on the horizontal axis and the percent of maximum capacity utilized on the vertical axis. Percent of maximum capacity is computed by dividing the number of CCS actually carried by the number of CCS of capacity represented by the number of channels available. In order to convert this to the number of subscriber lines, the network engineer must determine the average CCS load offered by each line.

If the concentration group is served by only one DS1 (capable of 24 simultaneous conversations), then, depending on the blocking tolerated, the efficiency of utilization of that resource ranges from 50% to 68%. On the other hand, if the concentration group is large enough to support 12 DS1s (288 simultaneous calls),

then the efficiency will range from 85% to 92% for the same range of tolerated blocking. The reason is that, as the number of lines increases, the probability of blocking goes down simply because there are more possibilities of finding an unused time slot somewhere in the group.

This efficiency speaks toward designing larger telephone nodes. Large nodes introduce other problems, however. First, large nodes allow more noise concentration, reducing the probability that the reverse path will work properly. Second, the larger the node, the more reverse spectrum will be required to operate the telephone system. A typical practical number for first-generation telephony systems is that a group of 24 to 30 channels will require about 2 MHz of both upstream and downstream spectrum.* In the preceding example, if only a single DS1 is required to service a telephony node, then 2 MHz of reverse spectrum will suffice (not including backup spectrum for use if the primary spectrum becomes corrupted). On the other hand, if 12 DS1s are used, then 24 MHz will be required. With some spectrum held in reserve (to allow for ingress), this could easily consume all the usable reverse spectrum. Network engineering on the HFC network involves trade-offs in the size of the nodes, the tolerated noise, the spectrum allowed for telephony, and the quality of service (percent blocking tolerated).

5.6 IP Telephony

Internet protocol (IP) telephony is currently generating a lot of interest in the cable television community as an alternative to conventional telephone service. Often it is seen as a low-cost long-distance service with a local interface, which uses the Internet as a transport mechanism. As such, some operators assume that it will not be a lifeline service, removing the obligation to provide for backup powering.

IP telephony began as several proprietary computer-based voice communications systems that used the Internet to avoid long-distance charges. Voice signals were digitized, frequently in a sound card, compressed in software, and sent as data packets on the Internet. Due to practical limitations in the computer, half duplex communications were employed. In half duplex communications, only one side may speak at a time, and when it is time to allow the other side to transmit, something must be done to enable the reverse transmission. The best-known example is conventional two-way radio communications, where each side places itself in the transmit mode at the appropriate time, and often indicates to the other side when it is ready to receive, with a word such as "over."

* If only a 1.544-Mb/s DS1 were carried, the theoretical RF bandwidth for QPSK modulation would be just over 0.75 MHz. As a practical matter, the data rate for 24 calls must be somewhat greater than 1.544 Mb/s to allow overhead for control signals. Further, as more excess bandwidth is used, the reliability of communications improves. Thus, though a smaller bandwidth is possible, 2 MHz has been used by several manufacturers as a basis for one "set" of 24 to 30 voice channels. Chapter 4 covers the bandwidth issues in more detail.

Later these early IP telephony systems were enhanced with the addition of gateways at which the IP telephone call could be connected to the public switched telephone network (PSTN). This allowed a computer-based user on one end to communicate with a telephone-based user on the other end. The next logical step was to enable telephone-to-telephone communications using the Internet as the communications medium between the two, at least for part of the route.

In order to save bandwidth, it is common in both Internet telephony and digital wireless telephony systems to compress the audio. A number of proprietary and standards-based compression systems have been used. As shown earlier in this chapter, uncompressed voice communications uses a data rate of 64 kb/s, and even at this rate, some modest compression, usually either mu-law or A-law nonlinear encoding, is used. A common compression algorithm is that specified for use with the GSM PCS system in North America, which uses a bit rate of 13.3 kb/s. Other compression algorithms involve data rates from about 4.8 kb/s to 16 kb/s. Generally, as we progress toward lower bit rates, voice quality suffers, and the complexity and delay involved in the compression process increases. Increased complexity generally corresponds to higher cost, at least to the extent that the compression algorithm relies on hardware. As delay increases, it becomes progressively more difficult to carry on a full duplex conversation, as is appreciated by those who have experienced international satellite-based long-distance telephone calls.

A unique issue involved in IP telephony is based on the fact that the Internet is basically a connectionless best-effort service in which there is no guarantee of when or even if packets will be delivered. The telephone system is a connection-oriented service with guaranteed latency (delay in delivering data). Chapter 4 describes the differences between the two types of service. The Internet protocol does include a type-of-service field that specifies how an upper-layer protocol (e.g., telephony) would like to have the packet handled. However, there is currently no guarantee that this request can be accommodated. Because there is no guarantee of when a packet will be delivered, it is necessary to buffer, or store, several packets at the receiver. Then if a packet is delayed, it will still have a good chance of arriving before it is needed. If this does not happen, the listener will hear an unnatural pause in speech or a speech artifact. The greater the variation in the time that packets are received (jitter), the more delay must be inserted to buffer the incoming signal. This delay is added to the delay that develops as a result of compression and error correction as well as the natural propagation speed of the network.

The applications for IP telephony in the cable industry are seen to piggyback nicely with the emerging DOCSIS standard for cable modems (see Chapter 4). They will also allow economical long-distance calling using the Internet as the transport backbone. In addition, some operators see such a service as not needing backup powering. It is possible to build a gateway to allow interconnection of an IP telephony system to the PSTN. In order for a service to be commercially successful, it will have to be based on telephone instruments, not computers, with the IP routing transparent to the user.

5.7 Summary

This chapter has presented a brief tutorial of the traditional telephone network, with emphasis on the interface-to-analog telephone instruments found in subscribers' homes. The chapter also presented a general description of a cable telephony system. The art of cable telephony is so new that no standard practices have evolved and, indeed, given the fast pace of the cable industry, may never evolve. A fairly generic TDM/TDMA system was chosen as a frame of reference. It was the intention to describe the technology, with no regard to one commercial system above another. The next chapters will move into the headend, describing equipment and techniques of importance to efficient headend operation.

Endnotes

1. Nortel, The Technology of Telephony. *Telephony 101,* November 1994, Northern Telecom, Richardson, TX.

2. Much of the information in this section is taken from Whitham D. Reeve, *Subscriber Loop Signaling and Transmission Handbook,* Piscataway, NJ: IEEE Press, 1992.

3. GR-499-CORE, published by Bellcore, Morristown, NJ, issue 1, December 1995.

4. Khalid Sayood, *Introduction to Data Compression.* San Francisco: Morgan Kaufmann, 1996, p. 294.

5. J. O. Farmer and E. J. Callahan, Issues in Handling Cable Signals Within Digital Optical Interconnect Networks, *1994 NCTA Technical Papers.* NCTA, Washington, D.C.

6. GR-499-CORE, published by Bellcore, Morristown, NJ, issue 1, December 1995, Chapter 10.

7. Adapted from Robert Eng, *RF Concentration: Planning/Engineering.* Customer training material, supplied by Arris Interactive/Antec Digital Systems, Duluth, GA.

Part Three
Headends

Chapter 6
Signal Reception

6.1 Introduction

This chapter is the first of three devoted to the headend. Chapter 6 covers the ways in which signals get into the headend. Chapter 7 will cover the electronics used in processing the signals once they get into the headend, and Chapter 8 will cover topics related to the signals after they have been processed.

Chapter 6 deals primarily with antennas of various types used to capture signals for the headend, and with propagation of signals from the source to those antennas.

Although cable operators usually use antennas to receive signals, antennas are sometimes described in terms of their operation when transmitting signals. The principle of duality applies, which states that an antenna is capable of operating as either a radiator or receptor of power. This principle is based on Maxwell's equations, which describe the co-existence of radiated fields and those contained within a conductor, without defining which causes the other.

6.2 Off-Air Reception

Cable television got its start as a community antenna service, in which a large antenna was erected in a favorable place and signals from it were supplied to the community. Today, off-air channels still tend to be some of the more watched though other channels are also popular. Some cities remain prime candidates for a community antenna service since propagation conditions don't lend themselves to good off-air reception at all locations. Cities located in mountainous terrain are good examples. As cable and broadcast television have learned to cooperate, it is becoming more common for cable headends to be supplied with a direct feed from the broadcaster. That way, the so-called perils of the radio path don't take their toll on the signals. Even so, it is still common for local and distant broadcasters to be picked up using off-air antennas.

6.2.1 Antennas for Reception of Off-Air Signals

Most, but not all, off-air antennas are of one of two types: Yagis or log periodic arrays. However, before we discuss them, it is necessary to discuss the dipole, which is fundamental to the operation of many other types of antennas. The dipole will be used to introduce certain concepts that are useful throughout the chapter.

The Dipole Antenna

A dipole antenna consists of a conductor broken in the middle to allow it to be fed from a balanced transmission line. The total length of the dipole is normally one-half wavelength though other lengths are used in certain cases.

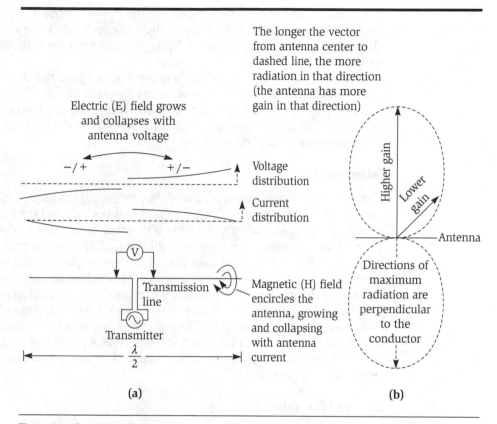

Figure 6.1 Operation of basic dipole antenna. (a) Half-wave dipole with current distribution and oscillating doublet. (b) E plane pattern.

Figure 6.1 illustrates operation of a basic half-wave resonant dipole antenna. Shown in Figure 6.1(a) is a dipole antenna excited from an RF source. The balanced nature of the ideal circuit ensures that identical current will exist on each side of the dipole. Current near the ends of the dipole is of necessity small, vanishing at the ends, because electrons, whose movement constitutes the antenna current, have nowhere to move at the very end of the wire. As you move from either end toward the middle, the RMS current magnitude increases until it reaches a maximum at the center. Similarly, the RMS voltage measured from one side of the antenna to the like position on the other side (if it could be so measured) is lowest in the middle and rises to a maximum at the two ends.

Instantaneously, the dipole is charged negatively on one side, beginning at zero and rising to a maximum charge proportional to the power supplied; then the charge decreases to zero, and that side of the antenna becomes charged positively on the next half-cycle of the exiting waveform. This process creates a rising and falling electric field from one side of the dipole to the other, which

moves away from the antenna. Similarly, the current in the dipole establishes a magnetic field encircling the dipole as shown, which also moves away from the antenna. The electric and magnetic fields together form the radiated electromagnetic field.

Maxwell's equations form the fundamental mathematics describing the action of antennas and the radiation of electromagnetic energy. They say that if an alternating electric field exists, then an alternating magnetic field must exist, and vice versa. The relative magnitude of each depends on the medium in which the electromagnetic field is propagating.

Polarization

Notice that the dipole, when parallel to the surface of the earth, radiates an electric field that is horizontal. Thus, it is said that the radiated field is *horizontally polarized*. The polarization ascribed to a radiated field is always that of the electric field. It is possible to position the dipole such that the electric field is vertically polarized. If two dipoles are placed at right angles and fed with the same signal, but 90° out of phase, then the electric field (along an axis perpendicular to the plane containing the two dipoles) will rotate between vertical and horizontal polarization, and the signal is said to be *circularly polarized*. Depending on which of the two dipoles is fed with the leading phase, the circularly polarized field will rotate either clockwise or counterclockwise (viewed with the field approaching the observer), resulting in a field that is said to be either *clockwise* or *counterclockwise* circularly polarized.

Near and Far Fields

Figure 6.2 illustrates a horizontally polarized wave front in the far field. Engineers often talk about *far field* and *near field* signals. Most computations assume far field conditions. Though there is no strict definition of far and near field, in order to be considered a far field signal, the electric and magnetic fields must be in a plane at right angles to the direction of travel of the wave, and the signal must be diverging according to the square law; that is, the signal power in a given cross-sectional area (the "power density") must be quartered every time the distance from the source is doubled. In general, these conditions are not met until the wave has traveled from the antenna by a few times the wavelength.

Note in Figure 6.1 that the E field close to the antenna is not quite flat (contained in a plane), and the H field is circling the antenna, so it cannot lie in a plane. As you go farther from the antenna, the diameter of the H field gets so large that it can be considered to lie in a plane for all practical purposes. Any magnetic field must close on itself, so the H field forms a circle. However, the dimensions of the circle become so large that for all practical purposes the H field lies in the same plane as that occupied by the E field.

Another concept of near and far field is that, in the near field, the electromagnetic field is influenced differently by different parts of the antenna, depending on the distance from the antenna part to the point of observation. That is, the an-

Figure 6.2 Plane wave traveling into page.

tenna cannot be assumed to be at one point with respect to the point of observation. As you move far enough from the antenna, the antenna becomes more of a point source so that each part of the antenna influences the field in the same proportion as does any other part.

Dipole Pattern—E Plane

Figure 6.1(b) shows a top view of the horizontal half-wave dipole antenna, with the radiation pattern it produces. The pattern is much different if the dipole is any length other than one-half wavelength.[1] The dashed lines illustrate the relative radiation in each direction. In a direction in which the line is farther from the antenna, the antenna radiates more signal. As shown, the radiation from a half-wave dipole is strongest in a direction at right angles to the direction of the dipole. Note that Figure 6.1(b) is drawn looking down on a horizontally polarized dipole, which has its E field parallel to the ground, and so is said to be horizontally polarized. Because the figure is drawn showing the radiation pattern such that the E field is in the plane of the pattern, this antenna pattern is said to be an *E field* pattern. Had the pattern been drawn looking at the end of the dipole, we would say that the pattern was the *H field* pattern.

Dipole Pattern—H Plane

In free space, a dipole would radiate equally in every direction when viewed end-on (the H field pattern). In the real world, dipoles are often operated close enough to ground that ground influences the pattern, and the radiation pattern would depend on how far above the ground plane the antenna was positioned. For example, a dipole located $\lambda/4$ (one-quarter wavelength) above a perfect ground has its maximum radiation at 90° (straight up). If the antenna is located at a height of $\lambda/2$, the maximum radiation is at about 30°. At a height of λ, the maximum

vertical radiation angle is about 15°. Above real ground, the effect on the pattern is different, depending on ground conductivity. The vertical pattern of the antenna has to be taken into account when using a dipole to estimate the field strength of a signal.[2]

Gain of a Dipole

Since an antenna cannot amplify a signal, gain of an antenna can come only from its directionality. It must rob power from one direction of radiation in order to transmit it in another. The pattern shown in Figure 6.1 shows some directivity associated with a dipole, which means that *with respect to an isotropic radiator,* a dipole does have gain. (An isotropic radiator radiates equally in all directions.) A dipole can be shown to exhibit gain in the direction of maximum radiation, of 2.15 dBi,[3] where the suffix i indicates that the gain is stated with respect to an isotropic radiator. Antenna gain is sometimes stated with respect to isotropic, and sometimes with respect to a dipole, dBd. Thus, the gain of a dipole may be expressed as either 2.15 dBi or 0 dBd.

Practical Dipole Antennas

Figure 6.3 illustrates some practical aspects of dipole antennas. This is of interest to the cable operator because he or she is sometimes called upon to use dipole antennas to make rough field strength measurements. The length of a half-wave dipole is given by dividing the velocity of propagation *in the antenna wire,* by twice (to make the dipole one-half wavelength) the frequency of the signal to be received and/or transmitted. The velocity of propagation in the antenna wire is a function of several things, but at lower frequencies with thin wire antennas, the length may be approximated by

$$l \approx \frac{468}{f} \tag{6.1}$$

where

l = antenna length (feet)

f = frequency (MHz)

As the frequency increases, the length-to-diameter ratio drops, and the resonant length of the antenna tends to decrease compared with that given in the formula. Capacitive loading may also shorten the antenna. Since a resonant dipole in free space has an input impedance very close to 75 ohms, the return loss of the antenna can be measured with common cable equipment, and the length of the antenna adjusted until the return loss is optimized for the frequency at which the dipole is to be used. At lower frequencies, the dipole will have to be suspended from its ends, so end insulators are used to attach support cables. Depending on the support material, capacitive coupling across the end insulators may lower the resonant frequency somewhat.

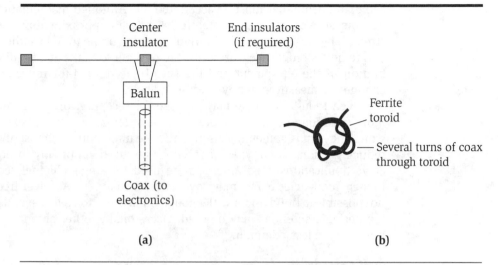

Figure 6.3 Practical dipole. (a) Practical dipole configuration. (b) One type of balun used to balance current.

In order to maintain the required balance (with respect to ground) between the two sides of the antenna, it must be fed from a balanced source (or must supply signal to a balanced load). Most equipment used in cable work is unbalanced, so it is necessary to convert from unbalanced to balanced using some sort of *balun* (*bal*ance to *un*balance). Several types of baluns are possible; the one shown is often useful at television frequencies. The coaxial cable closest to the antenna feedpoint is wrapped several times through a ferrite toroid. The resulting inductive reactance raises the impedance to any common mode current that may exist in the coax. (Common mode current is that which is in phase in the shield and center conductor of the coax.) Differential mode current, where the current in the center conductor and that in the shield are equal and out of phase, doesn't see the inductive reactance. Failure to ensure balance at the antenna driving point can cause distortion of the antenna pattern, and pickup on the coax, leading to errors in measured field strength.

A signal level meter and a half-wave dipole can be used to estimate the field strength of a radiated signal. The level of a signal is given by[4]

$$S \approx 0.021Ef \qquad (6.2)$$

where

S = signal level (dBmV)

E = field strength (µV/m)

f = frequency (MHz)

This relation is valid only for a resonant half-wave dipole in free space, oriented for maximum signal strength, and measured in a 75-ohm environment. Of course,

cable and balun loss must be accounted for. Although mathematically accurate, this measurement is represented to be only an approximation to the true field strength because a number of variables can combine to reduce the accuracy. Accurate field strength measurements are difficult to make and require careful calibration of the equipment and the site. Unaccounted-for reflections alone can change the measurement by several decibels.

Even though a dipole may be used for reception of television signals, it is more common to employ a more directive antenna, which will exhibit additional gain. The gain is sometimes useful, and the directivity is almost always useful to reduce ghosting. Two types of antennas are used commonly in professional receive applications. Yagi antennas are used for single-channel reception, or at higher frequencies a Yagi may cover several channels. At lower frequencies, it is not possible to build a Yagi with enough bandwidth to cover several channels, but log periodic antennas may be used. Occasionally, other antennas will be used, but they are less common.

Yagi Antennas

The Yagi (or Yagi–Uda) antenna is named after its inventors. It is a parasitic antenna, meaning that one of several elements is driven, and current in the driven element produces current in the others. Figure 6.4 illustrates a four-element Yagi. Yagis have at least two elements, and the number can go much higher, especially as the frequency increases. All else being equal, the more elements, the larger the structure, and the more directional and higher gain the antenna.

The driven element is essentially a dipole though the length is somewhat reduced owing to proximity effects of the other elements. A balun may be used to feed a Yagi, but other feed methods are more common. Shown is a gamma match, which is used frequently, especially at lower frequencies. The coax shield is connected to the boom, which is electrically attached to the center of the driven element. The center conductor is extended along one side of the driven element to the point at which the impedance is correctly matched to that of the coax.

A two-element Yagi typically has a driven element and one reflector, located behind the driven element, where *front* is the direction of signal radiation or reception. The reflector is slightly longer than the driven element. As more elements are added, they are added in front of the driven element and are progressively shorter. A Yagi allows a number of degrees of design freedom in the spacing between elements and the length of each. Because of this, Yagi characteristics may be modified, resulting in trade-offs between gain, bandwidth, and pattern.[5]

Log Periodic Antennas

Figure 6.5[6] illustrates the other type of off-air receiving antenna commonly used in cable television, particularly at lower frequencies. This is the log periodic antenna. It can be a very broadband antenna with good pattern characteristics. The

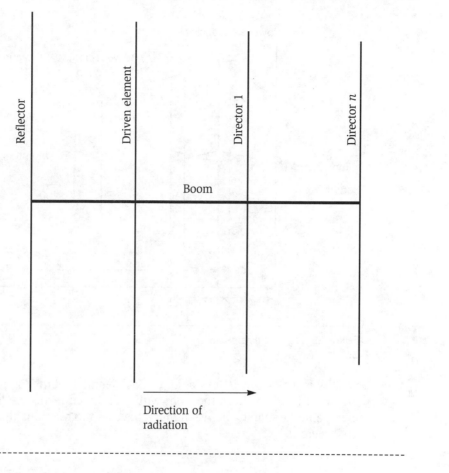

Reflector

Driven element

Director 1

Director n

Boom

Direction of
radiation

Move to best
impedance match

Driven element

Boom, with coax
shield attached

Gamma
match

Coax to electronics

Figure 6.4 Yagi antenna.

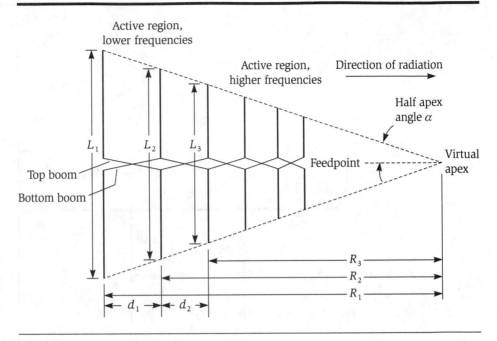

Figure 6.5 Log periodic antenna.

penalty is that it can get rather large. Each element of the log periodic antenna is driven. It is driven from the front with a balanced feed. The length of each element, and the spacing between them, have a ratio relationship, τ, called the *scaling constant:*

$$L_2 = \tau L_1 \qquad R_2 = \tau R_1 \qquad d_2 = \tau d_1 \qquad L_3 = \tau L_2 = \tau^2 L_1, \quad \text{and so on}$$

A second constant, the spacing constant σ, further describes the antenna:

$$\sigma = \frac{d_n}{2L_n}$$

The log periodic antenna exhibits an active region for every frequency at which it operates. At lower frequencies, the active region is toward the rear of the antenna. It moves progressively forward as the frequency increases.

The log periodic antenna used for television work is usually constructed using two conducting booms, one above the other. The booms are fed at the front with a balanced signal, which propagates along the two booms to the active region. In order to feed that antenna with an unbalanced coax line, the coax is

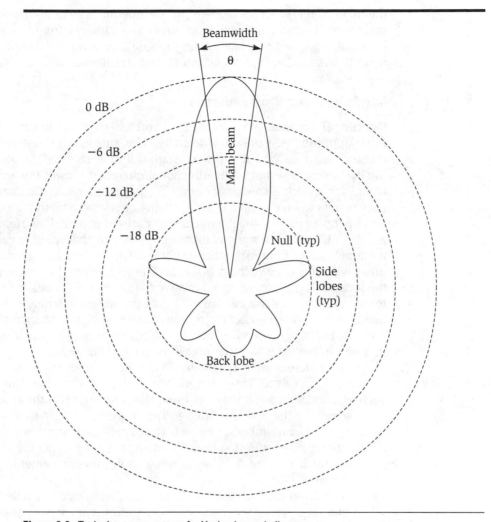

Figure 6.6 Typical antenna pattern for Yagi or log periodic antenna.

routed through one of the two booms, which then forms an infinite balun to convert from unbalanced to balanced.

Figure 6.6 illustrates a typical pattern for either a Yagi or log periodic antenna. As with all directional arrays, these antennas exhibit maximum radiation (sensitivity) in one direction. The antenna may be characterized by its *beamwidth*, the angle θ between the directions in which the power drops to one-half (−3 dB) of that in the direction of maximum radiation, or the *main beam* (sometimes called the main lobe). Ideally, the antenna would have no response in other directions, but as a practical matter, any antenna will have some response in other directions. Those responses are called *side lobes*, and an antenna will usually have several side lobes. The number of decibels that any side lobe is below

the main beam is a figure of merit for the antenna. The ratio of sensitivity in the main beam to that from behind the antenna is called the front-to-back ratio. This is usually specified separately from the side lobe level. Between the side lobes, there is often an angle of minimum radiation, called a *null*.

Interference and Ghost Reduction

Directional antennas are selected for two characteristics in cable television receive applications: gain and directivity. The antennas exhibit gain, which increases the signal level at which a distant station is received, possibly improving the carrier-to-noise ratio. The directional characteristics of the antenna can be useful for reducing co-channel and ghosting interference. Co-channel interference results when a desired station is being interfered with by a more distant station operating on the same frequency. In cases of potential interference, the FCC assigns offsets to the stations involved, such that their picture carriers will be nominally offset from each other by either 10 or 20 kHz. It is known that such an offset will reduce the visibility of a beat. The visibility of the beat is also related to the level of the interfering signal relative to the level of the desired signal. It is often possible to reduce a co-channel beat by using an antenna with better directional characteristics so as to reject the interfering signal. Alternatively, it may be possible to rotate the receive antenna slightly so as to place the interfering signal in a null in the antenna pattern, as illustrated in Figure 6.6.

The interference may not be due to another station, but it may be due to the station being received. This can occur when a signal is reflected from some object such as a building, water tower, or mountain, and arrives at the antenna after the main signal. By the time the reflected signal arrives at the antenna, the electron beam in the picture tube has moved, so the reflected signal is seen as a ghost. Again, it may be possible to move the antenna slightly to put the reflected signal in a null, or it may be possible to use a more directive antenna to reject the interference.

Figure 6.7 illustrates another technique that is used to reject interference when that interference is coming from a significant angle with respect to the desired signal. Two antennas are mounted horizontally, spaced by a distance *D*, which varies with the angle of arrival of the interfering signal. As shown in Figure 6.7(a), the two antennas are fed with equal lengths of identical coax, which supply the signals to a combiner. The lengths of the two pieces of coax are not critical, so long as they are exactly the same.

Figure 6.7(b) shows how the cancellation takes place. The phase of the signal coming from the desired source is the same at the two antennas, so at the combiner, the signals from each antenna add. However, distance *D* is selected such that the signal from the undesired source will phase cancel when the signals from the two antennas are combined. This condition is satisfied when *D* is chosen such that the undesired signal must travel one-half wavelength farther to the left antenna than to the right. The angle between the desired and undesired is θ, shown in right triangle ABC. The problem then reduces to defining *D* such that,

Figure 6.7 Use of multiple antennas to reject interference. (a) Antenna spacing. (b) Geometry showing why phase cancellation takes place.

given θ, side AC is an odd multiple of a wavelength at the frequency of reception (use the picture carrier frequency). This will cause the interfering signal to cancel. Substituting known relations, we can derive the formula for D:[7]

$$D = \frac{5{,}901n}{f \sin\theta} \qquad (6.3)$$

where

D = distance between antennas (inches)

f = frequency (MHz)

θ = angle between desired and undesired signals

n = an odd integer

Another technique that has been used in the past to cancel an undesired signal is a bucking array: a second antenna is pointed in the direction of the undesired signal. Its received amplitude and phase are adjusted such that it cancels the undesired signal being received from the main antenna. This technique was once popular, but seems to be less so now owing to the critical need for matching the level and phase between two (usually) unlike antennas.

Yet another method that has been used for ghost cancellation requires the cooperation of the television station. The station radiates a circularly polarized signal, and the cable system receives it on a circularly polarized receiving antenna. It may be shown that upon reflection, the polarization of a circularly polarized wave changes the sense of polarization, from left hand to right hand or vice versa. A circularly polarized receiving antenna can then reject the undesired polarity.

In the 1980s, the FCC permitted television stations to transmit circularly polarized signals so as to aid in rejecting ghosting. A few stations have elected to do so, but more have not. One problem is that a new and expensive antenna is required. The antenna must be installed and fed carefully to ensure truly circular polarization. Another problem is that, if all else stays the same when a station goes to circular polarization, the radiated signal in the horizontal plane traditionally used for television transmission is halved. This means that people receiving the signal with a horizontally polarized antenna will suffer a 3-dB loss of signal level. To recover this would require another 3-dB gain in the antenna or a doubling of transmitter power. Neither is a simple or inexpensive proposition.

Offsetting these objections somewhat is the observation that a circularly polarized signal is somewhat easier to receive on rabbit ear antennas, which respond to both horizontal and vertical components. This benefit may be obtained even with elliptical polarization, which is circular polarization in which the phase or amplitude of the vertical and horizontal components is not controlled carefully.

Yet another method that has been used to reduce reflection is the use of a ghost-canceling system. A training signal is transmitted in the vertical blanking interval of the television signal (line 19 has been designated by the FCC). This training signal is subjected to the same ghosting as is the video. A filter in the receiver is adjusted until its characteristics return the training signal to its known transmitted form. The same filter is applied to the video, removing the ghost. The system has been shown to be effective in reducing ghosts. Cooperation of the station is required to insert the ghost-canceling reference signal. This technique is simpler than going to circular polarization. Although it is being used in some places, the public has never bought a lot of television sets that have ghost-canceling circuits built in. Ghost-canceling hardware is available to the cable industry for headend use.

6.2.2 Propagation of Off-Air Signals

The best receiving antenna is of no use if signals are not present for it to receive. Signals propagate from the transmitting antenna to the receiving antenna in ways

that are not totally predictable though some characteristics are sufficiently constant that we can make some statements about the propagation. Signals are reflected from natural and man-made structures so that the resulting signal at the receiving antenna comprises direct and reflected *rays*. (A ray is a path taken by a signal from one point to another.) Direct and reflected rays may add constructively (in phase) or destructively (out of phase) or in between (some arbitrary phase angle). If sufficient delay is present between the signals, a ghost is produced. If the differential delay between paths is short, a distinct ghost may not result though the video waveform will be distorted.

Free Space Propagation

In the far field, the power varies inversely with the square of the distance from the source. This may be intuitively comprehended by remembering that the power from the source is spreading over two dimensions perpendicular to the direction of propagation. If so, then the area over which the power is spread quadruples every time the distance doubles. The same power is spread over four times the area, so the power density is quartered.

Signal strength is normally measured in microvolts per meter, and the units are often expressed in decibel terms, where the unit of signal level is decibels with respect to 1 microvolt per meter (dBμV/m). A particularly useful expression describing propagation of a signal in free space is[8]

$$E = 105 + P_t + G - 20 \log d \tag{6.4}$$

where

E = field strength (dBμV)/m)

P_t = power transmitted (dB above 1 kW, dBk)

G = gain of transmitting antenna (over isotropic) in the direction of concern

d = distance from transmitter (km)

Though useful in satellite work, free space propagation is more of a reference point in discussing terrestrial propagation. Reflections from the ground and other natural obstacles, and from man-made objects, will significantly modify the equation. At lower frequencies, reflections from the ionosphere will change the observed propagation. These ionospheric effects can sometimes extend well into the VHF TV band though they tend to show the greatest effect at lower frequencies.

Effects of Antenna Height

With the assumption that ground approximates a large flat surface, an approximation to field strength as a function of antenna height is given by Hall as[9]

$$E = 2E_d \sin\left(\frac{2\pi h_1 h_2}{\lambda d}\right) \tag{6.5}$$

where

h_1 = height of transmitting antenna

h_2 = height of receiving antenna

λ = wavelength

d = distance between transmitter and receiver

To get a feel for how receiving antenna height affects received signal level, consider Figure 6.8, which plots the relative strength of a signal from Hall's equation, for channels 2 and 40 (off-air), for a 1,500-foot transmitting antenna. At channel 2, the receive antenna is not high enough, even at 100 feet, to permit destructive interference between the direct ray and one reflected from the ground. At channel 40 (off-air, not cable), you can see clearly the effects of ground reflection.

We caution you against taking this illustration literally. It assumes a flat, perfectly conducting earth with no extraneous signal reflections. None of these conditions applies, except possibly to paths across open water. It does illustrate, however, that though increasing receive antenna height can be useful, there are times when you might expect a lower signal level as the antenna height is raised.

FCC Methodology for Prediction of Signal Strength

At television frequencies, radio waves are subject to shielding by the curvature of the earth though the effective radius of the earth for propagation purposes is larger than the physical radius. (See Chapter 12 for more information relating to the effective radius of the earth.) Also, one must account for reflection from the earth, which adds constructively or destructively with a direct signal. Thus, height of the transmitting and receiving antennas must be taken into account when predicting the strength of a received signal. For purposes of computing television signal strengths, the FCC uses a standard receiving antenna height of 30 feet (9 meters). The technical regulations for the broadcast industry, 47CFR section 73.699, feature graphs (Figures 9 through 11 of those rules) to be used in predicting the strength of a signal at various distances from a television station. All curves are based on a receive antenna height of 9 meters. For each frequency band, channels 2 through 6, and FM channels 7 through 13 and channels 14 through 69, two sets of curves are supplied. The first, F(50,50), is used to predict the field strength that is exceeded at 50% of potential receiver locations for at least 50% of the time. The second, F(50,10), is used to predict field strength exceeded at 50% of the potential receiver locations for at least 10% of the time. The curves are based on theory coupled with practical measurements. At higher frequencies, there may be modifications to the curves in the future.

Once a television station is on the air, it must make field strength measurements using procedures prescribed by the FCC to determine actual coverage and interference potential. These measurements are made using calibrated antennas and field strength meters not unlike the better selective voltmeters used in the ca-

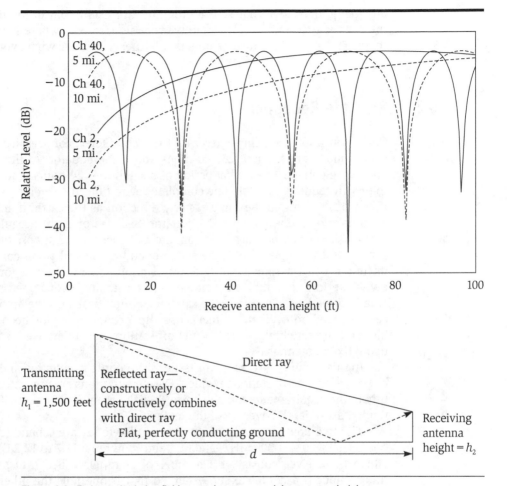

Figure 6.8 Estimated relative field strength versus receiving antenna height.

ble industry. Note that a selective voltmeter by itself is not capable of measuring field strength. The old name used by the cable industry, field strength meter, is a misnomer. The selective voltmeter is not a field strength meter until it is combined with an antenna having known gain-versus-frequency characteristics, and certain computations are made (Equation (6.2)).

The FCC sets standards for the minimum field strength for two grades of service. Grade A service is generally expected to be provided to the principal community served by a television station. Grade B service is a lower field strength, but one that should result in satisfactory pictures if good outside antennas are used. Of course, further improved service may be obtained by use of better antennas and lower noise receiving systems.

To aid in planning of cable reception, computer programs are available that use the FCC's database of station coverage to predict the received signal level at

any geographic coordinates. The programs are useful, but in the final analysis, the only way to know with certainty what signals exist where is to go look for them. The old cable industry axiom is still true: signals are where you find them.

6.3 Satellite Reception

Communications satellites used in the cable television service are in geo-synchronous orbit, about 23,500 miles above the equator. At this altitude and above the equator, the orbital period of a satellite is 24 hours, equal to the time the earth requires to rotate one complete revolution. If a satellite is launched to this orbit, moving in the same easterly direction as the earth, it appears to stay stationary over one point on the equator. Because of slight irregularities in the earth's shape, and because insertion into an exact orbit is impossible, it is necessary to track the position of the satellite, and command small corrections as it drifts out of its assigned slot. Tracking stations operated by or for the satellite owner perform this task. Fuel stored on the satellite is consumed making these small corrections. The life of the satellite is often based on the amount of fuel it can carry at liftoff. When the fuel is used up, orbital correction becomes impossible, and the satellite gradually drifts off station, becoming useless for reception using fixed antennas.

In this section, we discuss the issues that go into installation and operation of a receive-only earth station for video. Design equations are presented that will estimate the requirements for, and performance of, an earth station. This is not intended as in-depth coverage of the material.

The two popular bands for satellite service today are C band (downlink frequencies 3,700 to 4,300 MHz) and Ku band (downlink 11.7 to 12.2 GHz). The satellites themselves consist of a number of channels called *transponders*. Each transponder is a complete subsystem so that if one fails the others are not affected. A transponder consists of an input amplifier, bandpass filter, frequency translator, and output amplifier. It is usually a single conversion device. Most satellites today have 24 transponders on board though the number may vary.

Normally, one transponder will handle one analog television signal, which is frequency modulated onto a carrier. On occasion, two signals have shared a single transponder, in what is known as a *half transponder* mode. In the more common full transponder mode, the transponder is normally operated somewhat into saturation, meaning that it is slightly clipping the signal. This doesn't bother a frequency modulated signal and ensures maximum efficiency from the transponder. Each transponder is normally 40 MHz wide, including guard band between channels, with the actual signal occupying about 36 MHz.

In order to make the best use of the frequency spectrum, frequency reuse is practiced, in which signals of two polarities are transmitted. For example, some signals can be transmitted using vertical polarization, and others can be transmitted using horizontal polarization. The polarities are said to be *orthogonal*. The dictionary definition of orthogonal is that it has to do with a right angle, or some-

thing rectangular. An antenna that is horizontally polarized is theoretically insensitive to vertically polarized signals, and vice versa.

In U.S. satellite practice, the polarizations are linear, nominally vertical and horizontal, though the axes may be rotated by the time the signal travels from satellite to earth. In many international satellite applications, the orthogonal polarizations are right- and left-hand circular polarization. Either set may be used successfully.

Satellites are assigned orbital slots by the FCC for U.S. use and by the appropriate authority in other countries. Slots are assigned to each country by common consent under the auspices of the International Telecommunications Union. The orbital slots assigned to the United States are rather crowded, and the FCC has assigned orbital slots in two-degree increments. This places burdens on antennas to discriminate against another orbital slot two degrees away from the one to which it is aimed.

6.3.1 Antennas for Reception of Satellite Signals

Many different antenna types are used for reception of satellite signals. This discussion is restricted to those commonly used in cable television work today. Variations are common between different manufacturers. Almost all antennas are of the parabolic ("dish") type, in which a relatively large parabolic surface is used to collect signal power. That signal is focused on the actual receive antenna, known as the *feed*. At the feed, the signal is amplified by a low-noise amplifier (LNA) and usually block converted to a lower-frequency band. The device used to amplify and convert to a lower band has come to be known as a low-noise block converter (LNB). The most common band to which signals are converted is 950–1,450 MHz though other frequency plans have been used. The electronic hardware is described in more detail in Chapter 7.

The Parabola

Almost all satellite antennas are based on the parabola, whose geometry is shown in Figure 6.9. (Of course, practical antennas are realized in three dimensions rather than the two shown here for clarity. You can envision the third dimension by rotating the figure around the x axis.) The interesting characteristic of the parabola is that parallel rays approaching the parabola from the right, as illustrated, are all reflected to one point, called the *focus* or *focal point,* of the parabola. If the feed is placed at the focus, it will capture rays striking any point on the parabola, so long as those rays are coming from the direction shown. That direction is known as the *boresight* of the antenna.

The parabola as defined by the equation shown in Figure 6.9 extends to infinity along the x axis, with the slope of the parabola asymptotically approaching the horizontal. Real antennas are limited to some useful dimensions. The larger the area of the dish, the more power the antenna can collect. Antennas are characterized by their size, which is defined by the length of a line from one side of the dish to the other.

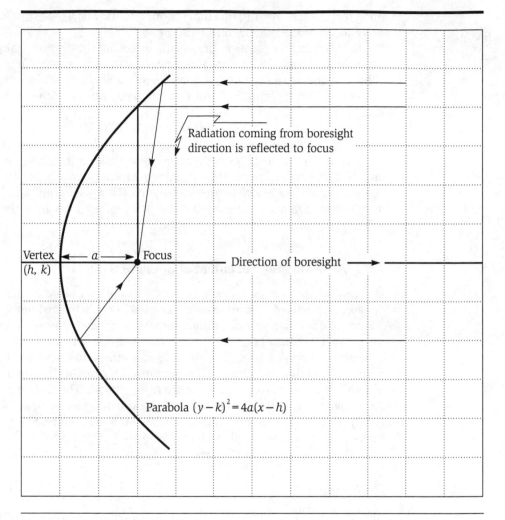

Figure 6.9 Parabolic surface.

An important quality of a real dish is the degree to which it conforms to the parabolic shape. Real dishes are constructed out of some sort of metallic surface, often of pieces bolted together in the field. A figure of merit for the dish is the RMS error in the conformity to the parabolic shape. The locations of many positions on the parabola are measured and stated as an offset, in the x axis for reflector orientation as shown in Figure 6.9, from the ideal location as defined by the parabola formula. The RMS error is then computed as follows:

$$\text{error(RMS)} = \sqrt{\frac{(x_1' - x_1)^2 + (x_2' - x_2)^2 + \cdots + (x_n' - x_n)^2}{n}} \tag{6.6}$$

where

x'_i = actual position

x_i = theoretical position

n = number of points measured

The error may also be measured as offset positions normal to the dish surface.[10]

As the operating frequency of the dish increases, the surface errors become more important. Thus, a dish that has low enough surface tolerances to be useful at C band (3.7–4.2 GHz) may not provide satisfactory performance at Ku band (11.7–12.2 GHz). Surface errors reduce the gain of the antenna and spread out the pattern: power radiated from the feed and striking an element of the reflector not positioned correctly will be reflected in the wrong direction. The effect is distortion of the pattern. Since some of the power is going in the wrong direction, less is available in the correct direction, so the gain is lower.

Prime Focus Feed

The most basic satellite antenna type is the prime focus feed, which places the feed at the focus of the parabola, as in Figure 6.10. Support spars hold the feed in position. The feed itself and the support spars offer blockage to some of the power coming from the satellite and also reflect power coming from other directions, somewhat compromising the side lobe performance of the antenna. Nonetheless, the prime focus feed is popular until the dish size becomes quite small, where the blockage and reflections are more serious.

Note that the pattern of the feed is critical for maximizing the gain and the directivity of the antenna. The feed pattern should ideally illuminate the entire surface area of the dish evenly, as shown in the dotted pattern. If the feed pattern is too narrow, the outer portions of the dish are not illuminated, so they cannot aid in focusing a transmitted signal. Similarly, a received signal will not be captured if it strikes the dish outside the feed pattern because the power will be rejected by the feed when it reaches the focal point. If the feed pattern is too wide, some of the power radiated from a transmitting antenna will miss the dish surface and go off in the wrong direction, spoiling the pattern of the antenna. Received power will not be collected efficiently because the feed antenna gain is lower than optimum in the direction of the dish surface.

Matching the feed pattern to the dish is an art in itself. Of course, no feed is perfect, but the closer you can come to the ideal, the better the antenna gain and pattern. Feed antennas include dipoles above ground planes, Yagi or log periodic structures, and microwave feed horns.

Prime focus feeds are relatively simple to build and align. A disadvantage is that the feed, which contains the active electronics of the LNB, is relatively inaccessible, being held several feet in front of a parabolic surface that is facing up at some angle. Some would argue that the feed in this location is subject to relatively more damage from electromagnetic pulses that occur when lightning strikes.

Extension of parabola
beyond antenna

Feed pattern
too wide

Proper feed
pattern

Support
spars

Feed and spars block
incoming signal and
reflect signal coming
from wrong directions

● F

Feed pattern
too narrow

Feed
and LNB

Figure 6.10 Prime focus feed.

Cassegrain Reflector

Another very common type of antenna is the Cassegrain reflector derived from
the telescope design by William Cassegrain. It is the same basic antenna as the
prime focus feed. The difference is that a second reflector is placed between
the vertex and the focus of the parabolic reflector. The most common type of
Cassegrain system is illustrated in Figure 6.11. A hyperbolic subreflector is used to
reflect signals from the main reflector to the feed. The feed is located either at or
in front of the vertex, depending on the design of the system. The hyperbolic
subreflector is located such that its first focus coincides with F1, the focus of the
parabolic main reflector. The second focus of the hyperbola is the location of the

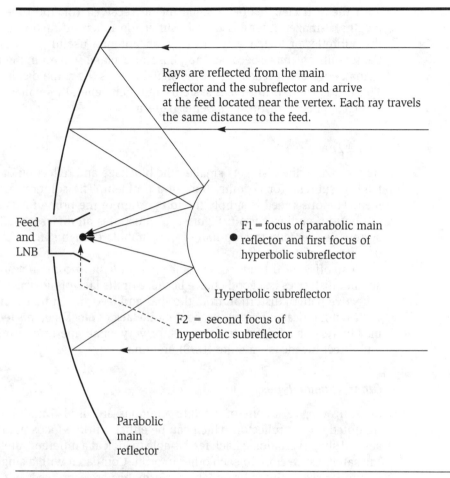

Figure 6.11 Cassegrain reflector.

feed. (The second focus is from the solution of the hyperbolic equation defining the subreflector.)

The Cassegrain reflector places the feed electronics (usually a low-noise block converter) in a position where it is considered to be more protected and easier to access. With a Cassegrain configuration, it is not necessary to route the signals along a spar, leading to somewhat lower cable loss. On the other hand, this is not especially critical with the block conversion systems commonly used in cable television. The subreflector must be supported by spars, and the subreflector and spars create blockage and pattern degradation, as in the case of the prime focus feed. The Cassegrain system is somewhat more difficult to align since the installer must align not only the feed but the subreflector as well.

In general, you would choose a prime focus feed where low side lobes are important and a Cassegrain where gain for a given size antenna is most important. This latter statement is true for antennas whose diameters are large compared

with the wavelength of the signal being received (on the order of 40 wavelengths); smaller antennas usually suffer too much diffraction on the edges of the subreflector to make the Cassegrain antenna useful. Gain is enhanced in Cassegrain antennas because the phase of a plane wave impinging on the antenna is preserved: each ray of the signal travels the same distance to the feed. This is not true of prime focus systems, but their generally smaller blockage of the primary reflector results in a better pattern.[11]

Offset Feeds

As the size of the dish gets smaller, the blockage and reflection of the spars and feed or subreflector become more of a problem. The solution can be an offset feed. The offset feed is simply the top portion of the prime focus antenna, with the bottom portion cut off. Figure 6.12 illustrates an offset feed. The illustrated system uses exactly the top half of the parabola though some offset feeds make use of less of the parabola.

The offset feed tends to be somewhat difficult to construct and align but in smaller dish sizes is a good choice because of the lack of blockage. There is some reflective pattern distortion from the spar and feed structure, but not as much as with symmetrical antennas. Because of the lack of blockage, this feed type is popular in very small aperture antennas, or very small aperture terminals (VSATs), and direct broadcast satellite (DBS) antennas.

Other Antenna Types

Several other variations of satellite antennas are used. Multiple feeds may be mounted on one reflector, which can be either a prime focus parabola or one of several shape variations. Each feed is able to look at a different satellite, so long as the satellites are close to each other in orbit. Compared with a single feed system of similar size, gain and side lobe performance are sacrificed, but the technique can be used to reduce costs and the amount of land needed for multiple dishes. A particularly interesting multiple feed antenna is a torus antenna. The reflector has a circular contour in the scan plane and a parabolic contour in the orthogonal plane. The scan plane must be carefully aligned with the orbital arc plane. However, once aligned, that antenna is capable of quite good performance over a wide section of the geosynchronous arc.

Shaped reflectors may be used rather than parabolas or hyperbolas to optimize performance of the antenna in some respect. Of course, the feed must be designed to properly illuminate whatever shape is used.

Shrouds are sometimes placed around the circumference of an antenna to reduce the magnitude of side lobes. These antennas are sometimes called ultra directive antennas (UDAs).

Horn antennas are sometimes used. They tend to be big and difficult to align, and require awkward supporting structures. However, their performance in terms of gain and side lobe suppression can be impressive. Horn antennas are used in certain high-density terrestrial microwave applications. They look something like

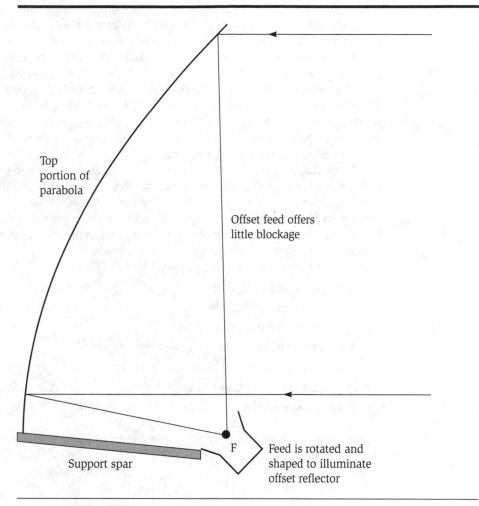

Figure 6.12 Offset feed.

gigantic cones with a dome on the wide end. A feed in the apex of the cone launches a signal, which is reflected from a shaped reflector at the wide end.

6.3.2 Earth Station Geometry

In this section, we deal with the computation of pointing angles from an earth station to a satellite, and the signal loss in the downlink. Following this, we present computations to illustrate the performance you can expect from an earth station.

Many different antenna mounts are in use. Each has advantages and disadvantages, proponents and opponents. The description here concentrates on the elevation over azimuth coordinate set, a traditional coordinate set that is relatively easy to understand and work with. Antennas designed for other than

communications with synchronous satellites often employ this set of axes. Antennas designed for communications with synchronous satellites may use different axes to permit the antenna to be moved between satellites by changing only one adjustment. Such systems tend to have proprietary elements, so you should consult the manufacturer for instructions on aiming the antenna.

The elevation-azimuth (or "el over az") coordinate system described here has its lower plane of rotation (azimuth) parallel to the local horizontal plane. A weather radar antenna, for example, rotates on the azimuth plane. The second axis, the elevation axis, is at right angles to the azimuth axis so that the antenna tilts up and down on the elevation axis.

Three pieces of information are of interest in defining how to point to the satellite: the azimuth (A) of the antenna, the elevation (E) of the antenna, and the path distance (D) to the satellite (also known as the *slant range*). These may be computed from knowledge of the satellite longitude (remember that the satellite latitude is zero, that being the only latitude for geosynchronous orbits). You must also know the earth station longitude and latitude.

Let the following definitions apply:

X = site latitude (degrees)

Y = site longitude (degrees)

Z = satellite longitude (degrees)

A = earth station azimuth angle (degrees)

E = earth station elevation angle (degrees)

D = slant range from site to satellite

C = difference between satellite and site longitudes

First compute the difference in longitude between the satellite and the earth station (the "site"):

$$C = Z - Y \tag{6.7}$$

Then the earth station azimuth angle is given by

$$A = 180° + \tan^{-1}\left(\frac{\tan C}{\sin X}\right) \tag{6.8}$$

The elevation, E, of the antenna is given by

$$E = \tan^{-1}\left(\frac{\cos C \cos X - 0.15126}{\sqrt{\sin^2 C + \cos^2 C \sin^2 X}}\right) \tag{6.9}$$

By convention, longitude is positive west of the prime meridian and negative east. Latitude is positive north of the equator and negative south. The earth sta-

tion azimuth angle is measured from true north, rotating clockwise as viewed from above the earth station. Use true north, not magnetic north, as the reference. The elevation angle is zero if the antenna is pointing horizontally and 90° if it is pointing vertically.[12] The constant 0.15126 is the ratio of the earth's radius to the satellite's orbital radius.

The slant range, or distance from the site to the satellite, is given by[13]

$$D = 26{,}199 \frac{\cos C \cos X - 0.15126}{\sin E} \text{ statute miles}$$

or

$$D = 42{,}164 \frac{\cos C \cos X - 0.15126}{\sin E} \text{ kilometers} \qquad (6.10)$$

6.3.3 Earth Station Performance Computations

From the slant range and certain other information, we can compute the expected performance of an earth station. The signal level to be expected at the antenna may be computed from the published effective isotropic radiated power (EIRP) of the satellite and the path loss, for which the slant range is needed.

Concept of EIRP

EIRP is the power that would have to be supplied to the antenna *on the satellite* if that antenna radiated equally in all directions. If the antenna on the satellite radiated equally in all directions, a typical transmitter output power on the satellite would be somewhat above 3,000 watts. This is hardly practical, but then it is not necessary for the satellite to radiate signal in all directions. It is necessary only to radiate signal in the direction of that portion of the earth that is to receive the signal. By concentrating power in that direction, the power required on the satellite is reduced drastically.

For each satellite, the owner publishes a contour map showing the EIRP at locations within the satellite's *footprint,* or intended area of coverage. This map is consulted to determine the EIRP at the location of the receive station.

Path Loss

The next quantity required is the path loss, sometimes called spreading loss because it is caused by the signal from the satellite spreading as it travels through space. Path loss may be computed from

$$L = 20 \log\left(\frac{4\pi D f}{c}\right) \text{dB} \qquad (6.11)$$

where

D = slant range (from Equation (6.10))

f = frequency (Hz)

c = speed of light

Slant range and speed of light must be in the same distance units. From the equation, path loss increases with increasing frequency. However, gain of the receiving antenna increases as frequency increases, keeping the required antenna size the same. If the slant range is expressed in miles and the frequency in GHz, Equation (6.11) reduces to the form shown in Equation (12.12):

$$\text{path loss (dB)} = 96.6 + 20\log(f) + 20\log(D)$$

where

f = frequency (GHz)

D = distance (statute miles)

Station G/T

The final quantity that must be known is the G/T of the earth station. G/T (pronounced "G over T") expresses the gain of the station, divided by its noise temperature. Noise temperature is a particularly useful expression of the noise contributed by the earth station from all sources. It is not a physical temperature, but rather is the temperature that will produce the same equivalent noise in a resistor. If a resistor is connected to a power meter, the power meter will read a certain power, given by

$$P_n = kTB \tag{6.12}$$

where

P_n = noise (watts)

k = Boltzmann's constant = $1.38 \times 10^{-23} \dfrac{\text{Joules}}{°\text{K}}$

T = temperature (°K)

B = bandwidth (Hz)

This noise power, P_n, is known as thermal noise, and is a basic law of physics: a resistor at a certain temperature will produce a certain amount of noise according to this equation. This doesn't have to be a physical resistor: it is often the terminals of an antenna, which looks electrically like a resistor. Real circuits, such as LNBs, add additional noise above and beyond the noise that would be predicted solely from the thermal noise. Noise temperature is a measure of the amount of excess noise, over and above thermal noise.

Other sources contribute to the noise temperature of a receive station. Space noise, which is produced by a number of objects in space, is a factor, but with

small earth stations, it is usually a small factor. An exception is the sun, which is a very significant noise-producing object in space: when the receiving station, satellite, and sun are lined up, communication from the satellite is lost, overpowered by noise from the sun. Also, the earth has a rather high noise temperature. When the elevation of the antenna is less than perhaps 20° or 30°, the earth's noise temperature can be significant. Of course, the contribution from the earth depends on the antenna pattern in the direction of the earth.

In addition, the LNB produces excess noise over and above what you would expect from thermal noise. The excess noise can be translated to an equivalent noise temperature by solving Equation (6.12) for T and substituting the equivalent excess noise at the input of the LNB for P_n. An equivalent concept somewhat more familiar to cable television engineers is that of noise figure. Noise figure expresses the same thing that noise temperature expresses, but in satellite work, we often work with noise temperature because certain computations are simplified.

The G/T of a station is a function of the gain of the dish, the pattern of the dish if the elevation angle is low, the noise figure of the LNB, and any losses in the structure. Technically, losses after the LNB must also be taken into account. However, they tend to have only secondary effects on the G/T of the station unless the loss between the LNB and the receiver begins to approach the gain of the LNB. The manufacturer of the station can supply appropriate G/T values.

Higher G/T will provide better carrier-to-noise ratio, and hence better baseband signal-to-noise ratio on the video. G/T can be increased by going to a larger dish size, improving the efficiency of the dish, or using an LNB with a lower noise temperature. Some slight improvement may be obtained by reducing losses following the LNB.

Carrier-to-Noise Ratio of Received Signal

Combining the preceding concepts, the carrier-to-noise ratio of the receive station can be expressed as

$$\frac{C}{N_o} = EIRP - L + \frac{G}{T} + 228.6 \tag{6.13}$$

where
$\dfrac{C}{N_o}$ = carrier-to-noise ratio (dB) per hertz

$EIRP$ = satellite effective isotropic radiated power (dBW)

L = path loss (dB) from Equation (6.11)

$\dfrac{G}{T}$ = figure of merit for earth station (dB/°K)

All quantities are expressed in decibels. The carrier-to-noise ratio expressed is per hertz. As shown in Equation (6.12) for noise power, the total noise is proportional to the bandwidth over which the measurement is made. The measurement bandwidth is the *effective noise bandwidth* of the receiver. Noise bandwidth is that

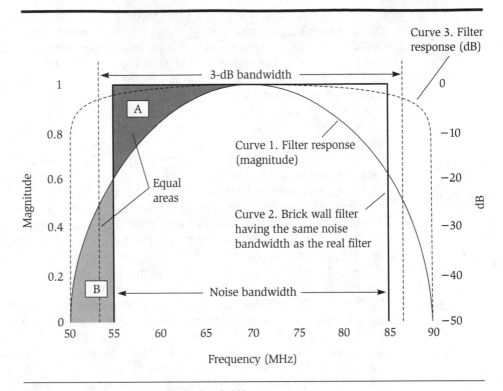

Figure 6.13 Filter response and noise bandwidth.

bandwidth a filter would have if it were perfectly rectangular and admitted the same noise as does the real filter.

Figure 6.13 illustrates the concept of noise bandwidth and how it relates to the actual bandwidth of a filter. The majority of the filtering comes from the IF filter in the earth station receiver. A filter response is shown expressed in decibels in dashed curve 3. The same filter response expressed as a magnitude ratio, [(power out)/(power in)], is shown as a thin solid line in curve 1. In determining the noise bandwidth of the filter, we work with the ratio quantity, not the decibel quantity. Filter bandwidth is usually expressed at the −3-dB points, corresponding to the half-power (10 log 0.5) response. The −3-dB bandwidth of this filter is 33.6 MHz. On the other hand, the noise bandwidth is a little narrower, 30.5 MHz. The noise bandwidth is computed by calculating the area under the magnitude response curve, then determining the bandwidth of an ideal "brick wall" filter that has the same area. Curve 2 is that of the ideal filter enclosing the same area as curve 1.

In order for curves 1 and 2 to enclose the same area, area A must equal area B. Area A defines the region where the actual response, curve 1, is below the brick wall response, curve 2. Area B defines the opposite condition, where the actual response exceeds that of the brick wall response.

Note that the noise bandwidth is an artificial quantity not representative of the actual filter response. It is merely a mathematical construct used to permit computation of the noise in the bandwidth of the receiver. This quantity is needed to compute carrier-to-noise ratio from the result of Equation (6.13).

Equation (6.13) is expressed in terms of carrier-to-noise ratio in a 1-Hz bandwidth. From Equation (6.12), the total noise is proportional to the bandwidth of the system, which is expressed by the effective noise bandwidth of the filter as shown. In order to correct Equation (6.13) for the noise bandwidth, a term must be added for the noise bandwidth. The manufacturer of the receiver will supply the noise bandwidth of the receiver. Equation (6.13) then becomes

$$\frac{C}{N} = EIRP - L_s + \frac{G}{T} + 228.6 - 10 \log BW_n \tag{6.14}$$

where

$\frac{C}{N}$ = carrier-to-noise ratio seen by receiver

$EIRP$ = satellite effective isotropic radiated power (dBW)

L = path loss (dB) from Equation (6.11)

$\frac{G}{T}$ = figure of merit for earth station (dB/°K)

BW_n = noise bandwidth (Hz)

FM Receiver Characteristics

Now consider the satellite receiver. The characteristic of interest is the baseband signal-to-noise ratio versus the RF carrier-to-noise ratio. By convention, baseband signals are characterized by a *signal*-to-noise ratio (S/N), whereas RF (and IF) signals are characterized by a *carrier*-to-noise ratio (C/N). At this point, we need to determine the receiver output S/N given the C/N computed in Equation (6.14).

Figure 6.14 illustrates a typical curve that relates baseband signal-to-noise ratio to IF carrier-to-noise ratio.[14] The measuring conditions must be stated and usually follow closely the transmission characteristics nominally used in satellite transmission (recommended by the CCIR):

Reference deviation: ΔF peak = 10.75 MHz

S/N per CCIR 421-3

525-line preemphasis

IF noise bandwidth: according to receiver (32.5 MHz in this example)

The solid line in Figure 6.14 represents the actual signal-to-noise ratio measured on the test receiver. At higher C/N, it is a straight line. As the C/N decreases, the S/N departs from the straight line, falling off more rapidly. If we were to extrapolate the curve back to a C/N of 0 dB (dotted line), we would find that the S/N was strangely high for a C/N of 0. The point at which the actual curve

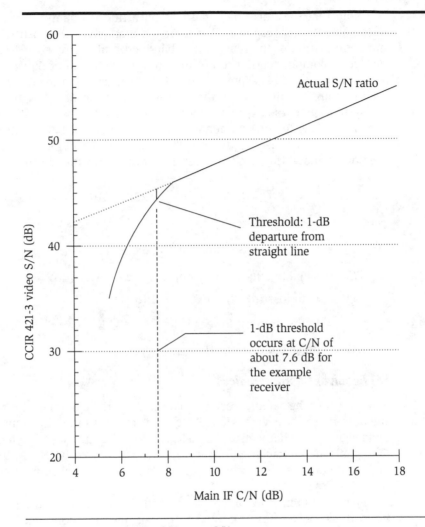

Figure 6.14 Typical receiver S/N versus C/N.

diverges from the straight line is called the *threshold* of the receiver. In order to provide a repeatable measuring point, the threshold is defined as the point at which the two curves depart by 1 dB.

In determining the needed characteristics of an earth station, it is necessary to establish a nominal operating point sufficiently above threshold so that we can tolerate real-world reductions in performance. Pointing errors can cause signal loss. Polarization discrimination can suffer because of Faraday rotation and rain, causing loss of discrimination against the cross-polarized signals. Signal can be attenuated because of rainfall: the more rain the signal travels through and the harder the rainfall, the greater the attenuation.

Terrestrial interference is a way of life at C band though arguably it may not be as severe as it was before many applications began migrating to fiber-optic cable. The C band is shared with terrestrial communications. Terrestrial interference has been known to make some receiving sites unusable on some transponders. However, in most cases, it can be controlled by erection of shielding fences or by putting the dish in a depression in the earth. If the interference is not too close to the center of the channel (the carrier), then sometimes a trap inserted in the IF of the receiver may prove effective at mitigating the interference. Many receivers include an IF loop-out feature to allow insertion of such traps.

Threshold Extension

In order to make use of smaller and smaller earth stations, which have lower and lower G/T, it is useful to find techniques that will enable the threshold to be reduced, enabling operation at lower C/N. Several proprietary techniques exist to provide a receiver with *threshold extension*. Regenerative techniques have been used, but a common threshold extension technique involves narrowing the noise bandwidth of the IF filter in the receiver. Some manufacturers have used a circuit that steers the filter, based on the luminance content of the scene. You can narrow only so far before the picture is adversely affected.

A common symptom of too low an IF bandwidth is streaking of colors in brightly colored areas of a scene. Red areas of the picture are particularly susceptible. This is a symptom of picture power that falls outside the passband of the IF filter of the receiver. If the filter was effectively widened, this effect would be reduced, but a higher carrier-to-noise ratio (higher station G/T) would be required.

Similarly, black dots ("sparklies") in the picture can indicate too narrow a filter bandwidth: as the carrier deviates with picture content, some power falls outside the filter passband, or the filter response is severely reduced at the frequency of the power. The effect is very similar to that observed when a station is operated too close to the threshold.

Pre- and Deemphasis

Frequency modulated signals have the characteristic that, when demodulated, the noise spectrum is triangular, increasing at frequencies farther removed from the carrier. In order to overcome this, many FM systems include preemphasis of the signal before modulation and deemphasis after demodulation. In audio work in North America, the traditional preemphasis network is a single break having a time constant of 75 µs. In satellite video work, the International Telecommunications Union prescribes a preemphasis curve that breaks upward at about 150 kHz, and breaks toward a flat response at about 860 kHz.

Figure 6.15 illustrates the recommended preemphasis curve. This shows that the color subcarrier is amplified nearly 15 dB with respect to low-frequency luminance information. Recall that this emphasis applies only between the uplink

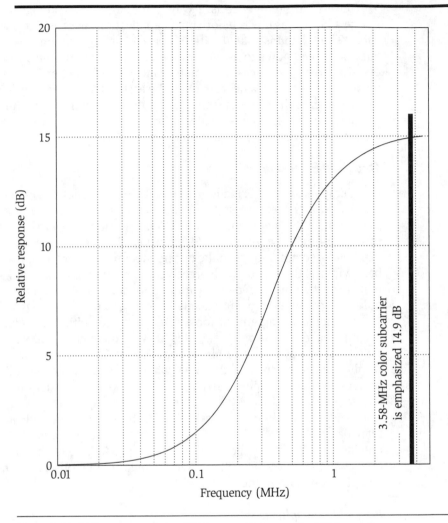

Figure 6.15 CCIR recommendation 405-1 preemphasis curve for 525-line systems.

exciter and the receiver at the headend. It does not apply to the signal out of the receiver. A deemphasis circuit in the receiver restores the signal to its flat frequency response.

However, this does illustrate that the majority of the frequency deviation is taken up by the color subcarrier in the satellite portion of the transmission path. In computing the relationship between RF carrier-to-noise ratio and baseband signal-to-noise ratio, the deemphasis must be taken into account. This is done in Figure 6.14.

Audio Transmission

In the early days of satellite transmission, signals were sent in the clear. To prevent theft of signals, it became necessary to scramble the signal. The scrambling system currently employed replaces the horizontal sync with digitized, encrypted audio. Enough bandwidth is available to provide for stereophonic transmission. If additional audio channels are required, as for cue tone and second language transmission, subcarriers are used. The subcarriers were used for transmission of the primary audio prior to the advent of scrambling.

A number of frequencies have been used for sound subcarriers, but the most common in North America are 6.8 and 6.2 MHz. The audio source is modulated onto the subcarrier (normally using the standard 75-μs preemphasis), then the subcarrier is combined with the video to form the complete signal transmitted through the satellite.

6.3.4 Satellite Frequency Plans and Interference Issues

The more established band for satellite reception is the C band, 3.7 to 4.2 GHz. This band is shared with terrestrial microwave links, and so interference becomes a significant issue. The issues are probably not as severe as they were a few years ago, however, since more terrestrial services have been migrating to fiber-optic cable.

Polarization Diversity and Frequency Offset

Figure 6.16 illustrates the frequency plan used at C band, effectively mandated by FCC rules.[15, 16] In order to use the spectrum more efficiently, two polarizations are used, as explained in the introduction to satellite communications. In North America, linear polarizations are used. In many other places, right- and left-hand circular polarization is employed. Polarization diversity allows transmission of two signals in the same frequency band without interference. However, as a practical matter, the discrimination that we can achieve is not perfect. In order to improve the performance of the system, signals using the two polarizations are also offset in frequency by one-half the channel width.

A channel is 40 MHz wide, with the occupied bandwidth of the signal approximately 36 MHz, leaving a 4-MHz guard band between channels on the same polarization. The opposite polarization is offset in frequency such that its picture carriers are in the guard band, as shown in Figure 6.16.

Polarization Rotation

Even if the signals leave the satellite polarized exactly vertically and horizontally, they don't necessarily arrive at the earth so polarized. Faraday rotation will cause both signals to rotate. Faraday rotation in the northern hemisphere is clockwise looking from the satellite toward the earth station for a downlink signal. Rain will

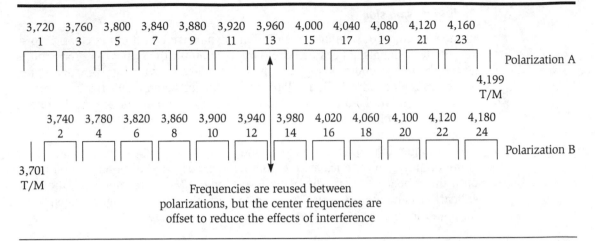

Figure 6.16 C band frequency plan.

also cause rotation of the polarization. The net effect is that the earth station antenna's ability to reject the undesired polarity is limited.

Faraday rotation is generated in the ionosphere. The ionosphere is a neutral gas containing charged particles called a plasma. In the presence of a magnetic field (from the earth), a plasma becomes an anisotropic medium, which will cause rotation of an electromagnetic signal propagating through it. Michael Faraday observed the same thing in light traveling through gyrotropic crystal with an applied magnetic field, so the phenomenon came to be named for him.

The amount of Faraday rotation depends on the electron density in the ionosphere, the magnetic field intensity, and the component of the wave traveling along the magnetic field vector. It will thus vary, as the concentration of electrons in the ionosphere varies with the season and with the time of day. The amount of Faraday rotation is inversely proportional to the square of the frequency, so it is more severe at C band than at Ku band. A typical peak value at 4 GHz is 4°. The loss of polarization discrimination at the earth station is given by the sine of the angle of rotation, so 4° results in polarization discrimination of about 23 dB. Of course, the feed may be adjusted for an average Faraday rotation, and if this is done correctly, the loss of polarization discrimination may be much less. Depolarization due to rain may be more severe in a number of locations.

Rain can cause depolarization because the raindrops are elliptical and cause amplitude and phase distortions in the two orthogonal linear polarization components of any incident polarization. Rain depolarization tends to be more severe at higher frequencies. Depending on the rain rate and the elevation angle, rain depolarization can cause the polarization discrimination to drop to between 30 and 10 dB (very high rain rates).[17]

Thus, though polarization can normally yield some 20 dB of discrimination against the undesired polarity, there are times that we wouldn't be able to expect

much more than 10 dB of discrimination, thus the need for frequency offset as well as polarization discrimination.

Numerical Observations Regarding Satellite Receiving Stations

The FCC uses a 2° separation standard for satellites serving the United States. Many satellites are so spaced, mostly near odd numbers of degrees. The physical separation is referenced to the center of the earth. Since real earth stations are located at the surface of the earth, about 15% of the distance from the earth's center to the synchronous orbit, the apparent spacing is going to be larger than is the specified orbital spacing.

Cook has presented an equation showing the apparent separation between two satellites from any place on the surface of the earth:[18]

$$\cos \alpha = \frac{2C_2 - C_3 \cos X (\cos \Delta L_A + \cos \Delta L_B) - 2C_1(1 - \cos \beta)}{2\sqrt{C_2 - C_3 \cos X \cos \Delta L_A}\sqrt{C_2 - C_3 \cos X \cos \Delta L_B}} \tag{6.15}$$

where

C_1 = 43.70532

C_2 = 44.70532

C_3 = 13.222

X = station latitude

ΔL_A = difference in longitude between first satellite and station (degrees)

ΔL_B = difference in longitude between second satellite and station (degrees)

β = satellite separation (degrees)

From this and other equations shown earlier, we can illustrate the geometrical issues for two satellites separated by 2° for several locations in the United States. Choose the eastmost positions in the orbital arc (84° and 86°) for which the satellites would be visible at all locations illustrated. The orbital slots shown are not occupied by actual satellites, but are intended to give a feel for the range of values we might encounter at moderate latitudes. Table 6.1 shows the azimuth and elevation for earth stations located in several places, for the satellite at 84°. The table also shows the apparent separation between that satellite and one at 86° and the slant range and space loss for the satellite at 84°.

The satellite locations are toward the east end of the orbit. They are roughly south of the Florida panhandle, one south of the eastern part and one south of the western part of the panhandle. The slant range from the farthest earth station (Alaska) is about 14% greater than the closest (Florida). However, the space loss, caused by spreading of the beam from the satellite, varies only by about 1 dB. A more dominant factor in determining station performance is the pattern of the antenna on the spacecraft. Typically, the EIRP in a particular direction will vary by a few decibels compared with that in another direction.

Table 6.1 Earth station geometry and space loss for representative locations in the United States for desired satellite at 84° and interfering satellite at 86°.

	Lat. (deg)	Long. (deg)	Az. (deg)	El. (deg)	App. Sep. (deg)	Slant Range (statute miles)	Space Loss (dB)
Equation:			(6.8)	(6.9)	(6.15)	(6.10)	(6.11)
Location							
Alaska	67	150	112	0.4	2.02	25,867	197
Hawaii	20	155	96.7	9	2.06	25,270	197
Washington State	47	120	135	26	2.16	24,239	196
Illinois	45	90	172	38	2.22	23,581	196
Florida	27	82	184	58	2.30	22,742	196

The apparent separation between the satellites is always at least as great as is the actual separation (as seen from the center of the earth). This may not be intuitive: you might expect that an earth station far from the subsatellite point (the point on the surface of the earth directly under the satellite), and near the equatorial plane, might see less separation than 2°. In this example, Hawaii is the station you might expect to experience the small difference in apparent separation. However, because the earth is round, Hawaii is "below" the normal plane at the subsatellite point, leading to more apparent separation.

6.3.5 Earth Station Design Example

A brief example of the parameters of an earth station design can be obtained from the preceding numerical examples. Assume a satellite at a longitude of 84°, as in Table 6.1 (no real satellite is assigned to this location). Compute the azimuth and elevation, which tell how to point the antenna. The space loss is computed once the slant range is determined. In order to compute the performance of the earth station, it remains to determine the carrier-to-noise ratio at the receiver. The first step is to obtain the EIRP of the satellite at the chosen location. This is obtained from the satellite operator. A typical C band number might be 34 dBW. Assume the Washington State location, with a path loss of 196 dB, and an earth station with an effective G/T of 20 dB/°K. Finally, assume that the receiver manufacturer specifies an IF noise bandwidth of 32.5 MHz. Equation (6.14) now becomes

$$\frac{C}{N} = EIRP - L_s + \frac{G}{T} + 228.6 - 10\log(BW_n)$$

$$= 34 - 196 + 20 + 228.6 - 10\log(32 \times 10^6)$$

$$= 11.55\,\text{dB}$$

The receiver will receive a signal having a carrier-to-noise ratio of 11.55 dB. If the S/N versus C/N curve is as shown in Figure 6.14, the expected video–signal-to-noise ratio will be on the order of 46 dB. The receiver is operating some 4 dB above threshold, before allowance for fade. Rain fade, antenna misalignment, polarization errors, foliage blockage, and other factors will reduce the actual carrier-to-noise ratio. Also, nothing has been allowed for terrestrial interference, interference from adjacent satellites, or the uplink (usually much better than the downlink). A prudent operator will allow several decibels for such.

6.4 Other Methods of Receiving Programming at Headends

By far the most common methods of receiving signals at headends is either by off-air antennas or by satellite. However, other methods are in use, and may even increase in importance in the future. These are covered briefly.

6.4.1 FM Terrestrial Microwave

In earlier days of the industry, it was very common to use terrestrial FM microwave to bring signals to the headend from distant television stations. Many of these links have now been replaced with fiber-optic cable or with closer television stations, but some links remain in operation. The equipment was the same as that used at the time by the telephone industry for microwave transmission, with the exciter changed to handle video. The standards used for modulation were very similar to those used for satellite work, so the concepts discussed earlier are valid for terrestrial microwave also. Propagation issues relating to microwave transmission are similar to those discussed in Chapter 12.

6.4.2 Direct Feed from Television Stations

Because of the perils of the radio path, it is sometimes not possible to deliver high-quality signals to a headend using pickup off the air. A moderate number of television stations now are using direct feeds from their studios to the cable headend. Typically, the baseband output of the master switcher, which supplies the broadcast transmitter, is taken as the input to the cable system.

A related activity today is cooperation between a local television station and one or more cable systems to produce local programs. Typically, the news department of a television station prepares a special cable edition of their newscast, or the off-air newscast is repeated on the cable. Other programming may also be involved, produced by either the television station or the cable operator.

Several methods are used to transport signals to the headend. Today, the most popular methods are analog using either coaxial cable or fiber optics. The trend is toward digital transmission using fiber optics. Some of the more common options are listed next.

Baseband Transmission

It is possible to transmit baseband signals several miles on coax, but the signal must be carefully conditioned to compensate for the considerable rolloff in response of a coax cable in the baseband spectrum. Also, hum pickup is a possibility that has to be dealt with. For stereophonic sound, the audio will have to be picked up off the air, brought down to the 4.5-MHz intercarrier frequency (or perhaps the 41.25-MHz sound IF), and combined with the video after modulation. Alternatively, the sound signal could be modulated onto a 4.5-MHz subcarrier at the TV station, and the sound subcarrier combined with the video for transmission. Most higher-end modulators can handle signals delivered in this fashion.

AM Supertrunk

It is possible to locate a cable television modulator at the TV studio and transport signals on a dedicated coax or fiber-optic path to the headend. At the headend, the signal is supplied to a processor of the same type used for off-air signals, which converts its frequency to the final cable channel and connects the signal to the outgoing cable plant. This is probably the oldest technique in use. If stereo transmission is to be employed, then a stereo encoder must be located at the studio so that BTSC stereo is transmitted from the studio. At the headend, most cable processors will handle BTSC stereo signals without modification. However, some operators have picked up the off-air signal using a second processor, and combined the sound IF from that processor with the picture IF from the processor handling the video coming from the studio. Engineers must be careful in doing this so that they don't exceed the frequency tolerance of the intercarrier frequency (see Chapter 7). The FCC requires that the intercarrier frequency error not exceed ±5 kHz on the cable (the broadcast tolerance, respected by high-end modulators, is ±1 kHz). If one processor is used to take an incoming picture carrier, convert it to a second channel, and combine with a sound carrier coming from an independent source, it is not too likely that the frequency tolerance will be met. Engineers must do custom-tuning of the oscillators involved in order to make this happen.

AM Sublow Band

Often if a cable operator wants to get programming originating in the community back to the headend, he or she will *backhaul* it on the sublow portion of a trunk cable. Many cable systems carry local sports events, government meetings, and government-sponsored information. The backhaul is frequently done by installing two-way amplifiers in the required portion of the plant. A sublow modulator is used at the point of origination, and at the headend the signal is converted to the downstream channel, again using a processor. This is essentially the same as the technique just discussed, except for the use of the sublow spectrum on the necessary portion of the plant.

Table 6.2 Sublow channel designations.

Channel	Picture Carrier (MHz)
T-7	7
T-8	13
T-9	19
T-10	25
T-11	31
T-12	37
T-13	43

By common convention, the industry has divided the return band into 6-MHz "channels," which are designated T channels. Table 6.2 shows the traditional designations used. Generally, the higher-frequency channels provide the higher-quality signal, due to noise on the return path. As shown in Chapter 14, frequencies generally below 15 or 20 MHz tend to experience a lot of interference on the return path. Thus, higher frequencies are preferred, especially for analog video, which tends to be more susceptible to interference than is digital transmission.

One factor, though, forces us to not put the return video too high in the spectrum. As shown elsewhere, the diplex filters used in two-way distribution amplifiers produce group delay near the crossover frequency. If the signal must be transported through too many amplifiers, the group delay may add to an unacceptable number. The usual countermeasure for this is to reduce the frequency of the analog transmission.

FM Supertrunk

A similar technique is to use FM transmission on a dedicated cable. This was popular a few years ago since the quality signal delivered to the headend could be quite good. The same conditions apply to audio as described before.

Digital Transmission with No Data Compression

Equipment is available today to allow noncompressed digital transmission to take place over a fiber-optic cable. The systems are low in cost but require a relatively high data rate for each channel. Data rates of 120–240 Mb/s per video channel are not uncommon in these applications. Several video signals can be multiplexed, with the total data rate being 2.4 Gb/s or higher. The technique is also used to interconnect headends, where it is desired to transport several high-quality signals

between headends. A dedicated fiber is usually used, though wave division multiplexing (WDM) is beginning to be practiced today.

Digital Transmission with Moderate Data Compression

Equipment is available to allow moderate compression in a transmission path. Generally, this technique will be more costly than use of noncompressed transmission, but it may be preferred if there is a scarcity of bandwidth. Compression is based on vendor proprietary compression schemes, most of which use *linear predictive coding* (LPC). In an LPC system, an algorithm is used to predict the content of the next picture (usually either a field or frame) based on past data. The encoder includes a decoder, which works identically to the one at the receiving end. The predicted output of the decoder is compared with the picture before encoding. The difference is an error signal, which is transmitted to the receiver. Because TV pictures tend to have a lot of commonality between frames, the amount of information transmitted is much less than if no compression were used. Data rates of around 45 Mb/s can yield excellent-quality pictures and are routinely used for backhaul of programs to broadcasters. Data rates in the neighborhood of one-half of this have met with limited acceptance.

Digital Transmission with Extreme Data Compression

This usually means some form of MPEG or MPEG-derived compression algorithm based on the *discrete cosine transform* (DCT) and possibly other aspects of MPEG compression. Because of the cost, no known applications exist today that utilize full MPEG compression for point-to-point transmission of video though this may happen in the future. However, it is not that expensive to implement just the DCT and maybe some other modest data reduction techniques. Data rates of under 20 Mb/s for broadcast quality have been reported. Chapter 3 includes more information on compression.

6.4.3 Local Origination

Finally, local origination (LO) is practiced in many systems. LO is done from studios operated by the cable company or sometimes by other parties. Typically, the studios will be similar to those at broadcast stations, except that the amount and quality of equipment is less.

Much programming originates on tape. Common tape formats in use include consumer VHS (very low end), super VHS, and 3/4 inch, in increasing order of quality and cost. Better operations employ time base correctors to compensate for the unavoidable timing errors that develop because of inconsistent tape transport speed. Computers may be used to control the tape machines, or they may be controlled manually.

The leading edge in local origination storage is the use of video file servers. Programs are digitized and compressed, usually using MPEG compression. The compressed program is stored on a large file server, not unlike those used in busi-

ness applications to store massive amounts of data. At the programmed time, the file server "plays" the file to an MPEG-to-NTSC converter (for analog channels), which converts the signal back to analog. This type of system is seeing its earliest successes in the commercial insertion (ad insertion) market, but likely will see future application in playing out long-form programs.

6.5 Summary

This chapter has discussed the underlying technology used to bring signals into the headend. The vast majority of signals are received today from satellites, with the next largest number being picked up off the air. Other techniques are used and are outlined briefly here. In the next two chapters, we will discuss headend-related concepts beyond getting signals into the headend. Chapter 7 will cover handling signals in individual pieces of equipment in the headend, and Chapter 8 will cover handling the signals outside those pieces of equipment.

Endnotes

1. Richard C. Johnson et al., *Antenna Engineering Handbook*. New York: McGraw-Hill, 1984, pp. 4–11.

2. Mark J. Wilson, ed., *The ARRL Handbook for the Radio Amateur*. Newington, CT: American Radio Relay League, 1988, p. 17.2. Regrettably, this figure seems to have been removed from recent editions.

3. Computed from Equations (12) and (14) of International Telephone and Radio Corporation, New York, *Reference Data for Radio Engineers*, 4th ed., 1967, pp. 676–677.

4. Scientific-Atlanta, Inc., Atlanta, GA, *CATV Data Book*, rev. 6, August 1990, p. 76.

5. Richard C. Johnson et al., *Antenna Engineering Handbook*. New York: McGraw-Hill, 1984, pp. 29–17ff.

6. Thomas Milligan, *Modern Antenna Design*. New York: McGraw-Hill, 1985, p. 383ff.

7. Scientific-Atlanta, Inc., Atlanta, GA, *CATV Data Book*, rev. 6, August 1990, p. 77.

8. Marty Barringer, Radio Wave Propagation. *Engineering Handbook*, 7th ed. National Association of Broadcasters, Washington, D.C.

9. Ibid.

10. Rick Barker and John Friesz, Earth Station Antennas, RF Considerations. Scientific-Atlanta Communications Symposium, 1983.

11. Ibid.

12. F. M. Fonda, Earth Station Geometry. Scientific-Atlanta Communications Symposium, 1983.

13. Jack Sanders, Satellites: The Birth of an Industry. *Installer/Technician,* May 1988, p. 15.

14. J. Searcy Hollis, Satellite Link Analysis. Scientific-Atlanta Communications Symposium, 1983, p. 12.

15. 47 CFR sec. 26.211. Scientific-Atlanta, Atlanta, GA.

16. J. Searcy Hollis, Principles of Satellite Communications. Scientific-Atlanta Communications Symposium, 1983, p 25.

17. Rick Barker, Re-Use Earth Station Antennas. Scientific-Atlanta Communications Symposium, 1983.

18. James H. Cook Jr., Interference Analysis. Scientific-Atlanta Communications Symposium, 1983. (The equation is in Appendix A.)

Chapter 7
Headend Signal Processing

7.1 Introduction

This chapter deals with the electronic devices most commonly used in headends to format signals for transmission on a cable plant. These devices include signal processors, RF devices used to take television signals off the air or from an incoming cable and put them on the cable. Also included are modulators, miniature transmitters that accept baseband audio and video signals and convert them to a channel. Demodulators do the opposite, converting an RF signal back to baseband video and audio. Stereophonic encoders convert left- and right-channel audio into the composite stereo format used in television transmission. Earth station receivers convert signals coming from communications satellites to baseband audio and video. Often these are used with descramblers that deal with the proprietary scrambling format currently used with most analog satellite transmissions.

Block diagrams and descriptions of the operation of the equipment are presented, along with information regarding the proper application of the hardware. The block diagrams are representative of commonly available equipment. They are intended to illustrate how the functions are accomplished; they are not intended to represent particular pieces of equipment. Variations will be encountered in the field. Some equipment may have controls that don't work as do the controls shown here, as a consequence of differences in block diagrams.

Chapter 6 covered inputs to the headend from various RF sources. Chapter 8 will cover methods used to couple signals from the equipment covered in this chapter to the cable. You may wish to review the fundamental composition of baseband and RF signals in Chapter 2.

7.2 Signal Processors

RF signal processors are normally used to transfer incoming VSB-AM signals from off-air antennas or incoming cables to the cable plant. The processor can change the channel (frequency) of the signal and the relative sound carrier level. The signal is bandpass filtered to remove any undesired adjacent channel power. Automatic gain control is employed to stabilize the output level. Optionally, a standby carrier can be substituted when the incoming signal goes away. Switching of IF sources can be done to maximize use of a channel.

7.2.1 Input Section

Figure 7.1 illustrates a typical block diagram of a processor. A number of variations are possible. Many modern processors are frequency agile. Differences between agile and nonagile processors are pointed out as appropriate. The incoming RF signal is supplied to a bandpass filter, BPF1, which is tuned to the incoming channel. This bandpass filter is typically of fairly low order and has only a little adjacent channel rejection. Its primary purpose is to reduce off-channel power to

Figure 7.1 Signal processor block diagram.

minimize distortion in the input circuitry. It, along with BPF2, provides image rejection, as described in Chapter 8.

The signal passes through a variable attenuator, AT1, which is part of the delayed AGC circuit. "Delayed AGC" is a somewhat unfortunate term because it has nothing to do with a time delay. Rather, the term *delayed* refers to the fact that AT1 is at minimum attenuation (the front end is at maximum gain) at low signal levels. As the signal level increases (above a predetermined threshold), it becomes necessary to reduce its amplitude before it reaches RF amplifier A1 and mixer M1. Failure to reduce the amplitude would result in unacceptable distortion. At lower signal levels, the gain of A1 is needed in order to overcome the noise introduced in the mixer and following circuitry.

7.2.2 IF Section

The signal is converted to the intermediate frequency (IF) in mixer M1. The commonly used North American IF is 45.75 MHz for the picture carrier and 41.25

MHz for the sound carrier.* Note that the spectrum is inverted at IF compared with the spectrum at RF, where the sound carrier is at a higher frequency than is the picture carrier. This is a consequence of using a high-side local oscillator (LO); that is, the frequency of the local oscillator is higher than is the RF frequency of the signal being converted. Use of high-side oscillators helps control spurious responses from the mixing process.

Following mixer M1 is another bandpass filter, BPF3, tuned to the IF band, normally 41 to 47 MHz for NTSC signals. Amplifier A2 provides gain to help overcome the insertion loss of later circuits. The IF AGC attenuator is AT2, which is adjusted such that the input to the AGC detector is constant regardless of the incoming signal level. Additional amplification may be provided, and at some point in the signal chain, a sample of the signal will be routed to the AGC detector. The output voltage of the AGC circuitry changes as the signal into the AGC detector increases. This change causes the attenuation of AT2 to increase, compensating for the increasing level of the incoming signal.

If the signal level increases sufficiently, the IF AGC voltage reaches the delay threshold, above which the gain of the input attenuator, AT1, is reduced to effect stabilization of the IF signal level. Depending on the design of the AGC system, the IF attenuator, AT2, may continue to drop in gain, or it may not. Usually, the best carrier-to-noise performance of the processor is achieved at input signal levels just below the delayed AGC threshold. If sufficient level is available, this is the preferred operating point for the processor.

Returning to the main IF signal path, as illustrated, the signal next is supplied to switch S2, which selects the standby carrier upon loss of input signal. The standby carrier was originally employed to prevent TVs from showing snow when a signal goes off the air. It is also employed if the output channel is used as a pilot for the distribution system. A pilot signal is used to set the gain of distribution systems, and if it disappears when a TV station goes off the air, the operating point of the distribution system will be upset.

Some processors include the ability to modulate the standby carrier with pseudosync pulses, a low-duty cycle 15.734-kHz square wave that simulates normal TV modulation. This is necessary to support distribution AGC systems that need sync tips to operate, rather than a CW signal.

Switch S2 is normally activated by a drop in the IF AGC voltage sufficient to indicate that the input signal level is lower than a minimum needed for useful performance. Following switch S2 is switch S3. This switch is optional and is used when it is desired to switch to an alternative IF signal. This is done, for example, when the primary signal leaves the air and it is desired to substitute another signal on the same channel. For example, the cable operator may carry a local television station that goes off the air at night. It is desired to substitute another signal on the channel when the local station goes off the air. The alternative signal, processed in another modulator or processor, is supplied at the "Composite IF in" connector. Optionally, a separate AGC circuit may be used to normalize

* PAL countries use 38.9 MHz for the picture carrier IF and 33.4 (PAL-B/G) or 32.9 (PAL-I) MHz for the audio IF. In China, 38 MHz is used for the picture carrier IF.

the level of the alternative IF signal. In this example, switch S3 would be connected to the IF AGC signal such that when the input signal drops, S3 is activated, selecting the alternative IF signal.

The picture and sound carriers are separated in the example processor. Aural IF, at a frequency of 41.25 MHz, is routed to a bandpass filter, BPF7, which eliminates any power associated with the picture signal or other channels. Following BPF7 is a limiter, A4, which functions identically to the limiter in an FM radio. Its output is constant regardless of the input signal level, within a reasonable level range. Following A4 is a manually adjusted attenuator used to adjust the sound carrier level. The sound carrier is normally adjusted to be −15 dB from the picture carrier level. A bandpass filter following AT3 removes harmonics of the sound carrier, which are introduced in the limiting process. The sound IF is then combined with the picture IF.

An alternative sound IF circuit uses a notch of variable depth to adjust the sound carrier level. This circuit works but cannot remove any residual AM on the sound carrier. Further, if not adjusted properly, it can add FM/AM noise, which hurts operation of many scrambling systems.

The visual signal at IF is supplied to BPF4, which provides the bulk of the selectivity of the processor. We describe the response of this filter later. After amplification, the visual IF is combined with the sound carrier.

7.2.3 Output Section

After combining, the composite IF signal is applied to output level control AT4. The location of this control in the signal chain allows it to simultaneously adjust the picture and sound carrier levels. After level control, the composite IF signal is split. One side of the splitter is supplied as the "Composite IF out" of the processor, to be used if needed as the alternative IF input to another processor or modulator. The other output of the splitter is amplified in A5 before being supplied to mixer M3, the output mixer. It is mixed with the output of local oscillator LO3. The signal is now on the assigned output channel. Bandpass filters BPF6 and BPF7 eliminate the image of the second conversion process and any spurious signals coming from mixer M3.

Amplifier A6 increases the signal level to the desired output level. Its performance is critical: any third-order distortion it introduces will generate frequencies too close to the output channel to be rejected by BPF7. The most common distortion produced is a third-order beat between the picture and sound carriers. This beat, appearing 4.5 MHz below the picture carrier, is derived from twice the picture carrier less the sound carrier. Another beat appears 4.5 MHz above the sound carrier but is lower in amplitude.

7.2.4 Local Oscillators

Normally, LO1 and LO3 are fixed, crystal-controlled oscillators in a fixed channel processor. If the processor is agile on the output channel, LO3 and M3 will be replaced with a dual conversion output converter, which converts the IF signal to

a high frequency (typically in the high hundreds of MHz), where it is filtered and down-converted to the output channel using yet another mixer and a phase locked local oscillator.

A processor with an agile input will have the input section replaced with an agile oscillator and tracking filter. In higher-quality processors, the front end will be double conversion, with the second IF at the normal TV band of 41–47 MHz.

If the processor is used as an on-channel processor, with the input and output on the same channel, then it is desirable that the output signal be phase coherent with the input signal to avoid beat products. To effect this phase coherency, local oscillator LO1 is not used for on-channel conversion. Rather, a sample of the output local oscillator, LO3, is supplied to mixer M1. Thus, the same local oscillator is used for both down- and upconversion. This may be shown to ensure phase coherency. Switch S1 is thrown to the alternative position for on-channel conversion.

If it is desired that the output be phase locked to an external reference signal at the frequency of the output channel, then LO3 will be replaced with a phase locked circuit.

7.2.5 Processor Frequency Response

Figure 7.2 illustrates the visual channel IF response of the processor, overlaid with the spectrum of signals that may be present. An important function of a processor is to provide off-channel rejection. This is important in cases where the processor is used to process signals coming in from another cable. For example, it is sometimes practiced to route signals by AM microwave or cable to a hub, where they are reprocessed and supplied to subscribers. In this case, adjacent channel power coming into the processor must be rejected.

Also, if a station is being received off the air along with an adjacent channel from another city, the adjacent channel must be rejected before the signal is supplied to the cable. It is accepted that, in order to be invisible, an undesired signal should be 60 dB below the level of a desired signal. For specification purposes, this is the desired adjacent channel rejection of a processor. In the case of incoming cable signals, adjacent signals will be at about the same level, so 60-dB adjacent channel rejection is adequate. If an adjacent channel has a greater signal level than the tuned channel, as could be the case in off-air processing, then more rejection may be required. This is considered to be handled best with external selectivity rather than to burden all processors with more rejection capability.

Figure 7.2 illustrates the ideal processor passband as a heavy line and a typical passband as a dashed line. The location of the relevant carriers is shown at IF, where the picture carrier is at a higher frequency than the sound carrier. Chapter 8 shows why the spectrum is inverted from that broadcast. Ideally, the visual IF passband would pass signals from the edge of the visual spectrum, 4.2 MHz from the picture carrier, to the edge of the flat portion of the vestigial sideband region, 0.75 MHz above the picture carrier at IF.

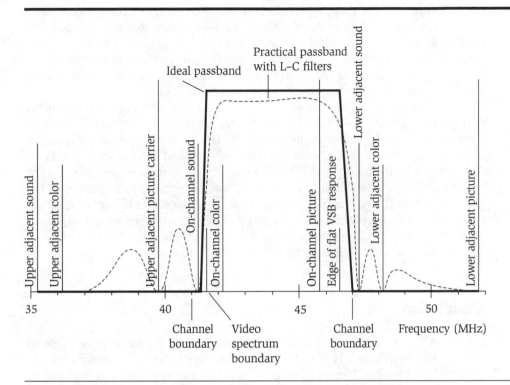

Figure 7.2 Visual channel IF response of processor.

This ideal spectral response may be approximated if a surface acoustic wave (SAW) filter is used as the IF filter (BPF4 in Figure 7.1). This filter sets the overall visual IF response of the processor for all practical purposes. Some processors employ L–C (inductor–capacitor) filters for this application. Practical L–C filters are unable to achieve the ideal passband shape. Bandpass sections are combined with traps, which provide high rejection over a very narrow band of frequencies. These traps are placed on the adjacent carrier frequencies as shown. Most of the power in an NTSC signal is concentrated at the picture carrier, the sound carrier, and the color carrier. The power density at other frequencies in a channel is relatively low. Because of this characteristic, it is possible to specify 60-dB rejection even if that much rejection is not available at all frequencies. The specification of 60-dB rejection is interpreted as meaning that power in an equal-level TV signal appearing in an adjacent channel is more than 60 dB below the on-channel picture carrier.

Even when a SAW filter is employed, it may be necessary to supplement its response with some traps. Practical SAW filters exhibit ultimate rejection of 40 to 50 dB, against a requirement of up to 60 dB depending on the frequency. Traps may be needed to achieve this last little attenuation.

L–C filters exhibit group delay. This means that signal components passing through the filter near either edge are delayed with respect to signal components passing through near the middle of the passband. The delay is a natural consequence of the attenuation of the stopband and is predicted from the mathematics describing the response of the filter. If the transition from passband to stopband occurs in a smaller frequency band, the group delay is greater.

Group delay causes components of the visual signal that occupy frequencies near the edge of the passband to propagate through the filter more slowly than do components near the center of the passband. This causes undesirable distortion of the video signal. To remedy this, another filter, called a *delay equalizer*, is incorporated with the bandpass filter. The delay equalizer must be designed for the bandpass filter with which it is to be used. It adds more delay to signals at the center of the passband, complementing the delay response of the bandpass filter.

A delay equalizer is necessary when an L–C filter is used as the bandpass filter, but not when a SAW element is used. Because of the way a bandpass filter is realized using a SAW device, group delay is inherently low.

7.3 Modulation

Figure 7.3 illustrates the video baseband, IF processing, and RF processing found in a typical modulator. (Figure 7.4 shows the audio circuitry.) Modulators are used with many more channels than are processors since the majority of channels are programmed from satellite-distributed programming or local sources. The output circuitry of the modulator is identical to that of a processor, but the front end is significantly different.

7.3.1 Video Switch and AGC

Video is introduced into the modulator in the top left-hand corner of Figure 7.3. An optional loop-through facility is shown on the input. Where a video source must be supplied to several places, it is possible to loop the video through from one piece of equipment to another by using a high input impedance in each. A termination is used only at the last piece of equipment in the signal chain. Section 7.8.1 covers this topic in more detail.

As shown, a video switch is optionally available to allow selection of an alternative video source. The switch may be activated manually, from an automation system, or automatically upon loss of the primary video.

Another option available with some modulators is a video AGC circuit. Typically, the amplitude of the sync tip is measured. If it is in error, attenuator AT1 is adjusted until the sync amplitude is normalized. It is assumed that the ratio of active video signal to the sync tip is correct. The proper level of video depends on the video content, which cannot be known at the modulator. However, the sync tip amplitude is well known. An error in sync tip amplitude is indicative of a gain

Figure 7.3 Modulator block diagram, baseband and RF.

error in video-processing circuitry. By correcting the sync tip amplitude, the active video amplitude is corrected.

Although the ideal correction strategy for the video AGC is to correct sync tip amplitude, a few modern modulators feature additional correction strategies. It is a fact of life that certain video signals handled in headends do not conform to the expected characteristics of amplitude and sync-to-video ratio. In order to deal with these signals, some video AGC circuits include a mode in which the amplitude of the sync tip to peak video signal is normalized rather than the sync amplitude. This mode should not be used except under certain conditions. If the video level drops because the picture intentionally goes dark, this mode would cause the sync level to increase and the video to be stretched to white unnaturally. However, the mode might be useful for video sources that are always expected to include peak white, but that are not always expected to be at the correct amplitude.

7.3.2 Sound Subcarrier Trap

Following the video AGC is a trap tuned to 4.5 MHz. In some cases, video is supplied combined with sound on a 4.5-MHz subcarrier. This is done, for example, when video is supplied via a terrestrial FM microwave link. The baseband video and subcarrier sound must be separated before modulation. The aural signal is coupled out through flag C (to Figure 7.4). The aural subcarrier is removed in FL1, a 4.5-MHz trap. Frequently, the trap is switchable because it does introduce some group delay. It may not be needed, and should not be used, if the incoming video doesn't contain significant power at 4.5 MHz.

Under some conditions other than the existence of a 4.5-MHz subcarrier, use of the trap may be desirable. If the video contains significant power at 4.5 MHz owing to distortion or other processing, that power would appear as noise to the sound subcarrier. This can adversely affect the sound–signal-to-noise ratio, and in certain circumstances, could affect descrambling (see Chapter 18).

7.3.3 DC Restoration and Peak White Clip

From the trap, video is supplied to a dc restoration circuit. Video signals may be interfaced between pieces of equipment with dc restored or ac coupled. As video content changes, the average signal level, and hence the dc value of the signal, changes. However, since it is required to transmit modulated signals with a fixed sync tip amplitude, the dc level must be reestablished before the signal is modulated onto the RF carrier.

A number of circuits have been used for dc restoration. One of the simplest is shown in Figure 7.3. If any portion of the video signal tries to drop below ground, diode D1 conducts, charging C1 positive at the right end. Since the video signal voltage is all positive with respect to the sync tip, when the charging of C1 is complete, the sync tip will be at ground level (ignoring the diode drop in C1), with all other elements of the video signal being positive with respect to it. This dc level is maintained to the actual point of modulation. Resistor R2 is used to discharge C1 so that if the sync tip rises too high the capacitor will discharge until D1 again conducts on sync tips.

After the video signal is clamped, it is applied to a peak white clipper. Visual carrier modulation is strictly limited to 87.5% in the NTSC system. If the carrier modulation is too deep, the carrier will at some point be cut off, or very nearly so. Because of the way sound is demodulated (see Sections 7.5.5 and 8.5), carrier cutoff will cause intolerable sound buzz. In order to prevent this, many modulators include peak white clip circuits. If the video voltage exceeds voltage +V in Figure 7.3, then diode D2 conducts, holding the voltage on the video line constant.

7.3.4 Modulation Process

The clamped and clipped video signal is coupled to modulator M1. It is no mistake that the modulator is shown modeled as a mixer with the multiplication sign.

Amplitude modulation is a multiplicative process, as is frequency mixing. The circuits used for both amplitude modulation and mixing are essentially identical and perform a multiplication process. Chapter 2 provides background on amplitude modulation theory. In that chapter, the mathematics describe suppressed carrier amplitude modulation as used on the color carrier. Here, we modify the expressions to preserve the carrier so as to be more reflective of how the video signal is modulated onto the visual carrier.

Conventional amplitude modulation of a carrier by a sine wave is expressed by the equation

$$V_0 = \cos\omega_c t(1 + M\cos\omega_m t)$$
$$= \cos\omega_c t + \frac{M}{2}\cos(\omega_c - \omega_m)t + \frac{M}{2}\cos(\omega_c + \omega_m)t$$

(7.1)

where

ω_c = carrier frequency

ω_m = modulating frequency

M = percent modulation of the signal (though TV signals use a different expression for modulation)

From the first form, we can see that the envelope (represented by the term inside parenthesis) never goes negative. (We can show that if the 1+ term is omitted, the carrier is suppressed, as in Chapter 2.) The second form is obtained by trigonometric expansion of the first. It shows that the frequency domain includes the carrier and two other frequencies, spaced above and below the carrier by the modulation frequency. These upper and lower modulation products are called *sidebands*. The information contained within the modulated carrier is carried redundantly by the upper and lower sidebands.

A video signal is much more complex than a sine wave. The single sinusoidal modulation term $\cos\omega_m t$ is replaced by a sum of sine waves of different frequencies, which taken together, define the video waveform (see Chapter 2). The resulting modulated signal has a series of sidebands above and below the carrier. One set may be removed without losing the information contained in the signal.

7.3.5 IF Processing

From modulator M1, the signal is often looped out for possible external signal processing. (The term *loop out* is applied to the routing of a signal from one piece of equipment to another, then back into the first piece for further processing.) Many common scrambling systems (described in Chapter 18) need access to the modulated signal at this point. They attenuate the picture carrier during the horizontal and, usually, the vertical blanking intervals by attenuating the IF signal. Though some older systems attenuated the sync by applying a sine wave attenuation, more modern systems use so-called gated sync suppression, achieved by

switching an attenuator into the signal path. The switching operation introduces high-frequency modulation components, which will interfere with adjacent channel signals if not eliminated. By arranging FL2 so that it follows the addition of the scrambling signal, the undesired modulation components are removed.

After scrambling (when used), the signal returns to the modulator and is supplied to vestigial sideband filter FL2. Vestigial sideband modulation and filtering were described in Chapter 2, and the precise shape of a VSB filter for NTSC is described later. It is used to eliminate one set of modulation sidebands. Modulation sidebands more than 0.75 MHz from the picture carrier are removed from one side of the picture carrier. If the signal has been scrambled using gated sync suppression scrambling, then FL2 will also filter those modulation sidebands.

After filtering, the IF signal is combined with the sound carrier signal. The combined signal may be looped out, at the composite IF loop, to provide for processing by an older sine wave sync suppression scrambling system. It is likely that the signal will be split, with a portion supplied to the composite IF out on the rear panel of the modulator. This signal is used with other processors or modulators in IF switching applications, as described earlier.

The composite (visual and aural) IF signal is up-converted to the RF channel in mixer M3, driven by local oscillator LO3. Attenuator AT5 controls the output level of the composite visual and aural signals. Remaining processing is identical to that described in the signal processor section of this chapter. The signal is filtered, amplified, and filtered again. A phase locking system can be used to phase lock the carrier to another source, and is described later.

7.3.6 Aural Modulation

Figure 7.4 illustrates the modulation process. Baseband audio is supplied to an optional switch S2. If video switching (S1 of Figure 7.3) is used, then audio switching is likely required, too. Normally, the audio switch will be driven by the video switch: when the video switch activates, it activates the audio switch. Sometimes a third audio input, intended for emergency override, is provided though it is not shown in the figure. Frequency response of the override input may be limited to voice band only. The reason it would be limited is to reduce any noise that may appear on an audio source that is outside the roughly 300–3,000-Hz band needed for understanding voice communication.

Preemphasis

Preemphasis is employed, as in FM radio broadcasting, to reduce the noise introduced by transmission. An FM demodulated signal can be shown to have rising noise density with frequency. In order to compensate, preemphasis is used. Above a certain frequency, the gain in the modulator increases 6 dB per octave (doubling of frequency). The demodulator employs deemphasis, which attenuates the signal by the same amount. In the process, the noise spectrum is flattened, and the total noise heard by the listener is reduced.

Figure 7.4 Modulator block diagram, aural.

In North America, the standard emphasis curve followed is called the *75-microsecond* emphasis curve because the break frequency, at which emphasis is 3 dB, is given by a 75-μs *RC* break:

$$e_0 = (1 + j\omega RC)e_i$$
$$= [1 + j\omega(75\,\mu s)]e_i \tag{7.2}$$

where

j = imaginary operator

ω = angular frequency ($\omega = 2\pi f$)

The corresponding 3-dB frequency, at which the imaginary component is equal to 1, is 2.122 kHz. In Europe and elsewhere, it is common practice to use an emphasis time constant of 50 μs, corresponding to a 3-dB break frequency of 3.183 kHz. This affords slightly less reduction in noise, all else being equal, while providing slightly more protection against overmodulation.

The issue of overmodulation with constant amplitude signals arises in the following manner. Consider that you are normalizing the audio level of a program using any of several accepted criteria. You should not set program level based on a preemphasized signal because that is not the way a listener hears the program. On the other hand, the program is transmitted with preemphasis. If the dominant program material occurs at a frequency significantly above the preemphasis break frequency, then overmodulation is possible with normal program level audio. Fortunately, most program material tends to be dominated by frequencies below either the 75-μs or the 50-μs break frequencies. However, sometimes this is not the case, and problems can develop. The 50-μs break frequency provides somewhat more immunity. In order to compensate for the reduction in signal-to-

noise ratio caused by the higher emphasis break frequency, countries using 50-μs preemphasis normally employ wider deviation of the sound carrier (50 kHz compared to 25 kHz for 75-μs countries).

When stereo audio is used, the preemphasis is more complex. The preemphasis circuit in the audio baseband circuits shown is disabled so that the input is flat. Stereo audio is covered in Chapters 2 and 7.

Aural Modulation Limit, or AGC

An optional modulation limiter may be provided. This circuit monitors the audio signal level, and if the level increases such that the deviation would be excessive, then the modulation limiter temporarily turns down the gain of the signal path by adjusting AT2 of Figure 7.4. After the overmodulation ceases to exist, the gain is returned to its previous value over some length of time, which reduces the audible impact of the change.

Some modern modulators include a second mode of operation, in which the audio level is measured and AT2 is adjusted to provide for a consistent sound level regardless of the incoming modulation. This mode would not be appropriate for a number of audio sources since it will make low-volume sounds the same level as high-volume sounds. (A whisper and a shout would be the same loudness.) It might be appropriate, though, if the program audio consists of normal speech only.

Some modern modulators also include audio clipping, in which audio that would result in significant overdeviation has its peaks clipped, eliminating the overdeviation. In the United States, the FCC requires that a television aural transmitter be able to deviate to at least ±40 kHz, so the appropriate clip point would be above this. The clipping process produces considerable distortion, so operation into the clipping region should never be done intentionally. However, it is a useful way to avoid gross overdeviation, which could cause interference into video or, worse, problems with scrambling recovery (see Chapter 18). It may be applied at all times, so long as the audio level is carefully set to avoid hitting the clipping level during normal programming.

Modulation Process

Audio is frequency modulated onto an aural carrier. Frequency modulation is usually accomplished by applying the audio directly to an oscillator, such that the frequency of oscillation depends in part on the instantaneous modulating voltage. As shown in Figure 7.4, the modulation is applied to LO2, a 4.5-MHz oscillator (other frequencies may be used with conversion).

Some method must be employed to stabilize the center frequency of the oscillator. Though not the only means to accomplish this, a commonly used stabilization method is a phase locked loop (PLL). It generates a frequency correction to the oscillator if the center frequency wanders. The bandwidth of the PLL is very low to avoid having the loop "fight" the modulation, which is intentionally forcing the frequency of LO2 to deviate instantaneously.

Audio from Other Sources

If audio is supplied already modulated on a 4.5-MHz subcarrier, then it is introduced into the audio section via flag C (see also Figure 7.3). At this point, the aural subcarrier may be combined with the baseband video, so bandpass filter FL3 is used to separate out only the sound subcarrier. It is passed to one input of switch S3, which selects either the audio from the modulated oscillator, LO2, or the external signal. Switch S3 may be actuated permanently by manual selection of the audio source, or it may be activated by some other event.

A common situation is to activate the switch upon loss of main video. Suppose a modulator ordinarily handles a signal delivered at baseband from a television station. The baseband link may be backed up by an RF link that picks up the off-air signal and demodulates it. The primary audio is delivered in baseband format, and so is applied to the main input to switch S2. The backup demodulator supplies sound on a 4.5-MHz subcarrier, combined with the video. If the video switch (S1 of Figure 7.3) detects loss of video, it activates to couple in alternative video from the backup demodulator. At the same time, it drives the 4.5-MHz switch (S3 of Figure 7.4) to select the external aural subcarrier of flag C.

Following the switch is a limiter whose purpose is to ensure that the sound carrier level is constant regardless of the source. After the limiter is a manual sound carrier level adjustment, AT4, which sets the level of the sound carrier relative to the visual carrier level. Finally, the aural carrier is up-converted from 4.5 MHz to the sound IF at 41.25 MHz using the same local oscillator as was used to generate the picture IF (LO1 of Figure 7.3). The aural IF is coupled at flag D back to Figure 7.3, where it is combined with the visual IF signal.

7.3.7 Vestigial Sideband Filtering

As described earlier, the majority of one sideband is removed by a device known as a vestigial sideband (VSB) filter (FL2 of Figure 7.3). Figure 7.5 illustrates the operation of the vestigial sideband filter in an NTSC modulator. Recall that the spectrum is shown at IF, intermediate frequency, where the picture carrier is higher in frequency than the sound carrier. This is reversed at RF, by the use of a high-side local oscillator.

The VSB filter has a response that is flat from 0.75 MHz above the picture carrier to at least 4.2 MHz below. The reason for the cutoff at 4.2 (really 4.18) MHz is that this is the maximum baseband video frequency that is handled in the NTSC transmitted signal. The color carrier and all its sidebands are contained within the passband of the VSB filter.

The high (vestigial) side cutoff is rather carefully defined. The response should be flat to 0.75 MHz above the picture carrier, then drop to zero at the band edge, 0.5 MHz higher. This frequency, 1.25 MHz above the picture carrier, is at the channel edge, 47 MHz at IF.

No filter has truly zero stopband response, so it must be decided how much attenuation is "enough" at the band edge and frequencies farther out. In the

Figure 7.5 Modulator vestigial sideband filter response.

section on processors, we argued that the power spilling into an adjacent channel should be at least 60 dB down from the picture carrier. This number is applied to the VSB filter also. The requirement does not mean that the filter must exhibit a stopband response of −60 dB because the power of the modulated TV waveform is not constant throughout the channel. Most of the power is concentrated at the picture carrier. Typically, other carriers are −15 dB or lower with respect to the picture carrier. The only frequency removed by the VSB filter that is likely to be that high is the color subcarrier image at about 49.3 MHz. Thus, if the VSB filter has 45-dB rejection, that is good enough. Even if the filter has only 40-dB rejection, that may be adequate given a small attenuation by other filters.

In some modulators, the sound signal may be supplied through the VSB filter, in which case a slightly wider response than shown is needed. However, with the modulator architecture of Figures 7.3 and 7.4, this would not be necessary.

The critical requirements for the VSB filter made it an early candidate for use of a surface acoustic wave (SAW) device, and in fact, this is the first known application of SAW devices in the cable television industry. SAW devices exhibit sharp cutoff regions, good temperature stability, and low group delay. However, they do exhibit higher propagation delay than do conventional L–C filters. Propagation delay is the absolute delay of a signal as it propagates through the device. The propagation delay is related to the construction of the SAW device and has decreased somewhat with improvements in the state of the art. All else being equal, the sharper the skirts (the narrower the rolloff region), the greater will be the propagation delay. Delay times of 3 to 5 microseconds tend to be typical of SAW filters built for this application. The propagation delay can be an issue when

events must be synchronized between those transmitted on the picture and sound carriers. Such a situation develops in the commonly used scrambling systems where synchronization data is transmitted on the sound carrier (see Chapter 18).

A SAW filter utilizes the propagation of an RF signal on the surface of a suitable material, such as a quartz or lithium niobate crystal. A pair of transducers is built onto the surface of the crystal, using the same techniques as are used to deposit metalization on the surface of integrated circuits. One transducer launches the signal into the crystal, where it propagates as a mechanical wave on the surface. The other transducer converts the wave back to an electrical signal. The shape of the transducers determines the passband of the filter. The loss of a SAW filter tends to be rather high, but the skirts can be made very steep without incurring the group delay that accompanies a sharp L–C filter.

7.4 Phase Locking of Carriers

The output carriers of both processors and modulators may be phase locked to external sources. Two primary reasons exist for phase locking the output. If the frequency is used for local off-air broadcasting, then there is a high probability that off-air signals will leak into some television receivers or, much less likely, into the cable plant itself. This leakage, or *ingress,* will produce a visible moving pattern on the screen if the on-cable picture carrier is not exactly equal to the ingress frequency. The industry commonly calls this pattern a *beat.* The processor or modulator output can be phase locked to the off-air signal to reduce the visibility of the ingress. This works because the phase locking process forces the frequency of the signal on the cable to be exactly equal to that of the ingress.

The other application for phase locking is to lock all picture carriers on the cable to a master reference frequency. Such techniques can be shown to reduce the interference caused by distortion in the distribution plant. Two techniques have seen fairly widespread use in North America. They are the harmonically related carrier (HRC) technique, in which all picture carriers are locked to harmonics of a master reference oscillator, and the incrementally related carrier (IRC), where all frequencies supplied from the master generator are offset 1.25 MHz. Neither HRC nor IRC techniques tend to be used as much today as in the past, but they will still be found in some locations. Chapter 8 describes the standard HRC and IRC frequency allocation plans.

Yet a third possible requirement for phase locking outputs is described in Chapter 8, in the discussion of node-specific programming. More than one program may exist in a headend on the same channel, and if leakage develops from one to the other, phase locking the two picture carriers will lessen the visibility of the beat.

The term *phase lock* refers to a technique of forcing one RF carrier to maintain a constant phase relationship with a second carrier. The second carrier is called the *reference carrier* because it is a reference to which the first is locked. If the two carriers have a constant phase relationship to each other, it follows that

they are identically at the same frequency. Indeed, a common technique for verifying a phase locked condition is to display one carrier on an oscilloscope that is triggered from the other. The displayed waveform should appear stable on the screen. Note that it is not necessary that the two carriers have a particular phase relationship to each other (though some applications require this), and in general, the phase relation might drift over time.

7.4.1 Phase Locked Circuitry for Processor or Modulator

Figure 7.6 illustrates one possible block diagram of a phase locked circuit. This can be used with either Figure 7.1 (processor) or Figure 7.3 (modulator). In either case, the output local oscillator, LO3, is replaced by the circuit of Figure 7.6. Mixer M3 is the mixer M3 of Figures 7.1 and 7.3.

The reference frequency, a sample either of the off-air signal or from the comb generator (for HRC and IRC applications) is selected in bandpass filter FL10. It is mixed to the common intermediate frequency of 45.75 MHz in mixer M5. The local oscillator used in this mixing process must be LO3, the same local oscillator used to mix the output channel.

After conversion to IF, the signal is filtered in FL12 before being supplied to limiter A11. This limiter must be used because, if the reference source is an off-air signal, it will have modulation on it, and the modulation must not interfere with the mixing process. From the limiter, the signal becomes one input to phase detector M4. The other input to M4 is a sample of the IF signal developed in the modulator or processor. This signal has been filtered (if needed) and limited in FL11 and A10.

Again, a multiplier is shown as the phase detector. A multiplier makes a good phase detector, just as it makes an ideal modulator or mixer. If the reference signal (at frequency ω_r) and the variable signal (at ω_v) are multiplied, the result is expressed as follows:

$$E_0 = \sin\omega_r t \, \sin(\omega_v t + \theta)$$
$$= \frac{1}{2}\cos 2\omega t - \frac{1}{2}\cos\theta \quad\quad (7.3)$$

where

$$\omega = \omega_r = \omega_v$$

This expansion uses the relation of Equation (2.5). The first term in the second line represents the second harmonic of the carrier frequency and is removed by the loop filter. The second term represents the phase difference between the two carriers, θ.

The output of the phase detector is supplied to the loop filter, which rejects the second harmonic and performs required filtering on the error voltage representing θ. These functions often include providing extremely high gain at dc, which forces $\cos\theta$ to be near zero. The phase locked loop will do whatever is nec-

Figure 7.6 Phase lock option for processor or modulator.

essary to the frequency of LO3 to keep θ constant. Note that, in order for the co-sine of the phase angle to go to zero, the two signals at the mixer will be held in quadrature, that is, with a 90° phase difference. This phase relationship is typical of multiplicative phase detectors.

If it is not intuitive that the output frequency is phase locked to (that is, the same frequency as) the reference, you are encouraged to write equations for the output frequency in terms of the frequency of the IF, of LO3, and of the reference frequency. Assume that the frequencies of the two inputs to M4 are equal (a condition forced by the phase locked loop), and solve for the output frequency in terms of the reference frequency.

Some modern agile modulators offer an alternative way to achieve phase lock between channels. If all relevant frequencies in all modulators are synthesized from the same reference frequency, then phase lock is accomplished.

Comb Generation Technique

Though not used extensively today, it is instructive to consider briefly how a comb of frequencies might be generated. The comb is used for phase locking processors and modulators in HRC and IRC frequency plans. Figure 7.7 illustrates an HRC comb generator in part a and the modification to make it into an IRC generator in part b.

HRC Comb Generation

The HRC comb is simply all the useful harmonics of a 6-MHz (for NTSC) generator. A simple way to produce the comb is illustrated in Figure 7.7(a). A precision 6.0003-MHz oscillator is used to generate the basic frequency. (The nominal

Figure 7.7 Comb generator block diagrams. (a) HRC comb generator. (b) IRC comb generator.

frequency is 6.000 MHz. See Chapter 8 for an explanation of why an oscillator at 6.0003 MHz is used.) After amplification, the signal is supplied to a fast recovery diode and to an inductor. When the signal from A10 goes positive, D10 conducts, setting up current in L10. When the signal from A10 begins to go negative, D10 becomes back biased and cannot conduct. The input impedance to A11 should be high so that no significant current exists in its input. When D10 is back biased, a current path to complete the circuit through L10 does not exist. If a closed circuit does not exist, then the current through L10 must stop. However, if you try to in-stantaneously interrupt the current in an inductor, the voltage across the inductor must go to infinity (less in a practical case, due to losses in the circuit). This pro-duces a voltage spike that approximates an impulse. An impulse may be shown to contain all harmonics of the fundamental frequency at equal amplitudes. After amplification in A11, the resultant may be applied to filters that remove unneeded harmonics. The small spectrum diagram near the output illustrates the comb.

IRC Comb

Figure 7.7(b) illustrates how the IRC comb is developed from the HRC comb of 6.000 MHz. The IRC comb is the HRC comb except offset up in frequency by 1.25 MHz. This is done by band limiting the HRC comb and using it as modulation on a carrier positioned at a picture carrier frequency of about one-half the maximum frequency. The effect is to generate a carrier at the center of the comb spectrum at one of the comb frequencies. The modulation sidebands are at the other frequencies required of the HRC comb. A double balanced mixer is used for the translation. Since it will try to suppress the frequency of LO11, a mixer balance adjustment is used, and the mixer is intentionally unbalanced to the translation oscillator, enough to cause the translation carrier signal to bleed through at the same level as the sidebands. The HRC comb elements should be completely suppressed, but if they are suppressed by only a few decibels, the phase locked loop in the processor or modulator should not have any trouble recovering the correct element.

7.5 Demodulation

A third type of RF product used in the headend is a demodulator. It is useful to understand the demodulation process in order to make sense of a lot of phenomena observed in cable systems. Also, demodulation is used for quality checks, for recovery of off-air signals for backup purposes, to put signals on a digital transportation route, and as interfaces for FM microwave systems. A few headends employ demodulation and remodulation instead of processing for off-air signals. This is advantageous if considerable signal switching is done. The use of a quality demodulator as a test tool is, alone, justification for studying the technique.

7.5.1 RF Input Section

Figure 7.8 shows the RF, IF, and video demodulation portions of a demodulator. The sound section is shown in Figure 7.9. The demodulator input is identical to that of the processor of Figure 7.1. Both processor and demodulator are used to recover signals from off the air or from an incoming cable. Thus, the input circuitry through attenuator AT2 is identical.

7.5.2 IF Processing Section

IF processing in a demodulator is significantly different from that of a processor. Power for demodulation of the sound carrier is taken at flag A. Sound demodulation will be covered later in conjunction with Figure 7.9. The AGC operates to hold the visual signal level after AT2 at a constant level regardless of the input level.

Zero chop switch S1 is optional but is often provided on professional demodulators as a way to measure visual signal depth of modulation. When enabled, the chopper momentarily interrupts the signal at IF, usually during the

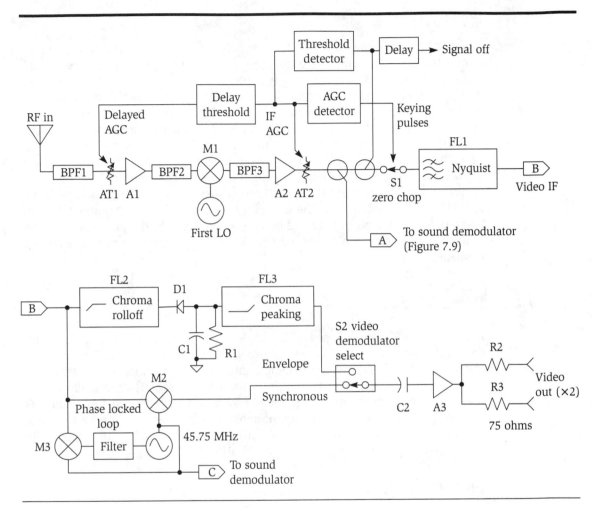

Figure 7.8 Demodulator block diagram, input and video.

vertical blanking interval. The interruption allows a waveform monitor or oscillo-scope to be calibrated to measure visual depth of modulation very accurately. Chapter 8 describes measuring depth of modulation using a zero chopper. The chopper is coupled to the AGC detector in the illustrated case so that the switch may be restored to the IF path during sync pulses. By doing this, the zero chop pulse may extend over several lines without interrupting sync.

From the zero chop circuit, the signal is supplied to the Nyquist slope filter, FL1, which provides the main filtering function for the visual signal. The filter is named after Harry Nyquist, a Bell Laboratories engineer who developed this tech-nique for recovering vestigial sideband signals. The operation of the Nyquist slope filter is described in Chapter 2.

7.5.3 Visual Carrier Demodulation

After the Nyquist slope filter, the signal is supplied to one of two demodulation circuits, an envelope detector or a synchronous detector. Home television receivers use one detector or the other, but it is advantageous to have both in a professional demodulator because the two will yield different information about a signal. Some of the differences are discussed in Section 7.5.7.

Envelope Detector

In the illustrated system, the envelope detector consists of a chroma rolloff filter, FL2, and following circuits. The chroma rolloff filter reduces the amplitude of the IF signal at the color subcarrier frequency. The purpose of the chroma rolloff filter is explained later. Diode D1 performs the actual detection process, and capacitor C1 filters the signal. The detector, except for component values, looks just like the detector for an AM broadcast receiver. Figure 2.21 provides another view of an envelope detector.

The direction of the diode is not arbitrary. Recall from Chapter 2 that the sync tip is the most negative portion of the video signal as the signal is normally interfaced. The sync tip also corresponds to the highest amplitude of the signal. With the diode in the direction shown, the voltage across C1 will be such that the highest amplitude of the RF signal (the sync tip) will be the most negative voltage across C1. Other voltages will be positive with respect to that. Of course, in a real demodulator, it is possible to detect the signal with the opposite polarity and then invert the video, but this is often not done for practical reasons.

Following the detector is the chroma peaking filter FL3, which restores flat chroma response, complementary to the operation of the chroma rolloff filter FL2. The signal is then applied to switch S2, which selects either the envelope detector or the synchronous detector.

Synchronous Detector

The synchronous detector is shown below the envelope detector in Figure 7.8. It consists of a phase locked loop used to generate a carrier phase coherent with the incoming picture carrier (at IF). In contrast to other applications described earlier, the phase of the locked carrier with respect to the input IF signal is of paramount importance: the two carrier phases must be the same. Because the phase of the carrier varies with modulation and, in addition, can vary owing to defects in transmission, the phase locked loop is normally keyed (closed) on either sync tips or the back porch portion of the modulated signal.

When a modulated carrier is multiplied with an unmodulated carrier at the same frequency, the result is recovery of the baseband signal. This is shown in Chapter 2. The advantage of the synchronous detector over the envelope detector is primarily associated with demodulation of signals having asymmetrical sidebands, which is the case for all analog television signals. An envelope detector introduces distortion that is avoided with a synchronous detector.

From the synchronous demodulator M2, the signal is low-pass filtered to remove the carrier and harmonics, and supplied to S2, the switch that selects either the envelope or synchronous detectors.

7.5.4 Video Output

Switch S2 is normally actuated manually to select either the envelope or synchronous detectors if both are provided. By comparing the results of demodulation using the synchronous and envelope detectors, we can deduce a lot about the performance of the RF components handling the signal. Under normal conditions, the synchronous detector will yield a more accurately demodulated signal than will the envelope detector.

A coupling capacitor C2 removes dc components of the demodulation process, as well as any bias voltages introduced in the demodulation process. It has been customary to ac couple the video output since this minimizes the power supplied or dissipated in the load and often simplifies circuitry. Today, both dc and ac coupling of the output are employed. If dc coupling is employed, capacitor C2 would likely be used anyway, with a dc restoration circuit in the output amplifier.

It is essential with video that the coaxial cable impedance be matched both on the source end and the load end. Failure to do so can cause unacceptable frequency response errors, due to reflections, if the interconnecting cable is long enough. A typical way to achieve multiple well-matched outputs is to provide a very low output impedance in amplifier A3 and multiple matching resistors R1 and R2 at the outputs.

7.5.5 Aural Carrier Demodulation

The aural demodulation process is shown in Figure 7.9. The sound IF is supplied from Figure 7.8 on flag A. The other input to the sound demodulation process is a sample of the signal from the 45.75-MHz phase locked oscillator, used in the synchronous demodulator (flag C, which is omitted if a synchronous detector is not supplied).

The intercarrier detection mode is the one used almost exclusively in consumer applications. The sound final IF frequency, normally 4.5 MHz, is derived from mixing the sound first IF, 41.25 MHz, with the picture carrier, 45.75 MHz. This final sound IF frequency is often called the *intercarrier frequency*. Use of intercarrier detection solves several problems related to television transmission. If any process in the transmission of the television signal introduces phase noise, then the phase noise will seriously damage the signal-to-noise ratio of the audio: phase modulation is a form of frequency modulation, so once phase noise is introduced, it cannot be removed. However, since phase noise will affect the picture and sound carriers identically, when the intercarrier signal is developed by taking the difference in the picture and sound carriers, phase noise is eliminated from the aural signal. The IF signal on flag A contains both the aural and picture IF car-

Figure 7.9 Demodulator block diagram, audio.

riers, so by applying this signal to both inputs of a mixer, intercarrier mixing results. (Additional filtering may be used.)

On the other hand, if phase noise is not a problem, then there are advantages to a direct detection process. The picture carrier may have some incidental phase modulation on it, and it certainly has visual modulation sidebands. These will cause interference with the aural signal if used in the intercarrier detection process. In direct detection, the CW signal locked to the picture carrier is mixed with the sound carrier to derive the intercarrier frequency. (In this mode, the sound carrier path may include filtering to remove the picture carrier though this filtering is not shown in Figure 7.9.)

After the intercarrier detector, M4, the signal is filtered in a 4.5-MHz bandpass filter FL4. Following the bandpass filter is a limiter, A4, which removes any residual amplitude modulation from the aural carrier. At this point, a sample of the aural carrier is usually taken. It is passed through a manually controlled attenuator AT3, amplifier A5, and a current source resistor R3. The purpose is to provide a sound subcarrier for external use. Often this signal is combined with the picture carrier to provide a signal to be transmitted by FM microwave link. The resistance of R3 is high compared with 75 ohms. If one of the video outputs on Figure 7.8 is paralleled with the subcarrier out connector, then R3 will act as a current source, adding the 4.5-MHz subcarrier to the video without significantly upsetting the impedance of the video circuit.

The other path from the limiter is through a discriminator, which recovers baseband audio. Note that the signal recovered at this point is either monaural (mono) or a stereo composite signal, as defined in Section 7.6. If the signal is mono, then a deemphasis filter FL5 is included. If it is desired to recover a composite stereo signal for subsequent demodulation or processing, the deemphasis is omitted in favor of more complex processing in the stereo decoder.

Finally, the signal level is adjusted at attenuator AT4, normally a manual control. It is amplified in output amplifier A6 and provided on balanced outputs. Balanced outputs are used in professional equipment to provide for reduction of pickup of 60-Hz ac components ("hum"). "Balance" refers to a transmission

technique in which the two sides of the output have equal impedance to ground, as opposed to unbalanced transmission, where one side of the circuit is at ground potential. Section 7.8.2 describes use of balanced interconnection for audio.

From a long-standing telephone specification, the impedance of the audio output is normally 600 ohms between the two balanced lines. In much equipment today, the source impedance is less than 600 ohms, but the performance of the amplifier is specified when the amplifier is working into a 600-ohm load. The lack of a defined 600-ohm output impedance is not a problem in most cases since the length of audio cable is very small compared with the wavelength of any audio frequency. If the output is to drive a line several miles long, then the impedance should be controlled more carefully.

7.5.6 Nyquist Slope Filter

The Nyquist slope filter shown as FL1 in Figure 7.8 has a shape shown by the solid line in Figure 7.10. At the low-frequency side (aural carrier at IF), the shape is fairly straightforward. The bandpass characteristic should pass the color carrier and its sidebands while rejecting the sound carrier.

The high-frequency side (near the picture carrier) is rather unusual. The purpose of this shape is to complement the shape of the vestigial sideband (VSB) filter used in the modulator, and shown in Figure 7.10 as a dotted line. The need is described in Chapter 2. The Nyquist shape is flat to 0.75 MHz below the picture carrier; then it begins falling off at a prescribed rate above that frequency. The rolloff must be linear with frequency on a voltage plot, not a logarithmic plot. The response of the filter must be 0.5 at the picture carrier, continuing to drop until the response is zero at 0.75 MHz above the picture carrier.

Other features of the Nyquist slope filter are emphasized in the next section, in which we describe the differences between synchronous and envelope detection. One of those features is not really a part of the Nyquist slope filter response. It is shown in Figure 7.10 as a dashed line and was mentioned in conjunction with the chroma rolloff filter (FL2 of Figure 7.8). This is the reduction of response near the color carrier applied in the case of an envelope detector. It is called a *haystack* response after traditional television usage: taken with the Nyquist slope around the picture carrier, the filter shape looks like a haystack. The haystack response is used to reduce the effects of luminance-to-chrominance crosstalk.

In consumer television receivers using envelope detectors, the haystack response is built into the Nyquist slope filter (FL1 of Figure 7.8). In professional receivers, the chroma rolloff may be realized independently of the Nyquist slope filter so that it is not in the signal path in the synchronous detector mode. Not that the haystack response would hurt the response of the synchronous detector if it were perfectly compensated at baseband. Rather, it is difficult to compensate perfectly, and for measurement purposes, it is better not to introduce a response error that must later be compensated.

In modern practice, the Nyquist slope filter is normally realized as a SAW filter in consumer receivers though the shape is compromised to allow the SAW filter to be built with good cost effectivity. In professional demodulators, the filter

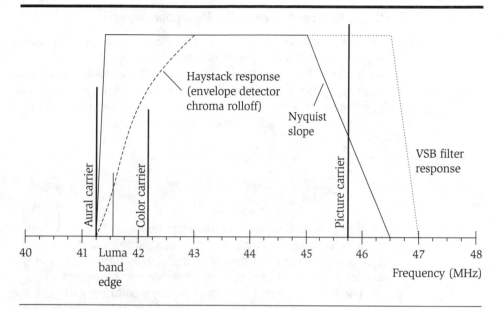

Figure 7.10 Nyquist slope filter response, with haystack.

may be realized either with a SAW device or as a conventional L–C filter with delay equalization.

7.5.7 Synchronous Versus Envelope Detection

Synchronous and envelope detection were covered in Chapter 2. Both synchronous and envelope detectors may be used for operational demodulation. Synchronous detection yields the higher-quality results absent a few distortions that prevent its proper operation. The most common problem that interferes with operation is phase noise, or phase modulation, on the picture carrier. A synchronous detector will not work as well as does an envelope detector under this circumstance.

Envelope detectors are susceptible to the practical limitations of diodes though there are ways to mitigate this problem. More seriously, envelope detection is susceptible to quadrature distortion errors that arise as a result of asymmetrical sidebands in the television signal. Quadrature distortion occurs when the upper and lower sidebands are not symmetrical. As shown earlier and in Chapter 2, the television signal presented to the detector does not have symmetrical sidebands at any frequency: at modulation frequencies above 1.25 MHz, only one sideband is transmitted. Below 0.75 MHz, symmetrical sidebands are transmitted but are turned into asymmetrical sidebands by the Nyquist slope filter. The lowest modulating frequencies approach sideband symmetry, but as the frequency increases, the symmetry is quickly lost.

Phasor Representation of the Modulated Signal

The mathematics of modulation were shown in Chapter 2. A useful graphical representation that shows the function of the two demodulators is to invoke the mechanism of phasors.[1] A phasor represents any steady-state sinusoid as a vector whose length is a function of the magnitude of the signal and whose angle is a function of the angle between a reference and the phasor. From Equation (7.1), an amplitude modulated signal can be represented as

$$V_0 = \cos\omega_c t(1 + M\cos\omega_m t)$$
$$= \cos\omega_c t + \frac{M}{2}\cos(\omega_c - \omega_m)t + \frac{M}{2}\cos(\omega_c + \omega_m)t \tag{7.4}$$

Figure 7.11(a) represents this equation in phasor form. The first term, representing the carrier, is defined as having a phase such that its phasor is straight up (this is arbitrary). The upper sideband has a phasor that rotates clockwise (with respect to the stationary carrier) at an angular frequency of ω_m. The lower sideband term, $\cos(\omega_c - \omega_m)$, is represented by a counterrotating phasor. The resultant phasor is shown to the right in Figure 7.11(a). The vector addition of the three phasors is always in phase with the carrier. An envelope detector responds to this resultant vector. Analysis of the vector at all different phase angles θ would show that the resultant traces out a sinusoid, which is the waveform used to modulate the carrier (the $M\cos\omega_m t$ term of Equation (7.1)). All is well.

Now consider the situation shown in Figure 7.11(b), a single sideband signal. The modulating waveform is the same sine wave as in Figure 7.11(a), but one sideband has been removed by the VSB filter (FL2 of Figure 7.3). Most notably, one sideband of the color subcarrier is removed, so the color subcarrier is transmitted as a single sideband signal. The vector sum of the carrier and single sideband phasor is shown to the right of Figure 7.11(b). Here the resultant is *not* in phase with the carrier. The resultant can be resolved into an in-phase component and a "quadrature" component at right angles to the carrier phase.

Response of an Envelope Detector

The envelope detector responds to the magnitude of the resultant carrier. By applying trigonometric relations to the resultant vector of Figure 7.11(b), we can compute the magnitude of the vector for various conditions. Figure 7.12 illustrates the result obtained using an envelope detector for two different relative levels of carrier and sideband.* The dashed line is the sinusoid with which the carrier was modulated. One of the two cases represented (the middle trace) results when the sideband is −6 dB (one-half the voltage) of the carrier. The third waveform represents the waveform produced by equal amplitude carrier and

* The figure is displayed as would be a television signal, with the lowest resultant amplitude at the top. The gain has been normalized in the three cases to produce the same peak-to-peak signal.

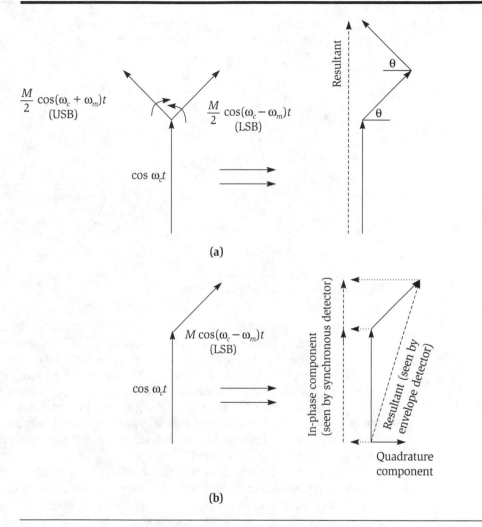

Figure 7.11 Phasor diagram, double and single sideband carriers. (a) Symmetrical sidebands. (b) Single sideband.

sideband. This is a realistic condition for a television signal: a scene with medium to high brightness has a relatively low carrier level during the lighter portions of the scene. If the object is brightly colored, the chroma carrier amplitude will be high, so when the camera is scanning the object, the chroma can be as high or higher in amplitude than the picture carrier. The resultant tends to be "scalloped," with the negative peak too wide and the positive peak too narrow.

From this illustration, it is evident why the chroma response of the Nyquist slope filter is reduced (haystacked) when an envelope detector is used (see Figure

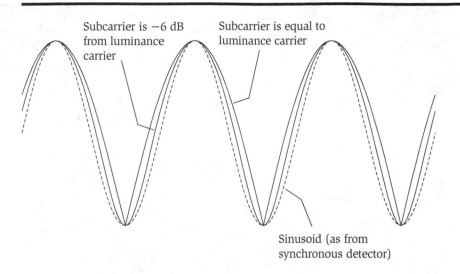

Subcarrier is −6 dB
from luminance
carrier

Subcarrier is equal to
luminance carrier

Sinusoid (as from
synchronous detector)

Figure 7.12 Demodulated results, envelope detector.

7.10). The chroma carrier is the highest amplitude sideband of the luminance carrier when brightly colored objects are being scanned. The higher the chroma signal is with respect to the picture carrier, the greater the distortion. By reducing the chroma amplitude before detection, the effect is reduced. Of course, the chroma amplitude must be corrected before the signal is presented for measurement purposes; hence the chroma peaking circuit FL3 of Figure 7.8. (In a consumer TV set, it is not necessary to use this peaking circuit because the chroma is separated from the luminance, and the correction can be approximated by adding more gain to the chroma channel.)

If the Fourier transform of the chroma waveform is taken, it will be seen that the amplitude of the 3.58-MHz fundamental component is less than what it would have been without quadrature distortion. This causes the chroma saturation (proportional to chroma amplitude) to be lower than it should be—the color appears washed out compared with the way it should look. An envelope detector that doesn't reduce the amplitude of the color carrier prior to demodulation will exhibit excessive differential gain and luminance-to-chrominance and chrominance-to-luminance crosstalk.

The average level of the chroma signal is no longer in the center of the waveform of Figure 7.12. Rather, it is lower than it should be. This means that the luminance level is lower than it should be, making the object appear darker than it should. Thus, a brightly colored object of relatively high luminance level (bright yellow is an example) will appear too dark and somewhat washed out owing to the quadrature distortion introduced by the envelope detector. Whether or not the distortion is objectionable depends on a number of factors.

One useful way to show the difference between envelope and synchronous detectors is to study the various sinusoids involved in the multiburst waveform. Switching back and forth between the two detectors will readily show the difference in waveform. Some bursts will show more distortion with an envelope detector than will others, depending on the sideband amplitude seen by the detector.

Another convenient test is to study the difference between a sine squared (2T) pulse when demodulated with synchronous and envelope detectors. The demodulated shape will be correct with the synchronous detector. With the envelope detector, the pulse will appear too narrow, and possibly too tall. This would have the effect on a picture of narrowing a vertical white line on the screen.

Response of a Synchronous Detector

The synchronous detector can be shown to respond only to the in-phase component of the phasor of Figure 7.11(b). In turn, it can be shown that the in-phase component, the projection of the sideband on the phase of the carrier, is an accurate representation of the modulation.

Application of Test Demodulators Having Both Detectors

Some demodulators have a second synchronous detector, in which the phase of the reconstructed carrier is 90° with respect to the incoming picture carrier (at IF). This quadrature detector is sensitive to the quadrature component of Figure 7.11(b). By displaying the quadrature component on the y axis of an oscilloscope and the in-phase component on the x axis, we can easily spot incidental carrier phase modulation (ICPM) in the modulator. Normally, a luminance ramp is supplied as video for this test. ICPM can be caused by improperly operated power amplifiers, and is a common problem with older broadcast transmitters, particularly UHF transmitters. It tends to be less of a problem with cable modulators because of lack of high power stages, but it can be introduced in an amplifier stage that is distorting. ICPM can cause excessive intercarrier buzz if it affects only the luminance carrier.

A problem with some older and lower-cost cable modulators is poor balance in the video modulator stage. This will distort the luminance level of bright portions of a picture. The presence of this may be deduced by measuring differential gain with the envelope and synchronous detectors. A good differential gain reading with the synchronous detector and a bad reading with the envelope detector indicate possible problems with the modulator's modulation stage.

Subtle amplitude and group delay problems in a piece of equipment can be masked when measured with an envelope detector. Amplitude and group delay problems can be evaluated by looking at the sine squared (2T) and modulated 12.5T pulses. These pulses are particularly convenient for use in evaluating problems because the results are presented in terms that directly relate to the visible distortion of a video signal. However, some problems produce waveform

distortions that are similar to the distortions produced by an envelope detector. Thus, a synchronous detector is more useful when studying response problems.

7.6 TV Stereo

At least three TV stereo systems are in use worldwide, with several variations to each. North America uses the BTSC system, named after the committee that adopted it. The Broadcast Television System Committee was formed under the auspices of the Electronic Industries Association (EIA). It was active from the late 1970s to the mid-1980s. The Japanese use a variation of BTSC stereo. In Great Britain, the BBC and IBC developed a digital system known as near instantaneous amplitude companding and modulation (NICAM). With variations, it is in use in a number of European countries and elsewhere. The German IRT earlier developed a two-carrier analog system, which is used there, in Korea, and elsewhere.

7.6.1 BTSC Stereo Generation and Application

The BTSC stereo system is described in Chapter 2. Figure 7.13 shows one possible block diagram of a BTSC stereo encoder. Left- and right-channel audio are supplied to identical low-pass filters, whose cutoff frequency should be as close to 15 kHz as possible, while providing a lot of attenuation at the horizontal sweep frequency of 15.734 kHz. Because of the sharpness demanded of these filters, their complexity is high. Failure to provide adequate reduction of power at 15.734 kHz will result in compromised stereo performance due to interference with the pilot carrier. Further, the difference channel approaches this frequency, and poor rejection could result in artifacts arising from spectral crossover between the sum and difference channels.

Note that the left- and right-channel program audio levels must be metered as shown to ensure the correct deviation. The monaural meter on the modulator is not usually suitable for this purpose.* In fact, it can be very misleading. Similarly, SAP level is monitored at its input.

From the filters, the signals are supplied to the matrix, which is simply a pair of networks that produce the sum (L + R) and difference (L − R) signals. The sum channel is supplied to a preemphasis circuit and a delay compensator, which compensates for the delay in the difference channel. See the discussion of emphasis in Section 7.3.6. It is necessary to use preemphasis in the sum channel in order to maintain backward compatibility with the mono system. From the preemphasis circuit, the sum signal is supplied to an adder that combines the different parts of the stereo signal.

The difference signal is supplied to the dBx encoder, which compresses it in order to reduce the effects of poor carrier-to-noise ratio in the transmission path.

* If the monaural meter drive signal is limited to a 15-kHz frequency response it may be marginally useful as a deviation indicator, but this is not recommended.

Figure 7.13 Block diagram, BTSC stereo encoder.

The compression is frequency selective and follows an algorithm proposed origi-
nally by the dBx Corporation. (For test purposes only, the dBx compression may
be replaced with a 75-μs preemphasis network.)

From the compression circuit, the difference signal is applied to the modula-
tor. It is double sideband suppressed carrier amplitude modulated (see Chapter 2)
onto a carrier at 2fh (twice the horizontal scan frequency), 31.47 kHz. The carrier
oscillator is phase locked to the horizontal sync rate, fh, derived from the video
signal. This need for phase lock is the reason that video must be looped through
the stereo encoder. Failure to lock to video could result in beats and separation
"pumping" from oscillators that are free running on almost but not quite the same
frequency. A portion of the fh signal, obtained by dividing the output of the 2fh
oscillator by 2, is filtered and supplied to the summing circuit as the pilot carrier.
This fh signal is also supplied to a phase locked loop, whose output adjusts the
2fh oscillator as needed to maintain the correct phase relationship between its

two inputs, one from the divider and the other from the sync separator. These three components—the sum, difference, and pilot signals—make up the basic stereo signal.

If the SAP is not used on a particular channel, that circuitry is not likely to be included in the stereo encoder. If it is, the circuitry at the bottom of Figure 7.13 is supplied. The SAP audio is compressed then frequency modulated onto an oscillator whose center frequency is locked to the fifth harmonic of the horizontal frequency, 5fh. This SAP signal is then summed with the stereo components. If the professional channel is used, then the blocks at the top of the page are supplied. It is rarely used in cable television.

Interfacing and Setting Levels in BTSC Service

A number of methods may be used to connect the BTSC encoder to the modulator. Selection of the interface method to use is based on inputs available on the modulator and outputs available on the stereo encoder. Of particular concern is the ability to set the deviation of the sound carrier correctly. Failure to do so will result in decreased stereo separation. For example, separation between the left and right channels will be degraded to less than 20 dB if there is a 1.9-dB error in setting the deviation.[2]

Several methods may be used to interface the encoder with the modulator. The most basic, but not necessarily the best, is to defeat the preemphasis in the baseband input of the modulator (because preemphasis is handled in the encoder) and supply the broadband signal to the baseband audio input of the modulator (Figure 7.4). This works, but you must be careful to set the deviation of the modulator correctly. Failure to do so will result in poor stereo separation, as shown earlier.

In order to use this interface, some modulator designers have arranged metering circuitry such that a modulation meter measures only the sum signal. The object would be to set the level using the left and right meters on the encoder, then set the deviation on the modulator using its meter. Setting deviation only with reference to the sum deviation is not at all accurate and will almost surely result in poor stereo separation. However, metering the sum channel at least makes a meter on the modulator make sense when handling a stereo signal.

In addition to metering the program audio, the encoder manufacturer may provide a test tone. When turned on, the test tone replaces the composite stereo signal supplied to the modulator. The modulator *deviation* adjustment (AT3 of Figure 7.4) is adjusted for the correct deviation of the test tone, usually 25 kHz. If this is done and the program audio levels are set correctly on the encoder, then the deviation will be correct. This method is far superior to metering only the sum channel program audio on the modulator.

A preferred interface method, if available, is to have a 4.5-MHz FMed oscillator built into the stereo encoder. An optional sound carrier modulator (oscillator) is shown in Figure 7.13. This way, the encoder manufacturer can ensure that the deviation is set correctly. The encoder output at 4.5 MHz is supplied to flag C of

the modulator block diagram, Figure 7.4. It is best to *not* combine the sound subcarrier with the video since doing so will reduce the carrier-to-noise ratio in the aural channel and will create more group delay problems in the video channel. The modulator must be capable of accepting a separate input to FL3.

Another possible but somewhat less desirable interface method is to locate a 41.25-MHz FM oscillator in the encoder and supply sound at this frequency to flag D of the sound IF loop at the modulator, Figure 7.3. This provides the same level calibration advantage as does the 4.5-MHz interface, but exacerbates the problem of maintaining accuracy in the frequency spacing between the picture and sound carriers. In any of the other methods described, sound carrier spacing depends only on a single 4.5-MHz oscillator. The FCC-required accuracy in spacing for cable television is ±5 kHz. Applied to the 4.5-MHz spacing, this represents an accuracy requirement of 1,100 parts per million, a trivial number for a crystal-controlled oscillator, and barely possible with an L–C oscillator (used in set top terminals but definitely not appropriate for professional use). If the interface is at 41.25 MHz, then two independent oscillators are involved, one at 41.25 MHz in the encoder and another at 45.75 MHz in the modulator. This means that each must be accurate to about 57 parts per million. This is not excessive for a crystal oscillator but is getting to the point that care must be exercised.

In addition, there is some possibility with a 41.25-MHz interface that ICPM could exist on the picture carrier and not on the sound carrier. This would cause a degraded signal-to-noise ratio in the intercarrier demodulator in the TV receiver. (See the discussions of ICPM and intercarrier detection in Section 7.5.)

Regardless of the interfacing method used, it is necessary to ensure that the deviation of the main sound carrier by the composite stereo signal is correct, then to set the level of the left and right channels using only attenuators AT1 and AT2 of Figure 7.13. The stereo encoder should provide metering to allow accurate setting of the program audio levels.

7.6.2 The NICAM Stereo System[3]

NICAM's formal designation is NICAM 728 because the bit rate is 728 kb/s. The system is a digital transmission system developed in the mid-1980s in the United Kingdom. It has been adopted for use in other countries as well. It is capable of carrying one stereo pair, or two independent monaural programs, or 352-kb/s data and one monaural program, or 704-kb/s data only.

Data is initially sampled at 14 bits per sample and down-converted to 10 bits (plus parity). The data is divided into blocks of 32 samples, and the scaling of the block is based on the highest amplitude sample in that block. A scaling factor is transmitted with the data. Transmission is divided into 1,000 frames per second. Each frame consists of 704 bits of data and parity, with the remaining 24 bits being overhead and synchronization.

Modulation is differential QPSK (see Chapter 4). As practiced with the CCIR-I television system (Great Britain), the modulator and demodulator both use a root raised cosine filter with an alpha of 100%. When NICAM is used in countries

employing the CCIR-B/G television format, the filter alpha is reduced to 40% to accommodate the narrower available bandwidth.

In the CCIR-I system (used in Great Britain), the NICAM carrier is located 6.552 MHz (nine times the 728-kHz bit rate) above the picture carrier. The normal analog sound carrier, 6 MHz above the picture carrier, is transmitted −7 dB with respect to the picture carrier, and the NICAM signal is set to −20 dB$_{PC}$. The power in the NICAM signal does spill out of the channel, slightly invading the upper adjacent channel. This has not been a problem.

In the -B/G system (used in much of continental Europe except France and Russia), the NICAM center frequency is slightly outside the channel, with energy spillage even farther out. At UHF (system -G), the occupied bandwidth falls in the 1-MHz guard band between the channels. At VHF (system -B), the power spills into the upper adjacent channel. By narrowing slightly the vestigial sideband filter for the upper adjacent channel, it is claimed that a cable system can successfully add the NICAM signal while employing adjacent channel transmission.[4]

7.6.3 The IRD Two-Carrier Stereo System

An earlier stereo system developed in Germany employed two analog carriers. It is used there with the CCIR-B/G television system. It has also been used, with modification, in Korea, where NTSC is the television system in use. The normal 5.5-MHz (352fh) sound carrier is modulated with a left plus right signal. A second carrier is added at 5.7421875 MHz (367.5fh). It is modulated with the right channel only. This second signal is demodulated and supplied directly to the right loudspeaker. The left signal is obtained by subtracting the right signal from the L + R signal on the main aural carrier. The reason for doing this rather than placing a difference (L − R) signal on the second carrier is that noise from the second, low-amplitude, carrier theoretically gets canceled by virtue of its phase being opposite on the two channels.

The main sound carrier at 5.5 MHz above the picture carrier is carried 13 dB below the picture carrier. The second sound carrier is carried −20 dB from the picture carrier and is also frequency modulated with a deviation of ±50 kHz. A pilot carrier of 64.6875 kHz (3.5fh) contains modulation that identifies the broadcast as mono, stereo, or dual sound (two mono channels, as for two languages). In addition, mode identification is carried in the vertical blanking interval of the video.

7.7 Satellite Earth Station Receiving Equipment

The modern cable television industry was enabled by the availability of communications satellites that could distribute many television programs simultaneously to all cable headends in North America. The watershed event was the September 30, 1975 (U.S. time), Ali–Frazier prizefight, the "Thriller from Manila," seen at the two or three headends in the United States that had received 10-meter earth station antennas by then.

Figure 7.14 Earth station electronics, block diagram.

Figure 7.14 illustrates the major electronic components of the earth station used at cable headends. Signal is coupled from the "dish" antenna (see Chapter 6) to a feed located at the focal point of the antenna. The signal is amplified and, typically, down-converted to a lower frequency, by a low-noise block converter (LNB) located at the feedpoint. The LNB is located at the feedpoint to minimize signal loss. Communication satellite downlinks today operate on two bands, C band (3.7 to 4.2 GHz) and Ku band (11.7 to 12.2 GHz). Coaxial cable loss is significant at these frequencies though older earth stations used large, high-quality cable to route C band signals into the headend.

The most common band to which signals are converted is 950–1450 MHz (roughly L band). Some older equipment used 270–770 MHz. At these frequencies, the signal can be routed into the headend using fairly small coaxial cable. The cable may also carry power to the LNB as shown. Fiber optics may also be used to bring signals to the headend.

Once in the headend, the signal is split to serve a number of receivers: today, up to 12 receivers may be connected to a single LNB. A second LNB located at the

same feedpoint can handle up to 12 additional channels coming from the same satellite in the same band, with different polarization.

At each receiver, the signal is again amplified by amplifier A2, filtered, and down-converted to another IF (dual conversion downconversion will be used if the receiver is frequency agile). The downconversion is done in mixer M2, using local oscillator LO2. After amplification and filtering, the signal is demodulated in FM demodulator DM1. As shown in the insert of Figure 7.14, the spectrum coming out of DM1 consists of baseband video (probably scrambled) and possibly one or more sound subcarriers. A number of subcarrier frequencies have been used for sound, but the two most common are 6.2 and 6.8 MHz. Today, these subcarriers are generally not used for primary program audio, that being carried digitally in the video. The subcarriers may be used for cue tones and supplementary audio. Cue tones are used to initiate local advertising inserts.

If an aural subcarrier is used, it is filtered in bandpass filter FL4, amplified, and detected in FM demodulator DM2. The audio is preemphasized using the standard 75-μs preemphasis, and so must be deemphasized in FL6. Attenuator AT4 allows the output level to be set, and differential output amplifier A4 supplies the output gain.

In the video path, a roofing filter FL5 may be used to remove power above the video spectrum. If not scrambled (rare today), the video passes from the roofing filter to deemphasis filter FL7. The video is emphasized to reduce the effects of noise on the signal, as is audio. A particular emphasis and deemphasis curve is used, as prescribed by the CCIR. After deemphasis, a dispersal removal clamp removes any dispersal signal that may be imposed on the video. Dispersal is used in C band satellite circuits because the same 3.7–4.2 GHz band used for satellite downlinks is also used for terrestrial microwave communications, primarily by the telephone industry. The fear is that if modulation is removed from the satellite signal, enough power could be concentrated at one frequency to interfere with terrestrial transmissions. A dispersal signal was required in the early days to prevent that from happening. The dispersal signal is a 30-Hz triangular waveform synchronized to the vertical blanking interval (VBI) and changing direction during the VBI. It is removed using a clamp or dc restorer, somewhat as illustrated by C1 and D1 of Figure 7.3. From the deemphasis filter, the video level is set using attenuator AT5. After buffering in A5, nonscrambled video is output from the receiver.

Dispersal waveforms are not always used in modern practice. If a transponder is used full-time for video, then the downlink power will not be concentrated at one frequency. For transponders having occasional video service, dispersal waveforms are still used, but they are frequently used only to replace video when it is removed but the uplink remains on the air (possibly for test purposes).

Almost all signals delivered to cable headends today are scrambled. Programmers went to scrambling during the 1980s to prevent theft of service by privately owned earth stations that had not paid for service. The scrambling system used almost universally in the cable industry today is a sync replacement system (simi-

lar to sync suppression—see Chapter 18), with the horizontal blanking interval (HBI) replaced by encrypted audio. The idea is to make the video scrambling hard to break. Then the audio is "hard" encrypted, such that it could not be decoded without authorization. Descrambled audio and video outputs from the receiver are amplified in A6 through A8, and supplied to the outside, normally to modulators and stereo encoders as described earlier.

Signals to be descrambled are passed to the decoder, which accepts authorization and then descrambles the signals. The system in use has the capability to deliver two audio signals so that a stereo pair can be transmitted. If a second language program is transmitted, it uses one of the FM subcarriers described earlier.

7.8 Signal Handling in Headends

This section includes information that should be useful to practitioners setting up and maintaining headends. Included are notes about video and audio signal handling. RF signal handling is covered in more detail in Chapter 8.

7.8.1 Video Loop-Through

Figure 7.3 included the illustration of video loop-through. This is a technique whereby you can supply video to several using devices such as modulators, video monitors, or switchers, without using amplifiers. It is normally recommended that no more than three devices be cascaded with a loop-through because there is adverse effect on the signal from each device. This section explains what loop-through is and the most common alternative to it.

Figure 7.15 illustrates video loop-through. Figure 7.15(a) illustrates video being looped through three modulators. The devices could also be a video switcher, a picture monitor, and a modulator. A common application in headends is to loop the video through a BTSC stereo encoder, a scrambler, and then terminate it at a modulator. Each modulator in the example has the ability to either terminate the video in a 75-ohm resistor or not. It is necessary to terminate video in the characteristic impedance of the coaxial cable carrying video so as to prevent reflections that would damage the frequency response of the picture.

If a signal is not terminated, it will be reflected toward the source. This reflected signal will combine with the direct signal, resulting in a frequency response that peaks at frequencies where the two signals are in phase and nulls where the two cancel.

So long as the coax is terminated at the last device in a chain of devices, there is not a problem. This is illustrated in Figure 7.15(a), where three modulators or other devices are in a cascade, with the video being terminated by the resistor in the last device. The modulators shown all have the ability to terminate or not. This is usually selected by a switch on either the back or front of the modulator.

Figure 7.15(b) illustrates the equivalent circuit of the arrangement of Figure 7.15(a). High input impedance amplifiers are used to sample the signal from the

Figure 7.15 Video loop-through technique. (a) Looping through three devices. (b) Equivalent circuit for (a). (c) Incorrect application of loop-through. (d) Video distribution amplifier alternative.

coax transmission line without affecting significantly the impedance of that line. At the last modulator, a 75-ohm resistor terminates the line. As illustrated in Figure 7.15(c), if the termination is placed other than at the last device, reflections will result. Signal power will be reflected from the last device in line, modulator 3. When it arrives back at modulator 2, several bad things happen. First, depending on the frequency and the length of cable between the two modulators, signal components at some frequencies will add, increasing the amplitude, whereas others will cancel. The frequency response at modulator 2 is thus corrupted.

Furthermore, reflected energy arriving at modulator 2 will see a double termination: the 75-ohm terminating resistor in modulator 2 and the 75-ohm impedance of the transmission line leading back toward the source. This impedance mismatch creates a 6-dB return loss, so one-half the signal propagates toward the source (to the left), and one-half propagates again toward modulator 3, upsetting the response there. Thus, it is essential that a looped-through signal be terminated at the last, and only at the last, device in the signal chain.

Similarly, it is unacceptable to parallel more than one signal path because reflections will cause frequency response problems. This would happen if you

branched off from device 1 and routed signals to devices 2 and 3 on different cables.

Each device the signal is looped through must use a very high impedance amplifier to bridge the video signal. As a practical matter, it is not possible to prevent some signal degradation as you loop video from one device to another. Because of this, the number of devices a signal is looped through should be limited. A rule of thumb is that three devices can be looped through without excessive deterioration in signal quality. Another limitation is that if a device the signal is looped through must be removed, then service will be interrupted to other devices in the loop-through chain. Finally, looping through may result in awkward headend wiring. An alternative, when video is to be distributed to many locations, is to use a video distribution amplifier (VDA).

Figure 7.15(d) illustrates a common form of VDA though most have more than three outputs. An amplifier provides a voltage gain of 2 to the video signal so that a 1-V p-p signal at the input becomes a 2-V p-p signal at the output. A number of outputs are provided, each with a series 75-ohm resistor to establish the source impedance of the path. This resistor and the 75-ohm terminating resistor in the following device form a voltage divider, reducing the voltage back to the video standard interface of 1-V p-p.

7.8.2 Audio Wiring

Many of the same factors that apply to video also apply to audio, except that audio is usually less critical in terms of signal wiring than is video because of the lower bandwidth. It is possible, and many times recommended, that audio distribution amplifiers (ADAs) be used for much the same reasons that VDAs are used. However, it is sometimes acceptable to wire a number of audio-using devices in parallel without regard for reflections since typical headend wiring distances are insignificant compared with the wavelength of audio. It is important, though, to provide one and only one 600-ohm termination in an audio chain.

Professional audio sources are usually specified to drive 600-ohm loads. Some sources may actually exhibit an output resistance of 600 ohms though it is not at all uncommon for the source resistance to be less. Since audio lines are usually of insignificant length compared with a wavelength, this lower driving resistance is not a problem unless it is desired to drive a signal several miles. In this case, more care should be taken to match resistance on both ends.

A major headache in audio wiring is hum elimination. Most hum is due to ground loops and other sources of potential difference. Ground loop mitigation is the most magic of all magic arts: no matter how many words are written about it, the problems encountered in the field will not be covered. However, some pieces of advice are sufficiently general to be worth mentioning.

Figure 7.16 illustrates options in audio interconnection between an audio source and load. Figure 7.16(a) represents the interconnection method strongly recommended and practiced almost exclusively in professional audio applications. Balanced transmission is used, in which the source consists of two signals

Figure 7.16 Options in audio interconnection. (a) Balanced audio. (b) Unbalanced audio with hum source. (c) Unbalanced source and load with transformer. (d) Unbalanced source and balanced load.

having equal resistance to ground (not shown) and exactly opposite phase. One way to produce such signals is to use two amplifiers, one for the + phase and another inverting that signal to generate the − phase, as shown. Another way is to use an audio transformer.

The signal is connected to the load on the right of Figure 7.16(a), using a balanced, shielded cable. Several cables made for this purpose are available. The two wires of the cable are terminated in a differential amplifier, whose output is an amplified version of the difference between the two input signals on the + and − inputs. By using balanced transmission, any hum pickup will be the same on the two wires of the balanced pair, so the hum will be subtracted out at the receiving differential amplifier (or transformer).

Notice that the shield is connected to ground at only one end. It is not correct to ground the shield at both ends since this can create a ground loop.[5] Ground loops exist any time there is more than one path for current between pieces of equipment. As a practical matter, this is hard to prevent: racks must be bonded together and to ground, providing a path. Individual pieces of equipment must have third-wire safety grounds, providing a second path. After that, consider all the interconnecting wiring, much of which involves grounds. A loop can form the secondary of a giant transformer, the primary of which is all the ac wiring in the headend. The result is current induced in the ground paths, which sets up potential differences between equipment. Besides the transformer model, it is likely that many headends will experience potential differences between racks because

of the various routes supply current can take to get back to the neutral of the primary power supply.

Figure 7.16(b) illustrates the unbalanced signal connection method used in most consumer equipment but not recommended for professional use. A single-ended output amplifier sends the signal to the receiving equipment, which receives it using a single-ended input. The shield of the cable serves as the return path. Also shown is a prime reason to not use this method in headends. If an ac voltage exists between the two ends, it will result in current on the shield. Because the shield has some resistance, a voltage drop will exist between the two ends. This voltage will be seen by the receiving amplifier, resulting in hum in the audio.

If the equipment at hand can accommodate only unbalanced audio, and if hum is a problem, then the solution of Figure 7.16(c) will have to be used. A high-quality audio transformer having a 1:1 turns ratio can be used at the receive side to isolate the ground on the two sides. The input side is connected to the center conductor and shield, but not to ground. The output side is connected to the amplifier input and ground. Transformers *must* be operated at their specified impedance. Small, low-cost transformers will distort the audio, particularly at lower frequencies, though high-quality professional transformers are available that do a good job.

Figure 7.16(d) illustrates an option where the sending equipment is single ended, but the receiving equipment can accommodate a balanced input. Balanced audio cable is used, with the sending ground connected to one of the wires of the pair. Again, note that the shield is connected at one end only.

In the end, though, we can say only that ground loops are so different from each other, and so difficult to resolve, that whatever works is the right thing to do.

7.8.3 Controlling Hum in Video

Hum in video systems using NTSC and in 60-Hz power grids will show up as horizontal bars moving slowly up through the picture. The upward movement is due to the slightly slower than 60-Hz field rate of NTSC. Just as ground loops can cause great anguish with technologists attempting to interconnect audio circuits, it can be a problem with video. Because of the frequencies involved, balanced (differential) coupling of video is not as predominant as it is in audio applications though it is possible to interconnect video signals using balanced transmission. It is far more common to use single-ended equipment to interconnect video. This makes resolution of ground loop problems even more difficult for video than for audio.

Figure 7.17 illustrates two methods that can be used to mitigate hum in video circuits. Shown are video source and terminating equipment. Both the signal and chassis grounds are shown, emphasizing that they are usually one and the same in video equipment. They may be kept separate internally to the equipment, but for the most part, they are the same outside the equipment.

Figure 7.17 Methods used to mitigate hum in video circuits. (a) Hum bucking coil. (b) Floating input connector.

Figure 7.17(a) illustrates use of a hum bucking coil. These may be purchased commercially and consist of a number of turns of 75-ohm coax through a high-permeability ferrite core. Typically, BNC connectors are provided on the case of the hum bucking coil. The connectors must be insulated from each other. To a current on either the center conductor or shield, the arrangement presents a high inductive reactance, which opposes the current. However, to a current traveling

in one direction on the center conductor and the opposite direction on the shield (correct differential mode signal transmission), the inductive reactance is not present: the inductance operates only on the net current in the coax, and in this case, the net current is zero.

So long as the inductive reactance at 60 Hz is high enough, it is not possible for significant ground loop current to develop through the coax. Thus, the hum bucking coil can offer significant reduction in hum, but only if it is not shorted between the input and output shields. The short would afford ground current a way to avoid the inductive reactance of the coil.

Some video equipment does offer a balanced input driven from an unbalanced source, as illustrated in Figure 7.17(b). This is analogous to the situation of Figure 7.16(d). Usually, a special BNC connector is used, which has its shield insulated from the chassis. A differential amplifier recovers the voltage difference between the center conductor and the shield. The resistor to ground is high and is used to prevent a static charge buildup on the shield if the other end is not terminated. A floating BNC connector as shown is used on some professional equipment, but the cost prevents its use in lower-cost lines of video-processing equipment.

7.9 Ad Insertion

Insertion of advertising in the programs transmitted by various programmers is a significant source of income for many cable systems. Many basic cable networks charge cable systems for the privilege of carrying the network. In turn, the networks make available "avails," or time each hour that is available for sale of advertising by the local cable systems. Though practices vary, it is not uncommon that two blocks of time, each 2 minutes long, are made available at predetermined times each hour for insertion of advertising sold by the local cable system. Control of insertion is done by the network, which transmits cue tones to initiate the play of local commercials and cue tones to transfer back to the network.

In the past, ad insertion was done using analog tape players controlled by a computer. Today, file servers have replaced the tape players in new systems. Figure 7.18 illustrates a file server used for this purpose. Differences will exist from one system to another. The file server itself is usually a computer workstation optimized for transfer of large digital files from its internal hard disk drives to the outside world. It interfaces with a traffic system, which provides the schedule for playing commercials. Through a second interface, commercials are loaded onto the hard drives. Typically, the commercials are compressed using MPEG-2 encoding. Because of the cost of encoding, the commercials are usually encoded centrally, either by an MSO or by an outside agency.

One file server is typically capable of supplying data fast enough to service a number of channels simultaneously. As shown in Figure 7.18, the file server "plays" the file out through a system interface, to an individual channel interface, which is shown in more detail in Figure 7.19. The channel interface receives

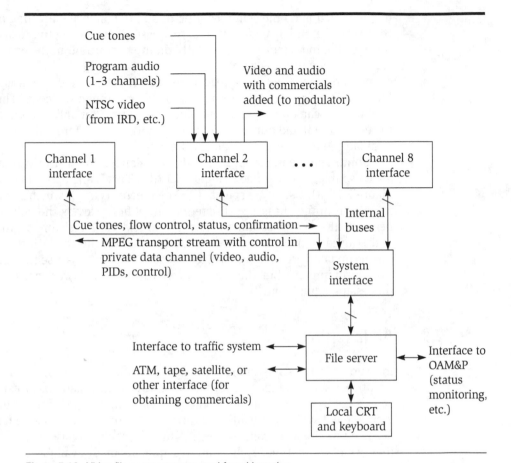

Figure 7.18 Video file server system used for ad insertion.

analog (in the future, digital) video and audio from the IRD or other source, along with cue tones. Usually, the cue tones are transmitted on analog sound subcarriers since the audio is transmitted digitally with the video. FM demodulators, such as DM2 of Figure 7.14, recover the cue tones. The cue tones are passed to the file server, where they initiate playing of commercials.

Figure 7.19 illustrates an individual channel interface. NTSC video is supplied to a switch, which allows replacement of the incoming video with the commercial. Sync is extracted from the video to allow the inserted video to be synchronized with the incoming video, a requirement for smooth switching between sources. An MPEG decoder receives video from the file server and converts it to NTSC video. The cue tones are supplied to a detector, which passes them on, ultimately to the file server. One to three (if SAP is included) channels of audio are switched with the video.

The video and audio from the ad insertion system are supplied to a modulator and, usually, stereo encoder, for transmission to subscribers. If the inserted

Figure 7.19 Individual channel interface in ad insertion system.

video is synchronized properly with the incoming video, and the actual switch-over between incoming and advertising video is done in the vertical blanking interval, the subscriber does not see a change that indicates the video is being originated locally. Cue tones cannot be heard because they are transmitted on a channel other than the one used for program audio.

7.10 Summary

This chapter has provided information about the common types of equipment found today in headends. This equipment included processors, modulators, demodulators, stereo encoders, earth station receivers, and advertising insertion equipment. We have demonstrated typical circuit sets and features, recognizing that a number of ways to do each job exist, and that not every piece of hardware will have every feature shown. The next chapter will cover what happens to signals after they exit the individual pieces of equipment and will also discuss other techniques related to headend use.

Endnotes

1. Donald G. Fink, *Electronics Engineers' Handbook*. New York: McGraw-Hill, 1975, pp. 2–26.

2. Christopher Bowick, *The Importance of Setting and Maintaining Correct Signal and Modulation Levels in a CATV System Carrying BTSC Stereo Signals*. This has appeared in several places, but a useful source is the *BCT/E Certification Program Reference Bibliography Reprint Manual*, Society of Cable Telecommunications Engineers, 140 Philips Rd., Exton, PA 19341-1318. Telephone 800-542-5040. Many other useful reprints are also found in this volume.

3. IBA, BREMA, BBC, *NICAM 728: Specification for Two Additional Digital Sound Channels with System I Television*, 1988. The current status of this document cannot be determined. However, the Instituto de Communicagoes de Portugal has published substantially the same information, which can be found at *http://www.icp.pt/legisuk/p316-93uk.htm*.

4. Televerket Radio Laboratory (Sweden), *Digital Multisound in Television— Feasibility of Filtered 728 Kbit/s NICAM with 5.85 MHz Carrier Frequency in CCIR System B. Adjacent Channel Operation in Cable Networks*. Paper is dated January 16, 1987, and was privately furnished to the author. Publication information is unknown.

5. Linc Reed-Nickerson, Audio Levels Scream for Attention. *CED*, October 1997, p. 54ff.

Chapter 8
Headend Operation

8.1 Introduction

Previous chapters in the headend section have covered the input devices to headends and the electronics that compose the bulk of the headend equipment. This chapter covers a number of issues related to headend operation. First, the accepted band plan for cable television is presented. This defines the accepted channel number versus frequency plan adopted by both the cable television and consumer electronics industries, and mandated by the FCC.

Management of optical and RF connections and signals is important in today's headends, so these topics are covered in some detail. This topic includes steering signals to and from nodes and organizing wiring such that performance is optimized. Finally, topics related to the assessment of the quality of signals in headends are covered.

8.2 Cable Television Band Plan

For reasons shown later, the frequencies used for transmission of television signals are not contiguous, as they are in AM or FM radio broadcasting. Therefore, television stations are identified, not by frequency as is the case with radio, but by channel number. The channel numbers are easy for viewers to remember and avoid the confusion of noncontiguous frequency assignment.

8.2.1 Concept of Image Response and Historical Background

The VHF television assignments are chosen to allow certain simplifications in tuner design. Recall from Chapter 2 that the NTSC television signal modulated onto an RF carrier has a picture carrier, a color carrier located 3.58 MHz above the picture carrier, and a sound carrier located 4.5 MHz above. The spectrum from 0.75 MHz below the picture carrier to 4.2 MHz above the picture carrier (at RF) is occupied with visual information. Engineers often talk of the channel frequency as the frequency of the picture carrier, understanding that the other carriers and the modulation sidebands are also present.

Inside a demodulator or television receiver, the incoming signal is quickly converted to an intermediate frequency (IF). This is a standard frequency at which a tuned signal on any channel will be amplified, filtered, and detected. The spectrum of the signal at IF is reversed, or inverted, from that at RF, due to use of a high-side local oscillator (LO) in the TV receiver. Using a high-side rather than low-side LO mitigates certain problems that would create major performance limitations. Local oscillators are used to convert an incoming frequency to another frequency and sometimes are used to convert from one internal frequency to another. When you tune a TV or radio, you are changing the frequency of the first local oscillator such that the difference between its frequency and the desired incoming frequency equals the IF frequency.

Figure 8.1 Operation of high-side local oscillator at channel 2.

Figure 8.1 illustrates the frequency conversion process, starting with the RF frequencies transmitted over the air or on cable and ending in the IF band normally used today. The top portion is taken from Figure 7.8 and shows the conversion of the received signal to IF. The lower portion illustrates what happens when a TV is tuned to channel 2. The local oscillator operates at 101 MHz for channel 2. The mixer takes the difference between the LO frequency (101 MHz) and the RF picture carrier (55.25 MHz). This difference is 45.75 MHz, the standard picture carrier IF used in NTSC television reception. Also shown are the conversions for the sound and color carriers. Notice that the spectrum of the channel at IF is

inverted from that at RF, due to subtraction of the RF frequencies from the local oscillator frequency.

If bandpass filter BPF2 is not present, and if there exist incoming signals at cable channels 17 and 18, then an "image" is produced in the IF. The desired IF signal is produced by subtracting the incoming frequency from the LO frequency, but the image is produced by subtracting the LO frequency from the incoming image frequency. This is shown in the IF spectrum, where the channel 18 picture carrier is converted to 44.25 MHz by the same local oscillator that converts the desired signal. The presence of these "spurious" signals in the IF passband will produce interference beats. No amount of filtering at IF (represented by BPF3) can remove the interference. The only way to remove it is to never allow the image-producing signal to reach the mixer. It can be removed by filter BPF2 before frequency conversion. This filter is tuned to the desired channel.

The image frequency is given by $f_{image} = f_{RF} + 2f_{IF}$. It is desirable that the image frequency be removed as far as possible from the desired RF frequency to simplify BPF2. The filter must allow adequate rejection of the image frequency so that the image does not adversely affect the received signal. If the IF frequency is raised, then the transition region of BPF2 can get larger, resulting in a simpler filter. The transition region is the frequency region between where the filter must pass signal (the passband) and the frequency at which it must provide a specified rejection (the stopband).

VHF off-air allocations don't include the image frequency for any channel. Thus, a TV receiver designed for off-air reception doesn't have to worry about a television signal being present on the image frequency. Furthermore, local oscillator frequencies don't fall within a channel. If a set radiates power at the local oscillator frequency, there is no danger of causing interference to TV reception. The UHF band does not include these desirable characteristics since channels there are contiguous over a frequency range of 336 MHz. However, the channel allocation plan the FCC uses includes "taboo" channels. When one channel is assigned in a city, several other channels become unavailable owing to image, local oscillator, and adjacent channel considerations. Assignment of channels for digital television broadcasting violates the taboo channels for the first time in the history of North American television broadcasting.

8.2.2 Introduction to the Modern Cable Television Band Plan

The need to use spectrum efficiently has driven the cable industry to include channels that violate the off-air plan and taboos. That this has worked is due in part to the fact that signal levels in cable are controlled much better than they can be controlled over the air. When all else fails, set top terminals are used to combat image problems. Set top terminals use a dual conversion frequency plan with a first IF most commonly at 608–614 MHz (converter frequency coverage to 550 MHz), so image rejection is not a problem. Set top terminals that tune above 550 MHz use higher IF frequencies.

Early on, the cable industry began using the so-called midband frequencies between the FM band and the VHF high band (channels 7–13). Initially, these were designated using letters, beginning with channel A (120–126 MHz). When need and technology opened up higher frequencies, the spectrum above channel 13 (210–216 MHz) was called the superband, and channel designations continued to use sequential letters. About 1980, set top terminals began using all-electronic tuning in order to permit the use of economical remote control. The low-cost numerical displays were not capable of displaying letters, so set top manufacturers began using numerical channel designations, beginning with channel 14 for what had been known as channel A.

Each manufacturer used his or her own allocation plan though to an extent a consensus developed: what had been channel A became channel 14, and sequential numbers were used from that point on. The problems developed when they got to channel 36, 298.25 MHz. This was the top frequency being used when mid- and superband numerical designations became popular. In order to get a little more capacity, cable systems were sometimes putting TV signals in the FM band, 88–108 MHz, and in the 108–120 MHz region previously avoided because of concern for potential aeronautical interference. In addition, HRC and IRC plans were becoming popular. As we will show, use of these plans opened up another 6-MHz block between channels 4 and 5. When set top manufacturers added all these channels to set top terminals, they took different directions in channel number assignment.

Shortly after, TV manufacturers began using tuners that tuned the cable channels. With such tuners and 75-ohm inputs, the sets were called "cable ready" and were sold as not needing set top terminals.* Lacking a standard channel assignment plan, each TV manufacturer copied the plan used on whatever set top terminal he or she looked at. This led to considerable confusion on the part of the subscriber, who had to be trained that, in order to get such and such a program, he or she would have to tune to channel 37 (for example) with one set top, channel 99 with another, or maybe channel 1 if he or she had a TV without a set top terminal. The cable television and consumer electronics industries began discussing a standard band plan, under the auspices of the joint EIA (Electronic Industries Association) and NCTA (National Cable Television Association) engineering committee. The first plan, produced in the mid-1980s, was assigned the designation IS-6, an EIA designation for an interim standard. It covered allocations to 300 MHz. Even as an interim standard, it received widespread support from both industries. Set top and TV manufacturers began applying it to

* In fact, the TV tuners often had to have set top terminals placed in front of them anyway. They were not well shielded against direct pickup. They didn't have adequate image rejection and were susceptible to overload from too many signals at the input. Further, the TVs often had inadequate adjacent channel rejection. These problems still required the use of a set top terminal, to the considerable frustration of the subscriber, the store that sold him or her a "cable-ready" set, the TV manufacturer, and the cable company. Eventually, the FCC got involved.

new product. (However, even today there is product in use that predates this standard.)

Before IS-6 could be adopted formally, the industry began using frequencies above 300 MHz. The committee extended the plan to cover channels through 1 GHz and to define how channels above this would be allocated. The new standard was designated IS-132. The prototype standard was rather quickly produced, but a number of years elapsed before it was officially approved and sent to the American National Standards Institute (ANSI). The final EIA-assigned standard number is EIA-542. This standard is described in the remainder of this section. Some seemingly strange violations of a straightforward channel number allocation plan exist for historical reasons: a new standard must not cause undue problems for equipment already in use when it is adopted.

8.2.3 The EIA-542 Band Plan

Outline of the Plan

Figure 8.2 shows the relation between the standard cable channels of EIA-542 and the FCC frequency assignments.[1] (Besides the standard assignments, EIA-542 defines HRC and IRC frequency plans, shown in Figure 8.4.) The boxes in Figure 8.2 contain the FCC channel numbers for off-air assignments plus the traditional letter designations for cable-only channels. The new channel numbers as defined by EIA-542 are indicated above the boxes. Below the boxes are representative frequencies for the channel boundaries. Finally, the traditional cable band designations are shown below the frequencies. These band designations are not used very often today.

Channels 2–13 are identical to the FCC assignments for off-air broadcasting. The midband comprises channels 14 (121.25-MHz picture carrier) through 22 (169.25 MHz). Channel 23 (217.25 MHz) is the first channel above channel 13. From channel 23, the channel numbers are progressive through channel 94 (643.25 MHz). Channels 95–97 lie in the FM band, and channels 98–99 define the frequencies between the FM band and channel 14. Channel 100 picks up at 649.25 MHz. From there, the plan progresses one channel number per 6 MHz, up to channel 158, 997.25 MHz, where the present revision of EIA-542 stops. Above this frequency, the plan specifies that the channel numbers shall continue to progress one channel number for each contiguous 6-MHz band.

Relation to the Off-Air UHF Band

Figure 8.3 shows the relation between the cable plan and FCC assignments in the UHF television band. The FCC plan is offset up 2 MHz from what we would expect if we were to keep adding 6 MHz to the VHF allocations. The cable plan fills the spectrum completely from above channel 13 to the UHF band. Consequently, the cable band edges are lower than the off-air band edges by 2 MHz. This can cause problems when these channels are used on cable systems. In the UHF re-

Figure 8.2 Relation between cable channels and FCC assignments.

Figure 8.3 Relation between cable and FCC assignments in UHF band.

gion, the channel offset can result in a local broadcaster rendering two channels less desirable on the cable. For example, assume channel 15 was used off-air. The picture carrier from channel $15_{\text{off air}}$ will fall in channel 66_{cable}. The aural carrier from channel $15_{\text{off air}}$ will fall in channel 67_{cable}.

If the field strength of channel $15_{\text{off air}}$ is high, the sound carrier will produce beats in channel 67_{cable}. On the other hand, if the field strength of channel $15_{\text{off air}}$ is not too high, the offset can help cable reception. Because the picture carriers are offset by 2 MHz, the visibility of any beat produced is less than if the offset were not present. The so-called W curve, described in Chapter 2, shows that an interfering carrier offset from the picture carrier by 2 MHz is not as visible as if it was closer.

Figure 8.4 Relation between standard, IRC, and HRC plans.

IRC and HRC Assignments

In order to control the effects of distortion in cable television distribution plant, some systems employ alternative frequency plans. Of a number of plans that have been tried over the years, only two alternative plans survive. Figure 8.4 contrasts these two plans with the off-air (standard) assignments. The low VHF band is shown since it is the most affected.

The incrementally related coherent (IRC) plan (lowest line of the figure) places all carriers, except for channels 5 and 6, on the same frequency as in the standard plan. All picture carriers are phase locked to a comb generator, whose output is derived from a common 6-MHz oscillator, but with the comb shifted 1.25 MHz high in order to place the picture carriers on their normal assignments. We can show that many third-order distortion products fall coherently on other carriers in this case, yielding lower beat visibility. The 4-MHz gap between channels 4 and 5 becomes 6 MHz because channels 5 and 6 must be shifted in order to accommodate IRC assignments. They are shifted up by 2 MHz, invading the FM band. This opens a 6-MHz gap between channels 4 and 5, which is assigned the designation "channel 1." (Some cable systems are using 48–54 MHz and calling it channel 1. This is not recognized in the plan.)

The other plan is the harmonically related coherent (HRC) plan (top row in the figure). All picture carrier frequencies are at harmonics of 6.0003 MHz. We explain the significance of the 6.0003-MHz frequency later. The picture carrier frequencies are derived by phase locking the picture carrier generators to a 6.0003-MHz comb. We can show that this causes all second- and third-order distortion products to zero beat other carriers, providing somewhat more immunity against distortion than does the IRC plan. All carriers are shifted low by 1.25 MHz, to fall on the harmonics of 6.0003 MHz. Again, the exceptions are channels 5 and 6,

which are shifted high by 0.75 MHz. The combination of shifting channel 4 lower by 1.25 MHz, and channel 5 higher by 0.75 MHz, again creates a 6-MHz opening that is assigned the designation "channel 1."

Operation in the Aeronautical Bands

Cable systems wishing to put signals in the frequency bands utilized in the aeronautical service (108–136 and 225–400 MHz) are required to periodically measure leakage from their systems and to repair any leaks that exceed a certain standard. The standard is much more stringent than is the standard to which consumer electronics manufacturers are held. In addition to controlling leakage, signals carried on the cable system are required to be offset in frequency from the nominal aeronautical carriers. The nominal offset puts the cable picture carriers at the edges of the aeronautical channels. The offset requirements apply to any carrier that exists in the distribution plant at a level exceeding 10^{-4} watts (+38.75 dBmV), measured in a noise bandwidth of 25 kHz and averaged over 160 μs.[2] With realistic distribution plant signal levels, this often includes picture carriers, but rarely includes sound or data carriers.

Channel Assignments in the Aeronautical Band

Figure 8.5 shows the relation between cable carriers and aeronautical channels. The aeronautical bands are subdivided into aeronautical radio navigation and radio communications bands. The navigation channels are assigned channels in 50-kHz increments, and the communications bands in 25-kHz increments. The aeronautical carrier frequencies fall precisely on the nominal TV picture carrier frequencies. Because of this, cable systems must offset their picture carriers so that they fall on the aeronautical channel edges. In the aeronautical radio navigation bands, this necessitates a 25-kHz offset. In the communications band, the offset is 12.5 kHz.* The FCC permits any aeronautical channel edge to be occupied by a picture carrier. In order to simplify the allocations, EIA-542 recommends picture carriers be placed at the next channel edge higher than the nominal frequency. This seemed to be the most common assignment used before the recommended standard was written.

Some cable systems, though, offset to the next lower channel edge, which is shown as the secondarily recommended picture carrier. One reason a system may want to use the negative offset is to help identify the source of leakage. Cable systems must formally measure leakage from their systems frequently, and must informally monitor leakage continuously. If two systems are adjacent to each other, it can be difficult to determine which one is responsible for a leak. When this condition exists, one of the systems will employ positive offsets, and the other negative offsets. Thus, by measuring the frequency of the leakage, it is possible to

* For a number of years, there has been talk of splitting the aeronautical channels into two, halving the separation between center frequencies. If this ever happens, the now-required cable offsets will fall on carrier frequencies rather than on band edges.

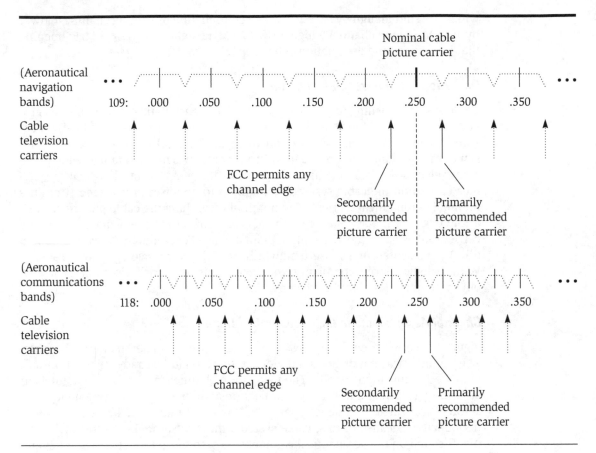

Figure 8.5 Relation between cable picture carrier frequency and aeronautical channels.

identify the system with the leak, allowing the correct repair crew to be dispatched. The recommended standard permits, but doesn't encourage, any offset up to ±250 kHz. It is felt that most TVs will accommodate this much offset without difficulty.

The tolerance on the carrier frequencies is specified by the FCC as ±5 kHz in the aeronautical band (elsewhere the tolerance is ±25 kHz). In order to accommodate HRC assignments within this limit, it is necessary to use a comb generator with a fundamental frequency of 6.0003 MHz and a frequency tolerance of ±1 Hz. Section 7.4.1 describes an HRC comb generator.

FM Band Usage

In some cable systems, spectrum is at such a premium that the operator has been driven to use the FM band for TV program distribution. This is not common, but is sometimes done. Also, some systems use the FM spectrum for digital music

channels. The recommended standard permits the use of the FM band frequencies, but does not require that the TV tune these frequencies. The cable operator is advised that many "cable-ready" sets will not tune to these channels. Many TVs include traps in the FM band to reduce distortion effects in the tuner. These traps preclude reception of the FM band.

Alternative Modulation Methods

Acknowledging the use of digital modulation in 6-MHz channelization, the recommended standard proposes that any signal delivered to the consumer should fit in channels having the edges at the frequencies of the NTSC channel edges. This does not preclude distribution of programming in the plant using some other frequency plan, as long as any signal handed to the subscriber at RF follows the plan. For example, digital signals might be carried in a 12-MHz band on the cable. This would be permitted so long as that signal was never handed to the subscriber. If a device supplied by the cable operator converts the signal to something contained within one 6-MHz channel having the band edges shown in Figure 8.2, that system would comply with EIA-542.

8.3 Headend RF Management

Until recently, headends were truly broadcast-oriented structures. They were built to collect various signals, format them, and send the same set of signals to all subscribers. This has been changing. The transition of the industry from tree-and-branch networks to nodal networks has enabled reuse of frequencies on different nodes. The industry has begun routing different signals to different nodes for telephony and data applications. Signals in these services are intended for individual subscribers rather than for all. In order to utilize bandwidth efficiently, it is desirable to determine the node location of an intended subscriber and to send signals bound for him or her to only that one node. This increases the efficiency of the cable network by allowing reuse of the same frequencies on each node, transmitting only information bound for someone in that node.

Besides the fairly obvious advantages of sending data and telephony signals to only the node requiring them, other services can benefit from this capability. For instance, the industry has talked about node-specific advertising, in which advertisements are selected based on the demographics of subscribers connected to that node. Yet another anticipated application is to supply customized video-on-demand (VOD) programming to individual subscribers.

RF management is the art of grouping signals bound for each node while maintaining level control in the most economical way possible. It involves physically segmenting the signals while maintaining isolation between them. It also involves handling return signals entering the headend, where it is desired to route signals to each of several return services, while minimizing the amount of hardware required.

Figure 8.6 Downstream RF management.

8.3.1 Downstream RF Management

Figure 8.6 illustrates a segment of a modern headend. Signals are received from satellite earth stations and other sources. As shown, each signal received from the earth station may have a number of pieces of hardware associated with it. The satellite receiver, described in Section 7.7, converts the signal from RF to baseband NTSC, usually in association with a decoder to descramble the satellite signal. These two pieces of equipment are often combined into an integrated receiver-decoder (IRD) in modern headends. If stereo is offered on the channel, video from the IRD is routed through a stereo encoder (see Section 7.6) and then to the scrambler if used (see Chapter 18). Finally, the signal is modulated onto an RF channel in the modulator. A number of such signals are acquired and combined in headend combiner CM2. Additional signals, such as digital audio broadcasts, may also be included. The output of combiner CM2 is the total set of broadcast signals, which are supplied to all subscribers.

Besides these broadcast signals, node-specific services are becoming popular. These commonly include telephony, data services, and in the future, node-specific programming. These services are duplicated for each node; different signals are transmitted to each node, but all use the same set of frequencies. Adequate input ports must be provided for addition of future services. Unused ports must be terminated. Services for node 2 are combined in combiner CM5, and those for node 1 are combined in CM4. These signals are combined with the broadcast signals before being supplied to the individual node optical transmitters.

The combining of broadcast and node-specific services is accomplished by first splitting the broadcast services in splitter CM3. One output of CM3 is supplied to directional coupler DC2, where signals bound for node 2 are coupled to the broadcast signals. Similarly, directional coupler DC1 is used to couple the broadcast signals with signals bound for node 1.

Level Management

A portion of the RF management issue involves physically arranging the combiners and directional couplers such that efficient use is made of the hardware, without causing problems for future expansion. Additional inputs are provided as illustrated at CM5 for services that may be added later.

Provision of additional inputs as shown does not come without penalty, however. Each time the number of ports provided at CM5 is doubled, the attenuation of signals from each existing service is increased by 3.5 to 4 dB. This means that at some point additional amplification may have to be provided.

Figure 8.7 illustrates a simple way in which the node 2 combining may be modified to reduce the level required of existing services, compared with Figure 8.6. In Figure 8.6, an eight-way splitter is provided at CM5, allowing eight services to be combined. The insertion loss of an eight-way combiner is approximately 12 dB (budgeting 4 dB per two-way combine). If this does not let the node 2 transmitters operate at comfortable levels, then Figure 8.7 can be used. Combiner CM5 is replaced with a four-way combiner, which has only 8-dB loss for each input. One of those inputs is from another four-way combiner CM6, which provides for additional services. By using two four-way combiners rather than one eight-way combiner, today's services can operate with 4 dB less level required at the respective specific services transmitters. The penalty is that future services will have to operate at levels 4 dB higher than they would otherwise (they see a 16-dB loss before directional coupler DC2), and one input is lost. Some cable operators consider this to be a good trade-off to allow the headend to be built simpler today—in some cases, using the Figure 8.7 configuration can save an amplifier in the combining network.

Signal Isolation

A major issue in both downstream and upstream RF management is isolation between nodes. This is illustrated in Figure 8.6, with the example of interference from node 1 into node 2. It is assumed that a similar service at the same frequency

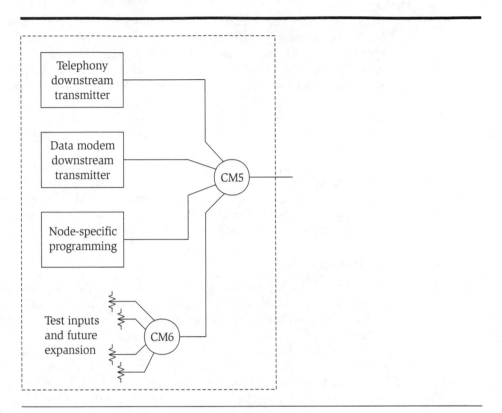

Figure 8.7 Alternative combining arrangement for node 2.

is used on each node. The problem is to avoid interference from node 1 into node 2, and so on. The heavy arrow of Figure 8.6 illustrates the example case. Signals from services specific to node 1 are coupled to the node 1 transmitter in directional coupler DC1. Theoretically, all the signal either is routed to the optical transmitter or is dissipated in the internal load resistor of DC1. If DC1 is a 17-dB coupler, then the power that appears at its output (connected to the node 1 optical transmitter), is −17 dB from the incoming signal.

In the real world, some power is coupled backward in the directional coupler and goes toward CM3. Theoretically, any power that impinges on CM3 from one of its outputs either is coupled to the input (connected to CM2) or is dissipated within CM3. In the real world though, some signal is coupled from one output of CM3 to the other. Thus, node 1 signal power from CM4 can wind up being combined with node 2 signal power, where it becomes an interfering signal.

It is assumed that the isolation of each directional coupler and combiner is 20 dB. In traversing the wrong way through DC1, the signal is attenuated 20 dB. In traversing from one output of CM3 to another, it is attenuated another 20 dB. Assuming then that it is combined with another signal of identical level

Table 8.1 Recommended headend isolation.

Interference	Recommended Isolation (dB)
Analog visual to analog visual	70 dB (phase lock recommended)
QPSK to QPSK	33 dB
64-QAM to 64-QAM	54 dB
256-QAM to 256-QAM	61 dB

in DC2, the signal from node 1 becomes a −40-dB interfering signal to a node 2 service.

The point at which interference becomes significant is a function of a number of factors. To provide a feel for the isolation necessary in headend combining, Table 8.1 lists some potential interfering situations and a recommended headend isolation number. You can see that as modulation density increases, more isolation is required. This is because higher-modulation density signals have less distance between modulation states and so can tolerate less interference (see Chapter 4).

The analog visual carrier recommended isolation is based on the traditional 60-dB threshold for interference, with 10 dB added to account for possible level differences between nodes. It is recommended that, even if this much isolation is provided, the two video sources be phase locked since this has been shown to significantly reduce the visibility of interference. Phase locking will afford additional margin to the threshold of visibility. It is known that the threshold of visibility can be more severe than −60 dB under optimum conditions, but this number, derived from work done at Bell Labs in the 1950s, is taken as a working number.

The recommended isolation for digital signals is derived from the crossing of a decision state threshold, taking into account the peak-to-average ratio of an unfiltered signal. No allowance is made for demodulation errors or filtering, but a 10-dB allowance is made for level difference, and a 20-dB allowance is made because other sources of signal degradation exist. The headend should not be a limiting factor.

8.3.2 Upstream RF Management

Some of the issues in RF management in the upstream direction are similar to those in the downstream direction, but upstream management introduces some unique issues. Signal isolation issues are similar to those in the downstream direction, but tend to be more complex owing to the need to share resources. Level issues become more complex because changes in the headend can easily affect operation of return components in the field, with operating personnel being unaware of the consequences of changes they make.

An ALC loop is set up between the headend and each home system individually. The headend senses the level of the return signal and adjusts the upstream transmitter such that the correct level is received at the headend receiver, Lvl 1. The most critical level is that at the reverse optical transmitter, Lvl 3.

Figure 8.8 Long loop ALC control of return transmission.

Level Management

Level management in the upstream direction is complicated because of the need to maintain accurate levels at the node (in the field) using long loop automatic level control (ALC). This issue doesn't have a close analog in the downstream direction and is not familiar to a lot of practitioners. Figure 8.8 illustrates a long loop ALC. The host digital terminal (HDT), a generic name for the headend equipment used with any two-way service, consists of a receiver for the upstream signals, a transmitter for the downstream signals, and logic to control those signals, as well as all the components required to operate the appropriate service. The topic of upstream level management is covered more fully in Chapter 14. The purpose of including this information here is to emphasize how the headend must be configured to allow it to function smoothly.

Downstream signals are transported to the optical node, where they are transferred to the coaxial network and to the home receiver, RX3. In the upstream direction, the home transmitter, TX4, supplies upstream signals, which are transported to the optical node and then supplied to the return optical transmitter,

TX2. From TX2, the signal is transported to the headend receiver RX2, then supplied to the RF receiver RX4 in the HDT.

Each HDT using the return path should establish a long loop ALC as shown, in which the return transmitter (TX4 at a subscriber's home) is commanded to increase or decrease its level until the level returned to the headend subsystem, at Lvl 1, is some target value. After the return system is aligned, it is imperative that the gain between the headend optical receiver, RX2, and the HDT not change. If a new service is added, then power will have to be tapped from optical receiver RX2 to supply the receiver of the new service. If this has not been accounted for initially, by provision of extra ports at CM1, the ports added will change the gain between Lvl 2 and Lvl 1. The existing subsystems will respond by increasing level at the home transmitter until the level at Lvl 1 has returned to its previous value. This will increase the level at Lvl 3, potentially overdriving the return optical transmitter, TX2. To avoid this, it is imperative that an adequate number of reverse ports be provided at CM1 when the headend is built.

Signal Isolation

Signal isolation issues in the reverse direction tend to be more difficult than in the forward (downstream) direction because many operators combine signals from different nodes for some services and not for others. For example, a telephony service may employ a separate HDT for each node. This would be done to permit good availability of service, with little chance of blockage. A modem service may combine two or more nodes to a common HDT to achieve better utilization of equipment—modem systems tend to degrade gracefully as more people use the service. Finally, a status-monitoring system doesn't need constant communications, so it is possible to combine signals from many nodes to minimize the number of status-monitoring HDTs required.

Figure 8.9 illustrates such a situation. The telephony system has a dedicated HDT for node 1. A data modem HDT is shared between nodes 1 and 2. Finally, a status-monitoring HDT is shared among many different nodes, of which three are shown. Splitter CM12 divides signal from node 1 to the telephony subsystem, the data subsystem, and the status-monitoring subsystem, and it has one or more additional ports for other services. As a minimum, a test point is needed, and ideally, the operator has made provision for future service expansion. Splitter CM13 provides the same service for node 2, as does CM14 for node 3.

Combiner CM10 has two inputs, from nodes 1 and 2, to allow them to share the same data modem HDT. Combiner CM11 has many inputs to allow many nodes to share the same status-monitoring HDT. All this sharing promotes efficient equipment utilization, but works against performance in that isolation is a problem. It is assumed that like equipment uses the same return frequencies on all nodes. This is usually the case, due to the scarcity of useful return spectrum.

In this limited example, three sneak paths exist to couple telephony signal power from nodes 2 and 3 to the node 1 HDT. Paths 1 and 2 both couple power

Figure 8.9 Return path signal coupling.

from node 2 to the node 1 HDT. Path 1 couples power from node 2 via CM10 and CM12. Path 2 couples power from node 2 to node 1 via CM11 and CM12. Path 3 couples power from node 3 to node 1 via CM11 and CM12. The power arriving at the node 1 telephony HDT adds from each sneak path.

In the example, if all combiners exhibit isolation of 20 dB, and all nodes have equal level return signals, then the total relative interference seen at the node 1 telephony HDT is three times the $10^{-40/10} = 10^{-4}$ provided by one sneak path. That is, each interference path traverses two combiners or splitters, each of which exhibits 20-dB isolation, so each interfering signal is −40 dB from the desired signal, and there are three such signals in the example. This works out to an interference level of $10 \log(3 \times 10^{-4}) = -35.3$-dB interference from the three sneak paths, assuming equal signal levels. In order to get back to parity with one sneak path, which would exist with no status-monitoring HDT sharing, the isolation of each combiner must improve by 4.7 dB. We can see the problem if the sharing of the status-monitoring HDT is spread over many nodes.

Isolation between nodes can be improved in several ways. Some operators characterize each combiner in the frequency band of interest and choose ports having the best isolation. This works for a limited number of ports in use, but as ports are filled up, the technique will ultimately break down. Sometimes attenuators in the combiner inputs can be used to effect improvement in isolation at the expense of more signal loss.

Filters can be used. For instance, suppose the status-monitoring HDT used a fixed return frequency of 8 MHz. If each input to CM11 had an 8-MHz bandpass filter, then many of the sneak paths used by the telephony subsystem (which probably uses frequencies above 20 MHz), would be broken. The problem, though, is that filters can limit frequency sharing between different services.

Finally, amplifiers may be used to effect isolation. A low-gain, high-isolation amplifier in each input to CM11 will significantly reduce the leakage. This, of course, is at the expense of more cost, possibly more distortion (though this can usually be controlled), and maybe a slight reduction of availability of the subsystem. An amplifier does not have unlimited isolation: many have as little as 15-dB isolation, but judicious selection of an amplifier and padding can often make the project succeed. Many operators use amplifiers as a last resort; nonetheless, the technique is gaining in popularity as headend RF management becomes more complex.

As a final note, it would be desirable to use some attenuation on each splitter or combiner output that goes to any receiver. Isolation of each splitter/combiner is enhanced by so doing. Power coming into one leg of CM10, for example, at the telephony receive frequency will probably see a poor return loss at the data modem upstream receiver. When the signal encounters a poor return loss, it must be reflected to the combiner, and one-half the reflected signal will be routed to the telephony upstream receiver. This mechanism reduces the isolation of a splitter compared to that isolation of which it is capable. An attenuator on each output port will improve the situation.

8.4 Headend Fiber-Optics Management

Almost all modern headends drive some amount of fiber-optic cable. The use of fiber-optic transmission has enabled many modern service offerings. However, fiber optics has its own set of requirements and limitations. Fiber-optic components and handling techniques have evolved rapidly. Chapter 11 describes the three methods of joining fiber-optic cables: fusion splicing, mechanical splicing, and remateable connectors. Optical connectors may be installed directly in the field, but it is much better to use factory-manufactured pigtails, which are cut and each end fusion spliced onto the incoming optical cables. Installation of optical connectors requires difficult polishing best left to a factory.

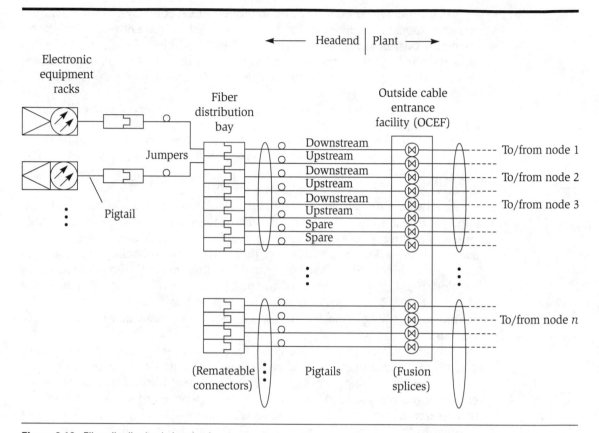

Figure 8.10 Fiber distribution in headend.

8.4.1 Fiber Distribution

Though many techniques are used to bring fiber cable into headends, a popular method is shown in Figure 8.10.[3] Outside plant fiber cables are brought to the headend at the right of the figure. The jumpers are fusion spliced in the outside cable entrance facility (OCEF), which typically is located just outside a headend wall. A pigtail brings signals into the fiber distribution bay inside the headend. The pigtails are often made by cutting a factory-manufactured jumper in two, yielding two pigtails. The jumper can be tested before it is cut; then two known good pigtails are produced when it is cut. Each pigtail is fusion spliced onto an outside fiber in the OCEF.

Connectors on the pigtails are permanently mounted to the rear of optical couplers (bulkhead connectors) in the fiber distribution bay. A fiber jumper is plugged into each active cable and couples signals to and from the electronic

equipment racks, where the optical transmitters and receivers are located. Optical transmitters and receivers are usually supplied with a pigtail that couples power from the laser diode or to the receiver diode. The pigtail normally terminates on a coupler (bulkhead) located on the chassis of the transmitter or receiver. The jumper from the fiber distribution bay plugs into that pigtail.

Of course, one could simply bring the outside plant cable into the headend, directly to the transmitters and receivers, bypassing the fiber distribution bay. The advantage of the fiber distribution bay and optical jumper is that they allow rapid recovery from plant damage and equipment failure with minimum sparing. The remateable connection at the transmitter or receiver is useful for recovering from equipment failure: if a transmitter or receiver fails, it can be replaced and the cable plugged into the new device. If a fiber fails, the connector at the fiber distribution bay permits rapid patching to a spare fiber strand.

Typically, the same bundle of fiber strands (individual optical conductors) will service several nodes, as suggested in Figure 8.10. Spares will be carried to each node. However, often it is desired that a spare optical strand will serve either a transmitter or receiver, depending on what fails. In some headend configurations, the transmitter and receiver servicing a particular node are not located close to each other. This means that it is not feasible to bring a particular strand up to, for example, a transmitter. If the strand were later needed for a receiver, that strand would have to be extracted from typically very large cable bundles in the headend and routed to the receiver. This would eventually destroy the organization of wiring in the headend, reduce reliability because of cable damage, and be time consuming. With the fiber jumper as shown, it is a simple task to move the appropriate jumper to the failed equipment on one end and to the appropriate outside plant strand.

The fiber distribution bay provides a convenient place to perform optical splitting when used. Optical splitting is used to route the output of one downstream transmitter to more than one node, and also to route light from one transmitter in two directions in a ring configuration. Ring configurations are used to provide path redundancy: if one path is damaged, such as from an overhead fiber being snagged, the second path takes over until the fiber is repaired.

In addition to covering the failure, use of the fiber distribution bay permits more rapid deployment of additional services that employ the unused ("dark") fiber running to a node. We do not have to worry about locating new equipment where the fiber is physically terminated; rather, a new jumper can be routed in the headend as desired.

It is also possible to omit the OCEF, bringing the outside cable directly to the fiber distribution bay. The outside cable will still need to be fusion spliced to a pigtail. Besides providing a convenient place to terminate outside plant fibers (if not done in an OCEF), the fiber distribution bay provides for labeling of the fibers. It also provides protection to the fiber, by shielding it from accidental contact, and provides for adequate strain relief and acceptable minimum bending radius.

8.5 Signal Quality Tests

In this section, we illustrate some simple yet effective basic quality tests that should be performed on video and audio signals in the headend. Also see Chapter 2, which contains information on the baseband video signal.

8.5.1 Visual Modulation Definition

Figure 8.11 shows a segment of the same video waveform used in Chapter 18 for illustration. The waveform is amplitude modulated onto an RF (or IF) carrier to the standard depth of modulation prescribed for the standard in use. As described in Chapter 7, the waveform is clamped on sync tips, such that the peak amplitude of the modulated waveform, which occurs during sync tips, is constant regardless of the modulation. The amplitude, A, during the sync tip is defined as *the amplitude* of the signal. As the video goes toward white, the instantaneous amplitude of the carrier drops. By definition, the peak white video modulates the carrier to 12.5% of the sync tip amplitude (that is, $B = 0.125A$, NTSC system). Video modulation is characterized by *depth of modulation* (DOM), which expresses the percentage of the distance the carrier transitions from peak sync amplitude toward zero amplitude:

$$DOM = 100\left(1 - \frac{B}{A}\right)\%$$ (8.1)

In most PAL systems, the depth of modulation is limited to 80%, whereas in NTSC, a depth of 87.5% is standard. The carrier is never totally cut off. This is important for several reasons. Many envelope detector circuits, as the envelope drops, tend to degrade in terms of their linear detection of the modulated signal. This is due to the diode curve, which exhibits a square law characteristic at low bias levels. This would tend to translate into differential gain (and differential phase due to capacitance modulation) if the carrier amplitude drops too far.

Another reason to maintain some carrier is the almost universal use of intercarrier sound detection in consumer receivers. The sound carrier at intermediate frequency is mixed with the picture carrier to produce the so-called *intercarrier* frequency, the difference between the two. Use of intercarrier detection is important to reduce the effects of incidental frequency modulation of the sound carrier. It also provides a convenient way to place the sound carrier in the middle of a passband that is narrow with respect to the accuracy of frequency conversions preceding it. However, if the picture carrier goes too low in level, the intercarrier detector will interrupt and/or phase modulate the sound intercarrier, with resultant buzz in the audio.

Thus, by standard and by FCC regulation, the depth of modulation of the visual signal is limited to 87.5%. Note that the color subcarrier is not included in

Figure 8.11 Definition of the modulated visual signal.

the depth of modulation issue: often depth of modulation is measured using a low-pass filter that eliminates the color subcarrier from the measurement. The issue of whether to include the color subcarrier in the measurement has long been controversial. For most programming, the issue is somewhat academic since most bright (high-luminance) picture elements tend to have little color saturation, so little color subcarrier is present near peak white. (A notable exception, and one that has caused some headache, is the famous Big Bird of PBS's *Sesame Street*. His bright yellow color has been known to cause clipping and out-of-channel emissions from older transmitters.)

If color information does cause overmodulation, there may be some color and luminance distortion, but it is in a portion of the color palette where the distortion tends to be somewhat less visible. Intercarrier problems in the audio tend to be masked because the color subcarrier is of such a high frequency compared with audio.

Measuring Depth of Modulation

Figure 8.12 shows a general setup for measuring depth of modulation (DOM). The measurement can be made in service using a test point on the modulator or an external directional coupler. Two DOM measurement methods are shown. A spectrum analyzer can be used as a linear receiver. Alternatively, a demodulator with a zero chopper can be used with an oscilloscope or waveform monitor. The waveform monitor is a specialized oscilloscope that is calibrated in IRE units and in depth of modulation.

Figure 8.12 Measuring DOM.

DOM Measurement with Spectrum Analyzer

A spectrum analyzer may be used by tuning it to the picture carrier. Suggested settings are shown in Figures 8.12 and 8.13. In order to measure DOM, two references must be established. Sync tips form one reference, and are set eight divisions above the bottom of the analyzer screen. The other reference is zero carrier, which by the nature of the analyzer display is on the bottom of the screen. Note that the video waveform is presented upside down from the way it is normally drawn: the analyzer displays increasing signal level toward the top of the screen. Recall that sync represents the highest amplitude of the signal. Also, note that most spectrum analyzers do not have a wide enough bandwidth setting to pass the color subcarrier, so the measurement won't include color signals.

DOM Measurement with Precision Demodulator and Oscilloscope

Figure 8.14 illustrates the waveform obtained with the lower configuration of Figure 8.12. As we noted, two points of reference are needed in order to establish a scale for measuring DOM. The sync tips at the bottom of the waveform are one reference. The other must be related to the white level. A demodulator can insert such a reference by interrupting the signal before the detector. Usually, the interruption is during the vertical blanking interval. This "zero chop pulse" is placed eight divisions above the sync tip. Then peak depth of modulation is adjusted until peak white is seven divisions above sync. Zero chop pulse generation is described in Chapter 7.

8.5.2 Aural Modulation

Equal in importance with setting visual modulation is the setting of aural modulation. One of the most frequent complaints against cable systems is inconsistent

Figure 8.13 DOM display using a spectrum analyzer.

audio level from one channel to another and from one program or commercial to another. This complaint has been leveled against broadcast stations as well. An objective measurement of audio level is difficult if not impossible owing to the subjective nature of human perception of sound. Much research has been done in this area, and a number of methods exist for making such measurements. This is not intended to be a thorough treatment of the subject. Rather, it is intended to impart some concepts that should prove useful in monitoring the audio being transmitted.

Deviation of the Sound Carrier

Television sound is frequency modulated onto the sound carrier in both NTSC and PAL television systems (it is amplitude modulated in SECAM). In frequency modulated systems, the amplitude is constant, but the frequency changes with respect to the center, or rest, frequency. The output of the detector is proportional to the deviation with respect to that center frequency. Note that the amplitude of the sound carrier is not a factor in loudness of the sound. The more the carrier is deviated, the louder the resulting signal. The art of setting the perceived loudness of the sound is often called "setting the deviation."

It is assumed that some facility for measuring deviation of the aural signal is available. This facility may be a meter in a modulator or a stereo encoder. It may

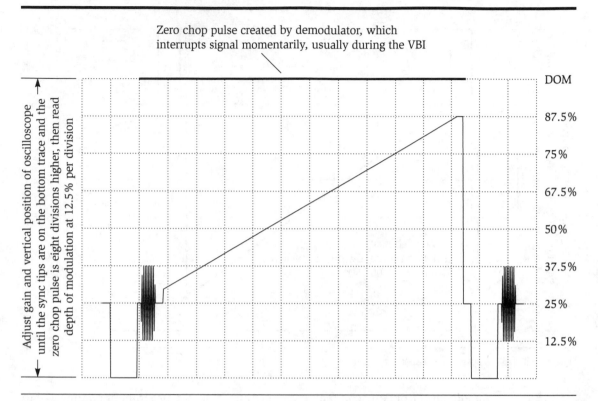

Figure 8.14 DOM display using a precision demod and oscilloscope.

also be an external demodulator with, or connected to, some sort of metering means. It may be a light that flashes at the prescribed peak deviation. This is undesirable as *the* modulation monitor because it gives so little information. However, many operators use it because of its low cost. The lights were intended to be overdeviation warning indicators only but have been used for setting deviation.

Confirming Accuracy of the Deviation Indicator

The modulation indicator must be properly calibrated. A particularly accurate method to ensure this, and one used in many labs, is the Bessel null technique. It utilizes the fact that, when frequency modulating a carrier with a sine wave, the carrier component goes to zero (all the power is contained in the sidebands) when the modulation index, M, is 2.4. This point is often called *carrier null*. The modulation index is defined as the ratio of peak deviation to modulating frequency:

$$M = \frac{\Delta f}{f_m} \tag{8.2}$$

In order to null the carrier component, we deviate the carrier by 25 kHz (the pre-scribed peak deviation for NTSC sound) using a modulating frequency that makes $M = 2.4$:

$$M = \frac{\Delta f}{f_m} = \frac{25 \text{ kHz}}{10.42 \text{ kHz}} = 2.4 \qquad (8.3)$$

The frequency of 10.42 kHz for the modulating signal is chosen to make the equation correct. When we modulate with a 10.24-kHz sine wave and find the first carrier null, we know that the deviation must be 25 kHz.

Carrier null is easily detected using a spectrum analyzer as a monitoring instrument and a sine wave generator as a source. Set the analyzer to distinguish between sidebands and the carrier. Set the oscillator frequency to 10.42 kHz. Begin increasing the amplitude of the oscillator output, starting from near zero, while watching the carrier component of the analyzer display. When the carrier component (not the sidebands) first goes to zero, the modulation index is 2.4, and the deviation should read 25 kHz on the indicating instrument. (Note that the carrier will go through zero for certain higher-modulation indices also.)

Exercise caution when making this measurement so as to properly account for preemphasis. The aural signal is preemphasized, as explained in Chapter 7, to reduce noise. At a modulating frequency of 10.42 kHz, preemphasis adds +14 dB to the low-frequency deviation. If the indicating instrument reads the nonpreemphasized signal, then it will read 14 dB low in the test described.

In countries using 50-kHz peak deviation and 50-µs preemphasis, the same procedure may be used, except that, at the indicated test point, the deviation should read −6 dB from peak (because the standard peak deviation is 50 kHz rather than 25 kHz). The correction for preemphasis, if needed, is 10.7 dB.

Often the actual modulation level used is backed off from peak by perhaps 6–7 dB if a traditional VU type meter ballistic is used so as to account for the dynamic response of such a meter. If a peak indicating instrument is used, this backoff is not needed.[4] Unfortunately, some meters that have been provided on cable modulators don't have dynamics that correspond to any particular standard. A meter that is accurate for a continuous sine wave may have significant error when measuring real audio.[5]

Routine Audio Monitoring

Figure 8.15 illustrates the setup for detecting the Bessel null using a sine wave audio oscillator and spectrum analyzer (switch down). This is the most accurate way to calibrate an audio deviation indicator. The figure also shows an audio demodulator and stereo decoder connected to an xy oscilloscope (switch up). This is a particularly convenient way to routinely monitor audio quality. Some broadcasters have such a monitoring facility permanently installed in their audio signal chain. An experienced engineer can obtain a lot of information quickly from such a display. One audio channel (the left as illustrated) is connected to the x input of

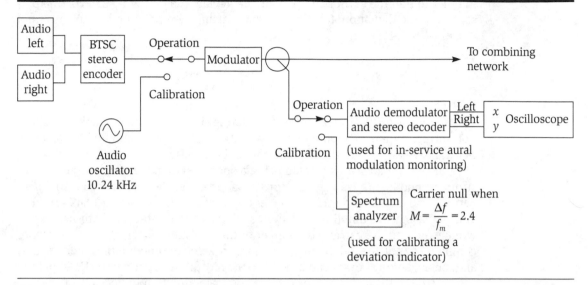

Figure 8.15 Audio monitoring options.

the oscilloscope, and the other is connected to the y input. The two channels must be calibrated for identical gain, and the position controls must be adjusted for zero crossing at the center of the screen.

Figure 8.16 illustrates a number of situations that can be detected using this pattern. Figure 8.16(a) illustrates a properly calibrated setup with a monaural signal. Since each channel of a mono signal contains the same voltage, the display should be simply a straight line from lower left to upper right.

Figure 8.16(b) illustrates a mono signal, but with the left and right channels out of phase. (An out-of-phase signal cannot be properly located by the human ear. When listened to on a mono TV, there will be no sound.) The length of the line is related to the peak deviation of the signal. Because a signal normally spends only a small fraction of the time at peak deviation, the line will become dimmer the farther you look from the center of the screen. Figure 8.16(b) also illustrates conventional quadrant designations: the majority of a proper in-phase signal will reside in quadrants I and III. An out-of-phase signal will largely reside in quadrants II and IV.

Figures 8.16(c) and (d) illustrate a stereo signal. The loci of all points occupied is "smeared" because the same signal is not being applied to each channel. The more the signal deviates from the mono axis of Figure 8.16(a), the greater the channel separation. (Compare Figure 8.16(c) with 8.16(e), which shows less separation.) Low or high separation in program material is often a fact, not necessarily an error condition. Once an engineer becomes proficient at looking at these waveforms, he or she knows whether the separation is normal for a particular type of program material.

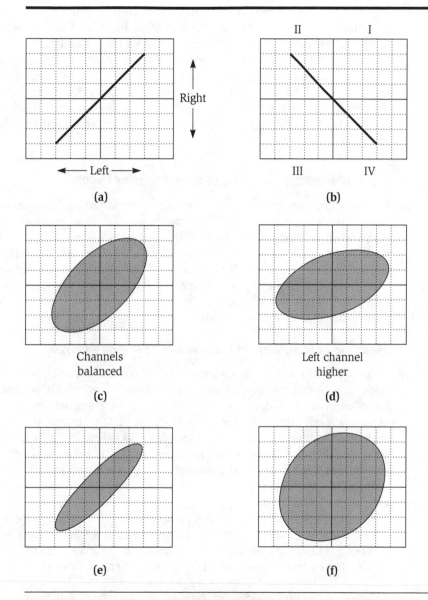

Figure 8.16 Oscilloscope patterns for stereo audio. (a) Mono, in phase. (b) Mono, out of phase. (c) Stereo, in phase, higher separation. (d) Same as (c), except left channel high. (e) Stereo, in phase, lower separation. (f) Synthesized stereo may show more occupancy of out-of-phase quadrants.

If the balance is not correct, then the waveform will be "tilted." The tilt will be down if the left-channel level is higher, as illustrated in Figure 8.16(d). If the right-channel level is higher, the waveform will be tilted up.

Notice that, though the majority of a stereo trace appears in quadrants I and III, some points appear in quadrants II and IV. This is to be expected with a stereo signal. Figure 8.16(f) illustrates a synthesized signal. Depending on the method of synthesis, more power may occupy out-of-phase quadrants II and IV. Some programmers use synthesizers when they are carrying programming that originates in mono.

8.5.3 Measuring Picture and Sound Carrier Levels

No measurements are as basic to setting up a headend as are measuring the picture carrier level and the sound carrier level. This may be done with a special-purpose signal level meter, or it may be done with a spectrum analyzer. The spectrum analyzer yields more information for the operator who understands it, but it is more expensive, more difficult to use, and capable of providing misleading results when used improperly. The signal level meter makes the reading without requiring setup, but it can be more difficult to use if many channels must be compared, and it gives less information.

Measuring Picture Carrier Level with a Spectrum Analyzer

Several points must be considered. The IF (resolution) bandwidth must be set wide enough to admit enough of the sync tip sidebands to allow the trace to reach its maximum. The video filter must also be set wide enough to allow the detected signal to reach full amplitude. As a practical matter, this means that both the IF and video bandwidths must be set to at least 300 kHz. The IF bandwidth must not be set so high that adjacent channel energy affects the measurement. Also, the sweep speed must be set low enough that you will see a number of horizontal lines on the waveform. With the correct setup, a spectrum analyzer can yield excellent results, though.

Scrambled signal level cannot be measured with any convenient setup. This is explained in Chapter 18, which makes the case for not trying to measure the level of a scrambled signal. Rather, to measure the level, turn off the scrambling, and make the level measurement normally. Then turn the scrambling back on.

Measuring the Picture–Sound Ratio with a Spectrum Analyzer

The ratio of the picture and sound carrier amplitudes must also be set. This may be done with a spectrum analyzer set to measure picture carrier level. Normally, in cable work, the picture–sound ratio is set to 15 dB. The off-air ratio is different, but cable uses a lower sound carrier to minimize adjacent channel interference.

The question often is asked about how to set the picture–sound carrier ratio when the signal is scrambled and pulses are used on the sound carrier. The pulses make the sound carrier level go high for a few microseconds to communicate in-

formation to the descramblers. Again, the proper way to set level is to turn off the scrambling, set the picture–sound ratio, then turn the scrambler back on. The scrambling manufacturer must ensure that the correct levels are maintained when the scrambling is turned on. Generally, the signal levels are set such that the picture–sound ratio is maintained at about 15 dB between pulses and goes to 9 or 10 dB at the peak of the pulses.

8.6 Summary

This chapter has dealt with a number of somewhat disconnected topics generally related to headend operation. Three general topics were covered: the standard band plan for cable television in the United States, management of outgoing optical and electrical signals in the headend, and monitoring of the quality of signals leaving the headend.

The television receiver and cable television industries have agreed on a band plan that defines the frequency versus channel number for television channels. The plan has been defined for three sets of frequency allocations: the standard plan, used in most instances today, and the HRC and IRC plans, chosen to reduce the visibility of distortion products. In the VHF band, the standard plan is identical to the off-air FCC channel assignments. In the old midband and superband regions, the plan follows traditional assignments with numerical channel designations. In the UHF band, the cable assignments are 2 MHz below the FCC plan because the FCC plan does not move from channel 13 (210–216 MHz) to channel 14 (470–476 MHz) in 6-MHz increments.

Signals in the headend must be managed in such a way that it is possible to carry different signals in different nodes but on the same frequency. This means that isolation must be maintained between unique signals bound to or from different nodes. On the other hand, certain signals are common to all nodes, and these have to be handled in common.

It is important to handle headend signals so that quality is maintained. Because the quality of signals exiting headends must be monitored, we have presented some fundamental ways to do so. Books have been written to cover the topic in detail.

We now move out of the headend and into the distribution plant. The next several chapters deal with the technology of distributing signals outside the headend.

Endnotes

1. J. O. Farmer, The Joint EIA/NCTA Band Plan for Cable Television, *IEEE Transactions on Consumer Electronics*, August 1994, p. 503ff.

2. Code of Federal Regulations, 47CFR sec. 68.610. U.S. Government publication, available at Government Printing Offices in major cities.

3. AT&T Corp., *LGX Broadband Fiber Management System Application Guide,* 1995.

4. IEEE Std 152-1991, *IEEE Standard for Audio Program Level Measurement* is one of many suggestions for measuring audio loudness and peak deviation. It includes an extensive bibliography on measuring audio, a controversial topic. IEEE, Piscataway, NJ.

5. J. O. Farmer, Setting and Maintaining Modulation Levels. *CED,* March 1987, reprinted in *BCT/E Certification Program Reference Bibliography Reprint Manual,* available from the Society of Cable Telecommunications Engineers, Exton, PA.

Part Four

Broadband Distribution Systems

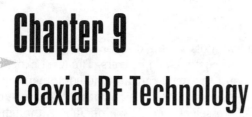

Chapter 9
Coaxial RF Technology

9.1 Introduction

Cable television's technical roots are the distribution of analog television signals to the antenna terminals of customers' receivers. The least expensive way to accomplish that was to avoid the need for in-home equipment by carrying each video stream on a different, standard television channel so that subscribers could use their existing TV sets to select and view programs. Although cable systems have evolved since then into sophisticated multiservice networks using combinations of linear fiber optics and coaxial cable for transmission, the essential characteristics have remained unchanged. In particular, the last portion of the network is still a linear, broadband coaxial network that simultaneously carries many modulated RF signals, each occupying a different band within the spectrum—frequency division multiplexing (FDM)—as discussed in several previous chapters. In modern systems, many of the modulating signals may be digital rather than analog; however, the network must still be linear to avoid generation of unwanted additional signals due to mixing among the desired carriers. This is discussed in detail in Section 9.3.3.

This chapter will treat coaxial network technology in detail, including cable, amplifiers, passive components, and powering systems. Basic linear network concepts will be introduced that will also apply to the linear fiber-optic links whose unique characteristics are discussed in Chapter 11 and to the microwave links covered in Chapter 12. Chapter 10 will discuss coaxial design practices and cascaded performance.

9.2 Coaxial Cable

Coaxial cable is not the only option for transmitting broadband RF signals. Indeed, many early systems were built using open, parallel-wire balanced transmission lines, and a few even used an ingenious single-wire cable known as G-line, which had only a center conductor and dielectric. Coaxial cable, however, offers the advantages of a high degree of shielding, coupled with relatively low cost and moderate signal loss. For that reason, most modern RF systems use coaxial cable as at least one of the physical media through which signals are transported.

9.2.1 Definition

Coaxial cable is constructed with a center conductor surrounded by a dielectric of circular cross section and by an outer conductor (shield), also of circular cross section. Signals within the normal operating bandwidth of coaxial cable have a field configuration known as transverse electric and magnetic (TEM). In the TEM mode, the electric field lines go radially between the center and outer conductor and are of uniform strength around a cross section of the cable, whereas the magnetic field lines are circular and perpendicular to the length of the cable (see Figure 9.1). In a cable with a continuous, perfectly conducting shield, no electric or

Figure 9.1 Coaxial cable basics.

magnetic fields extend beyond the outer conductor so that no signals can leak out of or into the cable.

9.2.2 Characteristic Impedance

Coaxial cables have a property known as surge impedance or, more commonly, characteristic impedance, which is related to the capacitance and inductance, per unit length, of the cable. The characteristic impedance is most easily thought of in terms of the effect on signals being transported: if a cable is connected to an ideal pure resistor whose value is equal to its characteristic impedance, a signal transmitted toward the resistor will be entirely absorbed by the resistor and converted to heat. In other words, no energy will be reflected up the cable.

The characteristic impedance, Z_0 (in ohms), is a function of the relative diameters of the center conductor and the inner surface of the outer conductor and of the dielectric constant of the dielectric:

$$Z_0 = \frac{138}{\sqrt{\varepsilon}} \log\left(\frac{D}{d}\right) \qquad (9.1)$$

where

D = the inner diameter of the shield

d = the outer diameter of the center conductor

ε = the dielectric constant

9.2.3 Attenuation as a Function of Frequency

Signal loss (attenuation) through coaxial cable can occur through any of four principal means:

- Radiation out of the cable due to imperfect shielding
- Resistive losses in the cable conductors
- Signal absorption in the dielectric of the cable
- Signal reflection due to mismatches between the cable and terminations or along the cable due to nonuniform impedance

Even when cables have perfect shields, exact impedance matches, and uniform construction, imperfect dielectrics and resistive conductors will cause loss. The general equation for this residual cable loss is

$$\alpha = 4.344\left(\frac{R}{Z_0}\right) + 2.774 F_p \sqrt{\varepsilon}\, f \tag{9.2}$$

where

α = the attenuation of the cable in dB/100 feet*

R = the effective ohmic resistance of the sum of the center and outer conductors per 100 feet of cable length at f

F_p = the power factor of the dielectric used

f = the frequency in MHz

The currents in the conductors are proportional to the strength of the magnetic fields at the conductor surface. If the conductors had no resistance, the RF currents would travel only on the surface. In real conductors, however, the current extends into the conductor, decreasing exponentially with depth. This property is known as *skin effect*, and the distance from the surface to where the current has decreased to $1/e$ (36.8%) of the surface amount is known as the *skin depth* at that frequency. It is related to frequency by the formula

* In accordance with common North American practice, all attenuation values will be given in decibels per 100 feet of length. This has the advantage that signal loss becomes a linear function of length and of simple conversion to metric units of length.

$$\delta = \frac{1}{\sqrt{f}}\, 2.60\, \sqrt{\frac{\rho}{\rho_C}} \qquad (9.3)$$

where

δ = the skin depth in mils (thousands of an inch)

ρ/ρ_C = the resistivity of the conductor relative to copper

f = the frequency in MHz[1]

The formula is valid only for nonferrous materials such as aluminum or copper.

Over a typical downstream frequency range of 54–750 MHz, the skin depth in copper will vary from 0.00035 to 0.00009 inch and will increase to 0.0012 inch at 5 MHz (the lower end of the upstream frequency range).

The effective resistance of the conductor is the same as if it were a tube of material whose thickness is equal to the skin depth with the current distributed equally throughout its volume. The resistance of this tube will be

$$R = 0.0996\, \frac{\sqrt{f}}{d}\, \sqrt{\frac{\rho}{\rho_C}} \qquad (9.4)$$

where

R = the resistance per 100 feet of the conductor

ρ/ρ_C = the resistivity of the conductor relative to copper

f = the frequency in MHz

d = the conductor diameter in inches

Taking into account the resistance of both center and outer conductors, Equation (9.2) can be rewritten as

$$\alpha = \frac{0.433}{Z_0}\left(\frac{\sqrt{\rho_d/\rho_c}}{d} + \frac{\sqrt{\rho_D/\rho_c}}{D}\right)\sqrt{f} + 2.774(F_p\sqrt{\varepsilon})f \qquad (9.5)$$

where

ρ_d/ρ_c = the resistivity of the center conductor material relative to copper

ρ_D/ρ_c = the resistivity of the shield material relative to copper

This has one term that increases as the square root of frequency and one that increases linearly with frequency. For most cable constructions (discussed later), the conductor loss will dominate the dielectric loss so that the overall cable attenuation will increase approximately as the square root of frequency. Figure 9.2 shows the specified variation in loss for several cables in comparison to an ideal

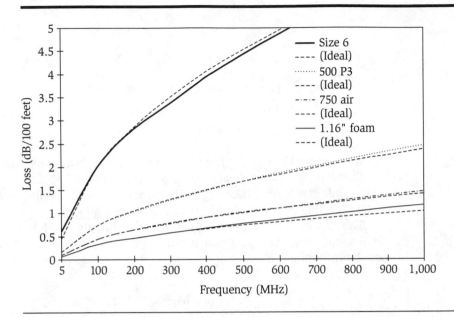

Figure 9.2 Cable loss: specified versus ideal.

square-root relationship (with the curves matched at 100 MHz). Note that the agreement is excellent for cables ranging from drop sizes to the largest trunk cables, indicating the relatively small contribution from dielectric losses.

A common cable construction, known as P3, utilizes a solid aluminum outer conductor, a foamed dielectric ($\sqrt{\varepsilon} = 1.15$), and a copper-coated aluminum center conductor. It is available in a number of nominal sizes (referenced to the *outer* diameter of the aluminum shield in inches). For this family of cables, the attenuation can be approximated using

$$\alpha = \left[\frac{0.036}{D_{\text{nom}}}\right]\sqrt{f} + 0.0002f \tag{9.6}$$

where

$$\alpha = \text{the attenuation in dB/100 feet}$$
$$D_{\text{nom}} = \text{the nominal shield outer diameter in inches}$$
$$f = \text{the frequency in MHz}$$

Other cable types using thinner shields and/or air dielectrics will have slightly lower losses for the same outer diameter, whereas solid dielectric cables will have higher losses.

9.2.4 Attenuation as a Function of Temperature

As Equation (9.2) shows, the attenuation of a coaxial cable, in decibels, is a linear function of its conductor resistance, provided that conductor losses are much larger than dielectric losses. As we will discuss in Section 9.2.10, the most common cable construction uses an aluminum shield and copper-coated aluminum center conductor. Copper has a resistivity at 68°F of about 0.68×10^{-6} ohm-inches, whereas aluminum has a resistivity of about 1.03×10^{-6} ohm-inches, and each has a temperature coefficient of resistivity of 0.22%/°F. Since conductor resistance varies as the square root of resistivity, the resistance of both conductors, and therefore the attenuation, would be expected to change about 0.11%/°F. In fact, the attenuation of commercial coaxial cables with hybrid aluminum–copper construction is typically specified to increase by 0.1%/°F.

9.2.5 Attenuation as a Function of Characteristic Impedance

For a given outer diameter (the primary determinant of the amount of materials used and, therefore, cost and weight), different impedance cables will be optimized for different characteristics. If the dielectric loss is assumed to be of only secondary importance and the usual construction is assumed (copper-plated center conductor with aluminum shield), then Equations (9.1) and (9.5) can be combined to find cable attenuation as a function of characteristic impedance:

$$\alpha = \frac{0.433\sqrt{f}}{D} \left[\frac{1.23 + 10^{\left(\frac{Z_0 \sqrt{\varepsilon}}{138} \right)}}{Z_0} \right] \tag{9.7}$$

As Figure 9.3 shows, the loss minimum is at about 80 ohms, for air dielectric (dielectric constant = 1.0) and decreases as the dielectric constant increases. Cables with air dielectric also have lower overall loss since the center conductors are larger, and therefore have less resistance, for the same impedance. Although the choice of impedances near the minimum of the curve was not important, 75 ohms may have been chosen because it is also close to the feedpoint impedance of a half-wave dipole antenna. In order to minimize the need for repeaters, wide area distribution systems universally use 75-ohm cables. They are also used for local interconnection of baseband video signals within headends and broadcast facilities in order to minimize the differential loss across the video baseband spectrum.

Although the losses are higher, 50-ohm cables are generally used by the broadcast and radio communications industries because the power-handling capability is higher, the cables are less fragile, and they are a close match to the feedpoint impedance of a vertical quarter-wave antenna.

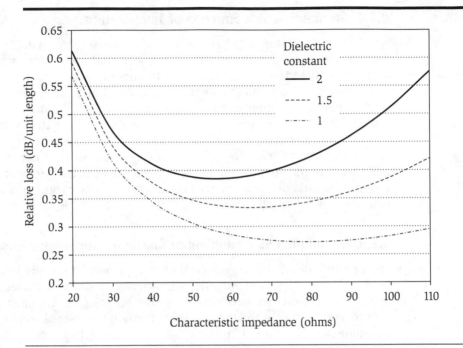

Figure 9.3 Coaxial cable loss versus characteristic impedance for various values of dielectric constant.

9.2.6 Wavelength

Along the length of the cable, at any one instant in time, the electric fields vary sinusoidally in strength. At some point, the center conductor will be at its most positive with respect to the shield. Moving along the cable, the voltage will decrease to zero, then become negative, then back to zero, and finally be at its positive maximum again. The distance between the maximum points of one polarity is known as the wavelength at the frequency being transmitted and is equal to the distance the signal transverses during one period of its frequency.

In free space, signals travel at the rate of 3×10^8 meters per second or, in more convenient terms, about 984 feet per microsecond. Signals in cable, however, travel more slowly due to the higher dielectric constant of the dielectric material. The ratio between velocity in cable and the free-space velocity is the *relative propagation velocity* V_P and varies from about 0.85 to 0.95 for common cable types. V_P is related to the dielectric constant by

$$V_P = \frac{1}{\sqrt{\varepsilon}} \tag{9.8}$$

The wavelength of a signal in a coaxial cable is thus

$$\lambda(\text{feet}) = \frac{984V_p}{f(\text{MHz})} = \frac{984}{f\sqrt{\varepsilon}} \qquad (9.9)$$

9.2.7 Theoretical Size Limitation

If the size of a coaxial cable is sufficiently large, it will support modes other than TEM. In particular, if the mean diameter exceeds one wavelength in the cable, it will support a mode in which the electric and magnetic fields are not uniform around the cable. This *cutoff wavelength* can be expressed as[2]

$$\lambda_c = \pi(D+d) \qquad (9.10)$$

where

D = the inner diameter of the shield

d = the outer diameter of the center conductor

Non-TEM mode generation is a problem since higher-order modes may not propagate at the same velocity and because coaxial components may not react the same to the higher-order mode. Standard practice is to limit cable sizes to those that do not support modes higher than TEM at the maximum operating frequency. The maximum size cable that will support a given frequency in TEM mode only can be found by substituting Equation (9.9) for λ in Equation (9.10), then solving (9.1) for d, substituting in Equation (9.10), and solving the resultant expression for D:

$$D_{\text{max}} = \frac{984}{\pi f \sqrt{\varepsilon}\left[1 + 10^{-\left(\frac{Z_0\sqrt{\varepsilon}}{138}\right)}\right]} \qquad (9.11)$$

Alternatively, we can solve for the maximum frequency that can be transmitted in only TEM mode through a given cable:

$$f_{\text{max}} = \frac{984}{\pi D \sqrt{\varepsilon}\left[1 + 10^{-\left(\frac{Z_0\sqrt{\varepsilon}}{138}\right)}\right]} \qquad (9.12)$$

For example, 75-ohm cables with foamed dielectrics (ε about 1.3) will support 1-GHz signals at up to 0.222-foot shield inner diameter, or about 2.6 inches. Since

this is more than twice the largest cable used by network operators today, higher-order mode suppression is not an issue with current operating bandwidths.

9.2.8 Precision of Match: Structural Return Loss

One measure of cable quality is how close it adheres to its nominal impedance. This has two aspects: the precision with which its *average* characteristic impedance matches the ideal value, and the *variations* of impedance along the length of a cable. As an alternative to attempting to measure the physical structure of the cable along its length, the most frequent measure of quality used is the percentage of incident power from a source that is reflected at the input of a cable being tested, when the cable's output is connected to a precision termination. If the source is calibrated using a precision, resistive terminator, then the result of the measurement is return loss. For cables, however, the variation of the reflection is often of more interest than the absolute match. In order to judge the variation, a variable bridge is used to match, as precisely as possible, the average surge impedance of the cable being tested. Then the reflection is measured over the full frequency range of desired operation, and the result is known as structural return loss (SRL). Its numeric value is the smallest measured ratio of incident to reflected power as the frequency is varied, expressed in decibels. Recommended worst-case SRL is 26–30 dB for trunk and feeder cable, whereas 20 dB is considered adequate for drop cable. The absolute impedance should be within ±2–3 ohms for trunk and distribution cables and ±5 ohms for drop cables.[3]

SRL is a compromise measurement, as the reflected power is a measure, at each frequency, of the composite effect of discontinuities at all distances. Because of the attenuation of the cable, signals traveling toward more distant discontinuities will be reduced in amplitude, and the reflections from those discontinuities will be further attenuated so that imperfections farther from the end of the cable being measured will have less effect than those closer. At higher frequencies, this effect will be increased. On the other hand, most cable imperfections cause a greater reflection at higher frequencies.

Where multiple reflections are spaced along a transmission line, the effects can add, cancel, or anything in between, depending on the length of line between them and the frequency. (Chapter 13 includes a more complete discussion of multiple reflections in coaxial transmission systems.) Absent line losses, small, identical discontinuities will approximately cancel when spaced odd multiples of a quarter wavelength, whereas they will reinforce when spaced multiples of one-half wavelength. This is especially important when considering multiple, equally spaced discontinuities arising, for instance, from an imperfection in a roller over which a cable passes during the manufacturing process. Each revolution of the roller produces an identical fault, spaced along the cable by a distance equal to the circumference of the roller. Testing such a cable over the full frequency range will show maximums and minimums of return loss as the effects of the discontinuities interact. This is why it is important to test cable over its full expected

frequency range with sufficient frequency resolution to detect sharply resonant discontinuities.

Though structural return loss measurements accurately assess the composite reflection from a cable, it is a poor method of determining the location of major defects. If a cable includes only a single major defect, its location can be estimated from the change in test frequency between successive maxima or minima of return loss; however, in most cases, cable loss plus a multiplicity of reflections will prevent accurate determination.

9.2.9 Precision of Match: Time Domain Reflectometry (TDR)

An alternative measurement method, also a compromise, is used to more accurately determine discontinuity location but averages the magnitude of the discontinuity as a function of frequency. In this test, a pulse is transmitted down the cable, and the distance is determined from the time required for the energy reflected from a discontinuity to return to the transmitter. The magnitude of the discontinuity is related to the amplitude of the returned signal. Secondary information about the frequency "signature" of the fault can be determined by comparing the shape of the returning pulse with the original.

Although TDR testing of short, low-loss cables is an accurate process, there are many compromises in testing of the relatively long cables used in typical cable system construction. With very short cables, the classic pulse waveform is a very fast rise time step function. A Fourier transform of this voltage waveform reveals that its frequency content is inversely proportional to frequency, and therefore, the power spectrum falls inversely as the square of frequency (see Figure 9.4(a)). When such a pulse is used to test long cables, it suffers from two problems: (1) most of the energy is concentrated at low frequencies, whereas most discontinuities primarily affect higher frequencies, and (2) the relatively higher cable loss at high frequencies further masks the reflections.

For that reason, long cables are most frequently tested using a narrow "impulse" (see Figure 9.4(b)). This waveform has an energy that extends more uniformly to a frequency equal to approximately $1/P_W$, where P_W is the pulse width (the exact shape of the curve in the frequency domain depends on the shape of the voltage pulse). The trade-off is that the distance resolution is limited to approximately cP_WV_P, or about 1 foot per nanosecond of pulse width for typical foam dielectric cables. By increasing P_W, it is possible to increase the total energy (and therefore sensitivity) at the expense of frequency and distance resolution. As with step function pulses, impulse function testing suffers in accuracy because of the variation of cable loss with frequency. Because of the compromises, TDR testing of CATV-type cables is primarily limited to field operations such as measuring the cable remaining on a reel or locating an underground cut so that the cable can be repaired.

An instrument that is able to simultaneously resolve both frequency and time (distance) domain information is a network analyzer with gated TDR and Fourier

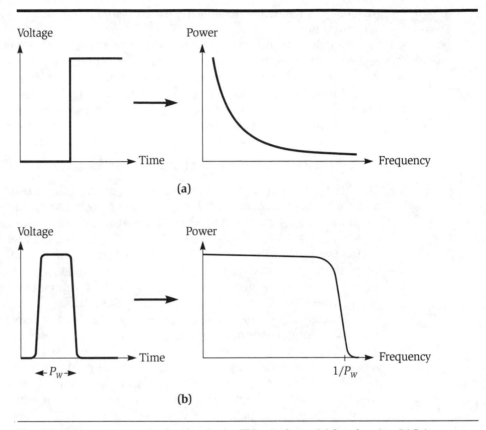

Figure 9.4 Frequency spectra of step and pulse TDR waveforms. (a) Step function. (b) Pulse.

transform capability. This instrument measures reflection coefficient, but "gates" the measurement so as to measure only reflections from a narrow range of distances at a time. A Fourier transform is then used to convert the waveform of the reflected signal into its frequency domain, resulting in a reasonably accurate measurement of the reflection coefficient of an individual discontinuity, even in the presence of other discontinuities at other distances. It is still, of course, affected by the cable loss and loss variation with frequency. As of the writing of this book, this technology has not been extensively used in field work in cable systems.

9.2.10 Practical Factors in Cable Selection

From an electrical standpoint, the ideal cable would have conductors with the highest possible conductivity, the largest diameter consistent with higher-order mode suppression, and a vacuum separating the conductors. Practical cables, however, are a compromise among several characteristics:

- Manufacturing cost (which will vary somewhere between being a linear function of diameter and as the square of the diameter)
- Weight (which affects the cost of the supporting structure for overhead lines)
- Diameter (which affects required conduit size for underground lines, visibility of overhead lines, and cost of connectors)
- Loss (which limits the distance between repeater amplifiers)
- Mechanical expansion coefficient (which affects the amount of sag that must be used with overhead lines)
- Connector cost (related to cable size)
- Shielding (although solid outer conductors provide nearly perfect isolation, braided shields may provide adequate shielding for some applications)
- Environmental protection (mechanical protection layers, compounds used to retard water ingress, integral support members, and so on)
- Handling characteristics (bending radius, reforming ability, crush resistance, pulling force resistance, and so on)
- Electrical resistance and current-carrying capacity at low frequencies

In oversimplified terms, the selection of optimum cable size for a distribution network design consists of comparing the cost of cable (and associated components) per decibel of loss with the cost per decibel of amplification to overcome that loss and choosing the least expensive combination. Of course, practical system design includes many additional factors that may affect the ultimate decision.

Trunk and Feeder Cable

Design optimization has resulted in several typical cable designs that are widely used. The most common distribution cable design in most of the world utilizes a solid aluminum shield, foamed polyethylene dielectric, and a copper-clad aluminum center conductor. The use of aluminum offers the best combination of cost, handling characteristics, weight, and strength, whereas the foamed dielectric offers a good compromise among loss, moisture ingress protection, and mechanical strength for the center conductor. Because of its rigidity, compared with cables that utilize braided wire shields, trunk and feeder cables using this basic structure are sometimes known as *hard-line* cable. The use of an aluminum core for the center conductor provides a match for the expansion coefficient of the shield, whereas the copper cladding provides a lower resistance to RF. Since the conducting area of the center conductor is much smaller than the shield, the total resistive losses are dominated by it so that copper cladding the center conductor is cost efficient though coating the inner side of the shield would not be.

The need to provide low resistance to the multiplexed power, as well as low RF loss, has led to the use of cables with outer shield diameters ranging from 0.412 to 1.125 inches, with larger cables being used where distance of coverage is paramount (trunk or express feeders), and smaller cables being used for shorter

Table 9.1 Common drop cable designations and sizes.

Size Designation	59	6	7	11
Approximate diameter over jacket in inches	0.24	0.27	0.34	0.40

distances and cases where a substantial amount of the loss is due to tap losses rather than cable losses (tapped feeder cables). This family of cables is generically labeled by the outer shield diameter in thousands of an inch; for example, "500 cable" refers to a hard-line cable with an outer shield diameter of 0.500 inch. Common diameters include 0.412, 0.500, 0.625, 0.750, 0.825, and 1.000 inch though others are also used.

Versions of these same cables are available with solid copper center conductors, for use where 60-Hz power losses must be reduced. Protection options include combinations of none (bare aluminum), polyethylene jackets of various designs and degrees of sun and fire resistance, flooding compounds (a substance inserted between the shield and jacket to retard the ingress of water), and additional steel shields for crush resistance. Since the added environmental protection layers can considerably increase the cost and weight, builders generally utilize several designs in a single system, optimized by application.

A variation of this design utilizes air dielectric with periodic center conductor supports. This design has slightly smaller diameter for the same loss per foot because of the lower dielectric loss and lower ratio of shield to center conductor diameter. Special design features are required to prevent water, which may enter the cable owing to a fault, from traveling down the center conductor/shield space. Other features are required to maintain the required concentricity and crush or bending resistance.

Though hybrid aluminum–copper cables are the most common, some network operators prefer solid copper for center conductors and, sometimes, shields.

Drop Cable

In a typical cable system, well over half the total cable footage is taken up by drop cables. For that reason, cost is an important parameter for that application, as well as weight and appearance. In North America, four common sizes of drop cable have evolved, as shown in Table 9.1. The approximate loss for each size, as a function of frequency, is shown in Figure 9.5.[4]

Drop cables are constructed using a copper-plated steel center conductor, foamed dielectric, a shield made up of at least one layer of an aluminum foil and one layer of an aluminum braid, and a plastic jacket. Additional layers of foil and braid are available as options for greater shielding effectiveness. Although size 59 cable was the historical choice of the cable television industry, size 6 cable is most frequently used in modern systems, due to the use of increasingly higher frequencies. The larger sizes are generally reserved for occasional longer drop cables.

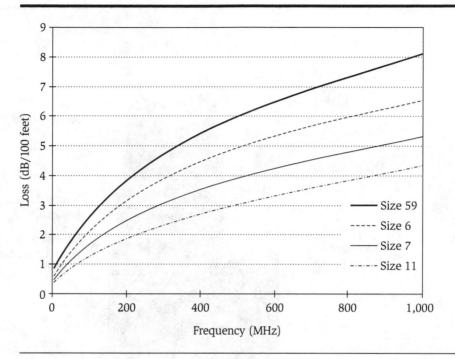

Figure 9.5 Drop cable loss versus frequency.

The jacket material for general use is usually PVC or polyethylene. Cables used for overhead drops often also include a separate steel strength member (the *messenger*) enclosed in the same plastic sheath structure. See Figure 9.6 for a composite drop cable structure that includes a messenger.

An array of drop cable variations are available for special purposes, including those with *flooding compounds* to retard water ingress in case the plastic coating is damaged, multiple cables within a common jacket, those with varying degrees of fire resistance for use inside buildings, silver-plated center conductors, copper braids, solid dielectric cables, and so on.

As with distribution cables, North American practice has been adopted by some countries, whereas others (for instance, Germany) have chosen to use semirigid drop cables with solid copper shields and center conductors for their underground drops.

9.2.11 Shielding Effectiveness in Drop Cables

Although distribution cables constructed with solid aluminum tube shields offer nearly perfect signal isolation from external fields, drop cables depend on the overlap in the aluminum wrap and the interaction between the aluminum braid and wrap for shielding integrity. It is a compromise that is justified on the basis of

Figure 9.6 Typical drop cable construction.

cost and cable-handling characteristics. The cable industry has long struggled to find nonambiguous testing methodology for evaluating shielding integrity. Although these are beyond the scope of this book, you may wish to refer to industry standard test procedures developed by the Society of Cable Telecommunications Engineers (SCTE) for this purpose.[5]

9.3 Amplifiers

In order to overcome the loss of the distribution cables and to provide power to drive end terminals, signal boosting amplifiers are utilized throughout the coaxial distribution system. These must be designed to add as little signal degradation as

possible, particularly noise and distortion, consistent with providing the required gain and total power output.

9.3.1 Broadband Random Noise

Because of the random motion of electrons in conductors, all electronic systems generate an irreducible amount of noise power. This noise is a function of the absolute temperature and the bandwidth in which the noise is measured. As discussed in Chapter 6, the minimum thermal noise power (the *noise floor*) can be calculated using

$$n_p = kTB \tag{9.13}$$

where

n_p = the noise power in watts*

k = Boltzmann's constant (1.374×10^{-23} joules/°K)

T = the absolute temperature in °K

B = the bandwidth of the measurement in Hz

Signal power in cable television systems is commonly expressed in decibels relative to 1 mV† across 75 ohms (dBmV), so the thermal noise power at 62°F is approximately

$$N_p(\text{dBmV}) = -125.2 + 10 \log B \tag{9.14}$$

Finally, although the allocated channel bandwidth for analog NTSC signals is 6 MHz, the effective noise bandwidth of receivers is less. For a variety of reasons, a 4-MHz bandwidth is used as the standard for noise measurements of analog NTSC signals and has been codified in the FCC's rules.[6] At that bandwidth, Equation (9.14) becomes

$$N_p = -59.2 \text{ dBmV} \tag{9.15}$$

This value will vary by a few tenths of a decibel across the temperature range normally of interest to cable operators.

* To avoid confusion, lowercase symbols will be used in this discussion to refer to scaler quantities (for example, watts), and uppercase symbols will refer to logarithmic quantities (for example, dBmV).

† In some countries, the reference is 1 μV rather than 1 mV. Levels in dBμV are therefore numerically greater by 60 than levels in dBmV. In either case, however, the reference is power, not voltage. Finally, for the convenience of those accustomed to working in systems referenced to 1 mW (dBm), 0 dBmV = −48.75 dBm. For consistency, this book uses dBmV as the units for RF power unless otherwise specified.

9.3.2 Amplifier Noise

Amplifiers generate added noise at various points in their circuitry. For convenience, however, the added noise is treated as if it were coming from an independent generator and summed into the input port (though it cannot be measured there, of course). The ratio of total effective input noise power to the thermal noise floor, expressed in decibels, is known as the *noise figure* of the amplifier. Thus, an amplifier with a noise figure of F_A dB will have a total equivalent input noise power of N_A, where

$$N_A(\text{dBmV}) = N_p(\text{dBmV}) + F_A(\text{dB}) \tag{9.16}$$

And the output noise power will be the input, increased by the gain, G, of the amplifier in decibels:

$$N_{\text{out}}(\text{dBmV}) = N_p + F_A + G \tag{9.17}$$

Similarly, a desired (noise-free) input signal, C_i, will be amplified by the same amount so that the desired signal output level $C_{\text{out}} = C_i(\text{dBmV}) + G(\text{dB})$, and the carrier-to-noise ratio will be*

$$C/N(\text{dB}) = (C_i + G) - (N_p - F_A + G)$$
$$= C_i - N_p - F_A \tag{9.18}$$

When evaluated relative to the 4-MHz bandwidth commonly used for NTSC C/N measurements, this becomes

$$C/N(\text{dB}) = C_i(\text{dBmV}) + 59.2 - F(\text{dB}) \tag{9.19}$$

Note that, in an FDM system, C/N is evaluated on each signal independently, with the level of individual carriers compared with the noise in the bandwidth that affects that signal. As will be seen, the C/N may be considerably different for different carriers in the spectrum. (The design of systems with cascaded amplifiers will be treated in Chapter 10.)

9.3.3 Distortion

Amplifiers are not perfectly linear. That is, if the instantaneous output voltage is plotted against the instantaneous input voltage, the graph will not be a perfectly straight line. There are two causes of this nonlinearity: the inevitable minor small-signal nonlinearities of the semiconductor devices used and the compres-

* As a reminder, the term C/N (carrier-to-noise ratio) is used in this book for RF modulated signals, whereas S/N (signal-to-noise ratio) is used for measurements of the noise level on baseband (unmodulated) signals. This is consistent with both regulatory terminology and common industry practice.

sion that takes place as the amplifier nears its saturation voltage. Though the small variations define a distortion "floor," most of the distortion is due to saturation effects. They can take any of three related forms: even-order distortion, odd-order distortion, and cross modulation.

Even-Order Distortion (Composite Second Order—CSO)

A perfectly linear amplifier has a transfer function that is a constant; that is, the output voltage is a linear function of the input voltage. If the transfer function is not linear, then, in general, it can be expressed as a power series; that is,

$$e_O = Ae_i + Be_i^2 + Ce_i^3 + De_i^4 + \cdots \qquad (9.20)$$

where

e_O = the output voltage
e_i = the input voltage
A, B, C, D, \ldots = the gains at the various powers of the input voltage[7]

Those terms with even-numbered powers are called even-order distortion, and those with odd-numbered powers are called odd-order distortion. They act in very different ways in cable systems utilizing the most common channelization plans.

If a single sine wave signal $[e_i = E_i \sin(\omega t)]$ is transported through an amplifier with second- and fourth-order distortion, the output voltage will be

$$e_O = A[E_i \sin(\omega t)] + B[E_i \sin(\omega t)]^2 + D[E_i \sin(\omega t)]^4 \qquad (9.21)$$

where

ω = the angular frequency
A = the linear component of gain
B = the amplitude of the squared nonlinear component
D = the amplitude of the fourth power component
E_i = the peak input voltge

Using standard trigonometric identities, this is equivalent to

$$e_O = \frac{BE_i^2}{2} + \frac{3DE_i^4}{8} + AE_i \sin(\omega t) - \frac{(BE_i^2 + DE_i^4)}{2} \cos(2\omega t) + \frac{DE_i^4}{8} \cos(4\omega t) \qquad (9.22)$$

Analyzing this expression, we find that

- The first two terms of this expression represent a change in the average dc voltage that may correspond to a change in the power dissipation of the active device, but is not transmitted into the ac-coupled transmission system.

- The third term represents the desired output signal. Note that this amplitude does not change as the result of the even-order distortion.

- The fourth term represents an additional signal at twice the frequency of the input signal and in phase with it $\omega = \pi/2$ radians. In real amplifiers, B is substantially greater than D. Thus, for small values of E_i, the amplitude of this factor will increase as the square of the input voltage. At some input level, the fourth power term will begin to dominate, and the second harmonic level will increase more quickly. Since the level of the fundamental increases linearly with input signal level, the ratio of the fundamental to the second harmonic, at low levels, is a linear function of input level. Expressed in logarithmic terms, the "second-order ratio" degrades by 1 dB for every decibel increase in operating level so long as the amplifier is "well behaved" (meaning the fourth power terms are insignificant relative to the second power terms).

- The final term is the fourth harmonic of the fundamental. Like the fourth power term of the second harmonic, this product is generally small at normal operating levels and at all levels is at least 12 dB (one-fourth the voltage) below the second harmonic term. Note that this term is also in phase with the fundamental at $\omega = \pi/2$.

If we plot the fundamental and first two even harmonics on a time scale (Figure 9.7), we see an interesting relationship: the distortion-caused products are *antisymmetrical* with respect to the fundamental; that is, whatever voltage they have at the positive peak of the fundamental, they have exactly the opposite during the negative peak. The usual case of the combined composite signal is shown in the bold line in the figure, which shows one peak of the composite waveform flattened relative to the fundamental, whereas the other peak is sharpened.

As discussed in Chapter 8, the North American Standard channel plan (ANSI/EIA-542) places most analog television visual carriers approximately 6 MHz apart and offset approximately 1.25 MHz upwards from harmonics of 6 MHz.* Each such carrier can be expressed as

$$\sin[2\pi(6n + 1.25)t] \tag{9.23}$$

where

n = the closest harmonic of 6 MHz

t = the time in microseconds

the angle is expressed in radians

When two such signals (on different channels) are transported through an amplifier with second-order distortion (for simplicity, and because the effects are often small, the fourth-order term has been dropped), the output is

* Variations from exact spacing are due to equipment frequency tolerances, broadcast station offsets of 10 or 20 kHz, and aeronautical band offsets of 12.5 or 25 kHz.

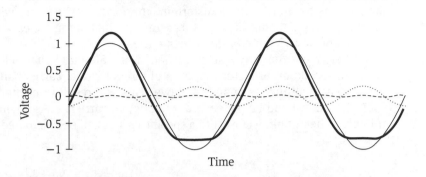

Figure 9.7 Even-order distortion.

$$e_O = AE_i(\sin[2\pi(6n+1.25)t] + \sin[2\pi(6m+1.25)t]) + \tag{9.24}$$
$$BE_i^2(\sin[2\pi(6n+1.25)t] + \sin[2\pi(6m+1.25)t])^2$$

where

the first term = the linear amplification of the input signals

n and m = the multiples of 6 MHz closest to the two visual carriers

E_i = the peak voltage of either of the input signals (they are assumed to have equal amplitudes)

The last term has a dc component plus the following intermodulation products:

$$- BE_i^2 \cos[2\pi(6[m+n]+2.5)t]$$
$$+ BE_i^2 \cos[2\pi(6[m-n])t]$$
$$- (BE_i^2/2)\cos[2\pi(12n+2.5)t]$$
$$- (BE_i^2/2)\cos[2\pi(12m+2.5)t]$$

The first term describes a product that falls 1.25 MHz above the visual carrier nearest the $(m+n)$th harmonic of 6 MHz, whereas the second product falls 1.25 MHz below the visual carrier nearest the $(m-n)$th harmonic of 6 MHz.

The third and fourth products will be half the voltage (−6 dB in power) and will each lie 1.25 MHz above another visual carrier (provided, in all cases, that the products fall within the occupied bandwidth of the distribution system). Note that these last products are identical in amplitude to those analyzed with a single carried signal and are, in fact, just the second harmonics of the two original signals.

The amplitude of each of these four distortion products, as with the single channel analysis, depends on the square of the input voltage so that the *ratio* of

fundamental to second-order products varies linearly with input level. A similar analysis for fourth-order distortion, the next even power, will show low-amplitude components at four times the original frequencies and additional components at twice the original frequencies.

When many similarly spaced carriers in the Standard channel plan are subjected to second-order distortion, each of the individual carriers and each of the carrier pairs will mix as analyzed above. This will result in clusters of products 1.25 MHz above and below each of the original carriers.* As a general rule, the number of beats falling closest to a specific carrier can be calculated using the expression

$$N_L = n - m - x + 1 \qquad N_U = \text{INT}\left[\frac{x - 2m + 1}{2}\right] \qquad (9.25)$$

where

N_L = the number of lower $(A - B)$ beats

N_U = the number of upper $(A + B)$ beats

n = harmonic number of highest carrier

m = harmonic number of lowest carrier

x = harmonic number of carrier being evaluated

INT indicates the expression is to be rounded down to the nearest integer

Since, in the ANSI/EIA-542 Standard channel plan, the carriers are not exactly spaced, the second-order products are not frequency coherent, but rather are clustered around the nominal product frequencies.

The usual measurement technique when evaluating this type of distortion is to use a filter bandwidth of 30 kHz, which is wide enough to contain all the significant products in a group and to measure the total apparent power within the filter bandwidth. The power level of the upper composite beat cluster, relative to the sync peak power of the visual carrier, is known as composite second order (CSO) and is sometimes expressed in units of dBc (meaning decibels relative to the carrier power level, which is always a negative number). Some engineers prefer the notation C/CTB, the ratio of the carrier to beats in decibels (a positive number). This book will use this notation. The absolute values are the same in either case. Although the lower CSO cluster will generally have greater magnitude than the upper cluster on lower-frequency channels, it falls at the edge of the video channel and is often ignored since it has a minor effect on video quality compared with the in-band upper cluster.

Since analog video carrier modulation results in average signal levels that vary with program content, equipment is specified and evaluated when loaded

* Actually, there are also some products spaced ±750 kHz from the visual carriers because two of the channels (5 and 6) are offset from the comb in the Standard channel plan.

Figure 9.8 Number of 1.25-MHz CSO beats: 550-MHz Standard channel plan.

with unmodulated carriers of a defined level in order to achieve repeatable measurements. When loaded with video carriers whose peak level is the same as the unmodulated levels used for the test, the average carrier power will be about 6 dB lower, resulting in an improvement of about 6 dB in C/CSO.

When the IRC channel plan is used, the visual carrier frequencies are derived from a common source so that all channels are phase coherent and offset from harmonics of 6 MHz by exactly the same amount. The result is that the second-order beats are also coherent.

When the HRC channel plan is used, the visual carrier frequencies are derived from a common source and *not* offset. Under those circumstances, the upper and lower second-order beats fall exactly on the visual carrier frequencies, and the difference frequency beats cannot be ignored.

The amplitudes of the individual products and the number in each cluster are a function of the nonlinearity of the amplifier, the levels of the carriers generating the beat, and the number of test signals used. Although the preceding analysis assumed a uniform carrier level, standard operating practice is to vary the carrier power in a controlled way as a function of frequency. This will result in varying individual beat levels.

When a comb of signals is subject to second-order distortion, the greatest number of products will fall on the highest and lowest channels. Figure 9.8, as an example, gives the total number of upper and lower 1.25-MHz second-order products in the Standard channel plan (in which channels 5 and 6 are offset by 2 MHz, and the 15–17th harmonics are not used).

The visual effect of the upper-side second-order products on analog NTSC television pictures is similar to that of a discrete interfering carrier—that is, closely spaced diagonal lines as discussed in Chapter 2. As discussed earlier, the lower CSO beats have relatively little effect on visual quality.

Many simple broadband RF amplifiers are single-ended class A devices, meaning that their output stages consist of a single active device that is biased to the approximate middle of its dynamic range. Such a device typically exhibits

an asymmetrical compression behavior so that second-order distortion products dominate. In recognition of this problem, the North American VHF off-air frequency plan designated channel boundaries such that neither second harmonic nor second-order beats from any combination of channels affect any other channels even if the drive levels to television receivers are high enough to result in compression. This was discussed in detail in previous chapters.

Early cable television amplifiers used single-ended designs for the same reasons. As the industry began using non–off-air channels, however, CSO quickly became the primary performance-limiting parameter.

Odd-Order Distortion (Composite Triple Beat—CTB)

In an attempt to reduce this problem, all modern cable amplifiers are "push–pull," meaning that their output stages contain two devices in a balanced circuit that assures symmetrical, or nearly symmetrical, compression. As will be seen, this will result in primarily odd-order distortion products.

If a single sine wave is transmitted through a circuit that has only third-order distortion, the output can be described by the following equation (derived from Equation 9.20):

$$e_o = Ae_i + Ce_i^3 \tag{9.26}$$

where

A and C = the coefficients of the linear and cubed products, respectively

$e_i = E_i \sin(\omega t)$

Using a standard trigonometric identity, this reduces to

$$e_O = \left[AE_i + \frac{3CE_i^3}{4} \right] [\sin(\omega t)] - \left[\frac{CE_i^3}{4} \right] [\sin(3\omega t)] \tag{9.27}$$

C may be either positive or negative. In the case where the nonlinearity is due to compression, it is negative so that the third-order distortion results in a reduction in the amplitude of the fundamental (unlike second-order distortion that does not affect the amplitude of the fundamental) plus the addition of a product at three times the original frequency whose peaks are in opposition to those of the fundamental. The level of the third-order products rises as the cube of the input voltage (e.g., the rise is 3 dB for every 1-dB rise in input level) so that the *ratio* of fundamental to third harmonic at the output decreases by 2 dB for every 1-dB rise in the input level *so long as the ratio between them is high.* When the reduction in the amplitude of the fundamental becomes material, the ratio will decrease more quickly.

Figure 9.9 shows visually the fundamental, third harmonic, and the sum of the two. As can be seen, third-order distortion is equivalent to a reduced ampli-

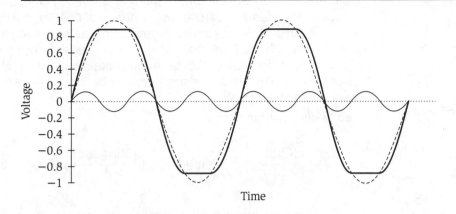

Figure 9.9 Effect of third-order distortion.

tude, symmetrically flattened waveform. If higher odd multiples of the original frequency are also plotted on the graph, it will be seen that all such harmonics will have exactly the same effect on both peaks of the original waveform.

If multiple signals are transmitted through a system exhibiting this type of distortion, the input can be described by the following general formula:

$$e_i = E_1\sin[\omega_1 t] + E_2\sin[\omega_2 t]) + \cdots + E_n\sin[\omega_n t] \qquad (9.28)$$

where

$\omega_{1,2,\ldots n}$ = the angular frequencies of the input signals

$E_{1,2,\ldots n}$ = the levels of each subcarrier (peak voltage)

If we insert this definition of e_i into Equation (9.26), it can be shown that the output spectrum will have signals at the fundamental frequencies of the input signals and all combinations of $3\omega_x$, $2\omega_x \pm \omega_y$, and $\omega_x \pm \omega_y \pm \omega_z$, where ω_x, ω_y, and ω_z are any of the input frequencies. In particular, single frequency terms, two frequency terms, and three frequency terms will be present:

- Single frequency terms:

$$\left[AE_x + \frac{3CE_x^3}{4}\right][\sin(\omega_x t)] - \left[\frac{CE_x^3}{4}\right][\sin(3\omega_x t)] \qquad (9.29)$$

Just as with a single sine wave, this indicates that the output contains, in addition to the linearly amplified input signals, two terms proportional to the cube of the input voltage level of each of the carriers, one at the fundamental carrier frequency and one at three times the carrier frequency. When the

input consists of carriers corresponding to the EIA Standard or IRC channel schemes (offset 1.25 MHz from harmonics of 6 MHz), the third harmonic term will fall 3.75 MHz above a harmonic of the root frequency (2.5 MHz above a carrier if one is present). With carriers that fall exactly on harmonics of a common root (HRC), the distortion products will also fall on harmonics of the root. The number of single frequency terms will be the same as the number of FDM carriers.

- Two frequency terms:

$$\left[\frac{3CE_x^2 E_y}{2}\right]\sin(\omega_y t) - \left[\frac{3CE_x^2 E_y}{4}\right]\sin([2\omega_x + \omega_y]t) + \left[\frac{3CE_x^2 E_y}{4}\right]\sin([2\omega_x - \omega_y]t) \qquad (9.30)$$

Each unique ordered pair of input signals mixes to produce a product at the fundamental of one of the input signals plus products at $2\omega_x \pm \omega_y$. The amplitude of the fundamental is twice that produced by the cubed term (assuming equal amplitude input signals). The voltage amplitudes of the other two terms are three times that of the cubed terms (+9.5 dB). When the input consists of carriers corresponding to the EIA Standard or IRC channel schemes, the $2\omega_x + \omega_y$ term will fall 2.5 MHz above a carrier assignment, while the $2\omega_x - \omega_y$ term will fall on a carrier assignment. With an HRC channel assignment scheme, all products will fall directly on channel frequencies.

When the input signals consist of an FDM spectrum of q input carriers, the number of ordered pairs is $q(q - 1)$. Since each ordered pair results in the generation of products at two new frequencies, the number of new products will be $2q(q - 1)$. Thus, when the number of carriers is large, the total number of products will increase approximately as the square of the channel count.

- Three frequency terms:

$$\left[\frac{3E_x E_y E_z}{2}\right]\left[\sin([\omega_x + \omega_y - \omega_z]t) + \sin([\omega_x - \omega_y + \omega_z]t)\right.$$
$$\left. - \sin([\omega_x + \omega_y + \omega_z]t) - \sin([\omega_x - \omega_y - \omega_z]t)\right] \qquad (9.31)$$

Each unique combination of three input signals mixes to produce terms at $\omega_x \pm \omega_y \pm \omega_z$. When the input signal amplitudes are equal, the voltage amplitudes of these terms are six times that of the cubed terms (+15.6 dB). When the input consists of carriers corresponding to the EIA Standard or IRC channel schemes, the $\omega_x + \omega_y - \omega_z$ and $\omega_x - \omega_y + \omega_z$ terms will fall on a carrier frequency, the $\omega_x + \omega_y + \omega_z$ term will fall 2.5 MHz above a carrier frequency, and the $\omega_x - \omega_y - \omega_z$ term will fall 2.5 MHz below a carrier frequency. Note that, though $\omega^x \pm \omega_y \pm \omega_z$ is sometimes negative, the only effect is the polarity of the magnitude since $\sin(-x) = -\sin(x)$.

When the input signals consist of an FDM spectrum of q input carriers, the number of unique three-signal combinations is $q(q - 1)(q - 2)/6$. Since each of these combinations results in the generation of four new products, the total number of new third-order products caused by three-signal mixing will be $4q(q - 1)(q - 2)/6$. Thus, when the number of carriers is large, the total number of products will increase approximately as the cube of the channel count.

In summary, when an FDM spectrum of q signals is transmitted through an amplifier exhibiting third-order distortion, the original signals will be amplified, and a number of new products will be generated:

- Some that affect the output amplitude of each original carrier.
- q new signals, one at the third harmonic of each of the original signals. With the ANSI/EIA-542 Standard or IRC frequency plan, these fall 2.5 MHz above visual carrier frequencies.
- $2q(q - 1)$ new signals at frequencies $2\omega_x \pm \omega_y$, where ω_x and ω_y are any two input carrier frequencies. Each of these signals is about 9.5 dB higher in level than the third harmonic frequencies. With the Standard or IRC frequency plan, half of these fall 2.5 MHz above, and half nominally on, visual carrier frequencies.
- $4q(q - 1)(q - 2)/6$ new signals at frequencies $\omega_x \pm \omega_y \pm \omega_z$, where ω_x, ω_y, and ω_z are any three different carrier frequencies. Each of these products will be about 15.6 dB higher in level than the third harmonic frequencies. With the Standard or IRC frequency plan, half of these will fall nominally on visual carrier frequencies, whereas one-quarter will be 2.5 MHz below and one-quarter will be 2.5 MHz above visual carrier frequencies.

The level of each of these products increases by 3 dB for every 1-dB increase in the levels of the input signals so that the ratio between the desired output signals and third-order products decreases by 2 dB.

Since the three-signal products ("triple beats") are the greatest in both number and magnitude, they quickly dominate other third-order effects. Since half of them fall nominally on visual carrier frequencies, those product groups are always the largest. Thus, composite triple beat (CTB) distortion is defined as the total power in a 30-kHz bandwidth centered on the visual carrier frequency without the visual signal present, relative to the peak level of the visual carrier in dBc (decibels relative to a reference carrier level). As with CSO, it can also be expressed as the ratio of visual carrier level to apparent beat power, or C/CTB, in decibels.

It is difficult to write a general expression that describes how many beats fall on any one channel, however, Figure 9.10 is a plot of the number of beats for some common, fully loaded, system bandwidths.[8] As can be seen, channels slightly above the middle of the spectrum are affected by the greatest number of beats. It does not follow, however, that the channel with the greatest number of

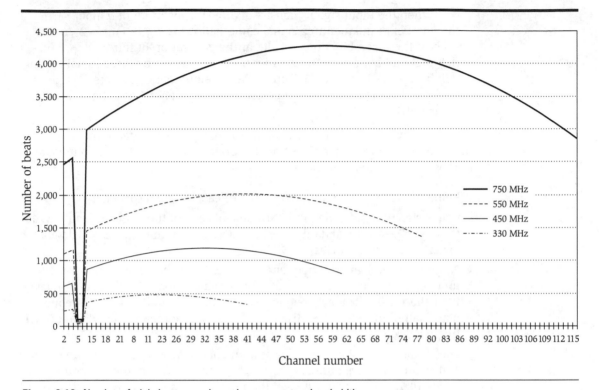

Figure 9.10 Number of triple beats per channel versus system bandwidth.

beats is the channel most affected by third-order interference. Amplifiers are not typically operated with equal power carriers, but rather with a uniform upward "slope" of levels with frequency (for reasons explained later in this chapter), causing some beats to be stronger than others. More important, since signal quality is affected by the *ratio* between visual carrier level and interference level, the lower carrier levels at the low end of the spectrum will sometimes cause the worst interference to occur on a lower-frequency channel than the one with the largest number of beats.

Since the third-order products are phase incoherent in most systems, their powers will add on a power, rather than a voltage, basis, and the measured CTB level would theoretically be proportional to $10 \log(n)$, where n is the number of beats in a channel if the levels of all carriers are equal. In actuality, owing to the unequal carrier levels and beat amplitudes, and possibly differences in distortion across the spectrum, CTB may increase as slowly as $5 \log(n)$ to $7 \log(n)$.[9]

The visual effect of CTB on an analog television picture depends on whether the signals are harmonically related or not. With unlocked carriers, the differences in frequency between the desired carrier and the various beats produces low-frequency interference that has been described as "streakiness" in the picture. With locked carriers, the various products will be frequency coherent with

the carrier and, like the fundamental frequency products, will alter the amplitude and phase of the original signals. The visual effects depend on many factors, including the amount of phase noise on each of the signals.

Since analog video carrier modulation results in an average signal level that varies with program content, equipment is specified and evaluated when loaded with unmodulated carriers of a defined level on each of the nominal visual carrier frequencies. This results in more repeatable measurements. When loaded with video carriers whose peak level is the same as the unmodulated levels used for the test, the average carrier power will be about 6 dB lower, resulting in an improvement of about 12 dB in CTB (though the improvement will vary over time depending on video modulation levels and timing and carrier phase relationships). Typical amplifiers provide a CTB of −70 to −90 dBc when loaded with unmodulated carriers at recommended operating levels, depending on design.

Cross Modulation (XMOD)

Another form of distortion occurs when amplifiers exhibiting third-order distortion are loaded with multiple amplitude modulated signals.

Although second-order distortion has no products that affect the gain of the primary FDM carriers, as Equations (9.27), (9.29), and (9.30) demonstrate, two classes of third-order products affecting every channel cause effective gain variations of the fundamental signal. In particular, the first term of Equation (9.30) demonstrates that the effective amplifier gain affecting each FDM subcarrier is modified by a series of terms that include the amplitude of every other signal in the spectrum carried. Thus, when the average levels of any other signals are varied because of modulation, the output level of the desired channel will also vary slightly. This effect is known as cross modulation.

If the modulation of the other signals were truly independent of each other and random, the effects of cross modulation from various channels could be expected to be noiselike. Unfortunately, video modulation is not random. As covered in Chapter 2, the change in amplitude from synchronizing peak to black level is identical for every channel and, furthermore, occurs at nearly identical and quite precise frequencies. Thus, if the signals modulating multiple channels happen to be synchronized (which happens when they are locked to a common timing reference in a multichannel generation facility), or if they drift into alignment occasionally, the level of cross modulation can become significant.

The standard test for cross modulation is to load the amplifier with a comb of simultaneously modulated carriers plus one unmodulated carrier on the frequency to be tested. The level of modulation is measured on the nominally unmodulated carrier. Cross modulation is defined as the difference between the cross modulation sideband level and the sideband level that would correspond to 100% modulation expressed in decibels.

Since it is caused by third-order distortion, the level of cross modulation varies in the same way as CTB; that is, the relative level varies 2 dB for every 1-dB

change in operating levels. Because of the difficulty in making the test and demonstrated inconsistency in the results, cross modulation has become less important than CTB as a means of characterizing third-order distortion. The inconsistency may be due to the many terms, whose effects may reinforce or cancel, affecting the cross modulation on any channel.

When the desired signal is an analog television program and the major cross-modulating signals are also analog television programs, the first visible indication of cross modulation takes the form of a single, faint broad horizontal line and another broad vertical line that may slowly move through the desired picture. These are caused by the horizontal and vertical synchronizing pulses (the highest carrier levels) of another television signal. In an improperly operated system, cross modulation can also be caused by a single very high amplitude channel that causes cross modulation onto other channels. In extreme cases, not only synchronizing bars but video modulation may be visible. The visual appearance of cross modulation is similar to that which results from direct pickup interference where an undesired signal "leaks" into the system and interferes with a cable signal on some channel.

Composite Intermodulation Noise (CIN)

The preceding analysis was based on the carriage of a spectrum of signals, each of which has a carrier that represents most of the energy in the channel. Though that is a valid model for analog television signals and many conventional AM and narrow band FM modulated signals, digital signals differ significantly. A QAM signal, for instance, has a suppressed carrier and, when viewed on a spectrum analyzer, looks like a flat block of noise occupying the entire communications channel (see Chapters 4 and 5 for detailed discussions of QAM modulation).

When such signals are subject to second- and third-order distortion, they do not produce single frequency products, but rather noiselike bands of energy whose amplitude frequency signature depends on the distortion mechanism. The products of this type of distortion are alternatively known as composite intermodulation noise (CIN), composite intermodulation distortion (CID), or intermodulation noise (IMN). This is an exception to our use of the term *noise* to refer to only thermal noise since CIN is actually a distortion product.

On the assumption that third order is the dominate distortion-producing mechanism and that, further, three frequency products ($\omega_a \pm \omega_b \pm \omega_c$) dominate, IM products involving digital carriers will be of three types:

- Products formed from the mixture of two analog carriers and one digital signal will appear as blocks of flat noise with a bandwidth equal to that of the digital signal.
- Products formed from the mixture of one analog signal and two digital signals will appear to have a symmetrical triangular-shaped amplitude-versus-frequency spectrum, where the triangle will have a spectral width equal to the sum of the widths of the two individual digital signals.

- Products formed from the mixture of three digital signals will appear to have a band-limited Gaussian shape with a spectral width equal to the sum of the widths of the three mixing signals.

If digital signals are carried lower in level than analog signals, then the preceding types of products are listed in decreasing order of amplitude.

Although standard operating practices have not yet been set for the cable industry, it has been reported that a 750-MHz plant, operating with a signal load of 77 modulated analog television signals (54–550 MHz) and 33 64-QAM digital signals (550–750 MHz) and typical operating conditions and levels, will suffer less than 1 dB of effective C/N degradation due to CIM with digital signal levels suppressed as little as 5 dB relative to analog signals.[10]

9.3.4 Group Delay Variation and Amplifier Stability

Group delay is a measure of the rate of change of phase shift through a device as a function of frequency. In equation form

$$\text{group delay(sec)} = \frac{1}{2\pi}\frac{d\gamma}{df} \tag{9.32}$$

where

γ = the phase shift through the device in radians

f = the frequency in Hz

In an ideal RF system, the transmission delay at all frequencies is a constant equal to the transit time through the length of line involved, and the group delay is zero.

If the group delay is not zero, signal components at one frequency will arrive shifted in time relative to other components. The degree of tolerance to group delay variation will vary according to the signal format. A particular case of interest to the analog NTSC format is chroma delay, a measure of the arrival time of the chroma signal relative to the luminance signal. As discussed in earlier chapters, the difference should not exceed 170 ns. Though modulators and demodulators are generally the primary contributors to chroma delay, channels close to forward or reverse transmission system band edges are also affected by amplifier design.

High-speed digital signals are also affected by group delay. For example, fast rise time waveforms will broaden if the higher-frequency components are delayed relative to the lower-frequency components. Occasionally, digital signals are carried below channel 2 in the downstream band and, thus, very close to passband edge.

The principal contributors to group delay variation in amplifiers are the diplex filters used at each of the signal ports to separate upstream and downstream spectra. Figure 9.11 illustrates the key components of a two-way amplifier. The input and output diplex filters each consist of a high-pass section

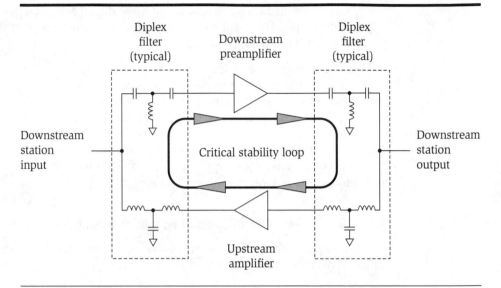

Figure 9.11 Bidirectional amplifier.

connecting the forward gain block to the external ports and a low-pass section feeding the reverse gain block. Each of the filter sections, as well as the amplifier modules, will have a gain (negative in the case of the filters) that is a function of frequency.

In order to assure absolute stability, the sum of the gains around the loop shown in the heavy arrow must be less than zero at all frequencies. This condition must be met not only for matched loads on the external ports but for any mismatch encountered in an actual field installation. With simple filter designs, this is most easily achieved by having the cutoff frequencies of the filters as far apart as possible and, thus, close to the edge of well-separated forward and reverse passband edges. Since group delay variation gets worse as the cutoff frequency is approached, there is an inherent conflict among (1) the separation between the top of the upstream spectrum and the bottom of the downstream spectrum, (2) the cost of the diplex filter, (3) the group delay variation at the extremes of the spectra, and (4) the stability margin of the amplifier.

Ensuring a nonoscillatory condition may not be adequate. Even if the loop gain is less than zero, frequency response variations may result unless there is adequate isolation. As a practical test, adequate performance may be deduced by observing the swept frequency response of an amplifier at all frequencies, including the diplexer crossover region. Peaking of the response indicates a possibility of a loop gain problem.

Figure 9.12 illustrates the downstream performance of a typical commercially available amplifier having a "42/52 MHz split" (upper edge of the upstream spectrum is at 42 MHz, whereas the lower edge of the downstream spectrum is at 52

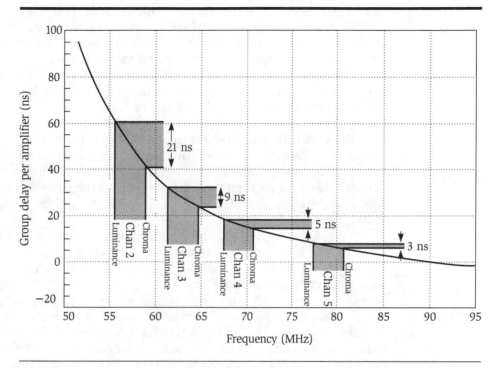

Figure 9.12 Downstream group delay due to diplex filters.

MHz). The curve illustrates the group delay, with the shaded regions indicating the frequencies and chroma/luma delay difference for the four lowest NTSC television channels in the Standard ANSI/EIA-542 frequency plan. The direction of the chroma delay is to advance the chroma with respect to the luminance. Serendipitously, this is the opposite delay sense (luminance is delayed with respect to chrominance) to that which is commonly encountered in the modulation–demodulation process, but any delay is not good.

The required end-to-end group delay requirements and typical network performance are discussed in Chapter 13, and upstream group delay is discussed in Chapter 14.

9.3.5 Hum Modulation

Although amplifiers that plug into conventional wall outlets are available (and used in headend and apartment house applications where convenient), most distribution amplifiers are powered via ac voltage that is multiplexed with the signals on the common coaxial distribution cable. Along with their signal-processing circuits, amplifier stations include circuitry for separating the power and signal voltages and for connecting the power to local power circuits that convert it to direct current to power the various electronic modules used.

Amplitude modulation of the transmitted signals at the power line frequency (hum) can occur in two ways: excessive ac voltage at the output of an amplifier's local power pack and parametric modulation of magnetic component properties. Power pack "ripple" can modulate amplifier gain or be coupled into other circuits such as the automatic gain control (AGC). This ripple can result from faults in the power pack or loss of voltage regulation due to excessive voltage drop between the power supply and the amplifier's power pack.*

Parametric magnetic modulation is less obvious. Consider the circuit segment in Figure 9.13. The current is tapped off the common coaxial input and sent through L to the amplifier power pack (and possibly inserted into the output coaxial cable(s) as well). Series capacitor C serves to pass the signals into the amplifier while blocking the supply voltage. Finally, T is the input winding of a typical signal-processing element, such as a directional tap, splitter, or part of a band-splitting filter.

One possible source of hum modulation is L. This component is required to exhibit a very high impedance to frequencies covering the entire forward and return spectra. If the magnetic core becomes partially saturated at the peaks of the supply current, then the impedance match of the amplifier will vary accordingly, resulting in a level variation of the transmitted signal at the line frequency rate.

T is a much more sensitive component. Typically, these transformers are constructed of tiny cores that must have constant magnetic characteristics over an extremely wide frequency range. Though C nominally protects T from the supply current, there is still a displacement current to contend with. In particular, an ac current will flow through C and the primary winding of T whose magnitude is

$$i_C = C\frac{dV}{dt} \tag{9.33}$$

When the power supply is a sine wave with a frequency of f_p and a peak voltage of E_p, the current will be

$$i_C = C\frac{d}{dt}[E_p\sin(2\pi f_p t)] = CE_p 2\pi f_p \cos(2\pi f_p t) \tag{9.34}$$

which has a peak value of $CE_p 2\pi f_p$. If the voltage is a square wave with fast rise and fall times, then the displacement current can be much greater for the same peak supply voltage. To illustrate the trade-offs in design of these circuits, assume that it is desired that C have no more than 1 ohm of reactance at the low end of the upstream spectrum (generally 5 MHz) so as to affect the input match as little as possible. This requires that C be at least 0.03 μF [$X_C = 1/(2\pi fC)$]. If the supply

* In accordance with standard industry terminology, the term *power supply* is used to designate the device that generates the system power for multiplexing on the coaxial cable, whereas the term *power pack* is used to designate the device within each amplifier that converts the multiplexed power to the appropriate regulated dc levels for the circuitry.

Figure 9.13 Power/RF separation in amplifiers.

voltage is 90-volts rms (127 volts peak) at 60 Hz, then the displacement current will be 1.4 mA. If, under square wave powering conditions, the rise time is five times that under sine wave conditions, then the current could be as high as 7 mA. The designer's challenge is to ensure that this level of additional current in the winding of T does not change the transformer's characteristics.

In the North American NTSC analog television system, the frame repetition rate is 59.94 Hz. Since the commercial power frequency is 60 Hz, hum modulation generally appears as a horizontal bar or brightness variation whose pattern slowly moves upwards with time (about 16 seconds to move the entire height of the picture). Typical coaxial components (both amplifiers and passive elements) are specified to create hum modulation at no greater than −70 dB, relative to 100% modulation, at their full rated supply voltage and pass-through current.

9.3.6 Frequency Response Variation and Impedance Match

Practical amplifiers do not amplify signals at all frequencies equally. Part of this variation is intentional and is intended to compensate for variations in cable losses and to optimize distortion performance, as we discuss in the next section. Any deviation from the intentional variation, however, is known as frequency response variation (quantified as $\pm x$ dB) or, more commonly, peak-to-valley (quantified as y dB peak-to-peak). Response variation is defined by the SCTE as the maximum gain deviation in either direction from a line drawn from

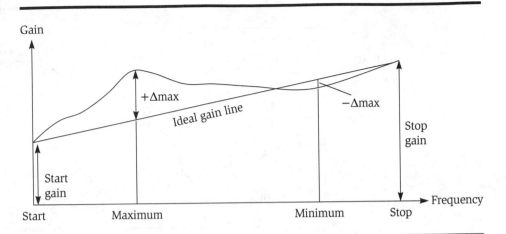

Figure 9.14 Definition of response variation.

the gain at the low end of the defined bandwidth to the gain at the high end of the defined bandwidth.[11] Figure 9.14 illustrates this definition; the peak-to-valley (P/V) is the sum of the absolute values of +Δmax and −Δmax. This definition, however, is somewhat controversial since it is highly affected by the endpoints. An alternative definition for frequency response is the variation above and below a line that is drawn so as to minimize P/V. For instance, if the lower end of the "ideal gain line" in the figure were raised slightly, the effect would be to lower +Δmax by more than the increase in −Δmax and thus reduce the total P/V.

The frequency response through a network may be affected by the imperfect impedance matching between the amplifier and attached transmission line. The effects of imperfect matches in coaxial components can best be understood in the context of their interaction. This is covered in detail in Chapter 13.

9.3.7 Practical Amplifier Design Choices

Complete amplifier stations may be simple or quite elaborate, depending on system needs. All share a common basic downstream structure consisting of several (generally two or three) cascaded gain stages, plus circuits that compensate for the loss variation of the cable before the amplifier (equalizers). Provisions are also made for adjusting the gain of the amplifier as needed to meet system needs. Amplifiers differ in how many stages are employed, internal splitting, redundancy provisions, where operating adjustments are in the chain, automatic gain control, and the performance of each gain block. Many have provisions for automatically adjusting amplifier gain to compensate for changing input signal levels (automatic gain control).

As will be seen, optimum operation of repeated coaxial distribution systems results when the levels of the FDM subcarriers at the output of amplifiers are not equal but increase with frequency. The variation in level across the operating range is known as *tilt*. The relationship between operating levels, amplifier response, and interconnecting transmission losses is explored more fully in Section 9.3.8.

In the upstream direction, all the amplifier designs are similar. Since cable losses are much lower in that portion of the spectrum, a single amplifier module is sufficient. Diplexers are used at each port to separate the upstream and downstream signals. All the upstream signals are passively combined and fed to the amplifier module. Unlike downstream, where all level setting takes place at the input or midstage with the aim of setting a consistent *output* level, the common upstream practice is to align the system so that the *input* level to the module is constant and to set the output gain and slope after the gain stage so that the input to the next amplifier station upstream is correct. Chapter 14 covers the alignment and operation of networks in the upstream direction.

To allow the most flexibility in application and system design, jumpers or switches allow power to enter the amplifier via any desired port and to be routed to any combination of output ports.

Early power packs used conventional transformer/linear regulator designs. Transformer taps were available to optimize the winding ratio of the transformer to limit required regulator dissipation. Modern designs almost universally use switching regulators (whose design is beyond the scope of this book) that have the property of operating at a constant high efficiency regardless of input voltage. To do so, they draw a constant *wattage* from the coaxial cable, and thus the current is inversely proportional to the supply voltage. Though this results in the greatest overall powering efficiency, it has some unexpected consequences when a number of amplifiers are powered from a common source, as will be seen in Chapter 10.

Beyond their basic functional elements, amplifier stations may be equipped with a variety of enhancements, including redundant gain modules and power supplies, status monitors and transducers, and remotely controllable upstream path disconnect. Often fiber-optic receivers and transmitters are combined with amplifier station components, creating a fiber-optic node.

Although the labels attached to different configurations are somewhat arbitrary and subject to change, the following classifications are typical:

- Trunk amplifiers are the workhorses of traditional all-coaxial cable television systems. They are optimized for long-distance, repeated signal distribution. As such, they are operated at output levels that cause only moderate distortion to allow long cascades. The gain is set as high as possible, consistent with low per-stage noise addition. Typically, the output levels of the highest channels are +35 to +40 dBmV, and the gain is about 22 dB. Trunk amplifiers are not used to supply signals to taps.

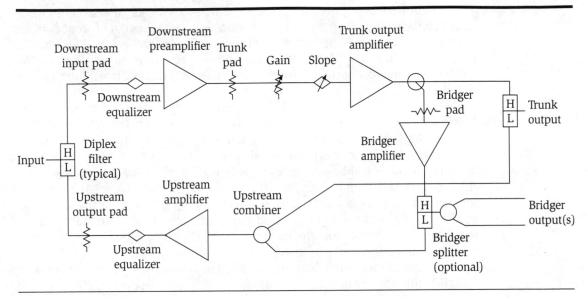

Figure 9.15 Typical trunk amplifier RF configuration.

- Where a local distribution circuit is also desired, a portion of the trunk signal is fed to a separate output stage known as a bridger. This amplifier generally runs at a higher output level (and distortion) to allow feeding as many customer taps as possible. Though generally configured as an optional module within the trunk station, bridgers may be contained in separate housings along with directional couplers (midspan bridgers). Figure 9.15 is a block diagram of the RF portion of a typical trunk station, including bridger.

- Distribution amplifiers are basically stripped-down trunk/bridger stations that are available in a wide variety of configurations. Typically, they are available with multiple output amplifier modules whose level can be independently adjusted. Figure 9.16 shows a typical configuration with three output gain blocks.

- Line extenders are used to allow feeding more taps than possible with just a bridger or high-level distribution amplifier port. Since they are generally operated at similar output levels, the cascade of such devices must be limited to keep distortions at an acceptable level. Line extenders are the simplest of the classifications, as shown in Figure 9.17.

9.3.8 Amplifier Operating Dynamics

The C/N degradation as a signal passes through an amplifier is related to the input signal level and the noise figure at any frequency (see Equation (9.18)). On the other hand, the loss of coaxial cable increases with frequency. Thus, the amplifier

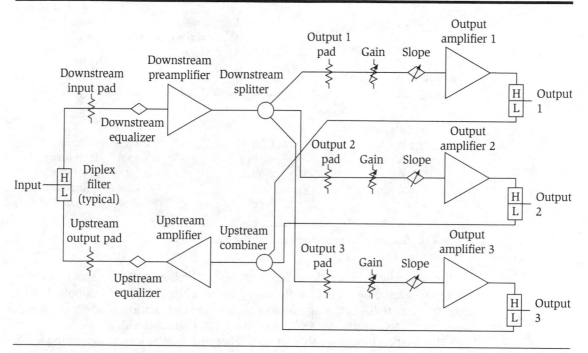

Figure 9.16 Typical distribution amplifier RF configuration.

Figure 9.17 Typical line extender RF configuration.

Figure 9.18 Downstream response and gain relationships and adjustments.

need not supply the same level for channels at the bottom of the spectrum as at the top to overcome cable loss. Reducing the levels of the lower-frequency channels, in turn, reduces the amplitudes of the intermodulation products to which those channels contribute, and thus CSO, CTB, CIN, and XMOD.

The relation between cable and amplifier characteristics and operating levels can best be illustrated by considering an amplifier and its driven cable as a system (see Figure 9.18). Only the downstream path will be discussed here; Chapter 14 will deal with the equivalent upstream issues.

The preamplifier will typically have a relatively uniform noise figure and gain across the spectrum. Between the preamplifier and output stage, a slope circuit is used to adjust the frequency response so that it uniformly increases across the downstream frequency range. Typical gain slopes are 6–10 dB for a 750-MHz amplifier. Additionally, pads are used to adjust the overall amplifier gain. The station noise figure and slope are illustrated in Figure 9.19(a).

Some combination of coaxial cable and passive RF devices is connected between this amplifier station and the input of the next station. Since both the cable and passive devices have a loss that varies across the spectrum (and, in the case of cable, increases as the square root of frequency), the variation in level reaching the next station will be reduced. Each amplifier input circuits are designed to accommodate a variety of equalizers that are optimized to compensate for the residual response variation and pads that are used to set the correct drive level to the preamp input. Occasionally, external in-line equalizers may be used also, for example, as a tool to reduce the total variation of levels fed to customers.

In station alignment, the equalizers and pads are used to align the *output* of each amplifier station as closely as possible to the optimum levels and "gain slope." Thus, the system gain and frequency response, as measured from the output of one amplifier to the output of any other similarly aligned amplifier in a

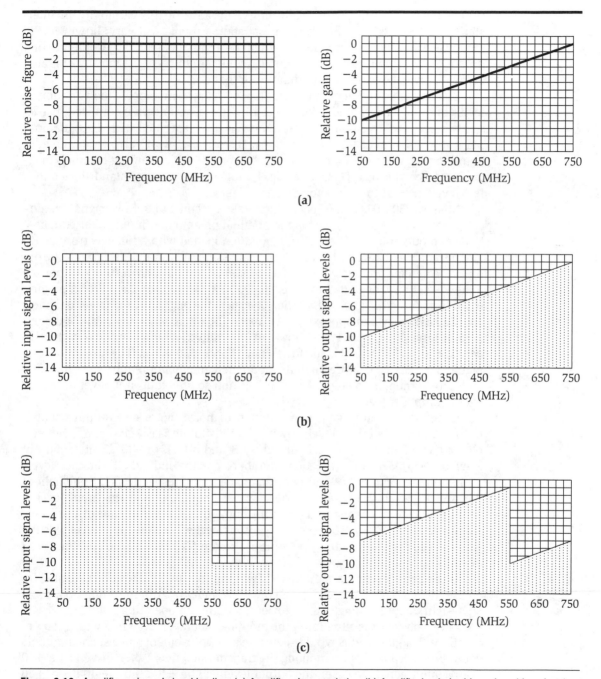

Figure 9.19 Amplifier gain and signal loading. (a) Amplifier characteristics. (b) Amplifier loaded with analog video signals. (c) Amplifier with hybrid analog–digital signal loading.

series-connected string, are ideally unity. Because of the variability in inter-station losses, the levels and response at station inputs will vary; however, the response at the internal preamp input will be approximately flat. Amplifiers that do not feed customer taps (for example, trunk amplifiers) are typically operated at levels several decibels lower than those that do.

By contrast to the frequency response of the system (slope), the variation in levels of the signals carried is known as tilt. The terms are often confused, particularly since modern amplifier performance is often specified with a tilt at amplifier outputs that matches the gain slope. Historically (with narrower bandwidth amplifiers), it was more common to use "block tilt" with groups of channels operated at different levels. This was easier for technicians to understand since only a few levels needed to be memorized.

Modern 750-MHz amplifiers are generally specified under two signal-loading conditions. Figure 9.19(b) shows the nominal preamplifier input (after equalization) and output levels of an amplifier station loaded with a full spectrum of 110 video signals that are tilted to match the gain slope of the amplifier. This is the optimum condition for carrying signals of similar format and results in approximately equal C/N degradation for each channel (because of the uniform preamp input levels) and the lowest distortion (because the outputs at the lower channels are reduced in level to approximately match the interconnecting cable and component losses). Figure 9.19(c) shows an alternative loading condition with a superimposed −10-dB block tilt for signals above 550 MHz. The assumption is made that these signals are digitally modulated and do not require the same C/N and C/distortion as analog video channels and so are reduced in level to reduce overall amplifier loading.

Optimum amplifier gain is a function of noise figure and output capability (for a given distortion level). By way of illustration, assume that a given amplifier has an effective noise figure of 9 dB and a CTB of −78 dB at an output level of +45/35 dBmV (the dual numbers are the individual channel levels at the top and bottom of the downstream spectrum, respectively). If the gain of the unit is set at 35 dB (at the top channel frequency) and the input level is +10 dBmV on all channels, then, from Equation (9.19), the C/N of this stage alone will be 60 dB (10 + 59 − 9). The output level will be the input level plus the gain or +10 dBmV + 35dB = +45 dBmV, so the CTB will be the specified −78 dBmV.

If the input level is varied, the C/N will improve by 1 dB for every decibel increase in input level, whereas the CTB will degrade by 2 dB. In setting up a coaxial system, input padding is used to set the optimum operating point.

Now consider the situation if the amplifier had been designed with a gain of 40 dB. At first glance, this would seem like an improvement since such amplifiers could be spaced farther apart along the transmission line. When driven by a +10 dBmV input signal, however, the output level will be +10 dBmV + 40 dB = +50 dBmV. At that level, the CTB will be only −68 dB, having degraded by 10 dB owing to the 5-dB increase in output level. If the input level is decreased by 5 dB to re-

store the distortions to the specified level, the station's C/N will decrease from 60 to 55 dB. In short, the excess gain forces the user to accept either higher distortion or higher noise.

Though line extender amplifiers often offer only an input padding option, some distribution and trunk amplifier designs also allow between-stage padding. This partially offsets potential excess gain problems by allowing the user to independently choose the operating level of each gain block though each gain block is still limited in performance by its individual noise figure, gain, and output capability.

The selection of optimum amplifier internal gains is very complex. It would seem from the preceding example that reducing gain would improve performance. Although that is true, more lower-gain amplifiers would have to be cascaded to reach the same distance, and that would lower reliability and increase power consumption. The industry's choice of operational gains ranging from 20 to 38 dB (depending on technology and application) is based on the best compromise among cost, reliability, and performance.

9.3.9 Amplifier Technology Choices

As previously discussed, modern broadband solid-state cable television amplifiers are all push–pull in nature, meaning that their output stages contain two matched transistors connected in a symmetrical configuration. To the extent that any nonlinearities are also symmetrical, they will produce only odd-order distortion products. Amplifiers whose output stages each contain only a single push–pull circuit are known generally as *push–pull amplifiers.*

An improvement in distortion performance can be realized by splitting the input signal, feeding two push–pull output gain stages in parallel, and then combining the outputs into a single port. This type of amplifier, generically known as *parallel hybrid,* offers about twice the output level at the same distortion levels compared with push–pull units. Since two output stages are involved, the power consumption and dissipation are also higher than with push–pull units. Parallel hybrid amplifiers are generally used where high output powers are beneficial, such as bridgers and line extenders.

The third dominant amplifier technology is known as *feed forward.* In a feed-forward amplifier, the input signals are amplified via a conventional inverting amplifier. Samples of the input and output signals, adjusted for gain, are combined and amplified in a separate inverting *error amplifier.* The output of the error amplifier is summed with the output of the original amplifier so that the errors approximately cancel.[12] Figure 9.20 shows this configuration. The delay lines are required to match the phases of the signals. Modern feed-forward amplifiers offer about 12 dB of CTB improvement and 6 dB of CSO improvement over power doubling units at comparable operating levels. Furthermore, as discussed later, the remaining distortions are not necessarily phase coherent among cascaded amplifiers, which improves the performance of a repeated system. Feed-forward

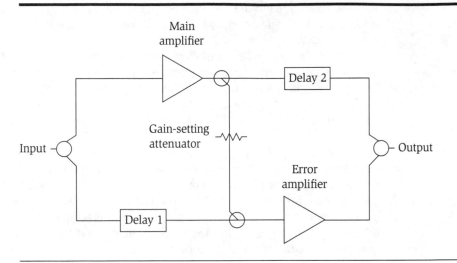

Figure 9.20 Feed-forward gain block.

amplifiers are generally used in long trunk lines where the lower cascaded distortion is beneficial.

9.4 Passive Coaxial Components

The nature of typical coaxial distribution systems is that the downstream path is split and resplit to create many endpoints from a common signal insertion point. In the upstream direction, the opposite is true, with signals from many potential insertion points combined at a common node. In the headend, signals from individual modulators and signal processors are combined to create the full downstream spectrum, whereas upstream signals are split to feed receivers for each service.

The signal splitting can occur either within amplifier housings or by use of stand-alone components. Those that split the signal equally are known as *splitters,* whereas those that divert a defined portion of the input signal to a side port are known as *directional couplers.* Finally, those that divert a portion of the input signal, then resplit the diverted signal to create individual customer drops are known as *taps.* All share certain characteristics and so will be treated in this section.

In general, all these are passive devices constructed using ferrite-loaded transmission lines and transmission-line transformers. They are bidirectional; that is, the same device can be used to split signals to feed multiple paths or to combine signals from multiple input ports. Important characteristics are signal loss, impedance match, and isolation between nominally isolated ports.

Figure 9.21 Directional coupler schematic. (a) Coupler plus power components. (b) Alternate RF configuration.

9.4.1 Directional Couplers

Figure 9.21 shows schematics of two directional coupler configurations. In both examples shown, the voltage transformer and current transformers have the same ratio so that, after combining, the ratio of voltage to current stays the same, meaning that the impedance of the side port is the same as the main line. In the main line, the relative polarity of voltage and current is reversed for signals traveling in one direction versus the other direction. The transformers sample voltage and current regardless of signal flow in the main line. The difference is that the sampled signal, after combining, will propagate either toward the termination resistor or toward the side port, depending on the relative polarity of voltage and current. This is what gives the coupler its directivity.

By varying the turns ratio of both transformers, various values of coupling are possible. Ellis[13] has published a tabulation of coupling values as a function of turns ratios, along with a good description of coupler operation. Table 9.2 lists some commonly used values and their theoretical signal loss values (practical units will exhibit 1–2 dB of excess loss due to imperfect components). The coupled loss is simply due to the voltage (or current) ratio defined by the transformers, whereas the input/output loss just reflects the loss diverted to the coupled port (or isolation resistor, depending on signal flow).

As is the case with amplifiers, added components in units intended for installation in trunk and distribution cables separate power from signal voltages and send them through different paths, as shown in Figure 9.21(a). Generally, means are provided to control transmission of power independently from RF signals so

Table 9.2 Theoretical directional coupler losses.

Turns Ratio (T1 and T2)	Input to Output Insertion Loss (dB)	Input to Coupled Port Loss (dB)
2.5:1	0.78	7.96
3:1	0.51	9.54
4:1	0.28	12.04
5:1	0.18	13.97

as to allow greatest flexibility in system design. Note that, if ac power is transmitted through all RF ports, the displacement current through the three blocking capacitors will add in phase through the primary of the voltage transformer. Thus, couplers generate parametric type hum modulation at levels as high as or higher than amplifiers.

The difference in sensitivity to downstream and upstream signals, as measured at the side port, is the *directivity* of the coupler. Sometimes, instead of directivity, the absolute upstream coupling factor, known as *isolation,* is specified. Numerically, directivity is simply the difference between coupling factor and isolation. As we will discuss in Chapter 13, directivity and return loss interact to cause additional system group delay and response variation.

The coupling value may not be uniform across the entire frequency range of the device. Even though the overall variation in coupling factor may not be desirable or intentional, the slope and response variation are measured separately and are defined for passive devices the same as for amplifiers.

9.4.2 Splitters

Two-way splitters differ functionally from directional couplers only in that the signal splitting is equal. Other common configurations include four-way and eight-way equal splits. Three-way unequal splits are internally constructed by splitting two ways, then resplitting one of the legs from the first splitter to create a 50%:25%:25% output ratio. Equal ratio three-way splitters are less common.

Figure 9.22 illustrates a two-way symmetrical splitter, with expansion to four way. Figure 9.22(a) is the schematic diagram of a two-way unit, and Figure 9.22(b) is the standard symbol, with terminology. Transformer T1 is a 2:1 impedance step-down transformer that creates a 37.5-ohm source to feed the center tap of transformer T2. T2 is a transmission-line transformer derived from a Wilkinson coupler (which is used for symmetrical splitting of high-power, narrow band signals for directional antenna arrays, among other applications).[14] The two outputs at ports 2 and 3 are nominally in phase and at the same level. Resistor R1 is required to provide isolation between them.

Because power is equally divided between ports 2 and 3, the theoretical loss from port 1 to either output is 10 log(0.5) = −3.01 dB. As with directional cou-

Figure 9.22 Splitters. (a) Schematic. (b) External power flow. (c) Multiple stages of splitting to create *m*-way splitter with enhanced isolation among some ports.

plers, the practical loss will be higher, due to core losses in the ferrite core material in the transformers and resistive losses in the transformer wire.

The dimensions of transformers used in high-frequency applications are constrained by the necessity for the winding lengths to be a small portion of a wavelength at the highest frequency of operation. This, in turn, mandates very small cores and fine-wire gauges whose resistance is a factor in excess signal loss. The excess loss over and above that due to coupling is typically 0.5 to 1 dB to each of the output ports, making the total loss typically 3.5 to 4 dB.

Loss in a splitter is called *flat loss* because it tends not to be a strong function of frequency, as is the loss in coaxial cable. However, there is some increased loss as the frequency increases, partially due to skin effects, which increase the effective resistance of the wire as frequency increases. At the higher frequencies, the ferrite core on which the transformer is wound ceases to have significant magnetic properties, but rather acts as a coil form. Some loss may be associated with it, but the ferrite material is primarily included to improve low-frequency

performance. At the lowest frequencies, loss also increases because the open-circuit reactance inductance of the transformer windings is no longer large relative to the circuit impedance.

When the splitter is used as a combiner, incoming signals on ports 2 and 3 are combined and appear at port 1. The theoretical loss from port 2 (or port 3) to port 1 is also 3 dB, the other half of the power being dissipated in resistor R1. Theoretically, no power from port 2 appears at port 3 (if all ports are properly terminated) and vice versa. The signal loss between the output ports, with all ports terminated in 75 ohms, is defined as the isolation. Typical values of isolation for commercially available splitters range from 20 to 30 dB.

Other important specifications of splitters include insertion loss, normally specified as the total loss from port 1 to port 2, with port 3 terminated (or vice versa), and the return loss at each port, with the other ports terminated.

Lack of termination drastically reduces the effective isolation. For instance, power injected into port 3 is attenuated −3 dB at port 1. Should that port be unterminated, the power would be reflected into the splitter dividing equally between ports 2 and 3. Thus, the isolation from port 3 to port 2 would be only 6 dB. Similarly, the effective return loss at port 3 would be 6 dB as a result of power being reflected to that port.

Splitters having more output ports can be fabricated by combining two-way splitters in tree fashion, as shown in Figure 9.22(c). Each leg of the leftmost splitter is connected to the input of another two-way splitter. At that level, four outputs exist. Each of those four can be connected to another two-way splitter, resulting in eight outputs, and so on. The practical loss at each stage will be 3.5–4 dB so that the four-way split loss is 7–8 dB. In an eight-way splitter, it is 10.5–12 dB and so on. In general, the loss in a multilevel splitter is given by

$$L_m = \left[\frac{L_1}{0.301}\right]\log(m) \qquad (9.35)$$

where

L_m = the loss of the m-output splitter chain in dB

L_1 = the loss in one two-way splitter (3.5–4 dB usually)

It is also possible to construct a splitter chain with a number of outputs not equal to 2^n by simply terminating the splits early on some of the chains. Equation (9.35) must be modified in this case.

Where output port isolation is important in multiport applications (such as headend combining) one alternative is to use ports in multilevel splitters that are not derived from common stages for critical applications. As shown in Figure 9.22(c), if two splitter outputs are connected to the same second-stage splitter, the isolation between them is merely the isolation of that stage. On the other hand, if two ports are not connected to the same second-stage splitter, then the isolation realized will include the effects of the first-stage splitter, as shown in the

figure. Unfortunately, manufacturers may not indicate internal splitter config-urations or specify nonadjacent port isolation.

Another higher-isolation alternative is to construct multiport splitters from cascaded directional couplers. Generally, side ports of adjacent, series-connected couplers will be better isolated than outputs from single-stage splitters.

9.4.3 Taps

The most common tap configuration consists of a directional coupler with the side arm signal split two, four, or eight ways to create customer drops. Although the input and through ports are typically provided with connectors appropriate for distribution cable, the side arms are equipped with connectors appropriate for drop cable. In North American systems, this means "5/8-24" ports for input and through ports and type F connectors for the drops.

A tap need not have a through port. At the end of the line, it is common to use a *terminating tap,* which is just a splitter whose output legs are equipped with drop connectors.

Until about 1995, taps supplied RF signals, but not power, to customer drop lines and were limited in main-line current-handling capacity to 7 amperes or less since this was adequate for entertainment-based cable television systems. With the active consideration of HFC architectures for telephony service came in-creased demands on tap capabilities:

- Bandwidth expansion to 5–1,000 MHz
- The ability to supply power to side-of-house active *network interface devices* (NIDs) that could provide an interface between baseband twisted pair in-home telephone circuits and the RF carriers in the distribution system
- Provisions for changing tap values and configurations without interrupting service to customers farther along the distribution cable
- Increased voltage and current capability, typically 90 Vrms and 10 amperes, in order to allow transmission of sufficient power for both network ampli-fiers and NIDs over the same coaxial cables formerly used only for video service

The new generation of taps that have evolved are known as *power-passing taps* or, sometimes, *telephony taps.* In addition to normal tap capabilities, they tap into the distribution cable power and selectively provide current-limited power along with RF signals to a NID on each home. Depending on the design, this power may be multiplexed with the signals in the coaxial drop cable or carried along a separate twisted pair of copper wires in a hybrid cable configuration. In general, these taps also provide a make-before-break connection between the in-put and through ports when the tap "plate" (which contains all the circuitry) is removed from the housing so that both RF and power continuity are maintained to customers downstream.

Figure 9.23 Power-passing tap using drop center conductor for power delivery to home.

A major concern for those using the drop center conductor to carry power to NIDs is that, if arcing occurs because of a faulty center conductor contact, the arc will transfer a very significant amount of RF power to the upstream plant, likely causing interference with all users of the reverse spectrum. On the other hand, an arc often causes healing of a bad contact and, thus, can be self-healing.

Figure 9.23 is a circuit diagram for a typical power-passing tap that uses the drop center conductor to pass power to the NID. The RF portion of the tap consists of a directional coupler and two-way splitter (the diagram is for a two-port tap). The directional coupler diverts a controlled portion of the downstream signal and passes it to the splitter, which divides the power to feed the subscriber ports.

Multiplexed ac power consideration in taps is similar to that on amplifiers and other network passives. Capacitors C1 and C2 isolate the multiplexed power supply voltage and current from the directional coupler RF components. Transformers T1 and T2 are built with very fine wire to permit operation at frequencies

up to 1 GHz and would be unable to withstand the currents that would result if the primary of T1 were required to handle several amperes of ac current or if the primary of T2 had tens of volts of 60 Hz across it. The multiplexed ac power is shunted across the tap through the large RF choke L1.

A choke with somewhat less current capacity, L2, provides a path for the multiplexed power that will feed the customer ports. Since this choke is shunt-connected across the main RF path, it must have a higher RF impedance than L1. The unavoidable RF loss due to this choke is one reason power-passing taps have a slightly higher RF loss than normal taps.

In order to permit the tap plate (which contains most, if not all, the circuitry) to be removed without interrupting either power or RF continuity in the hard cable distribution plant, many modern taps incorporate a mechanical bypass switch that engages before the tap-to-baseplate connections are broken to provide continuity from input to through ports. During the removal or insertion of a tap plate, therefore, there will be a brief interval when both the bypass and main connections are made. This will cause a transient condition during which the primary of T1 will be shorted, but the primary of T2 will function normally. This momentary mismatch will cause some attenuation of the transmitted signals. Whether or not it causes a functional problem to the services offered will depend on the characteristics of the signals used.

Continuing with the description of the circuit, capacitor C3 is used to ensure a good RF ground on the "cold" end of L2. Positive temperature coefficient (PTC) resistors PTC1 and PTC2 limit the ac current that can be drawn through the customer ports. At normal currents, the PTC resistor (commonly called just a PTC) exhibits low resistance. As more current is drawn, its temperature increases, causing a resistance increase and effectively limiting the total current. Electronic current limiters have been proposed as an alternative to PTCs, but to date have not been widely deployed because of cost. New National Electric Code (NEC) regulations, effective in 1999, require a more sophisticated current-limiting system if the drop cable does not meet certain construction and installation criteria.

Finally, inductors L3 and L4 couple power to the two tap ports shown, again isolating the RF signals from the ac supply. As with L2, the reactance of these components must be sufficient to minimize RF losses, and their presence leads to slightly higher overall RF losses in commercially available power-passing taps.

The relationship between passive specifications and system performance will be explored in Chapter 13.

9.5 Power Supplies

Power supplies convert commercial power to lower voltages that are multiplexed with RF signals in the coaxial cables. Although not part of the RF circuit, certain characteristics of power supplies have an effect on the transmission of signals.

Direct current, very low frequency (as low as 1 Hz) ac, and 60-Hz ac have all been considered for system power, as have voltages between 30 and 90 volts. Direct current (dc) has been rejected primarily because of possible galvanic corrosion. Early cable television systems used 30 volts, but the increased power demands of wider-bandwidth, higher-output amplifiers forced the industry to convert to 60 volts for systems constructed after about 1970. Two factors are forcing consideration of even higher voltages: the desire to power larger plant sections from a single power supply to improve reliability (see Chapter 17 on HFC reliability for a detailed explanation) and the possibility of powering some terminal equipment through the distribution plant.

Along with the increase in voltage has come a requirement to increase currents as far as practical, with requirements to pass 10–15 amperes being common. Since current from supplies can be transmitted both ways along the cable and/or inserted into more than one cable, power supplies with 40 or more ampere capacity are available.

Field power supplies are roughly divided among nonstandby, standby, and UPS/standby types. The basic component of a nonstandby, line frequency power supply is the ferroresonant transformer. Though its magnetic design is beyond the scope of this book, ferroresonant transformers have a number of properties of interest to operators: (1) output voltage regulation (a typical specification is ±5% output voltage change when the input voltage changes ±15%), (2) inherent current-limiting and short-circuit protection, (3) transient protection, and (4) 90% or better efficiency at rated output current.

Standby power supplies add a battery charger, a set of rechargeable batteries, an inverter, and a changeover switch. The charger assures that the batteries are always held at full charge, whereas the inverter starts and supplies square wave power to replace the sine wave commercial power when it is interrupted (see Figure 9.24). Depending on the application, sufficient battery capacity may be provided for 2–8 hours or more of standby operation.

An important parameter for operation in the standby mode is the rise and fall time of the square wave voltage. As Equation (9.32) shows, the amount of ac power supply current that "leaks" into the low-level signal path in amplifiers and passives is directly proportional to the rise time of this waveform. The trade-off, for power supply designers, is that the efficiency of an inverter is an inverse function of the time the transistors spend in their active region (that is, not in a saturated or cutoff condition). Also, the power lost due to slow switching is dissipated in the transistors and must be conducted away by heat sinks. Unfortunately, the industry does not have a consensus on the best compromise on rise time.

When commercial power fails, there is an unavoidable but brief interruption while the relay switches between the ferroresonant transformer output and the inverter transformer output. Typically, the time to switch is 8–16 ms. Most supplies synchronize their inverters with commercial power before switching so that phase shift transients in the power supplies of the amplifiers can be avoided.

Figure 9.24 Basic standby power supply functional diagram.

As will be discussed in Chapter 10, even with inverter phase synchronization, the short switching transient can be a problem when analyzed with respect to its effect on cascaded amplifiers. To avoid that, a third alternative is the uninterruptible power supply (UPS). Some UPS versions of field supplies avoid switching transients by utilizing the same ferroresonant transformer for both inverter and commercial power. During ac power outages, commercial power is disconnected, and the inverter continues to supply power via an additional primary winding.

9.6 Summary

This chapter has covered the technology behind coaxial distribution systems. The characteristics of the coaxial cable, amplifiers, passive components, and powering systems have been covered in detail, including both the theory behind operation and the performance of commercially available products.

Although other constructions are sometimes used, solid aluminum sheathed cable of 75-ohm impedance has been seen to offer a good combination of loss, cost-effectiveness, and handling characteristics and are in common use worldwide. Amplifier stations designed for cable systems are designed with unique characteristics that complement the characteristics of the interconnecting cable and passive devices. Integrated powering of active devices through the cable avoids the necessity of separately connecting each distribution system amplifier to the local utility, but requires special circuitry in every device to separate the power and signal paths.

Before the mid-1980s, cable television systems used primarily coaxial technology (sometimes supplemented by broadband microwave links) to serve their customers. Chapter 10 will cover the design of coaxial distribution systems, including both the shared portions and individual in-home "drop" wiring. As will

be seen, the performance of such networks is subject to certain unavoidable limitations that become more severe as bandwidth is increased. As a result, modern systems generally use fiber-optic "supertrunks" in place of the long, repeatered coaxial trunks to interconnect small, physically separate coaxial distribution networks with shared headends. Chapter 11 will cover the technology and performance of such links. Chapter 12 will then cover alternatives to fiber optics (primarily microwave). Chapter 13 will cover the end-to-end downstream performance of the broadband network, from headend to subscriber, and Chapter 14 will deal with upstream issues.

Endnotes

1. *Reference Data for Radio Engineers,* 6th ed. Indianapolis, IN: Howard W. Sams, 1977, p. 6-6.

2. Frederick Terman, *Electronic and Radio Engineering.* New York: McGraw-Hill, 1955, p. 121.

3. *NCTA Recommended Practices for Measurements on Cable Television Systems,* 2nd ed. NCTA, Washington, D.C., October 1993, Part Six.

4. The size 59, 6, and 11 cables were formerly known as RG59, RG6, and RG11 because of their physical size similarity to standard military-grade 75-ohm cables carrying those designations. Since CATV cables do not attempt to adhere to the military specifications, the RG notation has been dropped by the industry.

5. The SCTE is an ANSI-accredited standards-making body. It publishes a wide variety of standards covering both performance and test methods related to cable television products. In particular, SCTE-IPS-TP-011, *Test Method for Transfer Impedance,* and SCTE-IPS-TP-403, *Test Method for Shielding Effectiveness,* detail test methods for evaluating shielding integrity, whereas SCTE-IPS-SP-001, *Specification for Flexible RF Coaxial Drop Cable,* contains both mechanical and electrical minimum performance standards.

6. *Code of Federal Regulations:* C.F.R. 47 §76.5(w). Technically, the noise is defined as that occupying the band between 1.25 and 5.26 MHz above the lower band edge. Available from the Government Printing Office, Washington, D.C.

7. An excellent detailed mathematical treatment of amplifier distortion based on a power-series expansion is given in the *Technical Handbook for CATV Systems,* authored by Ken Simons and published by Jerrold Electronics Corporation, Hatboro, PA (the third edition was printed in 1968 and was still in print in 1978). His results were also published in the *Proceedings of the IEEE,* in July 1970.

8. Data taken from *Cable TV Reference Guide,* April 1992 edition, Philips Broadband Networks, 100 Fairgrounds Drive, Manlius, NY 13104.

9. Private correspondence from Archer Taylor based on his experiences with installed systems.

10. This treatment of CID is based on the paper by Jeff Hamilton and Dean Stoneback, The Effect of Digital Carriers on Analog CATV Distribution Systems, *1993 NCTA Technical Papers,* pp. 100–113. NCTA, Washington, D.C.

11. SCTE IPS-TP-201 *Test Method for Frequency Response and Gain.* SCTE, Exton, PA.

12. A basic discussion of feed-forward gain block principles is contained in the paper by John Pavlic, Some Considerations for Applying Several Feedforward Gain Block Models to CATV Distribution Amplifiers, *Technical Papers, Cable '83.* NCTA, Washington, D.C., June 1983.

13. Michael G. Ellis, RF Directional Couplers. *RF Design.* February 1997, pp. 33ff.

14. Robert P. Gilmore, Applications of Power Combining in Communications Systems. *Application Note.* Qualcomm, Inc., San Diego, CA.

Chapter 10

Coaxial Distribution System Design

10.1 Introduction

Distribution system design is, ideally, the process of specifying the most economical network that will provide the required bidirectional bandwidth to the required number of terminal points and still meet the performance goals. The performance goals, in turn, may be derived from internal quality standards, governmental regulations, the perceived requirements for reliable signal transportation, or all of the above.

Typical system specifications will include noise, distortion, response variation, delivered signal levels (and their stability), and hum. Reliability, another important consideration, is discussed in Chapter 17, and the effects of imperfect impedance matching among components are discussed in Chapter 13. The distribution system performance must also be considered as part of a chain that includes the headend channel processing and terminal equipment effects of signals. These are considered in separate chapters as well. Finally, the coaxial distribution system may well be cascaded with a signal transportation link such as microwave or, more frequently, a fiber-optic circuit. These, also, are treated in different chapters.

The final leg in the broadband distribution system is the drop wiring leading from tap ports to and throughout homes. By tradition, broadband networks are designed to meet certain goals from headend output to tap port, with an additional allowance made for the effects of in-home wiring. Modern drop systems, however, may vary widely in their complexity and quality, and in fact, may constitute minidistribution networks by themselves. The final section of this chapter deals with drops.

We begin by discussing basic principles and applying them to the "hard cable" portion of the plant.

10.2 Carrier-to-Noise Ratio

Noise is defined by the IEEE as "unwanted disturbances superposed upon a useful signal that tend to obscure its information content."[1] In many communications systems, that is a good working definition. In cable television, however, the term *noise* is generally applied to thermal noise, which is treated independently from disturbances caused, for instance, by intermodulation products, ingress, or other causes. One of the reasons for treating thermal noise differently from other disturbances is that the subjective effects of noise and various other types of interference are different for analog video signals. Another is that, unlike many communications systems, the sources and mechanisms by which interfering signals are generated are well understood, and a detailed independent analysis of each has proved useful. Therefore, in this book, unless otherwise stated, *noise* will be used to refer exclusively to thermal noise.

In general, thermal noise is *uncorrelated,* that is, unpredictable and non-repetitive. Noise generated in one device or at one time is totally unrelated to noise in another device or in the same device at another time although the *average* level of noise power may be quite predictable. Since that is the case, when the noise from two independent sources is combined, the total noise *power* will be the sum of the original noise powers.

Carrier-to-noise ratio (C/N) is defined as follows:

$$C/N(\text{dB}) \equiv 10 \log\left(\frac{c}{n}\right) \tag{10.1}$$

where

 c and n = the scalar power levels of the carrier and noise

For analog television signals, such as NTSC, the carrier power is defined as that rms power that occurs during the sync tip (the highest power level in the modulated signal). Noise is defined as the rms noise power measured in a defined bandwidth (4 MHz for NTSC). Equation (9.19) gives the C/N for a channel transmitted through an individual amplifier, given the input signal level and noise figure.

In order to design a coaxial distribution system, it is necessary to compute the C/N of cascaded amplifiers, each followed by loss equal to the gain of the preceding amplifier. For two amplifiers, the general expression for composite C/N is approximated by the expression

$$C/N_{ttl} = -10 \log\left[10^{\left(\frac{C/N_1}{10}\right)} + 10^{\left(\frac{C/N_2}{10}\right)}\right] \tag{10.2}$$

where

 C/N_1 and C/N_2 = the C/N of each of the amplifiers calculated independently using Equation (9.19) (for example, when driven by a noise-free input signal)

Although this general expression is useful when cascading two dissimilar network elements, a simpler approach is useful when cascading n identical amplifiers, separated by passive network losses equal to the amplifier gain:

$$C/N_{cascade} = C/N_A - 10 \log n \tag{10.3}$$

where

 C/N_A = the C/N of a single amplifier

n = the number of similar amplifiers cascaded and separated by losses equal to their gains

These equations are quite useful, but their limitations must be emphasized.

- Equation (10.3) is applicable only to a cascade of amplifiers and loss sections where the loss of each section equals the gain of the preceding amplifier, and each amplifier has the same noise figure.
- Equations (10.2) and (10.3) are accurate only where the gain of each amplifier is high (a minimum gain of 20 dB is adequate to assure reasonable accuracy).
- Equations (10.2) and (10.3) are accurate only where the amplifier output noise power is high when compared with the thermal noise "floor" (a noise figure of 2 dB or greater combined with a 20-dB or greater gain is sufficient to assure reasonable accuracy).
- The input to the first amplifier contains only thermal noise.

The preceding conditions are required because the equations do not rigorously account for the effects of thermal noise though they are adequate for analyzing typical amplifier cascades encountered in cable television systems. For other applications, a more precise set of equations is recommended.

In a simple repeated transmission line, the designer must "juggle" three parameters to reach a required end-of-line C/N: the noise figure of individual amplifiers, the input level to each amplifier, and the number of amplifiers to be cascaded. In more complex networks, the same variables apply, but the C/N "budget" may first have been allocated among several dissimilar network sections. As will be seen, noise considerations must be balanced against other requirements in reaching an optimal design.

10.3 Carrier to Distortion

Unlike the thermal noise, which is a function of only the transmission system characteristics and channel bandwidth, the distortion level depends on the signals carried. The distortion level of amplifiers is typically specified under one or two specific channel-loading conditions. Although it would be possible to simply specify distortion with just two or three carriers, the translation to expected IM products under the unique channel loading used in cable television is not straightforward, so vendors have resorted to testing under conditions that are easily related to actual field applications.

As discussed in Chapter 9, the channel loading under which amplifiers are typically tested and specified (for CTB and CSO) consists of independently generated, unmodulated carriers at each of the visual carrier frequencies specified in the ANSI/EIA-542 Standard frequency assignment list that falls within the bandwidth of the device. In addition, a portion of the bandwidth may be loaded with

signals that simulate the effect of digitally modulated signals (often just band-limited white noise at a controlled level). Testing of XMOD is similar except that the carriers are simultaneously modulated.

The results of such testing do not match the distortion when loaded with analog television carriers (whose sync peak carrier levels match the unmodulated carriers), but the results are much more repeatable, and there are standard assumptions regarding the approximate improvement when loaded with modulated carriers: 6 dB in the case of CSO and 12 dB in the case of CTB and XMOD, though the improvement will be somewhat less with short coaxial cascades.

The buildup of distortion as amplifiers are cascaded is less straightforward than in the case of noise. If the deviations from linear response are not predictable from amplifier to amplifier, then the distortions can be considered uncorrelated, in which case they will add in the same way as noise; that is, the distortion products will increase by $10 \log n$, where n is the number of amplifiers. On the other hand, if every amplifier has the same general distortion shape (for instance, a flattening of the curve at voltage peaks), then the distortion will be highly correlated, and the magnitude of the products will increase at a $20 \log n$ rate.

Since third-order distortion generally does result from symmetrical curve flattening and both CTB and XMOD arise from third-order distortion, most designs assume that these distortions increase at the $20 \log n$ rate although that may be overly conservative when evaluating cascades of amplifiers with dissimilar characteristics or those that nominally cancel third-order distortion (for example, feed-forward amplifiers). Second-order distortions, on the other hand, are nominally canceled in the universally used push–pull gain modules. For that reason, the residual second-order distortions are not highly correlated, and many designs are based on a compromise $15 \log n$ buildup.

As with cascades of identical amplifiers, the buildup of distortion when cascading different network sections depends on the degree to which the distortions are correlated.[2] In general,

$$C/dist = -x \log \left[10^{-\left(\frac{C/dist_1}{x} \right)} + 10^{-\left(\frac{C/dist_2}{x} \right)} \right] \tag{10.4}$$

where

$$C/dist = \text{the resultant carrier-to-distortion ratio in dB}$$

$$x = \text{from 10 to 20 depending on the degree of distortion correlation}$$

$$C/dist_1 \text{ and } C/dist_2 = \text{the distortion ratios of the individual sections}$$

Where identical amplifiers operating at the same output levels are cascaded, the expression for C/CTB can be reduced to

$$C/CTB_{cascade} = C/CTB_A - 20 \log n \tag{10.5}$$

where

C/CTB_A = the distortion of a single amplifier

n = the number of cascaded amplifiers

whereas the equivalent expression for CSO is

$$C/CSO_{\text{cascade}} = C/CSO_A - x \log n \tag{10.6}$$

where

C/CSO_A = the distortion of a single amplifier

n = the number of cascaded amplifiers

x = from 10 to 20 depending on the degree of second-order distortion correlation

10.4 Noise–Distortion Trade-Off

The interplay between noise and distortion can be readily illustrated by considering a cascade of n identical amplifiers. It can readily be understood that the loss between each pair of amplifiers must be made identical to the gain of one amplifier for optimum performance. If the loss is less than that value, then each amplifier's input level (and, hence, output level) will be greater than the previous amplifier, and the distortions will quickly build up to a high level. If the loss is greater than the gain, then the input of each amplifier will be less than the previous amplifier and thus contribute disproportionately to the overall C/N degradation.

Given equal gains and losses, let us assume a requirement that the cascade provide an analog television C/N of 50 dB and a C/CTB of 65 dB. (We could have considered any other distortion parameter as easily, but this will serve to illustrate the trade-offs involved.) Let us also assume that each amplifier has a noise figure of 8 dB, a gain of 22 dB, and a C/CTB of 80 dB at the required channel loading and at an output level per carrier of +40 dBmV. (Again, to make it simple, we will assume all carriers have the same level and the loss between the amplifiers is the same at all frequencies.)

If our cascade consists of just a single amplifier ($n = 1$), then we can calculate the minimum input level that will assure an adequate C/N. From Equation (9.19):

$$C_i(\text{minimum}) = C/N(\text{dB}) - 59.2 + F(\text{dB}) = -1.2 \text{ dBmV} \tag{10.7}$$

With a station gain of 22 dB, this corresponds to an output level of +20.8 dBmV.

Similarly, since CTB changes 2 dB for every 1-dB change in operating level, we can readily determine that we could increase the output operating level to

Table 10.1 Example of cascade effects on allowable amplifier operating levels.

Amplifier Cascade	Allowable Output Level Range (dBmV)	
	Minimum (C/N limited)	Maximum (CTB limited)
1	+20.8	+47.5
2	+23.8	+44.5
4	+26.8	+41.5
8	+29.8	+38.8
16	+32.8	+35.8
21	+34.0	+34.3
>21	Not possible and still meet both noise and distortion specs	

+47.5 dBmV and stay within the distortion limit. Thus, the output level could vary over a 26.7-dB range while maintaining the required performance.

If additional amplifiers are cascaded, the effective noise figure increases by 3 dB for every doubling of the cascade (Equation 10.3), so the minimum C_i (and thus output level) must increase by that amount to provide the same end-of-line C/N. Similarly, the distortion increases by 6 dB for every doubling of the cascade, so the maximum allowable output level must decrease by 3 dB to maintain the same end-of-line distortion levels. Thus the allowable range of signal levels decreases by 6 dB.

Table 10.1 shows the allowable output level range as a function of cascade for this example.

If we plot these general relationships as a function of cascade, we can see that there is a maximum attainable cascade and a unique operating level that allows that cascade to be realized. Figure 10.1 illustrates the usable operating range, along with the parameters that define the noise–distortion–cascade relationship. Complete system design requires that this calculation be performed for both second- and third-order distortion, as well as noise, and to pick operating levels that allow all specifications to be met. Generally, composite intermodulation noise (CIN) caused by digital signals is quantified as an equivalent increase in noise level that must be figured into the C/N calculation.

Another way of looking at the interaction between parameters is to consider the case of a hypothetical tapped distribution line that just meets end-of-line noise and distortion specifications, as well as required signal levels at subscriber terminal equipment. The operator desires to increase the bandwidth and channel loading of this system without degrading performance.

At the increased upper frequency limit, the cable loss is higher. The designer can compensate for that loss by using higher-gain amplifiers, spacing amplifiers

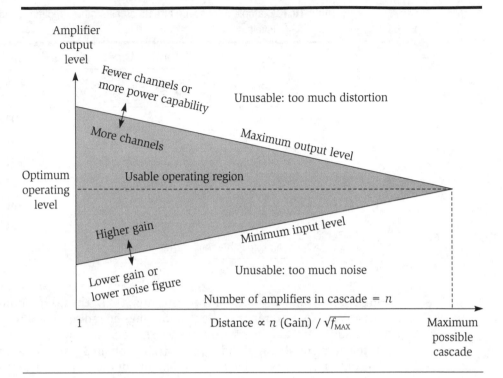

Figure 10.1 Relationship of cascade, noise, distortion, and levels.

closer together, or reducing the per-foot loss of the interconnecting cable. Additionally, the increased number of carried signals will increase the distortion products in each channel if the per-carrier level is held constant. Finally, the loss of subscriber drop wiring will increase, requiring higher tap levels or lower-loss cables.

If higher-gain amplifiers are used, then either output levels must be raised or input levels lowered. Absent a change in technology, the former will further increase distortion, whereas the latter will cause a greater C/N degradation in each amplifier station. If amplifiers are spaced more closely together, then more will be required to reach the same physical distance, thereby increasing both noise and distortion. Even if amplifier spacing is retained, either amplifier output levels must be raised or drop cables replaced to maintain adequate levels at subscriber equipment at the highest channels. Absent changes in technology, the designer is faced with an insurmountable obstacle.

Several techniques can be used to get out of this "box": (1) amplifiers with lower noise figures and/or distortions can be used; (2) both distribution and drop cables can be replaced with larger, lower-loss versions; and/or (3) methods can be used to transport signals to points closer to homes so that coaxial cascades can be shorter. Generally, upgrades to older all-coaxial systems employ elements of all three techniques. In particular, fiber optics or microwave is used as a *super-*

trunk to transport signals throughout large service areas with less noise and distortion than reasonably possible with coaxial techniques.

10.5 System Powering

As discussed in Chapter 9, coaxial distribution systems are usually powered by multiplexing 50- or 60-Hz power with the signals and transporting both through a common cable. Through use of separate signal paths through amplifier stations and passive devices, power routing is configured independently from the signal paths.

Coaxial distribution systems may be powered from one end (centralized powering) or via power supplies located along the cable route (distributed powering). In the latter case, it is common to locate the power supply midway through the section of plant it supplies so that roughly half the current can flow in each direction from the insertion point, reducing both the current capacity requirement for components and resistive losses in the interconnecting cables.

The reliability of distribution system power has a major effect on overall reliability. Therefore, systems designed for high availability are generally built with coaxial distribution networks that are centrally powered with high-capacity standby supplies. Chapter 17 covers reliability and availability.

Figure 10.2 is a simplified schematic of a section of cable and several amplifiers extending one direction from a power supply. From a network designer's point of view, the challenge is to minimize the number of power supplies required or, put another way, to determine the maximum number of amplifiers that can be powered from each supply.

Given a decision on system supply voltages, several parameters cannot be exceeded in making this determination: (1) the current rating of the supply, (2) the current rating of any coaxial component along the route, and (3) the minimum usable voltage at any powered component (typically around 40 volts for 60- or 90-volt systems).

As previously discussed, most modern amplifiers use switching power supplies. These draw current in approximately inverse proportion to the supply voltage so that the total power drawn is a constant. This allows them to maintain high efficiency and low power dissipation. It also complicates the power calculation.

Figure 10.3 is a redrafting of the first seven amplifiers in Figure 10.2 to show the power component interaction. Though it would be possible to set up simultaneous equations for each powering situation and solve for all the critical voltages and currents, a successive approximation procedure is generally used for both manual and computer calculations. The process is more or less as follows:

1. Determine the resistance of each cable leg at the maximum expected operating temperature.
2. Determine the voltage–current curve for each of the types of amplifiers to be used.

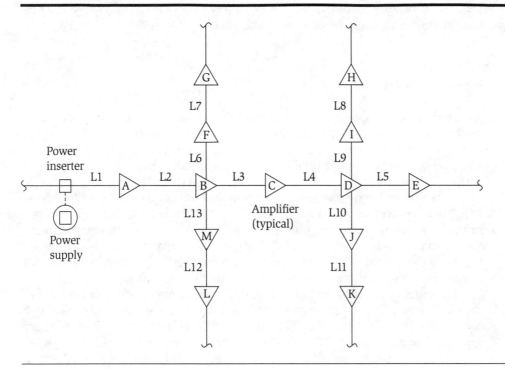

Figure 10.2 Example of amplifier powering.

where $I_n = P_n/E_n$
P_n = power consumption of amplifier n

Figure 10.3 Equivalent circuit of powered coaxial network.

3. Determine the pass-through current ratings for each of the system components.

4. Make an initial estimate of the supply voltages at each of the amplifiers and the number of amplifiers to be served.

5. From that data, determine the current draw of each amplifier and, therefore, the total current drawn in each cable segment.

6. Using the cable currents from step 5 and the resistances from step 1, determine the voltage drop in each segment.

7. Starting at the power supply, subtract the first segment drop from the supply voltage to determine the voltage at amplifier 1. If different from the initial estimate, reestimate the amplifier voltages and currents and loop back to step 6.

8. When the iterative process results in adequately consistent results (very small adjustments needed), confirm that each of the current ratings has not been exceeded and that the voltage at the most distant amplifier is adequate. If not, drop one amplifier and start over.

The actual calculation is very easy on a computerized system. The "black magic" is in strategically placing power supplies and determining network power boundaries. Not only is it important to minimize the number of power supplies, but from a reliability standpoint, it is important to minimize the number of cascaded supplies that affect any given subscriber. Finally, it is important to locate power supplies where the commercial power is most reliable.[3]

A technique sometimes used to overcome both the maximum component current ratings and the voltage drop in the high-current sections near the power supplies is to add a second cable extending from power supply locations to additional strategically located power insertion points. Since these cables do not carry RF, they can be of lower characteristic impedance and, thus, use larger center conductors. For instance, the loop resistance of standard 0.625-inch diameter P3 cable is 1.1 ohms/1,000 feet, whereas the same size "power feeder" cable is 0.290 ohms/1,000 feet.

10.5.1 Switching Transient Effects

A consequence of the interaction among switching power supplies located along a coaxial line is that anything that causes one amplifier to draw additional current will cause all other amplifiers along the same line to also draw additional current. This is because, regardless of where it is located, the additional current supplied to one amplifier will increase the current through the first coaxial segment, which will, in turn, result in lower supply voltages and therefore higher currents for all other amplifiers.

Though amplifiers normally draw a constant amount of power, consider what happens when a non-UPS system power supply switches between commercial and standby power. As discussed in Chapter 9, typical switching times are 8–16 ms. During this period, no power is supplied to the network.

When the power is restored, each of the amplifier power packs will attempt to draw additional current to make up for the missing 1–2 half cycles. The result is that the voltage at the most distant amplifier may well drop below operation range. In fact, one study has shown that interruptions as short as 8 ms can grow to effective interruptions of greater than 50 ms at the fifth amplifier away from a power supply.[4] It is incumbent upon network designers to evaluate whether such interruptions are unacceptable for anticipated services and signal formats and to consider UPS switching if appropriate.

10.5.2 Hum Buildup

The mechanism for hum modulation is similar in passives and the signal-processing components of amplifier stations. Since the period of the power signal is much longer than the transit time through the network (the wavelength of 60 Hz is about 3,000 miles), a conservative assumption would be that effects of the multiplexed power will be in phase throughout the system and that, therefore, hum modulation due to parametric effects should build up in a 20 log fashion. As a practical matter, however, much high-voltage utility distribution is three phase, and as a consequence, power supplies throughout a large service area are not likely to all be connected to the same power phase.

To illustrate the worst-case magnitude, assume that the cascade in a distribution line includes 5 amplifier stations, 4 splitters and/or couplers, and 30 taps. If each has a hum rating of -70 dB and all components are subjected to maximum voltage and current stress, the end-of-line hum would be $-70 + 20 \log(5 + 4 + 30) = -38$ dB, equivalent to 1.2%. Powering, however, will likely be distributed across several phases, and most components, especially end-of-line taps, are unlikely to experience anywhere close to their rated voltage and current so that total hum buildup should be well under 1% if all components are operating properly.

10.6 Signal Level Management

The requirements for level management are very different in the forward and reverse directions. In the downstream or forward direction, each amplifier receives signals from only a single source whose output is well defined. The design and operational requirements are constant amplifier output levels and adequate amplifier input levels. Additionally, to ensure that well-designed end-of-line receivers work satisfactorily, several parameters must be controlled: adequate signal strength to overcome receiver noise, avoidance of excess levels that could lead to overload, limits on adjacent channel level differences, limits on total instantaneous variation across the spectrum, and limits on changes with respect to time. These are more fully discussed in Chapters 19 and 20.

As discussed in Chapter 9, manufacturers typically specify performance with specific channel loadings and amplitudes. Since digital formats are generally more robust than analog television, it is common to specify performance under

two conditions: a full spectrum of analog television signals and a defined mix of analog television and digital signals, where the digital signals are operated at a relatively lower level (as illustrated in Figure 9.19). Note that the *net gain* of the network, as measured from the output of one amplifier to the output of the next amplifier, is 0 dB in either case. The depressed levels of the digital signals relative to the analog signals are set in the headend where they are combined.

To the extent that the positive gain slope built into amplifiers does not exactly compensate for the losses between amplifiers, *equalizers* are used to eliminate the remaining variability. To the extent that interamplifier losses (including any equalizer losses) are less than amplifier gain, attenuators (also known as *pads*) are used to set the gain block input level. Pads may be placed before the first stage (generally, the only option with line extenders) and sometimes also between stages. If the latter option is available, then in cases where excess signal level is available, operators can independently set the operating levels of input and subsequent gain blocks to optimize noise–distortion trade-offs. With amplifiers that offer multiple independent output stages, the interstage pads also allow setting of different output levels for different ports. Thus, a port that directly feeds a tapped line may be set to a higher level than one that will feed additional amplifiers and therefore needs to run at a lower distortion level.

In tapped lines, both distribution cable losses and drop cable losses must be considered. It is important that adequate and relatively uniform levels be delivered to terminal equipment (see Chapter 19). A 200-foot size 6 drop cable has about 8 dB greater loss at 750 MHz than at 54 MHz. This differential loss tends to offset some of the intentional tilt in amplifier outputs if connected to a tap that is close to a station. On the other hand, if it is connected to the last tap in a string, it will aggravate the situation if the slope is already negative (high-end channels below the low-end channels in level). To prevent excessive level variation, in-line equalizers are sometimes inserted in tap strings. As a general rule, operators try to limit analog video channel level variation across the spectrum to about 10 dB. Typical design rules call for +15 to +20 dBmV levels at tap ports on the highest channels.

Signal level management in the upstream direction is much more complex and must be treated as an overall system problem rather than as just part of the design. Signals may originate from anywhere in the system and may be of various types so that the channel loading and levels may vary continuously at any given amplifier.

The preferred approach to upstream level and gain setting is to align the system from headend out so that the *input* level to each upstream amplifier is consistent; that is, if a level of x dBmV is inserted into the input of any upstream amplifier, the level received at the headend will be the same. Pads (and, if required, equalizers) are placed in the *output* of the gain modules to set overall system gains properly.

This does not, however, mean that the required upstream transmitter output levels for terminal equipment will be consistent. Consider Figure 10.4, which depicts a typical tapped line. Following normal practice, tap values are selected to

Figure 10.4 Signal levels in tapped distribution line.

produce approximately the same downstream level at the highest channel at each home. The circuit losses depicted are based on typical manufacturers' specifications for four-port taps, the loss of 250 feet of .500-inch type P3 cable between the taps, and the loss of 100 feet of size 6 drop cable connected to each tap port. As can be seen, the downstream levels at the end of the drops are all within 1 dB.

If we calculate the insertion levels required to reach the return amplifier, however, we see a nearly 20-dB difference due to the differential loss of the cable and the taps. Rather than manually set the level of each upstream transmitter, most manufacturers sense the level reaching the headend and remotely control each device independently to set the appropriate level. Another approach is to equalize each drop to approximately compensate for the differential cable loss and equalize required levels. As of the writing of this book, there is no uniformly accepted practice in this area.[5] Chapter 14 deals with the numerous unique issues related to upstream transmission.

10.7 Signal Level Stability

As was discussed in Chapter 9, the attenuation of coaxial cable varies about 1% per 10°F. The losses of other components, as well as amplifier gains, also vary somewhat with time and temperature. Amplifier manufacturers offer two options in their equipment that are designed to compensate these variations: thermal compensators and automatic gain and/or slope control circuits (AGC/ASC).

Thermal gain compensation is *open loop,* meaning that the gain adjustment is only an approximation to the variation in interconnecting losses with temperature. AGC circuits work by sensing the level of one or more of the carried signals and controlling the gain of the station so as to maintain a constant output level for that signal. Automatic slope controls sense two signal amplitudes and control both the overall gain and the gain slope through the stage so that the levels of both signals are constant at the output port. Typical control ranges are 6–8 dB. Use of AGC and/or ASC circuits depends on the accuracy of headend level setting. Also, if one of the reference carriers fails, the entire distribution system can change gain sufficiently to cause serious distortion products.

The loss between amplifiers may be all cable or a combination of cable and passive devices. In the extreme case of an all-cable link whose loss is 35 dB, the loss will vary over a range of about 5.2 dB over a range from +110°F to –40°F (+43°C to –40°C). Thus every amplifier will require some sort of compensation in a trunk network.

In a tapped distribution line or one that is split several times, half or more of the loss may take place in passive devices. Thus, thermal compensation may be required only in every second or third amplifier. The trade-off is that compensation networks may have loss that reduces available gain, whereas level variations will affect C/N, CTB, and CSO.

10.8 The Service Drop

So far we have dealt primarily with the portion of the distribution network that lies between the headend output and the tap ports that serve individual customers. Generally, network designers design to specifications for this portion of the network, assuming certain average performance in the headend and drop systems such that the end-to-end performance will meet regulatory requirements as well as subscriber expectations. Chapter 2 dealt with subscriber quality expectations, and Chapter 13 will cover end-to-end performance. This section will deal with the variation in drop subsystems and suggest the performance degradation that might be expected in a "standard" drop that must be taken into account in calculating end-to-end performance.

As a practical matter, in the United States, as in many other countries, all or most of the drop system may not be designed, built, or maintained by the network operator. Thus, although designers must be aware of the variation in drop performance, little can be done to control quality or design within homes.

10.8.1 Simple Passive Architectures

At its simplest, the drop will consist of a flexible (either overhead or underground) cable from the distribution tap port to the building entry point, a connection to the building ground system at that point, and another flexible cable within the building that leads to a user's receiver, such as a television set or VCR. If there

are multiple receiver locations, a signal splitter is used to divide the signal to feed more than one outlet.

If the loss in the interconnecting cables plus splitters is too great to assure adequate signal level at receiver input ports, then either (1) higher tap levels must be provided, (2) lower-loss drop cable must be used, or (3) amplification must be provided in the drop system. As discussed earlier in this chapter, tap signal levels generally are difficult to increase because of the interaction among levels, noise, and distortion in the distribution network. That leaves the builder of the drop with two practical choices: reduce loss or provide gain.

In the early history of cable systems, almost all drop wiring was done with size 59 cable (originally known as RG59). As bandwidths have increased beyond 300 MHz, common practice has been to upgrade the portion of the drop between tap and bonding (ground) block (the point where the cable enters the dwelling) to at least size 6 and, as required, size 7 and even size 11. The shift from 59 to 6 is easily justified by savings in the hard cable plant due to lower required tap levels. The larger sizes, however, are considerably more expensive and, in the case of size 11, require special connectors that add to both hardware cost and installation labor. Even in systems with size 6 outside drop cables, the inside wiring is generally size 59 since it is easier to handle and less obtrusive visually where exposed. Common practice is to design the hard cable plant to have drop levels that will accommodate most situations with size 6 exterior and size 59 interior cables, and to use larger cable sizes, both inside and outside, only where required because of exceptional drop lengths and/or a high number of splits.

A Reference Drop Structure

Although there are many single-receiver households, it is increasingly common to find two (or more) television outlets in cabled homes, and therefore a splitter that may be mounted at the entry point to the dwelling and does double duty as a bonding block. Given that, we can define a *standard drop* (Figure 10.5) that will encompass the majority of current home wiring. It will include 25–100 feet of size 6 cable from tap to house, perhaps a splitter, and size 59 connections to outlets, with lengths varying from 10 to 50 feet.

Passive Drop Transmission Characteristics

In order to assess the transmission characteristics of this standard drop, we need to first predict the quality of the components used, then use the tools developed in Chapter 9 to calculate their effect on the signals.

Component Quality. A prime determinant of the degree of degradation signals experience in the drop system is the quality of the components used. Unfortunately, just as there are no mandatory standards on drop wiring or design, so there are none for the components used. Historically, drop wiring was installed and maintained by cable operators (much of it still is), and they purchased rea-

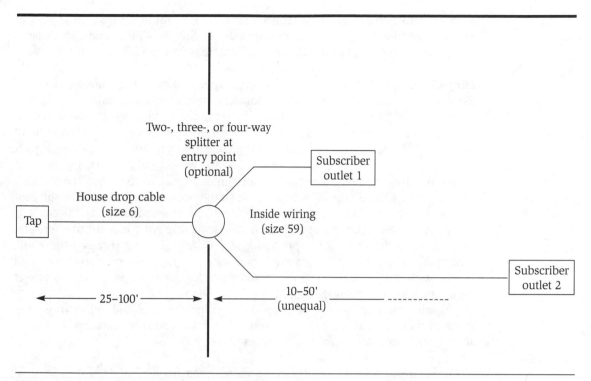

Figure 10.5 Typical drop configuration.

sonable quality components because it was less expensive than the trouble calls and customer dissatisfaction resulting from inferior materials.

As of the writing of this book, the Society of Cable Telecommunications Engineers (SCTE, an ANSI standards-making organization) is actively developing voluntary national standards covering virtually all drop components: cable, connectors, splitters, couplers, amplifiers, attenuators, filters, and so on.[6] Covered performance parameters include return losses, transmission losses, environmental requirements, shielding, surge protection, labeling, and a host of others. The specifications for amplifiers additionally include gain, noise, and distortion. Of importance in estimating drop performance, most devices must exhibit return losses of at least 18 dB, and excess loss (that is, loss exceeding that expected from the power division ratio) in splitters and couplers may not exceed 1.0 to 1.2 dB (for instance, the loss through a two-way splitter must be no greater than 4.0 dB) though in some cases, they can rise above 2 dB. Port-to-port isolation in splitters must be at least 20 dB. Cable-shielding requirements are consistent with good-quality double-shielded cables. There is no assurance, of course, that existing products all meet the proposed standards or that even conforming products will continue to do so after years of exposure to weather.

Furthermore, network operators can hardly be assured that components installed by a homeowner will meet these voluntary, future standards. In particular,

a conservative estimate is that return losses from components may be as poor at 10 dB, port-to-port splitter isolation may be as low as 10 dB, and cable losses may, after aging if not originally, exceed nominal values (Figure 9.5) by 10%.

Drop Cable. The characteristics of various types of drop cable were discussed in Chapter 9. Within the basic size options, however, are many quality and feature options. Chief among those are shielding integrity. Cables are available with a standard dual shield (foil and braid), triple shield (foil, braid, and foil) and quad shield (foil, braid, foil, and braid), with increasing effectiveness of shielding. Of serious concern to operators are dual shield (or worse) cables with sparse braid structures that are available in the retail market. Some of these poorly shielded cables do not adhere to SCTE tolerances for dimensions either, with the result that connectors are poorly attached, both electrically and mechanically, to the cable.

Aside from shielding options, cables are available with support messengers, with flooding compound (to prevent water from migrating through the cable structure), in a dual configuration (for dual cable distribution networks), and paired with one or more twisted copper pairs for carrying telephony or power along with RF in the same structure.

Finally, the nonmetallic portions of the cable are available with various degrees of flame resistance to meet electrical and fire codes when they are routed within some interior paths (such as plenums).

F Connectors. Whatever other components may be included in the drop, in North American cable systems, there will be at least four type F connectors—one at the tap, two at the bonding block, and one at the consumer's receiver input. This connector, designed for low-cost mass manufacturing, is at the root of many drop problems. The female connector uses a nonprecious-metal, two-point spring clip contact for the center conductor; the male cable connector is a *feed-through* type that utilizes the center conductor of the cable itself for a pin; and the mated pair depends on a threaded joint plus a butt contact for an outer RF connection. The result is a connector family that is extremely craft sensitive in installation. Despite ANSI standardization,[7] and detailed electrochemical studies,[8] F connectors remain the leading cause of trouble calls and signal leakage in many cable systems.

In cases where power is multiplexed with RF signals on drop cables (typically to power point-of-entry telephone or data terminal equipment), the limitations of F connectors are exacerbated. A less-than-perfect contact that has a reasonable capacitance may pass RF signals but will not pass power. Furthermore, a loose contact, whether in the center conductor or shield, may generate electrical arcing when cables move with temperature changes or wind. These transients generally will have frequency components extending through the upstream band and into the lower downstream channels. Since they occur *within* the cable, they are tightly coupled to the transmission line in both directions and can cause severe transmission problems. The effects of electrical transients on upstream transmission are discussed in detail in Chapter 14.

Table 10.2 Tap to receiver loss ranges for typical passive drop systems.

Frequency	Expected Loss Range	
	Minimum	*Maximum*
5 MHz	0.2 dB	9.5 dB
40 MHz	0.5 dB	10.5 dB
54 MHz	0.6 dB	13 dB
550 MHz	1.8 dB	18.5 dB
750 MHz	2.1 dB	20 dB

In addition to inherent problems with the quality of the mating surfaces, male connectors vary in how they are attached to the flexible drop cable. Until about 1990, the most common construction used a sleeve that was pushed between the foil and braid of the cable. It was retained by use of a separate hexagonal *crimp ring.* Improved designs used a heavier, integral crimp ring. With both designs, however, field experience showed that the retention strength of the connector on the cable reduced over time. More recent designs have featured a full annular press-fit connection that has proved to retain its strength.

The use of a feed-through design (where the cable center conductor forms the connector pin) with more than one size of cable has created problems with the mating connectors. Even with highly compliant female contact structures, once a female connector has been mated with a large cable center conductor, its ability to form a reliable contact with a small cable is reduced. The center conductor of size 11 cable is beyond the capacity of standard female ports so that special pin-type male connectors are used with it. Pin connectors are also sometimes used with size 7 cable to protect mating ports.

Traps. In addition to the normal drop components, some combination of bandpass, band-stop, low-pass, or high-pass filters are often inserted into the drop circuit at the tap to control which signals reach subscribers' receivers. The techniques used to restrict and tailor services are discussed in Chapter 18. However they are used, traps add loss. Generally, their out-of-band loss may be only a few tenths of a decibel, but their loss near cutoff frequencies can be as much as several decibels at the band edge of adjacent channels, coupled with a considerable slope through the channel. Where traps are used to control several channels individually, as many as four or five may be cascaded between tap and drop cable.

Transmission Loss. Given these variables, the total loss from tap to house will generally fall within the values given in Table 10.2, where the minimum losses result from short drop cables without a splitter, and the maximum losses are based on long drops, splitters, and two to three in-line traps.

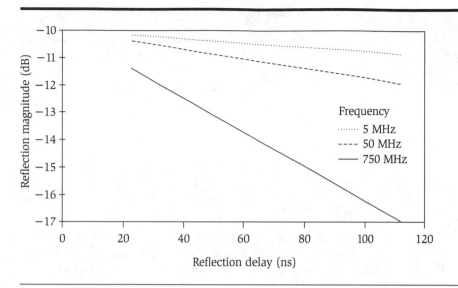

Figure 10.6 Worst-case microreflection magnitudes and delays in typical house wiring.

System designers must take into account this loss range in setting signal levels, and allowable level variation, at system taps. For example, in the "maximum" loss configuration, note that the *difference* in loss is 7 dB over the downstream bandwidth of a 750-MHz system. If the flatness of the signal at the tap port has a downward slope with frequency, the combination could result in widely differing signal levels being presented to receiver input ports.

Microreflections, Group Delay, and Frequency Response Variation. Characteristics of terminal equipment are dealt with in depth in Chapter 19. For now, it is sufficient to state that cable operators are required to deliver signals directly to consumers' television receivers under some circumstances (where analog channels are unscrambled and delivered on standard channels) and that the return loss of those receivers is essentially zero on untuned channels and may be as poor as 4–8 dB on the tuned channel. In any event, it is likely that an unused outlet will be left unterminated, so it should be assumed that 100% reflection occurs at each subscriber outlet. This will cause reflections whose delay and magnitude will depend on the isolation of the splitter ports and the length of the cables beyond the splitter. Using the methods discussed in Chapter 13, we can estimate the worst-case reflections. Figure 10.6 shows the results at various frequencies. Although operators should be prepared for occasional reflections of those magnitudes, a statistical study of actual cable systems showed 95% of all echoes, including those in the distribution plant, were more than 25 dB below carrier levels in magnitude.[9]

These reflections can lead to bit errors in digital transmissions since delayed signals cause successive transmitted symbols to overlap at the receiver and visible degradation of analog television pictures. As will be discussed in Chapter 13,

the effects also include group delay and passband response variations. Amplitude response variations exceeding 1 dB in each channel are quite possible.

Signal Leakage and Ingress. Most difficult to assess is signal leakage. In practice, most leakage and ingress is not associated with components or cable but rather is due to inadequate shielding in customer-owned terminal equipment and/or improperly installed or tightened drop connectors. Poor-quality connectors and poorly shielded cables may also contribute problems.

Given the statistical nature of both egress and ingress, there is no way of predicting their effects. The effect of shielding of consumer equipment is discussed in Chapter 19. The effect of ingress on upstream communications is treated, along with other upstream transmission issues, in Chapter 14.

10.8.2 Adding Amplification—The Trade-Offs

The alternative of providing gain is attractive when it is less expensive than drop cable replacement or where even the largest cables would not be adequate. However, many corollary issues are raised when amplification is added, including:

- Additional noise and distortion
- Signal reliability decrease owing to the amplifier failure rate combined with the availability of its powering source
- Gain stability and flatness, hum modulation, and all the other issues associated with active equipment
- Provision of a mounting location that has adequate environmental protection and available power, and yet is near where the cables to each of the outlets come together

If amplification is included in the drop system, the hard cable plant must perform at a higher performance level in virtually all respects in order to assure end-of-line quality. As a practical choice, drop amplifiers are seldom used except in multiple dwellings or commercial installations where they are, in fact, part of the distribution plant. One exception is in isolated single dwellings where there is no better solution to low levels.

Another even more rare instance is where the reverse isolation of the amplifier is needed to protect the system from ingressing signals that would otherwise travel up the drop and interfere with others' reception. This latter situation is discussed in depth in Chapter 19. Finally, there are homes where the need is not just to deliver cable signals to receivers but to interconnect multiple equipment.

10.8.3 Integrated In-Home Wiring Systems

In response to a need to have defined performance levels and a desire to interconnect home video equipment and multiple program sources, several industry groups have designed integrated home wiring systems. Although a complete discussion of these is beyond the scope of this book, two examples follow.

CEBus

The Consumer Electronics Manufacturers Association (CEMA) of the Electronic Industries Association (EIA) has developed a complex home wiring standard under the umbrella title CEBus. CEBus supports message and information transmission over a variety of physical paths within homes, each with its own special label: twisted pair wiring (TPBus), subcarriers on the power wiring (PLBus), coaxial cable (CXBus), infrared through the air (IRBus), over-the-air transmitted low-power RF (RFBus). A separate, but related, specification covers direct interconnection of home audio and video equipment using specially designed multiple twisted pair cables (AVBus). These standards have been under development since the mid-1980s.[10]

A common language is specified for medium-independent signaling of messages and control information, whereas routers interconnect different types of media. The language used for signaling is known as the *common application language* (CAL) and is defined by ANSI/EIA Standard IS-60.

Of greatest interest to broadband network operators, the CXBus distributes both externally supplied signals (such as from a cable system) and internally generated RF signals over dual, size 6 drop cables through the residence. Downstream signals from the external network are amplified, optionally combined with internal signals, then split to feed four or more "branch circuits" over one of the paired distribution cables. Each branch cable may feed a four-way splitter with shorter cables from there to individual termination points. Thus, one network may incorporate as many outlets as permitted by the cable losses and available RF power (see Figure 10.7). Note that the splitters may be internal to Node 0 and that this is, in fact, the preferred topology since all network elements and connectors are then accessible for maintenance or inspection, if required.

The downstream bandwidth of the distribution system (the "external" cable in CEBus parlance) extends from 54 to more than 800 MHz, so as to allow direct transmission of UHF signals from local antennas, if desired (switching between the external network and local antennas is provided in the design). Upstream signals are carried in the 5–30-MHz band and are combined and amplified before insertion into the external network. The bandwidth from 450 to 546 MHz may optionally be used for downstream transmission of video from local sources to television receivers. The second ("internal") cable has a more evenly split bandwidth and is used for routing of both messages and video among various consumer electronics devices. Data (information) channels are defined in 1.5-MHz bandwidth slices, with four contiguous channels required for internal transmission of normal analog video. Physically, the two cables are run together, with dual type F wall outlets provided for subscriber access.

All the signal processing is done at "Node 0." Figure 10.8 shows the signal processing within the node. Locally generated signals that will stay within the dwelling may be inserted into any internal cable outlet. The block converter is used to convert the frequencies of the internally generated signals from lower upstream channels to downstream channels for redistribution. The output cable

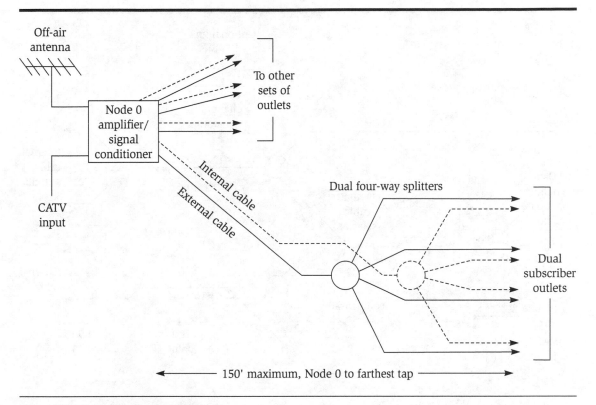

Figure 10.7 Basic CEBus coaxial topology.

selector determines whether the converted channels are distributed over the internal or external cables.

Locally generated signals that are intended for transmission through the external cable system can be inserted into any external cable outlet. Those signals will be amplified by the return amplifier and transmitted into the cable drop.

The control channel regenerator is an optional feature that allows control signals to travel both directions between coaxial and twisted pair distribution networks.

The CXBus is a well-designed specification that should assure proper level control, minimal signal leakage or ingress, and well-controlled noise and distortion degradation. Nevertheless, there are important issues when external network signals are transmitted over such a network:

- The network has fixed bandwidth boundaries, both downstream and upstream. If a network operator chooses to use frequencies outside those bandwidths, the signals will not reach subscriber outlets.

- As we discussed, the downstream and upstream amplifiers contribute both noise and distortion that must be accounted for in the end-to-end

Figure 10.8 CEBus Node 0 functionality.

performance budget. Since only a few homes may be equipped with CEBus systems, it is unrealistic to upgrade the entire external network to accommodate them.

- Upstream levels may vary widely in power level (see Chapter 14). Therefore, the return amplifier must have a very wide dynamic range.

Despite these concerns, it is expected that CEBus systems will generally interface with most current cable television network with minimal problems.

SmartHouse

The SmartHouse system is a privately developed alternative to the CEBus that is similar but different in some important aspects. SmartHouse is a project of the National Association of Home Builders (NAHB) and was developed at the same

general time as CEBus. Membership in the project includes many equipment manufacturers and utility companies.[11]

Key to the SmartHouse system, and a reflection of its parent organization, is the use of multipurpose cables and outlets. Various combinations of power wiring, telephone circuits, CATV cables, and twisted pair dedicated control cables are combined in bundles and connected to multifunction outlets, presumably with the intention of reducing the cost of wiring new dwellings.

The physical architecture of the coaxial portion is not unlike the CXBus network, with branch cables extending from a central service center to distribution interface units, where splitters will be used to create cables for routing to individual outlets.

One difference between the systems includes the use of separate control wires for SmartHouse-compliant equipment and appliances, whereas control messages in the CEBus system can travel over various media, including the coaxial network. Also, the CEBus network is a true "peer-to-peer" network (meaning that there is no central controller, and the control protocol provides for automatic system setup), whereas the SmartHouse system uses a central control system.

Other organizations and private companies have proposed various versions of integrated home video distribution networks, but none, including CEBus and SmartHouse, have been widely deployed.

10.9 Summary

Cable television coaxial distribution systems consist of a cascade of cable and amplifiers. The network typically branches in the downstream direction, and various passive RF devices are used both within and external to amplifiers to create the legs. The amplifiers are capable of amplifying signals flowing in both directions, with the downstream bandwidth extending from approximately the lowest broadcast television channel (54 MHz in the United States) to an upper limit determined by the bandwidth needs of the network, but typically 400–750 MHz. The upstream bandwidth typically extends from 5 MHz to somewhere between 30 and 42 MHz.

In the downstream direction, distribution networks with more than a few amplifiers in cascade are usually designed and operated so that the gain, measured from amplifier output to amplifier output, is unity. When systems carry only one signal type (for example, analog video signals), their levels are often adjusted so that they increase linearly with frequency, as measured at amplifier output ports. This usually results in amplifier input levels that are approximately the same across the spectrum and creates the optimum balance between noise and distortion. When different signal formats are carried, levels may be adjusted to a different pattern of levels. A common example is digital signals carrying packetized data or video, which are operated 6–10 dB lower in level than equivalent analog video signals would be at the same frequencies.

Amplifier gains must be sufficient to offset the loss of the intervening cable and passive devices at the highest operating frequency. A combination of attenuators, positive gain slope with frequency, and special response-shaping networks are used to just offset interamplifier losses. The last few amplifiers may be operated at higher levels to allow feeding more subscriber taps between gain blocks.

A distribution network will exhibit an overall downstream carrier-to-noise ratio that is inversely proportional to the logarithm of the number of amplifiers and proportional to the signal levels at the inputs to each amplifier. It will generate distortion products that are primarily determined by the amplifier output levels, as well as the number of amplifiers. The bandwidth, operating levels, noise, and distortion are interdependent and, as a set, describe the essential features of the system. Other important characteristics include hum modulation, reflections, frequency response, and gain stability.

Power is carried to amplifiers over the same cables used for signals. Most commonly, 60–90 VAC, 60-Hz quasi sine waves are used in the United States. In the event of commercial power failure, standby power supplies replace the sine waves with trapezoidal waves. The saturation of magnetic circuits used to separate 60-Hz power from signals at each device is the largest contributor to hum modulation.

Cable television drop systems can be as simple as a single coaxial cable from tap to one television receiver, or as complex as a full, integrated distribution system with multiple video sources and receivers. Today, the vast majority of cable drop configurations consist of a cable from tap to home, an optional splitter, and additional cables to one to four individual outlets.

The loss (and its variation) over this network, along with reflections from the termination points and shielding integrity, will have an effect on the overall transmission quality to consumers' receivers. In more complex networks that include amplification, the noise, distortion, gain stability, group delay, hum modulation, and so on must be included in end-to-end performance calculations. In the case of complex distribution systems, spectral usage by internal video sources may affect compatibility with the external network.

Endnotes

1. *IEEE Standard Dictionary of Electrical and Electronics Terms,* 2nd ed. IEEE Std 100-1977, Institute of Electrical and Electronic Engineers, Inc., New York.

2. Several excellent papers have been written on this subject including Doug McEwen et al., Distortion Characteristics of Integrated Fibre, AML and Cable Amplifier Systems, published in the *CCTA Convention Papers, 1990,* Canadian Cable Television Association, Suite 400, 85 Rue Albert Street, Ottawa, Ontario, K1P 6A4, Canada; and Doug McEwen et al., Distortion Accumulation in Trunks, published in the *1990 NCTA Technical Papers,* NCTA, Washington, D.C.

3. An excellent treatment of this subject, along with practical recommendations, is contained in CableLabs's 1992 *Outage Reduction,* in the section entitled Power Grid Interconnection Optimization. CableLabs, Louisville, CO.

4. Doug Welch, Cable Powering into a Distributed Load. Presented at the 1995 NCTA National Convention and published in the *1995 NCTA Technical Papers.* NCTA, Washington, D.C.

5. Dean A. Stoneback and William F. Beck, Designing the Return System for Full Digital Services, *Proceedings Manual, 1996 Conference on Emerging Technologies.* SCTE, Exton, PA, January 1996.

6. Society of Cable Telecommunications Engineers, Engineering Committee, Interface Practices Subcommittee. For current specifications and associated test procedures, contact the SCTE at 140 Philips Road, Exton, PA 19341.

7. ANSI/SCTE SP-400 specifies the mechanical dimensions of the female F port, and SCTE-IPS-SP-407 is the parent specification. Male cable F connectors are covered by SCTE-IPS-SP-408, and its dimensions are covered by SCTE-IPS-SP-401. SCTE, Exton, PA.

8. Brian Bauer, F-connector Corrosion in Aggressive Environments—An Electrochemical and Practical Evaluation. *1991 NCTA Technical Papers,* NCTA; and Brian Bauer, Hidden Influences on Drop Reliability: Effects of Low Level Currents on F-Interface Corrosion and Performance. *1992 NCTA Technical Papers,* NCTA, Washington, D.C.

9. Richard Prodan et al., Analysis of Cable System Digital Transmission Characteristics, *1994 NCTA Technical Papers.* NCTA, Washington, D.C., May 1994, pp. 254–262.

10. For current information on the CEBus standard, contact the Electronic Industries Association, 2500 Wilson Blvd., Arlington, VA 22201.

11. For current information on the SmartHouse system, contact SmartHouse, L.P., 400 Prince George's Boulevard, Upper Marlboro, MD 20772.

Chapter 11
Linear Fiber-Optic Signal Transportation

11.1 Introduction

Although coaxial cable distribution systems are used for the "last mile" delivery of broadband services to subscribers, other technologies are frequently used to cover most of the physical distance between headend and tap. One reason for this is that the interrelationship between bandwidth, signal levels, number of cascaded devices, noise, and distortion (discussed in Chapter 10) limits the size of purely coaxial systems. Second, large coaxial systems are best suited to delivery of a common spectrum of signals to all parts of the network, whereas many services such as data communications and telephony require a bidirectional bandwidth allocation to every customer. In order to provide this capacity in a network of attainable bandwidth, the network must be segmented, and this requires the use of small coaxial sections each connected by separate communications links to the headend.

In modern networks, these links almost always use linear fiber optics for at least the segment that interfaces with the coaxial network (though Chapter 12 will discuss alternatives). Fiber strands have a loss that, at the optical wavelengths used by cable television systems, is approximately two orders of magnitude lower than the best coaxial cables. Furthermore, the noise and distortion added by typical 5–15-mile-long fiber-optic links is less than that generated in comparable length coaxial trunk line sections.

This chapter will begin with a brief introduction to some principles of optics that are important to fiber transmission of light. Then optical fibers will be examined in detail, particularly the interaction between the fiber and the signals passing through it. Passive optical devices, comparable with their coaxial counterparts, will be treated next. Active devices will include transmitters, amplifiers, and receivers. Finally, the total performance of optical links will be examined with end-to-end performance calculations of noise and distortion.

11.2 Optical Basics

Light "waves" are just much higher frequency RF waves—about 300,000 times the highest frequency used in coaxial cable systems and 20,000 times the microwave frequencies sometimes used to relay cable television signals. Optical wavelengths are very short (remember that $\lambda = c/f$ in a vacuum). For instance, visible yellow light has a frequency of about 5.5×10^{14} Hz, and thus a wavelength of only 0.00000055 meter, more easily expressed as 550 nm (one nanometer is one billionth of a meter), considerably less than 1/10,000 of an inch.

At these frequencies, electromagnetic waves behave in ways that are sometimes counterintuitive. For one thing, physicists have long understood that the flow of energy is not continuous, but rather travels in "bundles" called *quanta*. In the case of light (and, for that matter, any radio waves), the carriers of the energy are massless subatomic particles called *photons* that travel, not surprisingly, at

the speed of light (about 3×10^8 meters/second or, if you prefer, about 11.7 inches per nanosecond, in a vacuum). The energy carried by each photon is directly proportional to the frequency of the signal, so photons at light frequencies are much more energetic than those at VHF radio frequencies. As a result, the "granularity" of RF power transmission is extremely fine, whereas it is sufficient at optical frequencies to add measurable noise to optical transmission links.

So, is light energy carried by electromagnetic radio waves or by particles called photons? Fortunately, we do not have to answer that question. In fact, they are both convenient mathematical models that explain the behavior of light under various conditions. Physicists use whichever model is most convenient to the situation at hand, and so will we.

Also, fortunately, we need to know but little of the broad subject called optics to understand the use of light as a carrier of signals through glass fibers. We do, however, need to understand the fundamentals of the interaction between light and materials.

11.2.1 Velocity of Propagation and Wavelength

RF signals propagate more slowly through coaxial cables than in a vacuum. Specifically (Equation (9.8)), the ratio of cable propagation velocity to that in free space is $1/\sqrt{\varepsilon}$, where ε is the relative dielectric constant of the cable dielectric.

Given that light waves are simply higher-frequency radio waves, we would expect that a similar phenomenon would occur in optical transmission lines, and it does. The difference is that the dielectric constant, at optical wavelengths, varies with frequency. At RF frequencies and below, the dielectric constant is dominated by molecular and atomic effects. At frequencies above the natural resonances of the molecular structures, they have less effect. The result is that the effective dielectric constant is both lower and wavelength dependent.

Rather than speak in terms of dielectric constant, the term *index of refraction* is used. The index of refraction of a material is numerically equal to the square root of the effective dielectric constant *at the optical frequency being considered.*

The speed of propagation through a material can be expressed as

$$v = \frac{c}{n} \tag{11.1}$$

where

v = the velocity of the propagated signal in meters/second

c = the speed of light in a vacuum (3×10^8 meters/second)

n = the material's index of refraction (≥ 1).

The slowing of the light waves through materials is the result of a phenomenon known as *forward scattering.* When light signals travel through material, it

interacts with the individual atoms. The wave model of what happens is that individual electrons get "excited" into a higher energy state, then give up that energy by generating a wave of their own. The particle model is that once in a while a photon slams into an electron, exciting it to a higher energy state. After a while, the electron returns to its former energy level, emitting a photon of equal energy in the process.

Whatever the model, the result is the same: energy from the transmitted wave is temporarily transferred to the material, then released in the form of new waves at the same frequencies. In the uniform density, almost pure glass used for optical fibers, most of the new waves are transmitted in the same direction as the original wave, but delayed by the period between excitation and reemission (when waves are retransmitted in different directions or with varying delays, the total transmitted energy in the main wave is reduced, as will be discussed later). The new waves vectorially combine with the original wave to form a composite transmitted wave that is slightly delayed from the original. This new wave then interacts with material further along its path, adding further delay. Since this delay is proportional to the distance the wave is transmitted through the material, the result is a uniformly slower transmission speed. The index of refraction is how we express the resultant velocity, relative to that in a vacuum.

The wavelength of the signal is related to the frequency, velocity of propagation, and index of refraction by

$$\lambda = \frac{v}{f} = \frac{\lambda_0}{n} = \frac{c}{fn} \qquad (11.2)$$

where

λ_0 = the wavelength in a vacuum

f = the frequency in Hz

11.2.2 Reflection and Refraction

When light strikes a smooth, nonopaque surface, we know from common experience that some may be reflected and some may be transmitted into the material. We also observe that when light strikes a piece of glass or a still water surface at a very low angle, it is nearly all reflected, and the material resembles a mirror. At greater angles, only a small portion is reflected, and it appears "transparent." These effects are due to the difference in index of refraction between glass and air.

Specifically, if we draw a general diagram (Figure 11.1) showing light propagating from a material on the left with diffraction index n_1 and hitting a perfectly smooth, flat interface with the material on the right (diffraction index n_2) at an angle θ from normal, we can express the equations governing the reflected and transmitted waves as follows:

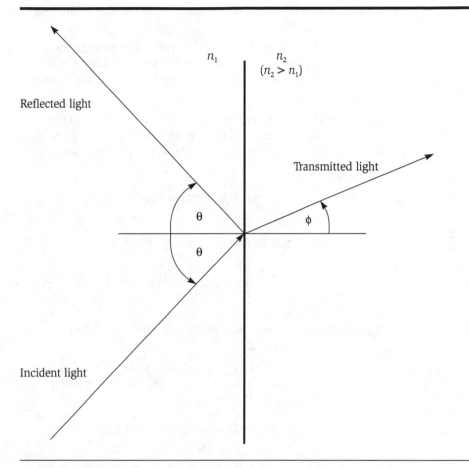

Figure 11.1 Light reflection and transmission at the interface between two materials.

- The transmitted wave will leave the interface at an angle ϕ, relative to the normal such that

$$\frac{\sin\theta}{\sin\phi} = \frac{n_2}{n_1} \tag{11.3}$$

Thus, if $n_2 > n_1$ (as in the figure), the transmitted wave will be closer to normal relative to the interface plane, whereas if $n_2 < n_1$, it will be more parallel to the interface. This relationship is known as Snell's law.

In particular, if $\sin\theta > n_2/n_1$, there is no transmitted wave, and all the energy is reflected (this is strictly true only if the n_2 material is thick). This is a crucial condition for some types of fiber transmission, as will be seen. Since the value of the sine function cannot exceed 1, total reflection is possible only

when $n_1 > n_2$. The angle where total reflection occurs (the "critical angle") is $\sin^{-1}(n_2/n_1)$.

- The reflected wave will leave the surface at an angle exactly opposite the incident wave, that is, at an angle $-\theta$ from normal (we will see later that this effect is used to advantage in certain types of optical fiber connectors).

The division of power between the reflected and refracted (transmitted) waves is a function of the refractive indices, the angle of incidence, and the direction of the electric fields of the incident signal. In the special case where the signal is normal to the interface, the ratio of reflected to incident power is

$$\text{relative reflected power} = \left(\frac{n_2 - n_1}{n_2 + n_1}\right)^2 \qquad (11.4)$$

and the ratio of transmitted to incident power is

$$\text{relative transmitted power} = \frac{4n_2 n_1}{(n_2 + n_1)^2} \qquad (11.5)$$

or in decibel terms more useful for transmission analysis:

$$\text{return loss(dB)} = 20 \log\left(\frac{|n_2 - n_1|}{n_2 + n_1}\right) \qquad (11.6)$$

$$\text{transmission loss(dB)} = 10 \log\left[\frac{4n_2 n_1}{(n_2 + n_1)^2}\right] \qquad (11.7)$$

so that perfect power transmission and zero reflected power require that $n_1 = n_2$.

As an example of less-than-perfect transmission, air has an index of approximately 1 (actually about 1.003) and common glass an index of about 1.5 at visible light frequencies, so about 96% of light striking window glass at right angles is transmitted and 4% is reflected, a transmission loss of about 0.18 dB. These relationships are important to the design of optical connectors.

11.2.3 Light Absorption

As with coaxial cables, some of the energy in a light signal propagating through materials is converted into heat. Though an exact understanding of the mechanisms is not essential, the effect is that an optical transmission path of uniform characteristics has a loss that, like coaxial cables, is linear (in decibels) as a function of distance.

Absorption, if it is due to atomic resonance effects, is very wavelength dependent. At the wavelengths used for signal transmission through fibers, reso-

nances effectively prevent the use of some of the optical spectra, whereas the effects of the resonances cause transmission characteristics at other wavelengths to be enhanced.

11.2.4 Scattering Loss and Rayleigh Scattering

Although on-axis (forward) scattering is what happens most of the time, photons may sometimes be emitted in different directions or with varying delays, leading to signal loss in the principal transmitted wave. Many mechanisms lead to this scattering, including impurities in the material, nonuniform densities, and nonlinearities due to extreme power levels. One such scattering phenomenon is known as Rayleigh scattering, after Lord Rayleigh who is credited with first demonstrating that the sun's light is scattered by nitrogen and oxygen molecules in the air. Rayleigh scattering is very wavelength dependent, varying approximately as the inverse fourth power of wavelength, which is why the blue (short wavelength) end of the solar spectrum is scattered much more than the red end so that we see a sharp-edged red-yellow sun surrounded by a diffuse blue sky.

The loss of energy in light waves transmitted through standard optical fibers is primarily due to scattering loss and secondarily due to absorption and several additional minor effects.

11.2.5 Wavelength Dispersion

Except for wavelengths that are close to the natural atomic resonance frequencies of the material, n varies with wavelength approximately as

$$n = A + \frac{B}{\lambda^2} + \frac{C}{\lambda^4} \tag{11.8}$$

where

A, B, and C = constants related to a particular material

This wavelength dependence is known as *material dispersion*. Near atomic resonances, n varies wildly, and a significant amount of absorption also takes place. Such a resonance happens to occur near the wavelengths used for optical signal transmission.

A common example of dispersion in the visible spectrum is the common glass prism (see Figure 11.2). When white light strikes a piece of glass at an angle, signals at the various colors bend *toward* the normal (right angle to the surface) at different angles due to the difference in n across the visible light spectrum (whose wavelengths range from about 400 to 650 nm). If the glass is a flat sheet, such as a window, the reverse bending takes place on the far side, and all signal wavelengths end up restored to their original angles, and we don't notice anything. If the sides are not parallel, however, then when the light impinges upon

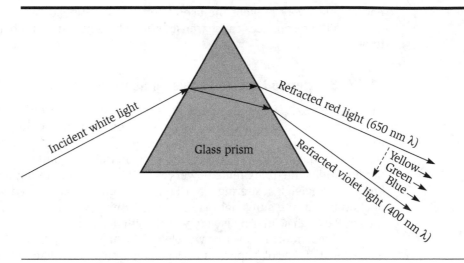

Figure 11.2 Chromatic dispersion of visible light in a glass prism.

the second interface, the angular deviation is made even greater as each wavelength bends *away* from the normal by varying amounts depending on the difference in n as a function of wavelength, and we observe the separated spectrum.

Dispersion is a major concern in optical communications cables, not primarily because of the bending effect, but because of the differing transmission speeds associated with changing n. If the components of a signal are, for some reason, carried over a range of wavelengths and transmitted through a cable whose index of refraction varies, then the components will not arrive simultaneously at the receiver. Depending on the signal format, this can lead to various kinds of distortion. Dispersion is, in fact, just optical group delay (see Chapter 13) and has the same types of effect.

In addition to the above, the interaction of light with very small glass fibers produces some unique interactions that will be discussed in the next section.

11.3 Multimode Optical Fibers

If we construct a long, smooth, narrow (but large compared with the wavelength of the light) circular cross-sectional bar of glass and shine light into one end, some of the light will travel along the length of the bar and exit the far end. That will be true even if the bar isn't perfectly straight. The reason can be seen in Figure 11.3.

Since the refractive index of glass is greater than that of air, light entering the bar at a sufficiently shallow angle will be completely reflected at the glass-to-air interface and will stay in the bar, simply bouncing along from side to side and eventually exiting the far end. Light entering at a greater angle will lose some of

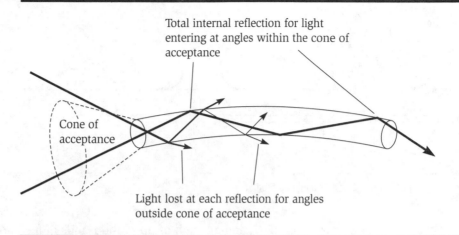

Figure 11.3 Light transmission through a bent glass bar.

its energy at each reflection and will quickly be dissipated. The maximum angle where all the energy is internally reflected defines the circular *cone of acceptance* for the bar.

If the bar is curved slowly (again, relative to the wavelength of the light), the light will still be totally reflected and will follow the curvature. If it is curved too tightly, however, some of the light will strike the outer surface at angles above the critical angle, and a percentage of the energy will be diffracted out of the bar.

The performance of this simple transmission line can be improved (though the cone of acceptance will be much smaller) if we coat the outside with a second layer of glass that has a slightly lower index of refraction than the core. So long as this coating is sufficiently thick, we can now handle the rod or place it next to other objects without interfering with its light transmission properties.

If the entering light is imperfectly collimated (that is, it enters at more than one angle), it can readily be appreciated that the light traveling straight down the core will go less distance than light that is near the critical angle and bounces frequently. The difference in transmission path length will depend on the length of the bar and causes another form of dispersion (known as *modal dispersion*) that is extremely important to signal transmission.

If we modify the index of refraction so that it decreases radially from the center, light that travels at a small angle with respect to the axis will slowly be bent back toward the core as it travels farther from the center. It will, however, have traveled more quickly than light transmitted straight down the core because it will have spent a percentage of its travel time in a medium with a lesser n and therefore greater velocity. Thus, by varying the index of refraction properly, the variation in transmission time with respect to angle can be reduced.

Now imagine that the "bar" of glass is round, with a core whose index of refraction is 1.49, and whose diameter is only 50×10^{-6} or 62.5×10^{-6} meters (50 or

Figure 11.4 Pulse degradation through multimode fiber.

62.5 μm or "microns"), and a length of up to thousands of feet, and you have a typical multimode optical fiber. The preceding analysis, with a few exceptions, explains how signals are transmitted in such cable. Fibers made with just a single coating are known as *step index* fibers, whereas those with a graduated index of refraction are known as *graded index* fibers.

Where the simplified analysis breaks down is that when the core of a fiber is not very large compared with the optical wavelength, only certain transmission paths are supported, whereas under the simplified analysis, light introduced at any angle within the cone of acceptance would be transmitted. The most common wavelengths transmitted through multimode fiber are 850 and 1310 nm. In glass with $n = 1.49$, these wavelengths shorten to 537 and 879 nm, respectively. Thus the diameter of a 62.5-μm fiber core is 116 or 71 wavelengths: large but not infinite.

Little or no multimode fiber is used for cable television long-distance signal transmission because of the angle-dependent dispersion effect discussed earlier. This form of dispersion is caused by the difference in axial transmission speeds of light traveling along different transmission paths or modes and is known as *modal dispersion*. How dispersion affects signals is illustrated by Figure 11.4. If a short burst of imperfectly collimated monochromatic (all one frequency) light is introduced into multimode fiber at time $T0$, the signal will be introduced into the fiber at a range of angles (relative to the fiber axis) and, thus, travel different total distances. If the pulse is examined after it has passed through a length of fiber, we will first "see" a little light from the signal traveling parallel to the core, then from the next fastest modes, and so on, until we finally receive light from the highest angles. Similarly, at the end of the pulse, the light from the fastest modes will disappear first, then gradually all the light will disappear. The effect is to shorten the top of the pulse and broaden the bottom ($T1$). If the pulse is sufficiently short, or the fiber sufficiently long, the top will disappear entirely ($T2$). For still longer fibers, the leading and trailing edges of the pulse will overlap, and we will see a longer, lower-amplitude pulse ($T3$). If we transmit a series of pulses (representing, for instance, a digitally encoded signal), then the maximum pulse rate will be determined by when the residual effects from one pulse interfere with the onset

of the next pulse. One way of expressing this maximum is the bandwidth of the fiber. If we drive the fiber with light that has been amplitude modulated at varying rates, the bandwidth is the point where the amplitude of the demodulated signal drops by 3 dB relative to its value at low modulation frequencies. Alternatively, the bandwidth is approximately equal to $0.35/t_r$, where t_r is the apparent rise time of the detected pulse when the input pulse has zero rise time.

Multimode fibers are rated according to the amount of this length-dependent bandwidth. Typical fibers with a 62.5-micron core diameter exhibit a bandwidth-length product of 300 MHz-km at 1310 nm and 160 MHz-km at 850 nm. Fibers with a smaller 50-micron core can reach 400 MHz-km. What this means is that a 50-micron core fiber that is 1 km long will have a bandwidth of 400 MHz, whereas one that is 10 km long will have a bandwidth of only 40 MHz.

Clearly, such fibers are not suitable for transporting 750 MHz of cable television spectrum over the long distances required in modern networks.

11.4 Single Mode Optical Fibers

If the core diameter is reduced sufficiently, fibers will support only light traveling collinearly with the axis (known as the LP_{01} mode), thereby eliminating modal dispersion. Such fibers are known as *single mode* and are almost universally used for long-distance, high-bandwidth applications.

11.4.1 Structure

The most common single mode fiber construction consists of an 8.3-micron diameter core, surrounded by *cladding* glass with a uniform, lower index of refraction and extending out to about 125 microns. This, in turn, is surrounded by a protective layer (the *coating*) extending out to 250 microns. This outer layer is generally colored to enable easy fiber identification in bundles. This construction is known as *matched clad, nondispersion-shifted* fiber.

The necessary condition for suppression of higher-order modes is

$$V = \frac{2\pi a \sqrt{n_1^2 - n_2^2}}{\lambda} < 2.405 \qquad (11.9)$$

where

$V =$ the V parameter or normalized frequency

$a =$ the core diameter in micrometers

$n_1 =$ the core refractive index

$n_2 =$ the cladding refractive index

$\lambda =$ the free-space wavelength in micrometers[1]

As can be seen, this requires very small core diameters and a very small difference in the refraction coefficients for the core and cladding.

A variant on the preceding construction is to use two layers of cladding: a lower n inner cladding followed by an outer layer with an intermediate n. This is known as *depressed cladding, nondispersion-shifted* fiber and is otherwise the same in construction to matched cladding fiber. With few exceptions, its performance as a transmission medium is very similar as well. Figures 11.5(a) and (b) show the refraction index profile and layers of construction of both types.

With either type, the light travels mostly in the core, but also partially in the cladding. The effective index of refraction is about 1.467 for either type at either of the common wavelengths used: 1310 or 1550 nm.

11.4.2 Attenuation

The total loss due to absorption and scattering is amazingly low. At 1310 nm, typical guaranteed losses for cables sold to the CATV market are only 0.35 dB/km, whereas at 1550 nm, they drop to 0.25 dB/km. Otherwise stated, at 1550 nm, only half the light energy is lost through a strand of glass 12 km (about 8 miles) long.

The curve of attenuation-versus-free-space wavelength is shown in Figure 11.6 (also from Ron Cotton). The local minima around 1310 and 1550 nm are known as the second and third windows of transmission (the first is at 850 nm). The high absorption peak around 1380 nm is known as the *water peak* and is due to absorption by hydroxyl ions (the peak is less pronounced in most modern production fibers than is shown in the figure). Fortunately, a knowledge of the physics behind the shape of the curve isn't important for system applications.

11.4.3 Chromatic (Wavelength) Dispersion

With the large modal dispersion eliminated, more subtle dispersion mechanisms become the limiting factors. *Chromatic dispersion* is a measure of the degree to which the effective propagation velocity changes as a function of wavelength. It is the sum of two factors: *material dispersion,* which is a measure of the change in refraction index of the glass with wavelength, and *waveguide dispersion.* In optical fibers, the signal travels partially in the core and partially in the cladding, and the total *mode field diameter* changes with wavelength. Since the refraction index is different in the core versus the cladding, a change in mode field diameter also results in a change in average dispersion index and, therefore, signal velocity. The ratio of wavelength change to velocity change due to this effect is known as waveguide dispersion.

As with modal dispersion, chromatic dispersion is a linear function of transmission system length. The units of chromatic dispersion are picoseconds per nanometer-kilometer; that is, for a one nanometer free-space wavelength change, this gives the number of picoseconds of delay change per kilometer of fiber length.

Standard fiber exhibits zero chromatic dispersion near 1310 nm (because the slopes of waveguide and material dispersion components are equal in magnitude

Figure 11.5 Cladding profiles for various fiber types. (a) Nondispersion-shifted matched clad fiber doping profile. (b) Nondispersion-shifted depressed clad fiber doping profile. (c) Dispersion-shifted fiber—one of several possible doping profiles.

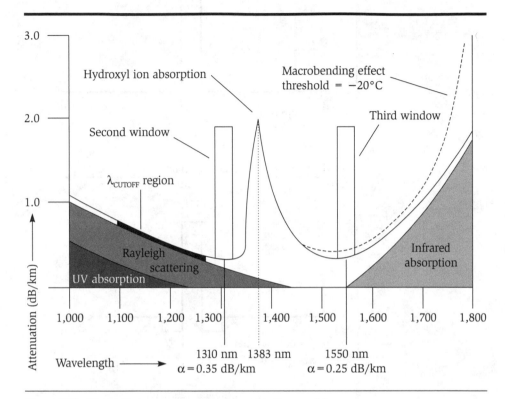

Figure 11.6 Fiber-optic loss as a function of wavelength.

and opposite in direction at that wavelength). In cases where it is important to have low dispersion at 1550 nm, the null point can be shifted upwards by altering the fiber doping profile and/or using several layers of cladding. Figure 11.5(c) gives one example of a doping profile for dispersion-shifted fiber.[2] Figure 11.7 shows a typical chromatic dispersion curve for both nondispersion-shifted and dispersion-shifted single mode fibers.

Typical commercial specifications for chromatic dispersion are 2.8–3.2 ps/nm-km from 1285–1330 nm wavelength and 17–18 ps/nm-km at 1550 nm for nonshifted fibers.[3]

Dispersion is very important in communications circuits because the optical sources used do not transmit on a single wavelength. The interaction between various source wavelengths and network performance will be discussed later in this chapter.

11.4.4 Polarization Mode Dispersion

Light transmitted through single mode fiber may be thought of as two separate signals (polarization modes) with their electric fields 90° apart, relative to the axis

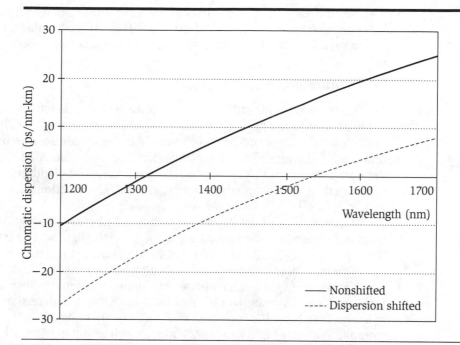

Figure 11.7 Chromatic dispersion for standard and dispersion-shifted fiber.

of the fiber. So long as the fiber looks exactly the same to both signals, they will have the same transmission time and will arrive in phase at the detector.

Two conditions, however, can detract from equal transmission velocity: dimensional irregularities and nonequal index of refraction.

- As discussed, waveguide dispersion is a reflection of the fact that signal velocity is a function of the mode field diameter because the signal travels partially in the core and partially in the cladding. Should the fiber core not be completely round, the ratio of core to cladding utilized for one polarity will be different from the other.

- More subtly, the glass may have a dimensional preference in its atomic structure that causes n to be greater in one direction than the other. This preference is typically nonuniform along the fiber and also modified by local stresses and temperature variations.

The situation is made more complicated because energy is exchanged between the polarization modes due to imperfections in the fiber, connectors, splices, and so on.

The net effect of all these mechanisms is called *polarization mode dispersion* (PMD). PMD is typically wavelength dependent and interacts with optical transmitter wavelength changes to cause system distortion. A typical commercial

specification for PMD is ≤ 0.5 ps/\sqrt{km} at 1310 nm in nondispersion-shifted cable,[4] though typical values are 0.1 to 0.2 ps/\sqrt{km}. At these values, PMD does not add materially to total system distortion in most cable networks.

11.4.5 Attenuation in Bent Fibers

As with the waveguides utilized for microwave signal transmission lines, single mode fibers may not be sharply bent without causing signal loss. Two classifications of bends are normally considered. Microbends are microscopic imperfections, often introduced in the manufacturing process, such as irregularities in the core material or in the interface between core and cladding. Losses due to microbending are included in manufacturers' specifications for cable performance.

Macrobends are bending of the completed fibers. If the radius of a macrobend is sufficiently small, significant amounts of the light will be transmitted out through the cladding and be lost. For that reason, manufacturers specify cable and individual fiber-bending radii that must be respected.

The fact that light is not constrained to remain within the fiber when bent has been exploited by manufacturers of light insertion and detection (LID) fusion splicers. They bend the incoming fiber sufficiently to insert light on one side of the splice location (light can "leak" into, as well as out of, fiber when it is bent), and then bend it on the other side and detect the leaked light. By manipulating the position of the fibers at the joint for maximum transmission before fusing, the splice loss can be minimized. One ambitious manufacturer even reached the field trial stage with a fiber-to-the-curb (FTTC) telephone and cable television distribution system that used bent-fiber couplers to extract and insert light at every node.[5] Their first version utilized multimode fiber, but later versions utilized single mode fiber.

A typical commercial fiber-bending specification is 0.05-dB maximum loss at 1310 nm and 0.10 dB at 1550 nm for 100 turns of fiber wound around a 75-mm (2.95-inch) diameter mandrel. The specified loss for a single turn around a 32-mm (1.26-inch) diameter mandrel is 0.5 dB at 1550 nm.[6] In general, bending losses are higher at 1550 nm than at 1310 nm.

11.4.6 Stimulated Brillouin Scattering

When the power level of monochromatic light transmitted into a long fiber strand is increased, the output power increases proportionately until a threshold is hit. Beyond that level, the received power stays relatively constant, and energy reflected toward the source increases dramatically. The signal-to-noise and signal-to-distortion ratios of the received signals both degrade. This phenomenon is known as stimulated Brillouin scattering (SBS).

SBS occurs because glass is electrostrictive—that is, an electrical field causes a mechanical stress on the material. When the field associated with the light signal reaches a certain amplitude, only some of the energy from excited electrons

is returned as forward-transmitted waves. The remainder is translated into an acoustic wave that propagates through the material. This acoustic wave, in turn, modulates the index of refraction due to the sensitivity of the index of refraction in glass to pressure. The varying index of refraction then causes the main light wave to alternately slow down and speed up, causing the detected signal to have increased distortion. A portion of the reflected light is re-reflected at acoustic wave fronts due to the change in the index of refraction, and the random phase relationship between the original and double-reflected signal, as received at the detector, causes increased noise.

SBS is a function of fiber geometry, loss, and length, as well as transmitted power. It is also a wavelength-specific process in that energy at light frequencies separated by as little as 20–100 MHz behave independently so that when light is transmitted at several wavelengths each experiences about the same Brillouin threshold.

The basic equation for the single-wavelength SBS threshold is

$$P_{SBS} = \frac{21A\alpha}{\gamma(1-e^{-\alpha L})} \qquad (11.10)$$

where

P_{SBS} = the threshold power of SBS in mW

A = the cross-sectional area of the fiber core (approximately $80 \times 10^{-12}m^2$ for nondispersion-shifted fibers and $50 \times 10^{-12}m^2$ for dispersion-shifted fibers)

γ = the Brillouin gain coefficient (2 to 2.3×10^{-14} meters/mW, depending on the degree of polarization randomness in the signal)

α = the fiber's attenuation in dB/km $\times 2.3 \times 10^{-4}$ (numerically equal to twice the attenuation in nepers/m)

L = the fiber's length in meters [7]

Figure 11.8 shows typical SBS threshold values for standard (Std) fibers at both 1310 and 1550 nm, and for dispersion-shifted (DS) fibers at 1550 nm.

Power higher than the monochromatic SBS threshold can be transmitted if divided among optical carriers whose frequencies differ by more than the SBS linewidth. Some vendors use spectrum-spreading techniques to allow transmitted powers as high as +17 dBm (50 mW) at 1550 nm in standard fiber without significant distortion. Clearly SBS extension becomes a balance between widening the spectrum sufficiently to avoid the Brillouin limit while keeping it sufficiently narrow to avoid significant chromatic dispersion. One method commonly used is to phase modulate the source with a sufficiently high amplitude and frequency to adequately spread the spectral power while creating added modulation sidebands that fall above the highest RF signal carried by the system.[8]

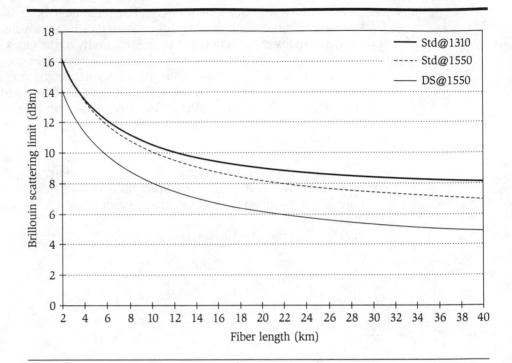

Figure 11.8 Brillouin scattering limit for monochromatic light as a function of fiber length, type, and operating wavelength.

11.4.7 Self-Phase Modulation

Another power-related effect that occurs in optical fibers is known as self-phase modulation (SPM). SPM. This is caused by the fact that the index of refraction, n, is slightly modified as a function of transmitted light intensity. Thus at the peak swings of instantaneous light intensity, n varies from what it is at the unmodulated average light level. This causes a different propagation velocity as the instantaneous power varies, and thus an effective envelope distortion, resulting in phase modulation as seen at the detector. The nature of the distortion will be affected by whether the wavelength is centered on the fiber's dispersion null or on one side of the null.

In general, the effect of SPM will be an increase in the level of composite second-order (CSO) distortion through the link. Although the analysis is not straightforward, the C/CSO will be degraded when

- The chromatic dispersion is higher
- The fiber is longer
- The RF carrier frequency is higher

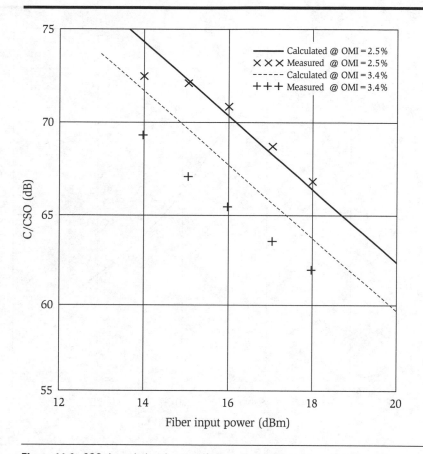

Figure 11.9 CSO degradation due to self-phase modulation.

- The wavelength of the light is longer
- The average light level is higher
- The optical modulation index (OMI, the intensity variation due to the modulating waveform) per carrier is higher
- The core diameter of the fiber is smaller
- The fiber attenuation is lower

Figure 11.9 shows the measured and calculated C/CSO degradation due to SPM as a function of average optical power level under the following conditions:[9]

- Length = 50 km of nondispersion-shifted fiber with a power coefficient of diffraction index equal to 2×10^{-20} m²/W, attenuation of 0.0576 nepers/km, and chromatic dispersion at 1553 nm of 17 ps/nm-km.

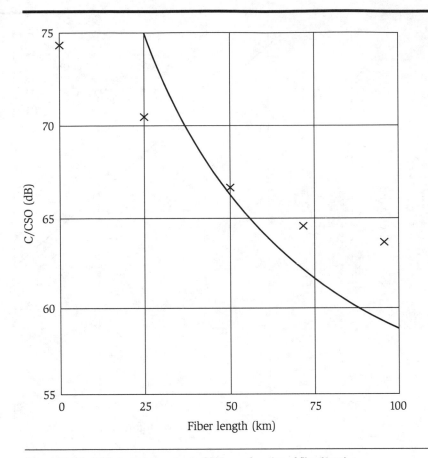

Figure 11.10 CSO degradation due to SPM as a function of fiber length.

- Channel loading = 77 RF carriers with OMI/carrier approximately 2.5% (top curve) and 3.4% (lower curve).
- C/CSO measured on top channel (547.25 MHz).
- The transmitter is an externally modulated DFB laser at 1553 nm (optical transmitter characteristics are discussed later in this chapter).

In Figure 11.9, the solid lines are the calculated results, and the discrete points are measured data. The referenced paper offers no explanation for the deviation between calculated and measured data at higher OMI values.

Figure 11.10 shows the variation as a function of length for a fixed +18-dBm optical power level. For lengths greater than 50 km, the signal was amplified after the first 50 km and transmitted through the additional fiber length. The solid line is the calculated performance, and the × marks are measured data. As can be seen in these graphs, at higher-optical levels, the degradation in C/CSO is significant.

In some cases, this may be a compelling argument for the use of dispersion-shifted fiber for long 1550-nm analog links.

11.5 Network Passives

In coaxial distribution systems, passive devices are used for various signal splitting and combining functions. Optical equivalents exist for many of the same functions, with the obvious exception of multiplexing of 60-Hz power with signals. Discussed next are the essential passive components used in linear optical links.

11.5.1 Connectors

Methods for joining fibers fall into three general classes: fusion splices, mechanical splices, and connectors. Fusion splicing consists of carefully preparing the end of each of the fibers to be joined, then bringing the ends carefully together and literally melting them. Automated and semiautomated fusion splicers have evolved that consistently produce splices with less than 0.1 dB of loss.

Mechanical splices generally involve similar fiber end preparation. The difference is that the fibers are brought into proximity and embedded in a gel-like material whose index of refraction closely matches that of glass. A mechanical sleeve holds the fibers in position. Though the loss of such splices is higher than fusion splices (typically 0.1 to 0.2 dB), they are quick to make and are favored for quick field repair tasks in cases where a fusion splice crew can later replace the temporary fix.

A wide variety of connector families have been developed by various vendors. These are used for interfacing between active equipment and fiber lines and for patch panels that allow test points and reconfiguring of systems. Typical connector losses are higher than mechanical splices (about 0.25 to 0.5 dB) and are variable with remating.

The difficulty in improving connectors lies in their imperfect fiber-to-fiber interface. As discussed at the beginning of this chapter, signals lose approximately 0.18 dB at each fiber–air interface. Unless the fiber ends are in intimate contact, a mated connector pair will have two such interfaces and a loss of 0.36 dB due to the two fiber–air interfaces alone. Worse, the return loss from each interface could be as bad as 14 dB.

A number of different connector families have been used over the years, and more will likely be developed in the future. The SC series is currently popular in North America. This is a small push-on–pull-off connector that is capable of low insertion loss and good return loss if the fiber ends are properly polished. A zirconia ceramic ferrule is used to align the two fibers. It has a central capillary hole into which the fiber is cemented, following which the end is polished. Connectors may be attached in the field or supplied preattached to fiber pigtails that can be fusion spliced in the field.

Ultra-Polished and Angle-Polished Connectors

SC connectors, as well as some other series, either come with flat, highly polished tips or have angle-polished tips. The flat (SC/UPC) version allows for somewhat better insertion loss when cleaned properly, but has typically a 5-dB worse return loss than the angled (SC/APC) version.

The reason for the lower reflections from the APC version is shown in Figure 11.11. Whatever light is reflected at the fiber end (due to alignment errors between the fibers, air gaps, imperfect polishing, or differences in the index of refraction) is returned at an angle equal to twice the end angle of the fiber, which is 8° to 12°. The reflected light is outside the cone of acceptance of the fiber, and so is not transmitted toward the source.

By contrast, the UPC connector requires a higher degree of polishing, which results in lower insertion loss, but with the disadvantage that whatever light is reflected will propagate back toward the source. Because of this difference, the return loss of an imperfectly cleaned UPC connector degrades more quickly than that of an APC connector.

The preoccupation with return loss levels in optical circuits, which are much lower than those required by coaxial networks (typically 16 dB, as discussed in Chapter 9), is because directly modulated laser diode transmitters are very sensitive to reflections, as discussed in Section 11.6. Generally, optical return loss values must be significantly better than 40 dB to avoid problems.

Connector Cleaning Issues

Proper cleaning of optical connectors is mandatory for good performance. Many problems in optical links are caused by dirty connectors. Proper installation and maintenance practices call for careful cleaning and inspection of all connectors before mating or remating, preferably using inspection microscopes made for that purpose. Recent experiments have shown that, even when connectors on new equipment were carefully cleaned and capped for protection, they were unacceptably dirty when uncovered for use after undergoing a simulated shipping environment.[10]

If dirty connectors are mated, it is possible that the foreign material, if abrasive, can damage the polish of the fiber end faces, permanently degrading the performance. In that case, the only remedy is to install a new connector.

Two methods of cleaning are currently accepted. In the first, a slightly abrasive tape is wiped across the connector end face. The tape is contained within a cartridge and is advanced after a set number of cleanings. This method has been found to be effective with SC/UPC connectors; however, its use on SC/APC connectors is questionable since excessive use can damage the angle polish.

The second method uses at least 91% pure isopropyl alcohol applied with a lint-free swab, followed by a dry swab, then blowing dry using canned air. This method seems to be adequate for any type of connector. Standard "rubbing alcohol" is not acceptable because of impurity levels that may leave a residue.

Figure 11.11 UPC versus APC connector.

11.5.2 Signal Splitters

Signal splitters can be made with virtually any splitting ratio. Excess losses (that is, the difference between input light power and the sum of the powers appearing at all output ports) are generally under 1 dB. Often multiple splitters are packaged to create multiple port splitters with 10 or more legs.

11.5.3 Wavelength Division Multiplexers

WDM couplers are the optical equivalent of RF diplex filters. They are used to route light to the correct port on the basis of the optical wavelength. The simplest versions divide 1310-nm light from that at 1550 nm. More complex WDM couplers can separate multiple wavelengths in the 1550 range that differ in wavelength by as little as 0.8 nm. Such systems are known as *dense WDM*.

11.5.4 Attenuators

Optical attenuators are used both as test equipment and for absorbing excess light in some transmission links. Both fixed and variable units are available. A variety of techniques, including controlled bending of an internal fiber, are used to create the desired signal loss.

11.6 Linear Optical Transmitters

Transmitters generate optical carriers, and intensity modulate those carriers with wideband RF spectra. In addition to linearity and noise requirements that are

analogous to those placed on RF amplifiers, optical transmitters must exhibit optical characteristics that are compatible with the interconnecting fiber strands.

11.6.1 General Characteristics

In general, optical transmitters use an electrical signal to modulate the power of a light source. An early, and still common, application is simple on–off modulation used to convert high-speed binary signals to light pulses. At the receiving end, detectors need only detect the presence or absence of light to accurately reproduce the original digital stream. Digital modulation is a conceptually simple and robust format, and relatively high noise levels can be tolerated. At the present state of the art, speeds in excess of 10 gigabits per second (Gb/s) are commercially deployed, whereas optical losses of 30 dB can be tolerated between transmitters and receivers.

Figure 11.12 shows theoretical bit error rate as a function of signal-to-noise ratio for a baseband, binary transmission system.[11] Quite moderate S/N ratios (compared with those required for modulated analog NTSC signals) yield very good bit error rates.

By contrast, transmission of an FDM analog signal is much more complex. In the usual cable television application, the composite headend output waveform (with spectral components extending from 50 to 750 MHz or more) is used to control the light intensity in a proportional manner. The modulation must be highly linear or, more correctly, must complement exactly the demodulation characteristic so that the total link is highly linear. Not only that, but the end-to-end C/N of each multiplexed RF signal must be compatible with signal-quality requirements. Typically, the fiber-optic link is required to provide a C/N of 50 dB or greater for each analog video carrier. Since stimulated Brillouin scattering and other nonlinear fiber effects limit the maximum transmitted power to +10 to +17 dBm, and receiver power levels of about 0 dBm are required to sufficiently overcome "shot noise" (see Section 11.8) at the receiver, this limits practical optical budgets for typical 80-channel fiber-optic links. These relationships will be further explored at the end of this chapter.

In fact, optical transmitters and receivers are not linear devices, but "square law" devices; that is, the instantaneous light output *power* of a transmitter is proportional to the input *current* and thus to the square root of the input signal power. At the other end of the circuit, the RF output power from the detector is proportional to the square of the optical power received so that the total link is nominally linear (predistortion is often used to overcome residual nonlinearities). As will be seen, however, the square law transfer function has an effect on noise and distortion addition. In particular, because of the square law detector transfer function, a change of 1 dB in optical loss will result in a 2-dB change in detected RF power, leading to the commonly stated, but incorrect, statement that "optical dBs are twice as big as RF dBs."

From a system standpoint, and regardless of the technology used to generate the signals, transmitters can be characterized by their wavelength, spectral pu-

Figure 11.12 Bit error as a function of signal-to-noise ratio for a binary digital transmission system.

rity, and stability, and by their bandwidth, linearity, and noise levels. These interact with the fiber and detector characteristics to determine the total link quality.

11.6.2 Directly Modulated Fabry–Perot Laser Diodes

Although a number of technologies can produce the roughly monochromatic light required for optical transmitters, few meet the dual requirements of sufficient power output along with a linear relationship between input voltage or current and light output power.

One that does is the Fabry–Perot semiconductor laser, named for the French scientists Charles Fabry and Alfred Perot. In a Fabry–Perot (F–P) laser, light is reflected and re-reflected between two "mirrors" at either end of a semiconductor material that has been biased electrically. The material and two mirrors form a resonant cavity that roughly determines the wavelength of the light produced. One of the mirrors is only partly reflective, allowing some portion of the light to "leak" out into an external fiber, whereas most is internally reflected. This is directly analogous to a "high-Q" resonant L–C tuned circuit where the circulating energy is much higher than that coupled into a load.

Physical processing limitations dictate the spacings between the mirrors, which form a resonant chamber long enough that oscillation can take place at any of several frequencies (longitudinal modes) for which the cavity length is an

integer number of half wavelengths. Though the total output power of F–P diodes is relatively constant (for a given bias current), that power fluctuates randomly among several modes, each with a slightly different wavelength. For typical 1310-nm F–P diodes, the spectral lines are spaced about 1 nm, and the energy is spread among several such lines. When this signal is transmitted through a fiber, light at each of these frequencies transmits at a slightly different speed due to chromatic dispersion. At the detector, the randomly varying transmission times translate into noise.

In addition to this *mode partition noise,* F–P diodes exhibit a certain amount of instability in their total optical power, known as relative intensity noise (RIN), and typically deviate slightly from an ideal square law transfer curve. For all these reasons, F–P diodes are not generally used for wideband downstream transmission although their moderate cost makes them attractive for upstream use, where typically, a single analog video channel or a limited number of digitally modulated signals are carried.

11.6.3 Directly Modulated Distributed Feedback (DFB) Laser Diodes

A modification of the basic F–P laser uses a diffraction grating (which serves as an optical tuned circuit) along the length of the cavity to restrict oscillation to a single mode. This is known as a distributed feedback (DFB) laser and is the most common light source used in downstream optical transmitters. High-grade DFB lasers may have linewidths (the width of the optical spectrum in the absence of modulation) as narrow as 1 MHz. As will be seen, source linewidth is an important factor for certain noise-generating mechanisms.

The key characteristics of F–P and DFB lasers are covered next. Although this discussion applies generally to either type, the typical values given will be applicable to DFB units optimized for direct modulation, unless otherwise specified.

Linearity as a Directly Modulated Source

The typical transfer curve showing the relationship between driving current and light output power for a DFB is shown in Figure 11.13. In modern diodes that are designed for analog modulation, a C/CSO of 60 dB or better and C/CTB of 65 dB or better are common for a signal loading of 80 unmodulated carriers and for signals that stay within the linear portion of the curve. The sharp discontinuity at low currents, however, causes abrupt clipping of applied waveforms that are too large. This is discussed later in this chapter.

Relative Intensity Noise (RIN)

The light from laser diodes has some random amplitude fluctuation. This is generally expressed in terms of the noise power in a 1-Hz bandwidth compared with the average optical power level. Typical values for quality DFBs are approximately −160 dB/Hz, provided there is no reflected light. Any light reflected from the load, however, will degrade this quickly. For that reason, most high-

Figure 11.13 Laser diode transfer function.

performance DFB sources are connected to fibers through optical isolators that serve to prevent reflected light from reaching the source.

In order to assess the effect of source RIN on link C/N, we need to compare it to the modulation of the light source by each carrier. The optical modulation index (OMI) is defined as the peak optical power variation from the unmodulated level divided by the difference between the unmodulated level and the threshold level (A/B in Figure 11.13). Since the device is highly linear, OMI can also be defined as the peak deviation from the bias current level relative to the difference between bias and threshold current levels.

We can speak of the total modulation of the transmitter by the sum of all the RF carriers or the modulation by some individual carrier. Though it is important that the total modulation not exceed 100% to keep the signal from being clipped, we are also frequently interested in the performance of individual channels. The peak modulation of the optical carrier by an individual carrier is abbreviated m_i,

which will be an important factor in calculating the per-channel noise performance of an optical link.

The relation between source RIN, m_i, and channel bandwidth is expressed by the formula

$$C/N_{RIN} = -RIN - 10 \log(BW) + 20 \log\left(\frac{m_i}{\sqrt{2}}\right) \tag{11.11}$$

where

C/N_{RIN} = the contribution of the source noise to the C/N of a signal, expressed in dB

RIN = the source noise level (relative to the unmodulated light power), expressed in dB/Hz

BW = the receiver noise bandwidth, in Hz, for the communications channel being evaluated

m_i = the peak modulation of the light source by the signal

As an example, if a source has an RIN of -160 dB/Hz and it is modulated 3% (typical for a 77-channel system) by an NTSC video channel whose noise is measured in a 4-MHz bandwidth, the C/N will be approximately 60.5 dB. This puts a length-independent ceiling on the performance of a fiber-optic link (which will be further degraded by detector and postamplifier noise, as discussed later).

Incidental Wavelength Modulation (Chirp)

When the current through a directly modulated laser diode changes as a result of modulation, there is also a small effect on the instantaneous wavelength of the produced light. This incidental FM is known as *chirp*. In typical laser diodes, the rate of frequency shift can vary over a range from approximately 50 to 500 MHz/mA with a total shift of ±1.5 to ±15 GHz, with DFBs exhibiting lower shifts than F–Ps and 1550 lasers exhibiting higher shifts than 1310 lasers.[12]

Chirp has several important effects on link performance. First, it spreads the energy over a range of wavelengths, which increases the effective Brillouin scattering threshold and thus allows higher optical power levels to be transmitted. On the other hand, it interacts with chromatic dispersion to cause a decrease in C/CSO. Finally, it interacts with reflected light to degrade effective source RIN.

Because the energy of a directly modulated DFB source is spread over a wider spectrum, it is possible to transmit higher power levels without reaching the Brillouin scattering threshold. Experience has shown that launch powers in excess of 25 mW into standard 1310-nm fiber are possible without a discernable increase in noise or distortion.

Over the relatively small wavelength shifts associated with chirp, the fiber's index of refraction will shift in an approximately linear manner. Thus, as the instantaneous modulating signal swings in one direction, the light will shift to a

wavelength that travels slightly faster, whereas at the other polarity, the generated light will travel slightly slower. Because the variance is approximately linear with modulation, the resultant distortion in the detected signal will affect one peak of the RF waveform exactly opposite of its effect on the other peak. As was discussed in Chapter 9, distortion with this characteristic causes only even-order products, of which the dominant is C/CSO. The relationship between chirp, RF frequency, fiber dispersion, and link length is given by the following formula:

$$C/CSO_{CHIRP} = -20 \log(DL\Delta\lambda f_{RF} 2\pi 10^{-6}) - 10 \log(N_2) + 3 \qquad (11.12)$$

where

C/CSO_{CHIRP} = the ratio of the level of the visual carrier in question to the level of the CSO products falling in that channel

D = the fiber chromatic dispersion in ps/nm-km at the optical frequency used

L = the length of the fiber in km

$\Delta\lambda$ = the laser chirp per channel in nm as a result of modulation

f_{RF} = the frequency of the CSO product being analyzed in MHz

N_2 = the number of second-order products in the composite beat that falls in the channel being analyzed (see Equation (9.25))[13]

The CSO affects channels differently because (1) a given amount of time shift represents more degrees of phase shift at higher frequencies, and (2) the number of second-order products falling into channels varies.

If the laser chirp per channel is known in terms of optical frequency, rather than wavelength, it can be converted to wavelength chirp using

$$\Delta\lambda nm = 0.0057\Delta f(GHz) @ 1310 \text{ nm} = 0.0067\Delta f(GHz) @ 1550 \text{ nm} \qquad (11.13)$$

For example, typical 1310-nm DFBs exhibit about 100-MHz/mA incidental frequency modulation. If such a laser were modulated at 2 mA/channel and the signal transmitted through 20 km of standard fiber with a chromatic dispersion of 3 ps/nm-km, the chirp-related C/CSO at 550 MHz (31 upper-side second-order products) would be 60.6 dB.

Directly modulated 1550-nm DFBs tend to have higher chirp values—typically 250 MHz/mA. If such a source were modulated at the same 2 mA/channel and the light transmitted through the nonshifted 20-km fiber described earlier, the C/CSO would decrease to 35.8 dB primarily due to the 18-ps/nm-km chromatic dispersion at that frequency. Thus, directly modulated DFB transmitters are suitable only for multichannel FDM video transmission when operated near the zero dispersion wavelengths of fibers.

Effects of Reflected Light

When light is reflected into a DFB (or F–P) laser from a connector, passive component, receiver, or from the fiber itself due to Rayleigh backscattering, it can have a significant effect on RIN. Because the wavelength of the laser changes with modulation, it is almost certain that its wavelength will be different from that of the reflected signal due to the round-trip transmission time from transmitter to point of reflection.

If the laser wavelength is close to that of the reflected signal, it will try to lock its present oscillation to the frequency of the returned light. However, this is not the frequency the laser "should" be on, based on the modulating signal. The result is similar to that of an electronic oscillator that is receiving energy at a frequency close to, but not exactly on, its own frequency. The oscillator "pulls" toward the incoming frequency, resulting in a noisy output. The same thing happens with a laser and results in spurious emissions and an overall degradation in effective RIN. Some protection is afforded where optical isolators (components that exhibit low forward loss but typically greater than 30 dB of isolation in the reverse direction) are installed in laser diode output fibers, but attention to reflections from the driven optical circuits is also required. The jumps in frequency also add a degree of uncertainty to the calculation of interferometric intensity noise (see Section 11.9.1).

Temperature Effects

The current-versus-light transfer function for diodes shown in Figure 11.12 is stable at any one temperature. As the temperature changes, the curve shifts horizontally and may also change slope. Since this could result in the operating point shifting from the optimum value, lasers used for wideband, downstream transmission are almost always connected to thermoelectric coolers (TEC) that act to maintain a constant chip temperature.

Though the use of TECs would be desirable in all linear modulation applications, the cost and power consumption has resulted in noncooled transmitters being used in many lower-performance upstream transmitters.

11.6.4 Externally Modulated Continuous Wave Sources

Externally modulated transmitters consist of a continuous wave light source whose intensity is varied through the use of an external device that is driven by the FDM waveform. This configuration offers several advantages, including:

- Freedom from the incidental wavelength modulation (chirp) associated with direct modulation of laser diodes (though modulators have some residual shift of their own)
- Improved isolation of the source from reflections in the transmission system, thereby decreasing effective source RIN

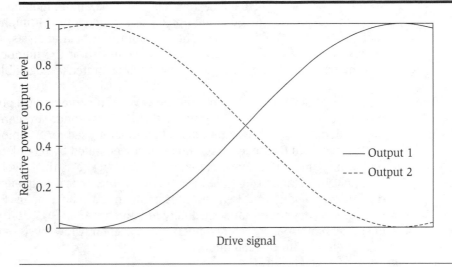

Figure 11.14 Transfer function, external optical modulator.

- The ability to use a variety of optical sources without regard to their ability to vary light output in proportion to an electrical modulating signal

The most common external modulator is a Mach–Zehnder (M–Z) device constructed on a lithium niobate (LiNbO3) substrate. The property of an M–Z modulator is that the incoming light is divided proportionately between two outputs in response to an electrical stimulus. The transfer function to each output is very nearly sinusoidal, as illustrated in Figure 11.14. When used as a linear transmitter for cable television distribution, the predictable symmetric curvature of the transfer function (which causes large odd-order distortion products) is compensated for by predistortion in the driver circuit.

The transfer function of M–Z modulators provides one important additional advantage over directly modulated sources: the distortion is symmetrical about the inflection point of the transfer function, offering significantly improved second-order distortion products (see Chapter 9 dealing with distortion in coaxial systems).

Externally modulated transmitters can be constructed at either 1310 or 1550 nm. At 1310 nm, the usual source is a relatively high-powered (80-mW typical) YAG oscillator. This device, whose wavelength is actually 1319 nm, exhibits a low RIN and optical linewidth.

At 1550 nm, it is more common to use a DFB diode for the light source, then amplify to the desired power level using optical amplifiers. As an unmodulated DFB has a spectral width of less than 100 MHz, this makes Brillouin scattering a concern.

Regardless of the light source, when an FDM spectrum extending up to between 500 MHz and 1.0 GHz is used to amplitude modulate the light, the optical

spectrum will have sidebands extending to the maximum modulating frequency on each side of the unmodulated optical center frequency. It might be expected that this would effectively raise the Brillouin limit. An analysis, however, has shown that 99% of the energy is still concentrated at the unmodulated wavelength, and actual measurements have demonstrated a negligible increase in scattering limit.[14]

High launch powers are possible, however, if the total optical power is spread more evenly over a range of wavelengths. Two methods are commonly used: phase modulation in the same external modulator used for AM modulation and, in the case of DFB sources, intentional chirp generated by modulating the DFB's operating point with a constant frequency. In either case, the wavelength spreading must be sufficient that the beats between the optical frequencies fall above the highest modulating frequency. Manufacturers who employ SBS extension techniques commonly claim the ability to launch +16 to +17 dBm into long single mode nondispersion-shifted fibers with acceptably low levels of CSO degradation at either 1310 or 1550 nm.

11.7 Optical Amplifiers

Optical amplifiers boost the level of amplitude modulated light without requiring the optical-to-electrical conversion, RF amplification, and remodulating otherwise required in a repeating location in a network. Understanding the process requires a knowledge of atomic physics that is beyond the scope of this book, but a broad explanation of the principles follows.

The most common form of amplifier is the 1550-nm erbium-doped fiber amplifier (EDFA). In an EDFA, the incoming signals are combined, in a WDM coupler, with a high-power unmodulated signal at 980 nm from one or more "pump" sources. The two signals are then sent through a special length of fiber that is "doped" with erbium (that is, a carefully controlled amount of erbium is added). The pump signal causes electrons in the erbium atoms to jump to a higher energy band. When they return to their normal state, the difference in energy boosts the signals at 1550 nm. Figure 11.15 shows the basic schematic of a dual pump EDFA.

The magic of EDFAs is that they have completely different characteristics than coaxial amplifiers with respect to noise and distortion. Even though they operate in a saturated mode (where a normal coaxial amplifier would produce pulses with a high harmonic and IM content), EDFAs add no measurable distortion to the modulated signal. Not only that, but they can be overdriven, if desired, so that their input noise level is small compared with the signal to be amplified. The C/N contribution from an EDFA is given by

$$C/N_{EDFA} = 86.2 + P_i + 20 \log(m_i) - NF_{EDFA} \tag{11.14}$$

Figure 11.15 Schematic of EDFA.

Figure 11.16 Noise figure variation in EDFA as a function of drive level.

where

C/N_{EDFA} = the carrier-to-noise per channel (measured in a 4-MHz bandwidth) in dB

P_i = the optical input power to the EDFA in dBm

m_i = the optical modulation index (OMI) per carrier

NF_{EDFA} = the noise figure of the amplifier

The noise figure of a standard EDFA depends somewhat on input power. Figure 11.16 shows typical variations in NF based on several manufacturers' specifications.

Commercial amplifiers are available with saturated power output levels in excess of 100 mW. The highest-powered units utilize two or more pump sources, injecting light into ends of the doped fiber (actually, four pumps are possible, two at each end, if they are cross polarized). Increasing pump power increases the available output power since more energy is stored in excited electrons that can be tapped to produce optical emissions. A modified technology uses erbium and ytterbium doping and an 800-nm pump wavelength, and achieves even higher powers, potentially in the hundreds of milliwatts.[15]

Unfortunately, the efficiency and performance of an optical amplifier is an accident of the properties of specific atomic elements. Although 1550-nm amplifiers are a mature technology, it has proved much more difficult to produce commercially practical units at 1310 nm. Recently reported results, though, include a +20-dBm unit with a 22-dB small signal gain and 6-dB noise figure. This unit uses praseodymium doping.[16]

11.8 Optical Receivers

Unlike transmitters, optical detectors are relatively standardized. The incoming light impinges on the active area of a photo diode that has a very precise square law transfer curve (that is, the output current is proportional to input optical power). Typical responsivity (the slope of the received optical power versus output current transfer function) is 0.8–1.0 mA/mW.

The noise performance of an optical receiver is limited by the noise current of the diode (shot noise due to the statistical variation in arriving photon distribution) and by the noise of the following coaxial amplifier. Since the inherent impedance of the detector diode is not 75 ohms, either an impedance-matching transformer or an active circuit is required to match it to the gain stages. Sometimes a *transimpedance amplifier* is used. Though these amplifiers have an excellent noise figure, they have limited dynamic range, with the result that the light level reaching a receiver must typically be controlled within a few decibels of 0 dBm for the best balance between noise and distortion for AM modulated links. It is common for manufacturers of analog broadband fiber-optic terminal equipment to specify link performance at or near 0-dBm received power.

The formula for the C/N contribution due to shot noise is

$$C/N_{SHOT} = P_r + 20 \log\left(\frac{m_i}{\sqrt{2}}\right) + 10 \log(R) - 10 \log(BW) + 154.94 \quad (11.15)$$

where

C/N_{SHOT} = the C/N of an individual carrier due to shot noise in the detector, expressed in dB

P_r = the received optical power level in dBm

m_i = the peak modulation of the light source by the carrier

R = the responsivity of the receiving diode in amperes per watt (or mA/mW)

BW = the noise susceptibility bandwidth of the channel in Hz

For example, if the OMI per carrier is 3%, the diode response is 0.9 A/W, the received optical power is 0 dBm, and the video bandwidth is 4 MHz, the noise contribution due to shot noise will be approximately 55 dB.

Added to that must be the contribution from the postdetector amplification system. If a transimpedance amplifier is used, its C/N contribution can be calculated using

$$C/N_{POSTAMPLIFIER} = 2P_r + 20 \log\left(\frac{m_i}{\sqrt{2}}\right) + 20 \log(R) - 10 \log(BW) + 10 \log(R_Z) - F + 137.91 \quad (11.16)$$

where

$C/N_{POSTAMPLIFIER}$ = the C/N of a signal due to thermal noise generated within the postamplifier, expressed in dB

R_Z = the postamplifier transimpedance in ohms

F = the postamplifier noise figure in dB

Note that the C/N varies 2:1 with received optical power (the factor $2P_r$) because of the square law nature of the detector as opposed to C/N_{SHOT}, which varies 1:1 with optical power because it occurs in the optical domain (before detection).

If, in the previous example, the detector diode is attached to a postamplifier having a 3-dB noise figure and 1,200-ohm transimpedance, the noise contribution due to postamplifier noise will be 65.3 dB. Transimpedance amplifiers, however, have very limited dynamic range. In commercial optical receivers optimized for cable television applications, a variety of means are used to couple photodiodes to postamplifiers, including both direct and impedance-matching transformers. A more general way to calculate postamplifier noise uses the total postamplifier equivalent thermal noise (which includes both the irreducible noise due to the impedance and the excess noise generated within the amplifier).

If Equation (11.16) is reconfigured to use equivalent amplifier thermal noise input current, the postamplifier C/N contribution can be calculated using[17]

$$C/N_{POSTAMPLIFIER} = 2P_r + 20 \log\left(\frac{m_i}{\sqrt{2}}\right) + 20 \log(R) - 10 \log(BW) - 20 \log(I_r) + 180 \quad (11.17)$$

where

$C/N_{POSTAMPLIFIER}$ = the C/N of a signal due to thermal noise generated within the postamplifier, expressed in dB

P_r = the received optical power level in dBm

m_i = the peak modulation of the light source by the signal being evaluated

R = the responsivity of the photodiode in amperes/watt or mA/mW

I_r = the postamplifier equivalent input noise current density in pA/$\sqrt{\text{Hz}}$

180 = a factor that results from converting the input power to mW from watts and the amplifier noise to pA from amperes

For example, if we use the same parameters as used in the last example, but assume instead that the amplifier input noise current density is 7 pA/$\sqrt{\text{Hz}}$ (typical values will be 6–8), we can calculate that the postamplifier C/N for an analog video channel will be 62.7 dB.

11.9 Interactions Among Transmitters, Fibers, and Receivers

In addition to the noise and distortion that takes place within transmitters and receivers, additional signal degradation takes place because of the interaction of fiber characteristics with the characteristics of the optical transmitter and receiver. These additional effects can have a significant effect on link performance, which is why it is important that fiber-optic equipment be specified and tested through actual fiber links, not just attenuators.

11.9.1 Double Rayleigh Backscattering: Interferometric Intensity Noise

Some portion of light that is scattered backward through the fiber due to Rayleigh scattering will be rescattered in the forward direction and will combine with the normally transmitted light. Since the sum of the double-scattered light from various lengths will not be coherent with the normal light, the effect when both reach the optical detector is that they mix to produce products at the differences in optical frequencies (and multiples thereof), and will show up as an effective increase in link noise. This interferometric intensity noise (IIN) is spread across a spectrum that is about twice the total effective linewidth of the transmitter (including the effect of chirp, if present). The effective link IIN is given by

$$IIN = 10\log\left[\frac{3.6\times10^{-14}\left[L - \frac{1}{2\alpha}(1 - e^{-2\alpha L})\right]}{\Delta f_{RMS}}\right] \qquad (11.18)$$

where

$$\alpha = 1 - 10^{-(\alpha_0/10)}$$

IIN = the level of double-scattering-caused noise relative to the unmodulated light level, in dB/Hz

L = the length of the fiber in km

α_0 = the loss of the fiber in dB/km

Δf_{RMS} = the total rms effective linewidth (in the case of directly modulated transmitters, this is the total wavelength spread, chirp, of the source as a result of modulation; in the case of externally modulated transmitters, this is the linewidth of the unmodulated source if not dithered or phase modulated)[18]

For affected frequencies, IIN can be related to in-channel C/N in the same way as transmitter RIN:

$$C/N_{IIN} = -IIN - 10 \log(BW) + 20 \log\!\left(\frac{m_i}{\sqrt{2}}\right) \tag{11.19}$$

where

C/N_{IIN} = the contribution of the IIN effects to the C/N of a signal, expressed in dB

IIN = the interferometric noise level (relative to the unmodulated light power), as calculated in Equation (11.18) and expressed in dB/Hz

BW = the receiver noise bandwidth, in Hz, for the communications channel being evaluated

m_i = the peak modulation of the light source by the signal

Figure 11.17 shows an example of the effect of IIN on video C/N as a function of fiber length for several optical linewidths at 1310 nm. The assumed OMI per carrier was 3%.

For externally modulated sources, absent SBS suppression techniques that cause considerable line spreading, the linewidth is so narrow that IIN noise density is very high and is concentrated at very low frequencies. In order to avoid severe IIN degradation to television channel 2, which starts at 54 MHz, the linewidth must be either less than a few MHz or artificially spread to levels at least comparable with those found with directly modulated lasers. Since a wider spectral line may be required as part of SBS suppression techniques, the latter approach is generally taken.

Finally, IIN may affect long upstream DFB links in a different way. The unmodulated linewidth of a typical DFB is approximately 10 MHz. Since an upstream link may not be used continuously, when it is idle, there may be a

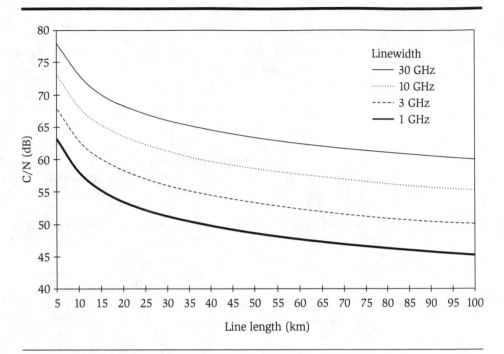

Figure 11.17 C/N degradation due to IIN.

significant level of IIN extending up to about 20 MHz much of the time. A possible answer is to apply a "dithering" modulation to the laser continuously to ensure a wider chirped linewidth.[19]

11.9.2 Phase Noise Contribution to Link Performance

Even though DFB diodes oscillate in only a single longitudinal mode, they still exhibit some residual frequency instability. As in F–P diodes, this variance interacts with dispersion to create link noise.

The relationship between source phase noise and link C/N is given by the formula[20]

$$C/N_{PM\text{-}AM} = 398.55 - 10 \log(\Delta v BW_{RF} D^2 \lambda^4 f_{RF}^2 L^2) + 20 \log\left(\frac{m_i}{\sqrt{2}}\right) \tag{11.20}$$

where

$C/N_{PM\text{-}AM}$ = the contribution to channel C/N (in dB) due to the phase noise in the optical source being converted to amplitude noise in the detector

Δv = the source linewidth in MHz

BW_{RF} = the noise susceptibility bandwidth of the channel in Hz

D = the fiber dispersion constant in ps/nm-km

λ = the operating free-space optical wavelength in nm

f_{RF} = the modulating frequency in MHz

L = the fiber length in km

m_i = the OMI per RF carrier at f_{RF}

The 398.55 factor is the result of converting all other values to common engineering units.

As can be seen, the noise affects the RF channels unequally, increasing as the square of the RF channel frequency. It also increases as the square of the fiber length.

Phase-related amplitude noise is generally not of concern in 1310-nm links using nonshifted fiber but may be significant in cases of long fiber links using nondispersion-shifted fiber at 1550 nm. As an example, assume a 750-MHz analog video channel used to externally modulate (OMI/ch = 3%) an amplified 1550-nm DFB whose optical linewidth is 1 MHz. The signal is then transmitted through a 60-km fiber whose chromatic dispersion is 17 ps/nm-km. The detected signal will exhibit a C/N due to PM-AM conversion of 53.8 dB.

11.10 End-to-End Fiber-Optic Link Performance

The characteristics of the transmitter, interconnecting fiber, receiver, and, optionally, any amplifiers work together to determine the link performance. In particular, the interaction between the signals and the transmission media is much more complex in optical than coaxial links.

11.10.1 Noise Performance

The total link C/N will have contributions due to transmitter RIN, detector shot noise, and postamplifier noise figure, as well as due to the IIN interaction among laser linewidth, double scattering, and detector mixing. These noise effects are noncorrelated and so will add on a power basis:

$$C/N_{LINK} = -10 \log\left[10^{-\left(\frac{C/N_{RIN}}{10}\right)} + 10^{-\left(\frac{C/N_{IIN}}{10}\right)} + 10^{-\left(\frac{C/N_{SHOT}}{10}\right)} + 10^{-\left(\frac{C/N_{POSTAMPLIFIER}}{10}\right)} \right] \quad (11.21)$$

where

C/N_{LINK} = the net per-channel C/N

C/N_{RIN} = the per-channel transmitter relative intensity noise (Equation (11.11))

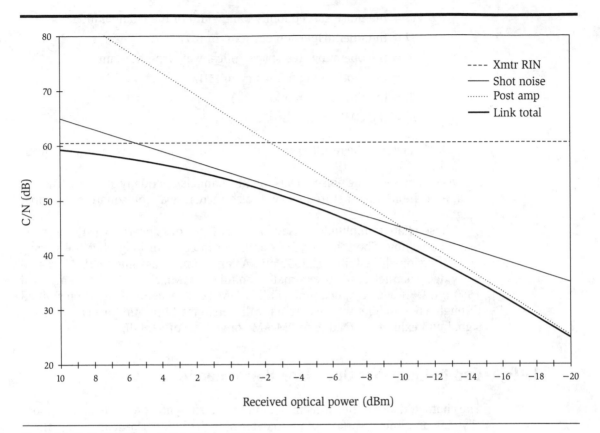

Figure 11.18 Typical fiber-optic link C/N contributions.

C/N_{IIN} = the per-channel interferometric intensity noise (Equation (11.19))

C/N_{SHOT} = the per-channel shot noise at the detector input (Equation (11.15))

$C/N_{POSTAMPLIFIER}$ = the per-channel postamplifier noise contribution (Equation (11.16) or (11.17))

If the end-to-end C/N of a simple DFB link is plotted as a function of optical loss, the curve will typically have three asymptotes, as shown in Figure 11.18 (the effects of IIN are not readily plotted on the same chart because they don't vary with received optical power, but with line length).

For high received power levels, the transmitter RIN will dominate, and the C/N will be independent of path loss; at lower received levels, the shot noise will begin to dominate, and the link C/N will change 1 dB for every decibel reduction in received power level. At some point, the postamplifier noise will begin to dom-

inate, and the link will begin to degrade 2 dB for every decibel of loss due to the square law receiver response. The range of receiver power over which each of these effects dominates depends, of course, on the quality of the individual components. For typical links, the slope of C/N versus the level near 0-dBm received power is about 1:1.

In addition to the factors shown in Equation (11.20), other elements and interactions may affect the end-to-end C/N. For instance, if an EDFA is used, its C/N must be included within the parentheses as an additional factor. In long 1550-nm links, residual phase noise may be a factor.

11.10.2 Small Signal Distortions

Directly modulated DFB transmitters behave similarly to single-ended RF amplifiers, generating both second- and third-order distortion whose amplitudes, as a function of RF drive level, increase roughly 1 dB/dB (in the case of second-order products) and 2 dB/dB (in the case of third-order products). As with their coaxial counterparts, CSO products tend to be the limiting distortion. As discussed earlier, typical specifications for such transmitters are 60 to 62 dB for C/CSO and 65 dB for C/CTB under normal operating conditions with a loading of 77 unmodulated carriers. These distortions arise from the combination of small transfer function nonlinearities and large signal clipping.

Externally modulated transmitters exhibit symmetrical distortion and are thus similar to conventional push–pull coaxial amplifiers in that regard. That, combined with the predictability of the nonlinearities, allows C/CSO and C/CTB values of 65 or better with channel loadings of 77 unmodulated carriers.

Distortion occurring in the optical receiver must be added to that generated in the transmitter. Like the transmitter, the detector–postamplifier combination behaves like a single-ended amplifier, with dominant second-order distortion. Often manufacturers specify link performance with a given received power, rather than individually specifying transmitter and receiver.

When the link loss includes fiber, the CSO degradations due to both chirp-induced dispersion and SPM must be included in the calculation of end-to-end performance. Since the CSO distortion mechanisms are primarily synchronized with the modulating waveform (with the exception of the small-signal DFB nonlinearities), the various CSO contributions could be expected to add on a voltage (20 log) basis rather than a power basis. Though this is true, the chirp and SPM effects may, in fact, be of opposite polarity and tend to cancel each other, rather than add. Nevertheless, an assumption of voltage addition is the most conservative approach to link design.

One technique for reducing the effects of chirp, SPM, and receiver-generated CSO is to transmit both outputs of a Mach–Zehnder type modulator through parallel networks to the receiver. If the two signals are separately detected, then combined out of phase at RF, these second-order effects will tend to cancel just as they do in push–pull RF amplifiers. C/CSO values of greater than 70 dB have been reported using this technique, known commercially as a Harmonic Link Extender.

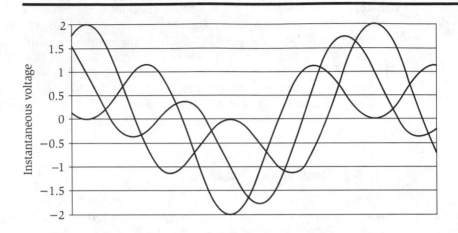

Figure 11.19 Addition of two sine waves with various phase differences.

As an added benefit, when the signals are combined, the noise will be largely uncorrelated (except for the contribution from transmitter RIN) and so will add on a power basis, whereas the signals will add on a voltage basis, resulting in a net 3-dB increase in C/N in the link. Link performance of 55-dB C/N over 50-km links at 1550 nm has been reported.[21]

11.10.3 Clipping Distortion

The ratio of peak-to-average absolute voltage in a simple unmodulated carrier is about 1.6:1. When two equal-amplitude nonrelated carriers are summed, however, the relationship is not as simple. Figure 11.19 shows three possible waveforms resulting from the simple addition of two sine waves when one is exactly twice the frequency of the other. Depending on their relative phases, the peak voltages can vary by nearly a factor of two.

The case when many signals are summed is much more complex. When many summed signals are randomly distributed in both frequency and phase, the peak-to-average level seldom exceeds 14 dB, as discussed in Chapter 14 (though with occasional higher peaks). The downstream spectrum of cable systems is not random, however. The most extreme case occurs when a large number of carriers are harmonics of a common root frequency (as in the ANSI/EIA-542 HRC channelization plan) and all the carrier phases are aligned. In that case, the time domain waveform resembles a string of impulses spaced by a time interval equal to the period of the common root frequency. This is not a surprising result since the Fourier transform of an infinite string of zero-width pulses is an infinite string of carriers.[22]

Figure 11.20 Instantaneous voltage from addition of 77 FDM carriers with standard frequency assignments and tolerances with phases aligned at $t = 30$ ns.

In the more commonly used EIA-542 Standard frequency plan, most of the visual carriers are offset by approximately 1.25 MHz from harmonics of 6 MHz and are neither frequency nor phase related. Despite this, the composite waveform has characteristics that are close to harmonically related carriers if the carrier phases do happen to align at some time. Figure 11.20 shows the resultant instantaneous voltage of the standard channels between 54 and 550 MHz, with each assigned a frequency at random within its tolerance range, but with the signal phases aligned at $t = 30$ ns on the horizontal scale. The peak-to-average voltage is approximately 34:1. If we look at the same waveform later in time, however, the phases slowly move out of coincidence, and the peak voltage decreases. Figure 11.21 shows the condition at $t = 15$ µs. By 20–30 µs, the pulses are no longer distinguishable. Figure 11.22 shows a typical situation with truly random phase relationships.

Observations of actual headends carrying NTSC modulated carriers have confirmed that peak-to-average ratios vary from about 3.5:1 to greater than 7:1. When some subset of carriers happen to phase align, the result is voltage peaks spaced approximately 167 ns apart (the inverse of the 6-MHz channel spacing), which build up to a peak over a few 10s of microseconds, then decrease again as the carriers move out of phase coincidence.

The importance of this occasionally occurring peak voltage condition is that, depending on the OMI, the optical transmitter may be driven into hard limiting (clipping) when a sufficient number of carriers are in phase alignment. This

Max = 39.1 Ave = 4.46 Max/Ave = 8.8 Start = 1.5 ns

Figure 11.21 Spike degeneration after 15 microseconds.

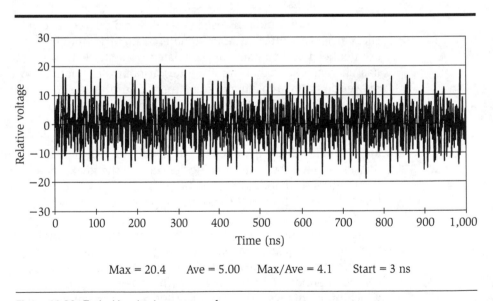

Max = 20.4 Ave = 5.00 Max/Ave = 4.1 Start = 3 ns

Figure 11.22 Typical headend output waveform.

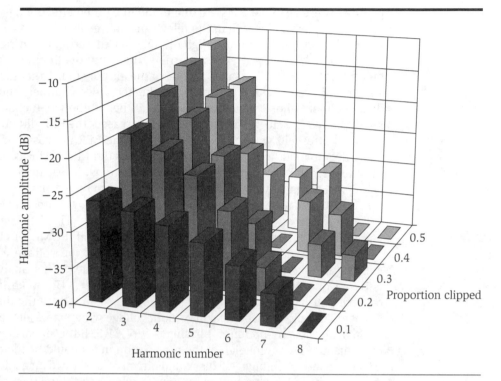

Figure 11.23 Harmonic content of clipped sine wave.

is particularly true in the case of directly modulated laser diodes, where a sharp knee occurs in the transfer function below which the light output is extinguished. When that occurs, intermodulation products are generated. As Figure 11.23 shows, both even- and odd-order distortion products are generated when one polarity of a sine wave is clipped off (though second-order products are always the largest). Thus, both CTB and CSO will show an increase when clipping happens with sufficient frequency to be statistically significant. Since clipping produces a string of pulses with the characteristic 167-ns spacing, the largest components produced will fall at harmonics of 6 MHz (which is at the position of the lower CSO beat when the EIA Standard or IRC channel allocation scheme is used and at the video carrier frequency when the EIA HRC channel scheme is used). With highly linear diodes, clipping may, in fact, be the principal contributor to IM products.

In the case of Mach–Zehnder external modulators, the nonlinearities are symmetrical, and the limiting is "softer," resulting in less clipping and primarily odd-order resultant distortion.[23]

The visible effect on analog video channels is that, as OMI is increased, "dashes" resembling electrical interference begin to appear on the displayed

picture. These visible effects are due to the integrated effect of a group of pulses (whose spacing is closer than the television can resolve).

In the case of high-order digitally modulated carriers sharing a linear link with a comb of analog television carriers, the result of clipping is an increase in bit error rate (BER). Data taken by one of the authors found that the BER through a fiber link loaded with 80 normally modulated video channels and one 64-QAM data channel (whose signal level was 10 dB below the video channels) was approximately 1.3×10^{-10}. As the OMI was increased, the BER did not change until a critical threshold was reached, whereupon the BER increased to 7.6×10^{-8}, a 600:1 degradation, with an OMI increase of less than 1 dB. This confirms findings of other researchers in this field. The observations of researchers using normal NTSC video signals were that the BER increased to unacceptably high values before any degradation was visible in displayed analog pictures.[24]

The crucial question is how often clipping occurs. This has been explored by a number of authors. David Grubb has suggested that if OMI per channel is held to $0.348/\sqrt{N}$ (where N is the number of channels), the voltage will exceed the clipping threshold only about 2.5×10^{-5} of the time.[25] That analysis, however, is based on unmodulated, randomly phased carriers. He noted, however, that optimal phasing of harmonically related carriers can reduce the theoretical peak voltage. The ratio of theoretical maximum voltage with phases aligned to the maximum achievable by optimum phasing is $(\sqrt{N})/1.5$. If the signals feeding an optical transmitter were so related, the modulation index could be increased to $0.67/\sqrt{N}$, or about 5.7 dB higher. This would directly translate to an achievable C/N increase in the link (whose exact amount is related to the relative contribution of transmitter, shot, and postamplifier noise, as discussed earlier in this chapter).

Another factor is video modulation. Since NTSC signals are only at maximum carrier level a small percentage of the time (during synchronizing peaks), the sum of the instantaneous voltages of a number of randomly timed video channels is almost always lower than unmodulated carriers. As with carrier phasing, however, the key word is "almost" since it is possible that the synchronizing pulses will occasionally drift into alignment. Stuart Wagner has measured clipping noise and its effect on BER of a digital signal carried along with 40 randomly timed analog video channels in a fiber link. What he found was that the peak OMI per analog carrier could be increased by approximately half the average difference in power level (in decibels) between the unmodulated and modulated carriers with no increase in noise or BER.[26] As discussed in Chapter 9, the usual assumption is that average power levels are about 6 dB below peak levels. In particular, the black level of a video signal is about 2.5 dB below sync so that with random video timing it should be possible, on average, to raise peak OMI per channel by about 3 dB compared with the unmodulated carrier case.

On the other hand, with optimum video timing, it should be possible to keep the total carrier power close to the average power by minimizing instances where carriers are simultaneously at peak levels. In that case, it should be possible to increase OMI by the difference between peak and average power, 6 dB,

without increasing clipping distortion.[27] That increase will lead to a similar increase in link C/N.

11.11 Summary

Linear fiber-optic links are capable of transporting the full spectrum of cable television services over distances exceeding 20 miles without amplification. Amplification, readily available at 1550-nm wavelength, can extend the attainable distance severalfold, with only a minor impact on end-to-end performance.

The basic C/N performance of optical links is limited by transmitter intensity noise, optical shot noise, the interaction between transmitter linewidth and doubly scattered light in the fiber, and receiver postamplifier noise. The interaction between transmitter residual phase noise and the fiber can further reduce the noise under some conditions.

Distortion is fundamentally limited by both small-signal nonlinearities in the transmitter and clipping caused by large signal peaks. Additionally, interaction between incidental transmitter FM and fiber dispersion can cause second-order distortion, as can interaction between peak power levels and the glass material itself.

Increasing the optical modulation index per carrier in an optical link will improve C/N at the expense of distortion, just as increasing operating levels in a coaxial network will have the same effect. Similarly, decreasing the number of carriers will allow higher C/N and lower distortion per channel, as in coaxial network.

Chapter 13 will discuss the total end-to-end performance of cascaded fiber-optic and coaxial broadband distribution networks, Chapter 16 will discuss various architectures, and Chapter 17 will discuss the reliability of various HFC architectures.

Endnotes

1. Ronald C. Cotten, *Lightwave Transmission Applications.* CableLabs, Louisville, CO, September 1993, pp. 80–84.

2. *AT&T Generic Specification: Fiber Optic Outside Plant Cable,* Issue 11, December 1995. AT&T Network Cable Systems, Norcross, GA.

3. *CommScope Fiber Optic Cable Catalog,* September 1995. CommScope Inc., Hickory, NC.

4. AT&T, op. cit.

5. Claude Romans and David Large, Optical Bus Architecture for Co-Deployment of Telephone and CATV Services in the FRG, *Conference Record,*

Globecom '89. IEEE Communications Society, November 1989, pp. 37.5.1–37.5.9. Institute of Electrical and Electronic Engineers, New York.

6. AT&T, op. cit.

7. David Grubb III and Yudhi Trisno, AM Fiber Optic Trunks—A Noise and Distortion Analysis, *1989 NCTA Technical Papers.* NCTA, Washington, D.C., May 1989.

8. F. W. Willems, W. Muys, and J. S. Leong, Simultaneous Suppression of Stimulated Brillouin Scattering and Interferometric Noise in Externally-Modulated Lightwave AM-SCM Systems, *IEEE Photonics Technology Letters.* Vol. 6, No. 12, December 1994. See also S. W. Merritt, G. J. McBrien, and E. R. Yates, Integrated Optic Modulators for 1 GHz HFC Systems, *CED,* February 1996.

9. Dogan A. Atlas, Fiber Induced Distortion and Phase Noise to Intensity Noise Conversion in Externally Modulated CATV Systems, *1996 NCTA Technical Papers.* NCTA, Washington, D.C.

10. AT&T Corporation, *LGX Broadband Fiber Management System Application Guide,* AT&T Network Systems, Norcross, GA, 1995.

11. Dan Harris, Primer on Baseband Digital Transmission on Single-Mode. *Communications Technology,* August 1996, pp. 92–98.

12. Ronald C. Cotten, op. cit., p. 148.

13. Henry A. Blauvelt et al., Optimum Laser Chirp Range for AM Video Transmission, *1992 NCTA Technical Papers.* NCTA, Washington, D.C., pp. 79–85.

14. X. P. Mao et al., Brillouin Scattering in Lightwave AM-VSB CATV Transmission Systems. Paper presented at OFC'92 Conference, San Jose, California, February 1992. Digested paper in *Technical Digest-Optical Fiber Communication Conference.* Published by Optical Society of America, Washington, D.C., 1992.

15. Roger Brown, Wave of New Products Set to Break. *CED,* September 1996, pp. 26–30.

16. Vincent Morin and Edouard Taufflieb, Improved Optical Fiber Amplifier for 1.3 μm AM-VSB CATV Systems. Scheduled for presentation at the 1998 NCTA National Conference and publication in the *1998 NCTA Technical Papers.* NCTA, Washington, D.C.

17. This equation is identical to that used by Ron Cotten in the previously referenced paper (his Equation 16, page 171), but with the input variables converted to more common engineering units and with the bandwidth kept as a separate factor to allow the formula to be applied to various bandwidth services.

18. Henry A. Blauvelt, op. cit.

19. Donald Raskin and Dean Stoneback, *Return Systems for Hybrid Fiber/Coax Cable TV Networks.* Upper Saddle River, NJ: Prentice Hall PTR, 1998, pp. 152–154.

20. Dogan A. Atlas, op. cit. Modified after consultation with Dr. Dan Harris of Corning Fiber to give the result per channel and to convert the units to engineering units more familiar to cable engineers.

21. Patrick Harshman, Ph.D., Applying Push-Pull Technology to Optical Networks. *CED,* September 1996, pp. 56–61.

22. Athanasios Papoulis, *Circuits and Systems: A Modern Approach.* New York: Holt, Rinehart & Winston, 1980, pp. 331–389. Fourier transforms are used to convert between time domain waveforms and their equivalent frequency content.

23. D. Raskin et al., Don't Get Clipped on the Information Highway, *1996 NCTA Technical Papers.* NCTA, Washington, D.C.

24. David Large and Rex Bullinger, Downstream Laser Clipping: Field Measurements and Operational Recommendations, *Proceedings Manual, 1997 Conference on Emerging Technologies.* SCTE, Washington, D.C., January 1997.

25. David Grubb III and Yudhi Trisno, Ph.D., AM Fiber Optic Trunks—A Noise and Distortion Analysis, *1989 NCTA Technical Papers.* NCTA, Washington, D.C., 1989.

26. Stuart S. Wagner, Thomas E. Chapuran, and Ronald C. Menendez, The Effect of Analog Video Modulation on Laser Clipping Noise in Optical Video-Distribution Networks, *IEEE Photonics Technology Letters.* Vol. 8, No. 2, February 1996, pp. 275–277.

27. David Large, Reducing Distortions Using Video Timing Techniques. *CED,* June 1989, pp. 114–119.

Chapter 12
Other Supertrunking Alternatives

12.1 Introduction

Although linear amplitude modulated fiber-optic links are the favored trunking methodology for most new systems, there are situations where they are simply not cost effective. For example, physical barriers, such as waterways, may make physical cable construction very expensive. Alternatively, the network may need to serve several small outlying communities that are widely separated so that the average home density is very low. In either of these cases, point-to-point or point-to-multipoint broadband amplitude modulated microwave links (AMLs) may offer a superior solution. This chapter will cover the basic operation of this equipment and the essential calculations and methodology required to engineer AMLs. No attempt is made to present a comprehensive manual on either internal designs of microwave equipment or the many subtleties of path engineering. Several standard reference books on the subject are available for those desiring to pursue the subject in greater depth.[1,2,3,4]

Before the commercial development of AM fiber technology, large cable systems were sometimes constructed with specially designed coaxial trunks that were optimized to cover long distances. These were known as *supertrunks*. Since these are seldom used in modern cable systems, we will discuss them only briefly.

12.2 Coaxial Supertrunks

Chapters 9 and 10 cover coaxial technology in general, along with the inescapable trade-offs among noise, distortion, bandwidth, and signal levels in a cascaded coaxial distribution system. Where cable systems found it necessary to cover large distances to interconnect local distribution cables in large systems, they constructed special supertrunk lines typically using some combination of large coaxial cables (typically 1.00–1.125 inches in diameter); feed-forward amplifiers (with high output power capability and low noise figures); and/or band splitting so that each of multiple parallel amplifiers carried only part of the entire spectrum.

There is no "magic" to coaxial supertrunks. They are just coaxial distribution lines that are designed to push the limits of performance. Since the loss of even the largest coaxial cables are about 100 times that of fiber and they are more costly to purchase and install, coaxial supertrunks are of interest for historical reasons only.

Occasionally, coaxial supertrunks have been used to carry video channels that were FM rather than AM modulated. Such trunks, however, require FM detection to recover the baseband video, then VSB-AM modulation to convert them into the format for transmission to homes. Thus, the amount of equipment required to convert many channels rivals that in a headend. FM transmission circuits, whether coaxial, fiber-optic, or microwave, are properly considered part of the headend interconnect network, not the broadband linear distribution network, and so will not be covered in this section.

12.3 Broadband Amplitude Modulated Microwave Links (AMLs)

The great advantage of microwave over any physical interconnecting link is that it doesn't require use of a right-of-way. As mentioned, this makes it the logical choice for signal distribution where, for whatever reason, construction of inter-connecting links is impossible or very expensive. It may also offer a solution to cases where signals need to pass through areas where the network operator does not have a right to construct plant.

Microwave may be used for relay of individual channels or for transport of the entire FDM spectrum. The use of channelized microwave (which often uses digital or FM modulation) is covered in Chapter 6. The discussion here will be limited to links that transport the entire spectrum and are part of the broadband distribution network.

12.3.1 U.S. Regulation of Microwave Transmission

Regulation of radio transmission will vary by country though broad spectrum planning is harmonized worldwide through international conferences. In the United States, the essential regulations pertaining to AML systems are contained in Parts 17 and 78 of the FCC's rules.[5] In particular, Part 17 covers towers, whereas Part 78 covers microwave licensing, channelization, and performance. Operators of microwave systems are required to have current copies of the rules on hand and to comply with detailed requirements regarding licensing, log books, post-ing of stations, periodic performance tests, and the like. Maintenance of tower lighting, where required, is a matter of critical concern because of aircraft safety issues.

The FCC has set aside the frequencies extending from 12.7 to 13.2 GHz for the Cable Antenna Relay Service (CARS). Some additional spectrum is as-signed at 18 and 31 GHz as well. Within the 12-GHz band, several acceptable channelization schemes are given for both FM and AM modulation: A, B, and K channels for FM, and C, D, E, and F designations for AML design techniques that will be discussed in this chapter. Each channel in each scheme is given a unique identifier.

In the case of the AML bands, each channelization group represents a direct translation of the FDM cable spectrum to a different portion of the assigned mi-crowave band as follows:

- Group C maps 54 through 300 MHz to 12,700.5 through 12,946.5 MHz.
- Group D maps 54 through 300 MHz to 12,759.7 through 13,005.7 MHz.
- Group E maps 54 through 300 MHz to 12,952.5 through 13,198.5 MHz.
- Group F maps 54 through 240 MHz to 13,012.5 through 13,198.5 MHz.

Transmission of a full 80-channel spectrum is possible through use of a combina-tion of Group C, D, and E channels (with some minor shifts from nominal micro-wave frequencies) occupying the full 500-MHz CARS band. Clearly, without some

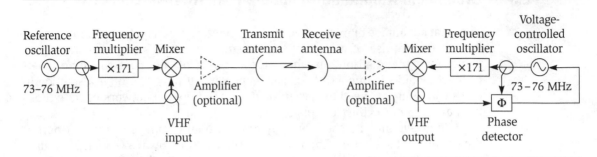

Figure 12.1 Basic AML principles of operation.

form of frequency reuse, no combination of available 12-GHz channels is adequate for a 750-MHz system.

In general, the FCC limits the microwave power level delivered to the antenna system to 5 watts per channel and the radiated power to +55 dBW EIRP.* Microwave frequencies must be accurate within ±0.005%, and the original cable channel frequencies must not shift by more than ±0.0005% through the link.

12.3.2 General Operational Principles

The basic principle of operation of AML equipment is simplicity itself. At the transmitter, the incoming cable spectrum is simply mixed with an unmodulated signal (the microwave local oscillator, or LO) whose frequency is equal to the difference between the original channel frequencies and the translated microwave frequencies. At the receiver, the microwave spectrum is simply mixed with another unmodulated carrier at the LO frequency, resulting in recovery of the original cable spectrum. In order to preserve the frequency accuracy of the VHF spectrum, as well as provide the required microwave frequency accuracy, a crystal reference oscillator operating at 1/171 of the offset frequency is used to phase lock the transmitter local oscillator directly. A sample of the reference frequency (which falls in the VHF spectrum between cable channels 4 and 5) is also upconverted and sent along with the remainder of the cable spectrum through the microwave link. At the receiver, the reference signal is recovered and used to phase lock the microwave LO there. Since both LOs are locked to a common reference, the recovered spectrum will be frequency coherent with the transmitted spectrum, and the accuracy of the microwave spectrum will depend on the accuracy of the reference oscillator. Figure 12.1 is a basic functional diagram of an AML. Table 12.1 summarizes the frequencies for various bands, and Table 12.2 shows the combination of AML channels used to transport a full 54–552 MHz cable spectrum.

* See Chapter 6 for a discussion of the relationship among antenna gain, transmitter power, and effective isotropic radiated power (EIRP).

Table 12.1 Frequency relationships in AML channel groups (in MHz).

FCC Channel Group	Cable Spectrum		Microwave Spectrum		Offset Frequency	Reference Oscillator
	Start	Stop	Start	Stop		
C	54	300	12,700.5	12,946.5	12,646.5	73.956614
D	54	300	12,759.7	13,005.7	12,705.7	74.302339
E	54	300	12,952.5	13,198.5	12,898.5	75.429824
F	54	240	13,012.5	13,198.5	12,958.5	75.780701

Transmitter designs vary in the way channels are handled and in their power capabilities. Some early AML transmitters used a separate transmitter for each cable channel, whereas most modern units are broadbanded. High-power transmitters use microwave amplification, whereas low-power units may use the mixer output directly. Finally, early units used klystron tubes, whereas most modern units use solid-state components though the highest-powered units still use traveling wave tubes (TWTs), which provide high power with both gain and phase stability.

Regardless of the difference in transmitters, all receivers are similar, differing primarily in their mounting, powering, and flexibility in connecting to the downstream coaxial distribution system. In some cases, microwave low-noise amplifiers (LNAs) are used to improve the noise figure of receivers and/or allow the receiver to be mounted at the base of a tower rather than at the back of the antenna.

12.3.3 Path Design

Microwaves travel approximately along "line-of-sight" paths, meaning that, unlike low-frequency AM broadcast stations, they do not bend significantly to follow the earth's curvature. Like light waves, however, they can be reflected from multiple intermediate surfaces between the transmitter and receiver. Under some atmospheric conditions, however, microwave signals can bend slightly, either toward or away from the earth's surface.

Designing a microwave path between two fixed points consists of determining the distance and direction between the points, then examining the path to make sure that there is adequate clearance from any objects near the direct line between transmitting and receiving antennas.

Path End Coordinates

Generally, the locations of the path ends will be described in terms of four parameters: the longitude and latitude, which define the location on the earth's surface, and the altitude of the base of the antenna tower and mounting height of the

Table 12.2 AML channel combination used to transmit 54–550 MHz cable spectrum.

AML Channel	Cable Channel	Channel Limits (MHz)		AML Channel	Cable Channel	Channel Limits (MHz)	
		Low	High			Low	High
C01	2	54	60	D33	37	300	306
C02	3	60	66	E01	38	306	312
C03	4	66	72	E02	39	312	318
C04	AML pilot	72	76	E03	40	318	324
C05	5	76	82	E04	41	324	330
C06	6	82	88	E05	42	330	336
C07	FM	88	94	E06	43	336	342
C08	FM	94	100	E07	44	342	348
C09	FM	100	106	E08	45	348	354
C10	FM	106	108	E09	46	354	360
C11	98	108	114	E11	47	360	366
C12	99	114	120	E12	48	366	372
C13	14	120	126	E13	49	372	378
C14	15	126	132	E14	50	378	384
C15	16	132	138	E15	51	384	390
C16	17	138	144	E16	52	390	396
C17	18	144	150	E17	53	396	402
C18	19	150	156	E18	54	402	408
C19	20	156	162	E19	55	408	414
C20	21	162	168	E20	56	414	420
C21	22	168	174	E21	57	420	426
C22	7	174	180	E22	58	426	432
C23	8	180	186	E23	59	432	438
C24	9	186	192	E24	60	438	444
C25	10	192	198	E25	61	444	450
C26	11	198	204	E26	62	450	456
C27	12	204	210	E27	63	456	462
C28	13	210	216	E28	64	462	468
C29	23	216	222	E29	65	468	474
C30	24	222	228	E30	66	474	480
C31	25	228	234	E31	67	480	486
C32	26	234	240	E32	68	486	492
C33	27	240	246	E33	69	492	498
C34	28	246	252	E34	70	498	504
C35	29	252	258	E35	71	504	510
C36	30	258	264	E36	72	510	516
C37	31	264	270	E37	73	516	522
C38	32	270	276	E38	74	522	528
C39	33	276	282	E39	75	528	534
C40	34	282	288	E40	76	534	540
C41	35	288	294	E41	77	540	546
C42	36	294	300	E42	78	546	552

Note: Channel D33 offset +0.8 MHz; channels E04–E09 offset +2.0 MHz from normal.

antenna on that tower, which together define the vertical location of the antenna relative to sea level.

Generally, the longitude is given as the sum of the number of degrees, minutes, and seconds that the location is east (negative) or west (positive) of the prime meridian, which passes through Greenwich, England. The latitude is given as the sum of the number of degrees, minutes, and seconds that the location is north (positive) or south (negative) of the equator.

Since there are 60 minutes in a degree of arc and 60 seconds in a minute of arc, we can convert the longitude and latitude of the transmitter and receiver antenna locations to decimal equivalents as follows:

$$\text{longitude or latitude (decimal degrees)} = \text{degrees} + \frac{\text{minutes}}{60} + \frac{\text{seconds}}{3,600} \quad (12.1)$$

For the remainder of this chapter, we will assume that this conversion has been done and that the transmitting location is at a west longitude of W_T and a north latitude of N_T, whereas the receiving antenna is at a west longitude of W_R and a north latitude of N_R, where all numbers are in decimal degrees. Minor modifications of the formulas will be required for links located in the southern hemisphere and/or eastern longitudes.

Path Length

Path calculations are based on an average earth circumference of 24,857 statute miles. Thus, each degree of arc along that surface represents $24,857/360 = 69.047$ statute miles.

We can approximate the path length by calculating the length of the diagonal of a hypothetical rectangle whose corners are located at the transmit and receiver locations, as shown in Figure 12.2. In the north–south direction, the height in miles is just the difference in the latitude values times 69.047. This is because latitude lines are all the same length. In the east–west direction, however, one degree represents only that distance at the equator. At any distance away from the equator, the distance is shortened by the cosine of the north latitude. Therefore, the width of the rectangle in miles is the difference in longitudes times the cosine of the average of the latitudes times 69.047.

The path distance, D, is the length of the diagonal of this rectangle, calculated by taking the square root of the sum of the squares of the height and width of the rectangle or*

$$D(\text{statute miles}) = 69.047\sqrt{(N_T - N_R)^2 + \left[(W_T - W_R)\cos\left(\frac{N_T + N_R}{2}\right)\right]^2} \quad (12.2)$$

* In all these formulas, the arguments for trigonometric functions are in degrees, not radians.

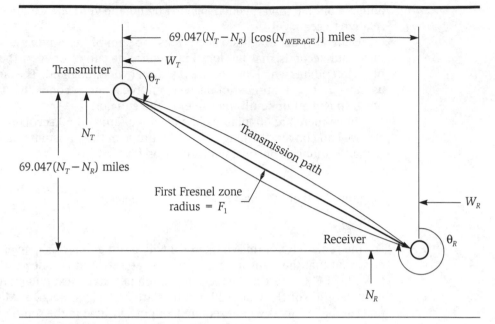

Figure 12.2 Microwave path horizontal parameters.

Though this gives results that are close for the path lengths likely to be involved in most AML applications, the precise formula for path length is

$$D(\text{statute miles}) = 69.047 \arccos[\sin(N_T)\sin(N_R) \\ + \cos(N_T)\cos(N_R)\cos(W_T - W_R)]$$

(12.3)

If the path length is desired in nautical miles for some reason, then use the correction factor 60 rather than 69.047 to convert degrees of arc to distance (nautical miles are based on an earth circumference of exactly 21,600 miles).

Path Azimuth

The next required value is the compass heading (azimuth) for the antennas in order to align the path. Compass headings are measured in degrees clockwise from true north. The first step is to calculate the angle θ using

$$\theta(\text{degrees}) = \arccos\left[\frac{\sin(N_R) - \sin(N_T)\cos(D/69.047)}{\sin(D/69.047)\cos(N_T)}\right]$$

(12.4)

If W_R is less than W_T, then the transmit antenna azimuth $\theta_T = \theta$. If W_R is greater than W_T, then $\theta_T = 360 - \theta$.

The azimuth for the antenna at the receive site is the opposite of that at the transmit site; that is, for θ_T between 0° and 180° $\theta_R = \theta_T + 180°$, whereas for θ_T between 180° and 360° $\theta_R = \theta_T - 180°$.

Path Clearances

Having determined the horizontal distance and azimuth, the next step is to check for adequate clearances along the path. In the horizontal direction, this requires a calculation of a factor known as the Fresnel zone, and in the vertical direction, it additionally requires a more complex analysis that includes the topology of the land along the path, the height of such features as trees that may be on the top of peaks between the antennas, and finally, the effective curvature of the earth along the path.

Fresnel Zone Calculations. The primary path from transmitting to receiving antenna is a nominally straight line. It is possible, however, for the signal to be reflected from objects that, though not in the direct path, are close enough that they are within the beamwidth of the antennas. If that happens, then those signals will arrive at the receiving antenna with a different phase than the direct signal, caused by the longer path length involved. The impact of this reflected wave will depend on the nature of the surface from which the signal is reflected, on the difference in path length, and on the pattern of the transmitting and receiving antennas.

The first Fresnel zone is defined as an imaginary ring surrounding the center of the direct path such that the distance from the transmitting antenna to this ring plus the distance from the ring to the receiving antenna is equal to one-half wavelength more than the direct path between the antennas. Subsequent Fresnel zones are defined as larger imaginary rings where the difference in path lengths is $n\lambda/2$, where n is an integer larger than 1, and λ is the free-space wavelength of the signal.

If the total path length is D statute miles, then we can calculate the radius of the first Fresnel zone at a point D_1 miles from the transmitting antenna using

$$F_1 = 72.1 \sqrt{\frac{D_1 D_2}{fD}} \qquad (12.5)$$

where

F_1 = the radius of the first Fresnel zone in feet
D = the total direct path length in miles
D_1 = the distance from the transmitting antenna in miles
D_2 = the distance from the receiving antenna in miles = $D - D_1$
f = the operating frequency in GHz

Figure 12.2 shows the width of the first Fresnel zone superimposed on the transmission path.

Although higher-order Fresnel reflections can cause problems with highly reflective surfaces that are oriented optimally (such as smooth, flat terrain or calm water), the general rule of microwave design is that it is sufficient to clear objects in the same horizontal plane by 0.6 F_1.

Additional Vertical Clearances. In the vertical plane, it is necessary to add the Fresnel zone clearance to the effective height of other objects that may be close to the direct path. Factors to be considered in the effective height determination include the curvature of the earth along the path, peaks in the terrain, and/or any objects, such as buildings or trees. Traditionally, such analyses were done graphically. Although it is more likely to be done with a computer today, Figure 12.3 illustrates how the terms add up.

The flat line at the bottom of the figure represents a straight line drawn from mean sea level at the transmit location to mean sea level at the receive location. At the transmit end, a vertical line represents the elevation of the ground at the base of the antenna tower, as surveyed or shown on a topological map. The units are feet *above mean sea level* (*AMSL*), and we have designated this elevation as $AMSL_T$ to identify which site is referenced. The elevation of the transmit antenna relative to the ground is *above ground level*, or AGL_T. The sum of the ground height and antenna mounting height we have designated as H_T. At the other end of the path, the equivalent values are $AMSL_R$, AGL_R, and H_R.

The curved line that intersects the ends of the flat bottom line represents the effective curvature of the earth between the endpoints. The effective curvature may not equal the actual curvature, however, since the microwave beam may be slightly bent due to atmospheric conditions. Though the beam most commonly bends toward the earth (thereby allowing longer distances without interference), occasionally the reverse is true. These effects are taken into account using a factor called the K factor. K is the ratio between the effective and actual earth radius. The effective height increase, h, at any point D_1 miles along a path of length D miles can be calculated using

$$h = \frac{D_1 D_2}{1.5K} \tag{12.6}$$

where

h = the effective height, in feet, due to effective earth curvature

D_1 = the distance to the transmitting antenna in statute miles

D_2 = the distance to the receiving antenna in statute miles = $D - D_1$

K = the K factor

Under normal atmospheric conditions, K is approximately 4/3. Conservative path design, however, frequently calls for calculating clearances using $K = 2/3$. K, for

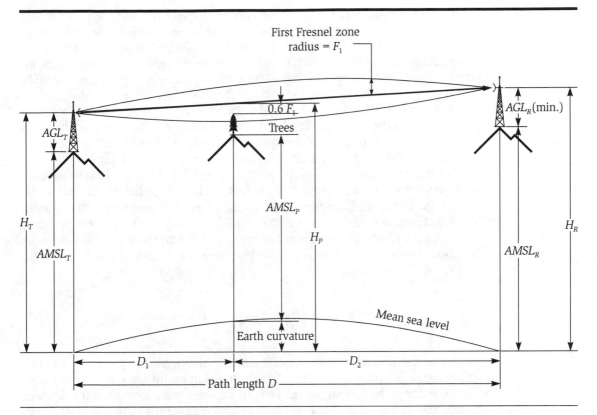

Figure 12.3 Microwave path vertical clearances.

any specific location, is determined from a factor known as the *august mean radio refractivity* and the path elevation above mean sea level.

In the special case of equal height antennas transmitting over a smooth surface (so that the maximum interference point is midway along the path), Equation (12.6) reduces to

$$h_{\text{max}} = \frac{D^2}{6K} \tag{12.7}$$

where

h_{max} = the effective height increase, in feet, at the midpoint of the path

D = the total path length in miles

Whatever the topology, the effective earth curvature line in Figure 12.3 represents the effective height of sea level along the path.

The next step is to examine the path profile. Generally, this is done using topological maps, such as those available from the United States Geological Survey (USGS). Rather than plot the entire profile, it is sufficient to plot selected points that represent elevation maxima. For simplicity, we have shown a single peak, located a distance D_1 from the transmit antenna, whose effective height is $AMSL_P$ plus the effective curvature of the earth at that distance along the path.

Added to this effective peak height must be anything that projects above the ground level at the peak. In wooded areas, for instance, it is common to add a number of feet that represents the maximum height of trees in that area. Finally, we must add $0.6 F_1$, the required clearance, to the first Fresnel zone.

Putting it all together, we get an effective height H_P, in feet, of the peak equal to the effective earth curvature at that point in the path, plus the ground elevation, $AMSL_P$, plus an allowance for anything extending above the ground level, plus $0.6 F_1$, where F_1 is the radius of the first Fresnel zone at that point in the path.

Using the height of the transmit antenna and the effective height of the peak and its distance along the path, we can calculate the minimum mounting height for the receive antenna using

$$AGL_R(\text{ft}) \geq (H_P - H_T)\left(\frac{D}{D_1}\right) + H_T - AMSL_R \qquad (12.8)$$

So long as the receive antenna mounting height above ground is greater than this value, the path should be acceptable from the standpoint of vertical clearances.

Note that in the special case of a flat landscape covering much of the distance between antennas, it may not be obvious where along the path the greatest potential interference occurs. A reasonable approximation is as follows.

First calculate the effective difference, ΔH, in elevation between the transmit antenna and the flat ground along the path:

$$\Delta H(\text{ft}) = AMSL_T + AGL_T - AMSL_P - Trees - 21.63\sqrt{\frac{D}{f}} \qquad (12.9)$$

where

$AMSL_T$ = the elevation at the base of the transmit tower

AGL_T = the mounting height on the tower

$AMSL_P$ = the elevation of the flat land between the antennas

$Trees$ = the allowance for foliage and objects on the flat land

D = the total distance between the antennas in miles

f = the operating frequency in GHz

The only approximation in Equation (12.9) is the last term, which calculates the Fresnel clearance on the assumption that the highest point will be roughly in the

center of the path. Obviously, to the degree that the total height of the transmitting and receiving antennas are different, this will be in error, though the effect on the total calculation will usually be minor.

Given the effective difference in elevation between the transmit antenna and the flat area, the maximum distance that a signal can travel from the transmitter and still clear the ground is

$$D_1 = 69.047 \ \arccos\left(\frac{R_E K}{R_E K + \Delta H}\right) \tag{12.10}$$

where

D_1 = the distance from the transmitter, in miles, to the point where the signal just clears the ground by the required amount

R_E = the radius of the earth in feet = 20,888,284

K = the K factor

ΔH = the effective elevation difference calculated earlier

D_1 can now be plugged into Equation (12.8), along with the other factors, to determine the required mounting height of the receive antenna.

12.3.4 Performance Calculation

In order to predict the performance of an AML, we first need to define the signal path from the transmitter to each receiver. Then we will start with the transmitter output RF power per channel and add up all the signal losses and gains (in dB) from the path elements to determine the nominal power level at each significant point. Knowing the performance of each circuit element, we will then calculate the cascaded noise and distortion of the link under normal weather conditions. Finally, using charts of expected weather disturbances, we will estimate the reliability of the circuit.

Signal Path Definition

The first step in performance calculation is to carefully list every circuit element in the path between transmitter and receiver. Since a single transmitter may feed more than one receiver (along different paths), the first element may well be a signal splitter. Following that will usually be sections of waveguide. Since elliptical waveguide is more lossy than circular, it is not uncommon to use circular waveguide to feed up tall towers, then transition to elliptical guide for the last few feet at both ends of the transmission line to allow flexibility for dish adjustments and equipment placement.

The dish will have a defined gain, followed by the path loss through the air, followed by the gain of the receiving antenna. Finally, there may be a low-noise

Table 12.3 Sample AML performance calculation.

Network Element	Quantity/ Property	Gain (dB)	Cumulative Power (dBm/ch)	C/N (dB)	C/CTB (dB)	C/CSO (dB)
Transmitter			+10.0	61.0	65.0	70.0
Splitter	2 way	−3.5	+6.5			
Circular waveguide	200 ft	−2.8	+3.7			
Elliptical waveguide	50 ft	−1.9	+1.8			
Transmit antenna	10 ft dia.	+48.8	+50.6			
Excess loss allowance		−2.0	+48.6			
Path loss	24 mi.	−146.5	−97.9			
Receiving antenna	10 ft dia.	+48.8	−49.1			
LNA	3 dB NF	+17.0	−32.1	55.9		
Circular waveguide	300 ft	−4.1	−36.2			
Elliptical waveguide	50 ft	−1.9	−38.1			
Receiver	13 dB NF			56.9	70.0	61.0
Total link performance				52.6	61.1	58.4

amplifier at the output of the receive antenna, followed by some combination of waveguides down the tower and finally terminating at the receiver input. All the applicable circuit elements should be separately listed for each section of the path and a separate list made up for each path fed from a transmitter. The left two columns of Table 12.3 show a sample list of circuit elements for a typical AML path at 13 GHz.

Power Budget Calculation

In the first row of the fourth column of the worksheet is the transmitter power output per channel in dBm. Unlike coaxial cable distribution networks where the reference for power measurements is 1 millivolt in a 75-ohm impedance (dBmV), the reference for microwave power is the milliwatt = 10^{-3} watts. Thus the power in dBm = 10 log (P), where P is the power in milliwatts. AML transmitters are available with rated power output levels per channel varying from about −12 to +33 dBm.

Next come any splitters used to feed multiple receive sites. Manufacturers' data sheets will give the loss of these. As a first approximation, the loss will typically be a few tenths of a decibel higher than the theoretical splitter ratio loss (for example, a four-way splitter has a theoretical loss of 6 dB and an actual loss of about 6.5 dB to each port). Enter the appropriate loss in the third column of

the worksheet as a negative number (representing "negative gain" in that path element).

After the splitter will be the waveguides connecting the splitter output port to the antenna. As a general rule, circular waveguide has a loss of about 0.014 dB/ foot at 13 GHz, whereas elliptical waveguide has a loss of about 0.038 dB/foot at the same frequency. The losses of transmission lines used at other frequencies will be given in the manufacturers' literature or in commonly available design tables. For each type of transmission line, multiply the line length by the loss per unit length, and enter the net loss into the third column, again as a negative gain number.

As a good rule of design, an extra loss allowance is added to account for aging, misalignment of equipment, and so on (sometimes known as the *field factor*). Generally, 2 dB is sufficient. Add that to the net loss at the input of the transmit antenna.

The transmitting antenna will exhibit a gain relative to an isotropic radiator. Parabolic antenna design and performance were treated in detail in Chapter 6. The antennas used for CARS band are almost always constructed with prime focus feeds. Diameters commonly range from 4 to 10 feet. Though actual antennas will vary slightly, a typical gain for a circular antenna with a parabolic cross section and a prime focus "button hook" feed is

$$\text{parabolic antenna gain (dBi)} = 20 \log(d) + 20 \log(f) + 6.5 \qquad (12.11)$$

where

d = the diameter of the antenna in feet

f = the operating frequency in GHz

Enter the transmit antenna gain on the appropriate line.

The loss suffered as a result of sending the signal through the air depends on both path length and operating frequency. Under normal atmospheric conditions, free-space loss is

$$\text{path loss (dB)} = 96.6 + 20 \log(f) + 20 \log(D) \qquad (12.12)$$

where

f = the operating frequency in GHz

D = the path distance in statute miles

Enter this loss in the third column.

The receive antenna will have a gain that will also be calculated in accordance with Equation (12.11). If an LNA is mounted at the back of the receive antenna feedhorn, as is common, enter its gain at this point. Now as at the transmitter end, enter all the feedline losses between the antenna or LNA and receiver input.

In the fourth column, calculate the power at each point in the circuit, adding power gain values in dB and subtracting losses in dB. The final number is the receiver input level, in dBm under normal atmospheric conditions.

Carrier-to-Noise Calculation

As discussed in Chapter 10, the noise power in a 4-MHz bandwidth is approximately −59 dBmV. 0 dBmV, however, is equal to −49 dBm so that the noise level referenced to 1 milliwatt is −108 dBm.

Thus, the equivalent input noise floor of any individual circuit element is equal to −108 dBm + F_A, where F_A is the noise figure of the device. Since C/N is the difference between the signal level and the noise floor, the C/N of any individual element, in decibels, can be calculated using

$$C/N_{Device}(dB) = 108 - F_A(dB) + P_{IN}(dBm) \tag{12.13}$$

where

$\qquad C/N_{Device}$ = the carrier-to-noise ratio contribution of the device

$\qquad\qquad F_A$ = the noise figure of the device

$\qquad\qquad P_{IN}$ = the driving power level in dBm

There are three principal contributors to link C/N: the transmitter, LNA (if used), and receiver. The transmitter C/N is generally given by the manufacturer for various channel-loading conditions. Since the input levels and noise figures of both LNA and receiver are known, their C/N contributions are calculated. All three C/N values are entered into column five of the worksheet.

The link C/N is calculated using the methodology of Equation (10.2):

$$C/N_{ttl} = -10 \log\left(10^{-\left(\frac{C/N_T}{10}\right)} + 10^{-\left(\frac{C/N_L}{10}\right)} + 10^{-\left(\frac{C/N_R}{10}\right)}\right) \tag{12.14}$$

where

$\qquad\qquad\qquad C/N_{ttl}$ = the carrier-to-noise ratio of the entire link

$C/N_T, C/N_L,$ and C/N_R = the carrier-to-noise ratios of the transmitter, LNA, and receiver, respectively

Distortion Calculation

The CTB and CSO distortion levels are given by the manufacturers for various channel loadings and power levels. In most cases, the LNA will contribute only slightly to the overall distortion levels.

Addition of second- and third-order distortion levels was discussed in Chapter 10. The link performance numbers in the worksheet were calculated using the most conservative assumption, which is

$$C/CTB_{Link} = -20 \log\left[10^{-\left(\frac{C/CTB_T}{20}\right)} + 10^{-\left(\frac{C/CTB_R}{20}\right)}\right] \qquad (12.15)$$

where

C/CTB_T = the transmitter CTB in dB

C/CTB_R = the receiver CTB in dB

A similar formula is used to calculate link CSO performance. Manufacturers should be consulted for the specific microwave equipment chosen to determine whether Equation (12.15) is appropriate for calculating cascaded distortion.

Atmospheric Effects on Performance—Multipath

The preceding calculation predicts the performance of the microwave link under normal weather conditions. Two significant atmospheric effects can degrade the signal, however. These are multipath and rain fade. An understanding of the general causes and methods for calculating their effects on link performance is important.

It will be recalled that generally a microwave beam bends slightly toward the earth (with the result that the earth looks "flatter" than it actually is). The principal cause of this bending is that signals are not transmitted through a vacuum, but rather through the air. Since the density of air usually decreases with altitude, there will be a slight gradient across the width of the microwave beam, causing a slight decrease in velocity of propagation at the bottom of the beam. This causes the beam to bend slowly toward the earth.

Unfortunately, the density and change in density of the air is not a constant, with the result that sometimes the beam bends more or less. Under extreme conditions, the beam can miss the receiving antenna altogether. Under other conditions, there may also be more than one path between transmitter and receiver, with slightly different transit times and, therefore, phase and amplitude variations at the receiver. These effects are called *multipath*.

Since multipath degradation is a statistically random effect, the usual way of quantifying its effect is to define a minimum usable condition for the link and then to predict how much of the time the link performance exceeds this condition (defined as the *availability* of the path). For analog video links, a common definition of the minimum usable C/N is 35 dB.

The first step in calculating path availability is to determine how much the path loss can deteriorate before the C/N reaches 35 dB (the *fade margin*). Realizing that the C/N of both LNA and receiver will degrade linearly with increasing path loss, while the transmitter's contribution will stay constant, we can write a single equation for fade margin:

$$M(\text{dB}) = 10 \log\left(10^{-\left(\frac{C/N_{min}}{10}\right)} - 10^{-\left(\frac{C/N_T}{10}\right)}\right) - 10 \log\left(10^{-\left(\frac{C/N_L}{10}\right)} + 10^{-\left(\frac{C/N_R}{10}\right)}\right) \qquad (12.16)$$

where

M = the fade margin in dB

C/N_{min} = the defined minimum usable link C/N, usually 35 dB

C/N_T = the transmitter's C/N

C/N_L = the LNA C/N under nominal signal conditions

C/N_R = the receiver C/N under nominal signal conditions

In the example given, the fade margin is 18.35 dB. Since the transmitter typically contributes relatively little to the C/N at the threshold, a good approximation to M is simply the difference between the link C/N under normal signal conditions and the defined minimum link C/N. In the example given, M determined by this method would be approximately $52.6 - 35 = 17.6$ dB.

Based on extensive field testing, the number of hours the path will be below the threshold can be estimated using

$$\text{multipath probability} = U_m = (a)(b)(2.5)(10^{-6})(f)(D^3)(10^{-M/10}) \quad (12.17)$$

where

U_m = is the average probability that the excess path attenuation will be greater than the fade margin due to multipath effects

a = a "terrain factor," varying from 4 for very smooth terrain or water to 1 for average terrain to 0.25 for very mountainous terrain

b = a "temperature/humidity factor," varying from 0.5 for very humid areas to 0.25 for normal interior or northern climates to 0.125 for very dry desert or high-altitude areas

f = the operating frequency in GHz

D = the path length in statute miles

M = the fade margin in dB, calculated using Equation (12.16)

When the terrain and temperature/humidity factors are not known, it is common to use 0.25 for the product of a and b for estimation purposes.

The unavailability probability can be converted into yearly hours of outage using

$$\text{unavailable hours(multipath)} = 8{,}760 U_M \quad (12.18)$$

Atmospheric Effects on Performance—Rain Fade

The second factor affecting path performance is rain. Water has a very high dielectric constant, compared with air (about 80 times as high), and therefore, when heavy rain cells move through the microwave path, the transmission can be seriously affected. The important parameters are the intensity of the rain, the

frequency of rain, the distribution of heavy rain peaks (that is, how big are the cells of downpour), and how fast cells of intense rain move through the microwave path.

As with multipath, these factors do not lend themselves to theoretical analysis as much as to field experience. Extensive research into rain fade effects at different frequencies and in different geographic areas has been carried out by many researchers. One approach to estimating rain fade uses the methodology outlined next.[6]

First, the available fade margin per kilometer, M', must be calculated by dividing the total fade margin M as determined in Equation (12.15) by the path length in kilometers. Since the path length, D, calculated in Equation (12.3), is in statute miles:

$$M'(\text{dB/km}) = \frac{M(\text{dB})}{1.609D(\text{mi})} \tag{12.19}$$

Second, it is necessary to determine the rainfall rate that will result in an attenuation per kilometer greater than M'. The relationship between signal attenuation and rainfall rate can be approximated[7] using

$$Y_R = kR^\alpha \tag{12.20}$$

where

Y_R = the attenuation along the signal path in dB/km

R = the rainfall rate in mm/hr

k and α = "constants" that are a function of frequency, polarization, and vertical path angle

k and α, in turn, are determined using the following formulas:

$$k = \frac{k_h + k_v + (k_h - k_v)\cos^2\theta\cos 2T}{2} \tag{12.21}$$

$$\alpha = \frac{k_h\alpha_h + k_v\alpha_v + (k_h\alpha_h - k_v\alpha_v)\cos^2\theta\cos 2T}{2k} \tag{12.22}$$

where

θ = the path elevation angle in degrees (zero for a horizontal path)

T = the path polarization tilt angle in degrees: 0° for horizontal polarization, 90° for vertical polarization, 45° for circular

k_h, k_v, α_h, and α_v are determined from Table 12.4.

Table 12.4 Rainfall attenuation constants as a function of frequency.

Frequency (GHz)	k_h	k_v	α_h	α_v
1	0.0000387	0.0000352	0.912	0.888
2	0.00154	0.000138	0.963	0.923
4	0.00065	0.000591	1.121	1.075
6	0.00175	0.00155	1.308	1.265
7	0.00301	0.00265	1.332	1.312
8	0.00454	0.00395	1.327	1.310
10	0.0101	0.00887	1.276	1.264
12	0.0188	0.0168	1.217	1.200
15	0.0367	0.0335	1.154	1.128
20	0.0751	0.0691	1.099	1.065
25	0.124	0.113	1.061	1.030
30	0.187	0.167	1.021	1.000
35	0.263	0.233	0.979	0.963
40	0.350	0.310	0.939	0.929
45	0.442	0.393	0.903	0.897
50	0.536	0.479	0.873	0.868
60	0.707	0.642	0.826	0.824
70	0.851	0.784	0.793	0.793
80	0.975	0.906	0.769	0.769
90	1.06	0.999	0.753	0.754
100	1.12	1.06	0.743	0.744

For relatively flat paths and horizontal polarization, $k = k_h$ and $\alpha = \alpha_h$, whereas for relatively flat paths with vertical polarization, $k = k_v$ and $\alpha = \alpha_v$.

The maximum allowable rainfall rate can be calculated by solving Equation (12.20) for R and substituting M' for Y_R:

$$R_{\max} = \left[\frac{M'}{k}\right]^{\frac{1}{\alpha}}$$

(12.23)

The probability that the rain rate exceeds this value in any given climatic region is given by[8]

Table 12.5 Rainfall variable by climatic region.

Climate Region	u	$R_{0.01}$
A	−0.2391	8
B	−0.0050	12
C	0.05669	15
D	0.10252	19
E	0.03726	22
F	0.04652	28
G	0.07652	30
H	0.04081	32
J	0.1082	35
K	0.0444	42
L	0.03368	60
M	0.04087	63
N	0.02924	95
P	0.02477	145

Note: Figures 12.4, 12.5, and 12.6 delineate worldwide climate regions.[9]

$$U_R = 10^{-4} \left[\frac{R_{0.01}}{R_{max}}\right]^b e^{u(R_{0.01} - R_{max})} = 10^{-4} \left[\frac{e^{uR_{0.01}+ln(R_{0.01})\cdot b}}{e^{uR_{max}+ln(R_{max})\cdot b}}\right] \qquad (12.24)$$

where

U_R = the time-averaged probability that the rainfall rate exceeds R_{max}

u depends on the climate region and is given in Table 12.5

$R_{0.01}$ = the rainfall rate in mm/hr that is exceeded in any given climate zone for 0.01 % of the time (0.86 hours/year), with an integration time of one minute; it is given in Table 12.5

b is determined from

$$b = 8.22(R_{0.01})^{-0.584}$$

Once the probability of rainfall exceeding the value needed to attenuate the signal below the specified performance threshold is determined, the annual number of hours of rainfall-caused outage can be simply determined by multiplying U_R by 8,760.

Figure 12.4 Rain zones: the Americas and Greenland.

Figure 12.5 Rain zones: Europe and Africa.

Figure 12.6 Rain zones: Asia and Australia.

Finally, once the rainfall and multipath unavailability probabilities are determined, the overall path availability can be calculated using

$$\text{path availability} = 1 - U_M - U_R \qquad (12.25)$$

Typical link specifications vary from 0.999 availability for noncritical video applications to 0.99999 or even higher for some telephony applications. 0.9999 availability (equivalent to 53 minutes per year average outage time) is a typical goal for a cable television trunking application.

12.4 Summary

Coaxial cable-based broadband trunking alternatives have been almost universally replaced by fiber optics in modern distribution systems. Microwave links, however, are still a viable and cost-effective way to transport signals where construction costs, physical, or regulatory barriers prevent cable construction.

Broadband amplitude modulated microwave links are simple in concept and straightforward to design. The performance of typical AMLs is comparable in noise and distortion to those constructed using directly modulated DFB laser transmitters but suffers from occasional outages owing to atmospheric conditions. More important, though fiber-optic links have almost unlimited potential bandwidth, current regulations limit 12-GHz AMLs to about 500 MHz of spectrum. The use of higher frequencies is limited by the dramatic increase in rainfall and atmospheric attenuation at higher frequencies.

The analysis of AML performance involves a combination of trigonometry and a lot of empirical data on climatic factors but is amenable to the use of straightforward computation using simple spreadsheet templates.

Endnotes

1. *Engineering Considerations for Microwave Communications Systems,* G.T.E. Lenkurt, San Carlos, CA, 1975.

2. *Engineering Information,* Rockwell International, Dallas, TX, 1979.

3. *The Handbook of Digital Communications,* Microwave Systems News, Vol. 9, No. 11, 1979.

4. *Transmission Applications Guide,* Harris Farinon Division, San Carlos, CA, 1987.

5. C.F.R. 47§17 and C.F.R. 47§78, published by the Office of the Federal Register National Archives and Records Administration and available from the U.S. Government Printing Office.

6. The authors are indebted to Dr. Francisco Bernues of CableAML, Inc. for providing this approach to estimating rain outage probability. Private correspondence in September 1997.

7. From CCIR Report 721-3, p. 229. Now published by the International Telecommunications Union (ITU), Geneva, Switzerland.

8. From CCIR Report 563-3, p. 132. Now published by the International Telecommunications Union (ITU), Geneva, Switzerland.

9. Figures 12.4, 12.5, and 12.6 are extracted from International Telecommunications Union Recommendation ITU-R PN.837-1, *Characteristics of Precipitation for Propagation Modeling.* These figures are reprinted with permission of the ITU, which is the copyright holder. The sole responsibility for selecting these extracts lies with the author and can in no way be attributed to the ITU. The complete volume(s) of the ITU material from which the figures were extracted can be obtained from International Telecommunications Union, General Secretariat–Sales and Marketing Service, Place des Nations, CH-1211 Geneva 20, Switzerland. They can be contacted on the Internet at *sales@itu.int.*

Chapter 13
End-to-End Performance

13.1 Introduction

The various parts of the broadband distribution system, the trunk, coaxial distribution, and drop, must work together to provide a transmission path whose performance allows delivery of adequate quality signals to the end user. This means that the total system performance requirements must be defined and that the allowable signal degradation must be allocated among signal acquisition, headend processing, distribution system, and terminal equipment.

This chapter will first discuss quality standards, then the question of allocation. Next, it will deal with the calculation of the cascaded broadband network sections. Finally, it will discuss the typical performance of a modern HFC distribution system, including factors that are not normally part of the cascaded network calculations, such as group delay and several classes of undesired signals that affect transmission channels.

13.2 Quality Standards and Requirements

In order to understand how performance standards are related, it is important to first determine the entire chain from signal generation to end user. The components of this chain will vary depending on the service, of course. Some signals may be generated in the headend or even closer to the subscriber, whereas others will be generated thousands of miles away. Figure 13.1 illustrates an example of the transmission elements through which an analog video signal may be sent.

The video source in this example is a television broadcasting station. The source degradation will include some level of white noise, crosstalk, transmitter distortion, and possibly quantizing noise if the video or audio were digitized in any part of the process. Often the (analog NTSC) video S/N is limited by the originating camera and further degraded by a videotape recording and playback process to result in a 55–60-dB input to the transmitter.* The broadcast transmitter diplexing filter (used to combine visual and aural signals) and antenna will affect the frequency response, whereas visual transmitter incidental carrier phase modulation (ICPM) limits attainable stereo audio performance.

The source-to-headend path, if the headend is receiving the signal via the normal transmission path, might include multipath distortion (echoes resulting from signals bouncing off objects to create more than one transmission path), frequency response variations due to receiving antenna characteristics, electrical noise interference, and a host of other factors.

Within the headend complex, the signal may be demodulated, digitized, switched, amplified, combined with other signals, modulated, and/or shifted in frequency. In some cases, signals may be received at a different location than where they are finally combined into the RF spectrum that is transmitted to cus-

* In this qualitative discussion, we are not distinguishing between baseband noise (S/N) and noise added to the RF modulated signal (C/N). Both affect the quality of the final product.

Figure 13.1 Various performance measures for a typical broadcast television channel.

tomers, adding additional layers of processing and transmission. In that case, the term *headend* is intended to include everything from first reception by the network operator to generation of the modulated FDM signal complex for transmission to customers. Each of the processing steps within the headend complex adds a measure of degradation.

Finally the complete frequency spectrum enters the broadband, linear distribution network, where, among others, noise, intermodulation distortion, cross modulation, hum distortion, and group delay are added.

Analog (and possibly digital) video signals may be directly connected to subscribers' receivers, or they may pass through a network termination device, such as an analog or digital set top terminal. If a signal passes through a set top terminal, it may be demodulated, descrambled, and remodulated—each an imperfect process.

Finally, the subscriber's own television receiver will add a measure of degradation. In addition to noise and distortion, it may be inadequately shielded so that off-air signals are mixed with the cable-delivered signals, creating a special form of crosstalk known as *direct pickup* (DPU) interference. DPU results in a variety of visible picture defects in analog signals and may reduce the bit error rate for digital signals.

As shown in Figure 13.1, the viewer sees and hears the composite effect of everything that happens to the signal from source to his or her screen and speakers (or other end application). Chapter 2 describes the results of tests undertaken to determine how overall technical performance in each of several categories is related to subscriber perception of NTSC signal quality.

Obviously, both the chain of network elements and the required quality of signal will be different for services other than analog video. In particular, it appears that various levels of QAM (from QPSK to 256-QAM) will be used to transport digitally modulated downstream signals, whereas lower levels of digital modulation, such as BPSK, QPSK, and possibly 16-QAM, will be used in the upstream direction. The requirements for the various digital signal formats are discussed in detail in Chapters 3 and 4.

Though no end-to-end formal standards have been proposed by the FCC for U.S. television distribution, the Deutsche Bundespost (DBP or German post office—the responsibilities for cable television have since been transferred to Deutsche Telecom) has published standards covering all segments of their entire network from camera to viewer.[1] The standard is very complete with 19 different quality measures covered. By contrast, the FCC regulates few aspects of television broadcast stations (primarily those that assure a lack of interference among stations) but many parameters of cable television systems (including drops if installed and maintained by network operators) and set top terminals (again, only if owned by the network operator). Performance standards for subscriber receivers vary according to how they are marketed. These issues are covered more thoroughly in Part 6. The FCC does not, as of the date of this book, regulate network performance for any signals other than analog video, nor does it regulate the quality of drop materials available for purchase by end users, though voluntary SCTE standards are being developed to cover drop components.

The FCC's standards applying to cable television systems generally specify the total allowable signal degradation from acquisition at the headend to delivery of the final signal to subscribers' receivers (though there are exceptions). This range of coverage is shown by the second horizontal line in Figure 13.1. The quality parameters are applicable (with a couple of minor exceptions) to every analog video signal at every subscriber equipment terminal under all normal operational conditions.

It is important to recognize that the FCC's requirements apply to performance when the network is carrying normally modulated signals. Although these parameters relate directly to subscriber picture quality perception, the nature of NTSC video is such that some parameters (for example, IM distortion products) are difficult to measure precisely under normal signal conditions because they vary depending on instantaneous levels, carrier phase relationships, and timing of video modulation. Since the total RF power in most cable systems is dominated by analog NTSC video signals, networks are designed and initially tested using unmodulated carriers in place of each expected visual carrier. This yields higher, but more consistent, levels of distortion. Various conversion factors are used to estimate performance under normal signal-loading conditions. In-service systems are tested using normal modulated signal loading, with precision sacrificed to avoid the subscriber service interruption that would be required to perform unmodulated signal testing.

The FCC's operational standards for television broadcasters are contained in the Code of Federal Regulations, Section 47, Part 73, and the cable television regulations are contained in Part 76 and are summarized in Table 13.1. Regulations pertaining to both set top terminals and subscriber receivers are contained in Part 15.[2]

In addition to the FCC's regulations, negotiations between the cable television and consumer electronics industries have resulted in supplemental voluntary standards applying to both cable systems and consumer receivers. Although these were not released as of the writing of this book (and are subject to change),

Table 13.1 FCC rules for analog signals in U.S. cable television systems.

Parameter/FCC Rule Paragraph	Performance Limit
Visual carrier frequency §76.612	No specific requirement, except that carriers within the specified aeronautical bands (108–137 and 225–400 MHz) must be offset from communications channels with a frequency tolerance of ±5 kHz (see rule for details)
Aural/visual carrier frequency difference §76.605(a)(2)	4.500 MHz ± 5 kHz
Minimum visual signal level §76.605(a)(3)	≥+3 dBmV at the end of a reference 100-foot drop cable connected to any subscriber tap port, and ≥0 dBmV at the input to each subscriber's receiver
Visual signal level stability §76.605(a)(4)	≤8 dB pk-pk variation in visual carrier level over the sum of two 24-hour tests, one taken in July or August and one taken in January or February
Adjacent visual signal level difference §76.605(a)(4)(i)	≤3 dB
Total variation in levels among all visual signals §76.605(a)(4)(ii)	≤10 dB pk-pk for systems with upper frequency limit of 300 MHz or less, plus 1 dB for each additional 100-MHz increment, or fraction thereof (e.g., for a 550-MHz system, the allowed variation is 13 dB)
Maximum visual signal level §76.605(a)(4)(iii)	Below TV overload point (level not specified)
Level of aural signal relative to visual signal §76.605(a)(5)	−10 to −17 dB (−6.5 dB maximum at set top terminal output)
In-channel visual frequency response variation §76.605(a)(6)	≤4 dB pk-pk from 0.5 MHz below the visual carrier to 3.75 MHz above the visual carrier; includes set top terminal response starting in 1999
Visual signal carrier-to-noise ratio §76.605(a)(7)	≥43 dB, measured in a 4-MHz bandwidth
Composite triple beat (CTB) and composite second-order (CSO) IM product levels, relative to visual signal level §76.605(a)(8)	≤−51 dB, time averaged, except ≤−47 dB for products that are frequency coherent with the visual carrier in HRC and IRC systems
Hum modulation §76.605(a)(10)	≤3% pk-pk modulation of the visual carrier (≤1.5% modulation by conventional definition)
Relative chrominance/luminance transmission delay (chroma delay) §76.605(a)(11)(i)	≤±170 ns, not including the contribution of any set top terminal
Differential gain §76.605(a)(11)(ii)	≤±20%, not including the contribution of any set top terminal
Differential phase §76.605(a)(11)(iii)	≤±10 degrees, not including the contribution of any set top terminal

key parameters in draft standard EIA-IS-23 affecting cable system signals delivered to the input terminals of consumers' receivers are as follows:[3]

- The visual level of any individual analog television signal delivered to a receiver should not exceed +20 dBmV, whereas the average of the power levels of all the visual carriers carried should not exceed +15 dBmV.

- When distribution systems use multiple cables, the levels of television signals on the same frequency, but different cables, shall not differ by more than 5 dB.

- The frequency of the visual (luminance) carrier of any analog television signal shall not vary more than ± 25 kHz from the nominal values specified in ANSI/EIA-542 (any frequency plan).

- The maximum levels of any non-analog-video signals whose frequencies lie between 0.5 and 30 MHz shall not exceed +42 dBmV, whereas the levels of signals between 54 and 1,000 MHz shall not exceed +20 dBmV (acceptable maximum levels for signals lying between 30 and 54 MHz have not been determined).

- The maximum total average power of nonvideo signals whose frequencies lie between 54 and 1,000 MHz shall not exceed +15 dBmV in any 6-MHz channel.

Regardless of their official status, these additional performance guidelines represent good engineering practice in assuring that cable television signals are compatible with most consumer electronics receivers.

Not included in either U.S. government regulation or interindustry agreements are specifications covering several aspects of the sound accompanying NTSC signals (such as loudness level or stereo separation) though these are included in the DBP model. An issue with both over-the-air broadcast and cable-generated video programming has been the perceived loudness variations among channels and within channels (for example, between the main program and commercials). Despite several attempts by both governmental and industry groups, it has never been successfully resolved.

Also not specified in any document are limitations on either microreflections or external signal ingress. Microreflections were discussed in Chapter 10 and result in visible ghosts in analog NTSC pictures if the amplitude and delay are sufficient. As will be seen later in this chapter, they also contribute to both group delay and frequency response variations. External signal ingress can occur in either the distribution network or within subscribers' terminal equipment. Signal pickup in subscribers' receivers, discussed in Chapter 20, can affect the network if ingressing signals are transmitted out the antenna terminals. It should be noted that, though the FCC does have strict standards covering signal *egress* from cable systems, where strong external signals are present, ingress will usually be visible to subscribers before the FCC egress limit is exceeded though the two effects both arise from inadequate shielding and are obviously related.

As of the writing of this book, there are several emerging industry standards for digitally modulated signals. The interindustry specification EIA-105 Decoder Interface, for example, includes several tuner performance requirements that will be necessary to successfully receive high-order VSB or QAM signals: the SCTE's Digital Video Subcommittee and the Advanced Television Systems Committee are developing digital video transmission standards, and the Multimedia Cable Network System (MCNS) data-over-cable service interface specification (DOCSIS) covers cable modems. Chapter 4 covers digital modulation in detail, along with the required transmission quality.

13.3 Performance Allocations Among Sections of the Cable System

Among the listed parameters, clearly some are determined exclusively (or primarily) in one element of the transmission chain, whereas others (such as carrier to noise) are cumulative as the signal travels through the sections. Specifically:

- The visual carrier frequency is initially determined at the point of RF modulation but may be modified by subsequent frequency conversions. In most set top terminals, the video signal is demodulated and remodulated so that the frequency of the signal delivered to the subscriber's equipment is not related to that on the network. In that case, the FCC's frequency standards apply to both the headend output and the output of each terminal.

- The difference between visual and aural carrier frequencies is also determined at the point of modulation and is unaffected by subsequent frequency conversions that modify the entire channel's spectrum (or the entire RF spectrum, for instance, in an AML transmitter). Similarly, the deviation of the aural carrier (and therefore the "loudness" of the detected sound) is determined at the last point of modulation, be that the over-the-air transmitter, headend, or set top terminal. Here again, the FCC's rules apply at the last point of modulation.

- The levels of signals delivered to subscribers' homes are primarily a function of the broadband distribution system but are also affected by changes in the headend. Where set top terminals are used, however, the signal levels reaching subscribers' receivers may have no relationship to system operating levels. The FCC rules on levels delivered to subscriber equipment apply both before and after any operator-supplied terminal equipment. In addition to the Part 76 rules covering cable television, specifications in Part 15 further limit the allowable output RF power range of terminals.

- In-channel frequency response variation, relative visual/aural carrier level difference, chroma delay, differential gain, and differential phase in a properly operating cable system are primarily determined in the headend and only slightly modified as a result of variations in the response of the

broadband network. As with visual levels, however, set top terminals significantly modify all these parameters.

- Intermodulation distortions (CTB, CSO, XMOD, and CIN) primarily occur in the broadband distribution network but can be further degraded by set top terminals.

- Hum modulation can occur at any place in the network though, in a properly maintained system, it will primarily occur because of the side effects of ac powering being carried over the same coaxial system as RF signals—a combination of power pack ripple and parametric modulation of magnetic components.

- Noise is added at all stages in the process though not in equal amounts.

In summary, the primary impediments that occur in multiple sections of cable systems are distortion (occurring in both broadband distribution network and terminal equipment) and noise (with contribution from headend, broadband distribution network, and terminal equipment). With this understanding, we can discuss allocations of the allowable degree of cable system signal degradation among the various major network sections and within the subparts of the broadband distribution section.

13.4 Noise and Distortion Allocations in Cable Systems

13.4.1 Carrier to Noise

Applying Equation (10.2) to the elements of a cable system, it can be seen that the total C/N measured from signal acquisition to the input of subscribers' receivers is

$$C/N_{ttl} = -10 \log \left[10^{-\left(\frac{C/N_h}{10}\right)} + 10^{-\left(\frac{C/N_s}{10}\right)} + 10^{-\left(\frac{C/N_d}{10}\right)} + 10^{-\left(\frac{C/N_t}{10}\right)} \right] \quad (13.1)$$

where

C/N_{ttl} = the system C/N considering only noise contributions from the headend input to the input of subscribers' receivers

C/N_h = the C/N of the headend, considered alone

C/N_s = the C/N of the supertrunk (generally fiber-optic or microwave link), considered alone

C/N_d = the C/N of the coaxial distribution, considered alone

C/N_t = the C/N of the terminal equipment (for example, set top terminal), considered alone

Although this formula appears somewhat complex, it reflects converting each section's relative noise level to a scalar value, adding the normalized noise levels, and reconverting to familiar logarithmic (dB) terms. The initial 10 and the 10 used to divide each of the original C/N values reflect the fact that the thermal noise generated in each section is noncorrelated and so adds on a power rather than voltage basis. A shorthand way of stating this relationship is to state that the "cascade factor" is "10 log."

One eminent cable engineer has suggested the use of a shorthand notation to express such equations, which we have slightly modified to make it more universal.[4] Using this notation, Equation (13.1) can more simply be written as

$$C/N_{ttl} = C/N_h \oplus C/N_s \oplus C/N_d \oplus C/N_t : 10 \log \qquad (13.2)$$

where the \oplus sign indicates the conversion to scalar values, addition, and reconversion to dB form, and the term after the colon indicates the cascade factor, which may vary from 10 log (totally uncorrelated power addition) to 20 log (voltage addition). We will use this notation throughout this section.

Typical set top terminals have noise figure specifications ranging from 10 to 13 dB. Since cable operators are only required to deliver analog television signals at levels of 0 dBmV (and typically deliver many channels at close to that level), we can use Equation (9.19) to determine that, with the noisiest terminals and minimum signal levels:

$$C/N_t = 0 \text{ dBmV} + 59.2 \text{ dB} - 13 \text{ dB} \approx 46 \text{ dB} \qquad (13.3)$$

At the other end of the chain, individual modulators and signal processors will generally exhibit 60-dB C/N or better. That is degraded, however, by the out-of-band noise from other modulators when many channels are passively combined. The various summing and isolation amplifiers add additional noise as well (see Chapter 8 for a discussion of required headend processing) so that the typical C/N of individual channels as measured at the input to the broadband distribution network is of the order of 57 dB.

Assuming that the goal is to just meet the FCC's required C/N_{ttl} of 43 dB, we can find the minimum composite C/N of the supertrunk and coaxial distribution together using

$$C/N_s \oplus C/N_d = C/N_{ttl} \ominus C/N_h \ominus C/N_t : 10 \log \qquad (13.4)$$
$$= 43 \ominus 57 \ominus 46 = 46.4 \text{ dB}$$

where the \ominus sign indicates that the scalar quantities are subtracted rather than added.

What this equation shows is that our worst-case terminal has effectively used up half the total noise budget of the system. Though most modern terminals have lower noise figures, set top terminals still contribute significantly to total noise.

Typical design specifications for the supertrunk plus coaxial distribution portion of the plant call for 48–49 dB C/N. The margin of about 2 dB provides an allowance for such operating variations as errors in setting operating levels at the headend, aging of components, and imperfect frequency response through chains of amplifiers. The variation in frequency response is known as peak-to-valley (P/V) and is a measure of the peak-to-peak variation, in decibels, from the ideal. Though fiber-optic links are typically flat within a few tenths of a decibel, the usual allowance for coaxial distribution networks is $N/10 + 2$, where N is the number of cascaded amplifiers.

The allocation of noise budget between supertrunk and coaxial distribution depends on both architecture and technology. Typical downstream fiber-optic links driven by directly modulated DFB transmitters will exhibit C/N values of 51 to 54 dB, as will broadband AML microwave links. In a simple architecture, a 49-dB supertrunk/coaxial C/N requirement might be met by cascading fiber-optic and coaxial sections, each independently providing 52-dB C/N. More complex architectures might require cascaded supertrunk links, requiring better performance in each section.

13.4.2 CTB and CSO

Since headend levels can be carefully optimized and premium amplifiers utilized, intermodulation distortion within the headend is minimized, and most distortion occurs in the supertrunk, coaxial distribution, and set top terminal. As discussed in Chapter 10, however, the calculation is not straightforward. Continuing the example of a simple HFC network, the distortion mechanisms in the directly modulated DFB laser, the balanced RF amplifiers, and the input stages of the set top terminal have some characteristics in common (all exhibit some form of limiting) and others that are different (the small-signal linearity variations and nature of the limiting when it does occur).

Using the same notation as in Figure 13.1, we can write a general equation for the composite distortion:

$$C/dist_{ttl} = C/dist_s \oplus C/dist_d \oplus C/dist_t : nlog \qquad (13.5)$$

where

$dist$ could indicate CTB or CSO

n = the cascade factor and can vary from 10 to 20

When the dominant distortion mechanism is similar in the cascaded elements, for instance, symmetrical compression in cascaded push–pull amplifiers running at the same output levels, the cascade factor will be close to 20, which was the number traditionally used for both CSO and CTB in large all-coaxial networks. Research in the performance of dissimilar cascaded sections, however, has shown that distortions often combine at less than a voltage addition rate (see endnote 2,

Chapter 10). In occasional cases, in fact, the distortion may build up at a less than 10 log rate when distortions partially cancel.

As discussed in Chapter 10, the broadband distribution network is designed and (often) tested using unmodulated carriers. This results in CTB and XMOD product design levels that are worse than average operational values by about 12 dB, whereas CSO products will be worse than operational values by about 6 dB, though the difference may not be that great for very short cascades or where the modulation on multiple channels is synchronized.

Although there are no standards in this area, common industry practice has been to design for about 53-dB C/CTB and C/CSO in the broadband network (supertrunk plus coaxial distribution) under test (unmodulated carrier-loading) conditions. Even when combined with typical set top terminal distortion levels (57–65-dB C/CTB and 60-dB C/CSO), the total distortion under operational conditions is well within FCC requirements, even with the most conservative cascade factor assumptions, as shown in Equations (13.6) and (13.7):

$$C/CTB_{ttl} = (53 \oplus 57 : nlog) - 12 \text{ dB}$$
$$= 60.8 \text{ dB for } n = 20 \tag{13.6}$$

$$C/CSO_{ttl} = (53 \oplus 60 : nlog) - 6 \text{ dB}$$
$$= 55.8 \text{ dB for } n = 20 \tag{13.7}$$

Typical specifications for directly modulated DFB optical transmitters include a C/CTB of 65 dB and C/CSO of 62 dB. Externally modulated transmitters are generally better, perhaps 70-dB C/CTB and 65-dB C/CSO, whereas broadband microwave links may generate slightly worse distortion.

The remaining distortion "budget" can be allocated to the coaxial distribution system or to additional supertrunk links, if required by the architecture. Fortunately, though fiber-optic links have higher CSO than CTB, coaxial amplifiers have higher CTB so that the cascade of the two technologies is somewhat complementary with regard to distortion.

Although we could do a similar calculation for XMOD, the results would generally track CTB since both are third-order effects. In practice, CTB testing has proved to produce more consistent results and so is generally preferred to XMOD evaluation. It should be noted that the FCC no longer specifies XMOD performance.

13.5 Typical Network Transmission Quality Under Operational Conditions

Having discussed how various signal imperfections arise in the network and how they are designed and tested, we will now consider, as an example, typical HFC network performance under the following conditions:

- 54–750 MHz total downstream frequency range, loaded with 78 analog NTSC channels extending up to 550 MHz, plus 200 MHz of digitally modulated signals extending to 750 MHz.

- A single fiber link extending from the headend to each node and feeding a coaxial distribution network that passes 500–2,000 homes.

- A nominally flat frequency response from the headend output through the fiber link and to the input of the coaxial distribution system; an interstage circuit to set an optimum gain/frequency slope within the first amplifier; and a cascade of amplifiers, interconnecting cable, and passive devices adjusted so that the frequency response from the output of any amplifier to the output of the subsequent amplifier is flat (see Figure 9.19 for a graphical illustration).

- Equal levels among the analog video channels leaving the headend and the levels of digitally modulated carriers set such that the total average power per 6-MHz channel is 10 dB below the levels of the analog video visual carriers during synchronizing peaks.

These parameters are typical of specifications, maximum signal-loading capabilities, and operational parameters of many cable systems built in the mid-1990s. Most coaxial and fiber-optic equipment designed for this application is specified under both the preceding carriage conditions and, alternately, under a load of 110 NTSC video signals.

This section should provide guidance on typical end-to-end transmission channel conditions to those designing products whose signals will be transmitted through cable systems, as well as general guidance on how operating performance is related to design parameters. An important caveat, however, is that although the cable television industry is moving toward widespread HFC architectures for most of its networks, many older all-coaxial networks are still in use, and new networks are being constructed with more complex architectures both with and without microwave supertrunks. These differences must be taken into account. As an example, microwave links are subject to such factors as fading, whereas all-coaxial networks will exhibit greater level instability and in-band response variations. Either technology, however, will outperform directly modulated DFB laser transmitters with regard to clipping distortion.

Each of the channel impairments summarized here will be expressed relative to two kinds of signals:

- Analog video signals that originate within or are received at the headend and that pass through the network and a set top terminal before delivery to customers' receivers

- Digital signals that are inserted at the last combining point in the headend, operate at 10-dB depressed levels (as defined earlier), and are delivered directly to terminating equipment

Given these two classes of signals, it is simple to estimate the effect on other signals of different characteristics.

13.5.1 Carrier-to-Noise Ratio

The FCC's video C/N specification is one of the most difficult to meet for the most distant (in terms of equipment cascade) cable system subscribers. For those subscribers who also use set top terminals, it is likely that the required 43-dB C/N is barely met on some channels. For customers who are located close to a fiber node and who directly connect their receivers to cable outlets, the C/N may be more than 50 dB (limited by the noise contributions of the headend, fiber link, and one coaxial amplifier).

A digitally modulated signal that is inserted at the input to the downstream laser transmitter (and thus avoids the noise contributions of headend amplifiers), which is operated at a level 10 dB below equivalent analog video channels and is directly connected to termination equipment, is affected by different factors. The C/N at the most distant subscriber tap for such signals is essentially that of the supertrunk and coaxial distribution network, corrected for the signal level difference between video and data signals, corrected for the noise susceptibility bandwidth of the digital receiver relative to 4 MHz, and with an added allowance for P/V, errors in setting levels, and aging of components. As with analog video, the total variation due to these factors may be of the order of 2 dB. Thus, we can write a generic equation for likely worst-case C/N for a non-analog-video signal:

$$C/N_{\text{digital}} = (C/N_s \oplus C/N_d :10 \log) - S - 10 \log\left(\frac{B}{4}\right) - margin \qquad (13.8)$$

where

C/N_s = the analog video C/N in the supertrunk

C/N_d = the analog video C/N in the coaxial distribution

S = the suppression of digital signal levels relative to analog video (in dB)

B = the noise susceptibility bandwidth of the digital receiver (in MHz)

$margin$ = the expected variation from design performance due to aging, P/V, and operational tolerances (in dB)

As an example, if $C/N_s \oplus C/N_d$ is 48 dB, the signals are run 10 dB below video, the bandwidth is 6 MHz, and we allow a 3-dB margin, the worst-case C/N_{digital} would be 33.2 dB.

13.5.2 Intermodulation Products from Discrete Carriers

As discussed earlier, typical broadband networks (supertrunk plus distribution) are designed to provide C/CTB and C/CSO levels of at least 53 dB under unmodulated signal test conditions. Combined with hypothetical set top terminal distortion levels and adjusted for the difference between unmodulated carriers and

normal television signals, worst-case C/CTB will be about 61 dB, whereas C/CSO will be about 56 dB.

As with C/N, a further allowance must be made for imperfect frequency response (P/V) and for operational variances and aging of components. Errors in setting signal levels will cause larger differences for CTB since the distortion product amplitudes change by 3 dB for every 1-dB change in desired levels, whereas CSO products change by only 2 dB. Furthermore, the levels of each video signal and the total peak RF power carried will change over time as average picture brightness changes and as the various video-modulating waveforms drift into and out of time synchronization. The result is that the observed product clusters affecting any one channel will vary by several decibels over time (the FCC allows time-averaged tests in evaluating operational systems). If we allow 5 dB for the total of these effects on CTB and 3 dB for the CSO effects, we can predict that the worst-case C/CTB on any channel at any time will be about 55 dB, whereas the worst-case C/CSO will be about 53 dB.

The improvement in performance experienced by subscribers who are near optical nodes is not as great as for C/N since the largest contributor to CSO is the optical transmitter, whereas the largest contributor to CTB is generally the last amplifier (or two) that drive each subscriber tap string.

In the case of digitally modulated signals sharing the cable network with analog television signals, some intermodulation products will fall into the digital spectrum. If the same channelization is used, then the products will appear in the same relative positions within each channel. That is, for ANSI/EIA-542 Standard channels, they will occur at harmonics of 6 MHz and 1.25 and 2.5 MHz above those harmonics, with the first and third of those being CSO products and the middle one being the CTB product. Scaling from the analysis just done for video signals, adjusting for the 10 dB of level difference, and eliminating the contribution from the set top terminal (approximately 69-dB C/CTB and 66-dB C/CSO with modulated signals), we would expect that the worst-case C/CTB might be 47 dB, whereas the worst-case C/CSO might be 45 dB.

Given the distribution of CSO products with frequency (Figure 9.8), it would be expected, further, that the only significant CSO products would be those occupying the highest frequency in each set (2.5 MHz above the harmonics of 6 MHz) when digital signals are placed at the top end of the spectrum. ANSI/EIA-542 Standard channels are further protected against lower-side CSO products because they occur only at the boundary between channels.

Though other ANSI/EIA-542 channelization alternatives are declining in popularity, it should be mentioned that with the HRC system, all the second- and third-order products will fall exactly on harmonics of the reference oscillator (generally 6.00003 MHz), as will the visual carrier, if present. Since the products are frequency coherent, they do not cause visible beats in analog channels and so are less visible. With the IRC system, the products have approximately the same frequencies as with Standard channelization, except that each product group is frequency coherent. Because the second-order products are not coherent with the visual carrier, second-order beats may become visible in analog channels carried in IRC systems before third-order beats in some cases.

13.5.3 Composite Intermodulation Noise (CIN)

As discussed in Chapter 9, intermodulation products among digital signals and between digital signals and analog visual carriers add "bands" of noiselike products, which add to thermal noise to raise the effective noise floor of the system. The frequency distribution of these products is a function of the distribution and relative power levels of both the analog and digital signals. Fortunately, when the average power density of digital signals is suppressed by 5 dB or more relative to visual carriers, the effective C/N degradation for the visual signals is less than 1 dB.

13.5.4 Laser Transmitter Clipping Distortion

As discussed in Chapter 11, downstream laser clipping distortion is a statistically rare event in a properly adjusted, directly modulated DFB transmitter. When external modulation methods are used, clipping is even rarer since other nonlinearities limit average modulation levels before clipping probability becomes significant. Thus, it can reasonably be expected that laser clipping will have no measurable, visible, or audible effect on analog NTSC channels under normal operating conditions. Should the optical modulation levels be adjusted too high, however, horizontal black streaks will appear in pictures.

Studies of clipping-induced errors in 64-QAM data signals that share DFB optical transmitters with a full spectrum (54–550 MHz) of analog video signals have shown that, with proper transmitter adjustment, bit error rates due to clipping are less than 10^{-9} even without forward error correction. If, however, the modulation is set too high, the BER can increase to the order of 10^{-7} or even lower before affecting subjective picture quality. Once the threshold of clipping is approached, the decline in performance with increasing modulation is very steep.

13.5.5 Hum Modulation

Hum modulation occurs in both powered equipment (due to imperfect power supply filtering) and as a result of parametric modulation of magnetic components in both amplifiers and passive devices. As discussed in Chapter 9, even though the FCC allows 3 % pk-pk modulation levels, systems seldom exceed half that in practice. Modulation due to power supply ripple should logically be more or less independent of RF frequency. To the extent that parametric modulation of magnetic components is a factor, however, its effects may vary with frequency. The primary mechanism is that magnetic cores may partially saturate at peaks of the ac current. This will affect their impedance at RF, and so create a change in return and transmission loss that will vary at a 60-Hz rate. Since the impedance change may vary with frequency, as may the effect of the impedance change, there is no way of predicting how the modulation percentage will vary across the spectrum. At a given frequency, however, the percentage modulation of a video signal should be the same as that of a digitally modulated signal though the latter may be able to tolerate a higher modulation percentage.

13.5.6 Microreflections

Microreflections can occur anywhere within the broadband coaxial network (or, for that matter, in the optical link) but are most severe where components are mounted within a few hundred feet of each other, namely, in the tapped coaxial feeder lines and drop structure. Even though component quality (and thus impedance match, isolation, and directivity) is higher in the "hard cable" plant (that is, the coaxial portion on the network side of subscriber taps), that is partially offset by the lower per-foot cable losses and greater reflection delays.

The mechanisms by which microreflections occur were analyzed in Chapter 9. Drops were analyzed in Chapter 10. In summary, occasional microreflections should be expected whose magnitude can vary from as large as −10 dBc at zero delay to a maximum amplitude that will decline in accordance with round-trip cable losses at the frequency in question.

Such large reflections, however, can only occur in the trunk and distribution network in the case of a major cable or component fault, poor-quality passive components, or a terminating tap, all of whose subscriber ports are unterminated. If any tap port is terminated, even by connection to a drop cable, the reflections will be significantly smaller. In the drop structure, reflections from receivers and unterminated outlets are partially compensated for by the larger attenuation of the interconnecting cables. As a practical rule, most delayed signals should be at least 20 dB below direct signals, with delays ranging from 0 to about 500 ns. In addition to reflections in tapped lines and within the drop structure, mismatches occurring in long, untapped cable spans (trunk or express feeder lines) can result in smaller reflections (≤ -40 dBc) with delays as long as 5 μs.

Microreflections can cause various picture degradations in analog-transmitted signals, ranging from subjective softening or overemphasis of vertical lines (depending on the relative polarity of the reflected signal) when delay times are short to visible ghosts when the delay time is sufficient that the reflected picture is distinguishable from the direct picture. Customer perception of microreflections' effect on picture quality is discussed in Chapter 2.

With data signals, the reflected signal may be interpreted in a number of ways. In the time domain, the effect is to create an amplitude uncertainty in detected amplitude and phase as the delayed and direct signals add at the detector. In the frequency domain, these same effects can be interpreted as group delay and in-band response variations.

13.5.7 Group Delay and In-Band Response Variations

To review, the return loss of a component is the ratio between the signal impinging on a port (with all other ports properly terminated) and the signal reflected from that port:

$$R(\text{dB}) = L_i(\text{dBmV}) - L_r(\text{dBmV}) = 10 \log\left[\frac{P_i}{P_r}\right] \qquad (13.9)$$

where

$$R = \text{the return loss in dB}$$

$L_i \text{ and } L_r =$ input and reflected power levels in dBmV (or any other consistent logarithmic units)

$P_i \text{ and } P_r =$ input and reflected power levels in mW (or any other scalar quantities)

Another way of quantifying the mismatch recognizes that the voltage due to the reflected wave will, at some points along the transmission line, add to the incident wave, whereas at others it will subtract from it. The ratio of maximum to minimum combined voltage is known as *voltage standing wave ratio* (VSWR). Note that although at any point the voltage will vary in a sinusoidal fashion at the frequency of the signal, the *ratio* of voltages along the line will not change, thus a "standing" wave ratio. The relationship between VSWR and return loss is

$$VSWR = \frac{(1+10^{-R/20})}{(1-10^{-R/20})} \tag{13.10}$$

Figure 13.2 illustrates the standing waves resulting when a transmission line is terminated in a pure resistance equal to half its characteristic impedance. When a line is terminated in a pure resistance of any value, the VSWR is Z_0/R or R/Z_0, whichever is greater.

As can be seen, for purely resistive terminations, the standing wave pattern has a maximum value at distances that are odd multiples of one-quarter wavelength from the termination and a minimum value at distances that are even multiples. The reverse would be true if R were greater than Z_0. If the termination is complex (not a pure resistance), then the distance to the first minimum or maximum will be other than a quarter wavelength. Regardless of the termination, however, the maxima and minima will be exactly one-quarter wavelength apart at the frequency of the signal.

The phase shift, ϕ, through a length of cable, in radians, is $2\pi L/\lambda$, where L is the length and λ is the wavelength in the cable. Substituting for λ from Equation (9.9):

$$\phi = \frac{2\pi L f}{984 V_P} \tag{13.11}$$

where

$\phi =$ the phase shift in radians

$L =$ the distance, in feet, over which the phase shift is measured

$f =$ the frequency in MHz

$V_P =$ the relative velocity of propagation in the cable

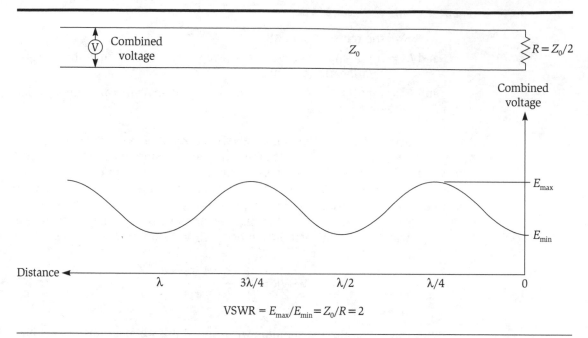

Figure 13.2 Standing wave pattern.

We would like to know what frequency shift, Δf, would be required to cause a relative phase difference of 2π radians through a fixed length L of cable. If we solve Equation (13.11) for f and let $\phi = 2\pi$, we get

$$\Delta f = \frac{984 V_P}{L} \tag{13.12}$$

If a direct signal and one delayed by being transmitted through a cable of length L are mixed, as through a reflection, then Δf is known as the ripple frequency.

Reflections Due to Mismatches in Trunk and Distribution Plant

To see how this affects amplitude response and group delay, let us examine a portion of a typical tapped distribution line (Figure 13.3). Signals exit the source, which might be an upstream tap or amplifier. They travel through cable 1 and enter the tap. Signals exiting the tap through-port travel through cable 2 to the next downstream component, labeled load, which might be another tap or the input to a following amplifier.

The following parameters will be used in the calculations:

Length of cable 1 in feet = L_1

Length of cable 2 in feet = L_2

Source return loss in dB = R_s

Figure 13.3 Sample of tapped line segment.

Load return loss in dB $= R_L$

Tap return loss in dB $= R_T$

Attenuation of cable 1 in dB $= A_{C1}$

Attenuation of cable 2 in dB $= A_{C2}$

Tap directivity in dB $= D$

Frequency in MHz $= f$

Reflected power will be coupled into the drop by two primary mechanisms:

- A portion of the signals that are reflected from the tap input will be re-reflected from the source and combined with the direct signals delayed by the transit time through $2L_1$ feet of cable. The difference in level between the desired and reflected signals, R_1 (in dB), will be $2A_{C1} + R_S + R_T$.

- A portion of the signals that are reflected from the load will be coupled into the drop due to the finite directivity of the tap. These signals will be delayed relative to the desired signals by the transit time through $2L_2$ feet of cable. The difference, R_2, in level between the desired and delayed signals in this case will be $2L_{C2} + R_L + D$.

In either case, the ratio of the scalar magnitudes of the desired, E_F, and delayed, E_D, signal voltages appearing at the drop port will be $10^{R/20}$.

The presence of the reflected/delayed signal affects the desired signal in three ways. First, the delay, if longer than the symbol period of a digital transmission, will reduce the error threshold by interfering with the following symbol. In the case of an analog television signal, the visible effect is a horizontally offset image whose visual appearance depends on the amplitude and delay time, as covered previously.

The second, the frequency-sensitive nature of the VSWR pattern between the direct and reflected signals will affect the apparent amplitude versus frequency response of the system. Using Equation (13.10), the peak-to-peak change in amplitude response can be calculated to be

$$\text{ripple(dB pk-pk)} = 20 \log\left[\frac{(1+10^{-R/20})}{(1-10^{-R/20})}\right] \tag{13.13}$$

This ripple will occur at the incremental frequency rate calculated in Equation (13.12), where L is either $2L_1$ or $2L_2$ depending on the dominant coupling mechanism.

Finally, the vector combination of the desired and delayed signals incrementally affects the phase shift through the system, and thus group delay. Figure 13.4 shows the relationship between the magnitudes of the desired and reflected signals, the angle between them, and the angle and magnitude of the resultant signal. In general:

$$\theta = \arctan\left[\frac{E_D \sin\phi}{E_F + E_D \cos\phi}\right]$$

$$= \arctan\left[\frac{10^{-R/20} \sin\left(\dfrac{2\pi Lf}{984V_P}\right)}{1+10^{-R/20} \cos\left(\dfrac{2\pi Lf}{984V_P}\right)}\right] \tag{13.14}$$

We can restate the definition of group delay (Equation (9.32)) in similar units:

$$\text{group delay(ns)} \equiv \frac{1,000}{2\pi}\frac{d\theta}{df} \tag{13.15}$$

where

θ = the incremental phase shift in radians

$d\theta/df$ = the rate of change of phase shift through the device as a function of frequency in units of radians per MHz

This allows us to solve for the peak-to-peak delay variation due to the presence of the reflected signal:

$$\text{group delay variation(ns pk-pk)} = \left[\frac{2.032L}{V_P}\right]\left[\frac{10^{R/20}}{10^{R/10}-1}\right] \tag{13.16}$$

where

L = the incremental cable length through which the reflected signal is transmitted in feet

V_P = the relative propagation velocity (dimensionless)

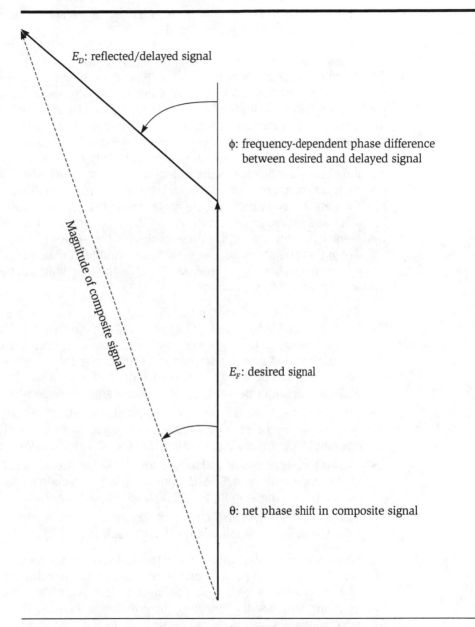

Figure 13.4 Vector addition of desired and reflected/delayed signals.

R = the ratio of the desired to reflected signal strength as measured at the drop port and expressed in dB

A practical example may be useful in quantizing the effects of reflected signals in typical coaxial distribution lines. Typical amplifiers and passives have return loss specifications of about 16 dB when properly terminated, with the worst performance near the ends of the passband, as can be expected. The specified directivity of taps varies as a function of both frequency and tap value. In general, the worst values are also at the extremes of the passband (especially at the low end of the upstream frequency range) and at high tap values, where the absolute coupling between ports limits the isolation. For most values, the directivity in the downstream spectrum is 10–15 dB, whereas the return band directivity is 6–9 dB. Finally, most distribution lines use cable whose loss is about 1.6 dB/100 feet at 750 MHz and varies, as we have seen, as the square root of frequency. The relative velocity of propagation for these cables is about 0.90.

Given these parameters, assume that an amplifier and two taps are spaced at 100-foot increments along a cable. At 750 MHz, the following will be the situation under proper operating conditions:

- The signal reflected from the first tap and re-reflected from the amplifier will be attenuated by $16 + 16 + 2(1.6) = 35.6$ dB and delayed by $984(0.2)/0.9 = 219$ ns. The signal transmitted through the first tap, reflected from the second tap and coupled into the drop due to a directivity (which we will assume to be 12 dB), will be attenuated by 31.6 dB and also delayed by 219 ns. We will use this larger delayed signal for calculating the effects.

- If a digital datastream is being transmitted, one effect of this "microreflection" is that if a pulse is transmitted, an echo of that pulse will create a voltage amplitude uncertainty of about 2.6% in the signal 219 ns later.

- The amplitude response will have a "ripple" versus frequency. The frequency of the ripple will be 4.42 MHz (meaning a full cycle plus in every 6-MHz channel), and the amplitude will be 0.45 dB peak-to-peak.

- The group delay will also have a variation versus frequency at the "ripple" rate. The delay variation will be 11.7 ns peak-to-peak.

Many cable operators, on the other hand, do *not* terminate unused tap ports, but only cover them to prevent undue weathering of the connectors. Their reasoning has been that a loose type F terminator acts as both a radiator and signal pickup point, reducing the shielding integrity of the system. The consequences for reflected signals, however, can be serious. In the case of a terminating 4-dB tap (actually just a two-way splitter), for instance, the return loss when unterminated is typically <5 dB. If the preceding calculations are made assuming a 5-dB return loss, the delayed signal will be only 20.2 dB below the desired signal, causing the delayed voltage uncertainty to increase to 9.8%, the amplitude response variation to increase to 1.7 dB, and the group delay variation to increase to 45 ns. Although none of the results from a properly terminated system would

Figure 13.5 Passband frequency response as a function of tapped line configuration.

significantly degrade most signal formats, the second set of values is large enough to be of concern.

If we do a similar calculation at 10 MHz in the upstream band, the principal differences are that the cable loss is much less, and tap directivity, and sometimes return loss, is degraded. Figures 13.5 and 13.6 show the amplitude and group delay variations for some typical situations based on an average of several manufacturers' component specifications.

Though the preceding calculations are simplified to a single reflection, in actual systems, many reflections occur. Since a tapped distribution line can easily contain a bridger, 2 line extenders, and 30 taps, many microreflections exist simultaneously and will add vectorially. Consider, for example, four taps equally spaced along a cable that has 0.5 dB of loss in each segment (a typical upstream value), as illustrated in Figure 13.7. Assume that each has a return loss of 16 dB at all ports and, for simplicity, that each of the first three has a through loss of 1 dB (including any mismatch losses). We can calculate the worst-case effective return loss and VSWR at the input to the first tap as follows (assume the last tap is a terminating type):

- The signal reflected from tap 4 is 16 dB below the incident signal and at the input port.
- Observed from the input of tap 3, tap 4's return loss will look 3 dB better because the incident and reflected signals are both attenuated by the through loss of tap 3 and the intervening cable loss. Using Equation (13.10), this corresponds to a VSWR of 1.253.
- The 16-dB independent return loss of tap 3 corresponds to a VSWR of 1.376.

Figure 13.6 Group delay as a function of tapped line configuration.

Figure 13.7 Cascaded reflections.

- The two mismatches will combine, in the worst case, to the product of the individual VSWRs, or 1.253(1.376) = 1.724.

- This can be converted back to an equivalent return loss of 11.5 dB using the reverse of Equation (13.10):

$$\text{return loss(dB)} = -20\,\log\left[\frac{VSWR-1}{VSWR+1}\right] \tag{13.17}$$

This, in turn, is attenuated so that, when observed from the input of tap 2, it is 14.5 dB. Converting to VSWR and combining with tap 2's own input match, we get a worst-case combined VSWR of 2.014 and return loss of 9.46 dB.

Repeating the process at the input to tap 1, we get a net VSWR of 2.24 and an equivalent return loss of 8.36 dB. If the string of components were longer, the equivalent match would continue to degrade though at a slower rate.

Since the phase shift through the interconnecting cables changes quickly as a function of frequency (rotating through a full 360° at the ripple frequency), the reflections from two mismatched components should be assumed to reinforce at many frequencies. This is the mechanism that causes structural return loss in co-axial cable to have sharp peaks at many equally spaced frequencies if the manufacturing process produces equally spaced physical defects. The reflections from multiple mismatches, however, will be less likely to reinforce if the spacings are random: a powerful argument against equally spaced similar components in co-axial distribution lines.

Reflections in Drops

As with the trunk and distribution plant, microreflections in the drop system lead to both group delay and frequency response variations. Equations (13.16) and (13.13) can be used to estimate the magnitude of these effects, whereas Equation (13.12) will provide the ripple frequency when applied to the typical drop configuration shown in Figure 10.5 as follows:

- For R in the equations, use the sum of the splitter isolation plus twice the loss of the cable from the splitter to an outlet.
- For L in the equations, use twice the length of the cable from the splitter to an outlet.
- For V_P in the equations, use the relative velocity of propagation in the drop cable (typically 0.85).

For example, at 550 MHz, a splitter isolation of 15 dB combined with a non-terminated 50-foot long, size 59 drop cable (one-way loss about 3 dB) will create a sinusoidal group delay variation of 21.5 ns peak-to-peak with a period of 8.4 MHz. The corresponding amplitude response variation will be 1.6 dB peak-to-peak with the same periodicity.

Group Delay Due to Diplex Filters

In addition to the group delay caused by reflections, frequencies close to cross-over frequency of diplex filters used in amplifiers to separate upstream and down-stream frequency signals will be subject to additional group delay. As discussed in Chapter 9 (see Figure 9.12), typical chroma delay (differential delay between the luminance and chrominance frequencies) per amplifier is 21 ns at channel 2, 9 ns at channel 3, and drops off rapidly at higher channels. Since this delay is cumulative through multiple stages, a typical five-amplifier HFC cascade will have a total chroma delay of 105 ns at channel 2.

Older all-coaxial cable systems and systems using longer cascades typically limit the upstream frequency range to 5–30 MHz so that the spacing between

upstream and downstream frequencies is greater and the group delay per amplifier is reduced accordingly.

As a general rule, the total channel 2 group delay through most systems will not exceed 100–150 ns and will meet the FCC's 170-ns requirement. Upstream group delays due to multiple reflections are often more severe than downstream due to the lowered cable loss and, frequently, worse component matches at the lowest frequencies. Upstream transmission is discussed in detail in Chapter 14.

Other In-Band Frequency Response Variations

In addition to response variations at the ripple frequency associated with multiple reflections, channels will experience in-band variations from three causes: the intentional gain/frequency slope of amplifiers (typically 10 dB across the full downstream spectrum), the variation in cable loss with frequency (in both distribution and drop cables), and the precision with which the various loss variations can be compensated within each amplifier (P/V).

In addition to the other effects discussed, individual drops may be subject to additional amplitude and group delay degradation when filters and traps are inserted to tailor the spectrum delivered to customers. Various combinations of high-pass, low-pass, bandstop, and bandpass filters are often used for this purpose, as discussed in Chapter 18.

It is rare for the total in-band response variation within any single 6-MHz channel to exceed 1.0 dB pk-pk in a well-maintained cable system, not including the effects of drop filters.

13.5.8 Local Oscillator and Other Interfering Signals from Receivers

The effects of the performance of consumers' receivers that are directly connected to the network are covered in detail in Chapter 20. In summary, however, local oscillator signals transmitted out the antenna terminals of receivers may have amplitudes as high as −10 dBmV, but are more likely to be −20 dBmV or lower.[5] These signals may be coupled into the downstream signal paths feeding other receivers via the imperfect isolation between tap or splitter output ports. Assuming 20 dB of isolation in such devices and short interconnecting cables, the likely maximum undesired carrier levels at the input to an adjacent receiver is about −40 dBmV, or 40 dB below the minimum acceptable analog desired signal.

The probability of such interference to any given channel, however, is slight. With normal television design, it occurs only when the receiver generating the interference is tuned exactly seven channels lower than the affected receiver, and is connected directly to the cable drop (as opposed to through a converter). When it does occur, the interfering carrier will appear 5 MHz above the lower band edge or 3.75 MHz above the desired luminance signal and very close to the chrominance signal, where the threshold of visibility is great, as discussed in Chapter 2.

Although the interfering signal magnitude is higher relative to depressed-level, digitally modulated signals, when those signals are placed above the analog

signals in the downstream spectrum, the potential for interference exists only in the first seven nonvideo channels above the highest analog channel, at least until digital standards are developed to the degree that digital descramblers are incorporated into consumer-owned receivers.

13.5.9 Antenna-Conducted Ingress from Receivers

Another potential source of interference is over-the-air signals that are picked up by the internal wiring of consumer receivers, by poorly shielded or faulty drop components (including cable and connectors), or coupled across antenna selector switches owing to finite isolation between ports. These mechanisms are discussed in Chapter 20. Such signals can affect the reception at the inadequately shielded receiver, but they can also be conducted out of the antenna terminals and back up the drop cable (thus, "antenna-conducted"). Assuming drops are properly installed using quality components, the largest effect is due to the finite shielding of consumer receivers.

If the source of the off-air signal is a VHF television broadcast station, then the interfering signal will be close in frequency to the corresponding ANSI/EIA-542 Standard scheme channel. Often, in fact, cable channels are phase locked to the corresponding strong off-air VHF stations to reduce the visibility of co-channel interference by eliminating the difference frequency "beat" lines that otherwise occur. A major disadvantage of IRC and HRC channelization plans is that this mechanism for improving subjective picture quality in the presence of ingress interference is unavailable since the visual carriers in each case are phase locked to a master headend frequency source. The potential for interference is especially severe in HRC systems, where the difference between off-air and cable visual carriers is 1.25 MHz and results in a highly visible diagonal line pattern on viewers' television screens when the interfering carrier levels are suppressed by less than about 55 dB below the desired signal.

If the interfering signal is a UHF television broadcast station, then its visual carrier will fall 3.25 MHz above the lower channel boundary of the overlapping Standard cable channel. (Cable channels 73–128 overlap the spectrum utilized for UHF broadcasting.) To the extent that other external radiators cause interference, the frequencies will differ. For instance, cable channel 19 is often affected by such services as paging transmitters.

Though the probability of destructive interference from external ingress is still relatively low, the interference, when it occurs, is likely to persist since it is not related to the tuning of the inadequately shielded receiver. Tests (discussed in Chapter 20) conducted on typical receivers that were immersed in an external field of 100 mV/m showed that most antenna-conducted egress levels from this cause were −20 dBmV or less.[6] Thus, like local oscillator interference, the probable maximum levels at adjacent receivers exposed to this field strength are about −40 dBmV. It must be pointed out, however, that significant areas in the country are exposed to much higher field strengths.[7] In some cases, well-shielded converters are connected between drop outlets to serve the dual purpose of converting all

channels off the off-air frequency in the downstream direction while isolating the cable system from antenna-conducted receiver egress in the reverse direction.

13.5.10 Signal Levels and Stability

Cable systems typically have little problem meeting the various signal level, spectral flatness, and stability requirements imposed by the FCC's rules. In part this is because it would be difficult to meet the noise and distortion requirements if signal levels varied far from optimum levels. It is unlikely, for instance, that total variation across the delivered downstream spectrum will exceed 10 dB while individual carrier levels will probably not vary by more than 5 dB with time and temperature.

The implications of the expected levels and their variation of required video receiver performance are as follows:

- The receiver front end may be exposed to as high as +35 dBmV total RF power across the downstream spectrum (100 channels at +15 dBmV), but more probably 5–10 dB less than that. Additionally, the receiver may be exposed to individual signals as high as +40 dBmV in the return band.

- The receiver must tolerate signals appearing at the image frequency (assuming a standard 41–47 MHz IF band) that exceed the desired signal level by 10 dB.

- The receiver must tolerate adjacent video channels whose levels exceed the desired channel by 3 dB.

- The level of the desired signal could be as low as 0 dBmV or as high as +20 dBmV but will likely not vary by more than 5 dB over time (the FCC allows 8 dB).

The requirements for nonvideo receivers are similar with regard to the total RF power presented at the input port and the stability of the desired signal; however, the desired signal will be lower by the degree of suppression of carrier levels with respect to analog video signals at the headend. Whether image frequencies are present at the receiver input will depend on the design of the receiver and the frequency of the data carrier.

13.6 Summary

In summary, typical HFC broadband distribution networks provide a broadband thermal noise floor at a level of approximately –45 dBc relative to normal video levels, as measured at the entry point to buildings and referenced to a 4-MHz bandwidth. The noise level in other bandwidths, after set top terminals, or relative to other signal levels is easily calculated.

Various signals may occur adjacent to or within standard 6-MHz channels as a result of network nonlinearities, normal signal loading, and/or attached subscriber receivers. These are summarized in Figure 13.8. For reference, the approx-

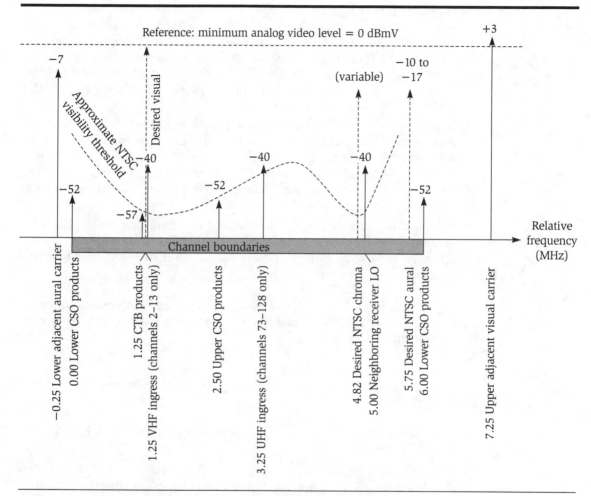

Figure 13.8 Maximum expected levels of extraneous signals in or near ANSI/EIA-542 Standard television channels as received at the input ports of terminal equipment.

imate level of suppression required to maintain a subjectively good analog NTSC picture is shown as a dotted line (see Chapter 2 for a detailed treatment of subjective picture impairments due to interfering carriers).

Finally, downstream signals transmitted through the network will be subjected to the following:

- Hum modulation that may be as high as 3% pk-pk, but is unlikely to exceed 1.5%.
- Microreflections that could be as high as −13 dBc, but are unlikely to exceed −20 dBc, with delays ranging from 0 to 500 ns, with smaller echoes having delays as long as 5 μs.

- Sinusoidally varying group delay (as a function of frequency) with an amplitude as high as 100 ns pk-pk, but unlikely to exceed 30–50 ns. The periodicity of the delay variation will typically be in the 2–5-MHz range. Additionally, channel 2 (in a standard low-split system) may be subjected to chroma delay as high as 170 ns, but more likely 100–150 ns owing to the diplex filters. Finally, individual subscriber drops may experience additional group delay due to the effects of any installed filters.

- In-band frequency response variations arising from microreflections, differential cable losses, intentional amplifier response slope, and imperfect frequency compensation. The total variation from all these causes is unlikely to exceed 1 dB pk-pk in any 6-MHz channel. As with group delay, however, the use of drop filters may add considerable response variations to some channels.

- Received signal levels (for signals launched at video-equivalent levels) ranging from 0 to +20 dBmV at subscriber outlets, with variations within that total range as large as 8 dB (but more likely less than 5 dB) over time and temperature.

As will be seen in the next chapter, signals transmitted in the reverse direction (from subscriber toward the headend) are subject to channel conditions that are less well defined.

Endnotes

1. Hans Stekle (Hrsg.), *Breitbandverteilnetze der Deutschen Bundespost* Heidelberg: R. v. Decker, 1988.

2. *Code of Federal Regulations, Section 47,* available from the U.S. Government Printing Office, Washington, D.C., in several volumes covering the entire FCC set of regulations, Parts 0 through 80.

3. *EIA-IS-23 RF Interface Specification for Television Receiving Devices and Cable Television Systems,* unreleased draft document of the Consumer Electronics Association, Arlington, VA, May 12, 1998.

4. Private correspondence from Archer Taylor of the Strategis Group, who doesn't claim to have invented it. We have modified Archer's suggested notation to add a variable cascade factor to broaden its applicability.

5. *Customer Premises Equipment Performance and Compatibility Testing,* CableLabs, Louisville, CO (undated, but published in late 1993). Part 4: Receiver Performance.

6. Ibid.

7. Ibid., Part 2, Potential Impact of Direct Pickup Interference. The report concludes that about 20% of U.S. households will be exposed to external field

strengths of at least 100 mV/m from at least one VHF television broadcast station, while 41% will be exposed to that magnitude of signal from at least one UHF station operating at or below 550 MHz. Television receivers, however, are usually placed within buildings that provide some attenuation of the external signal.

Chapter 14
Upstream Issues

14.1 Introduction

This chapter covers a number of topics of interest to those using return plant or needing to understand the issues involved in the operation of return plant. You may wish to review Chapters 9 through 11, which cover relevant topics in distribution systems. This chapter builds on the material in those chapters.

Return, or upstream, signal levels must be managed differently from downstream signal levels, and this topic is covered in some depth. Return laser characteristics are covered, along with methods being used to determine the proper level at which to operate return lasers. Finally, noise on the return path is dealt with. Several methods of analyzing the problem are presented, as are several alternative countermeasures.

Limited two-way cable plant has been built since the mid- to late 1970s. As early as 1972, the FCC required that plant be built to be "two-way ready." It was not until the mid-1990s, however, that the technology began to enjoy adequate applications to justify widespread deployment. Previously, the only applications were provision of a return path for impulse-pay-per-view (IPPV) set top terminals, some of which used an RF return path, and a few status-monitoring applications. Return analog video was and is practiced, but most links involve trunk runs only and are limited to local backhaul applications. Beginning in the mid-1990s, these applications were joined by HFC-based high-speed data transmission and telephony. More status monitoring came to be deployed in the plant though its required data rates remain modest. Interactive video continues to be a possible future application, though as of this writing, it has not enjoyed widespread acceptance.

The enabling technology that made the widespread use of the return path possible was the introduction of smaller nodal architecture in place of the older tree-and-branch architecture. This made feasible the control of noise buildup and also made it possible to realize adequate bandwidth to support marketable services. Thus, by the mid-1990s, the technology and applications had both arrived to make return plant practical and economically justifiable.

14.2 The Two-Way Node

In earlier fiber-optic deployment, the device that mounted on strand (or in a pedestal), and that received optical signals from the headend and converted them to electrical form, was called a *receiver*. As two-way services came to be, a return optical transmitter was added to the optical receiver. Use of the term *receiver* to describe the device became confusing, so the device came to be known as a *node*.

Figure 14.1 illustrates the basic optical node. It consists of a forward (downstream) receiver, a reverse (upstream) transmitter, and a diplex filter to separate signals moving in the two directions on the coaxial cable. The diplex filter, or *diplexer*, is a matched set of low-pass and high-pass filters, which separate signals

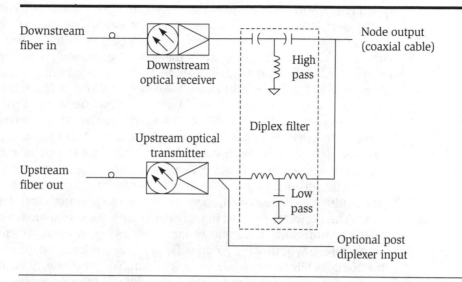

Figure 14.1 Optical node.

by direction (due to their difference in frequency). Diplex filters are described in Chapter 9.

Depending on the node design, the signals at the common point of the diplexer (the node output in this case) may be split, or several diplexers may be supplied to serve different routes from the node. Also shown is an optional post diplexer input, which may be used when a block converter translates multiple input paths to different frequencies. We discuss this technique later.

14.3 Downstream and Upstream Frequency Partitioning

A number of different plans are used in assigning downstream and upstream frequencies. In North America, the downstream band usually begins at 54 MHz, the low end of channel 2 (see Chapter 8). However, some operators prefer to use the frequency band from 50, or 52, to 54 MHz for narrow band services, and so use diplex filters that cut off below 50 MHz. The maximum downstream frequency is usually between 450 and 870 MHz, depending on when the system was built and the economic viability of various bandwidths in a particular system.

The traditional upstream band was 5 to 30 MHz, but recently, the maximum frequency of the upstream band has been increased to either 40, 41, or 42 MHz in order to allow more usable return bandwidth. Caution is advised when using frequencies above 40 MHz, and especially when using frequencies above 41 MHz. The common IF band used by NTSC television receivers is 41–47 MHz. Not all television sets exhibit good IF rejection to signals entering the antenna terminals because this has not been a requirement. Since return signals cannot be

completely isolated from television receivers, the possibility for interference is significant. Issues relating to subscriber susceptibility are covered in Chapter 19.

A common practice is to specify two numbers to indicate the high end of the upstream band and the low end of the downstream band. A partitioning having a maximum upstream frequency of 40 MHz and a minimum downstream frequency of 54 MHz would be called a 40/54 split. Other splits in common use include 55/70 (Japan) and 65/82 (Australia, New Zealand). Various European systems have used 26/45, 33/50, 42/54, 50/70, and 65/85. Splits tend to be somewhat customized. In certain applications, the split tends to yield more symmetrical spectrum, with larger upstream bands. These are not popular for residential service due to the need to place TV signals beginning at channel 2.

Useful upstream bandwidth may be somewhat less than the entire return bandwidth. On the low end, ingress often is the limiting factor below 15 or 20 MHz. The subject is covered in more detail later. As you approach the upper end of the return band, group delay increases owing to the characteristics of the diplex filters. Whether the group delay is a problem or not depends on the number of diplex filters in cascade, the rate of cutoff of the diplex filter, and the type of modulation used. Thus, the usable bandwidth of the return plant may be lower than the reverse passband.

14.4 Group Delay of Diplex Filters

Of concern with two-way plant is the group delay of the diplex filters used to separate the upstream and downstream frequency bands. The narrower the crossover region, the worse the problem becomes. In the upstream direction, the effect is to introduce group delay to the data signals that are typically carried in that direction. Group delay can produce intersymbol interference, the error in a signaling symbol produced by adjacent data symbols. If bad enough, the bit error rate of the signal is adversely affected. In the upstream direction, it is generally not feasible to equalize for group delay, as is done when necessary in the downstream direction, because the equalization would have to be different for each of several transmitters that may be using the same upstream bandwidth. Group delay is discussed in more detail in Chapter 9.

14.4.1 Upstream Group Delay

Figure 14.2 illustrates the upstream group delay of an amplifier having a 42/52 split.[1] The left vertical scale shows the absolute group delay (defined in Equation (9.32)), and the right scale is the differential group delay in a 1-MHz frequency span. (If we were concerned about carrying NTSC video, we would measure the differential group delay over a frequency span of 3.58 MHz, and we would call it chroma/luma delay.) Notice that at the high end of the return path, the differential group delay increases significantly, as the frequency is approaching the low-pass cutoff of the diplexer. If the differential group delay is too great, then some

Figure 14.2 Upstream group delay due to diplex filter.

spectrum at the high end of the band may not be usable. "Too much" group delay is a function of the modulation type and the particular system being used. Higher-density modulation systems (that is, those having a higher bandwidth efficiency) are more susceptible to group delay and other impairments than are lower-density modulation systems.

The group delay increase at the low end of the band is a result of a high-pass filter included in the amplifier measured. The high-pass filter is advantageous in that noise below the 5-MHz low cutoff is not as much of a factor in loading the return system.

Notice that the chart in Figure 14.2 reflects the group delay per amplifier. Multiply the group delay shown by the number of amplifiers in cascade. Optical nodes will exhibit somewhat less group delay because they have only one diplex filter (see Figure 14.1). Figures 9.15 through 9.17 show that two diplex filters are cascaded in an amplifier. Furthermore, amplifiers need more crossover attenuation than do nodes to avoid the stability issue described in Chapter 9.

14.5 Block Conversion

It is sometimes desirable to increase the return capacity of a node. The limited return bandwidth that can practically be used limits the number of services that can be accommodated in the reverse direction. Also, noise buildup may limit the performance of the upstream bandwidth. Capacity may be increased, and noise buildup reduced, by splitting the return into several sectors (frequently based on multiple cables branching from the node) and using a separate return transmitter

Figure 14.3 Block conversion at a node.

for each. Rather than using multiple return transmitters, block conversion has been used as an alternative. A block conversion system takes several return paths and converts all except one to a unique block of frequencies. The blocks are combined and transmitted to the headend using one return optical transmitter.

Figure 14.3 illustrates block conversion in a common configuration that uses a 200-MHz return optical transmitter. Typically, this circuitry would be housed in a node though it might be housed in an ancillary housing. Up to four return paths may be accommodated. The first is coupled directly to the optical transmitter. The other three are up-converted to other frequencies up to about 200 MHz and combined with the unconverted spectrum to supply signals to a return transmitter. At the headend, the process is reversed, developing four individual 5- to 40-MHz spectra to be supplied individually to receivers. Alternatively, it is possible to build return receivers that tune to 200 MHz, eliminating the need for downconversion in the headend.

Single conversion is used for economy. For other applications, double conversion might be used if it is necessary to put many blocks close together, but the cost can be higher. The single conversion results in inversion of the spectra of the three converted bands so that what was the low end of the band becomes the

high end of the converted band. The conversion back in the headend will reinvert the spectrum. Inversion is brought about by the practical need to use high-side local oscillators.

In some cases, it is necessary to provide very accurate frequency conversion in the node and headend, such that the frequency to which a signal is restored in the headend is very close to the frequency at which it existed before conversion at the node. If this is required, then most commonly, the oscillators in the node will be phase locked to those in the headend, or vice versa. A pilot tone is transmitted between the node and headend to allow phase locking.

Block conversion increases the bandwidth available per customer and reduces noise funneling without requiring more optical transmitters and receivers. However, since the one optical transmitter is loaded with more signals, the performance of that transmitter must increase, or the OMI must be reduced. This reduces the carrier-to-noise ratio available in the optical portion of the return path. The use of block conversion may force an operator to use a DFB return laser rather than a lower-cost F–P laser.

14.6 Return Signal Level Issues

Of paramount importance to the operation of the return path is the proper handling of signal levels. The issue is far more complex in the upstream direction than it is in the downstream direction. In the upstream direction, you deal with many transmitters located on or in subscribers' homes, with each signal path having a different gain between the home and the headend. Several conditions must be addressed in setting the level of the transmitters. It is as important to design and balance the reverse path as it is the forward. If you balance the forward plant and let the reverse fall where it may, reliable return performance will not be obtained.

14.6.1 Measurement of Return Signal Levels

In Chapter 4, the definition of "signal level" of a digital signal was presented as the level, expressed in dBmV, at which a sine wave would produce the same heating in a resistor as does the signal being measured. That is, the level of a digital signal is taken to be the average power. Most digital signals have the characteristic that the amplitude of the signal changes for different transmitted states, so the peak level is greater than this average. The peak-to-average ratio is small with QPSK, and indeed, would be zero were it not for transmitter filtering. Higher-order modulation exhibits a greater peak-to-average ratio, as shown in Chapter 4. Peak-to-average ratio is rather difficult to measure, so the *NCTA Recommended Practices for Measurements on Cable Television Systems,* "Upstream Transport Issues" section, recommends that each manufacturer state the peak-to-average ratio of his or her modulation.[2]

Many signals in the upstream direction employ time division multiple access (TDMA), a technique in which many transmitters transmit on the same frequency at different times. An explanation is presented in Chapter 5. Measurement of individual elements can be accomplished by triggering the measuring instrument (usually a spectrum analyzer) on some reference time, allowing one of the signals in the return "parade" to be identified. This triggering signal would normally be supplied by the system being analyzed. First-generation equipment does not necessarily have the capability to provide such a trigger, however.

If you set a spectrum analyzer to the zero span mode, it can be used to measure the signal level on only one frequency, with the bandwidth of the analyzer determining the bandwidth of the measurement. You may be restricted as to the measurement bandwidth due to adjacent carriers, and this will vastly complicate the measurement.

Figure 14.4 illustrates the measurement issues for a TDMA signal in the return path. Figure 14.4(a) shows the preferred test configuration: received signals are supplied to the using system and also to a spectrum analyzer used to measure signal level. Ideally, the using system would supply a triggering signal to the spectrum analyzer, which would allow identification of the return signal to be observed. Again ideally, the spectrum analyzer would have a trigger input and would have a time base setting that allows separation of the return signals into the individual elements. These ideal situations do not exist with all first-generation hardware, but it is hoped that future systems will allow for this type of observation.

The spectrum analyzer is set to the zero span mode, in which it operates as a fixed tuned receiver, plotting amplitude versus time of the signal, as shown in Figure 14.4(b), (c), and (d) for three different resolution bandwidths of the analyzer. If the resolution (IF) bandwidth of the analyzer is wider than the occupied bandwidth of the signal being measured, as shown at the left of Figure 14.4(b), the waveform might look as illustrated in (b). The TDMA parade consists of a number of responses from different transmitters, each being received at a slightly different level at the headend. Ideally, the long loop automatic level control (ALC) would bring all elements to the same level. However, there will be differences due to the limited resolution with which the transmitters can adjust level.

Issues Regarding Making Accurate Measurements of TDMA Signals

Particularly with higher-density modulation formats (bandwidth efficiency of greater than two bits per hertz), the amplitude will change within one time slot since a number of bits are transmitted in each time slot, and different bit combinations correspond to differing amplitudes. This is illustrated to the extreme right of Figure 14.4(b) and further complicates the job of accurately measuring signal level.

Yet another difficulty is that, if the spectrum analyzer resolution can be set wider than the occupied bandwidth, it will tend to track envelope peaks, measuring the peak, and not the average, signal level. This will necessitate adding another correction term.

Figure 14.4 Measurement of TDMA return signals. (a) Instrumentation to observe return levels. (b) Wideband measurement of amplitude. (c) Measurement with restricted bandwidth. (d) Measurement with severely restricted bandwidth.

Figure 14.4(c) illustrates a situation in which the spectrum analyzer bandwidth is slightly narrower than the occupied bandwidth of the signal. Analogous to measurement of the amplitude of an analog television signal, if the bandwidth is too narrow, the individual packets will not be present long enough to allow the analyzer output to stabilize at the correct level for each packet. Figure 14.4(d) illustrates a further reduction in bandwidth of the analyzer, showing even more difficulty in accurately reading the bandwidth of any element.

One reason that the spectrum analyzer bandwidth may be too narrow for the signal being measured is that the signal may be wider than the widest bandwidth of the analyzer. It is not uncommon to deal with return signals having a 2-MHz occupied bandwidth, and some analyzers are not equipped with IF filters wider than 1 MHz. Also, if adjacent channels are occupied, it may be necessary to reduce the bandwidth of the analyzer in order to avoid interference from those adjacent signal(s).

Some spectrum analyzers use IF filters with relatively high shape factors (the ratio of bandwidth at high attenuation to bandwidth at lower attenuation). Such filters tend to ring less and can allow accurate measurements to be made faster. However, it does mean that the resolution bandwidth may need to be less than the occupied bandwidth in order to reject adjacent channel power. Other

analyzers use filter shapes that are more squared (somewhat as illustrated in Figure 14.4). These filters, if available in the correct bandwidth, do a better job of rejecting adjacent channel signals, but they tend to ring more, introducing yet another source of error.

Making the measurement with a significantly narrower spectrum analyzer bandwidth can yield a fairly accurate average channel power measurement *if* return signals are present substantially all the time. This can be useful in checking that a long loop ALC system is set up to operate at the desired point. It cannot tell anything much about the amplitude of a specific element in the return signal. If there are gaps in the received signal, the results can be quite misleading. Consider a telephony system that multiplexes 24 telephone lines into a single TDMA stream. If the node has fewer than 24 lines, there will be some gaps in the received signal. These gaps can provide misleading results when you attempt to measure the signal level of the stream by looking at average power.

In summary, it is difficult to make accurate measurements of the amplitude of TDMA signals in a real network. Gross errors can often be spotted fairly easily, but very accurate measurements are hard to make. The *NCTA Recommended Practices* document, referenced earlier, defines the "level" as being that which would be measured if the signal was on continuously. It further requires the manufacturer of a signal-measuring device to publish a valid procedure to allow a user to translate the level indicated to the defined level. However, this is not always easy to do.

Measurement of the level of return signals using more advanced techniques, such as code division multiplexing and multiple carriers, is not covered here. The vendor of such systems should provide guidance for making the required measurements.

14.6.2 Long Loop Automatic Level Control

Figure 14.5 illustrates the critical points in the return signal level control issue. It is similar to Figure 8.8, which addresses the same issue from the headend. As described in Chapter 8, the return service includes a long loop automatic level control (ALC), wherein the signal level is measured in the headend (Lvl 1 in Figure 14.5), and a signal is sent to the home transmitter, TX4, to adjust its output such that the correct level is received at the headend. This corresponds to some level at the tap, Lvl 5, and to other levels at each RF amplifier (Lvl 4) and at the node, Lvl 3. The point is that, once the system controls the signal level at Lvl 1, all the other signal levels are set according to the design of the system.

The most critical level is that at the return optical transmitter, Lvl 3. The dynamic range (from maximum signal level down to noise level) tends to be relatively low for return path optical transmitters, both due to technology and cost. Dynamic range is explored in more detail later. See Chapter 11 for a detailed discussion of optical technology.

The system must be set up to optimize the level at Lvl 3. However, Lvl 3 cannot be directly observed by the adjusting system. Instead, the gain between the

An ALC loop is set up between the headend and each home system individually. The headend senses level of the return signal and adjusts the upstream transmitter such that the correct level is received at the headend receiver. The most critical level is that at the reverse optical transmitter, Lvl 3.

Figure 14.5 Pertinent points in return signal level control.

point marked Lvl 3 and that marked Lvl 1, where the level observation is made, must be known and must be stable. The target value of Lvl 1, to which the system is adjusted, is set based on the desired value of Lvl 3. At the same time, it is necessary to set the gain in the system such that the input and output signal levels at all amplifiers, represented by Lvl 4, are correct. Attenuators and equalizers are often provided in the return amplifiers for this purpose and must be properly selected during system alignment. "Correct" is based on the manufacturer's specified operating levels for the conditions under which the amplifier is being used. When these levels are set (by design and alignment of the system), then the level at the tap, Lvl 5, will be set by the tap value and the gain from there to the node.

A few systems have attempted to use automatic gain control (AGC) in the return path. This is not recommended for a number of reasons. First, it fights the setting of levels as described earlier. Second, at times it may be normal for the number of carriers in the reverse direction to change. For example, some services employing time division multiple access (TDMA) may permit signal dropout as a normal condition at times. Further, as return services are added and deleted, it is not desirable that gain in the return path change. Next, real return paths often

branch and have a different number of amplifiers in each branch. AGC can significantly complicate operation of such systems.

In some cases where reverse AGC has been employed, it has been done with a pilot tone generated somewhere near the end of the system and detected by all amplifiers and the node. This may work, but raises many questions when downstream signals branch after leaving the node. With branches having a different number of amplifiers in cascade, where do you put the pilot tone generator? You would logically place it at the end of one of the branches, but that leaves the other branches with no coverage.

14.6.3 Thermal Gain Control

It is often desirable to provide thermal control of return path equipment. Even at return frequencies, cable exhibits some variation in loss with temperature, and it may be desirable to also compensate for variations in the operating point of lasers. This is often done by using thermistors (temperature-sensitive resistors) to sense temperature level inside the return optical transmitter and adjust gain based on temperature variation. It is also practiced at times to cool the laser, as mentioned in Section 14.7.

14.6.4 Return Signal Levels at the Tap

Of interest to those designing or using systems needing the services of the return path is the issue of the signal level at the tap, Lvl 5, needed in order to produce the desired level at Lvl 3. Intuitively, we might assume that the difference between the lowest and highest levels would be about the same as the difference between the lowest and highest tap value in the system. Tap values are chosen such that the downstream level at the tap is more or less constant, fixed by system design. This calculation is done at the highest frequency of interest, where cable loss is maximum. At the return band, cable loss is much lower, so the level required at the tap, Lvl 5, would be lower at lower tap values. Generally, those lower-value taps are located farther from an amplifier (in the downstream direction).

Analysis of the return gain characteristics of real nodes, however, shows that this line of reasoning is oversimplified. Sometimes low-value taps are located at the end of a long cable run, so there is little upstream loss at the low frequencies composing the return band. Other times, though, low-value taps are located after a directional coupler and near an amplifier. At those locations, the loss is primarily flat loss (not a function of frequency), and the required output at Lvl 5 will be higher.

Figure 14.6, from the analysis of one real node, shows the levels found to be required at the tap (Lvl 5 of Figure 14.5) in order to produce a level of 0 dBmV back at the node (Lvl 3). Each dot represents a tap of the value shown on the x axis, which requires the y-axis level at any tap port. Notice that low tap values exhibit a wide variation in level required, with the level range tending to compress slightly as you move to higher tap values. This slight compression is a result of

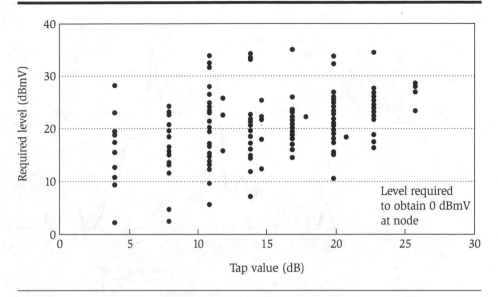

Figure 14.6 Required signal level as a function of tap value.

having few places to use high-value taps, other than shortly after an amplifier. This data is from one node, but examination of other nodes owned by the same and other cable operators shows the same tendency.

It is not necessarily desirable to have a node that requires low signal levels at the return input. It is desirable to require the highest level that can be supported by the reverse amplifiers that supply signals to the node. This will help minimize the effect of noise entering the plant (primarily from homes). Of course, you must ensure that the requisite signal levels are available at the home transmitter. The use of 0 dBmV as the input to the node is not intended to be a real case but to provide a convenient normalized reference point.

Figure 14.7 illustrates the signal level issue another way. The population of taps having certain level requirements is plotted for the node of Figure 14.6. The distribution of levels appears to be a very crude approximation to a normal distribution. It illustrates that, though there is a clumping of levels, a significant number of taps will require a higher or lower level. The product designer must plan for the required level range, also allowing for variations from system to system, errors, and variations with temperature. Note that the level of the signals at the home will vary because of temperature-dependent changes in gain in the system. When the signal level received at the headend changes, this will cause the signal level commanded from the home to change, due to operation of the long loop ALC.

Reducing the required tap return level range would be desirable, in that a reduced level range specification would provide some reduction in cost and would enhance performance in at least some cases. Homes with the lowest signal level

Figure 14.7 Distribution of tap signal levels.

requirement are also the homes that have the greatest potential to introduce noise into the return path. A wide level range implies high output level capability from the device, which implies possible high noise emission when the level is set low. It also implies rather larger current drain since the output amplifier must have the capability to handle large signals with minimum distortion.

14.7 Optional Ways to Specify Return Lasers

As of this writing, the industry has not standardized the method of specifying return lasers, so we can show only the methods being used or proposed. At least three types of lasers are in common use at this time: cooled distributed feedback (DFB) lasers, uncooled DFB lasers, and Fabry–Perot (F–P) lasers. The list is in the order of decreasing performance (and decreasing cost). It is required to select the correct laser for each application, bearing in mind the eventual loading, not just the initial loading. Refer also to Chapter 11, which discusses laser technology.

14.7.1 Data- or Video-Grade Classification

An early method of classifying return transmitters (and receivers) is to classify them as to data grade or video grade, with the implication that video-grade lasers are better. This method has proved unsatisfactory because it doesn't take into account the real requirements placed on the system. A return path with many data signals requires at least as good a laser as does a return path with several analog

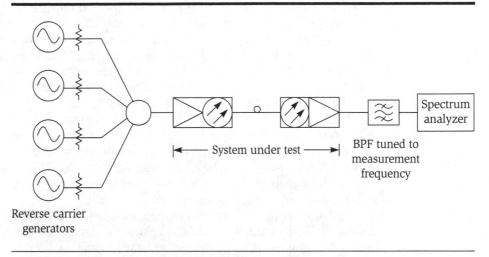

Figure 14.8 Four-tone measurement of a return optical system.

video signals. Further, there is no standard criteria for classifying lasers: in at least one case, a manufacturer interpreted "data grade" to mean capable of handling one FSK return signal only. When it was used with multiple QPSK return signals, it was incapable of handling the job.

14.7.2 Discrete Carrier Testing and Classification

An older method of measuring the performance of return components is taken directly from downstream measuring practice. Either two or four CW carriers are used to excite the laser, and the results are measured on a spectrum analyzer. The second- and third-order beat products are recorded.

Figure 14.8 illustrates the method. Either two or four oscillators are combined, with identical output levels, to produce a spectrum, usually on the highest T-channel carrier frequencies used by the system. The performance of the system is reported as the composite second- (CSO) and third- (CTB) order beats and the cross modulation, as if this were a miniature downstream system. Typically, lasers tend to be more second-order than third-order limited, so CSO becomes the dominant specification.

The attenuators and bandpass filter are shown for practical reasons. The attenuators help isolate the signal sources so that you don't suffer intermodulation distortion in the output circuits of the generators. The bandpass filter (BPF) tuned to the frequency of measurement prevents overload of the spectrum analyzer.

In evaluating the performance of the system for carrying a limited number of signals, some practitioners have tried to predict the frequencies at which the maximum distortion is present. These frequencies are then avoided in order to

maximize the performance of the link. The cost of doing this is that many poten-
tially useful frequencies are excluded. You are left with lower capacity and less
ability to move carriers to avoid noise. Another problem with this approach is
that, as we will show, when the laser goes into clipping, the normal model used
for predicting the frequencies of distortion components is invalid. Also, this tech-
nique fails to stress the laser the same as it would be stressed with a number of
real digital signals. Again, this is explained later.

Yet another problem encountered in comparing specifications between man-
ufacturers by using discrete test tones is that some manufacturers, following
downstream practice, quote specifications with equal level carriers. Others as-
sume that some number of the carriers represent analog signals, and others repre-
sent digital signals that are carried at a lower level. They reduce the level of some
of the signals when making measurements.

Still another problem in comparing specifications is that manufacturers may
choose different channels on which to place the carriers. Depending on the place-
ment, differing numbers of composite beat products will exist on the measure-
ment frequencies. Only the most careful reading of the data sheet can reveal the
exact conditions under which the measurements were made.

14.7.3 Noise Power Ratio Testing and Classification

A newer method of characterizing systems is coming into common practice. It
comes much closer to testing the laser as it will be operated in service. The tech-
nique is not really new: it has been used in the telephone industry for years to
characterize linear systems, such as FDM (frequency division multiplex) coax
and radio systems. HFC systems are FDM systems because they employ multiple
frequencies to carry different information.

Figure 14.9 illustrates the so-called noise power ratio (NPR) testing tech-
nique. A broadband white noise generator is used to simulate a spectrum filled
with data signals. Data signals generally have a flat spectrum over the occupied
bandwidth of the signal (see Chapter 4), so simulating them with a white noise
signal is reasonable. ("White noise" simply means that the spectrum is flat over
the frequency range of interest.) A bandpass filter limits the spectrum to that for
which the system under test is rated.

An attenuator is used to isolate the bandpass filter from a notch filter that fol-
lows. The notch filter removes the white noise in some measurement bandwidth.
The total power in the signal is important, and would be measured using, ideally,
a thermocouple type power meter. The variable attenuator is used to adjust the
signal level supplied to the system under test.

A spectrum analyzer, protected if necessary by a bandpass filter tuned to the
notch frequency, is used to observe the output. The notched frequency will be
somewhat filled in by the system under test. At low signal levels, the filled in
noise is due to noise introduced by the system under test. At higher signal levels,
it is filled in by distortion introduced by the system under test. It is not possible to
look at the filled in level and tell whether the fill is due to noise or distortion or

Figure 14.9 Noise power ratio testing of a return path.

both. The ratio of noise on the flat portion of the spectrum, and that in the notch, is called the noise power ratio (NPR). Some practitioners have called it carrier to noise plus intermodulation noise, or C/(N + IMD).

Use of NPR to Set Operating Level

If we were to plot the NPR against the power supplied to the laser, we would obtain a plot similar to that of Figure 14.10. The total power supplied to the laser is plotted on the x axis. Often the quantity plotted will be the power spectral density, which is the total power divided by the bandwidth over which the power exists.

At low signal levels, the NPR is noise limited. One would expect the NPR to increase 1 dB for every decibel increase in total power into the laser. In practice, there will often be a region on the left of the diagram in which the NPR increases faster than decibel for decibel.[3] This is especially true for distributed feedback (DFB) lasers feeding long fiber-optic cables. The reason is that at low signal levels, lasers are susceptible to reflections from the optical cable, due to Rayleigh backscatter. The reflected light "pulls" the laser wavelength, creating noise. As the drive to the laser increases, the laser tends to "chirp" (change wavelength with instantaneous drive) more, reducing the pulling by reflected light. The reflected light is a function of the length of optical cable used: the longer the cable, the more reflection. DFB lasers tend to be more susceptible than are F–P lasers because their wavelength is chirped less by modulation, making them more susceptible to wavelength pulling by reflected light.

Wavelength pulling by reflected light is a phenomenon similar to the tendency of an oscillator to oscillate off of its normal frequency if a signal at a close frequency is supplied to the circuit. The more the separation between the two fre-

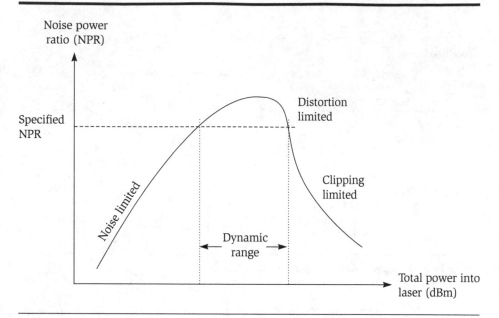

Figure 14.10 NPR as a function of the total power into the laser.

quencies, the less the tendency to pull. Chirp refers to the changing of frequency (wavelength) of an oscillator (or light source) with applied modulation.

At some drive level, the NPR is maximized. Above this level, the NPR drops quickly. The initial drop may be due to distortion, primarily second order, but you quickly enter a region in which the primary effect is from laser clipping. Clipping noise tends to be the dominant mechanism by which the NPR is reduced on the right side of the graph, and is covered in greater detail later in this chapter.

NPR should not be confused with carrier-to-noise ratio, C/N. C/N represents the ratio of a signal to the noise lying under that signal. It is a function of signal level and the total noise provided from thermal sources plus the excess noise generated in real electronics. On the other hand, the noise in the notch of an NPR measurement either is a function of the thermal and excess noise or, at higher levels, is caused by distortion, which transfers noise from other frequencies into the notch. The mechanism may be the familiar second- and third-order distortion, or much of the noise filled in the notch may be clipping noise from the laser (or electronic amplifiers). Do not look for a simple and universal relationship between NPR and C/N. Later in this chapter, we show how to compute C/N from NPR.

One convenient way to establish the level at which to operate the laser is to determine the range of input signal levels across which some specified minimum NPR is obtained, as indicated in Figure 14.10. The laser should be operated with total power within this "dynamic range." The engineer might be tempted to oper-

ate the laser at the total power that represents the highest NPR. This will work, but because the dropoff in performance is so steep on the right side, he or she may want to operate with slightly less power (to the left of the maximum NPR). (In some lasers, the slope on the right side has a region of gentler slope, where it is limited by second-order distortion. In other lasers, the slope tends to be quite steep just to the right of the peak. This indicates that the laser performance is limited more by clipping than by second-order distortion.)

A noise power ratio test has several advantages over a four-tone test. Noise has an amplitude probability distribution that is different from that of four carriers, but is similar to that of a number of digital carriers added together. The central limit theorem from statistics states that, when a number of independent variables having any arbitrary distribution are added together, the resultant distribution is normal. A normal distribution also describes the amplitude distribution of noise.

Further, the measurement is independent of the bandwidth of measurement, so long as the bandwidth is significantly less than the notch width. It is also independent of the characteristics of the spectrum analyzer detector, so repeatability is good. The only other instrument (besides a spectrum analyzer) needed is a thermocouple power meter, which is low in cost and very accurate.

Use of BER to Set Operating Level

An alternative way to look at the optimum laser operating point recognizes that the vast majority of signals likely to be carried in the reverse direction use low-density digital modulation. The return spectrum is loaded much as shown in Figure 14.9, except that a modulated (usually QPSK) signal is added into the notched frequency band. The bit error rate is plotted against the signal level.[4]

Figure 14.11 illustrates the principle. At low signal levels, the bit error rate (BER) is poor owing to noise corruption. As the signal level increases, the BER gets better (lower on the y axis is better BER). In the central part of the diagram, the BER is so good it is below the chart. Toward the right side of the graph, the BER again begins to deteriorate, due to clipping. The operating total signal level into the laser is set to be somewhere in the middle of the range where the BER is extremely good.

14.8 Characteristics of Return Lasers

Two basic types of laser diodes are used in return path service today: Fabry–Perot (F–P) and distributed feedback (DFB) lasers. DFB lasers may be either cooled, with an integral thermoelectric cooler, or uncooled. F–P lasers tend to have lower performance but are lower in cost. F–P lasers are rarely cooled but have been a few times. Since lasers perform somewhat more poorly at high temperatures, a cooled laser can offer better performance at higher optical output levels. The

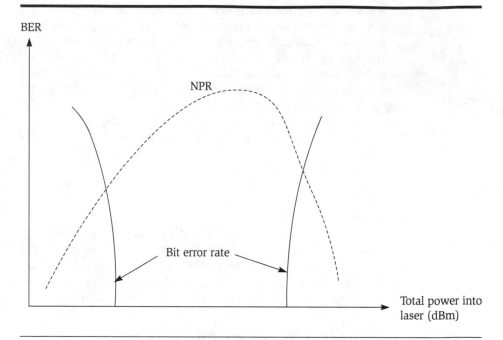

Figure 14.11 BER versus signal level.

cooler can also reduce the shift in transfer function of the laser with temperature. However, the cooler increases power consumption and usually increases cost.

14.8.1 Fabry–Perot Lasers

Fabry–Perot lasers have their frequency controlled by the spacing of mirrors at each end of the laser. The frequency control mechanism is such that the laser can oscillate simultaneously, or jump in rapid succession, to several wavelengths that are close to each other. Each wavelength propagates through the fiber at slightly different velocity because of chromatic dispersion in the fiber.[5] As shown in Chapter 11, this results in restricted performance in terms of carrier-to-noise ratio or NPR on the low side.

Figure 14.12 illustrates the performance of an older real F–P laser diode using the NPR test described earlier. On the left side of the graph, the performance is limited by inherent noise in the device. It increases almost precisely 1 dB for every decibel increase in signal level. Above the level of peak NPR performance, the drop in performance is precipitous as the laser rapidly enters the region in which the signal is clipped.

Note that Figure 14.12 is the same plot as Figure 14.10, except that Figure 14.12 is for a real laser. The horizontal axis of Figure 14.10 is the total power supplied to the laser, whereas the horizontal axis of Figure 14.12 is that total power

Figure 14.12 NPR performance of an F–P return laser diode.

divided by the bandwidth over which that power is spread. For example, if the power supplied to the laser in Figure 14.12 is spread over 35 MHz (5 to 40 MHz), then the total signal delivered to the laser will be that level on the *x* axis of Figure 14.12 plus 10 log(35,000,000) = 75.44 dB. That is, at an operating point at −35 dBmV/hertz, the total signal into the laser is −35 + 75.44 = +40.44 dBmV. A strategy for setting the level into the laser is to apportion the total power desired, 40.44 dBmV, over the actual occupied bandwidth of all signals. This subject is covered more fully later in the chapter.

14.8.2 Distributed Feedback Lasers

DFB lasers operate in a similar manner to F–P lasers, except that a diffraction grating restricts the operating wavelength of the laser primarily to a single mode (wavelength) of oscillation. This means that, relative to an F–P laser, the DFB laser exhibits a lot less noise and distortion as a result of multiple propagation speeds in the cable.

Figure 14.13 illustrates the NPR of a DFB return laser diode on the same NPR scale as the F–P laser of Figure 14.12. (The absolute values on the *x* axis are irrelevant since they are separated from each other by only an attenuator or amplifier.) Both diodes were measured using the same 6-km signal path, with no splitting loss. The increase in NPR with increasing signal level on the left side is only

Figure 14.13 NPR performance of a DFB return laser diode.

slightly greater than dB for dB because the path loss is low, meaning the path is short. Because the path is shorter, less Rayleigh backscatter reflected signal arrives at the laser, and it arrives with less time delay, so there is less wavelength pulling, and therefore less excess noise.

As shown in Figure 14.13, the NPR performance of the laser is somewhat better at room temperature and below than it is at high temperature. For this reason, cooled DFB lasers are sometimes used in return path operation and are necessary where higher power levels are required. However, the added cost and power drain (which adds heat to the node) often argue against adding a cooler. The performance of uncooled DFB lasers seems to be considered adequate for many purposes today, though long, heavily loaded paths may benefit from a cooled laser. The laser of Figure 14.13 was embedded in a temperature-compensated transmitter module. Had it not been, the performance range with temperature would have been worse.

Compare Figure 14.13 with Figure 14.14, which shows three different return DFB lasers at 25°C. These lasers are of one type from one manufacturer, so the chart illustrates the difference between lasers that are as nearly identical as possible. Although the laser of Figure 14.13 was driving 3.55 dB of fiber, the lasers of Figure 14.14 are driving 10 dB of fiber. Because of the longer fiber length, more Rayleigh backscatter can develop, so we see a greater slope of the NPR on the left

Figure 14.14 NPR performance of three DFB lasers at room temperature.

side. The slope is around a 4-dB increase in NPR for every 3-dB increase in signal level. This is rather typical of lasers, particularly DFB lasers, driving a long fiber.

14.8.3 Return Path Combining at the Headend

In order to save the cost of multiple RF receivers in the headend, it is a frequent practice to combine several reverse path signals into one. This can work if adequate total data capacity (called "bandwidth" by data communications engineers) is available to service multiple nodes as one. This combining is usually done electrically so that there is one optical receiver for each node. RF combining (in the headend, after the optical receiver) can be done satisfactorily as long as adequate carrier-to-noise ratio is maintained, considering both thermal noise and ingress. The transfer gain from the electrical reverse input to the node, to the point of combining in the headend, must be the same for all return paths that are combined. Otherwise, the return operating level of at least one of the nodes will be incorrect.

However, some operators like to do optical combining to further save the cost of optical receivers. Optical combining must be done carefully, if at all, because if the lasers operate at close to the same wavelength then interference (beats) between the lasers will develop in the receiver. These can increase the noise level

much more than would be expected from the addition of incoming noise from each transmitter. Also, the received power from each return laser must be the same. Optical combining is not a recommended practice.

Since DFB lasers have relatively narrow ranges of operational wavelengths, it may be possible to combine them. You must ensure that the wavelengths will stay sufficiently separate over life, temperature, and modulation. Because F–P lasers have a less well-defined wavelength operating range, optical combining when F–P lasers are being used is particularly dangerous. Optical combining may work when it is first implemented, but if at a later date the two lasers drift to the same wavelength, the link could fail.

14.9 Spurious Signals in the Return Path

Of paramount concern to operators of two-way plant is noise induced into the return path. Anecdotal and numerical evidence point to the house as being the prime contributor of noise, with the drop contributing the next largest amount. Most practitioners contend that the hard-line plant is not responsible for many noise problems, with the exception of common path distortion. A few practitioners have reported noise problems in the hard-line plant, however.

Considerable investigation has gone into this problem of noise on the return plant. Figure 14.15[6] illustrates a fairly typical result. This is a three-dimensional plot of the frequency and time of interference in the return band. Frequency is plotted along the bottom of the chart, progressing from right to left. Along the left edge is plotted time, over a period of about nine hours. The vertical axis represents the relative level of signal observed. This data was taken with a special-purpose receiver, which can sample signals sequentially at a number of frequencies. The data is stored on a computer for later analysis. Other investigators have used a spectrum analyzer interfaced to a computer. A third method for obtaining the data is to capture the return spectrum on a high-speed digital oscilloscope and do a fast Fourier transform (FFT) on the resultant waveform. (An FFT is a mathematical way of computing the frequency spectrum of a complex waveform.)

The figure shows considerable interference at lower frequencies. The frequency spectrum below about 7 MHz is relatively invisible in the upper right-hand corner of the box, but shows less occupancy (by interfering signals) in the morning than in the afternoon. This is consistent with typical off-air usage patterns on these frequencies, which exhibit better propagation characteristics during the later parts of the day and at night. The large energy content between about 22 and 30 MHz is an intentional return signal. The other energy content represents ingress. Above about 15 MHz, this node was essentially ingress free. Below this frequency, interference was substantial and had the distinct probability to cause serious interference with communications.

That the spectrum below about 15 or 20 MHz is relatively "dirty" and the higher spectrum tends to be much cleaner is a fairly typical result. Traditional services, such as status monitoring and return signals from set top terminals us-

Figure 14.15 Interference versus frequency and time.

ing so-called RF-IPPV technology, use very slow and robust signaling methods, such as FSK and BPSK, at low data rates. These are transmitted in this lower portion of the band, where interference is worse. However, the nature of the transmissions is that if one transmission doesn't get through, the system will try again and nothing is lost. (RF-IPPV is the industry acronym for impulse-pay-per-view, IPPV, services that use an RF return channel as opposed to a telephone channel.)

14.9.1 Plant Unavailability Analysis Based on Undesired Signals

Measurement of the quality of the upstream signal path is complicated by the absence of a spectrum of steady-state known signals and by the varying nature and presence of the undesired signals.

Generally, the dominant interfering signals are common path distortion (CPD), discrete ingress signals, and electrical transients. The first two result in

signals that may be present for relatively long periods, but occupy narrow frequency bands, whereas transients are present for a very short duty cycle, but may have broad spectral content.

It is important to be able to measure these two classes of signals separately since their effects and the required countermeasures are different:

- Because discrete signal interference varies with time and frequency, the most effective countermeasure is often the ability of a service to shift the upstream frequencies of field transmitters as required to avoid these signals. In order to do that, the headend equipment must continually monitor the spectrum from each node and keep a list of available frequency bands.

- Because electrical transients are typically very brief in duration, the most effective countermeasure is often some combination of interleaving (to prevent one transient burst from affecting more bits than can be corrected by the error-correction capability of the FEC), combined with adequate FEC techniques to correct isolated errors.

The following two measurement techniques are offered as examples of emerging industry practice in this area. The first measures the entire spectrum and range of possible operating levels, but includes only discrete signals, whereas the second measures multiple types of interference, but on only a single channel at a single defined operating level.

14.9.2 Discrete Interfering Signal Probability (DISP)

An NCTA recommended practice for quantification of discrete interfering signals is discrete interfering signal probability (DISP).[7] This procedure was developed by Large and Bullinger as a result of extensive field testing.[8]

Continuous spectrum analyzer measurements are taken over the entire upstream spectrum. The operating parameters are set up to reject transients so that the measurements reflect only discrete signals. The level reading at each frequency on each sweep is downloaded to an attached computer for analysis. Each reading is compared with several predefined thresholds to create a three-dimensional matrix whose axes are frequency, time, and level, and whose values are the probability that a signal is present at a given frequency and time whose amplitude exceeds a given threshold.

From this matrix, postprocessing is used to derive overall discrete signal/CPD performance of the system as a function of threshold, system availability as a function of frequency and operating level, channel availability as a function of time and operating level, and other data as required.

14.9.3 Plant Unavailability Analysis Based on Threshold Boundaries

Studies have been made of the availability of return plant using a CW tester developed at Cable Television Laboratories (CableLabs).[9] The technique uses a CW carrier in the return band and analyzes the characteristics of the signal received at

Figure 14.16 CW tester block diagram and constellation.

the headend to predict the error rate of a signal at that frequency and with that bandwidth. From that data, we can predict the percent availability of a channel. Whereas the DISP measurement was based on interference amplitude measurement, the CableLabs technique analyzes the errors in location of a point in the data constellation (see Chapter 4 for an explanation of the data constellation).

Figure 14.16 illustrates the CW tester block diagram and the resulting constellation. A CW carrier is added somewhere in the node to be studied. At the headend, the carrier is up-converted to the normal TV IF of 44 MHz and passed through a SAW filter. After AGC (not shown), the signal is supplied to a phase locked loop, which generates quadrature demodulating carriers (see Chapter 4).

The outputs of the demodulator are supplied to a digital-sampling oscilloscope and to a window comparator that determines when the signal is outside the accurate decoding window. The accurate decoding window is shown in the figure; as shown, it is the decoding window for 16-QAM. While the constellation point is in the window, the correct state is decoded. While it is outside the window, a totalizing counter counts pulses from an 11-MHz clock. The output of the totalizing counter is accumulated by a computer, along with output data from the oscilloscope.

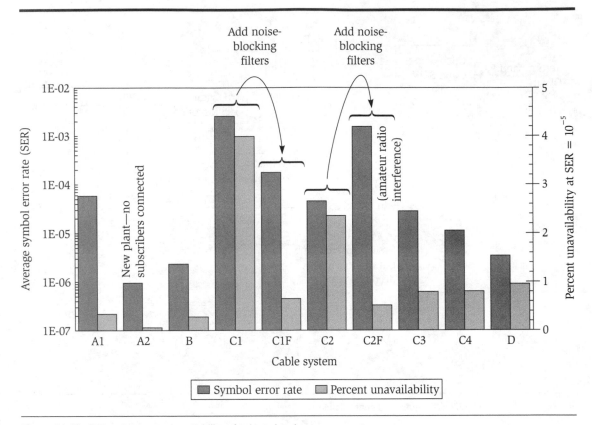

Figure 14.17 SER and percent unavailability of selected nodes.

The constellation diagram at the bottom of the figure illustrates the data point crossing the threshold to the region occupied by another state. Assume for a moment that the study concerns use of 16-QAM modulation on the return carrier. The box defining the threshold region, in which the signal will be correctly demodulated, is shown. If a disturbance of any kind causes the carrier to go outside the box, then the totalizing counter is started, and it measures the total time the signal is out of the box. From this information, we obtain the symbol error rate (SER) equal to the proportion of time out of the accurate decoding region. If we assume that the channel is available when the SER is below some level (10^{-5} in the example shown), then we can chart the percent of unavailability of the channel.

Notice that the parameter measured is called the symbol error rate rather than the bit error rate. This is because it is the symbol that is subject to error, and typically each symbol comprises two or more bits. Chapter 4 explains the differences between bits and symbols in the context of data transmission.

Figure 14.17 illustrates the symbol error rate and consequent percentage of unavailability of nodes in several cable systems based on one set of tests. Each

was observed for about one week. To the left is a logarithmic scale showing the symbol error rate, and to the right is a scale showing the percent of unavailability. Note node A2, which was a new node with no subscribers. Its good performance is evidence of the observation made by others—that drops and house wiring tend to be responsible for many return path problems.

Nodes C1 and C2 were studied with and without noise-blocking filters. Noise-blocking filters are high-pass filters placed on taps to preclude in-home–generated noise from reaching the return path. In node C2, interference developed from an amateur radio operator during the testing; nonetheless, the percent of unavailability dropped significantly when the filters were added. The symbol error rate and unavailability improved dramatically in node C1 when the filters were added. This observation again lends credence to the hypothesis that the home is responsible for most of the problems in return path operation. It also attests to the efficacy of noise-blocking filters (covered in more detail later in the chapter).

14.10 Characteristics of the Composite Reverse Signal[10]

A sinusoid (CW signal) has a well-known peak-to-average ratio given by $20 \log \sqrt{2} = 3$ dB. A composite signal composed of n carriers of equal level and randomly phased, has a total average power n times the power of one carrier, or $10 \log n$ dB greater than one carrier. The peaks add on a voltage basis (because at some time they will all add in phase), so the peak power is $10 \log(2n^2)$ dB greater than the average power of a single carrier. As more carriers are added, the peak power compared with the RMS power increases though the peak is reached for progressively shorter times.

This peak-to-average ratio argues again for the characterization of return lasers using NPR techniques rather than discrete carriers. When a number of carriers are combined, as for transmission on a return optical transmitter, the composite signal is noiselike in its amplitude and frequency characteristics. A noiselike signal has a well-defined and measurable RMS power, but it doesn't have a distinct peak value. Very occasionally, a high peak comes along, which will cause the laser (and electronic amplifiers) to clip.

As we add more carriers to the signals applied to a laser, the more noiselike the signal becomes in terms of peak-to-average ratio. Figure 14.18 illustrates the addition of carriers. Plotted on the x axis is the probability (percentage of time) that the waveform will exceed the RMS value by the number of decibels shown on the y axis. For example, look at the x-axis value of 10%. Following it up to the 3-carrier line shows that a signal consisting of three (sinusoidal) carriers exceeds its RMS value by just over 4 dB for 10% of the time. If the total signal comprises more than three signals, it will exceed its RMS value by at least 4.5 dB for 10% of the time.

As the graphs level off (toward the left side of the figure), it is proper to speak of the y axis as representing the peak-to-average ratio. For three carriers, the peak-

Figure 14.18 Peak-to-average ratio of multiple CW carriers.

to-average ratio is about 8 dB, and for five carriers, it is about 10 dB. We cannot see the peak-to-average ratio for a larger number of carriers because it occurs for a smaller fraction of the time than 0.001 %, the leftmost value on the graph.

As the number of carriers is increased, the curve approaches that of random noise, which has a distribution equal to the infinite curve. This is a confirmation of the central limit theorem from statistics. Stated in engineering terms, the central limit theorem may be expressed as follows:

> The sum of independent variables X_j will be normal for n→∞, even if the X_j each have a different distribution, provided that . . . the variance of any one term is negligible compared to the variance of the sum. If this is not so, the term whose variance is not negligible will predominate in the sum and its distribution will still be apparent in that of the sum.[11]

This demonstrates one of the shortcomings of evaluating the return laser with four CW carriers, as described earlier. The peak-to-average ratio of four carriers is around 9 dB, much less than the peak-to-average ratio of random noise or, equivalently, a large number of independent signals sharing the return spectrum. A composite signal consisting of 10 carriers will have a peak-to-average ratio of about 13 dB, within 2 dB of the peak- (probability 10^{-8}) to-average ratio of ran-

dom noise, so this signal may be acceptable for testing purposes.[12] However, in terms of peak-to-average ratio, random noise remains the most appropriate test signal.

14.11 The Reaction of Active Components to Signal Characteristics

The "peaky" nature of the signals has caused a great deal of concern within the engineering community regarding the reaction of laser transmitters to the peaks in the signal. At least two differing ways to analyze the problem have been used. These are summarized here.

14.11.1 Effect of Laser Clipping—Frequency Domain View

One can analyze the effect of laser clipping in the frequency domain, as done by West.[13] Figure 14.19(a) illustrates no laser clipping. The laser transfer characteristic curve shows that, with laser bias current (the x axis), there is no light output until a threshold current is reached, above which the light output power is proportional to the current through the laser. The input signal is shown below the transfer curve, and the output is shown to the right. So long as the input signal remains such that the light output never drops to zero, changes in the output light are proportional to changes in the input current.

As shown in Figure 14.19(b), if the input signal current is such that the light output ceases for a portion of the input signal, then the light output "clips" the signal. The portion of the signal represented by the dotted line is not really transmitted, so the signal is distorted. One way to model this is to assume that the signal was converted to light with no clipping, but then a signal was added, exactly equal in amplitude and opposite in phase, to the portion of the signal clipped. This added signal is shown in Figure 14.19(c). It represents a series of near impulses, which have a frequency spectrum that extends from some very low frequency equal to the lowest beat note between any elements of the input signal to very high frequencies. Thus, the spectrum of the clipping energy extends from very low frequencies to frequencies generally exceeding the bandwidth of the return path. The result is noise degradation, as represented by the degradation in NPR on the right side of Figure 14.14 and earlier figures.

14.11.2 Effect of Laser Clipping—Time Domain View

Kenny[14] has demonstrated the effect of clipping on a single data signal being carried in the reverse path. The effect is extended to more than one carrier by the central limit theorem. He found that, with a variety of lasers, the effect of clipping is to compress the signals, which compromises bit error rate, as shown by West.

As shown in Figure 14.20, the test setup is similar to that for the NPR test of Figure 14.9. A signal from CW generator G2 is placed in the noise notch, and in one test the noise source, G1, is replaced by a second signal generator. The

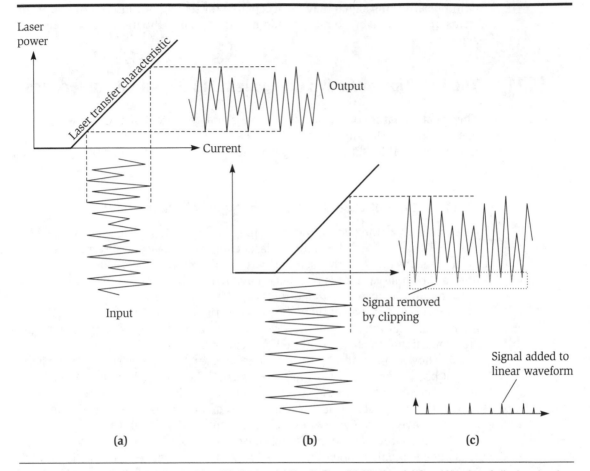

Figure 14.19 Laser in linear region and in mild clipping. (a) No clipping. (b) Clipping. (c) Quasi impulse of clipping signal.

observation of test results follows a technique developed by Cable Television Laboratories, which is described in Section 14.9.3. A CW carrier may be thought of as "1-QAM" modulation: a single state in a constellation diagram. This is not useful for communicating information, but much can be learned about the communications channel with this test. (For a discussion of levels of QAM modulation, see Chapter 4.)

Generator G1 is adjusted to sufficient amplitude to cause the laser to clip. The "signal" is demodulated by quadrature detectors, which produce in-phase (I) and quadrature (Q) components. (Chapter 4 describes the resultant constellation display in detail.) The phase of the two channels is adjusted to place the recovered "information" in the first quadrant, as shown on the oscilloscope screen. As expected, the spot that should represent the state of the modulation is smeared

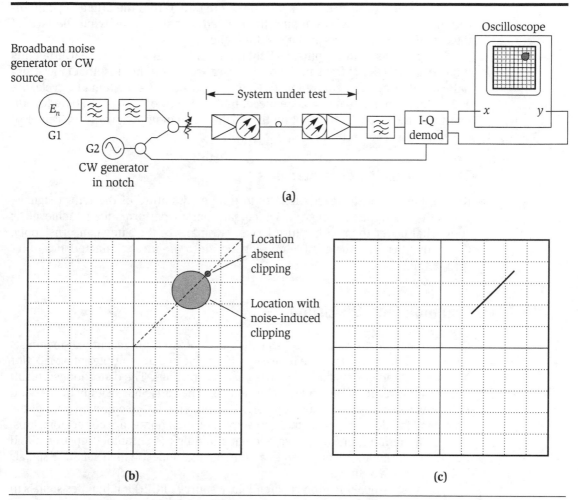

Figure 14.20 Time domain analysis technique. (a) Test configuration. (b) Constellation with noise causing clipping. (c) Constellation with CW signal causing clipping.

owing to the noise that fills in the notch, as shown in the description of NPR test-ing. The smear is circular, indicating phase effects as well as amplitude effects, due to clipping noise falling on the frequency of the test carrier. The location of the smear is toward the origin, indicating a reduction of signal level as expected.

In a second test, the noise source was replaced with a CW generator, whose frequency was chosen such that distortion products did not fall within the pass-band of the receive filter (the frequency of the test carrier). The laser was forced into clipping by increasing the amplitude of the CW generator that replaced the noise generator, and the resultant constellation spread is as illustrated in Figure

14.20(c). The spot turned into a line along the radial from the origin, again indicating that the laser was clipping the desired signal, but showing the lack of intermodulation products at the test frequency.

When a laser is in clipping, its light is cut off for some short time. During this time, all signals being transmitted by the laser are cut off, not just that largest signal. Since the signal is cut off for some time, typically for a fraction of a cycle, the amplitude of that signal will be reduced. In the display of Figure 14.20, reduction in amplitude results in the spot on the oscilloscope moving toward the center of the screen, as shown in (c).

14.11.3 Amplifier Characteristics

RF amplifiers tend to react similarly to the peaky nature of the return signals. Typically, the dynamic range in which satisfactory performance is achieved is somewhat better than with return lasers, but many of the same concerns apply. Currently, there is considerable investigation into the ideal characteristics of return amplifiers.

14.12 Common Path Distortion

Common path distortion (CPD) has gotten a lot of attention recently. It manifests itself as a series of beats in the return spectrum, located at multiples of 6 MHz (for North American systems using 6-MHz channel spacing). The distortion is a result of intermodulation of the downstream signals. The cause has usually been found to be mechanical connections at passive components in the common path of the upstream and downstream signals. Examples of such points are connectors, taps, drops, and terminators. These components handle the combined upstream and downstream signals, as opposed to amplifiers and nodes, which separate signals bound in the two directions.

Any mechanical junction is potentially a source of CPD if it handles signals in both directions. If the junction exhibits any nonlinearity, then intermodulation develops. Corrosion on the mating surfaces will typically create such a nonlinearity. Mitigation consists of careful selection of materials on mating surfaces to minimize corrosion and to ensure that the connections are airtight, meaning that they are held in place with so much force that there is a region that is always not in contact with air so that corrosion cannot take place.

This latter criterion is not easy to achieve. Often, because of temperature fluctuations or mechanical vibrations, a point on a contact that is airtight at one time will at another time be exposed to air. Corrosion can develop whenever water (in the air) is present. This mechanism has been called *fretting corrosion*.

There have been reports of components that were assembled with small contacts between, for example, ground and active components. Small contacts with relatively little force holding them together are particularly susceptible to fretting corrosion, especially if inappropriate materials are used. However, the majority

of devices using mechanical connections work quite well today: manufacturers have had extensive experience with the problem and are able to control it well.

14.13 Return Path Interference Mitigation Techniques

A number of proposals have been made for noise mitigation on the return path. Some operators who have reasonably new and properly installed plant feel that noise mitigation of any sort is unnecessary. However, it is not possible for the cable operator to maintain control over in-house wiring over time since the subscriber has the legal right to provide his or her own wiring. Traditionally, the availability of poor-quality coaxial cable and connectors in the consumer market, coupled with lack of proper training and installation tools, has resulted in degradation of in-home wiring over time. Some success has been achieved by operators who make available installation kits having high-quality components and, perhaps, instructions for proper installation. Some operators have also worked with local retail establishments to ensure that they carry quality components.

Figure 14.21 summarizes a number of proposed mitigation techniques, all of which are now being considered. They involve putting something in the hard line between amplifiers or at the drop (or in some cases, at a splitter in the house). Several approaches have been proposed, and they are summarized here.

14.13.1 Option 1: Drop Filters

Drop filters can be used to remove return band energy that might be coming out of the house. Two alternatives, 1a and 1b, are shown in Figure 14.21. Option 1a is a high-pass filter that cuts off below about 54 MHz, the low end of the forward band. It is intended for use at homes not taking any services requiring two-way facilities. With typically 25 to 40 dB of attenuation in the return band, it does not allow return band signal power in the home to reach the hard-line plant.

Filter option 1b is similar to 1a except that a bypass is added to allow signals in a narrow band, often between 10 and 15 MHz, to pass. These are used when the home has an RF-IPPV converter, which must communicate back to the headend on these frequencies. Interference from the home that occurs in the passband obviously appears in the hard-line plant, but interference coming out of the house at other frequencies is blocked.

Filters are usually placed at the tap though they may be placed at the side of the home in some cases. They are sometimes seen as a stopgap measure because they don't remove the root cause of the problem, and they interfere with provision of modern two-way services in the home. When it becomes the practice to allow the customer to purchase two-way hardware at retail and install it himself or herself (a desirable way to sell data services), the filter can cause particular difficulties. Also, the filter is seen by some as an undesirable expenditure at homes not paying for enhanced two-way services. Countering this last argument is the

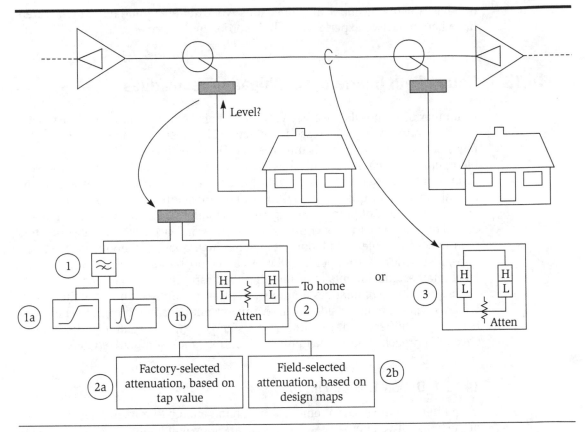

Figure 14.21 Options for interference mitigation.

thought that use of the filters postpones the day that homes not taking two-way services will have to have their internal distribution wiring improved.

Even when two-way services are added to the home, it may be possible to split signals at some point, with a new cable supplying only the device requiring return facilities. The filter can then be placed in the path to all services not requiring two-way facilities.

14.13.2 Option 2: Return Attenuation

A second interference mitigation option is to provide, in the tap or externally, two diplex filters, which separate the upstream frequency band (5–40 MHz) from the downstream band (54+ MHz). An attenuator is inserted into the low side of the diplex filter chain so that the loss is increased in the return direction while not changing the loss in the forward direction. The rationale is that tap values are selected to yield the correct signal level at the highest downstream frequency. Since cable loss is roughly proportional to the square root of the frequency, the loss in

the upstream direction is considerably lower than in the downstream direction. Thus, whereas the *downstream received* level is the same at all houses (within a certain range), the *upstream transmit* level is significantly different between homes.

Figures 14.6 and 14.7 illustrate the range of levels that could be expected in a particular node that was analyzed for return level. Generally, as you move downstream from an amplifier, the tap value is reduced to compensate for signal loss at the high end of the downstream band. Since return signals suffer much less loss, the required upstream transmit level is reduced as you move downstream from an amplifier. Since the taps farther from an amplifier require lower transmit level, they also admit more potential interference from the house.

The thought behind adding the attenuator in the return direction is that this will bring taps requiring a low signal level up to requiring a higher signal level. In the process, it will attenuate interference coming out of those homes. It would not deter return services from being installed, as might filters. However, the diplexers will add some group delay.

Option 2a: Factory Selection of Attenuation Based on Tap Value

Within the spectrum of options for the return band attenuators is the possibility that the value of return attenuation is selected at the factory, based on the tap value. This would not place additional burden on field installation personnel, who treat the tap as they would any other tap. The problem with this approach, however, is that the range of transmit levels required at any particular value tap (Figure 14.6) is so great that factory selection of an attenuator value would be very difficult.

Figure 14.22 illustrates an attempt in one real node to add reverse attenuation. In this example, all tap values 23 dB and below had reverse attenuation added such that the total attenuation in the reverse spectrum equaled 23 dB. Dots show the required levels before addition of the reverse attenuation, and + marks show the required levels after. Notice that after adding attenuation, some of the low-value taps now require the highest return levels. The reason is that some low-value taps are placed after a lot of cable attenuation, so they benefit from a lot of added attenuation. Other low-value taps follow a directional coupler so that the loss is primarily flat (frequency insensitive) loss, and the return signal level required after addition of the attenuation is high. This is not a good condition since it could require excessive output level from a transmitter.

Preliminary analysis has raised questions about the efficacy of factory selection of attenuator values. If the amount of attenuation added is low enough to prevent possible excessive level requirement, the benefit derived is minimal.

Option 2b: Field Selection of Attenuation Based on Design Maps

The other option for attenuators in taps is to provide a plug-in attenuator that is selected in the field, based on computed gain back to the previous amplifier. This technique shows considerable promise to improve the ingress problem while

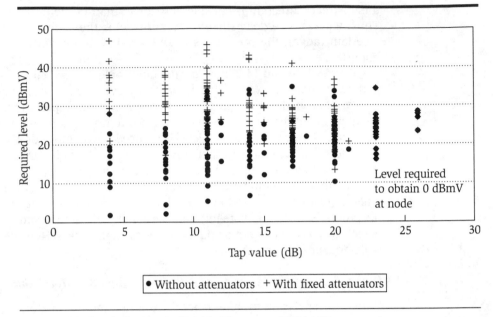

Figure 14.22 Required tap level with and without reverse attenuation.

permitting return transmitters to be designed with a very narrow return signal range. The problem is that additional burden is placed on field installation personnel, who first have to select the tap based on system design and then have to select the correct attenuator value to plug in. After the tap is installed, it is not possible to inspect the attenuator for the correct value. If the tap value is changed later, the new tap may not be equipped with the optimum attenuator value.

14.13.3 Option 3: Move Diplexers and Attenuator to Midspan

Others have suggested that the diplexer be moved from the tap to approximately midspan between amplifiers, and the attenuation used be selected to make the transmit levels required downstream of the diplexer to be roughly equal to those before the diplexer. This would reduce the number of diplexers required by somewhere around a factor of three or four, compared with putting them in taps. It does necessitate inserting another device into the hard-line cable. Co-locating the diplexers with an equalizer has been suggested, but this does not seem to be optimum.

Preliminary analysis indicates that such placement, combined with optimum attenuation selection in the field, would significantly reduce the signal level range required of transmitters. However, the estimated improvement in ingress is not particularly good, raising questions about the efficacy of the technique.

Notice that the industry has not firmed up its philosophy for applying these ingress countermeasures. Future work may reveal advantages that are not obvi-

ous at this time and may result in newer techniques that yield more benefit. Also, it is possible that the industry will learn how to improve performance of problem homes without having to use any of the techniques shown in this section.

14.3.4 Option 4: Frequency Hopping

Many services that use return plant service have provisions for frequency hopping. They can move from one frequency to another if their original frequency develops interference problems that make communications unreliable. This subject is covered in some detail in Chapter 5.

14.13.5 Option 5: Error Correction

Error correction is often added to digital datastreams to allow correction of any bits that are corrupted in transmission. The effect is to sharpen the curve of error rate versus carrier-to-noise (or interference) ratio. As noise increases, the recovered signal retains its quality longer, but at some point, the quality will collapse rapidly with worsening carrier-to-noise ratio. Chapter 3 describes error correction as it is being applied in the cable industry.

14.14 Examples of Signal Apportionment

We showed earlier that an optimum operating level for a return path laser exists. It is desirable to set the actual operating level as close to this as possible without going over the limit where clipping gets to be a significant issue. The operator must take into account temperature effects, errors in level setting, and laser loading due to ingress. He or she must also account for future services that may be added: it is not good to go back and readjust operating levels to accommodate a new service. After the operating point has been determined, it remains to apportion power among multiple services intelligently. This section gives two examples of how an optical transmitter could be operated under different conditions.

14.14.1 Few Return Services, F–P Laser

The first example illustrates the apportionment of signal power for a fairly simple node having only a few return path services and using a Fabry–Perot (F–P) laser. The laser is the one illustrated in Figure 14.12. The chosen operating point is −35 dBmV/Hz, where the laser exhibits about a 42-dB NPR at room temperature. In practice, we would operate lower on the curve to allow more room for error, but we take this point for illustration.

The services assumed include the following:

1. A status-monitoring carrier using BPSK modulation at a center frequency of 7 MHz, with an occupied bandwidth of 200 kHz (total, not plus and minus).

2. An RF-IPPV carrier at 13 MHz, with an occupied bandwidth of 300 kHz, BPSK modulation.

3. A telephone service using QPSK, centered at 23 MHz, with an occupied bandwidth of 1.8 MHz.

A significant omission from this list, done to keep the example simple, is an allocation for future expansion. It is essential that all future services be allocated the amount of power they will need when added. Failure to do so can result in extreme laser clipping later, and possibly a situation where a node that previously worked, no longer does. Also, we have not allowed for ingress. It is advisable to allow some power for ingress.

The level on the x axis of Figure 14.12 is measured at some convenient reference location. It may be the input of the return laser transmitter (Lvl 3 of Figure 14.5), which is likely a convenient place. It may be at any other convenient place in the network. The important thing is to make sure that the laser is characterized at the same point as that where operating levels are set, or that the gain between the two has been properly accounted for.

The technique used to apportion signal level among the services is to assign the individual signals levels such that the total signal level is equal to the total level at the operating point of the NPR curve, Figure 14.12 (x axis). That total signal level is given by the input in dBmV/Hz, plus 10 times the logarithm of the bandwidth over which the transmitter was excited when the NPR curve was plotted. In this case, the laser was excited with power spread over 35 MHz, so the total signal level into the laser was

$$\text{signal}_{total} = -35 \text{ dBmV} + 10 \log(35 \times 10^6) = 40.4 \text{ dBmV}$$

Note that, though signal level is expressed in dBmV as is customary in the cable television industry, we are actually manipulating signal power, expressed in dBmV in a 75-ohm system. For computation, we express this level in watts because it is necessary to add the drive power contributed by several signals. We cannot add power when that power is expressed in decibels. This conversion can be made as follows.

$$
\begin{aligned}
40.4 \text{ dBmV} &= 10^{\frac{40.4}{20}} \\
&= 105 \text{ mV} \\
&= 0.105 \text{ volts} \\
\text{power} &= \frac{E^2}{R} \\
&= \frac{0.105^2}{75} \\
&= 147.6 \times 10^{-6} \text{ watts}
\end{aligned}
\qquad (14.1)
$$

This much signal power is imposed on the laser during test and will be imposed by the three signals of the example. Since the total bandwidth occupied by the three signals is less than 35 MHz (by a lot), then it is possible (and, indeed, necessary) to increase the level per hertz of each. One philosophy popular in the industry today is to apportion power on a constant power-per-hertz basis, in which each service receives the same power per hertz. The total signal power is then given by adding 10 log(bandwidth) to the power per hertz, as was done here. This assures that each signal operates at the same carrier-to-noise ratio.

In the apportionment philosophy shown here, this is modified to give more power to services requiring a higher carrier-to-noise ratio. See Chapter 4 for a discussion of the carrier-to-noise ratio (C/N) required to meet a certain bit error rate (BER). The power ratios for different modulation formats are based on the distance between state boundaries in the constellation diagrams, with the average level of the signals held constant. In order to exhibit the same BER, a QPSK signal must have a 3-dB higher C/N than that needed by a BPSK signal. Thus, in the example, the QPSK signal is allowed 3 dB per hertz more level than is the BPSK signal.

It is possible to construct a table that will allow assignment of levels to each of the signals. Then we can express the power of each signal as follows. Let p represent the power per hertz in the BPSK signal. Table 14.1 illustrates how power is added. We multiply the power per hertz, p, by the occupied bandwidth of the service. The QPSK telephony service is assigned twice the power because it is twice ($10 \log 2 = 3$ dB) as sensitive to noise as are the signals of the BPSK services. With this weighting factor, the total power as shown at the bottom of the chart is $4,100,000\,p$. Equating this to the total power to be distributed among the services, 147.6 microwatts, we obtain

$$4,100,000\,p = 147.6 \times 10^{-6}$$

$$p = 36 \times 10^{-12} \text{ watts per hertz} \tag{14.2}$$

This may be converted back to dBmV:

$$p = \sqrt{PR}$$
$$= \sqrt{(36 \times 10^{-12})(75)}$$
$$= 51.96 \times 10^{-6} \text{ volts per hertz}$$
$$= 51.96 \times 10^{-3} \text{ mV per hertz}$$
$$= 20 \log(51.96 \times 10^{-3}) \text{ dBmV per hertz}$$
$$= -25.7 \text{ dBmV per hertz} \tag{14.3}$$

Recall that this is the signal level per hertz that we are allocating to the BPSK signals, based on distributing the total power we wish to supply among the several using systems. We can then determine the total power of each BPSK signal by

Table 14.1 Power assigned to each service, node with F–P laser.

Service	BW (Hz)	Total Power (W)
1. Status monitoring	200,000	200,000 p
2. RF IPPV	300,000	300,000 p
3. Telephony	1,800,000	2 × 1,800,000 p
Total power		4,100,000 p

adding 10 log(bandwidth) to the signal level in dBmV. For the status-monitoring signal (200 kHz wide), this gives a total power of $P_{sm} = -25.7 + 10 \log(200{,}000) = 27.3$ dBmV.

We can estimate the carrier-to-noise ratio as follows. From Figure 14.12, the NPR at the selected operating point is 42 dB. This means that the noise per hertz is 42 dB below the signal level (per hertz), selected to be –35 dBmV per hertz (from Figure 14.12). Thus, the noise level is $-35 - 42 = -77$ dBmV per hertz. To convert this to the total noise, we must correct for the actual bandwidth of the receiver's filter. For the status-monitoring signal, we shall assume that the noise bandwidth of the receive filter is equal to 1.1 times the occupied bandwidth of the signal. Thus, the actual noise level under the status-monitoring signal is given by $-77 + 10 \log(1.1 \times 200{,}000) = -23.6$ dBmV. But the carrier level was found earlier to be +27.3 dBmV, so the carrier-to-noise ratio is

$$\frac{C}{N} = 27.3 - (-23.6) = 50.9 \text{ dB} \tag{14.4}$$

Thus, we have so far established that the amplitude (when the entire signal is measured) of the status-monitoring signal is +27.3 dBmV, and the carrier-to-noise ratio is 50.9 dB (much more than is needed for a BPSK signal). It remains to determine what we would see on a spectrum analyzer if we looked at this signal. In order to determine this, we must make some assumptions concerning the way the analyzer works and how it is set. For the purpose of illustration, we assume that the analyzer is set to a resolution bandwidth of 100 kHz, narrower than any of the real signals being observed. We further assume that the noise bandwidth of the analyzer is 1.2 times the resolution bandwidth, or 120 kHz. A detector correction factor is not used though the analyzer manufacturer may have one that should be taken into account; each analyzer model can be different in this respect. From the preceding, we have the level of the status-monitoring signal as -25.69 dBmV per hertz, so if it is observed in a bandwidth of 120 kHz, the level we should see is given by

$$\text{level}_{analyzer} = -25.69 + 10 \log 120{,}000 = 25.1 \text{ dBmV} \tag{14.5}$$

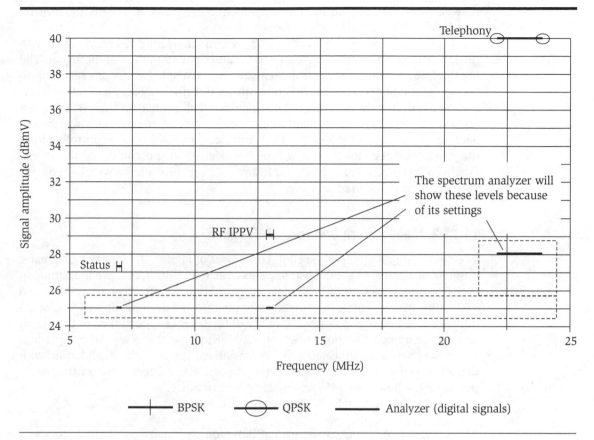

Figure 14.23 Signal level apportionment, F–P laser, limited services.

In a similar manner, we can determine the operating point of the RF-IPPV signal and the telephony QPSK signal. When we compute the amplitude of the telephony signal, we must recall that it is twice what we compute for the BPSK signals, as a result of adding 3 dB (that is, multiplying by 2) to the level of the BPSK signals. This was done in order to account for the lesser distance between decision thresholds of the QPSK signal. The results are shown in Figure 14.23. Notice the level indicators enclosed within dashed lines near the bottom. They represent what the spectrum analyzer described earlier would indicate. For the two BPSK signals, the analyzer indicates the same level because it measures signal level in less than the occupied bandwidth. In order to correct the analyzer reading for the actual signal level, we would have to add to the reading 10 log(ratio of occupied bandwidth to measuring bandwidth).

Further, note that the analyzer reading for the telephony signal is substantially below the actual signal level because the telephony signal is so much wider than the analyzer's measurement bandwidth. Also, note that the difference

between the analyzer readings for the QPSK and BPSK signals is 3 dB, the level by which they differ on a per hertz basis.

Figure 14.23 is not to be taken as a spectrum analyzer display: the lines with the labels next to them show the frequency and total amplitude of the signal. The lines below show what a spectrum analyzer would display *under one set of conditions*. Under other conditions, the spectrum analyzer will display a different level.

Also note that the example did not include an allowance for ingress, nor did it include an allowance for errors in level of the various return signals. As shown in Section 14.6.2, the signal levels at the return laser transmitter are controlled from the headend. Inevitably, there will be some error in the control of the levels. Finally, allowance was not made for future services; it would be wise to allow for such future expansion of offerings.

14.14.2 Many Return Services, DFB Laser

The next example involves the DFB return laser of Figure 14.13 carrying a greater number of signals. We'll choose to operate at −65 dBmV/Hz. We'll assume an NPR of 40 dB, which gives a dynamic range of about 15 dB. Power is apportioned first to the analog signals to allow them to operate at a desired carrier-to-noise ratio. Next, the remaining signal capability of the laser is apportioned to the digital signals using a constant power-per-hertz philosophy within a type of modulation. The power per hertz allocated to a signal is offset based on the modulation, with denser modulation formats receiving a proportionally larger share of the power per hertz to allow for their greater sensitivity to noise.

The following signals are included in this example:

1. A status-monitoring carrier using BPSK modulation at a center frequency of 7 MHz, with an occupied bandwidth of 200 kHz (total, not plus and minus).

2. An RF-IPPV carrier at 13 MHz, with an occupied bandwidth of 300 kHz.

3. A telephone service using QPSK, centered at 23 MHz, with an occupied bandwidth of 1.8 MHz.

These first three signals are identical to those illustrated in the previous example, Figure 14.23. Several additional signals are added:

4. An analog video return signal from the local high school, with picture carrier at 29.75 MHz. Its amplitude is set to yield a 47-dB carrier-to-noise ratio from the laser path (noise in other portions of the path will reduce the carrier-to-noise ratio).

5. An analog video return signal from city hall, with a picture carrier at 35.75 MHz. Its amplitude is set to yield a carrier-to-noise ratio from the laser contribution of 49 dB.

The digital signals "get" their allocation of power from the power left after the analog signals receive enough to yield the desired carrier-to-noise ratio:

Figure 14.24 Signal apportionment, DFB laser, many services.

6. A DOCSIS-compliant QPSK return datastream centered at 20 MHz, with an occupied bandwidth of 800 kHz.

7. An "old" (meaning pre-DOCSIS) QPSK return datastream centered at 18 MHz, with an occupied bandwidth of 1 MHz.

8. A DOCSIS-compliant 16-QAM return datastream centered at 26 MHz, with an occupied bandwidth of 1.6 MHz.

Figure 14.24 summarizes the signal level of all signals using the procedure outlined earlier. The level offset from QPSK to 16-QAM, taken to account for the relative sensitivity of the two to noise, is 4.1 dB, taken from Figure 4.13.

Though not shown on Figures 14.23 and 14.24, the carrier-to-noise ratio of the signals retained from one example to the other drops even though the signals are transported using a better laser. The reason is other services that are also taking a portion of the power available with the better laser. The BPSK signals in the

node with the F–P laser had a C/N of 51.3 dB, whereas they had a C/N of 45.7 dB in the node with the DFB laser. Of course, both numbers are much, much greater than that required for quality service. It would have been possible in both nodes to add more carriers. Alternatively, in the DFB node, the analog carriers could have been set higher to yield a better carrier-to-noise ratio.

In both of these examples, notice that the carrier-to-noise ratio was computed based on the assumed operating parameters. Since those parameters will not necessarily be followed in practice, it is prudent to allow several decibels for operating at other points. Commonly, you will simply assume a given NPR, such as 40 dB, even though there is a range over which better performance may be obtained. Then look at the range over which this minimum level of performance may be obtained (see Figure 14.10).

14.15 Summary

The chapter has dealt with a number of items of interest in the operation of return HFC plant. It began with a description of the components unique to a two-way plant. We showed key characteristics of those components. Level setting in the return plant received a lot of attention because it is crucial to operation of reliable return plant, just as it is to proper operation of downstream plant. We examined lasers used for return services and gave practical methods of evaluating the lasers. We discussed important characteristics of return signals and their interaction with the laser transmitters.

Noise in the return plant is important and received considerable attention, along with proposed mitigation methods. Finally, we gave examples of ways to set up the levels of various return services and showed how to estimate key performance parameters.

The next chapter deals with system architecture. Methods of plant design are described that optimize performance using the principles shown in this chapter.

Endnotes

1. Data courtesy of H. Carnes, R. Oberloh, and team. Antec, Norcross, GA.

2. National Cable Television Association, Upstream Transport Issues, *NCTA Recommended Practices for Measurements on Cable Television Systems*. The "Upstream" portion was added to the *Recommended Practices* in late 1997. The document is available from the NCTA, Washington, D.C.

3. John Kenny, Characterization of Return Path Optical Transmitters for Enhanced Digitally Modulated Carrier Transmission Performance, *Technical Papers of the NCTA Annual Convention*, NCTA, Washington, D.C., 1997.

4. Lamar West, Analysis of Reverse Path Laser Loading in HFC Networks. *HFC'97 Conference,* Phoenix, AZ, September 1997. SCTE, Exton, PA.

5. H. Blauvelt et al., Return Path Lasers for High Capacity Hybrid Fiber Coax Networks, *1995 NCTA Technical Papers.* NCTA, Washington, D.C.

6. Data courtesy of D. Junghans of Arris Interactive, and T. Mitchell, formerly of Arris Interactive. Private communication.

7. Supplement on Upstream Transport Issues, *NCTA Recommended Practices for Measurements on Cable Television Systems,* 2nd ed. NCTA, Washington, D.C., October 1997, pp. 57–63.

8. D. Large and R. Bullinger, A Proposed Method for Quantifying Upstream Ingressing Carriers, *1997 NCTA Technical Papers.* NCTA, Washington, D.C.

9. Richard S. Prodan et. al., Results of Return Plant Testing, *1997 NCTA Technical Papers.* NCTA, Washington, D.C.

10. This section is adapted from the *NCTA Recommended Practices for Measurements on Cable Television Systems.*

11. Petr Beckmann, *Probability in Communication Engineering.* New York: Harcourt, Brace & World, 1967, p. 107.

12. National Cable Television Association, op. cit.

13. West, op. cit.

14. John Kenny, Characterizing Return Path Transmitters. *CED,* May 1997, p. 26ff.

Part Five
System Architecture

Chapter 15
Service-Related Architecture Requirements

15.1 Introduction

The architecture of a cable system is a description of how the major elements—signal acquisition, signal processing, signal transport, and subscriber terminal equipment—are interconnected. The first step to selecting or analyzing a specific network architecture is to analyze the services to be delivered and their requirements. It is that analysis that will be covered by this chapter.

First, we discuss the categories of service-related parameters that are affected by architecture, then the requirements of various specific service types. The emphasis will be on the process to setting standards and on alternative means of meeting those standards rather than on suggesting specific values.

Chapter 16 will deal with specific architectures and how they differ relative to the parameters discussed here, and Chapter 17 will cover the calculation of network reliability and availability, given a specific equipment configuration.

15.2 Performance Parameters

Network performance can be measured by many "yardsticks." Signal quality in the broadband distribution system, including noise, distortion, levels, frequency response, and other related parameters, is extensively discussed in Chapters 9–14. Other architecture-affected topics are discussed in this chapter.

15.2.1 Information Capacity

The information capacity of a network is a composite of many factors, including the RF bandwidth, the efficiencies of the processes by which raw information is encoded for modulation onto RF carriers, spatial reuse of the network, and efficient sharing among users with varying needs for information transmittal rate.

Downstream RF Bandwidth

One of the fundamental measures of the broadband, linear distribution network is the range of frequencies used to transport signals toward customers. In cable television systems, the lower limit of downstream frequency range in coaxial distribution networks is constrained by the fact that upstream signals are almost always carried below downstream signals in shared cables.* The bands must, furthermore, be separated sufficiently that reasonable-cost diplex filters can be deployed in amplifier stations to properly route the different spectra for processing without causing amplifier instability or excessive group delay at the band edges. Additionally, systems that deliver analog television programming to resi-

* One experimental network has been built in which the upstream signals are carried in the 900–1,000-MHz band, thereby constraining the upper, rather than the lower, end of the downstream spectrum.

dential customers must include the lowest television channel within the downstream spectrum, whereas those not carrying such programming are sometimes designed with different crossover points between the bands.

In both North and South America, the lowest television channel starts at about 53 or 54 MHz, depending on channelization plan, with the result that the lower downstream frequency limit is typically near 50 MHz (see Chapter 9 for a discussion of amplifier station internal RF circuitry and Chapter 8 for a discussion of typical channelization plans). These are called *low-split* networks. Common lower downstream band limits for nonresidential networks are 150 and 234 MHz, commonly called *midsplit* and *high-split* networks, respectively. For networks requiring more RF bidirectional bandwidth, separate cables and amplifying stations may be used for each direction of transmission with dual coaxial cables, thereby freeing constraints on bandwidth in both directions. Occasionally, dual-cable low-split networks are used for residential cable television networks as a means of increasing both downstream and upstream bandwidths while retaining conventional channel plans.*

At the top end of the spectrum, the bandwidth of coaxial amplifiers is primarily limited by amplifier technology. Chapter 10 contains a discussion of how total RF bandwidth can be balanced against other broadband performance parameters, such as operating levels, noise, and distortion. The maximum downstream frequency has steadily increased as technology has improved with 550-, 750-, and 860-MHz amplifiers commercially available. In some areas, 1,000-MHz systems have been built.

Most installed fiber-optic links use separate fibers for upstream and downstream communications and thus there are no restrictions on RF bandwidth occasioned by the need to share spectrum space. As with coaxial technology, however, there is a trade-off in linear fiber-optic links among bandwidth, signal loading, noise, and distortion.

Upstream RF Bandwidth

Historically, the majority of all-coaxial cable television systems were not enabled for two-way communication with subscribers for the very practical reason that large, all-coaxial plants contained so many amplifiers and branches that the combination of noise funneling and ingress made upstream communications problematical. Among those with two-way capabilities, some were used for point-to-point applications in which the system had few, if any, upstream signal combining points. Others were used to carry low data rate signals, such as data from two-way set top terminals or status monitoring, where only a single box transmitted at a time, upstream levels could be high, and rugged frequency shift keying (FSK) modulation was used. In the few systems that offered true sub-

* A number of dual cable systems were built in the 1980s as a means of getting more channels than allowed by the technology current at the time. The cost of maintaining two networks, however, has led operators to convert most to single cables as they are upgraded.

scriber interactivity (such as the Columbus, Ohio, system), electronic switches were employed at field branching points to dynamically reduce the effective upstream combining. Two-way operation is practical in modern HFC systems due primarily to the smaller, electrically separate, broadband distribution subnetworks. Chapter 14 discusses the unique problems with upstream communications.

The low end of the upstream spectrum is limited by the need to separate signals from the harmonics of the powering that is multiplexed with the RF signals. It is also effectively limited by the practical frequency range of the magnetic components used in amplifiers and passive elements, such as power inserters, taps, splitters, and directional couplers. The most common lower frequency limit is 5 MHz.

The upper end of the return spectrum is limited by the need to provide an adequate crossover band between upstream and downstream signals, and also by the need, in residential networks, to avoid feeding high-level signals into the input terminals of television receivers in the common intermediate frequency (IF) band, which is 41 to 47 MHz. This is discussed in more detail in Chapter 20.

Typical upstream band limits in older amplifiers were 5–30 MHz, whereas modern designs are typically 5–40 MHz or even a little higher. For nonsubscriber midsplit and high-split networks, the upper upstream frequency limits are typically 112 and 174 MHz, respectively. These three common band plans are shown in Figure 15.1.

Bandwidth per Customer

The distribution system RF bandwidth is a good measure of information capacity when a common set of television signals are to be delivered to all customers, which was exclusively the case in the early history of cable television and true for much of the programming on most cable systems today. Increasingly, however, new services need be delivered only to small groups or individual customers. To the extent that the network can be logically or physically divided into independent transmission systems, we can use the same frequencies to simultaneously deliver differentiated services to subscribers in different network sections. This is known generically as *space division multiplexing* (SDM).

A measure of the network's ability to deliver customized services is the bandwidth per customer (or, alternatively, bandwidth per home passed). In the downstream direction, this is simply equal to the RF bandwidth devoted to some service divided by the number of homes in the smallest portion of the network that can be fed unique signals multiplied by the penetration of that service among homes passed. In the upstream direction, it is the bandwidth devoted to that service divided by the number of customers sharing that bandwidth at any point along the path. The upstream and downstream bandwidth per customer may be highly asymmetrical.

The independence of upstream and downstream bandwidth in optical links has led to designs in which the upstream signals from each of several coaxial distribution lines emanating from a fiber node are frequency translated to nonover-

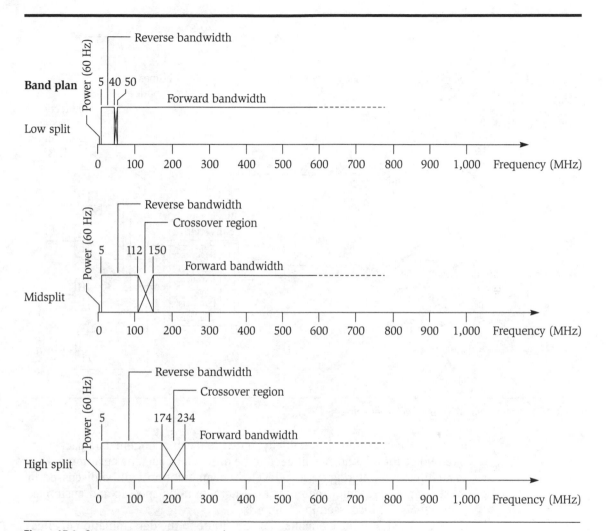

Figure 15.1 Common two-way spectrum plans.

lapping bands and combined before feeding the upstream optical transmitter. This is known as *block conversion* and is discussed in Chapter 14 and illustrated in Figure 14.3. The major benefit is increased average upstream RF bandwidth per customer since signals can use the same upstream frequencies on different legs simultaneously without mutual interference.

It is common, at least in residential system planning, to divide the available downstream bandwidth among common signals, those directed to large groups of customers, and those directed to individual customers. For example, it is efficient to distribute the most popular television channels systemwide, whereas local educational or governmental access channels may vary across a large system that

Figure 15.2 Factors controlling bandwidth per home.

covers several communities, and finally, individual cable modem data packets or digitized voice for telephony will need to be directed to specific customers. Network architectures offering various levels of segmentation will be discussed in the next chapter. Headend requirements to create node-specific and customer-specific signals are discussed in Chapter 8.

Figure 15.2 shows the elements that contribute to the determination of effective bandwidth per home passed for any given service. The effective downstream bandwidth per customer for any given service in that configuration is

$$B_c(fwd) = \frac{B_s(fwd)}{P_s RH} \tag{15.1}$$

where

$B_c(fwd)$ = the downstream bandwidth per subscriber to that service

$B_s(fwd)$ = the total downstream bandwidth assigned to that service

P_s = the penetration of that service among homes passed

R = the number of nodes served from one downstream optical transmitter

H = the number of homes passed by the coaxial distribution lines extending from each node

In the reverse direction, the effective bandwidth per customer is given by

$$B_c(rev) = \frac{nB_s(rev)}{mP_sH} \tag{15.2}$$

where

$B_c(rev)$ = the upstream bandwidth per subscriber to that service

$B_s(rev)$ = the total upstream bandwidth assigned to that service

P_s = the penetration of that service among homes passed

H = the number of homes passed by the coaxial distribution lines extending from each node

m = the number of nodes whose signals are combined into each data receiver input

n = the number of independent (and equally sized) coaxial distribution lines emanating from each node whose signals are combined using block segment converters at nodes

Note that the upstream equation is valid only if the data receiver contains an independently functioning input module for each upstream frequency block simultaneously in use.

For example, a network consisting of multiple fiber nodes, each fed by an independent optical transmitter and each providing signals to a coaxial network serving 400 customers, can provide different programming to those customers than to those fed from other similar nodes. If the downstream bandwidth allocated to wired telephone service is 6 MHz and the service penetration is 30% of homes passed, then the bandwidth per customer would be 50 kHz. If the upstream bandwidth were 2 MHz, block segment conversion used to isolate the signals from each of four coaxial legs, and no node combining used at the headend, then the return bandwidth per customer would be 66.7 kHz. Whether this is adequate depends on whether that bandwidth is all usable, how efficiently it is used, and simultaneous usage rates among telephone subscribers. It can be increased, however, only by choosing a smaller node size, further subdividing the node, or increasing the bandwidth assigned to telephony service. Such considerations have been a major factor in choosing node sizes where extensive subscriber-specific services have been contemplated.

Bandwidth Efficiency

Having determined the available bandwidth, the next important parameter is how much information can be carried in the available spectrum. This comprises two parts: information coding and modulation efficiency.

Nonscrambled analog television and FM radio signals are universally modulated in the same format as over-the-air broadcast signals so that direct connection to subscribers' television receivers and/or connection to those receivers after processing through simple frequency conversion set top terminals is possible. In the United States and Canada, in fact, most cable systems are required to carry the programming making up their lowest level of service in this format. The signal formats and modulation are discussed in Chapters 2 and 8 of this book.

Where it is desired to scramble analog programming to control access, the signals are modified in such a way as to render decoding difficult, as discussed in Chapter 18. Depending on the scrambling method, such signals may decrease the effective power loading on the cable system, relative to nonscrambled channels (for instance, by removing the high-power synchronizing pulses), but will still occupy the same bandwidth—6 MHz for NTSC and PAL-M formats, and 7 or 8 MHz for PAL-B/G formats.

When video signals are converted to digital form, the resultant data rate is a measure of the information required to adequately describe the pictures and sound for the specific application. For entertainment-quality television, the Moving Picture Experts Group (MPEG) has defined a number of levels of data compression, which are described in Chapter 3. These cover resolution levels comparable to analog television as well as higher resolutions and different screen aspect ratios that will be used for high-definition television (HDTV). Since the information needed to describe a video program depends on the rate of change in the picture from frame to frame, the data rate can vary widely. On average, however, it appears that approximately 3 Mb/s is required for NTSC-quality pictures, whereas 19 Mb/s will be required for the format chosen for U.S. HDTV broadcasting. The difference in data rate between the digitized-but-not-compressed video stream and the output of the compressor is the *coding gain,* which can easily be 30 or more. For nonentertainment applications, such as videoconferencing and surveillance cameras, standards are available that use much lower data rates and higher coding gains.

Nonvideo data signals include packetized data for cable modems, system-monitoring signals, digitized telephony of several varieties, utility control and monitoring applications, digitized audio services, interactive games and video-related services, and so on. Some of these require permanent dedicated virtual channels, whereas others are "bursty," with peak rates far exceeding average rates. Average data rate requirements per customer for these may vary from tens of bits per second to hundreds of thousands of bits per second.

Once the datastream is defined, it must be prepared for transmission. Baseband optical digital modulation is the process whereby a single datastream (which may include the time-multiplexed content of more than one information stream) is used to turn the output of an optical transmitter on and off. This technique is used in such formats as SONET and FDDI. Though such formats have the advantage that very high data rates can be supported (of the order of 10 Gb/s) and long transmission lines are possible, all signals must be both digital and able to time-share a common datastream.

Where digital signals are to be transmitted over a shared wideband analog network, they must first be modulated onto RF carriers. Although simple on/off modulation of the RF carrier could be used, that would result in a usable bandwidth efficiency of only about 0.6 b/s/Hz. Since cable networks must provide a very high quality signal path for analog television, more bandwidth-efficient modulation schemes are possible.

Over-the-air digital television broadcasting will use an 8-VSB format, whereas cable will likely use 16-VSB and/or 64- or 256-QAM modulation for digitized high-quality video. (Chapter 3 covers digital video modulation.) Because each transmitted symbol represents more than one bit of information, high-order modulation schemes have greater bandwidth efficiency. For instance, 64-QAM has a theoretical bandwidth efficiency of 6 b/s/Hz and a practical efficiency of about 5 b/s/Hz. Note that, after modulation onto an RF carrier, we no longer have a strictly digital signal, but rather a digitally modulated analog signal that will share the analog transmission network with other analog signals.

Similarly, nonvideo digital signals can be modulated on RF carriers, as discussed in Chapters 4 and 5. The more "aggressive" modulation methods offer greater bandwidth efficiency, but are less tolerant of channel impairments. Though there is no universal agreement on a common method, the DOCSIS standard provides for up to 256-QAM for downstream transport and up to 16-QAM for upstream modulation, depending on the service tolerance for errors and channel conditions. Some upstream technologies are able to adjust their modulation order to suit channel conditions. Several proposed advanced standards, such as code division multiple access (CDMA) and orthogonal frequency division multiplexing (OFDM) would allow a more efficient use of impaired upstream bandwidth.

In summary, the instantaneous information capacity of any part of the broadband network is equal to the RF bandwidth multiplied by the bandwidth efficiency of the signals occupying it, whereas the information capacity per customer is equal to the information capacity of each separate network segment divided by the number of subscribers served from that segment. In the special case of digital video, digital music, and certain voice signals, the effective instantaneous channel capacity is further related to the coding efficiency used to convert the analog signal to a bitstream. Note that system operators may not realize all the theoretically available information capacity if, for instance, they use part of the bandwidth for systemwide or regional common signals.

Time Sharing

The third leg of network information capacity is how efficiently the available information capacity is shared among users and applications. As mentioned previously, traditional cable systems delivered all video channels to all homes in an analog format. In a system with many channels and small nodes, however, it is likely that some channels are not being watched in some nodes some of the time. If less popular channels are delivered to nodes only when desired by a subscriber, it is possible to offer more program choice without increasing the total available

information capacity. A network able to do this incorporates a form of time sharing known as *switched virtual circuit,* whereby a user is granted full-time use of an RF channel for the duration of his or her use, after which it is freed up for other users.

This is the same principle used in the design of central office telephone switches, where use of a "path" through a switch is granted exclusively to someone making a call, but then freed up for the next caller. Depending on the statistical usage pattern, switched-channel time sharing can be a very inefficient use of available information capacity. Most telephone users, for instance, spend roughly the same amount of time listening as talking, and even when talking, do not talk in a continuous stream but are silent for a surprising percentage of the time (though we all know exceptions!). Thus, they use only a small fraction of the system resources dedicated to their use during the call.

Sometimes the time sharing is on a much shorter scale. For instance, a T1 telephony circuit at 1.544 Mb/s is made up of 24 digitized voice streams, each of which requires 64 kb/s (plus a total of 8 kb/s of "overhead"). Thus, 24 telephone calls are carried over one datastream, with the time slots allotted to each call on a rotating basis. This is known as *time division multiplexing* (TDM), and hierarchies are available up to multiple Gb/s rates. TDM methods that grant a fixed percentage of the total data rate to each user are known as *constant bit rate* (CBR). In the T1 example, each voice channel is a 64-kb/s CBR virtual circuit.

Where user needs are variable, more efficient sharing of a data channel results when the capacity is divided among datastreams as needed, a method known as *statistical multiplexing.* For instance, where many telephone calls share a common circuit, it is possible to greatly increase the total call capacity by simply not transmitting any bits when a caller is silent and using that capacity for another conversation. Another example of statistical sharing is the Ethernet local area network, where computers send data packets to other computers on the network only as needed. A final, and very important, example is digital video, where the bit rate required for each channel varies greatly depending on the degree of change from frame to frame of a video source. It is possible to squeeze more video programs into a given datastream if we take advantage of the statistics of combining sources with varying data rates. In the case of MPEG-2 compressed, standard resolution video, it may be possible to combine as many as 12–15 programs into a single 30-Mb/s datastream, resulting in a significantly increased program capacity compared with analog video. In general, data channels whose capacity is varied in response to users' requirements are known as *variable bit rate* (VBR) services.

Where many nonsynchronized users must share a channel, a time-sharing scheme known as *time division multiple access* (TDMA) may be used. The upstream communications from cable modems is an example of such a situation.

A key advantage of statistically sharing a digital datastream is that each application can have access to peak data rates that are much greater than needed on average. This is particularly important in the case of highly bursty data services such as cable modems used for Internet access.

Summary of Information Capacity

In summary, the concept of effective information capacity per customer needs to be expanded to include statistical sharing of available capacity on two levels: simultaneous service usage rates and the instantaneous circuit data rate required to support simultaneous users. Predictions of *simultaneous service usage rates* are done to predict how many subscribers to a given service will use it simultaneously. This is generally known as traffic engineering and is well known and documented with regard to voice telephony (see Chapter 5). Usage of switched video channels and packetized data channels is less well studied. *Instantaneous circuit data rate* is a measure of the average and peak data rates needed to support simultaneously active users of VBR services.

Thus, the total information capacity of a network can be seen to be a complex that includes the following:

- The total RF bandwidth in each direction of the broadband distribution system
- The logical segmentation of the distribution network into subsections that can support independent signals (space division multiplexing)
- The division of the available bandwidth in each subsection among common and switched services (frequency division multiplexing)
- The coding efficiency with which signals are prepared for transmission and the bandwidth efficiency with which they are modulated onto RF carriers
- The statistical usage of services among subscribers
- The time-shared use of available datastreams among active users (time division multiplexing)

Figure 15.3 shows the elements that determine effective information capacity. Not shown in this simplified diagram are such elements as forward error correction and network layer overhead, both of which detract from net capacity.

15.2.2 Network Reliability and Availability

Service availability, in the most general sense, is a measure of the likelihood that a given service will be available for use by customers. Some factors that affect that availability are discussed here.

Equipment Reliability

The first requirement to build a reliable network, to state the obvious, is that its components must be reliable. Though the discussion of internal equipment reliability is beyond the scope of this book, in general the overall reliability (usually expressed in terms of mean time between failure, MTBF, in hours) will be affected by such factors as operating temperatures, the complexity of the unit, the environmental protection, the mechanical strength of interconnections,

Figure 15.3 Elements that determine effective network information capacity.

protection from powering transients, and the long-term aging characteristics of materials and components. Typically, manufacturers provide MTBF data for their network components based on theoretical analyses and sometimes also based on test data taken under controlled environmental conditions.

Experience has shown that classical failure analyses, based on the internal electrical parts and how they are interconnected, typically yield reliability predictions that have a limited relationship to the performance of field-installed equipment. On the other hand, accurate field failure data is difficult to gather, given that field crews are often more focused on restoring service than performing postevent failure analysis.

Network Failure Rate

The failure rate of the network is determined by the number of components, their individual failure rates, and how they are interconnected, as well as certain outside factors. Chapter 17 is devoted to the calculation of network failure rates and availability. In general, however, networks that have more components that can

independently cause a signal transmission failure will be less reliable. Thus, a 40-amplifier cascade will be less reliable than a 4-amplifier cascade. Not only electronic components, but the power to those components contribute to failures and must be included. Service-related reliability requirements are one factor leading to the choice of serving area size.

Network reliability can be improved by architectures in which the failure of some network elements does not lead to a communications outage. A simple example is the provision of two diversely routed fiber cables between the headend and a node, with separate optical transmitters and receivers on each link and the node configured to automatically select the best signal. The probability of failing to get signals from the headend to that node then becomes the product of the probabilities of failure of each of the redundant links taken independently. For example, if each of the links has a probability of failure in some period of 1%, then the probability of failure of the redundant network is only 0.0001 in the same period. The gain from redundancy, however, is reduced by the failure rate of the added equipment that must decide which of two redundant paths to use and switch between them. Redundant structures are part of many high-reliability architectures.

In most existing cable television networks, service outages are more likely to be due to commercial electric power interruption than actual equipment failures. Thus, an important factor in distribution system design is the degree and type of backup powering provided and the placement of field power supplies at points in the utility's power grid that are less likely to fail.

Network Availability—Failure Related

Given a network failure rate, whether due to component failure or powering failure, the next important parameter is the percentage of time that the network (or a specific service) is available for use. Although this is treated in detail in Chapter 17, the general equation for availability is

$$\text{availability}(A) = \frac{1}{1 + \dfrac{MTTR}{MTBF}} \tag{15.3}$$

where

> $MTBF$ = the mean time between failure
>
> $MTTR$ = the mean time to restore service when failure occurs

MTBF is governed by the reliability of network elements and how they are interconnected as discussed earlier; however, MTTR is both an operational practices and network design issue. Such issues as field response time and field crew training are operational issues. System-monitoring capabilities, however, are part of the network design and control how soon an outage is known to repair crews.

Finally, more subtle design issues such as the number of fibers in shared cables may affect the time required to repair a cut.

Network Availability—Signal Quality Related

Even when the physical network is operating properly, individual services may experience outages due to temporarily degraded transmission conditions. For example, an ingressing signal may interfere with an upstream data transmission (or, more rarely, a downstream transmission). If the terminal equipment operates on a fixed frequency, the service will be lost until the signal stops transmitting or the point of ingress is located and repaired. If, on the other hand, the headend equipment is equipped to sense such interference and the upstream transmitters are frequency agile, the service can be restored by merely shifting the frequency of operation. Although such considerations are not affected by distribution system physical architecture (except, of course, that smaller node serving areas are less likely to be affected by the interference in the first place), they are affected by the selection of terminal equipment and the allocation of available operating spectrum—in other words, the architecture of the specific service.

Several means of quantifying data circuit availability include signal-quality issues. Bit error rate (BER) is a measure of the percentage of individual bits of information that are lost, whereas errored seconds, severely errored seconds, and errored minutes all measure lost information on related, but different, scales. These are discussed in more detail in Chapters 4 and 5.

Service Availability—Traffic Loading Related

Where transmission paths are statistically shared among users, there is always the possibility that demand will sometimes exceed available capacity. Thus, even if the network is working properly and the transmission channels are unimpaired, the service may sometimes be unavailable to a specific user. This happens in the U.S. telephone network every Mother's Day, for instance, when, at some hours of the day, more people attempt to call their families than there are available long-distance circuits.

As with signal-quality issues, capacity-limited blockages are not affected by the reliability or architecture of the transmission network, but rather by the architecture of the individual service. In essence, they are designed in as an economic choice since providing sufficient capacity for the largest possible traffic load would not be economically justifiable if that load level occurs only rarely.

The design of an appropriate load capability for a given service is known as *traffic engineering*. This is a mature art in the case of voice telephony networks and a less well-developed art in the case of such services as cable modems and video on demand.

In summary, overall service unavailability is approximately the sum of unavailabilities due to network failures, signal degradation, and capacity-related blocking (the approximation is due to the possibility of simultaneously occurring unavailabilities in two categories).

Single Points of Failure

A final topic related to network reliability and its consequences is the management of single points of failure. Failures of components at various points in the network will affect different portions of the service and differing numbers of customers. For instance, the failure of a terminating tap will, at most, affect only the customers directly connected to it, whereas failure of the optical receiver at the node may affect hundreds of customers. Failure of a single headend channel processor may affect only 1 channel throughout the network, whereas failure of the satellite antenna low-noise amplifier may take out 12 channels or more.

One measure of the quality of an architecture is how immune the system is to single points of failure that affect many customers and how quickly such failures can be detected and repaired. Note that this is not necessarily the same as providing high reliability and availability. For instance, a single headend outage that takes out all services for several minutes will not affect overall yearly availability greatly, but will likely overload the customer service telephone lines with complaints, something that will not happen with a random distribution of small outages. For this reason, if no other, systems are designed to eliminate as many large single points of failure as possible. That is one reason for providing redundant structures closer to the headend, where many customers are affected, but not deeper in the network.

15.2.3 Signal-Processing Equipment Distribution

In a simple network designed to distribute a common spectrum of signals to all terminals, the signal-processing equipment can all be concentrated at the headend. This has the advantage of easy service, and it becomes simpler to design for reliability and secure powering. Unfortunately, with HFC systems and switched services, use of centralized equipment may not be possible and, even if technically feasible, may not result in the best performance or reliability. The distribution of signal-processing equipment is an important part of architectural design. Some of the options are discussed below.

Common Versus Switched Information Capacity

Given a need to deliver a defined range of services to a group of customers, operators have the option of providing sufficient networkwide information capacity so that all signals can be delivered throughout the network, or alternatively providing switching and additional modulators and signal processing so that a reduced instantaneous information capacity combined with a customized signal load for each network segment accomplishes the same end. In multilevel architectures, this switching can take place at the headend, at intermediate hubs, or at some combination of levels.

Where the number of offered video programs exceeds the instantaneous analog channel capacity of the network, systemwide video information capacity can

be increased through the use of digital video compression. Thus, the cost of video compression and in-home digital reception and processing must be balanced against the cost of switching and additional modulators (and sometimes additional fiber-optic transmitters) required to make more efficient use of a bandwidth-limited network.

Single Point Versus Distributed Signal Processing

At some point, signals in an HFC network are modulated onto RF carriers and frequency multiplexed for delivery to homes. That point may not be entirely at the headend, however. In some architectures, only commonly distributed signals are generated at the headend, whereas hub, node, and/or customer-specific signals are generated and received at lower levels in the architecture.

When the conversion from baseband digital to analog FDM signals occurs at lower levels in the architecture, the digital signals can be transported using baseband binary digital signaling to the point where they are modulated onto RF carriers. In the case of fiber-optic transport, this allows longer spans that are free of the types of mutual interference that limit the performance of linear distribution techniques.

Another example that illustrates how the choice of processing location affects architecture and effective information capacity is the general area of Internet access services. Operators may choose to develop shared downstream packetized datastreams containing data for all the customers served from a hub or node, then modulate that datastream onto an RF carrier for transmission through the broadband network to customers' cable modems, which extract packets destined for their own computers. Upstream packets in such systems are typically generated at each modem, with the upstream transmissions controlled by the headend equipment to avoid mutual interference. With this delivery method, the data for more than 100 customers of the service can share a single 6-MHz downstream channel and even narrower upstream channel, but at the cost of a separate cable modem for each customer.

At the other extreme, a group of statistically shared computers may reside at the headend. When a customer is using one of these to access the Internet, the computer's display is converted to analog video format, scrambled, and sent to only that customer. A simplified keyboard and navigation device in the home sends commands upstream to control the computer.[1] The advantage of this method is that the in-home equipment is very inexpensive (just the customer's normal television set with a standard set top descrambler; a keyboard; and a simple, low data rate, upstream transmitter) and the service offerings can be easily upgraded and changed at the headend. Channels may even be shared among this service and other on-demand video services. Obviously, such a service is limited, compared with the increased display resolution and flexibility offered by a home computer, but the service can be offered to all homes, whether or not they have computers. A significant difference is that more bandwidth is required to send video than to send the unprocessed data from the Internet. The bandwidth

required for such a service depends on node size, service penetration, and the rate of simultaneous usage among users.

In between these two extremes is a third option whereby the shared computers are located at the headend, but transmission to homes is through the unused capacity in the vertical interval of analog channels. In that case, digital signals are sent to specially equipped in-home set top terminals that convert the data either for analog television display or for connection to a home PC. As can be seen, the result is somewhat more signal processing in the home but greater system information capacity.[2]

All three of these Internet access service offerings have been offered commercially and are presented here to illustrate how the distribution of signal processing can affect net system information capacity. Figure 15.4 illustrates the downstream service architecture differences.

Point-of-Entry Versus User-Terminal Signal Processing

At the customer end of the link, it is possible to distribute broadband signals directly to terminals at the point of use (and to generate any required upstream signals from those same points). Doing so, however, has several distinct disadvantages. These are most apparent in two-way systems:

1. There may be unnecessary duplication of functions among receivers and transmitters within a single building.
2. Electrical transients or over-the-air signals can leak into the system anywhere within the building and pollute the upstream spectrum of the entire node.
3. Although equipment from multiple manufacturers and transmission standards can be accommodated, these must all compatibly use the available spectrum.
4. Network powering of terminal equipment, though possible, is much more difficult since many buildings are wired by other than network operators, and many drop passive components are not designed to handle multiplexed power.
5. Owing to the many possible in-building distribution system topologies, the available downstream signal strength and required upstream transmitter levels will vary widely.

The alternative is to terminate and generate some or all network signals at a single *point-of-entry* box located where the drop cable enters the house. This is sometimes also known as a network interface device (NID) or network interface unit (NIU). Typically, but not always, analog video signals are passed through the NID to customer terminals, whereas digital signals of all types are terminated at the NID and reprocessed for distribution through the house. Similarly, all upstream signals share a single upstream transmitter at the NID.

The advantages to use of a NID include (1) isolation of the network from the effects of in-building wiring, both losses and signal ingress; (2) simplified network powering, particularly for wired telephony services; and (3) cost savings in multiservice installations through sharing of common equipment. Balanced

Figure 15.4 Effect of placement of signal-processing equipment in Internet access application.

against these advantages are the generally higher costs of the first installed service due to the flexibility required in the NID to support incremental features or services. Also, the use of a common NID implies, in most cases, that all two-way services must be interoperable. Often in the cable industry, this means a common and sole source for equipment.

In summary, network architecture includes not only the physical network topology, but the distribution of signal-processing equipment for various services. Network information capacity can vary by service and includes the RF bandwidth, how efficiently that bandwidth is used and shared, and the effects of equipment distribution on the information required to support the service.

On an operational level, architecture has an effect on the reliability and availability of services, on the size and frequency of outages, and on how difficult it is to recover from those outages.

15.3 Requirements by Service Type

A key factor in designing a network architecture is a clear understanding of the services that will be offered since each has unique requirements. The signal-quality requirements for various analog and digital services are covered in detail in Chapters 2–5 and will only be summarized here. Additionally, specific services require different levels of reliability and availability, switching, per-customer data capacity, and links between network customers and networks that are external to the cable system. Although many service categories may not be covered in the following discussion, the general process of assessing requirements should be similar.

15.3.1 Broadcast Video Services

The U.S. cable television industry was founded on the retransmission of analog, NTSC, over-the-air broadcast service to improve reception for viewers compared with what they could get with private antennas. In the late 1970s, those channels were supplemented by satellite-delivered networks. In the context of this chapter, *broadcast services* will refer to all programming that is delivered to all subscribers in common regardless of its origin. Today, most of the bandwidth of typical residential networks is still dominated by programming that is delivered in nonscrambled, broadcast analog format to subscribers. The requirements for such networks are that they have sufficient downstream bandwidth for all the channels, coupled with a quality and availability that is acceptable to subscribers (and that conforms with regulatory requirements, as outlined in Chapter 13).

With the practical implementation of digital video compression, the theoretical program capacity of such networks has been increased by 6 to 10 or more times for standard resolution programming and by 2 times for high-definition programming, compared with over-the-air broadcast formats. Digitally modulated video signals, whether in broadcast (VSB modulation) or cable (QAM modulation) format, have a greater tolerance to noise and intermodulation distortion than do analog video signals (of the order of 30-dB C/N for 64-QAM modulation), with the result that they can be run at lower power levels in the network and, therefore, provide less loading of the active devices.

As opposed to the lifeline telephony situation where network reliability is related to emergency services access and is, therefore, a governmental and public concern, most video services consist of nonessential entertainment, and therefore, network reliability is primarily a marketing issue. In other words, the programming must be delivered with sufficient reliability that customers do not

prefer alternative service providers. It can be argued that the transmission of Emergency Alert System (EAS) warnings through cable networks creates an overriding public interest in their reliability, but no federal standards have been issued (though local governmental authorities sometimes negotiate customer service standards with franchised cable operators that include outage measurements).

CableLabs' Outage Reduction Task Force reported on industry research to determine customer tolerance of various outage rates and durations in their landmark 1992 publication *Outage Reduction*.[3] Based on its findings, that group proposed the following performance target:

> No customer should experience more than two outages within a three month period. While less accurate, because the cumulative effect of outages over time is lost, a secondary standard could be set at no more than 0.6 outages per customer per month.[4]

This guideline is based on research that showed that, when outage frequency exceeded the target rate, it became a major factor in customers' perception of service quality, whereas below that rate, the overall satisfaction rating was largely independent of outage frequency.

Another important result of the research showed that, though there was a difference in perceived quality for outages of various lengths, the dissatisfaction level for 10-minute outages is nearly as high as for all-day outages, and the dissatisfaction at 1-minute duration is about one-half the 10-minute rating. The task force, however, did not put forward a recommended standard for service restoration, or network availability percentage, though they state that "A[n availability] figure in a good month is 99.99 percent or greater. In a poor month, the figure might be as low as 99.95%."[5] If we use an average service restoration time of 4 hours for all outages, then 0.6 outage per month corresponds to a network availability of 99.7%, a frequently quoted cable industry availability goal.

In the same document, the task force has proposed the following definition for an outage:

> An outage is defined as any event in which two or more customers experience loss of reception on one or more channels arising from a common cause, regardless of the cause. Loss is defined as an interruption rather than degradation of signal. Loss of a single channel at the headend or hub site is included.[6]

This is an interesting definition from a network design standpoint. For one thing, it discusses only outages that are "experienced" by subscribers. Thus, outages that are detected and repaired before they are noticed by subscribers do not count. The relationship between actual failure rate and subscriber-experienced failure rate is discussed in Chapter 17.

Second, by excluding single subscriber outages, it ignores all outages occurring because of faults in drop wiring or terminal equipment, whether furnished by the network operator or the subscriber. Given that many, if not most, sub-

scriber outages arise in either the drop or terminal equipment, this is an important exception.

By excluding single subscriber outages, the definition also eliminates from consideration any single-channel outages that affect only single customers. Though this is not important for broadcast services, it is important in the case of interactive services such as video on demand, where each video stream is utilized by only a single subscriber.

Clearly, the outage definition is not consistent with the research on subscriber tolerances. Subscribers are likely to be just as dissatisfied with an outage occurring because an individual set top terminal failed to descramble as one occurring because a headend modulator failed. It would be prudent for those designing networks to include all elements involved in service delivery in setting acceptable outage rates. Given the relative independence of outage length on dissatisfaction rating, it would also be prudent to set separate goals for outage rate and network availability (or service restoration time).

A final parameter in determining reliability parameters for delivery of broadcast video services is outage size. When customers experience an outage, a certain percentage of them will attempt to contact the network operator to report the event. If the outage affects many customers, this can cause the telephone call volume to swamp the customer service department and its incoming telephone trunks. Since many franchises (as well as voluntary industry guidelines) contain language on telephone responsiveness, it is incumbent upon operators to design networks that have as few common elements as possible and to tailor repair efforts to respond most quickly to those outages affecting the largest number of customers.

Though individual network operators must set their own service standards, a reasonable goal for broadcast video service levels is an outage rate not exceeding 0.6 per month, but including single, as well as multiple, customer outages. Independently of the outage rate, network availability should be at least 0.997, corresponding to an average restoration time of 4 hours combined with 0.6 outage per month. Service restoration times should be managed so that they are inversely proportional to the number of customers affected.

15.3.2 Narrowcast Video Services

Narrowcast services are defined, for purposes of this discussion, as those that are not available to all subscribers, but whose generation is not under direct control of viewers. Included in this category are the services known as pay, pay per view (PPV), and near video on demand (NVOD). In each case, the services are delivered, in accordance with a fixed transmission schedule, to a subset of the subscribers receiving the broadcast channels.

The three categories blend into each other. Pay services historically refer to those like HBO that are typically subscribed to on the basis of monthly fees. PPV refers to a fixed schedule of movies and events that are sold on a per-event basis,

with different orderable events transmitted sequentially on one or more channels. NVOD is a logical extension of PPV, whereby movies are typically displayed with sufficient frequency that the average wait time for a particular movie is short enough to discourage subscribers from seeking other means to access the programming (such as renting a tape).

The parameters specifically related to narrowcast services include the means used to control access, ordering mechanisms, and the efficiency of network information capacity usage. Access can be controlled by either signal denial or signal scrambling, as discussed in Chapter 18.

Ordering mechanisms include the following:

- Manual advance ordering by telephone (person-to-person to a CSR)
- Manual advance ordering by telephone using the telephone touch-pad dialer to enter the information into an audio response unit (ARU)
- Manual or automatic advance ordering using automatic number identification (ANI) technology
- "Store-and-forward" systems whereby advance access is granted to programs with postevent accounting of purchased events
- Real-time ordering of programs

Store-and-forward systems typically use a downstream data carrier to poll individual descramblers on purchases. Depending on the model, the individual descramblers may use a telephone modem and the public switched telephone network to upload information to the computer or, in the case of a two-way cable system, may transmit the information upstream through use of an RF transmitter. Real-time ordering systems using two-way transmission capacity have an advantage over use of the telephone since they allow the operator to avoid paying for multiple telephone trunks and modems and to also avoid problems when customers' phone lines are busy.

Whether store-and-forward or advance authorization is used, the upstream data requirements are modest—just sufficient information to identify the subscriber and programs involved in the transaction. Transmission latency requirements for store-and-forward systems are very lenient—hours or even days being acceptable in some cases. In real-time ordering systems, customers may tolerate a few seconds to a couple of minutes between requesting access to a program and its enablement.

It has been shown through numerous marketing studies that the ease with which programming can be accessed is a critical parameter in controlling average buy rates. It has also been shown that, despite incentives for advance ordering, subscribers tend to order programming immediately before needing it so that order-processing rate is also a critical parameter.

As systems offer an increasing number of optional video services, the effective information capacity of the network devoted to those services must be increased. There are basically three ways to do that: increasing downstream

bandwidth, greater information density (for example, digital video compression), and/or network switching. The first two are obvious, but the last mentioned has not been widely exploited as an alternative.

Switched delivery of narrowcast services requires that the network keep track of which optional services are requested by subscribers attached to each node and deliver only the immediately required services. It can easily be shown that with typical node sizes of 500–2,000 homes passed, a reasonably wide selection of programming, and typical North American usage and viewing statistics, a considerable savings in bandwidth can be realized. The trade-off in network architecture is the savings in RF bandwidth (and possibly terminal equipment for digital compression) against the cost of implementing the switching and spectrum management required, with node size as an independent variable.

As an example, consider an NVOD service that offers the top 10 movies, each starting every 15 minutes. Given a 2-hour average movie length, this would require 80 program streams, if every event were to be available to every subscriber. Given that few, if any, cable systems have 80 analog video channels that could be devoted to such a service, the usual way of approaching such a service offering would require digital video compression. At a 10:1 net compression ratio (that is, 10 digitized, compressed movies transmitted in the spectrum generally used by a single analog channel), 8 RF analog channels would be required.

If this offering is made using headend switched channels, then we need to make some traffic assumptions. Assume that the distribution plant fed from each node passes 1,000 homes. At a 70% basic service penetration, the node will serve 700 customers. If 50% of those choose to lease a digital set top terminal, the potential market for NVOD services in each node will be 350 customers. Assuming the monthly movie buy rate among those customers is 100% (1 movie, on average, per subscriber per month), the daily buy rate will be about 12. If one-third of those daily buys overlap in the busiest evening hour, then four RF channels would be required to deliver the service using standard, analog modulation. Thus, even if 50% more channels (two more) were provided to take care of statistical variations, there is savings relative to digital video compression. If the switching has the capability to share available modulators across nodes, then the statistical variations may further reduce the amount of required equipment though not the number of channels.

In analyzing which is the better solution for a given service and operator, it is necessary to compare the cost of the digital compression and receiving hardware with the cost of the headend switching equipment and node-specific modulators used to deliver a switched service. Also, a full analysis must include other revenue streams that might be enabled by the digital terminal versus possibly higher movie buy rates in the switched case enabled by the greater percentage of subscribers who can access the programming.

The U.S. cable industry has not suggested reliability standards for optional video services, as opposed to broadcast services. The research done by CableLabs, discussed earlier, did not distinguish among service levels and the

relative expectation of subscribers who subscribed to premium services. Since customers pay more per channel for such services, though, it is logical to expect that they would expect a higher reliability for such services.

15.3.3 On-Demand and Interactive Video Services

On-demand video services are defined as those delivered to specific subscribers, generally in response to real-time requests. Such programming might include more than entertainment categories. Generally, this implies a degree of real-time interaction between the subscriber and the program generation and control system through a two-way cable system. As an example, a subscriber might use a multilevel menu system to select from among audio, video, or print material, or might search a database of articles in a research project. Alternatively, the subscriber might select a movie and start time from an available library, then be granted VCR-like control over its playback (known as video on demand or VOD).

From an architectural standpoint, on-demand services will differ primarily in the acceptable latency times and in the nature of the information source. In the case of video games, for instance, acceptable latency times may be of the order of a few milliseconds, whereas database searches may tolerate several seconds, depending on the application.

Whatever the nature of the communication, the downstream requirement is essentially the exclusive use of a video stream per simultaneous user per isolated network segment. As with narrowcast services, this is affected by service penetration rate, simultaneous usage rates, and node size. The upstream requirement is for a channel with appropriate bandwidth and latency. Generally, the upstream bandwidth per user, even for the extreme case of interactive games, is relatively modest.

With respect to reliability of service delivery, it can be assumed (because of its date) that the CableLabs research on subscriber tolerance of outages was largely based on broadcast and monthly services. As with narrowcast services, it is reasonable to assume that subscribers who are sole users of channels, and who may be paying far higher per-channel costs for that exclusivity, will be more demanding of service reliability. On the other hand, to the extent that narrowcast services are viewed for a shorter time than, say, broadcast services, the statistical probability of failure will be lower, a balancing factor. To the extent that these narrowcast, on-demand services share the same physical cables as the broadcast video channels, but also use relatively complex server devices, their inclusion in a network operator's offerings may place a more stringent reliability requirement on the broadband delivery system.

15.3.4 Packet Data Services—Internet Access

For purposes of this discussion, packet data services are defined as nonvideo digital transport in which data is packaged in short bursts and sent through a time-shared channel on a demand basis. The most common noncable example of such

a network is an Ethernet, in which variable length packets are sent between any two terminals on the network. Included within this broad definition are both TCP/IP and ATM.

Packet data services may be offered through a one-way cable network (typically with telephone modems used for the upstream packets) or through a full two-way network. The network quality requirements are those that will result in an acceptable service quality. In the case of TCP/IP, this means that the net loss of packets (after any provided forward error correction) must be low enough that the net throughput does not materially suffer. Although industry standards for transmission quality are still being developed, typical, well-maintained HFC networks will provide uncorrected bit error rates in the range of 10^{-9} to 10^{-10} in the downstream direction with 64-QAM modulation, and 10^{-5} to 10^{-7} in the upstream direction with QPSK modulation.

In the downstream direction, a 6-MHz RF channel containing a 64-QAM modulated signal will provide a data rate of about 30 Mb/s. The total number of such channels required will be a function of service penetration, simultaneous usage among subscribers, average data rate per simultaneous user, acceptable blocking rates, and the size of network segments that are fed independent data signals (equal to node size unless data transmitters or optical transmitters are shared among nodes).

In the upstream direction, the requirements are more complex. In addition to the preceding factors, the upstream capacity includes the efficiency with which users are controlled so as to share each RF channel. A common scheme requires modems to request use of the channel, then to be allotted time slots on a specific upstream RF channel, as discussed in Chapter 4. In order for the system to accommodate subscribers who might be located at different distances (and therefore different transmission times) from the headend and at different tap ports, upstream transmitters must be level-agile and able to synchronize their transmissions in time.

Though average data rates will change as usage patterns and applications develop, a typical cable modem traffic analysis might include the following assumptions:

Each simultaneously active user's session is assumed to include

	6	downstream messages/minute
each consisting of	100	packets
with an average length of	1,600	bytes/packet
multiplied by	8	bits/byte
divided by	60	seconds/minute
equals	128	kb/s *average* data rate per active session

On the assumption that the dominant use of the modem is for file downloading, the upstream traffic will consist primarily of 64-byte acknowledgments of the

1,600-byte downstream packets, a 1:25 ratio. Thus, the upstream data rate to support the above downstream rate is 128 kb/s ÷ 25 = 5.12 kb/s.

The relationship among node size, number of nodes sharing a datastream, usable downstream data rate, service penetration (measured against homes passed), and simultaneous usage rate is given by the following:

$$\frac{\text{usable data rate}}{\text{average data rate per active session}} = \text{HP/node} \cdot \text{nodes} \cdot \text{penetration} \cdot \text{simultaneous usage}$$

where

usable data rate = the net available user data rate in the datastream

average data rate per active session was calculated earlier

HP/node = the number of homes served from each node

nodes = the number of nodes that share a downstream datastream

penetration = the percentage of homes in each node that subscribe to cable modem service

simultaneous usage = the number of data subscribers who are simultaneously using the service

For example, if we assume that the net usable data rate from a 64-QAM, 6-MHz bandwidth signal is 20 Mb/s and that the average downstream data rate per active user is 128 kb/s, then the system will support 156 simultaneous users. If 2 1,000-home nodes share a common downstream signal and we allow for up to 35% simultaneous usage among data subscribers, then the system will support a 22% penetration among homes passed.

In the upstream direction, total usable data rates, data rates per session, and circuit sharing will differ; otherwise, the calculation is similar.

Since the upstream spectrum is subject to ingressing signals (and also malfunctioning modems that transmit continuously), many modem systems are also upstream frequency agile so that the system can "sidestep" a degraded frequency.

Network reliability and availability standards have not been set for packetized data services. To the extent that such services compete with those delivered by dial-up telephone modems, the local public switched telephone distribution network arguably may meet the 0.9999 availability requirement for local telephone service. Balancing the possibly higher cable network failure rate, however, is the ability of cable modem equipment to "ping" (test) each modem on an ongoing basis and thus to quickly determine if there is a communications or equipment problem. Even a relatively low penetration of such devices will ensure that virtually the entire cable network can be continuously monitored for integrity. By contrast, the telephone company knows if there is an individual line problem only when the subscriber detects it and manually calls it in.

Although telephone company physical networks are generally reliable, the accessibility of the local Internet service provider (ISP) may be far less certain. A 1997 survey by Inverse Network Technology, for instance, found that the call-in failure rate for all measured ISPs was 22.4%, whereas the largest provider in the country had a failure rate of more than 60%![7] By providing a permanent virtual connection between users' PCs and the Internet, network operators using cable modems will be able to offer a considerable availability enhancement over dial-up alternatives.

For another competitive comparison, AT&T's Digital Data System (DDS), which offers duplex data channels to users at speeds ranging from 2.4 kb/s to 1.544 Mb/s, guarantees a BER of at least 10^{-7} and an availability of 0.9996, with the local circuits on each end of the network (T1 carrier plus baseband loop) assumed to have an availability of 0.9999.[8]

15.3.5 Voice-Grade Telephony Services

Voice-grade telephony comes in several forms, including conventional wired telephony, PCS telephony, cellular telephony, and IP telephony. Wired telephony and IP telephony provide a physical connection between the user's terminal equipment and the distribution system, whereas PCS and cellular systems use bidirectional radio transmission. Although there are distinct legal and frequency usage differences between cellular and PCS telephony, from a network perspective, there is almost a continuum of terms of cell sizes, cell interconnection requirements, and available services. Generally, cellular base stations and portable telephones both use higher power than PCS and correspondingly greater spacing between base stations. Also, PCS is a purely digital service, whereas cellular can be either analog or digital in the air link.

Wired telephony service can certainly be provided over an HFC network, as can the interconnection between PCS transceiver antenna locations though the network requirements are considerably different. IP telephony is an emerging competitor that "rides" on a TCP/IP data circuit and thus can be accessed transparently over a cable modem network.

Wired Telephony

Among the telephony options, wired telephony—the service most people take from their local phone company—places the greatest demands on the cable system. By definition, the service requires that the cable systems support two-way communication to and from every home. Since the greatest percentage of upstream ingress problems occur in the drop system and internal building wiring, the communications equipment used must be designed to work in relatively "dirty" channels. One architectural choice that eliminates many in-home problems is the use of an external NID on each home that contains the data receivers and transmitters for telephony, with standard twisted pair copper used to transport baseband analog telephony within the building. By using a NID, the network

is protected from in-home ingress (see Section 15.2.3 for a general discussion of the advantages and disadvantages of the use of NIDs).

The requirement to have the service survive localized power outages also favors the use of a NID, which can be directly powered from the coaxial plant, which in turn, can be equipped with standby power. In other cases, operators may choose to equip each NID with batteries and a local charger. Either choice involves compromises. Network powering, unless the NID has very low power consumption, can significantly increase the total power consumed by the network and, for that reason, make the design of reliable standby power more difficult. The use of a home-powered charger and batteries creates the need to maintain that equipment and to dispose of worn-out batteries on a regular basis. Also, the power for the charger within the home is likely to be less reliable than distribution plant power. Finally, though it is common for telcos to plan for long-term outages by equipping standby plant power supplies with generators or very high battery capacity, that is not economical on a home-by-home basis, with the result that individuals will be protected only for short power outages.

Telephony signals are sent to and from individual subscribers. Thus, the RF bandwidth used will be directly related to node size, modulation efficiency, service penetration, and simultaneous usage among subscribers. As an example, assume that a system is built with 1,000-home fiber nodes. If the network operator is a local exchange carrier (LEC) and expects to serve virtually all residences with at least a primary telephone and a 50% chance of a second phone, then it must plan on a capacity of 1,500 lines per node. Assuming the company's traffic engineering studies predict 20% maximum simultaneous usage (this is known as *line concentration* in the telephone industry), then the network must be able to handle 300 simultaneous telephone calls per node. At 64 kb/s per voice stream, the network must be able to transmit 19.2 Mb/s per node. Finally, if QPSK modulation is used on the RF carriers (with a bandwidth efficiency of 1.5 b/s/Hz), the required bandwidth will be 12.8 MHz in each direction. In the downstream direction, this will require the allocation of about two standard 6-MHz television channels. In the upstream direction, however, this will require the usage of a significant percentage of the usable bandwidth of most standard low-split systems, especially if the noisier frequencies below about 12 MHz are avoided.

As the primary communications link into homes, the local telephone service is expected to provide very high reliability and availability, and to work when the power is out. All these are greater than the demands required to deliver nonessential entertainment video services. On the other hand, some feel that there is a viable business in providing lower-cost secondary lines whose requirements are less strict.

Bellcore, the research consortium of the telephone industry, has established reliability objectives for the telephone industry. Those relating to the subscriber loop are contained in TR-NWT-000418.[9] The basic criterion is that "The two-way service availability objective of a narrowband transmission channel should be a minimum of 99.99% (0.9999 probability) for the subscriber loop." This corre-

sponds to an average yearly unavailability of 53 minutes. "Narrowband transmission channel" is further defined to mean services, including voice, that are transmitted at speeds lower than DS1 (1.544 Mb/s).

It is important to understand what the Bellcore criterion includes and what it does not. In terms of applicability, everything in the central office on the subscriber side of the switch is included (for HFC systems, this would include the digital multiplexers and RF modulators, for instance), as is the distribution network and, if used, the NID. The subscriber-provided in-building wiring and telephone instrument are not included.

Equipment failures and the time to repair them are included in the calculation. Not included, however, are service outages due to both primary and backup powering. Since the majority of outages in HFC networks are power related, actual subscriber unavailability may far exceed 53 minutes per year and yet meet the Bellcore criterion. By comparison, the CableLabs guideline quoted earlier includes power outages. Also, Bellcore's definition defines the restoration time from when the operator becomes aware of the outage, which may be long after it actually starts in the case of an outage affecting single customers. Finally, the unavailability experienced by an individual customer is further affected by his or her usage pattern and by failures of his or her own terminal equipment. Despite its limitations, the Bellcore criterion is the accepted criterion for measuring availability of residential lifeline telephone service.

The relationship of subscriber-perceived outage rate and unavailability to network reliability and availability is complex and involves usage patterns, as well as the depth of network monitoring. This subject is further discussed in Chapter 17. It is interesting that, though the added complexity of HFC networks using NIDs may *increase* the actual hardware failure rate, the addition of automatic subscriber line monitoring may allow a *decrease* in customer-perceived outages relative to lower failure rate, but unmonitored, networks.

PCS Telephony

PCS telephony, like cellular service, uses a direct radio link to subscriber's terminal equipment. Typically, the portable telephone is owned and maintained by the subscriber. The subjective difference between them is that, since radiated power levels are much lower, PCS requires many more radiation/reception locations to cover the same area. Thus, though it is economically practical for cellular providers to use point-to-point microwave links or leased data lines to interconnect radio modems with a centrally located switch, PCS requires many more such links.

Where a broadband HFC network operator happens to serve the same area as a PCS operator, there is an opportunity to provide the interconnection service over the network. This is far less demanding, particularly in the upstream direction, than provision of wired telephony service since the service drop, inside wiring, and much of the distribution plant is not involved in the signal transmission, thus eliminating most sources of interfering signals and simultaneously

improving network reliability. Since individual telephones are battery equipped and charged from user-provided local power, the constraint to provide network powering to homes is also eliminated.

One PCS configuration that takes advantage of the broadband network's ability to carry RF modulated signals of any format is known as remote antenna driver (RAD) or cable microcell integrator (CMI). The technology was developed by Rogers Cable, Cox Cable Communications, and Cable Television Laboratories. Rather than putting a full base station at each RF modem location and interfacing with the network at the baseband digital level, a RAD system simply down-converts the as-received 1,900-MHz PCS signals to a range of frequencies within the upstream bandwidth of the cable network. At the headend, signals from as many remote locations as desired can be combined to feed shared base-station radios. Similarly, the base station's output signals are carried in the downstream cable spectrum to the remote antenna locations, where they are up-converted to feed the radiating antennas.[10]

This system has many advantages. Primarily, the cost of the remote frequency converters is much lower than full base stations so that sharing a station across several antennas is cost effective. Second, cell sizes can be shaped to meet traffic patterns by selecting which antennas share use of each base station. As service penetration grows and changes, the cells can be redefined as required by merely changing the RF combiners at the headend. Thus, a low penetration service may use only a few base stations and many antennas to get full coverage, with additional base stations added as service demand increases. Finally, since several neighboring antennas are connected to the same base station, there is no "handoff" required until the user moves to antennas connected to another base station. This considerably reduces the network overhead required to handle frequent handoffs.

Against these advantages must be considered one major disadvantage. The coding level and modulation method required to survive the uncertain over-the-air transmission environment limits bandwidth efficiency. Because of the wide range of received signal levels (preserved when the signals are merely frequency translated), the rugged coding is also required in the cable environment since the C/N will vary linearly with signal level. For those reasons, the typical RF bandwidth required to transport 28 voice streams to an antenna is 2–4 MHz in each direction, whereas at least one manufacturer claims that the same information, optimally coded for cable transmission, can be fit into 600 kHz of bandwidth.[11]

Given the uncertainty in the radiated portion of the transmission path, both with transmission to a given cell and the handoff to another cell, the reliability and availability of radio-based telephony is described in terms of grade of service (GOS). Good service quality is defined as the probability that the sum of blocked calls plus 10 times the number of dropped calls is less than 0.5%. Based on traffic engineering, the 28 voice channels that occupy the bandwidths discussed earlier are sufficient to support approximately 200 customers with a blockage probability of 1%.[12] Both the availability of PCS service and the probability of successfully completing a call obviously include an allocation for the network tying the an-

tenna locations to the base stations; however, the relative allocations to the air link, transmission system, base-station reliability, and traffic-blocking probabilities are not specified.

IP Telephony

IP telephony began as a means of bypassing the long-distance carriers. Computers at each end of an IP telephone link were equipped with microphones, speakers, and audio coders/decoders (CODECS), and could use voice communications provided the two-way data rate was sufficient to support the digitized voice.

Although the initial results were crude in comparison to the normal public switched telephone network, hardware improved to the point that major carriers have taken an interest in it. One crucial development has been the emergence of facilities in numerous cities that serve as gateways between the Internet and local exchange carriers (LECs). This allows calls between standard phones and IP phones. Both the legal and technical framework for this service is still under development.

Should IP telephony become commonplace, it will change cable modem service usage patterns, including session timing and length and the degree of upstream–downstream data rate symmetry.

15.3.6 Utility Monitoring and Control Services

A final example of services that can use the broadband infrastructure are various low-average-data-rate monitoring and control functions. Examples are electric utility customer load control and meter reading; burglar, fire, and medical alarm monitoring; and traffic light monitoring and control.[13] These services can be characterized generally as relatively low-speed data circuits with varying latency and transmission reliability requirements. Most require highly accurate end-to-end data integrity.

The two applications with the greatest apparent hope of commercial success are electric meter reading and demand-side management (DSM). Since electric power companies must provide capacity based on peak, rather than average, demand, DSM offers them a means of trimming peak electric loads by selectively turning off high-powered appliances, such as hot water heaters and air conditioners, during the hottest hours of the day. By turning off rotating groups of such appliances for a short time, they can materially reduce load while minimizing inconvenience to users.

Although the data rates are not high, the overall reliability by which the DSM process functions must be exceptional since the result of failure could be a widespread power outage. On the other hand, only the common equipment must meet this high standard since the failure of a single node or the equipment in a single home will not have a material effect on overall load management. Nevertheless, it has been suggested that network availabilities in excess of 0.99997 might be

required by some utilities.[14] By contrast, the requirements for meter reading are less stringent since timing is not critical and data can be retransmitted if necessary to assure accuracy.

A similar situation exists for medical, burglar, and fire alarm monitoring. The average data rates are quite small, and transmission delays of a few seconds are tolerable, but network availability is crucial, as is transmission accuracy, since lives could potentially be lost if alarms are not promptly and accurately reported.

15.4 Summary

Networks can be described by a number of parameters, including noise, distortion, and other signal degradation measures; instantaneous RF bandwidth; various means, such as segmentation, that increase the effective bandwidth; and reliability and availability.

Services that may be delivered over those networks place varying demands on the distribution system: (1) the quality of the RF channel occupied by its signals; (2) the occupied bandwidth, both forward and reverse; (3) the availability of the network to carry signals; and (4) network powering for unique service-related processing equipment.

The architecture of the network must be chosen so that the network provides adequate performance and effective bandwidth for all the contemplated services. Chapter 16 will discuss various architectural examples and how they meet the needs of certain services, and Chapter 17 will deal with reliability and availability modeling.

Endnotes

1. W. Leo Hoarty, Less is More—The Case for Concentration, *Montreux 1997 Record—International Television Symposium and Technical Exhibition,* Montreux, Switzerland, June 1997, pp. 29–37.

2. David Wachob, WorldGate Communications . . . Connecting the Unconnected. Eighth Annual Digital Audio/Video Workshop, Philadelphia, PA, October 1996. IEEE Consumer Electronics Society/Electronic Industries Association, Rochester, NY.

3. *Outage Reduction.* September 1992, CableLabs, Louisville, CO.

4. Ibid., p. I-18.

5. Ibid., p. V-10.

6. Ibid., p. I-8.

7. As reported in David Plotnikoff's Dialing up Answers from the FAQ File. *San Jose Mercury News,* June 15, 1997, p. 6E.

8. Bernhard E. Keiser, *Broadband Coding, Modulation, and Transmission Engineering.* CEEPress Books, Washington, D.C., 1994, pp. 96–97.

9. *Reliability Assurance for Fiber Optic Systems System Reliability and Service Availability.* TR-NWT-000418, Issue 2, December 1992. Bell Communications Research, Piscataway, NJ.

10. George Hart, Cost Effective Cable Television Transport for PCN, *1992 NCTA Technical Papers.* NCTA, Washington, D.C., pp. 148–156.

11. Douglas E. Holulin and Robert W. Hammond, Integrated Wireless PCS—Hybrid Fiber Coax Network Architecture, *1996 NCTA Technical Papers.* NCTA, Washington, D.C., pp. 10–18.

12. Hart, op. cit.

13. Rainer Kochan, Dipl. Ing., Value-Added Services and Functions for MATV and CATV Systems, *Symposium Record, CATV Sessions, 16th International TV Symposium,* Montreux, Switzerland, June 1989, pp. 524–535.

14. Dean Ericson and Dan Carter, Demand Side Management Considerations in Advanced Network Deployment, *1994 NCTA Technical Papers.* NCTA, Washington, D.C., pp. 8–14.

Chapter 16

Architectural Elements and Examples

16.1 Introduction

Network architecture is the logical and physical interconnection of all the elements between a signal's generation and its termination. The previous chapter dealt with ways of measuring architecture-related parameters and with the needs of various types of services. This chapter will first delineate each of the elements from which an architecture is constructed, then present some examples of high-level distribution networks that have been proposed or constructed. The relationship between network architecture and service reliability and availability will be dealt with in Chapter 17.

16.2 Architectural Elements

Although network architecture is often thought of in terms of the physical distribution topology, everything from the internal headend (and preheadend) processing through network termination equipment affects how services are delivered. It is important, therefore, to understand the implications of available options at all levels of the network. We discuss in this section only those structures that are part of the cable television network; we do not include information flow outside that system.

16.2.1 Terminal Equipment

The network terminal equipment provides the interface between the user's end application devices and the shared broadband distribution system. Depending on the service, it may perform format conversion, testing, or security functions, as well as protect the network from disruption. One example is set top terminals, which perform tuning and selective descrambling functions for video services and may also provide menuing or ordering capabilities. In the case of digital video channels, the terminal also converts the signal to standard analog video format. In two-way systems, the terminal communicates user's service requests to the headend. Another example is a telephony point-of-entry network interface device (NID), which converts between digitally modulated RF signals and analog baseband telephone signals for routing within a home. Choices in terminal equipment can have a significant effect on overall cable system design and performance.

In-Home Equipment

Historically, the full cable spectrum (possibly minus some channels blocked by traps) has been delivered throughout subscribers' homes. In many cases, subscriber-owned television receivers have been directly connected to cable outlets, whereas in others, converters or converter/descramblers have been placed between the drop and subscriber's receiver.

This scenario is consistent with U.S. regulations that allow customers to wire their own residences and that require cable operators to deliver all channels in the lowest programming tier in unscrambled, analog broadcast format.

One advantage to delivering the full cable spectrum throughout homes is that any service can be accessed from any outlet by connecting the proper equipment. Second, no home equipment investment is required in advance of service delivery. Finally, the cost and responsibility of equipment powering is borne by subscribers.

Balancing those advantages are the relatively high and variable loss of drop wiring in homes, which makes it more difficult to assure that terminal equipment receives the proper signal levels. Additionally, home wiring has been a major ingress point for both signal and electrical transient ingress in two-way systems. Although upstream transmitters may be located anywhere within the home drop system, the full bidirection bandwidth must be preserved, and upstream plant is therefore exposed, on average, to more frequent interference. Often cable operators reduce this risk by installing filters to block upstream ingress on portions of the drop system that will not be used by two-way devices. Finally, and depending on the services being offered over the cable system, there may be a duplication of some functionality among different pieces of terminal equipment.

Point-of-Entry (POE) Equipment

Terminating the broadband linear network at a common point exterior to the residence provides a means to isolate two-way cable networks from signals generated or picked up within the home. Any upstream signals to be transmitted into the distribution system can be generated by a common transmitter at the POE device. Similarly, multiple downstream services may share some of the common POE circuitry. In some cases, for instance wired telephony, the POE device may generate a standard baseband signal for transmission over twisted pair cables to telephone instruments. The same equipment may drive an Ethernet bus serving multiple computers.

Where several services are provided and can be coordinated, the use of POE equipment can result in both cost savings and improved operation. The addition of automated status monitoring may be justified by expanding the maintenance capabilities of network operators. On the other hand, the cost of equipment for the first service delivered to a home is often more expensive than a single-purpose in-home version. Finally, lacking interindustry standards, use of a multi-service POE device restricts the ability of operators to buy equipment from multiple sources.

There has been a long-standing debate regarding the relative advantages of network versus home powering of POE devices. Home powering significantly reduces the power consumption of the distribution network, but it is less reliable. Also, it is much simpler to maintain large shared standby powering systems than individual in-home units. The goal to provide 99.99% availability for basic

telephone service certainly tends to favor network powering. The proper decision is less obvious for nonessential services.

Shared Terminal Equipment

In some cases, terminal equipment is integrated with multiport taps so that some functions (for instance, power supplies and data transceivers) can be shared among multiple homes. The most common examples, for video services, are addressable disconnect functions, interdiction, and "whole-house" descrambling systems (these are discussed in detail in Chapter 18).

Two obvious advantages of shared terminal equipment are easier maintenance access and, potentially, improved security. Deployment of such a system, however, requires either that the network provider completely equip the plant with such equipment or that the plant be modified each time a new subscriber is installed. In other words, the operator is forced to choose between a high initial capital cost and an even higher evolutionary upgrade cost accompanied by service interruptions.

As with POE equipment, shared terminal equipment has the further disadvantage of limiting future flexibility—it is hard to upgrade program scrambling security when the descramblers are part of the distribution system. These problems have limited the widespread adoption of shared interdiction technology by the cable television industry.

The greatest commercial success with shared equipment has been in multiple dwelling units (MDUs), where the rate of connects and disconnects exceeds the average, signal theft is high, or physical access is difficult.

Shared terminal equipment is central to the fiber-to-the-curb (FTTC) architectures proposed by some telephone companies. In an FTTC architecture, the fiber distribution network extends to curbside shared terminal equipment, whose functionality is shared among a few homes or businesses (typically 6 to 25 or so). Individual homes are fed from the terminal equipment through twisted pair lines, coaxial drop cable, or both. An example is discussed in Section 16.3.4.

Hybrid Arrangements

Most commonly, where either POE or shared terminal equipment is used, it is coordinated with some in-home equipment. For example, one manufacturer offers a POE network interface for telephony and data but passes the bandwidth containing video signals to set top descramblers. The upstream bandwidth is also divided, with data and telephony signals generated by the interface equipment and signals related to ordering of television signals originating in the set top terminal. Filters block only a portion of the upstream bandwidth from the home wiring. Though such arrangements are arguably more expensive than totally integrated solutions, they preserve the network operator's flexibility to deal with different vendors and formats, and work well where new services are to be installed without disrupting older service platforms. Figure 16.1 illustrates terminal equipment location options.

Figure 16.1 Choices for network termination equipment location.

16.2.2 Coaxial Distribution Network

The generic design of coaxial distribution systems was covered in Chapter 10. Within the general framework of amplified, split, and tapped networks, however, are many architectural variables.

RF Bandwidth

The most basic parameter of any coaxial network is the forward and return bandwidth. Since these are primarily determined by the amplification equipment, some operators are proposing the use of sufficiently small coaxial distribution networks that they can be totally passive between the output of the fiber node and the user. Such networks have the advantage that the division between upstream and downstream bandwidth can be altered at will, as can the upper frequency limit. As a practical matter, however, these freedoms are restricted by the spectrum usage issues discussed in Chapters 8 and 15, whereas the upper frequency will be constrained by the characteristics of the passive devices used, the increasing loss of the cables, and the practical limits on signal levels in the network.

Coaxial Serving Area Size

The next most basic decision in the design of an HFC network is the size of the serving area (measured in homes passed) to be served by one physically separate coaxial network. Since the use of the RF bandwidth is shared among all the users in the serving area, this determines the bandwidth per home passed.

Since reliability is inversely proportional to the number of series-connected critical components, the required network availability and failure rate for the services to be offered should also be a factor in determining serving area size. The calculation of network reliability and availability is discussed in detail in Chapter 17.

Typically, three to four separate coaxial distribution legs will be served from each fiber-optic node, offering opportunities for subsequent node size reductions. Some operators, in fact, design nodes for subsequent subdivision.

Coaxial Powering

The average reliability of commercial power is often lower than the total end-to-end required network performance. Thus, systems are required to employ a powering strategy that minimizes the effects of power interruptions. Conventional cable television standby power supplies typically provide 1–2 hours of operation from self-contained batteries. Although this is sufficient for short interruptions, some percentage of interruptions will exceed this length. Thus, in systems where customers are served by distribution networks that are powered by more than one such power supply, reliability will decrease due to the typically noncorrelated nature of those interruptions.

Where the individual coaxial distribution legs are limited to approximately 150 homes passed (assuming typical suburban, single-family home densities), it is practical to power the entire network from a single location, even when network-powered POE interface devices are used. This is, in fact, the strategy typically used by U.S. telephone companies that have constructed HFC networks. Single point powering of this size network, however, generally requires 90-volt powering, with maximum currents on the order of 15 amperes. Voltage limitations as well as electrical safety issues are controlled in the United States by the National Electrical Safety Code.[1]

In order to achieve telephone-grade reliability, the telcos have generally equipped these common power supplies with 8-hour battery capacity and/or backup generators. The generators, in turn, can be powered by self-contained fuel supplies or tied to available natural gas mains.

In order to avoid the problems and costs of installing these large, unsightly power supplies at each node, one telco (Southern New England Telephone) proposed to power nodes from the central office. To accomplish that, they developed a hybrid cable containing optical fibers for signals plus copper conductors to carry 440-volt, three-phase power. At each node, the power was converted to acceptable voltages for multiplexing onto the coaxial distribution systems.[2]

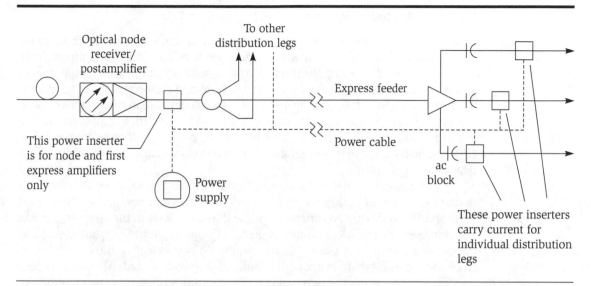

Figure 16.2 Bypass powering to increase reach of node-located power supply.

Where the current-handling capabilities of components or the voltage drop through distribution cables might otherwise limit application of single point powering, the output from one power supply is sometimes divided among several power distribution cables whose resistance per unit length is less than the distribution cables. These power bypass cables may be routed alongside the signal cables past the first split point, where lower currents can be inserted into each of several network legs. This is illustrated in Figure 16.2. Since these power distribution cables do not carry RF, lower impedances and nonstandard construction are sometimes used to reduce power loss.[3]

When battery backup supplies are used, major improvements in outage rate and availability can occur if the supplies are equipped with monitoring systems that alert network operators of commercial power outages so that crews with portable generators can provide interim power before the batteries expire.

16.2.3 Fiber Node

The individual fiber nodes provide the interface between the linear fiber trunking system and the coaxial distribution legs. At the simplest level, such a node may consist of a single optical receiver whose output is amplified to feed the downstream amplifier and, optionally, an upstream optical transmitter whose input is driven by the output of a combiner whose inputs are the upstream signals from all connected coaxial distribution legs. There are, however, many architectural variations.

Split-Band Multiple Fiber Inputs

Some nodes provide the option for multiple downstream input fibers that are fed to separate optical receivers whose outputs are then combined in a diplex filter. There are basically two applications for this capability: improved signal quality and network flexibility. In the first case, the C/N and distortion of the fiber-optic links can be improved by simply dividing the total signal load between two transmitters. This allows greater optical modulation index and thus improved C/N. At the same time, fewer distortion products will be produced. The relative gains in C/N and distortion can be exchanged, within a limited range, by adjusting the optical modulation index.

A more common use for split-band optical inputs is to allow those signals that are common to multiple nodes to be carried on one fiber, and those that are node specific to be carried separately. As will be seen later in the chapter, this can result in overall network flexibility and cost savings. It should be cautioned, however, that a system that uses split-band inputs cannot easily adjust those bands since the diplex filters in every node would have to be changed. Also, signals near the crossover between the two bands will inevitably suffer some degradation: increased noise, group delay, or both.

Redundant Multiple Fiber Inputs

An alternative use for multiple downstream input fibers is redundancy of the fiber-optic trunking network. In nodes equipped with redundant inputs, each fiber is routed to its own full-bandwidth receiver. A switch at the output of the receivers selects the better (or surviving) signal in the case of failure or degradation of one path. If the fibers feeding the node are diversely routed (so that the probability of simultaneous cut is low), then significant gains in reliability can result. As a practical matter, the gain in reliability is limited by the reliability of the switch (and its control circuit), which is an added critical element in the signal path. As Figure 16.3 shows, the configurations are very similar, and in fact, some manufacturers offer either as an option in the same node platform.

Redundant Signal Processing

For applications where availability is critical, nodes are available with additional redundant elements. Power supplies, downstream and upstream amplifier modules, and upstream optical transmitters are all available in dual configurations. In the last-mentioned case, dual upstream optical outputs are provided, enabling diversely routed external cables.

Handling of Multiple Coaxial Distribution Legs

Generally, optical nodes are used to feed more than one coaxial distribution leg. Although this is a design convenience that allows shorter coaxial cascades by placing the feedpoint near the logical center of the node serving area, electrically separating the coaxial legs also offers opportunities for effective bandwidth

Input A: lower channels

Input B: upper channels

Separate optical receivers

Diplex filter

L

H

Common downstream amplifier

Configuration A: split-band inputs

Input A: primary feed

Input B: redundant feed

Redundancy switch

Configuration B: redundant fiber inputs

Figure 16.3 Fiber node multiple input options.

expansion and improved upstream performance. One option, described in Chapter 14, is block conversion whereby the upstream signals from different legs are converted to nonoverlapping frequency ranges. Thus the bandwidth per home passed is limited by the coax serving area rather than the larger node serving area. Offsetting the per-subscriber bandwidth gain is the increased performance required of the upstream optical transmitter, which must handle potentially more signals and certainly increased RF bandwidth. In some cases, this will require upgrading from FP to DFB transmitters at a considerable increase in cost.

An alternative way of achieving the same upstream bandwidth expansion is to use separate upstream transmitters and fibers for each distribution leg. This reduces the performance requirements for each transmitter, but requires more transmitters and fibers. A third alternative is to use separate upstream transmitters operating on different optical wavelengths, and combine their outputs using a wave division multiplexer (WDM). Whether separate fibers or WDM is less expensive will depend on the length of the fiber transmission path and the relative cost of the transmitters and multiplexers.

Figure 16.4 Options for segmentation of upstream coaxial inputs.

These three options are illustrated in Figure 16.4. Note that optical splitting or use of additional laser transmitters can be used with any option to create dual outputs for feeding diversely routed fibers for higher upstream reliability, though the incremental cost is least for configuration B.

16.2.4 Fiber Interconnects

In most systems, linear fiber-optic links provide the connectivity between the point where the FDM spectrum is created and its delivery to coaxial distribution legs. Fiber links are also commonly used at higher levels in the network to distribute selected signals of all types, often in digital form. Although fiber-optic links are used among headends and between headends and outside information sources (such as broadcasters or Internet service providers), this section will dis-

cuss only those links that form part of the network extending from headends to and from end customers.

Although this book will treat interconnect architectures as if they were all constructed with fiber optics, it should be remembered that AM modulated broadband microwave radio technology (AML) is still used for both individual signal and broadband distribution in those cases where it is more cost effective than construction of physical optical cables. Clearly, however, microwave links can replace only point-to-point fiber links and not more complex fiber topologies. Also, the available microwave spectrum may not permit as many channels as does a fiber-linked system. For instance, the U.S. 13-GHz CARS band will support only 80 analog television channels.

We discuss some basic architectural structures next.

Star

The simplest architecture, and the most commonly deployed for transmitting analog FDM signals from headends or hubs to and from nodes, is the star, defined as separate paths from a common point to multiple termination points. When constructed with pairs of optical fibers driven with separate transmitters for each node (in each direction), the star architecture allows complete independence in the signals delivered to each node (as opposed to the most common type of microwave broadband star network that splits the output of a common transmitter to feed multiple antennas and paths). In fact, many cable systems optically split fiber transmitter outputs to feed two to three nodes, but do that splitting in the headend so as to allow easy separation of nodes in the future when greater network segmentation is appropriate.

Though the failure of an individual fiber feeding a receiving point affects only the subscribers on that node, practical cable routing often results in one cable sheath holding the fibers for many nodes, and thus creating a possible single point of failure that is much larger (see Figure 16.5).

Sheath Ring

The exposure to outages caused by fiber cuts or failures can be reduced significantly by sending the signals between the common point and each termination via physically diverse routes. When multiple termination points can be arranged in a ring configuration, considerable savings result from sharing of common construction costs, as shown in Figure 16.6. This architecture is known as a *sheath ring*. Note that the diagram shows only a single line in each direction around the loop from the common point to each node though in fact there may be several (at least one upstream and one downstream). Analysis of the figure will show that the cable can be cut at any point without destroying continuity from the common point to each node.

Although the sheath ring architecture preserves the signal independence of the star and adds a considerable degree of protection against fiber cuts, it does so at the expense of a considerable increase in total *fiber* footage. Depending on the

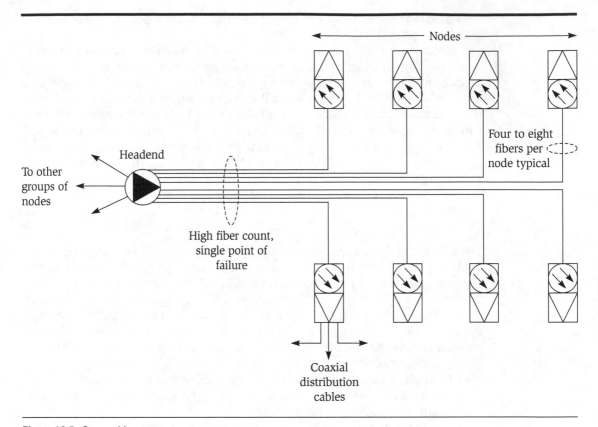

Figure 16.5 Star architecture.

physical layout of the ring, however, the total *cable* footage may be only slightly greater than a simple star. Since construction costs typically are much greater than fiber costs, the incremental cost of the installation may be only moderately higher.

Sheath rings are a common architecture among competitive access providers (CAPs) that offer highly reliable data transport to commercial customers.

Analog Shared Ring

Where a common set of signals must be sent to multiple nodes, an economical structure is the shared ring, which is illustrated in Figure 16.7. The common signal spectrum is transmitted both clockwise and counterclockwise from the common point using separate fibers (and often separate transmitters to increase the degree of redundancy). At each node, a portion of the signal is coupled from both rings and fed separate optical receivers. A redundancy switch, as described in the optical node options, selects the surviving path in the event of a cable cut or transmitter failure.

Switches automatically
select surviving fiber
(may be placed after
dual optical receivers)

Nodes

Headend

To other
groups of
nodes

High fiber count
but no single
point of failure

Figure 16.6 Sheath ring architecture.

Though shared rings can be constructed at 1310-nm wavelength, the maximum available optical budget of approximately 17 dB (see Chapter 11) limits the number of receiving nodes. At 1550 nm, however, optical amplifiers are available that can considerably increase the available loss budget (1310-nm amplifiers are expected to be available within a few years).

A major advantage of the shared ring is obviously a considerable savings in optical fiber usage and number of transmitters compared with either a simple star or sheath ring while preserving the protection against fiber cuts or transmitter failures. Its application, however, is limited to distribution of a common signal spectrum. As will be seen, shared analog rings are effective when combined with other structures.

Digital Repeating Ring

A ring structure that is well suited to digital signals is the repeating baseband digital ring, an example of which is shown in Figure 16.8. As discussed in Chapter 11, baseband digital optical transmission allows much greater optical budgets (on the order of 30 dB) and very high modulation rates.

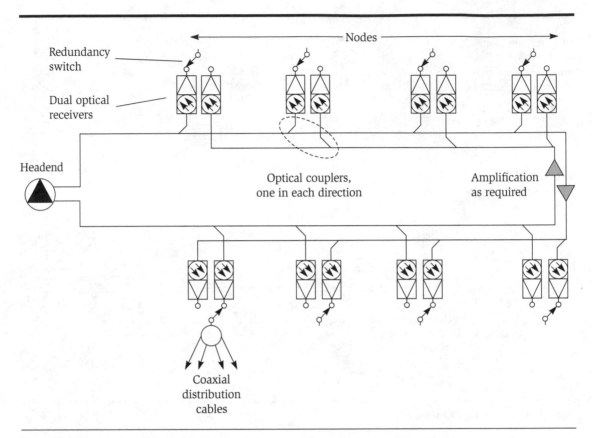

Figure 16.7 Analog shared ring architecture.

In a repeating ring of the form shown, the information destined for all nodes is time multiplexed into a single datastream, which is transmitted both clockwise and counterclockwise around the ring. At each node, the signals destined for customers served from that node are separated from the stream, and signals destined for the common point are inserted. Should a cable be cut or any piece of terminal equipment fail, the remaining equipment automatically "loops back" so that it communicates with the common point via the surviving path.

As with the shared ring, an advantage to the architecture is the small number of fibers required while preserving protection against cuts or equipment failures. A limitation is that its application is restricted to the distribution of information that can be converted to the common digital transmission format. Several such formats are available, including proprietary time division multiplexed (TDM) schemes, asynchronous transfer mode (ATM), and synchronous optical network (SONET). Compared with analog, digital transmission requires much more costly signal processing at the point where the signals are converted to RF since each

Figure 16.8 Baseband digital repeating ring.

must be demodulated to a baseband digital stream, converted to an analog baseband signal, remodulated to the desired RF channel, and finally, combined into the desired FDM spectrum for delivery.

A variant that is lower cost than SONET and designed specifically for the cable television market uses pulse code modulation (PCM). Rather than starting with a baseband digital signal, it encodes the modulated IF signal for transmission. Then, at the destination, it is necessary only to reverse the process, then convert the reconstituted IF signal to the desired RF output channel.

Depending on the format, digital terminal equipment may take the form of switches, routers, or add-drop multiplexers (ADMs). Where standards-based formats such as SONET are used, network capacity can be upgraded by simply using terminal equipment that supports higher data rates on the network side. Rates up to 10 Gb/s are commercially deployed.

16.2.5 Switched Video Headend Structures

In traditional entertainment-only cable systems, the headend served the function of receiving signals from various locations and processing them into a common spectrum for distribution throughout the distribution network. As discussed in

Chapter 8, considerable complexity is added by the requirement to generate custom spectra for each node containing both regional differentiation and information to be transmitted to and received from individual customers.

An important example that illustrates the architectural trade-off between node size, RF bandwidth, and frequency reuse is the use of headend switching in the offering of near video on demand (NVOD)—typically a selection of movies played on a fixed schedule with frequent start times—or narrowcast—meaning low viewership—video services. The characteristics of both service categories are that the video sources operate on a fixed schedule that is not controlled by the viewer.

If a common spectrum of signals is distributed throughout a cable system, then the number of program offerings is obviously limited by the spectrum devoted to that service classification and the efficiency with which that spectrum is used. In particular, if the number of offerings exceeds the RF bandwidth needed for analog transmission, then the only options are to rebuild the system for greater bandwidth or to use digital video compression and accept the increased headend and set top terminal cost.

However, when the usage of video channels is very nonuniform, as is the case with entertainment, it becomes likely that some channels are not always being watched in all nodes. For example, if a network is built with 1,000 home nodes, subscriber penetration is 60%, and simultaneous usage (percentage of homes watching television at any one time) is 67%, then 400 viewers will be watching at the peak hour of usage in any one node. Based on current viewing patterns in North America, the three dominant television networks command approximately 50% of the viewing audience. Including those, the total percentage of customers tuned to any of the 20 most popular channels on any cable system is likely to be more than 90%. Thus, regardless of the number of channels offered, the maximum number of additional channels that will be watched at any one time in any node is probably fewer than 40, and will approach that number only if the viewers of less popular programming all want to view something different from one another. Put another way, any channel whose audience share is less than 0.3% has a less than 50% likelihood of being accessed in any given node at any given time.

These statistics suggest an alternative way to provide channels of limited interest. If viewers' viewing requests could be communicated to the headend, then a video switch and separate set of modulators for each node could be used to provide less popular channels only as required. If this structure were put in place, it would remove any limitation on the number of programs offered except for the economic limitations on switch size. Such an architecture has, in fact, been tested successfully in a U.S. cable system though it has not been widely deployed.[4] The principles are illustrated in Figure 16.9. Although the diagram shows use of an audio/video baseband switch driving a set of modulators for each node, alternative approaches include IF switching and baseband digital switching.

Note also that the switched NVOD or narrowcast headend differs from a true video-on-demand (VOD) headend in the assumption that all available NVOD

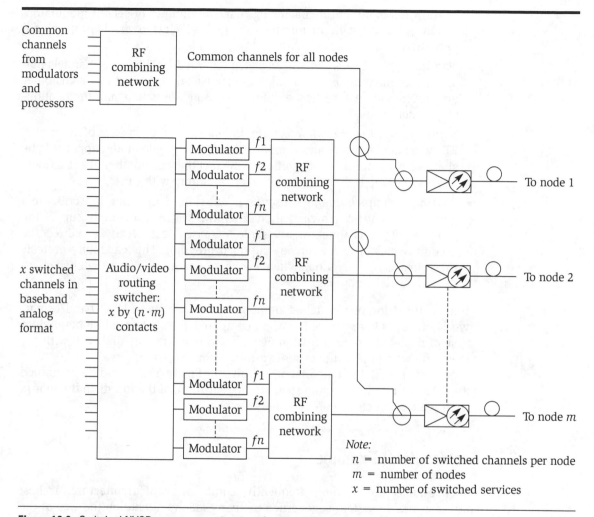

Common channels from modulators and processors

RF combining network

Common channels for all nodes

Modulator $f1$
Modulator $f2$
Modulator fn

RF combining network

To node 1

Modulator $f1$
Modulator $f2$
Modulator fn

RF combining network

To node 2

x switched channels in baseband analog format

Audio/video routing switcher: x by $(n \cdot m)$ contacts

Modulator $f1$
Modulator $f2$
Modulator fn

RF combining network

To node m

Note:
n = number of switched channels per node
m = number of nodes
x = number of switched services

Figure 16.9 Switched NVOD or narrowcast headend functional diagram.

and narrowcast channels are always present at the input of the matrix switch, whereas in the VOD case, channels are generated only by the server in response to (and under control of) individual subscriber requests. In the application described, video is not *generated* in response to subscriber request but only *switched* so that it is available to subscribers fed from specific nodes. Thus, the result is switched NVOD, not VOD.

The trade-offs in deploying a headend switching system for NVOD or narrowcast services are complex:

- *Node size.* Smaller nodes will each need fewer modulators since the number of switched channels required will statistically be less though the

relationship is not linear. Balancing that, the number of sets of modulators increases linearly with the number of nodes, which is inversely proportional to node size.

- *Division between common and switched bandwidth.* The more channels that are offered in common, the smaller the statistical likelihood that a customer on a given node will request a different channel though, again, the relationship is not linear.

- *Statistical sharing of modulators among nodes.* At the expense of additional RF switching hardware and complexity, frequency agile modulators could be shared among nodes in recognition of the fact that the number of subscribers requesting switched channels will vary randomly with time.

- *Sharing of channels among subscribers.* When more than one subscriber in a given node requests a switched channel, the system can simply supply the program on two channels or, at a slight increase in complexity, can direct the second requestor to the already supplied program. This capability protects the system against overload when an occasional offering on a switched channel is unusually popular.

Whether utilization of a switched architecture for delivery of NVOD and/or narrowcast channels is cost effective when compared with increased RF bandwidth or use of digital video compression depends on many factors that vary with the system and with time as the cost of various technologies matures.

Note that the NVOD architecture shown in Figure 16.9 could be combined with VOD services in the same spectral range if some of the inputs to the matrix switch were connected to on-demand servers.

16.3 Architectural Examples

Which architecture or, more frequently, combination of architectures is best suited for a cable system is governed by many factors, including the size of the system, the services to be offered, the density of homes, the cost of converting existing plant and headends, and the cost of building new plant. Discussed next are examples of hybrid architectures that have been built or proposed to suit specific situations and strategic plans.

16.3.1 Cable Area Network (CAN)

One of the earliest HFC architectures was the cable area network (CAN), conceived as a rebuild strategy for older coaxial networks. The principle is illustrated in Figure 16.10. Fiber-optic links from the headend are used to transmit the channel lineup to selected nodes inserted at regular intervals in a former all-coaxial distribution network. This effectively shortens the cascades, which allows the system to be upgraded to a greater RF bandwidth. Reliability, which is improved

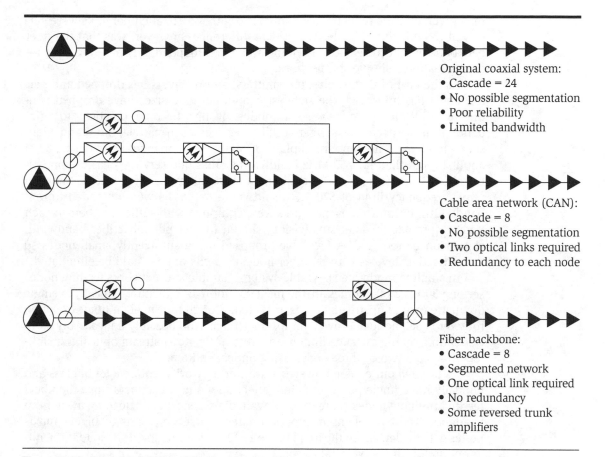

Original coaxial system:
- Cascade = 24
- No possible segmentation
- Poor reliability
- Limited bandwidth

Cable area network (CAN):
- Cascade = 8
- No possible segmentation
- Two optical links required
- Redundancy to each node

Fiber backbone:
- Cascade = 8
- Segmented network
- One optical link required
- No redundancy
- Some reversed trunk amplifiers

Figure 16.10 Principles of CAN and fiber backbone architectures.

by shortening the cascade, is further enhanced by leaving the coaxial network between the nodes in place as a backup in case a fiber strand or transmitter fails.

Although a number of systems were upgraded using the CAN architecture, they were limited by being constrained to a common channel lineup. In fact, the architecture was useful primarily as a mechanism for minimum-cost bandwidth expansion and reliability improvement.

16.3.2 Evolutionary Star Networks

The first step toward truly innovative fiber architectures is similar to the CAN architecture, except that instead of just inserting nodes into an existing amplifier cascade, greater cascade reduction was realized by feeding both directions from the nodes. This was originally known as a fiber backbone and, like CAN, was proposed as a cost-effective system upgrade strategy. It became favored over CAN because, although it lost the redundancy feature, far fewer fiber links were needed

for a given cascade reduction, as shown in Figure 16.10, and fiber links were perceived to be highly reliable. The critical difference, however, was that the fiber backbone architecture had created the ability to independently program nodes—thus, creating a simple star network.

As the cost of fiber cables, transmitters, and receivers has dropped and performance has improved, the smallest economic node sizes have dropped continuously. With sufficiently small nodes, it has become practical to offer bandwidth-intensive, subscriber-specific services such as telephony and high-speed modems. The announced plans of many cable operators and telcos, as of about 1996, called for 750-MHz bandwidths and node serving areas of approximately 500 homes. By 1998, equipment was commercially available with an upper frequency limit of 870 MHz, and some networks were upgraded to that bandwidth. In some cases, networks were equipped with sufficient fibers to each node to provide for later subdivision, should further per-subscriber bandwidth expansion be required. A few have proposed using sufficiently small nodes so that all active devices between fiber nodes and customers could be eliminated.

In small to moderate-size cable systems, simple star networks feeding nodes serving 500 to 2,000 homes are the most common architecture. Typically, a single downstream and upstream fiber is activated, though four to eight fibers are often routed to each node to provide for future services and possibly node subdivision. Depending on the distance from headend to node, downstream optical transmitters may feed two or three nodes using optical splitters.

As pointed out earlier, however, pure star networks require a lot of glass and typically have single points of failure affecting many subscribers since the fibers feeding multiple nodes share the few available physical paths leading away from headends. These problems are proportionately worse as systems get bigger and as nodes get smaller. Given the trend in the 1990s toward regional consolidation, different architectures were called for.

16.3.3 Double Star Networks

One way to decrease the required fiber footage is to create intermediate hubs between the headend and the more remote nodes. The signals can then be transmitted to the hubs, received, split, and retransmitted to nodes. Since there are fewer hubs than nodes and the distances between hubs and nodes are shorter than between nodes and the headend, a substantial reduction in total fiber footage can be realized. This is known as a double star architecture (Figure 16.11).

An example of a double star is the network that was constructed for one of the major telephone companies in several cities. In that system, a single headend sent a common set of video signals to each of the local telephone switching offices, which were utilized as hub locations. The signals were transmitted from there to nodes in each neighborhood of approximately 500 homes.

Though the double star is less expensive to construct, it has several disadvantages. For one thing, although it is possible to customize signals on a hub level, no

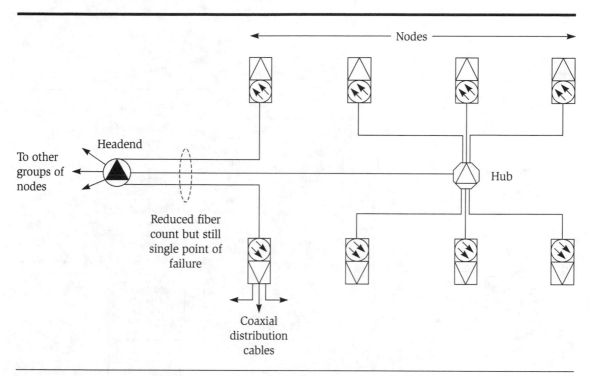

Figure 16.11 Double star architecture.

mechanism exists for creating individual node signals at the headend. Thus, per-subscriber signals must be generated at each hub. Second, the performance of each fiber-optic link must be better than in a single star since the noise and distortion will add. In one design, this forced a change from cascaded 1310-nm DFB transmitters to the use of an externally modulated 1550-nm transmitter with optical amplifiers and splitters at the hubs so as to get the required end-to-end performance and optical budget.

In the case of the telco system described earlier, the use of switching offices for the hubs provides a future expansion option since digital telephone signals can be generated by host digital terminals at each switch, modulated onto RF carriers, and inserted into the spectrum transmitted to each hub. Where DFB transmitters are used at the hub, this combining can occur at the RF level. Where optical amplifiers are used, separate DFB transmitters can send the signals over separate fibers to nodes using a split-band system.

16.3.4 Star/FTTC Hybrid

An interesting variant on the double star system was prototyped by another major telephone operator. In that case, a common spectrum of analog

Figure 16.12 A hybrid HFC/FTTC network for telephony, VOD, and broadcast video.

television channels was sent to 500 home nodes using a simple double star similar to that described earlier. Multiplexed telephony and switched digital video-on-demand signals were sent via independent fibers directly to shared terminal equipment in an FTTC architecture, with those fibers sharing the same rights of way as the HFC plant's components. The curbside network interface devices (NIDs) were powered from the HFC network, thereby solving one of the most difficult problems of FTTC networks. In fact, this was a dual cable distribution system—one coaxial and one fiber. At the interface unit, the telephony signals were separated from the digitized video and sent to homes via conventional twisted pair cables. The digital video signals (whose spectrum all lay below 50 MHz) were combined with the analog video signals and sent into homes using conventional coaxial drop cables. A set top terminal was required to decode the digital signals and to generate programming requests back to the

switched headend facility. Figure 16.12 is a simplified diagram of a portion of this network.

16.3.5 Hybrid Sheath/Shared/Repeating Digital Ring

Most current architectures utilize a star architecture, at least at the lowest level of fiber trunking. One U.S. cable operator, however, has elected to extend the reliability benefits of ring technology to the node level. Using an architecture it dubs "ring-in-ring," Cox Communications connects video services to nodes using a counterrotating sheath ring configuration (with a minimum of two fibers per node in each half of the ring).[5] This allows for the reliable distribution of node-specific programming.

In some cases, the backup fibers, rather than being dedicated to a specific node, carry only the downstream common signal lineup and are optically tapped to serve several nodes. This is a shared, analog ring configuration. Its use is an economic and reliability trade-off since it reduces the required fiber footage, as compared with a full sheath ring, but means that all node-specific programming, as well as direct RF return signaling, is lost if the primary link is broken. Where the shared ring is used, two fibers are used (one clockwise, one counterclockwise), with the optical taps in each case only in the second half of the ring to provide the required diverse routing.

An additional pair of fibers in the ring carry time multiplexed baseband digital signals and are used with SONET or similar terminal equipment to form a repeating digital ring.

Figure 16.13 shows a composite of the two ring-in-ring architecture choices. The primary FDM feed to each node is shown as light solid lines, whereas the dedicated backup feeds are shown dotted. The broadcast backup option is shown as a heavy, solid line, with optical couplers to feed nodes. Finally, the baseband, digital SONET ring fibers are shown in a dot-dash pattern. To illustrate both options, the backup for node 1 is from the shared analog broadcast feed, whereas the backup for node 2 is via dedicated, redundantly routed fibers. In each case, switches at the output of the node optical receivers select primary or backup feeds. If required by the length of the ring, the signals in the tapped, shared ring can be carried at 1550 nm to reduce losses and possibly amplified to reach all nodes (though it is not clear that Cox chose to do so). A pair of fibers serves SONET add-drop multiplexers in each node, which interface with host digital terminals and RF transceivers for telephony and data services.

The ring-in-ring name results from the operator's desire to limit the number of nodes per ring so that multiple, logically concentric, rings are typically constructed in large metropolitan areas. Cox suggests that the optimum configuration is 6–12 nodes per ring.

Although more fiber cable footage is required than for a simple star network, only two dedicated fibers per node are required throughout the ring so that the total fiber trunking costs are expected to be lower than for a star network with eight dedicated fibers per node.

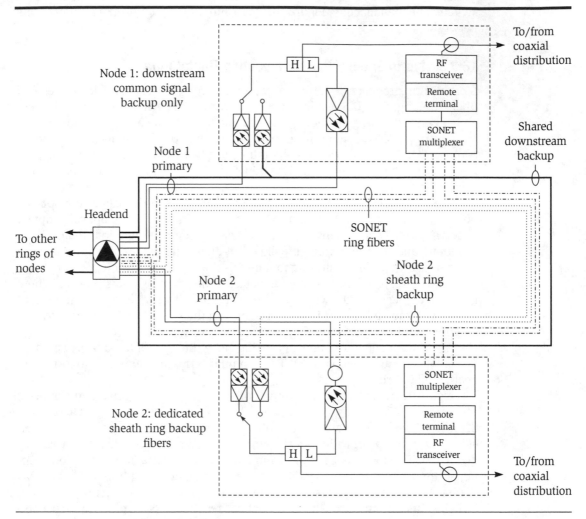

Figure 16.13 Cox ring-in-ring architecture.

16.3.6 Ring/Ring/Star with WDM

A double fiber ring architecture has been proposed that allows segmentation of broadcast signals on a regional (but not node) basis without requiring individual channel processing at major nodes.[6] The basic topology is shown in Figure 16.14.

At the headend, a separate combining network is used to create the FDM spectrum of analog video and QAM-modulated digital signals for the region served by each subring and master hub. Each of these spectra is used to modulate an optical transmitter operating on a slightly different wavelength in the 1550-nm region. The outputs from the transmitters are optically combined us-

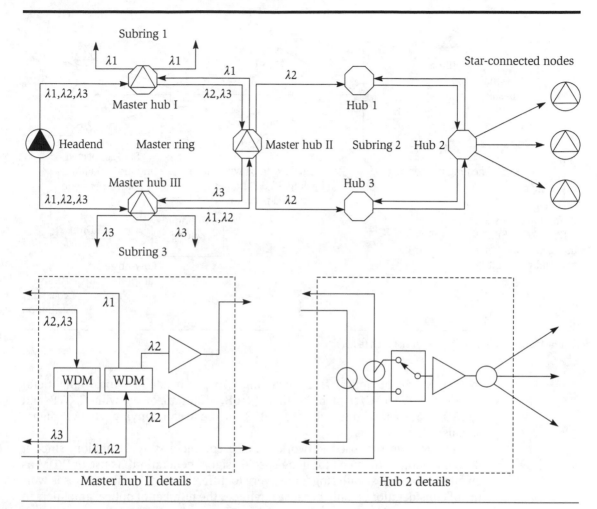

Figure 16.14 Ring/ring/star architecture using WDM.

ing a wave division multiplexer (WDM). The combined output is transmitted both clockwise and counterclockwise on the main ring using separate fibers. Depending on the degree of redundancy desired, the single optical output could be split, or separate optical transmitters could be used for each direction.

At each major hub, another WDM filter is used to separate the wavelength containing signals for that hub, which is then transmitted to secondary hubs via the subring. Depending on the degree of redundancy desired, this ring could be single or dual directional, as well. Finally, at each secondary hub, a portion of the signals are split off and sent to individual nodes in a star configuration.

Because the entire fiber network is broadband, the only postheadend active equipment required is optical amplification of the combined multiwavelength, modulated complex light signal. Single wavelength tests conducted through a

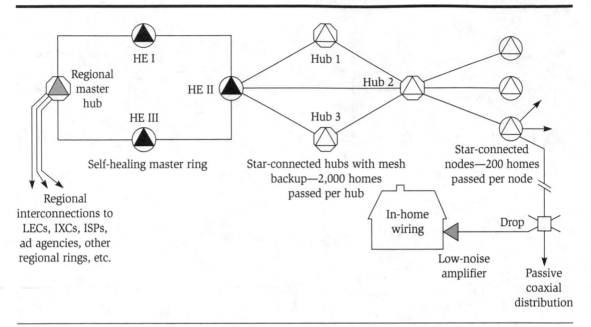

Figure 16.15 CableLabs' ring/mesh/star architecture.

total of 35 dB of optical loss (including two postheadend optical amplifiers) achieved a video C/N of more than 49 dB, simultaneously with C/CTB and C/CSO of at least 65 dBc and $<10^{-9}$ BER for the accompanying 64-QAM digital signals.

Though the proposed network is not a complete system solution, since it does not address the need for node-specific signal generation, its use of WDM as an alternative to a sheath ring for delivery to differentiated signals to hubs is worthy of consideration, if only because it reduces the number of optical amplifiers in physically long rings.

16.3.7 Ring/Mesh/Star Regional Architecture

In very large metropolitan areas, networks not only must distribute signals to end users but also must interconnect cable systems with studios, broadcasters, local and long-distance telephony carriers, and so on. The economies of scale allow the development of highly sophisticated multiservice headends in fully redundant configurations.

In 1992, CableLabs introduced a prototype architecture designed specifically for these situations,[7] which was refined in 1993.[8] The high-level structure, shown in Figure 16.15, consists of a master ring interconnecting a regional hub with formerly independent headends, mesh-connected hubs serving approximately 2,000 homes connected to each headend, star-connected nodes serving approximately

200 homes each, passive coaxial distribution to groups of homes, and amplified drops to each home.

The concept behind the master ring is that it is simply more economical to concentrate certain functions (such as advertising insertion, telephone switching, large servers, and digital video compression) at one location in each metropolitan area. This allows for very high quality and a high degree of redundancy with the costs spread across a large customer base. The architecture contemplates that the regionalized functions may not all exist at the same location in the master ring so that, for instance, one large cable operator may provide video server or advertising insertion services for smaller operators within the region, whereas another location may contain the telephone or PCS switch.

The master hub or any of the headends also serve as interconnection points to outside networks and sources of signals, such as other regional rings, local telcos, long-distance carriers, Internet service providers, and ad agencies.

The connections from the individual headends to hubs is basically a star arrangement, with cross connections between hubs providing the reliability enhancement of path-diverse routing of signals. Nodes are connected, using a pure star topology, from the nearest hub. By limiting the number of homes per node, it is proposed to use only passive coaxial distribution. This is made possible, in part, by the use of low-noise amplifiers in each home's drop so that lower than normal drop levels can be used and still provide sufficient signal strength to serve multiple user terminals in each residence.

Few details are given about detailed signal and transportation formats, except that both analog and digital distribution are shown at all levels in the network. It is contemplated that at least some of the signals in the master ring may be transported in an ATM-over-SONET repeating ring.

16.3.8 Hybrid Ring/Star

A final example is the architecture proposed by one of the authors to meet the needs of a particular client. This client served an area of fewer than 200,000 homes, including several small communities surrounding a central city. It was felt that regionalized advertising was not of benefit in this service area. The network was to have the capability to support broadcast video, NVOD, data modems, and possibly PCS, but not VOD or wired telephony. Four major hubs were connected to the headend using three ring structures:

- An analog, shared ring using 1550 nm and optical amplification is used to provide reliable delivery of the common signals to each hub, where they are optically amplified, split, and sent to nodes using a star. Use of separate optical transmitters at the headend and optical amplifiers at each hub provides a high degree of reliability. Given the size of the system and a lack of interest in eventually upgrading to VOD or other interactive services, a logical choice for NVOD services was the use of a 750-MHz bandwidth and digital video compression. This network layer is shown in Figure 16.16.

Figure 16.16 Shared hub ring/star node common signal distribution.

- An overlaid SONET ring carries the signals from the headend Internet high-speed gateway to and from the data-terminating equipment in each hub for the cable modem signals. This same ring may also carry the PCS telephone traffic from the headend-located telephone switch. A SONET add-drop multiplexer (ADM) separates the signals for each hub from the backbone stream. The PCS base stations and cable modem terminal equipment each create node-specific RF signals. These are combined at RF and sent via individual low-cost linear optical transmitters to each node on separate fibers from the common signals, as shown in Figure 16.17. The nodes are equipped with dual input optical receivers and are configured for split-band input (see configuration A of Figure 16.3).

 The return signals from each node will be received at the hub, and the PCS and data signals routed to data terminal equipment or PCS base-station transceiver, as appropriate. The network-side signals from that equipment, containing the multiplexed data from all nodes, will be connected to the SONET backbone using the same ADM.

- Finally, a sheath ring was provided to carry nonstandard, hub-specific signals, such as telemetry or upstream signals from set top terminals. In this

Figure 16.17 SONET ring/star node switched digital signal distribution.

case, it was less expensive to use separate fibers than to convert these signals to SONET format. The use of the sheath ring also preserves future options to send other signals to and from each node in non-SONET format.

Changing from a single star network to this configuration eliminated several very large single failure points and reduced significantly the total fiber footage required by the network. Furthermore, it allowed the system to defer most of the capital investment for nonvideo services until they were deployed.

16.4 Summary

The architecture of a system determines what services it can deliver. It controls effective bandwidth, reliability, flexibility, and distribution of signal processing. This chapter has described some of the architectural elements and how each is related to essential network characteristics. The examples given illustrate choices made or proposed to meet specific service requirements in specific service areas.

The most common architecture in use today is the single star with nodes connected directly to a single headend. In large regional systems, formerly indepen-

dent headends are often linked by either digital or 1550-nm analog fiber-optic links to a large master headend and become hubs. Single coaxial cable, low-split band plans are used in the vast majority of systems, with individual nodes serving 500 to 2,000 homes. Coaxial amplifier cascades vary from 2 to 10 in typical new upgrades.

A combination of unscrambled, analog scrambled, and sometimes digital encoding and scrambling is used to deliver basic, premium, and NVOD programming. Simple traps are still in common use to create multiple tiers among the unscrambled channels. Few systems offer true VOD or switched video programming. Point-of-entry and shared terminal equipment is still the exception, with addressable set top descramblers and individual cable modems the most common terminal equipment.

The relationship between architecture, service reliability, and service availability is a major topic in itself. The methodology for estimating reliability-related parameters is given in Chapter 17.

Endnotes

1. *National Electrical Safety Code.* Published by the Institute of Electrical and Electronics Engineers, New York, and approved by the American National Standards Institute as C2-1997.

2. Tom Osterman, Broadband Powering Issues, *Cable-Tec Expo '96.* Nashville, TN, June 1996, pp. 469–473. SCTE, Exton, PA.

3. Dan Kerr, Power Distribution in a Lifeline Network. *Communications Technology,* May 1997, pp. 64–70.

4. Leo Hoarty, Interactive Television—Architecture Under Test, *Symposium Record.* International Television Symposium, Montreux, Switzerland, June 1995, pp. 221–239.

5. Cox Communications corporate engineering staff, Rings, Clamshells and Spurs. *CED,* February 1995, pp. 24–34.

6. Dr. Shlomo Ovadia, Dr. Hongxing Dai, and Dr. Chinlon Lin, EDFA-Based Video Lightwave Trunking Systems. *CED,* June 1997, pp. 32–44.

7. Stephen Dukes, Photonics for Cable Television System Design. *CED,* May 1992, pp. 34–48.

8. Stephen Dukes, Next Generation Cable Network Architecture, *1993 NCTA Technical Papers.* NCTA, Washington, D.C., pp. 8–29.

Chapter 17
Network Reliability and Availability

17.1 Introduction

Chapter 15 described the network requirements of certain classifications of services that can be offered over broadband FDM distribution networks. Among those are transmission quality, effective bandwidth, *service reliability* (the probability that a system will survive without interruption for a defined period), *outage rate* (the average rate at which service interruptions occur), and *availability* (the percentage of time that service is available). Transmission quality was treated in detail in Chapters 9–14, and effective bandwidth was covered in Chapter 16. This chapter will deal with the calculation of network reliability and service availability. It will also discuss how those parameters vary as a function of the topology of the distribution system, among other factors. Finally, it will explore the difference between true availability and that experienced by users of a particular service.

Service interruptions can result from a variety of causes, including equipment failure, commercial power problems, interfering signals, or blocking due to inadequate circuit capacity, as discussed in Chapter 15. This chapter will deal with outages caused by either equipment or powering. Chapter 4 dealt with data traffic engineering, while Chapter 14 dealt with upstream interference issues.

17.2 History and Benchmarking

Logically, service availability should be just that—the net availability to the end user, irrespective of the cause of any interruptions. Historically, however, different industries have used different measures of their own service integrity.

17.2.1 Telephony

As we discussed in Chapter 15, the most widely quoted service availability requirement is 99.99% for local, wired, voice telephony service. It is important to review, however, how the exchange carriers define this parameter:

> The service availability objective for the subscriber loop is 99.99%, which corresponds to 0.01% unavailability or 53 min/yr maximum downtime. This objective, incorporating all network equipment between the local switch and the network interface (excluding the local switch and customer premises equipment), refers to the long-term average for service to a typical customer.[1]

This is not the same as end user service availability. It excludes, for instance, availability of dial tone from the switch, for whatever reason, or availability of interswitch or long-distance circuits. It also excludes any wiring within the home or terminal equipment problems. Though this may be understandable from a local network operator's point of view, the history of the cable television industry has been that in-home problems are the largest single cause of service outages.[2]

Second, this is a goal for performance that is averaged over both the entire customer base and over time. Thus, a significant minority of customers may have consistently unreliable service, yet the carrier meets the service goal if the majority of customers have few problems.

Third, carriers do not count "unavailability due to the loss of both primary and back-up powering"[3] in cases where there is powered equipment between the central office and home, such as fiber-in-the-loop (FITL) configuration. The importance to this exception depends, obviously, on the integrity of the field powering systems. It has been the experience of cable television network operators, however, that utility power interruptions cause more outages than all equipment failure classifications combined.

Finally, availability includes not only failure rates but also the time required to restore service. The problem is that, though a cut telephone drop wire (as an example) causes service to be unavailable, the network operator will typically not be aware of the outage until the customer tries to use the telephone and discovers the line is dead, then reports the outage to the phone company. The phone company calculates the length of the outage, for purposes of measuring compliance with the availability goal, from when it is *reported,* not from when it actually begins.[4] Arguably, therefore, the actual outage times may far exceed the reported times. As the traditional telephone network evolves by deploying fiber and performance-monitoring equipment closer to homes, this disparity will presumably decrease.

17.2.2 Cable Television

Although the cable television industry has not historically had formal goals covering availability or outage rates, CableLabs undertook a study of field reliability and subscriber tolerance of outages under the auspices of its Outage Reduction Task Force. Its report, published in 1992,[5] is discussed in detail in Chapter 15. In summary, the report set as a failure rate target that no customer should experience more than 2 outages in any 3-month period, with a secondary standard of no more than 0.6 outage per month per customer.

As with the telephony goal, it is important to review what is included in the definition of an outage:

- Only multicustomer outages are included, thereby excluding all drop, NID, set top terminal, or in-house wiring problems.
- Single-channel outages are included, thereby including headend processing equipment.
- Only signal outages, not degraded reception, are included.
- All elements of signal distribution between headend and tap are included, including the effects of commercial power availability.
- Although the language refers only to outages that are "experienced by customers," the software distributed with the report calculates all outages, and therefore, it can be assumed that customer awareness of outages is not

important to the definition, even though the goal was based on customer experience.

- The "no customer shall experience" clause in the goal clearly indicates that a worst-case, not average, analysis is called for.

Based on an estimated mean time to restore service of 4 hours, CableLabs translated 0.6 outage per month into a minimum acceptable availability of 99.7%.

The equivalence between customer-perceived individual outages and distribution system outages is tenuous at best. On the one hand, some outages may be detected and repaired without some subscribers being aware of the outage; on the other hand, individual subscribers may have in-home problems that will not be included in the preceding definition. Nevertheless, CableLabs chose 0.6 multicustomer outage per month as its benchmark for distribution system outage modeling.[6]

In the field measurements portion of its study, the task force gathered actual field failure analysis reports from some of its members and used those to predict failure rates of certain classifications of components (discussed further later) and to develop models that could be used to predict the failure rates of arbitrary network architectures. The accuracy of the raw data used in this type of analysis is limited by the analytic skills of the field technician and by its completeness. For instance, it is not obvious that a technician would think to report a brief outage caused by a routine maintenance procedure or by a brief power outage that did not cause a formal trouble call to be entered into the dispatch system. Second, the gathered data is necessarily representative of typical cable systems of that time and did not include modern HFC systems. Finally, cable systems (at least in 1992) routinely reported only *distribution system outages,* not *individual customer outages,* in their reports.

17.2.3 Cable Television Versus Telephony Reliability and Availability Definitions

Clearly, the CableLabs and Bellcore definitions of reliability and availability are not equivalent, and the numbers should be not compared directly. Table 17.1 illustrates the approximate inclusion of each industry's outage definition compared with those elements that actually affect customer service.

As a final note, *availability* is traditionally the most important parameter applied to telephone services though CableLabs found that the *rate at which outages occurred* was most important to television viewers, with the length of the outage (and thus availability) of only secondary importance.[7]

17.3 Definitions and Basic Calculations

It is important to have a clear understanding of the terminology used in calculating reliability-related parameters. Thus, the following definitions will be used throughout this chapter.

Table 17.1 Comparison of Bellcore and CableLabs outage definitions.

	Factors Included in Definition of Outage						
Outage Definition	*Centralized Processing/ Switching*	*Distribution Network*	*Distribution Power*	*Drop*	*Network Terminal Equipment*	*In-Home Wiring*	*Analysis*
Bellcore		X		X	X		Average customer
CableLabs	X	X	X				Worst-case customer
Customers' Experience	X	X	X	X	X	X	

- *Component failure rate.* The failure rate of a component or system is defined as the statistical probability of failure of any one of a similar group of components in a given time interval. The measured period should be contained within the normal service life of the device—that is, failure rate is not intended to measure time to wear-out, but rather "midlife" random failures. For the purposes of this chapter, we will generally express failure rates on a yearly basis.

 Failure rates of individual electronic components (capacitors, resistors, semiconductors, and so on) are generally determined by their manufacturers using carefully designed tests that are conducted on a sample of products selected at random from the production line.

 Failure rates of system-level components such as amplifiers, taps, connectors, and set top terminals can be determined theoretically based on the failure rates of their internal electronic parts and the way in which they are interconnected, or they can be determined based on actual failure data from field-installed units. We have a strong preference for the latter since the installed failure rate is influenced not only by design but also by craft issues, environmental conditions, and how the unit is powered.

 When field observations are used to determine failure rate, the appropriate formula is

$$\lambda = \frac{k}{nt} \tag{17.1}$$

where

λ = the annual failure rate

k = the number of observed failures

n = the number of items in the test sample

t = the observation time in years

The failure rate can be converted to other units as required. For instance, the failure rate per hour is the annual failure rate divided by 8,760.

- *System failure rate—simple nonredundant connection.* Where n components are connected in such a way that the failure of any one of them will lead to a failure of some defined system or function (such as a subscriber service outage or a single-channel failure), and the failures do not have a common cause, the resultant failure rate of the system is given by

$$\lambda_{system} = \sum_{i=1}^{n} \lambda_i \qquad (17.2)$$

where

λ_{system} = the net system failure rate

λ_i = the failure rate of component i

n = the number of components, any of whose failure will cause a system failure

Note that this formula is not exact since it assumes that none of the failures overlap in time. Thus, it is accurate only when the failure rates and repair times are such that multiple, independent, simultaneously occurring outages are statistically insignificant. Generally, this is a good assumption for well-designed and managed cable systems.

- *System reliability—simple nonredundant connection.* System reliability is the probability that the system will not fail during some specific period. In general, reliability can be calculated once the failure rate is known using

$$R(t) = e^{-\lambda t} \qquad (17.3)$$

where λ and t are in consistent units (that is, failure rate per year and years or failure rate per hour and hours).

The result of this calculation may be less than intuitive. For instance, if the annual failure rate of a given system were 1 (meaning that if we measured a large number of similar systems, we would detect a number of failures per year equal to the number of systems), the probability of any single system surviving one year without a failure is $e^{-1} = 0.37$.

Where components or subsystems of known reliability are connected in a configuration where the failure of any component leads to a system failure, the net system reliability can be calculated using

$$R_s(t) = R_1(t) \; R_2(t) \dots R_n(t) \qquad (17.4)$$

where

$R_S(t)$ = the net system reliability over period t

$R_1, R_2, \ldots R_n$ = the reliabilities of the individual components, measured over the same period.

- *Mean time between failure (MTBF)*. The predicted time between failures is simply the inverse of the failure rate, or

$$MTBF = \frac{1}{\lambda} \tag{17.5}$$

where *MTBF* and λ are in consistent units.

Generally, this chapter will express MTBF values in hours. When manufacturers provide predicted MTBF data, it is also generally given in those units. As with failure rate, MTBF is not a measure of useful service life. In fact, many components have an MTBF that exceeds their wear-out time.

- *Mean time to restore (MTTR)*. When a failure does occur, the mean time between failure and restoration of the defined function is the MTTR. For the purposes of this chapter, MTTR values will usually be expressed in hours.

 Note that MTTR is sometimes used to express mean time to *repair* rather than mean time to *restore*. We use the latter definition here to distinguish between when service is restored and when the failed component is actually repaired. Since it is more common in cable television to restore service after a failure by swapping a good for a failed network component, this is more accurate when our goal is to determine service reliability.

- *Availability (A)*. The ratio of time that a service is available for use to total time is known as availability. It can be calculated from MTBF and MTTR (expressed in consistent units) using

$$A = \frac{MTBF}{MTBF + MTTR} \tag{17.6}$$

- *Unavailability (U)*. Unavailability is the ratio of time that a service is unavailable to total time and is, by definition, equal to $1 - A$.

- *Outage time (T_U)*. Outage time is the amount of time that the network is unavailable during a defined period. In general, the outage time during some period t is equal to the unavailability multiplied by t. It is common to express outage time in minutes per year, which can be calculated as follows:

$$T_U(\text{min/yr}) = \text{minutes/year}(1 - A) = 525{,}600(1 - A) \tag{17.7}$$

17.4 Effects of Redundant Network Connections

The simple calculations above apply to situations in which the failure of any single component leads to the failure of the defined system or function. In a

redundant connection, either of two (or more) components can provide a function should the other(s) fail. Thus, the system user experiences a failure only if the redundantly connected components simultaneously fail. As discussed in the previous chapter, a common example in fiber architectures is a ring configuration where signals can travel by either of two cable routes between head-end and hubs.

- *Unavailability.* When two network sections are connected in a redundant configuration, the equivalent unavailability is the product of the unavailabilities of the two individual sections. Since, for each section, the unavailability is simply the number of outage minutes per year divided by the total minutes in a year, the equivalent unavailability is

$$U_R = \frac{(T_{U1})(T_{U2})}{(525{,}600)^2} \tag{17.8}$$

where

$$U_R = \text{the net unavailability}$$

$$T_{U1} \text{ and } T_{U2} = \text{the yearly outage times (in minutes) of the redundantly connected sections}$$

Converting this back into equivalent yearly outage minutes for the combined section:

$$T_{UR} = 525{,}600 U_R = \frac{(T_{U1})(T_{U2})}{525{,}600} \tag{17.9}$$

- *Failure rate.* The equivalent failure rate is the sum of the failure rate of the first feeding segment multiplied by the probability of the second section failing before the first can be restored, plus the failure rate of the second section multiplied by the probability of the first section failing during an outage in the second section.

 The probability of section one failing during an outage in the redundantly connected section two is

$$P_{F1} = 1 - e^{\frac{-(\lambda_1)(MTTR_2)}{8{,}760}} \tag{17.10}$$

where

$$P_{F1} = \text{the probability of failure of section one}$$

$$\lambda_1 = \text{the failure rate of section one per year}$$

$$MTTR_2 = \text{the period (in hours) over which the probability of failure is to be measured, namely, the MTTR for a failure in network section two}$$

Thus, the net failure rate when two sections join in a redundant feed is

$$\lambda_R = \lambda_1 \left[1 - e^{\frac{-(\lambda_2)(MTTR_1)}{8,760}} \right] + \lambda_2 \left[1 - e^{\frac{-(\lambda_1)(MTTR_2)}{8,760}} \right] \qquad (17.11)$$

17.5 Absolute Versus User-Perceived Parameters

The preceding equations yield reliability-related performance parameters for the network, independent of usage. For applications that require continuous use of the network, the calculated failure rates and availability of the network are the same as those experienced by a user. For most services, however, usage of the network by individuals is only partial. For instance, most families watch television about 4 hours per day, whereas telephone users may access the public switched telephone network for only a half hour daily.

The importance of partial usage is that some outages will occur and service be restored before all potential users are affected. Thus, the average network reliability, as experienced by occasional users, will be higher than for full-time users. However, the relationship between the network outage rate and availability and that experienced by occasional users is complex and affected by not only average aggregate usage time but also by the pattern of usage.

As an example, if a television subscriber watches 4 straight hours of television every evening, but at no other time, and the network MTTR is 1 hour, the probability is that he or she will experience slightly more than one-sixth of all outages. On the other hand, a person who uses the same network to make a 1-minute phone call every 6 minutes (say, a 24-hour business using a telephone modem to do credit card verifications) will experience every outage (and, in fact, will perceive a greater outage *rate*, since the restoration time for each outage will affect 10 attempted phone calls) even though his or her total daily usage time is the same.

A further complication is in the user's response to a failure and how we quantify the modified usage pattern. For example, assume that a user normally averages one phone call every 4 hours. If the user attempts to make a call and finds the phone not working, he or she may try again every few minutes until successful. If the outage lasts an hour, the user will have experienced many more failures than during his or her normal usage pattern. At the other extreme, a television viewer who would normally watch an evening's programming may, upon discovering an outage, abandon viewing for the entire evening and thus not be aware that the outage lasted only a few minutes. Again, the viewer's perception of the network's availability is affected by his or her changed usage pattern. A legitimate question for reliability modeling is whether these failure-induced retries are counted as independent failures.

For a hypothetical fiber-to-the-curb, fully status-monitored network, one researcher has calculated that a telephone user making 10-minute calls spaced 100

minutes apart (roughly a 9% utilization factor of the network) would experience 30.2 of the network's 53 minutes of unavailability every year.[8] If any portion of the network is unmonitored, the outage time experienced by the customer will increase and could, in fact, exceed the Bellcore-defined outage time for the network!

For cases where usage patterns are not modified as a direct result of outages, we can estimate perceived outage rates and service availability as follows.[9]

First, define usage pattern for a service consisting of an average *call length* as T_C, and an average *call cycle* time (from the start of one call to the start of the next call) as T_S. To keep things simple, all times will be expressed in hours.

The probability of a service failure during a call period T_S can be calculated from Equation (17.3) (the probability of failure is 1 minus the reliability):

$$P_f = 1 - e^{-\lambda T_s} \tag{17.12}$$

where

λ = the service failure rate per hour

T_S is in hours

If we multiply the probability of failure per call cycle by a scaling factor that takes into account both call length and service restoration time, we can calculate the probability of failure per attempted usage of the network:

$$P_C = (1 - e^{-\lambda T_s}) \left(\frac{T_C + MTTR}{T_S} \right) \tag{17.13}$$

where T_C and $MTTR$ are both in hours.

The total number of attempts to use the service per year can be calculated using

$$B = \frac{8,760}{T_S} \tag{17.14}$$

from which we can calculate the user-experienced yearly failure rate:

$$\lambda_{EXP} = P_C B \tag{17.15}$$

The average length of experienced outages is

$$T_O = \frac{(T_C)(MTTR)}{T_C + MTTR} \tag{17.16}$$

so that the total experienced outage time per year is

$$T_{EXP} = (P_C)(B)(T_O)$$ (17.17)

Finally, the experienced availability (that is, 1 minus the ratio of outage time to attempted usage time) can be calculated using

$$A_{EXP} = 1 - \left[\frac{(P_C)(T_O)}{T_C}\right]$$ (17.18)

As an example, assume that a network fails once every 3 months (MTBF = 2,190 hours, $\lambda = 0.000457$ per hour) and that the MTTR is 2 hours. Based on this data, an application that used the network continuously would experience an outage rate of 4 per year, a yearly outage time of 8 hours, and an availability of 0.999088.

A television viewer using the network 4 hours per day ($T_C = 4$, $T_S = 24$) would experience a yearly outage rate of 0.9953, a yearly outage time of 1.327 hours, and an effective availability of 0.999091. Thus, although using the network only one-sixth of the time, the viewer experiences roughly one-fourth the outage rate but nearly the same effective availability as a 24-hour user.

A telephony user who makes a 10-minute call every 4 hours ($T_C = 0.16667$, $T_S = 4$) would experience a yearly outage rate of 2.1665, a yearly outage time of 0.333 hour, and an effective availability of 0.999087. Thus, even though the telephone user makes use of the network for a lower percentage of the time, he or she experiences about the same effective availability but a higher failure rate than the television viewer because of his or her more frequent attempts at access.

As can be seen, network effective availability—the probability that service is available—is quite insensitive to usage patterns, whereas experienced outage rate is closely tied to both call length and call rate.

17.6 Network Analysis

Once the basics of reliability are understood, they can be applied to the analysis of actual and proposed networks. The steps to analysis include breaking the network into manageable sections, properly assigning the components to the sections, determining the failure rates of individual network components, estimating the service restoration times for failures at various network levels, and calculating the various reliability parameters. Though the principles of the process will be described here, in practice, the calculations are much more easily performed using a computer, which also allows the engineer to assess the effects of changes in topology, operating practices, and component reliability.

17.6.1 Logical Subdivision of the Network for Analysis

The first step is subdividing the network into segments. The degree to which this is required depends on the level of analysis required. For instance, if the only parameter of interest is the failure rate and availability as experienced by the most distant subscriber, then the many branches in the coaxial distribution system can be ignored and only the longest cascade analyzed in a single calculation.[10] As a minimum, however, any network branches that form redundant structures must be separated, analyzed separately, and then used as variables in the equations in Section 17.3 to determine the effective performance of the redundant structure.

For those desiring a more thorough analysis, a separate analysis of each physically separate branch and calculation of reliability data at every customer port will allow additional data such as total failures and the performance as experienced by the average as well as by the most disadvantaged customers. It also allows analysis of the contributions of various network sections to overall failure rate and unavailability. The analysis given at the end of this chapter was done using a large spreadsheet that allowed a very detailed analysis.

In order to limit the amount of data to be analyzed, it is common to analyze only a typical subsection of a large network. In the case of an all-coaxial network, this might be a single trunk, whereas in the case of an HFC network, it is likely to consist of a single fiber-fed node, along with the fiber trunking structure feeding that node.

Figure 17.1 shows a typical example. In this case, a sheath ring fiber structure is used between the headend and the node receiver, while coax cascades of up to five amplifiers are used beyond the node. Should only the performance to the last user be required, then analysis of all but the longest cascade can be omitted. Only the downstream optical structure is shown in the figure. See Section 17.6.6 for a discussion of the calculation of two-way structures.

Figure 17.2 shows the logical division of this network into segments for analysis. Note that the number assigned to each section identifies a unique structure, whereas the letter denotes the instance of that structure. Thus, all four of the segments numbered five are similar, and only a single analysis of that structure is required. Since a single headend optical transmitter feeds both downstream fibers between the headend and node, it is part of segment 1A. Separate optical receivers are used in the node, however, so each of the segments numbered 2 includes both the fiber and connected node receiver.

17.6.2 Assigning Network Components to Subdivisions

Sometimes, network components' effect on transmission integrity may be determined by other than their physical placement in the network. Power supply 1 in Figure 17.2 is physically located in section 5A, but powers the node and the first amplifier in both sections 5A and 5B. Therefore, for purposes of calculating reliability parameters, it should be included in section 3A. Similarly, power supply 2 is physically located in section 6A, but must be included in section 4A since it

Note: Only downstream F/O network shown for simplicity. Taps omitted for clarity, untapped cable shown solid, and tapped cable shown dashed.

Figure 17.1 Topology of typical HFC node and sheath ring trunk.

Figure 17.2 Division of network into segments for analysis.

powers the amplifier that feeds sections 7A and 6A, and its failure would affect the signals in all the downstream branches.

With the components properly assigned to network segments, based on their effect on operation, the components in each section should then be listed in a column as a first step to segment analysis.

17.6.3 Determining Component Failure Rates

In order to estimate the performance of the defined sections, the field failure rate of each network component type must be determined. Manufacturers' design analysis, based on the MTBF of the individual components used and their temperature/voltage stress levels, often have limited relationship to failures of field-installed equipment where such factors as voltage transients and the skill levels of field personnel may dominate inherent failure rates. Also, a failure of a series-connected component may or may not cause a loss of service to all downstream subscribers. For instance, a tap port may fail, causing a loss of service to one customer but not disrupting the through-signal to subsequent taps.

With those caveats, it is instructive to examine failure rates of a few key components.

- *Commercial power.* CableLabs' task force recorded average commercial power failure rates of 30% (all failure rates are on a yearly basis unless otherwise noted), meaning that the probability of a commercial power interruption at any given power supply is 0.3 each year. Since this data is based on outages caused by power failures, it is to be presumed that this figure relates to *unprotected* commercial power outages rather than all power outages. Also, given that their data was derived from reported system outages, it can be presumed that many short interruptions were not included in the data.

 By contrast, the Network Reliability report quotes from a study showing average commercial power *unavailability* of 370.2 minutes per year.[11] They did not report on the rate at which outages occurred, however. The net unavailability with varying durations of standby power capacity is shown in Table 17.2.

 From this data, it is readily apparent that even standby battery capacity of several times the 2 hours normally designed into cable television standby power supplies will result in unprotected outages well in excess of Bellcore's 53-minute availability guideline. Thus, normal standby powering will need to be supplemented by other means (such as built-in generators or status monitoring combined with deployable mobile generators) for high-availability services.

 It is the experience of one of the authors, however, that the reliability of commercial power varies widely from location to location so that Table 17.2 can be taken only as a measure of average conditions. Accurate network reliability predictions must depend on locally viable commercial power data

Table 17.2 Average commercial power unavailability in the United States.

Average Service	Zero Backup	4-Hour Backup	8-Hour Backup	12-Hour Backup
Unavailability (minutes/year)	370.2	192.4	153	128.5

unless networks are so hardened against commercial power outages as to make the issue irrelevant.

- *Power supplies.* Independent of the availability of commercial power is the reliability of the power supplies themselves. CableLabs' data has suggested a failure rate of only 3%. Though such a rate may be achievable by a non-standby power supply, it is the industry's common experience that standby power supplies are often much less reliable (though the overall reliability of commercial power plus standby power supply will exceed that of a non-standby supply by itself). Standby supply reliability will be dramatically affected by the age and condition of the batteries, as well as by the level of routine maintenance in the system.

- *Amplifiers.* CableLabs found field failures occurring in trunk amplifiers at the rate of 10%, whereas failures in the simpler line extenders occurred at a 2% annualized rate. "Distribution" amplifiers used in modern HFC networks have a complexity somewhere between these extremes, and 5% is often used in modeling HFC networks though upgraded designs may justify lower values. Werner of TCI has quoted "empirically derived" failure rates for distribution amplifiers as low as 1.75%.[12] Merk has suggested that operators use failure rates of 0.5% for trunk amplifiers and 0.15% for line extenders, based solely on warranty repair data.[13] This range of values illustrates how difficult it is to be confident of failure rates and how competent engineers can arrive at rates that vary by more than 10:1.

- *Optical transmitters.* Werner and Merk are consistent in reporting 2.3% downstream transmitter failure rates. If CableLabs' failure rates for other headend equipment of 20–30% are to be believed, then these values are suspiciously low. On the other hand, the base of equipment that went into the latter study varied in age, whereas optical transmitters are likely to be much newer given the relatively recent widespread rollout of HFC systems.

 Werner rates upstream transmitters identically to downstream, whereas Merk suggests that upstream transmitters experience a failure rate of only 0.9%. Given that most upstream transmitters operate at much lower powers, do not require a thermoelectric cooler, and do not require a high-power driver, there is some logic for believing a lower upstream value.

- *Optical receivers.* When evaluating optical receivers, it is important to distinguish between the receiver itself and the entire optical node. Since nodes often include full trunk amplifier stations, it is advisable to rate the receiver independently of its host station unless it is a stand-alone unit. Werner's

value here is 1.7%, whereas Merk uses 0.7%, with both authors rating upstream and downstream receivers essentially identical.

- *Passive devices (taps, power inserters, splitters, directional couplers).* MTBF data supplied to CableLabs by manufacturers suggested that passive device failure rates should be on the order of 1%. CableLabs' field data suggested that actual numbers were closer to 0.1%—a rare case where actual experience was superior to predicted. Werner's data is even more optimistic at 0.04%. Merk suggests the use of values between 0.07% and 0.26%, depending on the device.

- *Coaxial connectors.* Neither CableLabs nor Merk suggests failure rates for coaxial connectors though CableLabs' "cable span" failure rate may be a composite of the actual cable and the attached connectors. Werner suggests a failure rate of 0.01%. Although this may seem insignificant, it must be remembered that the signal must pass through at least two connectors for every series-connected device in the system so that a cascade of 5 amplifiers, 1 power inserter, 2 splitters, and 25 taps will include 66 connectors.

 The preceding values are for the connectors used for solid-sheathed aluminum cable. From experience, the failure rate (as measured by outages and associated trouble calls) for drop connectors is much higher. Based on actual experience operating numerous systems, it is suggested that 0.25% be used pending better industry statistics. As with distribution connectors, a typical drop system will include several series-connected F connectors. Among all the components considered, the greatest actual field variation may be in the performance of drop connectors owing to both craft sensitivity and the wide range of available connector quality choices.

- *Fiber-optic cable.* The modeling program provided with CableLabs' report suggests the use of 3% failure rates for fiber cable spans (regardless of length). Merk suggests the use of Bellcore data of 0.44% per mile, whereas Werner suggests the use of 0.44% (presumably per span). Hamilton-Piercy suggests that Bellcore's data on fiber failures is equivalent to 0.3% per span per year but uses his company's experiential value of 1.5% per span.[14] Clearly, these are widely varying numbers that reflect several variables.

 For one thing, the Bellcore data was based on analysis of "FCC-reportable" outages only, meaning those that affect at least 30,000 customers.[15] Thus, most instances of fiber damage on the subscriber side of switches went unreported, whereas the failures occurred in telephone interoffice cables that are placed in well-protected trenches because of the high volumes of traffic being carried. Fiber serving small pockets of customers is more economically installed and more likely to be damaged as a result. For example, most of the damaged cables reported by telephone companies were located 4–5 feet underground, well below the depths used in most cable television construction.

 Second, the lengths of the spans considered vary from 5 miles or less for typical suburban HFC architectures to 50 miles or more for long-distance

telephone links. Finally, available failure data may have been difficult to coordinate to the total length of the failed cable.

Given that most fiber cable failures are due to dig-ups and should increase proportionally with distance, a value of 0.1% per mile is suggested pending better data. This is roughly equivalent to Hamilton-Piercy's value if 15-mile average link distances are assumed, or Werner's value if 4.4-mile spans are assumed.

- *Coaxial cable.* Coaxial distribution cable might be thought to suffer similar failure rates. CableLabs' field data, however, suggests 3% annual failures. Even if it is assumed that spans are a half-mile long (about the longest expected between trunk amplifiers), this would be the equivalent of 6% per year per mile, or 40 times the optical cable failure rate. Werner's 1996 data suggests the use of 0.23%—less than one-tenth the failure rate. Based on the fact that fiber cable is generally armored and therefore both stronger and not as subject to rodent damage, it seems reasonable to use a value of 0.2% per mile for distribution coaxial cable or double the risk factor for fiber cable.

 Drop cable is not as well protected, nor is it made to the same quality standards as distribution cable. Analysis of trouble call rates and causes suggests that annual failure rates of drop cables may be in the 1–2% region.

- *Set top terminals and other operator-supplied terminal equipment.* The failure rates of terminal devices will obviously vary considerably with complexity and age. New, nonaddressable terminals will have very low failure rates, whereas first-generation addressable products often failed at several percent per month. Few authors who have modeled cable television failure rates have included terminal data since CableLabs' definition excludes individual subscriber outages. Based on field failure data from several systems over a number of years, we suggest that a default value of 7% be used, lacking more device-specific data. For purposes of analysis, this failure rate is taken to include nonhardware failures such as the failure of an addressable terminal to receive an enabling command, data entry errors that result in unintended deauthorizations, and power failures that cause the terminal to lose its authorization status for some time after power restoration. From the subscribers' viewpoint, these are still service outages.

- *Active network interface devices (NIDs).* Where services terminate at the outside wall of residences, the equipment is subjected to greater weather exposure but may be rigidly mounted and is not subject to handling damage. Bellcore suggests that NIDs containing active termination equipment, whether for video or telephony service, should have a net unavailability of about 26 minutes per year.[16] If it is assumed that individual customer outages are restored in a mean time of 8 hours, this is equivalent to a 5.4% yearly failure rate.

- *Headend/rack-mounted equipment.* CableLabs' initial data indicated very high (20–30%) failure rates among headend equipment. In general, however, a failure rate of 5%, averaged over all classifications of headend equipment,

is more in line with operators' experience and will be used in our example calculations.

In the case of telephony, headend units are required to interface between the ports on the switch and the HFC distribution system. Although the term often has a more precise meaning in different contexts, host digital terminal (HDT) will be used to refer to all the headend equipment between the switch and broadband distribution system, considered as a subsystem. Although Bellcore has not predicted the failure rate of HDTs, they have assigned 10 minutes of annual unavailability to this subsystem. On the basis of an assumed 0.5-hour MTTR, the HDT is assigned an annual failure rate of 33%.

In summary, there is reasonable industry agreement on some average failure rates but widely diverging data on others. As the cable industry gains more experience with reliability as an essential element of network design, better data will likely become available for predicting performance. We have a bias toward carefully collected field data derived from systems operating in conditions similar to the system to be analyzed (regional differences in such factors as lightning strikes and the availability of laws requiring utility locates before commencing underground construction can have a major effect on outages). For the purposes of the example discussed later in this chapter, the annual failure rates shown in Table 17.3 will be used.

17.6.4 Estimating MTTR

Although the elapsed time between failures is a function of network architecture and the quality of the components used, the time required to restore service after an outage occurs is almost entirely within the operator's control and is a function of staffing levels and training in addition to performance-monitoring systems.

Total time to restore service consists of four time segments:[17] (1) the time between the start of the outage and when the operator is aware of it, (2) the time for crews to respond to the outage location, (3) the time for the technician to identify the cause of the outage, and (4) the time to repair or replace the failed item and restore service.

Werner quotes average MTTR values varying from 66 to 243 minutes, depending on the diligence of the system operator. Though CableLabs did not put forth an industry goal, they did quote from a Warner Cable guideline calling for 2-hour maximum MTTR for damaged fiber-optic cables.[18] Data provided by the telephone industry to the FCC on fiber repairs showed that, in most cases, service was restored within a 2–4- or 4–6-hour window.[19]

Historically, many cable systems used manual trouble call routing, staffed during business hours only (except for major outages, storms, and so on), and operated many headends without any regularly present technicians. As larger, regional systems offering a wide variety of services have evolved, 24-hour, 7-day staffing of headends is no longer uncommon, and field service staffs often work 2 shifts, with on-call technicians for late-night situations.

Table 17.3 Typical individual network component annual failure rates.

Component	Failure Rate	Component	Failure Rate
Commercial Power (unprotected)	300%	Optical Receivers	1%
Power Supplies (nonstandby)	3%	Passive Devices	0.1%
Power Supplies (standby)	6%	Coaxial Connectors (distribution)	0.01%
Amplifiers (trunk or distribution)	4%	Coaxial Connectors (drop)	0.25%
Amplifiers (bridger modules or line extenders)	3%	Fiber-Optic Cable	0.1% per mile*
Optical Transmitters (downstream)	3%	Coaxial Cable (distribution)	0.2% per mile
Optical Transmitters (upstream)	1.5%	Drop Cable (length independent)	1.5%
Redundancy Switches	0.5%	Set Top Terminals/Modems	7%
NIDs	5.4%	Headend Equipment	5%
HDTs	33%		

* Since the major cause of failure is dig-ups and cuts, the failure rate applies to the entire cable and all the fibers within it. Therefore, no redundancy gain is realized by using separate fibers to deliver signals to a node, for instance, if they share a common cable sheath.

As an economic choice, the response to outages is proportional to the number of customers affected. Large outages inevitably clog the customer service telephone lines and may lead to bad publicity since customers already affected by loss of service are doubly inconvenienced by being unable to report the situation.

Based on experience from a number of systems, the MTTR values in Table 17.4 will be used in the example discussed later in the chapter. To the extent that operators do better or worse than these values (which may be overly pessimistic for a well-organized system), the availability will vary from that calculated.[20] Factors such as traffic conditions, dispersal of personnel, training, availability of repair materials, and communications systems and utilization will have a major impact on MTTR.

17.6.5 Special Considerations Within Headends and Hubs

When Bellcore set up the availability standards for telephony, they specifically excluded the switch and everything on the network (nonsubscriber) side of the

Table 17.4 Typical cable system MTTR values.

Failure Level	MTTR
Headend (24-hour, 7-day staffing)	0.5 hour
Fiber trunk cables	2 hours
Optical node components	1 hour
Distribution system coaxial components	4 hours
Individual customer drop or components	8 hours with remote monitoring 12 hours without remote monitoring (customer call-in delay)

switch. CableLabs, on the other hand, specifically included headends to the extent that a single-channel outage is counted the same as a total system outage.

Those wishing to model cable system reliability factors may wish to consider some interim approach that accounts for all critical common equipment (such as combining amplifiers and headend power) but only some percentage of individual channel-processing equipment. Without such an approach, it can readily be shown that the failure rate of a modern 750-MHz fully loaded headend will be theoretically very high, yet the average viewer will be unaware of those failures if they happen to occur in channels he or she is not watching.

In previously published studies, one of the authors has proposed that viewers, in a typical viewing session, are exposed significantly to 10 channels (that is, they access and view programming on 10 channels in an evening of typical viewing), and he therefore modeled headend reliability based on the sum of the critical common equipment, plus all the equipment required to create 10 channels (satellite antennas, receivers, modulators, and so forth).[21] In the example discussed later, it has been assumed that the viewer is exposed to the potential failure of the approximately 30 channel-specific pieces of equipment needed to process the 10 channels viewed, plus 3 common combining amplifiers. For other services, different calculations of headend equipment reliability might be appropriate. As with equipment failure rates, there is little statistically valid field information on the number of channels viewed by typical subscribers.

A complete analysis of video headend-caused subscriber outages should include not only signal-processing equipment but control equipment, for instance, the telemetry generators that send enabling signals to terminals, addressability control computers, and (for two-way terminals) upstream data receivers.

17.6.6 Special Considerations for Two-Way Systems

Most of the distribution system is shared between upstream and downstream signals, including amplifier stations (though a detailed analysis might assign different failure rates to upstream and downstream modules within the station), power supplies, passive devices, and the cable itself. The optical portion of the plant,

however, almost always uses separate fibers for upstream and downstream communications, and always uses separate upstream and downstream optical transmitters and receivers.

When modeling the optical trunking segments, therefore, both upstream and downstream transmitters and receivers are critical to two-way applications and must be included. Assuming the fibers share a common cable, however, the reliability of a pair of fibers will not differ significantly from a single fiber, and so the number would be the same as for one-way transmission. When calculating the failure rate of the headend for a two-way application, it is also necessary to include equipment that processes both incoming and outgoing communications.

17.6.7 Calculation of Failure Rates and Availability

Once the structure of the network is defined and the field failure rates and repair times of the equipment used are known, the next step is to calculate the total failure rate of each segment. This is simply the sum of the failure rates of each of the elements in the segment (Equation (17.2)). Next, using Equations (17.6) and (17.7), it is possible to calculate the availability of each section and the total yearly minutes of outage.

If the only result desired is total system failure rate and availability to the most affected subscriber, the failure rates and outage minutes of each of the critical series-connected sections can be added. The net availability can then be calculated using Equation (17.7). Additional data can be derived from the analysis, if desired, by, for instance, calculating the number of subscribers affected by each outage, calculating the outage rate and availability to each subscriber, and/or calculating the cost of repair for each class of failure. Although this level of analysis is much more complex, it yields a significant amount of additional information about the performance of the network, including

- Reliability and availability to the average, as well as most affected, customer
- The total number of outages and the total number of customer outages (the summation of each outage multiplied by the number of customers affected)
- The total cost of outage repair
- The distribution of outages by size

This additional information is very helpful in making design trade-offs that include not only the cost of construction but also the required customer service staffing levels and predicted repair costs.[22]

17.7 Analysis of a Typical HFC Network

There have been many published analyses of the reliability and availability of specific network configurations. One of the most comprehensive compared 13 different architectures and found availability values ranging from 99.7% to

Figure 17.3 Simplified schematic diagram of analyzed system.

99.99% based on 0.3% fiber link failure rates and 10% amplifier failure rates (failure rates for other components were not given). Using 1.5% fiber link failures and 5% amplifier failures, two architectures were found to exceed 99.98% availability: a ring/ring/star and a double star.[23]

As an illustration of attainable reliability factors, as well as the degree of information available from a comprehensive analysis of failure rate data, we give a complete analysis of a typical HFC cable system.

17.7.1 Architecture

The analyzed system is typical of an early 1990s upgrade. The downstream bandwidth extends to 550 MHz, whereas the return bandwidth is limited to 30 MHz. Although it is logically a simple single star network, several nodes are served from each large fiber-optic cable leaving the headend, and the analysis accounts for this shared risk. Figure 17.3 is a simplified diagram of a typical node. As with Figure 17.1, the tap configuration is not shown though each of the dashed lines contains taps (and sometimes splitters and/or directional couplers). Even though

not shown in the figure, the taps and branching are included in the analysis. The distribution system extending from each node passes approximately 2,000 homes, with a basic penetration rate of 70%. The total number of homes served from the headend is 150,000, split among 75 similar nodes.

The coaxial amplifier cascade beyond the node is limited to 4, and the entire node distribution system contains 53 amplifiers. Three power supplies are required to power all the active devices, with a maximum power supply cascade of two. The total number of series-connected taps in any one distribution leg is about 20.

The initial analysis is based on the use of a generator, but no uninterruptible power supply (UPS), at the headend and field standby power supplies with 2-hour battery capacity. It is assumed that this results in a 30-second headend outage every time the commercial power fails (three times per year) until the generator "kicks in," and that the field standby power supplies have the effect of reducing the field failure rates to 50% per year in each location, with 1 hour of unprotected outage when the batteries do run down (based on dispatching a crew with portable generators as a result of customer-reported outages). It is assumed that there is no status monitoring of power supplies that would have allowed crews with portable generators to back-power supplies before the batteries expire.

17.7.2 Video Services Performance

The network was analyzed for delivery of conventional cable television video services. It was assumed that the viewer was "exposed" to the reliability of 30 pieces of headend channel-processing equipment plus 3 headend combining amplifiers during a typical viewing session, as discussed in Section 17.6.5.

A typical home drop system was assumed, including a set top terminal. No monitoring of the home equipment was assumed so that subscribers were exposed to all failures occurring beyond the tap (since they wouldn't be known to the network operator until reported by the user), but only a portion of the failures earlier in the network, as discussed earlier under user-perceived reliability factors.

The outage rate per year for the most affected customer is 6.11. Though there are no specific standards for telephony service, this is close to CableLabs' suggested maximum of 0.6 per month for video service. (Even though the CableLabs' guideline did not include drop and premises elements, this analysis does since subscriber satisfaction was measured on overall outage rate independent of cause.) Figure 17.4 shows the relative contribution of various network sections to this rate. As can be seen, powering is by far the largest contributor though headend outages also have a major effect than for unavailability. The large power contribution reflects the use of a generator, but no UPS at the headend, so that customers experience a short outage each time the power fails at the headend.

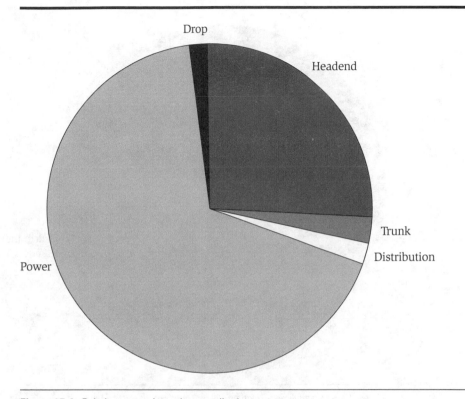

Figure 17.4 Relative network section contributions to outage rate.

The network availability to the most affected customer was calculated to be 0.99951, equivalent to yearly service outage time of about 4.3 hours. This unavailability is only one-sixth CableLabs' initial goal for cable television service.

As Figure 17.5 shows, the largest contributor to unavailability is also powering, followed by the drop, headend, coaxial distribution network, and fiber trunking. The large contribution from powering reflects both the limitations of battery capacity and the lack of status monitoring. (In Section 17.7.4, we will see the effect of adding power supply status monitoring.) Although the drop hardware is reasonably reliable, its contribution to total outage time is disproportionately large since the MTTR is much higher than for outages affecting multiple customers. The size of the headend contribution is affected primarily by the large number of moderately reliable units involved in producing a multichannel video lineup. Nevertheless, the headend contribution is proportionately less than for outage rate since the MTTR is much less.

A third useful result of the failure analysis is the distribution of multi-subscriber outages on the basis of the number of customers affected. This analysis is important because it allows the network operator to size his or her customer

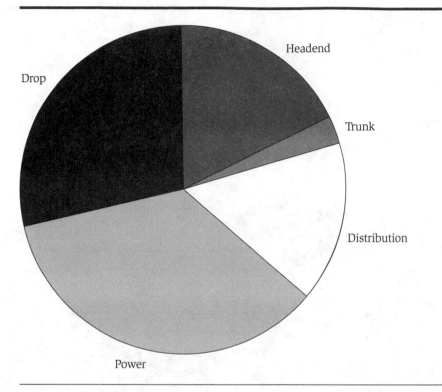

Figure 17.5 Relative network section contributions to outage hours.

service operation knowing the expected volume of outage-related customer calls. Figure 17.6 shows the outage size distribution for the network analyzed.

17.7.3 Wired Telephony Services Performance

The network was also analyzed as a delivery means for wired telephony. In that case, there was no exposure to headend video equipment, and in accordance with Bellcore methodology, the switch reliability was not included though the interface between switch and distribution system (the host digital terminal) was. In the case of telephony, the drop system included only the drop itself, plus a network interface device (NID), on the assumption that the in-house wiring and analog instruments should not be included in accordance with Bellcore practice. The failure rates assigned to the HDT and NID resulted in 10 and 26 minutes per year of downtime in accordance with Bellcore's allocations in TR-NWT-000418.[24]

When analyzed in this telephony configuration, the outage time is 26% less than for video services, resulting in an availability of 0.999636, whereas the number of outages is 27% less (4.82 per year). Analysis of the causes of the remaining outages shows that power-related issues dominate all others. Although the

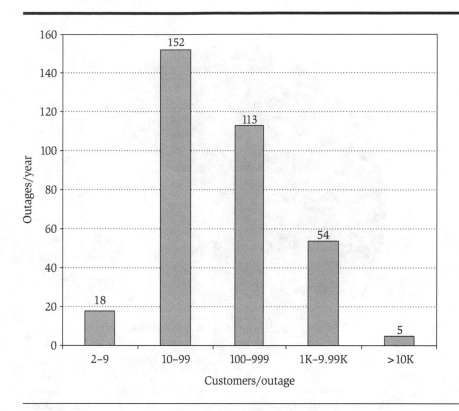

Figure 17.6 Outage distribution by number of affected customers.

unavailability is about 3.6 times worse than Bellcore standards for wired tele-
phone service, they are not comparable because of the factors not included in
the Bellcore outage definition (this analysis, for instance, includes powering and
measures outages from actual occurrence, not from when they are reported).

17.7.4 Effect of Improved Powering on Telephony Performance

If a UPS is added to the headend and status monitoring to the field supplies (so
that crews can deliver temporary generators before the batteries lose their
charge), then all power-related outages can be eliminated except those due to the
reliability of the powering equipment itself. In that case, the telephone service
availability improves to 0.99976, and the outage rate declines dramatically to
0.82 per year.[25] Having said that, however, it must be pointed out that the cable
industry's experience with early status-monitoring equipment was often that it
was less reliable than the equipment being monitored.

 As Figure 17.7 shows, powering is still a significant contributor to network
unavailability but is less important relative to the drop (dominated by the NID
failure rate and slow repair time) and the distribution network. Figure 17.8 shows

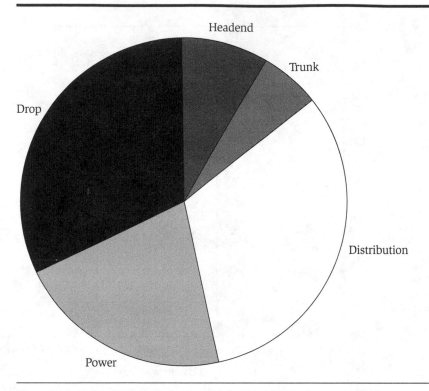

Figure 17.7 Reduced effect of powering with headend UPS and status monitoring.

that the number of large outages has dropped dramatically as a result of improved powering.

It is clear from this analysis that achieving a true 0.9999 availability to the consumer, should that be required, requires several fundamental changes in this network:

- Improved HDT and NID reliability—their Bellcore-assigned outage time allocations consume 68% of the 53-minute total allowed.

- Hardened and more reliable powering. The field power supplies installed by most telcos for their HFC networks were extremely reliable (and expensive) by traditional cable television standards.

- Shorter cascades of both coaxial equipment and power supplies. In general, telco-rated HFC networks are powered by a single power supply located at the node and have active equipment cascades beyond the node of only two to three.

- Reliable status monitoring throughout the network. Since the time required to be aware of a failure, then to analyze its cause, is a material part of total

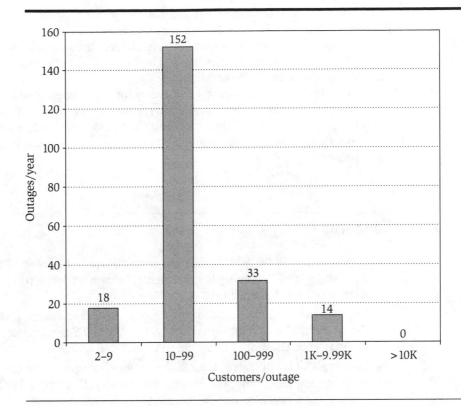

Figure 17.8 Reduction in number of large outages with improved powering.

MTTR, extensive status monitoring plays an essential part of reducing repair times.

Note that the fiber trunking portion of the network was a minor contributor to unavailability in all cases, even in this simple star network. Thus, use of various ring topologies will reduce the instance of large outages (sometimes an important consideration) but will not significantly reduce network unavailability in this network. As other factors are brought under control, the fiber contribution will become more important.

17.8 Summary

The estimation of network availability, failure rate, and other reliability-related factors is very straightforward, if somewhat tedious, given a topology, known component failure rates, and repair times. Unfortunately, a great deal of the effort needed to produce accurate reliability predictions is involved in developing believable component and commercial power failure rates. Once a logical model of

a proposed network is constructed and entered into a spreadsheet for analysis, it is simple to test the effect of different component choices and operating practices.

The preliminary analysis done in this chapter illustrates that modern HFC networks are easily capable of achieving low customer-experienced outage rates and unavailability for video services.

HFC networks that are designed properly for wired telephone services are capable of achieving the required availability, whereas lower-cost networks may be entirely adequate for various levels of video and data services. It is the function of the network engineer to design the most cost-effective network to carry the desired services.

Endnotes

1. Bellcore TA-NWT-00909, Issue 2, Generic Requirements for FITL Systems Availability and Reliability Requirements, December 1993, p. 13-1. Bell Communications Research, Piscataway, NJ.

2. Brian Bauer, In-Home Cabling for Digital Services: Future-Proofing Signal Quality and Minimizing Signal Outages, *1995 SCTE Conference on Emerging Technologies*, Orlando, FL, pp. 95–100. SCTE, Exton, PA.

3. Bellcore TR-NWT-000418, Issue 2, Reliability Assurance for Fiber Optic Systems System Reliability and Service Availability, December 1992, p. 31. Bell Communications Research, Piscataway, NJ.

4. Network Reliability Council, *Final Report to the Federal Communications Commission, Reliability Issues—Changing Technologies Working Group, New Wireline Access Technologies Subteam Final Report*, February 1996, p. 13.

5. CableLabs, *Outage Reduction*, September 1992. CableLabs, Louisville, CO.

6. Ibid., p. II-5.

7. Ibid., p. V-10.

8. Network Reliability Council, op. cit., p. 15.

9. Private communication from Andrew Large. The formulas are valid for cases where $T_c + MTTR < T_s$.

10. This statement is not true when considering *signal* as opposed to *hardware* reliability in the upstream direction since ingress and noise from all branches will affect the reliability of communications on any given branch. Also, as will be seen, we are talking about logical, not physical, cascade here, and the powering configuration may require the modifications to what we would normally consider the cascade order. Finally, electrical short circuits in network branches that are not part of the signal cascade can nevertheless affect communications.

11. Network Reliability Council, op. cit., p. 16. The figures are quoted from the paper authored by Allen L. Black and James L. Spencer entitled, An Assess-

ment of Commercial AC Power Quality: A Fiber-in-the-Loop Perspective, and published in the *Intelec '93 Proceedings.*

12. Tony Werner and Pete Gatseos, Ph.D., Network Availability Consumer Expectations, Plant Requirements and Capabilities of HFC Networks, *1996 NCTA Technical Papers.* NCTA, Washington, D.C., pp. 313–324.

13. Chuck Merk and Walt Srode, Reliability of CATV Broadband Distribution Networks for Telephony Applications—Is It Good Enough?, *1995 NCTA Technical Papers.* NCTA, Washington, D.C., pp. 93–107.

14. Nick Hamilton-Piercy and Robb Balsdon, Network Availability and Reliability. *Communications Technology,* July 1994, pp. 42–47.

15. Network Reliability Council, *Network Reliability: A Report to the Nation, Compendium of Technical Papers.* National Engineering Consortium, Chicago, June 1993, pp. 1–32.

16. Bellcore TA-NWT-00909, Issue 2, Generic Requirements for FITL Systems Availability and Reliability Requirements, December 1993, p. 13-5-6. Bell Communications Research, Piscataway, NJ.

17. Werner, op. cit., p. 318.

18. CableLabs, op. cit., p. V-16.

19. Network Reliability Council, *Network Reliability: A Report to the Nation, Compendium of Technical Papers,* National Engineering Consortium, Chicago, June 1993, p. 22.

20. One large MSO reports systemwide MTTRs ranging from 1.64 to 2.39 hours for multiple customer outages. The measured systems are all HFC with various sizes of nodes and wide deployment of both fiber node and coaxial amplifier status monitoring. All headends have UPS plus generator backup. Private correspondence from Nick Hamilton-Piercy.

21. David Large, User-Perceived Availability of Hybrid Fiber-Coax Networks, *1995 NCTA Technical Papers.* NCTA, Washington, D.C., pp. 61–71.

22. David Large, Reliability Model for Hybrid Fiber-Coax Systems, *Symposium Record, 19th International Television Symposium and Exhibition.* Montreux, Switzerland, June 1995, pp. 860–875.

23. Nick Hamilton-Piercy, op. cit.

24. Bellcore TR-NWT-000418, Table 4, Loop Downtime Allocations—FITL Systems, December 1992, p. 37. Bell Communications Research, Piscataway, NJ.

25. The average monthly video availability among cable systems comprising one large MSO (that has widely deployed status monitoring) varied from 0.9992 to 0.9996 over a 10-month period. Private correspondence from Nick Hamilton-Piercy.

Part Six
Customer Interface Issues

Chapter 18
Program Denial Technologies

18.1 Introduction

To a large degree, the modern cable television industry is founded on the ability to deliver premium programming for which subscribers pay, either on a monthly subscription basis or on a per-program basis. Key to offering premium programming is the ability to deny the program to anyone who has not paid for it. The denial technology must offer an adequate level of program denial combined with a high-quality recovered signal and reasonable cost. It must be transparent to the paying subscriber, and it must offer reasonable immunity to defeat. These requirements conflict and have led to many advances in technology.

The cable industry, in contrast to the telephone industry, tends not to be standards oriented though this is now changing. Rather, individual manufacturers have developed proprietary program denial schemes. This has led to equipment incompatibility, but it has also resulted in a degree of resistance to pirating: the industry has not been a single large target, but has been a number of small targets, making it somewhat less attractive to pirates. Nonetheless, theft of premium signals has been a major concern of the industry for many years. In this section, we describe the major program denial technologies that are in use though we stop short of describing technology that may aid in the development of pirate devices.

The word *scrambling* is used to mean modification of the video (and sometimes the audio) signal such that it cannot be viewed on a television receiver. A *descrambler,* which has some unique piece of information, is able to convert the signal back into a form visible on a television set. This usage of the word *scrambling* is contrasted to the unrelated usage in the data transmission field. In data transmission, the word refers to the process of modifying a datastream to remove any repetitive bit patterns for the purposes of equalizing the spectral density of the signal and aiding in data recovery. It is common in the data communications field to provide for separate access denial, the technology of which is called *encryption.*

We use the word *pirate* to refer to anyone who attempts to receive programming for which no payment has been made. Usually, it is applied to those who either develop or sell the technology to receive such programming. Sometimes, though, the word is used to refer to a subscriber who purchases equipment to receive unauthorized programming.

18.2 Non–Set Top Resident Program Denial

The first class of program denial technologies to be discussed are those that normally don't reside in set top terminals. These have also been known as *whole-house* systems because they operate on signals before they enter the subscriber's home. A signal enabled with any of these technologies is available to all TV sets in the home that are connected to cable. Some of these denial technologies represent the oldest technology still in use, whereas others represent much more re-

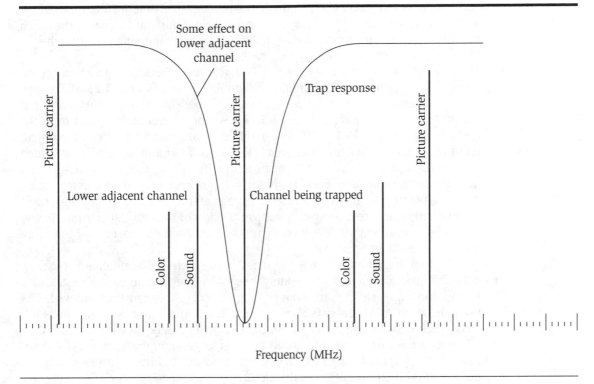

Figure 18.1 Operation of a negative trap.

cent efforts. Today, the efforts in this direction are driven by the need to allow the subscriber to avoid using a set top terminal if he or she doesn't need it for tuning.

18.2.1 Negative Traps

The first technology widely used for denial, and one that still has many adherents, is called *negative trapping.* Signals are sent over the cable in the clear, and a trap is installed at each home that does *not* take the program. Negative traps are frequently employed for a service with a high take rate, that is, with a large percentage of subscribers who buy a subscription. By protecting such services with a negative trap, the operator needs to supply a trap only for the minority of customers who don't take that service.

Figure 18.1 illustrates the functioning of a negative trap. The filter has a notch, or trap, at the picture carrier of the channel to be denied. By removing the picture carrier (and lower-frequency sidebands), the TV has no signal to recover. Since no signal enters the home, the technology is rather secure. The biggest security threat is unauthorized removal of the trap from a drop. This may be discovered by physical audit in many cases though in some apartment complexes

auditing may be a problem. The other issue in security is that the trap may fail, allowing signals to pass. Traps may drift with time or temperature so that the notch moves off the picture carrier. Enough signal power is then delivered that the TV can recover a picture, although a degraded one.

A disadvantage of a negative trap is that it will often have an effect on the lower adjacent channel. The frequency to be removed is only 1.25 MHz away from the channel edge and 1.5 MHz above the adjacent channel sound carrier. If the trap is used on channel 3, the adjacent sound carrier is separated from the notch frequency by 2.4%, so it is relatively easy to control the rise of the notch. However, if the notch is on channel 13, the adjacent channel sound is separated only by 0.7%, making the notch much more difficult to realize. The nature of a filter is that the transition band is proportional to frequency all else being equal. Also, temperature drift is a constant percentage of the filter frequency. Thus, traps are normally restricted to use in the lower portion of the cable spectrum. As you go to higher frequencies, the effectiveness of the notch decreases, and the problems it causes increase.

A trap will also introduce group delay in the lower adjacent channel. Group delay is introduced near the transition region of a filter: signal components closer to the band edge of a filter are delayed with respect to those further removed. The lower adjacent color carrier is closer to the transition region than is the lower adjacent picture carrier so the color information will be delayed with respect to the luminance information. If the group delay is large enough, it produces the so-called funny paper effect, in which color information is shifted with respect to the black-and-white information in the picture.

18.2.2 Positive Traps

A similar technology is that of positive trapping. The shape of a positive trap is nearly identical to that of a negative trap, but the trap frequency and method of use are different. An interfering carrier is inserted at the headend, and a trap is used to remove the interfering carrier at the home of a subscriber who takes the service. Positive traps are popular for services having low penetration since they are installed only at the homes of the people taking the service.

Figure 18.2 illustrates the operation of a positive trap. One or two jamming carriers are inserted at the headend, usually 2.25 MHz above the picture carrier. The peak amplitude is usually the same as, or slightly higher than, the picture carrier amplitude. The jamming carrier(s) may be modulated with various signals. In a mixing process, such as the picture demodulator in the TV, the second harmonic of the jammer (with respect to the picture carrier) is generated and falls on the sound carrier, affording a large degree of scrambling for the sound carrier.

The positive trap is located at the jamming carrier frequency so as to remove the jammer. Assuming the jammer is completely removed, the TV will be able to recover the signal. Since the trap cannot be made infinitely narrow, it will remove some desired modulating sidebands as well as the jamming carrier. Except at the jamming frequency, this may be at least partially compensated for by

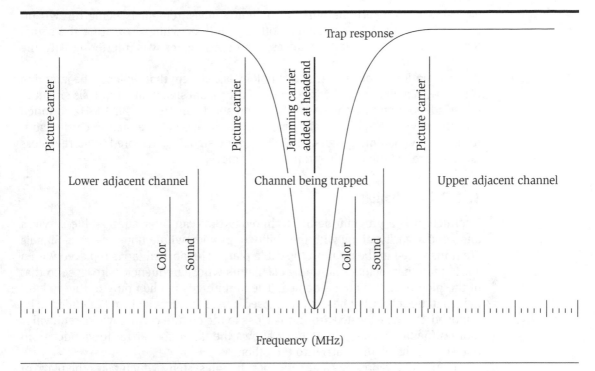

Figure 18.2 Operation of a positive trap.

preemphasizing the midband frequencies at the headend. Preemphasis accentuates the frequencies that will be partially removed later such that the net result is an approximation of normal sideband levels. Preemphasis may also be applied to the group delay introduced by the trap. Because the amount of power around 2.25 MHz in a typical video signal is low, the picture damage is often considered acceptable. Nonetheless, positive traps have a reputation for "softening" a picture.

Since the positive trap notch frequency is toward the center of the channel, we would expect less effect on the lower adjacent channel than that caused by a negative trap. Since the trap can be located inside the subscriber's home, the temperature range over which the trap must function is reduced.

A jamming carrier generator is connected to each modulator carrying a signal to be positively trapped. This jamming carrier usually interfaces with the modulator using either the picture IF loop or the composite IF loop of Figure 7.3. In addition, the baseband video signal may be predistorted to compensate for the amplitude and delay effects described earlier. This can be done either at baseband or at IF.

Engineers must be cautious when measuring the level of a signal that includes a jamming carrier. Depending on the instrument used to measure the level, the jamming carrier may or may not interfere with the measurement. The

easiest way to determine whether or not a measurement is being thrown off by the jamming carrier is to turn off the jammer while measuring the signal level. If the measured level changes, then the jammer was interfering with the measurement.

A variation on positive trap technology is a system that replaces the jamming carrier with a highly emphasized copy of the visual signal: an emphasis circuit at the headend amplifies the baseband video spectrum around 2.25 MHz so much that the power at that frequency is about as great as at the picture carrier, so it forms its own jamming signal. This "jamming signal" is restored to more-or-less normal level by the positive trap at the home.

18.2.3 Interdiction

Interdiction is a program denial technology that can offer addressable services and good security, along with the ability to provide whole-house service. Signals are transmitted in the clear on the cable plant. The normal tap is replaced with a special tap that includes jamming oscillators whose frequencies are close to that of the picture carriers to be denied. The jamming is applied only to those channels that the subscriber has not purchased. Since the jammer is not intended to be removed, it can be located arbitrarily close to the picture carrier, where removal is more difficult. One system actually allows the jamming carrier to wander from one side of the picture carrier to the other.

Figure 18.3 represents a generic interdiction system, which takes the place of the normal tap. The interdiction system includes a directional coupler DC1 and an amplifier A1. After amplification, the signal is split in CM1 so as to be routed to all tap ports. Directional coupler DC2 routes signals to the data receiver or transceiver of the addressing system. Directional coupler DC1 and splitter CM1 compose a normal tap. (In this text, splitters are referred to using the designation CM for combiner because the same device is used for either task.) Each port has another amplifier, used mainly to provide isolation for the jamming carriers, to prevent them from leaking to other outputs. A series of jamming oscillators is provided, which can be switched in and out under command of the addressing system. When an oscillator is connected to a port, it is set to a frequency close to, and sometimes equal to, the picture carrier. The signal received in the home is rendered useless because of the jamming.

The advantage of the interdiction system is the ability to control a number of channels, allowing the operator to provide advanced programming options such as can be provided using a set top terminal. At the same time, interdiction provides excellent security and no equipment in the home. It does not interfere with the use of VCRs or picture in picture, and the signals are available to all TVs in the house.

The disadvantages include the need to supply equipment for every subscriber, regardless of his or her purchase of premium services. Powering is also an issue: the equipment must be mounted away from a subscriber's reach so that he or she doesn't have access to the unscrambled signals, which exist on the plant

Figure 18.3 Interdiction system.

side of the interdiction equipment. This normally means mounting at the tap location. Powering is from the plant, so the cable operator assumes the power bill rather than its being assumed by the subscriber. Home powering has been proposed, but the problem is delivering significant useful power to the unit through the drop cable while observing Underwriters' Laboratories guidelines for the maximum voltage and power delivered in certain service categories.

18.2.4 Broadband Descrambling

In a broadband descrambling system, signals are scrambled at the headend, and descrambled enmasse at the home though only the signals purchased are descrambled. One technology employs carrier suppression as the root scrambling mechanism. If the carrier is suppressed on a television signal, we can show that the signal demodulated with a conventional demodulator will have sync in the correct relation to the darker portions of the picture. But as the signal becomes whiter, at some point the carrier vanishes, and as the signal continues to go toward white, the demodulated signal will "turn around" and go back toward the sync level.

At the home, special circuitry recovers the phase of the carrier, reinserting it with the desired signals and rendering the signal descrambled. When this type of system is deployed, it is necessary to treat the scrambled signal much as one would for other common types of headend scrambling. This subject is covered in more detail in the next section.

18.3 Set Top Resident Program Denial

The dominant program denial methods involve use of set top terminals in the home. Although many technologies have been proposed to effect scrambling, only two are in common use in North America at the time of this writing: sync suppression and video inversion (others have been used elsewhere and are covered later). The security afforded with these program denial methods is not total. However, the industry has decided, at least for analog signals, that other methods are not cost effective. Indeed, a review of the plethora of attacks on signal security in the last 15 years reveals that the most successful attacks against reasonably modern scrambling methods are those commonly called "spoofing." In spoofing systems, a modified set top terminal is "spoofed" into "believing" that it is authorized to descramble a particular signal. The attack is through the authorization system, not the scrambling system. Though there have been attacks against the scrambling system, they usually either have been minimally successful or have yielded to signal modifications in the headend.

18.3.1 Sync Suppression Scrambling

The most common scrambling method in North America is, by far, sync suppression. Variants of sync suppression have been employed since the earliest practical scrambling systems, dating back at least to the mid-1970s.

Recall from Section 8.5 that the sync tip corresponds to the highest amplitude of the modulated RF envelope. If the sync is somehow modified by moving it to amplitudes occupied by active video information, then a television set will not be able to synchronize. The effect is a screen full of random, more-or-less horizontal, lines.

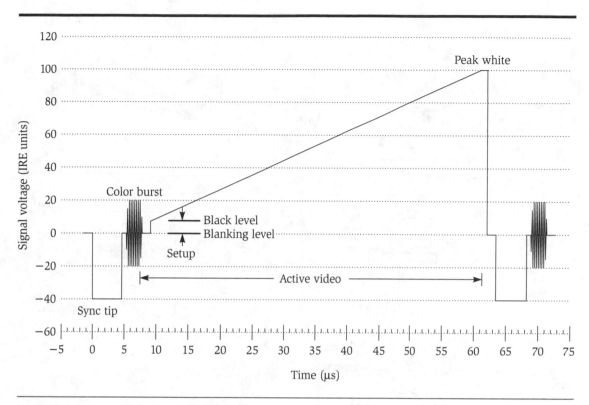

Figure 18.4 Reference video waveform used to illustrate scrambling systems.

The Scrambling Process

In order to illustrate certain features of scrambling systems, a reference waveform is used, as shown in Figure 18.4. Scrambling operations will be illustrated with reference to this waveform. (Refer to Chapter 2 for more information concerning the video waveform.) The reference waveform shown in Figure 18.4 is a ramp, which rises from black, at the left side of the screen, to white, at the right. The setup shown is a feature of NTSC video that is left over from the early days of television. Color burst is shown though no color information is shown during the active line. In general, we would expect additional color subcarrier power during the active line.

Figure 18.5 illustrates a typical relationship between a modulator and an RF sync suppression scrambler. The portions of the modulator shown are identified as they are in Figures 7.3 and 7.4. Flags B and D are as shown in those two figures, where they were used to route signals from one figure to the other. Baseband video is looped through the scrambler, where a high-input impedance amplifier couples video to a sync separator. The scrambler must have access to the video sync so that it can synchronize the suppression and restoration information. The

Figure 18.5 Sync suppression scrambling circuitry and waveforms.

process of routing video through a piece of equipment is often called *loop-through* and was described in Section 7.8.1.

From the loop-through input of the scrambler, the video is routed to the modulator, where it is processed as described in Chapter 7. In the scrambler, the sync separator extracts vertical and horizontal sync, which are passed to timing circuitry. The timing circuitry operates a switched attenuator, AT10, in the visual IF path, and an amplitude modulator, M10, in the aural IF path.

The top left of Figure 18.5 shows the reference waveform of Figure 18.4 modulated onto the IF carrier. As explained in Chapter 2, the modulation format prescribed for NTSC (and PAL) video is negative modulation, in which the sync tips correspond to the highest amplitude of the RF (or IF) signal, and peak white cor-

responds to the minimum amplitude. Sync can be suppressed simply by switching in an attenuator, AT10, during the sync time. This results in the waveform shown at the top right of the figure. During the horizontal (and, usually, vertical) blanking intervals, AT10 is switched in, and during the active video, it is switched out. Now the sync tips no longer correspond to the highest amplitude of the modulated waveform. Two values of attenuation are commonly employed: 6 and 10 dB. The waveform shown has been scrambled using a 6-dB attenuator. Had a 10-dB attenuator been used, the horizontal blanking interval would be of even lower amplitude.

From attenuator AT10, the signal is returned to the modulator, where it passes through the vestigial sideband filter FL2. The process of switching AT10 effectively applies amplitude modulation to the visual IF. This modulation produces sidebands that would extend outside the television channel were they not filtered in FL2. After the vestigial sideband filter, the signal is combined with the aural IF, with the remaining signal processing being as shown in Figure 7.3.

Inspection of the scrambled video waveform at the top right of Figure 18.5 shows that the peak amplitude of the scrambled signal is not constant. The peak amplitude depends on the darkest portion of the picture. Because of this, it is not possible to use a sync-suppressed signal as a pilot to operate the AGC of a coaxial or microwave distribution link. Furthermore, it is not straightforward to measure the signal level of a sync-suppressed signal. Manufacturers normally recommend turning off scrambling before measuring signal level. The manufacturer must then ensure that the scrambled signal level remains correct when scrambling is turned on.

Some sync suppression systems take advantage of the reduced peak amplitude of a sync-suppressed signal to raise the signal level about 2.5 dB when scrambling. This returns the peak amplitude of the scrambled signal to the peak amplitude of the nonscrambled signal, thus improving carrier-to-noise ratio by 2.5 dB during active video.

Synchronizing the Descrambler

Several methods are used to communicate descrambling information to the descrambler. One of the more common, though not the most recent, is to place pulses on the sound carrier, as shown in the lower portion of Figure 18.5. The aural IF is routed out to the scrambler at flag D. Amplitude pulses are modulated onto the sound carrier as shown by the envelopes in the bottom of the figure, where the left envelope is unmodulated, and the right envelope has the modulation impressed on it.

Recall that the audio is frequency modulated onto the sound carrier. Theoretically, the carrier may be amplitude modulated with an independent signal, such as the scrambling synchronization waveform shown, and the two will not interfere with each other. In practice, a number of mechanisms can cause crosstalk between the amplitude and frequency modulation components. The

manufacturer must take pains to control these sources of so-called AM/FM and FM/AM conversion. It is possible, though not common, that the distribution plant can introduce undesirable AM/FM or FM/AM conversion.

The sound carrier occupies a relatively narrow spectrum (see, for example, Figure 7.2). You must carefully control the spectrum that results from impression of the AM modulation on the carrier. For this reason, filter FL10 is used in the path between the timing circuitry and modulator M10. The filter controls the spectrum of the modulat*ing* signal and, hence, of the modulat*ed* signal. It reduces the rise and fall time of the modulating waveform, resulting in the slow transitions shown in the right envelope. A fast transition would be preferable to allow accurate recovery of timing information from the waveform because noise can distort the recovered waveform. The slow rise time means that such distortion could translate into timing errors. However, fast rise times mean the spectrum will spread, possibly causing interference with either the on-channel or adjacent channel video. The manufacturer must walk a fine line between timing recovery accuracy and spectrum spreading.

Descrambling is effected with an attenuator in the descrambler that operates backward from attenuator AT10: the descrambler attenuator must be IN during active video, and OUT during blanking intervals when AT10 is IN. This operation must be timed very carefully to coordinate the operation of the two attenuators to within tenths of microseconds. Besides adjusting the value of attenuator AT10 to effect variations of scrambling in order to frustrate pirates, some systems also vary the timing of the aural carrier pulses with respect to the video.

A common variant of the scrambling system described here is baseband sync suppression scrambling. The sync is suppressed before the video is modulated. Because of the way the systems most logically work, baseband sync suppression is accomplished by offsetting the sync level rather than multiplying it. Baseband sync suppression systems don't require the visual IF interface shown, and since commercial systems don't rely on pulses on the sound carrier, they don't use the aural IF interface either. However, because the sync is not the most negative portion of the waveform as it enters the modulator, the clamp in the modulator (see Figure 7.3) has to be modified to clamp at the modified sync tip level, and a signal must be supplied to the modulator to define the clamp time.

Characteristics of the Scrambled Waveform

Figure 18.6 illustrates the demodulated waveform that would be processed by the baseband circuits of a TV receiving the scrambled signal. Also shown, for reference, is the nonscrambled waveform in dashed lines. Note that the sync, which should be the most negative part of the waveform, has been moved up to a voltage range normally reserved for picture information. The darkest portion of the picture is now the most negative portion of the waveform, assuming the picture is not all "whiter" than the modified sync tip level. Sync circuitry in the TV will try to synchronize to the wrong place in the line.

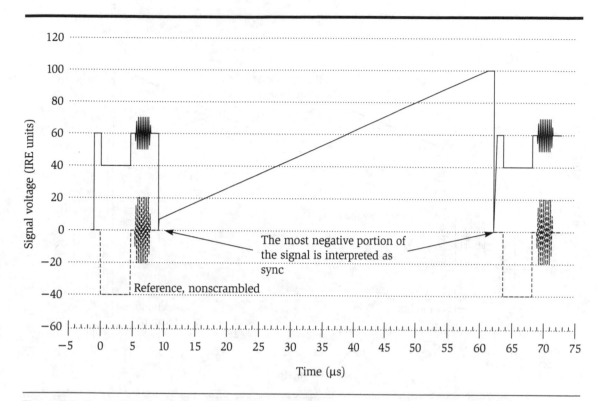

Figure 18.6 Scrambled signal recovered at television receiver.

As illustrated, the most negative portion of the waveform occurs at both the beginning and end of the line, where the picture content was black as scrambling ended and started. With most real video patterns, the most negative portion will be somewhere else along the line and will vary from one line to another.

Notice that the peak-to-peak amplitude of the sync tip and the color burst, which is also suppressed, is reduced to one-half that of the unscrambled signal. This is a characteristic of RF sync suppression systems. Since attenuator AT10 of Figure 18.5 reduces the amplitude of the signal by one-half (for 6-dB suppression), all portions of the suppressed signal are reduced in amplitude by one-half. Baseband sync suppression, on the other hand, operates by shifting the signal level during the HBI using a voltage added to the video signal. The peak-to-peak amplitude of the sync and the burst is not reduced with baseband scrambling.

The Descrambled Waveform

Figure 18.7 illustrates the signal as it would appear in a TV that has received the signal after descrambling. The horizontal (and usually vertical) blanking intervals have been suppressed in the scrambler and expanded in the descrambler.

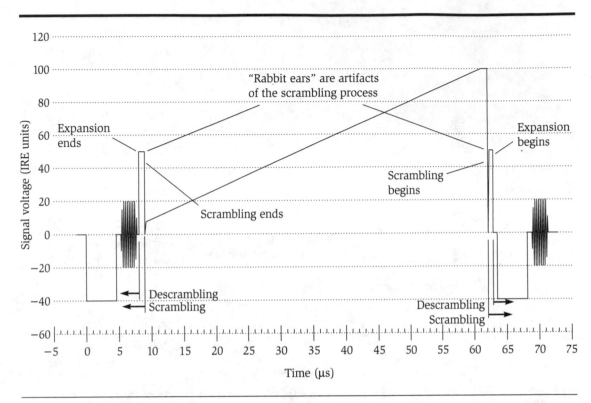

Figure 18.7 Descrambled signal at authorized TV receiver.

The order of events is very important. If the descrambler ever expanded (by switching out attenuation) a signal that had not been suppressed, the result would be a sliver of signal more negative than sync. This could cause unacceptable operation of the TV receiver.

Because some time must be allowed between suppression and expansion, a small sliver of time remains in the descrambled signal, during which the waveform was suppressed but not expanded. This artifact of the scrambling and descrambling process is characteristic of sync suppression systems, unless removed by another signal-processing step. The artifact is sometimes called a "rabbit ear."

The rabbit ears of Figure 18.7 can have deleterious effects on the signal. One early closed caption decoder required all the run-in pulses transmitted at the beginning of the closed caption line (line 21). In a particular cable system, the scrambler was not set properly, resulting in the rabbit ear at the beginning of line 21 impinging on the first closed caption run-in pulse. The problem was fixed by moving the timing in the scrambler so that the run-in pulse was spared. (Run-in pulses appear at the beginning of the line of closed caption data and are used to synchronize the decoder.)

The rabbit ears are normally in what is known as the *overscan* region of the TV picture. That is, the electron beam in the TV scans off the edge of the picture tube screen and does not show the rabbit ear. Occasionally, a TV is encountered that does not exhibit normal overscan: the picture does not cover the entire screen plus a little spilling off the edge. In this case, a white bar may be visible on one side of the screen.

Sometimes electrons emitted during the rabbit ear reflect from the side of the picture tube and strike the phosphors on the screen. The electron beam is badly defocused but can result in a slight lightening of the picture on the edges. Usually, the effect is not noticeable because it is always present. However, in scrambling systems that change scrambling modes such that the rabbit ear changes, the effect can be noticeable at a scrambling mode change. For this and other reasons, scrambling modes are normally changed at scene changes.

Yet another issue relating to the rabbit ear is the minority of television sets that perform dc restoration (clamping) on the back porch rather than on the sync tip. If the set tries to clamp when the rabbit ear occurs, the clamp level will be in error, leaving the picture darker than it should be. A back porch clamp can be identified by changing the timing of the back porch rabbit ear and noting if the picture brightness changes. Such a test is best done using video from a test generator.

The suppression interval is determined solely at the scrambler, which will usually contain some sort of adjustment to set the suppression start and finish times. The expansion interval begins at a time related to the time of a pulse on the sound carrier (or some other time marker depending on how the system works). The end of the expansion interval is then timed from the beginning of expansion. Because of the criticality of the sound carrier pulse timing, it must be set at the scrambler/modulator. Reference to Figure 18.5 will show that the sync-suppressed picture carrier is routed through the vestigial sideband filter FL2. The sound carrier, with the synchronizing pulses, is frequently not routed through FL2 (though some manufacturers may do so).

Modern vestigial sideband (VSB) filters are almost always realized using surface acoustic wave (SAW) technology. A characteristic of this technology is unavoidable delay to the signal traversing the filter. The amount of delay is related to the steepness of the filter transition region. To an extent, it is also related to the present state of the art, but in every practical case, there will be a rather large delay to the signal in the SAW filter. Practical filters for professional VSB applications have 3 to 5 µs of delay, which the sound carrier does not experience. This is long, compared with the 100 ns or so timing accuracy desired, making imperative the careful setting of the relative timing of the scrambling and sound carrier pulses. This timing is done in the scrambler, which must be mated with the modulator with which it will be used. If the SAW filter or the scrambler is changed, timing must be reset because a different SAW filter in FL2 may have a different delay.

The scrambler manufacturer will provide a procedure to adjust the scrambler. Usually, the most accurate method involves detecting the video signal and the

aural carrier pulses, possibly using a spectrum analyzer, whose output is applied to an oscilloscope. The time between the leading edge of sync and the 50% point of the pulse on the sound carrier is measured and set to a specification supplied by the manufacturer. This technique, though very accurate, is tedious and therefore prone to error. Often a substitute method, using a certified descrambler, is used. This certified descrambler is subject to drift in descrambler timing but is adequate if sufficient care is exercised.

18.3.2 Scrambling by Video Inversion

Most modern scrambling systems used in North America employ video inversion in addition to sync suppression. In video inversion systems, the active video is inverted with respect to its normal polarity. The produces a negative picture image. In addition, since the HBI is rarely inverted, color information is reversed as well: the phase relation between the active color and the burst is changed by 180°.

Figure 18.8 illustrates video inversion of the reference waveform. The nonscrambled waveform is shown as a dotted line. The video in this case is inverted with respect to an axis of 50 IRE. One way to think of the inversion is that the video (but usually not the sync) is rotated about an axis, in this case, of 50 IRE. Mathematically, the operation performed on the active video is given by the following equation:

$$V_{out} = 2V_{axis} - V_{in} \tag{18.1}$$

This axis information, V_{axis}, must be communicated to the descrambler in some manner. If it is not communicated, then gain errors will create an error in the axis about which the video is reinverted (reinversion applies the same equation), resulting in a picture that is too dark or too light. Some early video inversion systems suffered from this problem, but the issue was quickly recognized, and more modern systems include a means for communicating the axis information to the descrambler. One method is to transmit a level in the vertical blanking interval that is related to the axis of inversion. Another method is modification of the sync pulse in some manner to communicate the axis information. Either method can give satisfactory results.

The illustrated scrambled waveform communicates axis of inversion information to the descrambler by shifting the last half of the unsuppressed sync tip to the level of the axis of inversion. Some scrambling systems use an axis of inversion of 50 IRE, and others use 30 IRE or some other axis. Use of 30 IRE would yield a picture in which the white level is translated to the normal unscrambled sync tip level.

Video inversion is often accompanied by sync suppression since inversion by itself doesn't necessarily yield a completely unrecognizable picture. By itself, the effect of video inversion is much like that of looking at a photographic negative. Combined with sync suppression, it makes life more difficult for the pirates. Frequently scrambling manufacturers will combine sync suppression and video in-

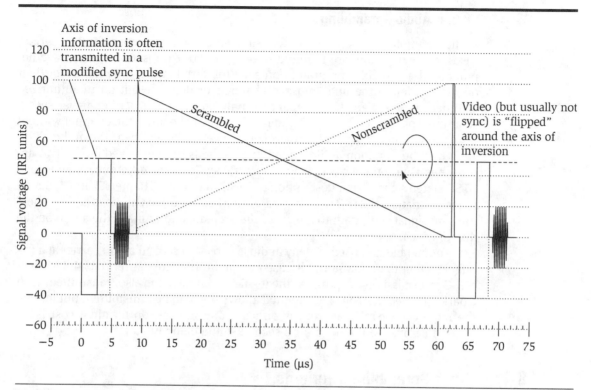

Figure 18.8 Video inversion with the axis of inversion at 50 IRE.

version, using several variations on each, to yield a number of scrambling modes they can switch between so as to cause more problems for the pirates.

18.3.3 Phase Modulation of the Picture Carrier

A variation of sync suppression scrambling employs a special surface acoustic wave (SAW) filter to shift the phase of the picture carrier. The filter is switched in during the HBI and VBI. It has the characteristic that the amplitude is reduced at the carrier much more so than at the color subcarrier. Also, the phase of the picture carrier is shifted by 180°, whereas that of the color subcarrier is shifted very little.

The scrambled video is sync suppressed, and the picture waveform is distorted by virtue of the varying phase shift with frequency between the picture carrier and the color subcarrier. When demodulated using an intercarrier detector (see Chapter 7), the aural carrier would include rapid 180° phase modulation, effectively masking the sound. The decoder operates by switching in and out a complementary SAW filter, with the times of the switching based on detecting a sudden 180° phase shift of the picture carrier.

18.3.4 Audio Scrambling

Audio scrambling is used at times as a supplement to video scrambling. In certain cases, cable operators are required to deny sound to subscribers who have not purchased a particular program. Some set top terminals, by the nature of their operation, don't tune unauthorized channels or don't permit an unauthorized channel to appear at the terminal's output. Thus, the subscriber never sees a scrambled picture nor hears the audio unless he or she tunes the channel without a set top terminal. If the subscriber does so, he or she will hear the audio.

The more common audio scrambling methods have involved one of two approaches. One approach is to place monaural audio on the difference channel of a BTSC stereo signal. For a description of BTSC stereo, see Chapters 2 and 7. Sometimes a variation of BTSC will be used. A "barker," or teaser, audio is added where the sum signal goes, both to advertise the program and to mask the audio on the difference channel.

Another method used to deny audio is to place a second audio carrier at a frequency other than that normally used for audio (picture carrier plus 4.5 MHz in NTSC transmissions). Normally, the regular aural carrier is also transmitted, with a barker carried thereon. This method allows stereo to be transmitted, but finding the required "clean" spectrum is difficult, and the set top terminal cost is increased because of the need for a second sound demodulator.

18.4 Other Scrambling Methods

Besides sync suppression and video inversion, a number of other scrambling methods have been proposed and even applied to analog television signals. Figure 18.9 illustrates four alternative scrambling methods, which we describe in the following sections. Most of the work on these have been done in Europe though some has been done in North America. Many of these other methods have the advantage of not affecting the sync. Such methods are useful in that they pass through normal signal-handling equipment without having to modify the hardware. Typically, they have not enjoyed success in North America, though, because of the cost of recovering pictures.

18.4.1 Line Dicing

Figure 18.9(a) illustrates line dicing, also known as line cutting, or sometimes, cut and rotate. Each line is cut, usually at one of many predefined locations. As illustrated, the line is cut after segment C. Segments D through I are transmitted at the beginning of the line, and segments A through C are transmitted after them. On the next line, the cut may be made after segment H, then on the next line, after segment E, or elsewhere. The picture-hiding capability of this system is excellent, but some problems exist in recovery by a legitimate descrambler. In the figure,

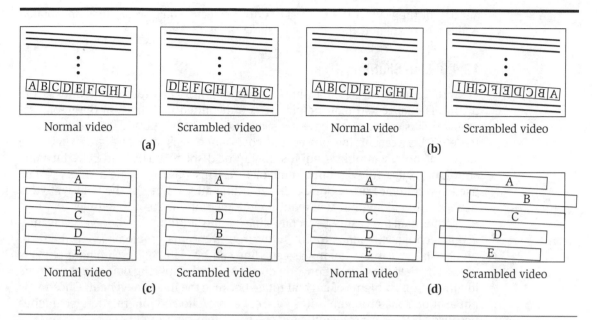

Figure 18.9 Alternative scrambling methods. (a) Line dicing (cutting). (b) Line rotation. (c) Line shuffling. (d) Time shifting lines with respect to sync.

segments C and D are transmitted on opposite sides of the picture, but when the signal is recovered, they will abut each other. Any line tilt will be visible at the point they are spliced together, and care must be taken that no pixels are omitted or repeated. Some practitioners claim to have overcome this problem.

Another problem occurs in over-the-air transmission. If a ghost is introduced during transmission, the ghost of the sync tip (and video segments) will not be stationary as it is with normal signals. Rather, the sync ghost will "hop" around the screen depending on where the cut point is on each line. This is somewhat more annoying than is a simple static ghost.

18.4.2 Line Rotation

Figure 18.9(b) illustrates line rotation scrambling, in which certain lines are reversed in the transmitted sequence. Many variations are possible by selecting different sets of lines to rotate (reverse). Arguably, picture hiding may not be quite as complete as with line cutting, but it is probably adequate so long as only a few contiguous lines are transmitted either rotated or not rotated. Ghosting can also be a problem with this system. For example, if a ghost of the sync tip normally would show up in segment B, then on the rotated line, it will show up in H. When the line is returned to its normal left–right orientation, the ghost will be seen

almost all the way to the right side of the screen. Since some lines are rotated while others are not, the effect will show up as flicker.

18.4.3 Line Shuffling

Figure 18.9(c) illustrates line shuffling scrambling. The order of the lines in the picture is changed so that the picture is transmitted with lines in the wrong sequence. This method can have excellent picture-hiding power though there is a trade-off between hiding power and cost. Since each line of a group must be stored at the descrambler until it's time to be displayed, cost can be reduced by shuffling groups of fewer lines in the picture. This way, the same memory can be used several times during one field. Shuffling all the lines of the field will yield better picture hiding but will require more memory in the descrambler.

Other than the cost issue, scrambling by line shuffling can be a very advantageous technique. It tends to not suffer from the same problems as do the techniques shown earlier, so long as each line can be clamped to eliminate artifacts caused by field time distortion. You do not suffer the splicing point level and time identification problems of cut and rotate because the line is never cut. Ghosting is not as much of a problem as it is with cut and rotate and line rotation techniques because all lines are transmitted in the same left–right direction.

18.4.4 Time Jitter

Figure 18.9(d) illustrates time jitter scrambling. In this technique, each line is transmitted at a different time than it should be. This can be done as long as the advance in time is offset by an equal retardation in the same field. In order to achieve significant time offset, this technique is often used in conjunction with sync replacement techniques, in which sync is transmitted in some unconventional manner. This frees up the time during which sync is normally transmitted so as to allow more shifting of the video line. Even though between any two lines there may not be much if any time for sync, on a net basis over a field, the sync time still remains. Whenever it occurs, digital audio may be inserted. Because of sync replacement, time jitter scrambling doesn't qualify as a scrambling technique that will pass through unmodified video hardware.

The technique has been used in private video networks in North America, with the sync being replaced with digitized audio. The application for which this technique has seen use in North America is in conjunction with a system of video transmission called multiplexed analog component (MAC) video, which was developed in England. In MAC systems, the chroma information is transmitted separately from the video, with each being time compressed to allow both components to be transmitted during the time of one video line. The advantage accrues mainly in FM systems, such as are used in analog satellite video transmission. In FM systems, differential gain can create significant video problems and is mitigated by use of MAC systems. Also, by not combining monochrome and color information, an improvement in signal recovery is possible.

18.5 The Set Top Terminal

For many years, the set top terminal has been a part of subscribers' lives, first as a device that enabled more channels than what a TV could tune. Later, set top terminals added remote control before that feature was standard in TV sets. Set top terminals are also commonly used as a vehicle for program denial. Most scrambling systems rely on a descrambler built into a set top terminal. A very recent ruling by the FCC requires descrambling to be separated from the set top terminal early in the 21st century.

18.5.1 Set Top Terminal Classification

Set top terminals are frequently considered to come in three fundamental types. The simplest is the plain converter, which performs no functions except channel tuning and perhaps volume control. This type of converter is the one traditionally used for expanding the tuning of a TV that cannot tune all the channels available on the cable. Though still available, plain converters tend to be sold more at retail today, for use with the dwindling number of TV sets that don't tune all cable channels.

The next class of set top up from the plain set top is the descrambling set top terminal. It is similar to a plain converter, except that it includes some sort of descrambler. Authorization of programs must be embedded in the terminal because the cable operator has no way to communicate with the terminal once it is installed in the home. Older descrambling terminals had programmable read-only memory (PROM) devices installed, which controlled what could and couldn't be descrambled. Later models usually had a facility to be programmed either by plugging a programmer into a socket on the terminal or by using the infrared remote control link. In every case, the terminal manufacturer provided the programming facility. This type of terminal is rarely sold today though many remain in use.

By far, the most popular class of set top terminals today are addressable terminals, in which the cable operator maintains some sort of communications with the set top when it is in a subscriber's home. This communication can be accomplished in several ways, the most popular of which we describe briefly here. The communications link is used to authorize the terminal to descramble services according to what the subscriber has purchased.

The addressable data sent traditionally has included authorization information for the terminal to specify what channels or programs should be descrambled. Some set top terminals tune only authorized channels. Others tune all channels and search for program identification information ("tags") to determine descrambling authorization.

Recently addressable terminals have been upgraded to utilize other types of information. Data "channels" are possible, which look to the subscriber like TV channels with characters and graphics on the screen. The information displayed may be sent specifically to that terminal, or it may be broadcast to all terminals.

Electronic program guides (EPG) are popular now. In this application, a program schedule is transmitted to the set top terminal, which formats it for display to the subscriber. The subscriber typically moves a cursor on the screen until finding a program he or she wants to view. By pressing a "select" button, the set top is directed to the relevant channel.

In yet another newer application, information from the World Wide Web is transmitted to the set top terminals via the addressability channel. The terminal displays the received data on the TV when commanded to do so by the subscriber. The data available is selected by the cable operator and broadcast continuously and cyclically (a "data carousal"). If a two-way connection is available, then data may be requested directly from the Web. Advanced terminals today include some facility to download software to customize what the terminal does.

Out-of-Band Addressability

Out-of-band addressability uses a separate data carrier to communicate data to the set top terminal. A separate data receiver is contained in the set top terminal. The advantage of out-of-band addressability is that the operator can communicate with the set top terminal at any time. Instructions to change the terminal's state, such as when a pay-per-view event is ordered, can be processed at almost any time, assuming that the terminal is powered and connected to the cable. Consequently, a message does not have to be transmitted much more than one time to ensure receipt.

In-Band Addressability

Out-of-band addressable terminals must bear the expense of a separate data receiver. In-band terminals use addressing information carried in one or more TV channels. This data may be carried in the vertical blanking interval, as additional pulses on the sound carrier, or using some other method. One advantage is the reduction in cost by not having a separate data receiver. Some practitioners feel security is enhanced by using in-band data. The problem usually cited with in-band addressability is the inability to communicate with the set top except when it is tuned to the proper channel(s). To overcome this problem, some manufacturers place data on many different channels, even if the channel is not scrambled. They also creatively divide the data among channels to maximize the probability that a terminal will receive intended data. However, the operator can never be sure that a particular terminal has received all the data sent to it.

Alternative Authorization Procedures

Several methods may be used to authorize programs or channels to be descrambled. This section describes the most common.

Subscription programming is the oldest method of authorizing pay programs. In a subscription system, the subscriber pays a monthly fee and receives a pre-

mium channel all the time. This requires that a terminal be addressed only when the subscriber changes the channels to which he or she subscribes. Only one-way communications is required.

Pay-per-view (PPV) programming is sold on a per-program basis, but the subscriber must call the operator in advance to order the program. Ordering may be done by talking to a customer service representative (CSR) with the cable company, who adds the price to the subscriber's bill and causes an authorization signal to be transmitted to the terminal. Alternatively, the customer may call an automated system that uses automatic number identification (ANI, popularly known as caller ID) to identify the caller. The caller punches in a number corresponding to the program desired, and his or her set top is automatically sent an authorization.

Impulse-pay-per-view (IPPV) programming is often considered to be the preferred method of selling individual programs since the subscriber doesn't have to preorder: he or she can order a program on impulse, even after it has started. IPPV systems require two-way communications with the terminal because the terminal must inform the headend of programs viewed so as to allow billing and payment for the programs.

Commonly, IPPV systems work in what is called a store-and-forward manner. When the subscriber requests a program by actuating a prescribed key sequence, the terminal allows the program to be viewed without checking first with the headend. (If the headend must provide preauthorization, the resulting time delay is considered unacceptable.) At a later date, the terminal is polled from the headend and reports what programs have been watched. From the time the program is viewed until the time the terminal reports the viewing, the viewing information is stored in nonvolatile memory. If the subscriber interferes with communications to the headend, the operator will not know to bill for the viewing. To minimize losses in such a case, the terminal limits the number of programs (or the cost thereof) that can be viewed before it communicates with the headend.

Video-on-demand (VOD) programming is perhaps the ultimate in pay programming. VOD systems, which have not reached significant penetration as this is written, allow the subscriber to select what he or she wants to see, when he or she wants to see it. Programs are stored on a video file server similar to those used to store local advertising (see Section 7.9). VOD requires the use of RF return signals to allow the subscriber to request a program from some sort of menu. Advanced systems include so-called VCR functions of pause, rewind, and fast forward. For this to work, the RF return path must exhibit low delay (latency) so that the experience is truly similar to that of using a VCR.

Near-video-on-demand (N-VOD) is a compromise system used when an adequate number of channels are available. Selected movies are transmitted, each movie on several channels. On each channel used to transmit a particular movie, the start time is staggered by 10–30 minutes. This way, a subscriber gets convenient starting times. By tuning to a channel with a later start time, it is possible to simulate "pausing" the movie.

18.5.2 Elements of a Modern Set Top Terminal

This section describes the general elements of a modern set top terminal. Of necessity, our discussion is not all-encompassing since each manufacturer does things differently, and techniques are continuously evolving as market needs change. We present this material to allow those not familiar with set top terminals to understand what is in one.

The Signal Path

Although variations abound, Figure 18.10 illustrates a typical block diagram of a modern addressable set top terminal. RF from the cable enters at the top left. A directional coupler, DC1, or some other arrangement, is often used to take some power in the downstream direction so as to supply signals to a data receiver if out-of-band (OOB) data is being used. Also, if an RF return path for impulse-pay-per-view service is used, it will be coupled to the cable through DC1.

Following DC1 is a low-pass filter, FL1, used primarily to remove any energy above the highest frequency the set top can tune so as to reduce noise pickup and local oscillator leakage into the distribution plant. An attenuator, AT1, performs a delayed AGC operation. At lower signal levels, AT1 is at its minimum attenuation point, permitting the set top to exhibit the best noise figure it can. As signal levels increase above a certain threshold, AT1 increases in attenuation to protect the amplifier and mixer following it from being overloaded. Chapter 7 describes the delayed AGC function.

From AT1, the signal passes to a broadband preamplifier, A1, which largely sets the noise figure of the converter. The signal is up-converted in mixer M1 to the first intermediate frequency. For terminals that tune no higher than 550 MHz, a common intermediate frequency is 608–614 MHz. This would be channel 37 off-air, but no TV stations in North America are assigned to this channel, which is reserved for radio astronomy. Thus, the manufacturer can assume little likelihood of signal pickup off the air. For terminals that tune to higher frequencies, several intermediate frequencies in the neighborhood of 1,000 MHz have been used.

Upconversion is accomplished by mixing the incoming signals with local oscillator LO1, a high-side local oscillator. Use of a high-side LO results in the signal at the first IF being spectrum inverted (see Chapter 8). The local oscillator frequency is changed to effect tuning to the desired channel. A phase locked loop is used to ensure that the local oscillator remains tuned to the proper frequency. After filtering in FL2, the signal is down-converted either to the normal IF (NTSC picture carrier 45.75 MHz) or to channel 3 or 4 if the converter is an RF-only terminal (rare today). The second LO, LO2, may be fixed in frequency, may have limited tuning range, or may be controlled by an automatic fine-tuning (AFT) circuit, as shown here. The functions just described generally compose the tuner in the set top terminal.

The signal next encounters AT2, which does sync restoration for RF sync suppression scrambling systems, as described earlier in this chapter. This is a

Figure 18.10 Modern addressable set top terminal.

switched attenuator, which offers, usually, 6 or 10 dB of attenuation during active video, and is switched out during horizontal and vertical blanking intervals. Operation of the descrambler is described in Section 18.3.1.

After sync suppression restoration at IF (for the common RF sync suppression systems), the signal is supplied to video and audio demodulators. At this point, the video has had sync suppression removed, but if it is inverted, that still must be corrected. The video inversion circuitry handles this task. As in the case of sync suppression descrambling, recovered in AT2, video inversion is controlled by the controller circuitry, as appropriate to the scrambling system being used.

Most set top terminals today have an on-screen display (OSD), which is used to indicate the channel tuned, time, the program service on that channel, and other functions. In ordering pay events, the OSD is normally used. It is becoming popular to include electronic program guides in set top terminals, and the OSD will also be used for this.

Audio is normally demodulated to just the composite BTSC stereo signal, as described in Chapters 2 and 7. It is not common, but is possible, to include a BTSC decoder. It is not likely that a BTSC encoder would be included owing to its cost. Volume control is effected at attenuator AT3, which operates on the composite BTSC stereo signal. This does not yield a truly accurate volume control

mechanism since you cannot retain satisfactory stereo separation when changing the amplitude of the composite signal. The technique is used as a compromise to keep the cost low. Because it is critical to maintain the correct gain for proper recovery of BTSC stereo, most set top terminals include some means for normalizing the volume control setting such that the deviation out of the set top equals the deviation into the set top (unity gain). This setting is often called "best stereo," or something similar. It is the only setting at which stereo can be recovered with optimum separation.

Volume control of the NICAM digital audio signal (see Chapter 7) is not possible without bringing it back to an uncompressed format (either digital or analog). This operation tends to be expensive, and the expense is greatly worsened if it is required to restore the NICAM signal to transmit it to the TV. Since the NICAM signal cannot be volume controlled, it is normally separated in the set top terminal using a filter much like the one used to separate out the analog sound IF signal. After the analog video and audio are processed, and just before the RF modulator, the NICAM signal is recombined with the analog video and audio.

Terminal Control

The intelligence of the terminal is embedded in a block labeled "Control" in Figure 18.10. This block must receive addressable information from whatever means are provided by the manufacturer. If out-of-band data is transmitted, a data receiver as shown will be used to recover the data and pass it on to the intelligence of the terminal. If in-band data is used, then it must be coupled to the control section. Some terminals use data transmitted as pulses on the sound carrier using the same technique used to transmit descrambling information. Others embed control information in the VBI.

The control also interfaces with some sort of nonvolatile memory, which can retain information even when the terminal is not powered. Several technologies are in common usage. The control must receive user inputs, usually from a remote control receiver (infrared or RF), and usually from a limited function keypad on the terminal. Some sort of output must be provided to the subscriber. This output could be some combination of a numerical display on the terminal and, more likely today, an on-screen display.

Finally, the terminal must be able to communicate back to the headend. This communication is necessary to facilitate impulse-pay-per-view functions as described earlier. One way of effecting this communication involves using the subscriber's telephone line (telephone IPPV), in which case, the terminal includes a simple telephone modem. Alternatively, if two-way cable plant is available, RF-IPPV can be used, in which case, an RF transmitter sends signals upstream to the headend.

TV Display Issues

A problem with display of data on a TV set is that the TV system is optimized for displaying pictures, not small text. The compression methods used, even in analog TV to save bandwidth, don't allow for transmission of small characters, espe-

cially colored ones. The methods used to send display data from a computer to a monitor are so optimized, however. This difference in function is not always recognized, leading to frustration on the part of people who try to place too many characters on a TV screen. Generally, you should not use normal TV transmission techniques to place more than 32–40 characters on one line of text. This is contrasted to the 80 or so characters routinely used in computer work.

18.6 The Back-of-Set Interface

As a result of consumer problems involving set top terminals, the Cable Act of 1992 required an alternative to set tops and directed the FCC to develop a standard. The FCC turned to an existing joint engineering committee of the NCTA and CEMA (Consumer Electronics Manufacturers Association), a division of the Electronic Industries Association (EIA), which was already considering such a device. The product of this work is embodied in two specifications promulgated by the EIA, IS-105.1 (physical) and IS-105.2 (logic and control).[1] These documents standardize the interface between equipment but do not place any restrictions on what the equipment can do or how it is built. The anticipated device is sometimes called a set back terminal because it is intended to be placed behind the TV. Operation should be invisible to the consumer.

Figure 18.11 illustrates the principles embodied in the standard. The idea is to move security functions from the set top terminal to a box that sits behind the TV and does nothing until the subscriber tunes a scrambled channel for which he or she has paid. At that time, IF is routed from the TV tuner to the back-of-set box. There the signal is descrambled and demodulated, and passed back to the TV as baseband audio (if required) and video.

Besides accommodating analog video and/or audio, the specification is able to accommodate digital signals contained within a 6-MHz channel, tuned by the TV and delivered to the decoder. In addition, other functions not related to program security may be added. These include electronic program guides, front door video monitors, telephone caller identification on the TV, and any other functions that can be envisioned. The additional functions may be included in the same decoder as the program security functions, or another module may be added. The standard provides for adding multiple modules, which typically will be supplied by different manufacturers. Each module may or may not require the IF signal from the TV. A VCR may be added to the setup if desired. Expansion and switch interfaces are defined to allow multiple devices to avail themselves of the IF signal if necessary.

A common control channel, described in IS-105.2, arbitrates between different devices needing and supplying different resources. The control channel is based on the EIA Homebus protocol.[2] A central intelligence is not required to arbitrate among users: the protocol allows each device to automatically learn what other devices are present and their capabilities.

Four audio interfaces and four video interfaces are provided. Video can be interfaced as composite analog or S-video. S-video, also known as supervideo,

Figure 18.11 EIA-IS-105 back-of-set interface.

transmits the luminance and sync on one wire and the chroma on a second. The chroma is modulated on a 3.58-MHz subcarrier just as in normal NTSC. However, since the chroma is carried on a separate path, it is possible to provide wider luminance and chrominance bandwidths than with NTSC. Furthermore, chroma/luma crosstalk artifacts (sometimes called "cross color") are avoided.

Following the standard of the EIA-693 AVBus audio–video interface,[3] each audio and video interface is transmitted using a balanced, unshielded twisted pair. The cable consists of 13 such twisted pairs: 4 for audio, 4 for video, 1 control pair, 1 common mode reference pair, and 3 for future expansion. Each signal is superimposed on a 5-volt common mode voltage to accommodate single polarity power applications. The video cable pairs must be terminated on each end and are limited to a maximum length of 10 meters. Sources and loads bridge the appropriate pair. Inputs have a high impedance so as not to upset the cable's nominal 120-ohm characteristic impedance. Sources exhibit a 120-ohm impedance while transmitting and a very high impedance when not transmitting. This permits the audio cable to act as a proper balanced and terminated transmission line, thereby controlling reflections.

18.7 Program Denial for Digital Transmission

The preceding methods of program denial apply to analog transmission of video signals. In the digital world, program denial is done somewhat differently. All the video signals that make up the program stream in a given channel are multiplexed together using (usually) MPEG transport streams. The data is then encrypted using any of a number of public or private encryption algorithms, as chosen by

the manufacturer. The basic method of encryption usually involves exclusive ORing the data signal with a pseudorandom bitstream generated in some manner not known to the would-be pirate. The pseudorandom bitstream is communicated to the decryptor using a method not known to the would-be pirate. At the decryptor, the incoming datastream is again exclusive ORed with the pseudorandom bitstream. The resulting output is the MPEG transport stream that existed before encryption.

Some differences in terminology exist between program denial in the analog and digital domains, which are a result of the evolution of usage of different terms by the cable television and data transmission communities. For this reason, the terms *encryption* and *decryption* are used in the preceding paragraph to mean the same thing that is meant by the terms *scrambling* and *descrambling* in the analog sections of the chapter. The basis for this difference is explained in the introduction to this chapter.

18.8 Summary

In this chapter, we have tried to show the most important features of scrambling systems that are in use or proposed. We have intentionally avoided, however, discussing the details of the methods used to communicate descrambling information to the subscribers' premises. The most common analog scrambling technique in use in North America is sync suppression. It modifies a normal signal in ways that hide the sync. In the process, it becomes very difficult to measure signal amplitude, so this is usually done with the scrambling turned off. Video inversion is often used in combination with sync suppression to provide a signal that is more difficult to restore. Other scrambling techniques have been used in Europe but have not enjoyed much success in North America.

An older technology, which is particularly friendly to subscribers but has limited capabilities, is trapping—either positive or negative. Interdiction has been used as an off-premises addressable technology. Most scrambling systems today are integrated into set top terminals, which were also described. We introduced important characteristics. Chapter 19 will deal with compatibility issues between cable plant and consumer equipment.

Endnotes

1. EIA-IS-105.1, Decoder Interface Standard, Part 1 and EIA-IS-105.2, Decoder Interface Standard, Part 2. Documents are available from the EIA's literature contractor, Global Engineering Documents, Englewood, CO, 800-854-7179.

2. EIA-IS-60, Home Automation Systems, October 1992. Available from Global Engineering Documents.

3. EIA-693, Audio/Video Bus (AVBus) Physical Layer and Media Specification. Available from Global Engineering Documents.

Chapter 19
Consumer Electronics Interface

19.1 Introduction

Cable services are enjoyed via the subscriber's consumer electronics products, the most important of which are television receivers and videocassette recorders. Stereo and surround-sound systems are also important to home theater installations. Coordinating the use of all the features of these products with cable services causes complications. Add-on devices such as video games, Web appliances, digital video disc (DVD) players, personal computers, home security and control systems, and other consumer electronics products add still more challenges. The cable–consumer electronics interface issues are complex, and satisfactory solutions have been elusive even with extensive committee work and cooperation between the industries. This chapter will survey the issues and some of the attempts to alleviate the problem.

There are two principal aspects to the issues concerning the interface between consumer electronics products and cable systems. One aspect is the way in which this interface affects the usage of cable services and of consumer electronics products and their features. The other aspect involves matters of law and politics. Both aspects will be considered, but separately.

Five major constituents have interests in these issues: (1) the subscriber/consumer, (2) the service provider, (3) the service provider's equipment manufacturer, (4) the consumer electronics and related industries, and (5) government agencies creating laws and rules related to these issues.

19.2 The Nature of the Interface

19.2.1 Cable Services

For purposes of this discussion, cable services can be categorized as (1) basic, (2) expanded basic, (3) pay, (4) pay per view, and (5) other.

Basic services are not scrambled as a matter of law. This is a requirement of the 1992 Cable Act, which we discuss later. Signal delivery is accomplished merely by connection—usually by a technician physically connecting the drop cable to the cable plant. In a few cases, a switched tap is used to connect the subscriber under electronic control without the need for a technician to visit the tap location. Signal denial for those who no longer subscribe is done by disconnecting the subscriber from the cable plant. The subscriber who takes only the basic service and has a consumer electronics product that can tune all the channels of interest experiences the same amount of complexity as found when connected to an antenna.[1]

A converter with appropriate tuning capability is used in situations where (1) the subscriber's consumer electronics product does not tune all the channels of interest, (2) the product suffers from inadequate tuner performance, or (3) the product does not include remote control and the subscriber desires that feature. A converter is an extended capability tuner with a channel 3 or 4 output. It usually

also includes a remote control. Converters were once commonly leased by cable operators. That is much less common today because converters are rarely needed since few very old receivers are still in service. Converters are available at retail.

Expanded basic services involve additional channels and an additional price. Signal denial for nonsubscribers is usually accomplished by traps or, less commonly, by scrambling. Alternative signal denial possibilities include interdiction and broadband descrambling. These alternatives have not found wide acceptance. All these technologies are described in the previous chapter. When traps, interdiction, or broadband descrambling is used, the subscriber experiences the same amount of complexity as found when the TV and VCR are merely connected to an antenna in the usual way recommended by the TV and VCR manufacturers. This assumes that the consumer electronics products tune all the channels of interest; otherwise, a converter is required.

Pay services, also called premium services, are sold by the month on a channel-by-channel basis. Pay services feature movies and special events and require a substantial additional fee. Signal denial for nonsubscribers is usually accomplished with traps for services with stable acceptance, that is, that have subscribers who keep their subscriptions for extended periods. There are two major types of traps: negative and positive. Negative traps allow subscribers to the premium service to receive it "in the clear" without any signal processing. With this technique, basic subscribers who do not take popular pay services have the negative trap installed in their connection to the cable system. The negative trap filters out the channel that is not taken. Popular services are those where more than half of basic subscribers take the service. Positive traps are installed at premium service subscriber locations when less than half of the basic subscribers take the service. A factor in making the choice between positive traps and negative traps is the need to minimize the number of units required. Alternative signal denial methods include interdiction and broadband descrambling. Conventional descrambling is used for services with higher rates of turnover or where additional security is desired. When traps, interdiction, or broadband descrambling is used, the subscriber experiences the same amount of complexity as when the TV and VCR are connected to an antenna (assuming the consumer electronics products tune all the channels of interest).

Pay-per-view services are sold one at a time and almost always utilize an addressable descrambler. In the past, disposable traps that timed-out after the pay-per-view event were used, but with only moderate success.

Other services are generally taken by more sophisticated subscribers who have little or no difficulty with the interface between their equipment and the cable system.

19.2.2 Relevant Cable Set Top Terminal Features

Cable set top terminals have a couple of important features that have been designed specifically to help with the interface between cable systems and consumer electronics products.[2]

The *auto bypass switch* is a feature that bypasses the entire cable spectrum to the TV (or VCR) when the set top terminal is turned off. This is an option that either is built into set top terminals or is available as an add-on. It is intended to allow the TV's (or VCR's) remote control to operate the TV's (or VCR's) tuner when the set top terminal is not used for scrambled channels. Though the auto bypass switch was available before the 1992 Cable Act's Consumer Electronics Interface amendment, the act motivated wider availability of the feature. Cable operators are required to make these switches available to subscribers but are allowed to charge for their costs. This feature has had little consumer demand.

The *dual descrambler* feature puts two tuners and descramblers in one package so that subscribers are not limited to unscrambled channels for some of the modes of operation of consumer electronics products. Cable operators are required to make these units available to subscribers and are allowed to charge for their costs. This feature also has had very little consumer demand.

The cable industry and independent entrepreneurs have developed a number of switching appliances to facilitate configuration of the interconnections for various purposes. Other devices intended to ameliorate these difficulties have been offered for sale. None of them has found long-lasting marketplace acceptance.

19.2.3 Relevant Consumer Electronics Characteristics and Features

TVs and VCRs are the primary consumer electronics equipment affected by cable service. These products come with a wide variety of features. The most important features for these issues relate to the remote control and the tuner. The remote control is used to turn the device on and off, change channels, and operate other features. Another feature of the VCR is its timer, which is used to commence and terminate recording and to change channels at appropriate times to capture programs of interest. Other TV and VCR features of interest include parental control, last channel tuning, sleep timers, and favorite channel access.[3]

Premium TV receivers include a feature called *picture-in-picture,* (PIP), which allows a small picture to be inserted into the large picture. Most often, the signal for the second picture comes from another source such as the tuner in a VCR or a television camera that might be placed in a baby's room or at the door to monitor visitors. Often the two signals may be interchanged with the press of a button. Another variation of the PIP feature includes the scanning of the channels with the capture of a set of miniature still frames providing a glimpse of what is on the various channels.

Most modern television receivers and VCRs have a feature, called *auto channel setup,* that automatically searches for available channels when the unit is first plugged in (and after every extensive power outage). The TV or VCR takes several minutes to search all the channels it can tune and loads into memory those that contain programming. Many (but not all) of these receivers will skip over most scrambled channels. When the viewer uses the channel up or down button, these channels will be skipped.

The *antenna and cable/converter switch* feature (usually operated by remote control) switches between two or more inputs to the TV. One input is intended to be from the converter and the other directly from the cable (or an antenna). The intention here is to allow the TV's tuner to do the work when nonscrambled channels are watched and just use the converter/descrambler's tuner when scrambled channels are watched. This feature is similar in function and purpose to the auto bypass switch option in converter/descramblers.

The *TV/VCR switch* feature on a VCR determines whether the entire cable spectrum is presented to the TV or just the channel tuned by the VCR or the tape playback.

19.2.4 Consumer Frustration

The two principal aspects of consumer frustration are the installation and the utilization of consumer electronics products with cable services. Both of these frustrations come from the variety of services and features found in these products.

19.3 Connecting Cable, TVs, VCRs, and Set Top Terminals

In 1987, the NCTA Engineering Committee formed a Subcommittee on Consumer Interconnection and developed a set of 27 diagrams indicating the most likely methods of connecting a TV, VCR, and one or more set top terminals along with an assortment of switches and signal splitters. Many of these diagrams are also found in consumer's manuals that come with new TVs and VCRs. All the interconnection schemes have advantages and disadvantages. None solves all of the problems. Reviewing some of these diagrams results in a better understanding of the complexity of the problem.[4]

19.3.1 Basic Cable and Trapped Premium Service Connection

Figure 19.1 is illustration 12 of the NCTA connection diagram set. The cable is connected to the VCR, which is connected to the TV. This diagram is for the basic-service-only subscriber and the subscriber who has premium services controlled by traps. No set top terminal is used. Since less than half of cable subscribers have set top terminals and since most subscribers connect more than one TV or VCR to cable, the majority of cable connections use this simple configuration.

Advantages

1. Least expensive.
2. TV and VCR operate as with antenna connections.

Figure 19.1 In-home equipment and simplest connection without set top terminal.

3. Can consecutively record different unscrambled channels automatically.

4. Can watch one unscrambled channel while recording a different unscrambled channel.

5. PIP works with unscrambled channels when VCR's baseband output is connected to TV's baseband input.

Disadvantages

1. No scrambled services available for either watching or recording.

2. No interactive services available.

3. Have to remember to put TV to channel 3 when watching recorded tapes.

4. Need to use two remote controls or a "universal" remote control.

5. TV's parental control is inoperative when using VCR's tuner.

6. Consumer's tuner subject to overload and damage from lightning strikes on cable.

7. Signal leakage possible from inadequately shielded tuners.

Cable input

Figure 19.2 Most common cable connection with set top terminal.

19.3.2 A Simple Start

Figure 19.2 is illustration 1 of the NCTA connection diagram set. The cable is connected to the set top terminal, which is connected to the VCR, which is connected to the TV. This is by far the most popular connection method involving a set top terminal because it is the least expensive and simplest to implement and use.

Advantages

1. Inexpensive and easy to install.
2. Easy to understand.
3. Can record scrambled channels.

Disadvantages

1. Can't watch one channel and record a different channel.
2. Can't consecutively tape different channels automatically because VCR tuner does not do channel selection.
3. Neither the TV tuner nor the VCR tuner has any control.
 (a) TV's parental control, favorite channels, and last channel recall features are inoperative. The TV is just tuned to channel 3 (or 4).

(b) VCR's timer, parental control, favorite channels, and last channel recall features are inoperative. The VCR is just tuned to channel 3 (or 4).

4. Need to use three remote controls or a universal remote control to turn on the three devices and put the TV and VCR to channel 3 (or 4).

5. Makes the TV's picture-in-picture feature useless.

Special Case

1. If the converter/descrambler has an auto bypass switch, then when the converter is turned off, the entire cable spectrum passes through to the VCR, and the situation is exactly the same as in Figure 19.1, which is the case where no set top terminal is used. When again using a scrambled channel, the subscriber must remember to put the VCR to channel 3 (or 4). If the set top terminal has the channel-mapping feature and the TV or VCR does not, the same programming may appear on two different channel numbers. One channel number will be found on the set top terminal and the other on the TV or VCR. This may be confusing.

19.3.3 Two Set Top Terminal Case

Figure 19.3 is illustration 2 of the NCTA connection diagram set. The cable signal is split and connected to two set top terminals. The left set top terminal is connected to a VCR, which is connected to one input of an A/B switch, which is connected to the TV. The right set top terminal is connected to the other input of the A/B switch.

This is one of the least popular installations because of its expense and complexity. It would be used by "techies" who want the most flexible system and can both visualize how it works and handle multiple remotes with ease. It will be very frustrating to others. The dual descrambler feature could be used to simplify both the installation and operation of this connection. On-screen displays could further assist the subscriber. Despite the flexibility of this connection and the potential to make it relatively easy to use with on-screen displays, its usage is very limited.

Advantages

1. Most flexible.

2. Can watch any authorized scrambled or unscrambled program.

3. Can record any (other or same) authorized scrambled or unscrambled program.

4. PIP can be used with both channels if the TV has baseband inputs that are connected to the VCR's baseband outputs.

5. Useful in systems that have both analog scrambling and digital television.

Figure 19.3 Connection with two set top terminals.

Disadvantages

1. Can't consecutively tape different channels automatically because VCR tuner does not do channel selection.

2. Expensive hardware and monthly fees.

3. Complex to install, many cables, many opportunities for signal leakage.

4. Complex to operate; need to use at least three remote controls.

5. Neither the TV tuner nor the VCR tuner has any control.
 (a) TV's parental control, favorite channels, and last channel recall features are inoperative. The TV is just tuned to channel 3 (or 4).
 (b) VCR's timer, parental control, favorite channels, and last channel recall features are inoperative. The VCR is just tuned to channel 3 (or 4).

6. Need to use three of four remote controls or a universal remote control.

7. Most brands of set top terminals will operate simultaneously from one remote control. Consequently, the subscriber may have to operate one terminal

as manual. If the set top terminal supports two control codes, four remote controls may be used with this configuration.

8. If both analog scrambled and digital channels are being used, may need to interchange set top terminals depending on which is to be recorded and which is to be watched.

9. Switch at TV is usually manual, requiring getting out of the chair unless the switch is built into the TV and remote controllable.

10. Splitter cuts signal in half and may introduce noticeable noise.

Special Cases

1. If the left converter/descrambler has an auto bypass switch, then when the converter is turned off, the entire cable spectrum passes through to the VCR, and the situation is exactly the same as in Figure 19.1.

2. If the right converter/descrambler has an auto bypass switch, then when the converter is turned off, the entire cable spectrum passes through to the TV, and the situation is similar to that shown in Figure 19.1 but without the VCR.

3. If the set top terminal has the channel-mapping feature and the TV or VCR does not, the same programming may appear on two different channel numbers. One channel number will be found on the set top terminal and the other on the TV or VCR. This may be confusing.

19.3.4 Minimal Scrambled Channel Usage

Figure 19.4 is illustration 5 of the NCTA connection diagram set. The cable signal is split into two paths. The left path is connected to a set top terminal, which is connected to an A/B switch, which is connected to a TV. The right path is connected to a VCR, which is connected to the other input of the A/B switch.

This connection could be used by someone who wishes to continue the use of the features of their TV and VCR as enjoyed before cable, and only occasionally watches, but never records, scrambled channels. This is especially convenient for someone with only one scrambled channel or who uses the converter/descrambler only for pay per view but doesn't record that programming. The right-hand path is identical to Figure 19.1. Alternatively, since the path with the converter can be used to watch either scrambled or unscrambled channels, the only purpose for the switch is to watch recorded tapes. That could also be accomplished with the VCR's baseband output and the TV's baseband input.

Advantages

1. Easy to use if most of the usage is for unscrambled channels.
2. Can watch any channel while recording any unscrambled channel.
3. Can consecutively record unscrambled channels automatically.

Figure 19.4 Connection for few scrambled channels.

4. Especially easy if only one scrambled channel is taken.

5. PIP can be made to work as long as the second picture is unscrambled.

Disadvantages

1. Can't record scrambled channels.

2. On the left side, TV's parental control, favorite channels, and last channel re-call features are inoperative. The TV is just tuned to channel 3 (or 4).

3. Switch must be manually operated unless it is built into the TV.

4. Need to use three remote controls or a universal remote control.

5. Splitter cuts signal in half and may introduce noticeable noise.

Special Cases

1. If the converter/descrambler has an auto bypass switch, then when the con-verter is turned off, the entire cable spectrum passes through to the

Figure 19.5 Modified most common connection.

TV, and the situation is similar to that shown in Figure 19.1 but without the VCR.

2. If the set top terminal has the channel-mapping feature and the TV or VCR does not, the same programming may appear on two different channel numbers. One channel number will be found on the set top terminal and the other on the TV or VCR. This may be confusing.

19.3.5 Improved Simple Connection

Figure 19.5 is illustration 4 of the NCTA connection diagram set. This configuration inexpensively overcomes some of the difficulties of Figure 19.2 and especially makes sense if the TV has a built-in A/B switch.

Advantages

1. Inexpensive and easy to install.

2. Easy to understand.

3. Can record scrambled channels.

4. PIP feature can be used when one of the channels is always unscrambled and the VCR's baseband output is connected to TV's baseband input.

Disadvantages

1. Can't consecutively tape different channels automatically because VCR tuner does not do channel selection.

2. Need to use three remote controls or a universal remote control.

3. On the left side, neither the TV tuner nor the VCR tuner has any control.
 (a) TV's parental control, favorite channels, and last channel recall features are inoperative. The TV is just tuned to channel 3 (or 4).
 (b) VCR's timer, parental control, favorite channels, and last channel recall features are inoperative. The VCR is just tuned to channel 3 (or 4).

4. Switch at TV is usually manual, requiring getting out of the chair unless the switch is built into the TV and remote controllable.

5. Splitter cuts signal in half and may introduce noticeable noise.

Special Cases

1. If the converter/descrambler has an auto bypass switch, then when the converter is turned off, the entire cable spectrum passes through to the VCR, and the situation on the left side is exactly the same as in Figure 19.1.

2. If the set top terminal has the channel-mapping feature and the TV or VCR does not, the same programming may appear on two different channel numbers. One channel number will be found on the set top terminal and the other on the TV or VCR. This may be confusing.

19.3.6 Problematic Connection Example

Figure 19.6 is illustration 7 of the NCTA connection diagram set. This figure is almost never used with modern set top terminals because they usually require a separate frequency for receiving control signals. The VCR would block those signals, and the set top terminal would turn itself off after a preset time. Also, this configuration would block upstream signals, making impulse pay per view (IPPV) and other two-way services impossible.

Advantages

1. Attempts similar functionality to Figure 19.4 but without the need for the A/B switch.

2. PIP can be used with the second picture unscrambled if the TV has baseband inputs connected to baseband outputs from the VCR.

Cable input

Figure 19.6 Connection with set top terminal difficulties.

Disadvantages

1. Many set top terminals can't be used in this configuration.
2. Impulse pay per view and other two-way services are impossible.
3. Can't record scrambled programs.
4. TV's parental control, favorite channels, and last channel recall features are inoperative. The TV is just tuned to channel 3 (or 4).
5. Need to use three remote channels or a universal remote control.
6. Converter/descrambler's parental control is inoperative when using VCR's channel 3 (or 4) output.

Special Cases

1. If the converter/descrambler has an auto bypass switch, then when the set top terminal is off, the entire output of the VCR passes through to the TV, and the situation is exactly the same as in Figure 19.1.
2. If the set top terminal has the channel-mapping feature and the TV or VCR does not, the same programming may appear on two different channel numbers. One channel number will be found on the set top terminal and the other on the TV or VCR. This may be confusing.

19.3.7 Improvements

One of the main disadvantages of the configurations shown in Figures 19.2, 19.3, and 19.5 is that the VCR is not able to consecutively record different programs except if the set top terminal has an auto bypass switch and no scrambled programming is to be recorded.

This disadvantage can be overcome, at least in theory, with set top terminals that include a programming function similar to that found in VCRs. This programming function can be set up to turn on and change channels at appropriate times. Usually, an on-screen menu is used for this purpose. The subscriber then uses the same principles to program the set top terminal as are used to program the VCR, with the VCR programmed only to turn on and record the output channel of the set top terminal at the appropriate time. In practice, this procedure meets with limited success by the occasional user since there are additional opportunities to make mistakes and thereby fail to record the desired programs. Nonetheless, the careful subscriber can master these techniques and successfully record different channels consecutively.

An alternative approach to overcoming the deficiency that inhibits the recording of consecutive programs on different channels is found in certain premium VCRs. These products have an infrared (IR) light emitter on a short wire. This device is sometimes called an IR blaster. The IR control is set up to use the IR control codes of the brand set top terminal employed by the cable operator. The VCR can then change the channels of the set top terminal and thereby record different channels, scrambled or unscrambled, consecutively. Since the set top terminal is now controlled by the VCR, the set top terminal's remote control is not necessary. Only two remote controls are needed by the subscriber, one for the TV and one for the VCR.

In the case of Figure 19.3, this type of VCR overcomes yet another difficulty. That is, if the two set top terminals respond to the same IR codes, the VCR's IR controller can be placed in such a way that the set top terminal used with the VCR is isolated from the IR signals used with the set top terminal connected to the TV. Then only three remote controls are necessary, and both set top terminals can be controlled remotely.

It is possible to use the same IR blaster technique with the TV controlling the set top terminal (if the TV is so equipped). Also, the set top terminal may use an IR blaster to control the VCR or TV. In that case, the subscriber programs the set top terminal and uses its remote control.

An unfortunate complication with the IR blaster stems from the fact that the usual on–off control on consumer electronics products and set top terminals is a toggle function. That is, there is not a separate on and a separate off control. Rather, if the unit is on, the control turns it off, and vice versa. This uncertainty over the current state is a source of error in setup and usage of these techniques.

There are numerous complications in these configurations. Most subscribers have difficulty visualizing what is happening with the multiple remote controls or the multiple modes of use of the universal remote control.[5]

19.4 Consumer Electronics Interface Issues

19.4.1 The Importance of Compatibility

The cable–consumer electronics interface establishes the manner in which a cable system, a consumer, and consumer electronics products function together to provide a useful result. Compatibility means the convenient use of the features of in-home equipment that are of interest to the subscriber. It also means a minimum of hurdles to subscribing to advanced cable services. From the perspective of the cable operator, compatibility is an issue of customer service and satisfaction. Effective compatibility can be a strong competitive advantage over other service providers. It can mean the difference between subscriber retention and the loss of subscribers to other media.

19.4.2 The Participants and Their Priorities

To better understand the consumer electronics interface issues, it is helpful to consider the problems from the perspectives of the various participants. These participants have differing goals and desires that are not always in harmony.

There are five major participants: (1) the subscriber/consumer, (2) the service provider, (3) the service provider's equipment manufacturer, (4) the consumer electronics industry, and (5) government entities making laws and rules.

The subscriber/consumer is the ultimate objective of all the efforts. If the subscriber/consumer is not happy and if there are alternative entertainment and information choices, the result will be a loss of business. Of course, there always are alternatives. The simplest alternative is to simply not subscribe to premium services and just use an off-air antenna with videotape and video disc rentals as supplementary video.

The next participant to be considered is the service provider who has a love–hate relationship with the in-home hardware. Following is the service provider's equipment manufacturer who has legitimate business concerns. The participant that is most anxious to become an active player in this arena is the consumer electronics industry. Lastly, the Congress has established laws that require the FCC to regulate these matters.

19.4.3 The Subscriber/Consumer

The subscriber/consumer participates in two ways. The subscriber/consumer buys equipment from the consumer electronics industry and subscribes to services that play on that equipment. The ultimate goal is the services. Additional programming choices make the TV and VCR even more useful.

The best reason for a large screen and a surround-sound system with multiple speakers and high-wattage amplifiers is the premium programming that comes with pay services and pay-per-view events. Likewise, having a quality receiver and sound system makes movies and premium services more enjoy-

able. Given all this, it is clear that the cable and consumer electronics industries are highly motivated to work closely together to reinforce this symbiotic relationship.

The consumer wants good programming, good pictures and sound, cost-effectiveness, and simplicity. The subscriber does not care which medium makes the programming available. The consumer is frustrated with product features that are difficult to use and understand. The blinking "12:00" on the VCR has become emblematic of consumer technophobia—many consumers do not even know how to set the correct time. The maze of wires behind the TV and VCR are bad enough with just off-air reception, but they become beyond comprehension for many consumers when a premium service provider adds a set top terminal of some sort.

19.4.4 The Service Provider

There are two elements to the service provider participant: the cable operator and the program supplier. Although this is an interdependent relationship, there are differences in their priorities.

The service provider's love–hate relationship with the in-home terminal concentrates on its major disadvantage, the cost, and its major advantage, the facilitation of new services. The capital cost of the in-home terminal is a significant portion of the total capital investment required for service provision. This is especially burdensome with digital set top terminals.

A rule of thumb used in the cable industry for a decade and a half is that for every $50 of in-home hardware, $1 a month is required just to cover the cost of the equipment. This rule of thumb is based on the assumption that the money to buy the set top terminal is borrowed from some capital source that requires repayment. The capital source may be internal or external, but capital is always a scarce commodity with multiple demands, each of which must justify its use. The capital must be repaid over the period during which the set top terminal remains functional, and the interest must meet corporate hurdle rates. Over the years that set top terminals have been used, the interest rate has varied over a wide range. Using the usual financial equations that are available in spreadsheet programs, Figure 19.7(a) shows a table of the amounts that must be paid back monthly to cover a $50 loan for various interest rates over a number of years. Figure 19.7(b) graphs this data with the number of years across the top, the dollars of outflow on the vertical axis, and a separate curve for each percentage value.

In the graph, the curves for the various percentages cluster closely. Clearly, the number of years over which the loan must be paid back is the more important parameter. Going from 3 years at 5% interest to 3 years at 10% interest increases the monthly cost by only 7%. Going from 3 years at 5% to 4 years at 5% reduces the monthly outlay by 23%. Long life of the equipment is important to low monthly costs. Conversely, high-tech equipment that is soon obsolete is very expensive on a monthly basis. In addition to repaying the capital for the set top terminal, there is a monthly maintenance cost.

Interest rate in percentage

Years	4	5	6	7	8	9	10
3	$(1.48)	$(1.50)	$(1.52)	$(1.54)	$(1.57)	$(1.59)	$(1.61)
4	$(1.13)	$(1.15)	$(1.17)	$(1.20)	$(1.22)	$(1.24)	$(1.27)
5	$(0.92)	$(0.94)	$(0.97)	$(0.99)	$(1.01)	$(1.04)	$(1.06)
6	$(0.78)	$(0.81)	$(0.83)	$(0.85)	$(0.88)	$(0.90)	$(0.93)
7	$(0.68)	$(0.71)	$(0.73)	$(0.75)	$(0.78)	$(0.80)	$(0.83)
8	$(0.61)	$(0.63)	$(0.66)	$(0.68)	$(0.71)	$(0.73)	$(0.76)
9	$(0.55)	$(0.58)	$(0.60)	$(0.63)	$(0.65)	$(0.68)	$(0.70)
10	$(0.51)	$(0.53)	$(0.56)	$(0.58)	$(0.61)	$(0.63)	$(0.66)

(a)

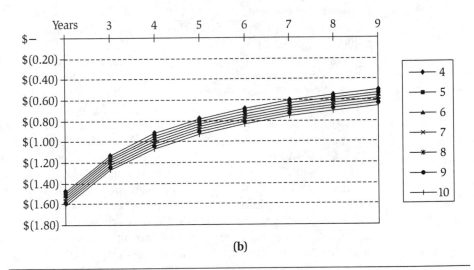

(b)

Figure 19.7 Set top costs. (a) Monthly cost of $50 loan at various rates paid back in multiple years. (b) Graph of data.

The benefit of the in-home terminal to the service provider is that it facilitates services. The in-home terminal does this in two ways. First, it provides the electronics necessary for the service. For example, a Web appliance provides the signaling, memory, and display electronics needed to browse the Web and display the results on the TV. Second, the in-home terminal is the instrument of conditional access. Unless access can be restricted, a price cannot be demanded. A wide variety of ways of protecting signals was discussed in the previous chapter. Nearly all of them require some in-home device.

Experience shows that consumers respond to the ability to make impulse purchases of programming. The easier the mechanism of purchase, the more is bought. This capability also requires some sort of in-home equipment.

The in-home terminal is one of the fundamental determinants of the quality of the video and audio experienced by the subscriber. Technical specifications such as noise figure, linearity, and the tightness of the shielding are critical to good pictures and sound. The need for a quality terminal is paramount.

Lastly, there is a need to be able to experiment with new services. In the past, this has been done by the system operator providing the in-home hardware necessary for the marketing experiment. If the service proves compelling, additional equipment is procured, and the service is rolled out across the population. The cable industry is very good at copying success and applying it widely. If the experimental service is disappointing, the equipment is abandoned and forgotten. In neither case is the consumer required to purchase new hardware that might be later orphaned. This is just an extension of one of cable's major marketing advantages. The salesperson can offer a low-cost or even a free trial hookup to the consumer without asking for the purchase of any equipment. Consumers know that if they are dissatisfied with the service, they may discontinue at any time without getting stuck with hardware that has no other use. This ability to offer a risk-free trial is an important advantage.

The program supplier's principal concern is protection against theft. Though this is important to the cable operator, it is even more critical to the program suppliers. The framers of the U.S. Constitution realized the importance of protecting the rights of authors, artists, and other creative workers when they provided for the copyright. Protection encourages the creative contributor. Without the fruits of these contributors, there would be no cable industry and no consumer electronics industry. The consumer would have minimal interest in what little would be left. Although the public may view theft of programming as a "victimless crime" or one of little significance, this is a misperception. All suffer from the loss of incentive to produce quality programming. Legitimate consumers pay higher prices because those who steal do not contribute their fair share to covering the costs of operation. Fortunately, the 1992 Cable Act, the 1996 Telecommunications Act, and the actions of the FCC have recognized the importance of preventing signal theft.

The program supplier is also interested in minimizing the impact of the copying of programs on other means of distribution. Businesses in the rental and sale of videotapes and video discs are undermined by easy copying of programming delivered on cable.

19.4.5 The Service Provider's Equipment Manufacturer

The priorities of the service provider's equipment manufacturer are perhaps easiest to understand. Its goal is to profitably sell systems involving large quantities of

in-home hardware and the headend equipment needed to supply signals. Since cable systems are individual closed systems of limited geographic area and since the in-home hardware has traditionally been leased, adjacent cable systems can utilize totally incompatible equipment and still well serve the TVs and VCRs owned by subscribers. The subscriber who moves from system to system is not inconvenienced by these incompatibilities since the system that serves the new residence will supply the necessary in-home equipment while the system serving the previous residence takes back its equipment.

A major advantage of this approach is that new technology is easily accommodated. When a cable system is ready to upgrade its hardware, it can purchase the best of the newest equipment available without concern about making the consumer's equipment obsolete. From the manufacturer's perspective, this keeps the market from becoming saturated. New services and new technologies can recapitalize the consumer's set top while bringing excitement to the service. Imprudently implemented consumer ownership of this equipment would interfere with this process of innovation.

19.4.6 The Consumer Electronics Industry

The consumer electronics industry has two main elements: the manufacture and the retail sale of products.

The nature of the consumer electronics business is one of high competitiveness. The manufacturer is interested in making more products and in selling upscale models. Almost 100 million residences receive television signals in one form or another in the United States. There are more than 250 million television receivers and 150 million VCRs in those homes. About 25 million new TVs and 15 million new VCRs are sold annually. The consumer electronics industry has the capacity to produce even more. Each year, new efficiencies make possible the production of more units from existing production lines. There simply is massive production overcapacity. The consequence of this is that it is impossible to maintain prices, much less raise them. Margins are tight. Only the full-featured products make money. The simple products now sell for something like $10 per inch of screen size. The cable set top terminal has the potential to interfere with attractive features in the TV and VCR. This drives the consumer toward the profitless low end of the product line. Clearly, this is a survival problem for the consumer electronics industry.

The retailer does not have the option of selling set top terminals to cable operators. Those are large-scale industrial sales that are seen by the retailers as a bypass of their business. The retailer's desire is either to totally prevent the use of set top terminals or to get into the business of selling them. After a few years of attempting to outlaw set top terminals, the retailers have come to realize that the age of interactive television, convergence, and advanced services requires set top terminals. The emphasis is now on requiring these devices to be available at retail.

19.4.7 The Two Extremes

After the 1992 Cable Act, discussions between the cable and consumer electronics representatives began in earnest. Some on each side adopted extreme and simple-minded positions. The extremists in the consumer electronics industry said, "Why don't they just not scramble?" Whereas the extremists in the cable industry said, "Why don't they just make monitors?"

Both of these outlooks are life threatening to the industries concerned. Without scrambling, cable cannot protect signals and offer premium services. This would spell the end of the cable business. Without premium features, the monitor business would degenerate into a profitless commodity business that would cause chaos to the consumer electronics sector. Neither alternative is workable. Neither alternative considers the consumer. Something else is required.

19.4.8 The Origins of the Problem

The history of the situation provides some help in understanding its complexity. The first "cable-ready" TV was almost an accident. A couple of decades ago, television tuners were analog devices with mechanical rotary turrets. A multitude of contacts switched coils, changing the frequency tuned and the local oscillator frequency. Remote control was implemented with a motor driving the shaft of the tuner. The tuner contacts wore out or oxidized, and the tuner became the major source of service problems. The digitally controlled varactor tuner was an innovation that tuned the receiver with electrically variable capacitors (the varactors) instead of screwing a ferrite core into a coil of wire. The tuning voltage for the varactors was derived from a digital-to-analog converter driven by digital electronics. The voltage data was stored in semiconductor memory. The same digital electronics served to implement the remote control. The advantage to the consumer was a significant increase in reliability because of the elimination of moving parts and the vulnerable electrical contacts. However, the cost of TVs with electronic tuners was around $100 more than comparable units with mechanical tuners, making these advanced receivers difficult to sell.

An engineer recognized that memory comes in fixed sizes and that most applications do not use all the memory available on the chip. He proposed filling the memory chip with additional data that allowed the varactor tuner to tune some nonbroadcast frequencies used on cable. With the addition of a cable television F connector to accommodate connection to a cable, the product could be sold as "cable ready." The additional cost was near zero—just better utilization of the memory that was already there. This was a feature consumers could understand. They were willing to pay more for "cable ready" because it promised to free them of the set top terminal.

As time went on, subscribers noticed that though the TV gave them access to some cable channels, it did not give them access to all. A "horsepower race" developed where the number of channels tuned was part of the advertising. As

more channels were added, a larger size memory was needed; this added a small increment of cost. Further increases in channel capacity required slightly more complex tuners. But this still was a minor cost increment.

19.4.9 Other "Cable-Ready" Issues

There are a number of important design features included in most cable set top terminals that very few "cable-ready" TVs and VCRs incorporate. Most cable set top terminals have a feature called channel mapping. Channel mapping decouples the channel display number from the frequency tuned. Thus, if there is a direct pickup problem in the cable system because of the presence of a strong broadcast signal, the broadcast channel can be placed at a different frequency to avoid this interference while allowing the subscriber to request it by its familiar channel number. The remote control command and the channel indicator all provide the familiar number while the tuner is "mapped" to another frequency. The set top terminal has downloadable control commands to change the mapping as channel lineups change and to suit the needs of the cable system using the unit. A consequence of the absence of this important feature in "cable-ready" TVs and VCRs is the annoying need for multiple channel numbers in many local program listings. The familiar channel number is used for those subscribers using a cable operator-provided set top terminal. An alternative channel number is used for "cable-ready" TVs and VCRs. This problem exists in the same home when some TVs have set top terminals and others do not or when the auto bypass feature is used with a set top terminal. Receivers built in compliance with the EIA-608A standard may include automatic channel mapping as an option.

A more important application of channel mapping is the manner in which scrambled channels are handled in a set top terminal. When a subscriber using a set top terminal attempts to tune a channel for which he or she is not authorized, the set top terminal avoids that channel. Either it goes to another channel that has an appropriate message or it simply presents a blue (or blank) screen and muted audio. The alternative channel may include advertising for premium cable services. In that case, it is referred to as a *barker channel.* The term comes from the circus "barker" who hawked sideshows and attempted to attract an audience. If a blue screen is used, it may include a text or graphics message indicating how the subscriber could upgrade the subscription. In either case, the video and audio are kept from the receiver. This feature allows for practical video scrambling that has relaxed constraints. The video scrambling does not need to completely hide the video. There is no concern that "objectionable" video might "bleed" through to offend sensitive viewers. Additionally, audio does not need to be scrambled since, in most cases, scrambling the video is sufficient to require a subscription. Children are protected from unintentional exposure to the audio of objectionable programming and do not accidentally see the "random lockup" of a scrambled video because of the set top tuner's mapping to another channel. This protection is lost when the "cable-ready" product does not include the same channel-mapping feature.

The television audio signal is complex. It includes a stereo component, a second audio program (SAP), usually for monaural second language, and auxiliary signals. Since the human ear is very sensitive to audio distortion, scrambling the complex analog television audio signal in a manner that results in quality sound when descrambled is expensive and technically challenging. This expense would be unnecessary if channel mapping were universally used.

Set top terminals have other important features as well. Parental control features are required by some franchises. These franchises often specify the manner in which the parental control feature must function and the level of security it must provide. "Cable-ready" TVs and VCRs usually have simplified parental control if they have the feature at all. Additionally, set top terminals often have an emergency alert feature that allows local government officials to alert citizens of hazardous conditions. "Cable-ready" TVs and VCRs do not include this capability.

19.5 Frustrated Product Features

Adding a set top terminal either makes certain features and functions of TVs and VCRs redundant or interferes with their operation. In most situations, the feature or function is not blocked, but its use requires additional effort that the consumer finds inconvenient. It is this frustration that motivated the compatibility section of the 1992 Cable Act.

19.5.1 Remote Control

Remote controls were included in even early set top terminals. The first remote controls were "wired." A long wire led to a separate unit positioned near the viewing chair and allowed channel changing without having to go to the TV. Cordless remotes based on ultrasonic tones, infrared light waves, or radio links were introduced later on. Consumers were able to add the remote control feature to existing nonremote-controlled TVs by utilizing the cable set top terminal. This was seen as an important ancillary advantage to a cable subscription. The consumer electronics industry was not pleased with the slowing impact this had on replacement sales. However, a benefit accrued in that consumers were presold on the advantages of remote control. It became almost impossible to sell a new TV or VCR without remote control.

A problem arises when both the TV and the set top terminal are remote controlled. The customer may feel he or she paid for a remote control feature on the TV that is not needed because the set top terminal has the remote control. Worse yet, the consumer may need to juggle two remote controls. He or she may have to use the TV's remote to turn it on, put it to the output channel of the set top terminal, and adjust volume. The consumer then uses the set top terminal's remote to change channels. The set top terminal often includes a switched convenience power outlet. This works well with older, nonremote-controlled TVs. But there

are problems with remote-controlled TVs. These products usually have a toggle on–off push button, which means that simply powering the line cord does not result in the TV going on. Worse yet, if the TV has an automatic channel setup feature, it will go through the setup procedure every time it is repowered. Since this takes several minutes, the result is unacceptable. The net result is the juggling of two remote controls. Note that the same situation exists in a noncable application where the VCR is connected between the antenna and the TV.

19.5.2 Timers

When a set top terminal is used with a VCR, further complications are encountered. Since the VCR timer controls only the VCR, it cannot turn the set top terminal on and off, and it cannot change the terminal's channel. Early on, this was recognized, and set top terminals with built-in timers have been available for well over a decade. The very same techniques used for programming the VCR are used for programming the set top terminal. The consumer who can do one can do the other as well. Unfortunately, most consumers do this function so infrequently that they must relearn it every time they wish to record a program. The chances for error are great, and the success rate is low. This leads to frustration. The majority of consumers—cable and noncable—use VCRs for playback of prerecorded tapes and just avoid these headaches.

Some set top terminals have the ability to set their clocks using signals coming from the cable headend. Thus, their time is always correct, even in spring and fall when daylight savings time changes occur. Some newer VCRs implement this feature with signals from the broadcaster. Another common timer found in TVs is the sleep timer. Although a cable set top terminal does not interfere with the TV's sleep timer operation, some set top terminals have a sleep timer that makes one of the two timers redundant.

19.5.3 Picture-in-Picture

Picture-in-picture (PIP) is the insertion of a small picture within the larger picture. The purpose is to be able to monitor a second program while watching and listening to another program. If the action in the second program becomes compelling, the two images can switch places with the push of a button. Most TVs above the basic models include PIP.

There are a number of implementations of PIP. The most common embodiment has only one tuner, which is used for the main picture; the image for the second picture comes from an external device—usually a VCR's tuner. Other video source alternatives include a baby-monitor camera or a front-door camera. More expensive TV receivers have two tuners. An advanced form of PIP can be operated to scan through channels and put up a mosaic of small still pictures from a number of channels. As such, it is a kind of program guide that allows owners to scan the channels prior to making a selection.

In a TV with two built-in tuners, a problem arises if an internal splitter is used and the set top terminal is placed ahead of the single antenna input. In that case, the output of the set top terminal appears on both the large and small pictures. If on the other hand, two antenna inputs are provided, the use of the set top terminal is noninterfering except when the customer wishes both pictures to come from scrambled channels. In that case, two descramblers are required.

The least common PIP implementation is the one that presents a mosaic of pictures from various channels. This system is interfered with since the program selector is in the set top terminal and is not under the control of the TV receiver.

19.6 Early Attempts

There were a number of early attempts to solve or mitigate the compatibility problems between consumer electronics and cable systems. All proposed solutions were partial. The 80/20 rule applies. Eighty percent of the problem can be solved with 20% of the effort (and expense). However, solving the remaining 20% of the problem takes 80% of the effort.

19.6.1 Universal Remotes

Universal remote controls sought to alleviate the need for using two or more remote controls with the set top terminal and TV or VCR. The universal remote control is configured for the brand TV, the brand VCR, and the brand set top terminal. Simple universal remote controls have a hidden row of switches that are configured according to printed instructions that come with the device. An alternative approach is to enter appropriate setup codes from the printed instructions. Most models direct the channel change commands to the set top terminal, the volume commands to the TV, and the on–off commands to both units. VCR control can also be accommodated. Either a slide switch or a mode button governs which device is primarily controlled.

A problem arises with simple universal remote controls when a new model of TV, VCR, or set top terminal becomes available or if a favorite feature of the TV or VCR is not included in the universal remote control. "Learning" universal remote controls solve this problem by allowing the universal remote control to learn the signals emitted by the remote control that comes with the product. A disadvantage is the complexity; some consumers seem unable to master the instructions. Another type of learning universal remote control uses acoustic signals from the telephone to enter the data requested by the customer.

Advanced universal remote controls include *macros*. A macro is a collection of commands emitted in response to the pressing of just one button. For example, a common macro is the advanced on command. First, the set top terminal is turned on, then the TV receiver is turned on, then the TV receives a command to tune to the output channel of the set top terminal. It is important that all the

devices "see" the complete IR emission sequence and respond appropriately, or the result may be disappointing.

Another model of advanced universal remote control includes a clock and a timer. The remote control can be positioned so that it simultaneously communicates with both the VCR and the set top terminal. The universal remote control is programmed with the VCR control instructions. The remote turns on the VCR at the appropriate time, ensures that the VCR is tuned to the output channel of the set top terminal, then tunes the set top terminal to the desired channel, and finally starts the VCR recording. Consecutive recording of different channels is made possible.

A few models of two-way remote controls have a small screen on the remote control that displays signals received from the set top terminal. Titles of songs, purchase order forms, interactive games, and other instructions can be viewed.

19.6.2 Electronic Program Guides

Electronic program guides (EPGs) have been introduced with features that alleviate some of the interface problems. The EPGs are available as stand-alone devices, built into cable set top terminals, or built into TVs or VCRs. Several of these devices include one or more infrared light-emitting diodes (LEDs) on short wires that are meant to be attached near the remote control window of the VCR, the cable set top terminal, or the TV. The device with the EPG controls the other devices to accomplish the desired result. If the EPG is built into the cable set top terminal, the LED controls the VCR so that it turns on and off at the correct times and is tuned to the correct channel. If the EPG is built into the VCR, the LED controls the channel on the cable set top terminal. Of course, none of these approaches can ensure that sufficient blank tape has been put into the VCR!

A related device, called VCR Plus, uses numeric codes published in printed guides (or conveyed electronically in a later version) to the VCR or to a form of universal remote control. The codes are decoded using a proprietary algorithm to derive the start and stop times and the channel number. Infrared emissions control the VCR and cable set top terminal.

19.6.3 The TV Integrating Device

In the mid-1980s, the research department of a cable operator, American Television and Communications in Englewood, Colorado, began development of an approach the engineers called the TV integrating device (TVID). It was intended to be included in a cable set top terminal and to use IR LEDs to control the TV and one or more VCRs. It had a built-in EPG and an on-screen display. The goal was to provide a simple menu that would allow subscribers to indicate their desires. The subscriber would indicate what he or she wanted recorded and what he or she wanted to watch. The TVID would take care of switching the descrambler into the correct part of the system and turn things on and off at the right time to implement the subscriber's desires. If a conflict arose, such as wanting to watch a

scrambled channel while recording another scrambled channel, the subscriber would be advised of his or her choices and allowed to act upon them.

The TVID included a "mood guide," which gave the subscriber a way of indicating his or her mood. Those programs that were appropriate to his or her mood would then be displayed for selection to watch or record. The microprocessor could keep track of the programs watched and deduce the subscriber's interests. The subscriber would then be alerted to future programs that fit his or her previous viewing habits. The project reached the early prototype stage but was not completed because of other priorities.

19.7 Management of Expectations

Most of the subscriber's problems have solutions. Some solutions are more complex than others. The crux of the compatibility problem seems to be the expectation that the term "cable ready" brings. To consumers, this means that the TV or VCR can be connected to cable and all features and all services are readily available. When operation of features is complicated or unexpected results are obtained, frustration and anger set in. If instead, expectations were properly managed and the subscriber/consumer had a correct understanding of what was possible, the likelihood of disappointment would be reduced. The fundamental problem is the management of expectations.

19.7.1 Complexity Overload

A major contributor to the difficulties of the consumer electronics interface is the wide variety of choices available in features and functions, the way in which the features and functions are operated and controlled, and the diversity of services offered by cable operators. The number of combinations and permutations is almost infinite. Just the simple matter of entering channel numbers has several methods. Some products use an enter button. Others prescribe the way in which the digits must be entered. For example, on some products, requesting channel 7 requires pressing 07. Others can require waiting for a fixed time after entering the command. The names for common functions are brand specific. The number of complex features boggles the mind. It may simply not be possible to preserve all the product functions and features while offering access to all the available cable services.

19.8 Legal Aspects of Cable–Consumer Electronics Compatibility

19.8.1 Compatibility Definitions

The word *compatibility* should mean that two things work well with each other. In the case of compatibility between cable service and consumer electronics,

compatible should mean that the subscriber should be able to enjoy the cable services of interest and the features of the consumer electronics that he or she finds appealing. The subscriber should be able to conveniently do both.

Note that compatible is a situational definition. If a subscriber takes only basic (unscrambled) service from a cable system with a modest number of analog channels and is located far from strong interfering broadcast signals, he or she may enjoy all the functions of his or her "cable-ready" TV and VCR. If the subscriber then decides to add a scrambled service or moves near a broadcast transmitter, the situation changes significantly. He or she may lose compatibility between the TV or VCR and the cable system.

19.8.2 The 1992 Cable Act's Definition of Compatibility

Senator Patrick J. Leahy (Democrat from Vermont) drafted and introduced a bill in 1991 pertaining to the operation of cable systems and the design of consumer electronics products so as to improve compatibility. The bill did not pass. In the following year, it was introduced as Section 17 to the 1992 Cable Act.[6] Section 17 requires the Federal Communications Commission (FCC) to create rules to improve the compatibility between cable systems and consumer electronics products. The effect of this section of the law was to define compatibility as having three principal aspects:[7]

(i) to watch a program on one channel while simultaneously using a video cassette recorder to tape a program on another channel;
(ii) to use a video cassette recorder to tape two consecutive programs that appear on different channels; and
(iii) to use advanced television picture generation and display features

Although this definition of compatibility may not be complete, it does define a requirement for the FCC's rules.

The bulk of the 1992 Cable Act pertains to economic issues such as rate regulation, the rules for buying and selling systems, and the permissible business relationships between programmers and operators. A small, but very important part of the Cable Act concerned the interoperability of cable systems and consumer electronics products. Though the 1992 Cable Act gave the FCC a difficult assignment, it also laid down some guidelines. The FCC was told to consider the costs and benefits to consumers of its rules and the provision of effective protection against theft. But the ultimate goal was clear: the minimization of interference with certain special functions in consumer electronics devices when connected to cable systems.

The FCC was given some new authority to help it achieve these goals. The FCC's primary influence over the cable industry was granted in the following clause: "the Commission shall determine whether and, if so, under what circumstances to *permit* cable systems to *scramble or encrypt* signals or to *restrict* cable systems in the manner in which they encrypt or scramble signals."

The authority given the FCC over the consumer electronics industry is much more limited: "to specify the technical requirements with which a television receiver or video cassette recorder must comply in order to be sold as 'cable compatible' or 'cable ready.'"

To ensure that the goals of compatibility continued to be pursued in the future, the FCC was given continuing responsibility:

> The Commission shall periodically review and, if necessary, modify the regulations issued pursuant to this section in light of any actions taken in response to such regulations and to reflect improvements and changes in cable systems, television receivers, video cassette recorders, and similar technology.

Further requirements imposed on the operation of cable systems included obligations for the FCC to:

> promote the commercial availability . . . of converter boxes and of remote control devices to the extent technically and economically feasible, to offer subscribers the option of having all other channels delivered directly to the subscribers' television receivers or video cassette recorders without passing through the converter box.

The FCC was required to provide a report to Congress within a year of the passage of the act. This voluminous report was based on the extensive responses to the FCC's initial notice of inquiry on the subject. The FCC issued a notice of proposed rule making on November 10, 1993, and released its first Report & Order on May 4, 1994. In those documents, the FCC indicated its directions:

> [W]e clarify that the compatibility regulations being adopted herein apply only to cable systems and do not apply to . . . video dialtone services.

> Signal thieves [have been] notoriously successful. . . . We will not require cable systems to allow their subscribers to own descrambling equipment.

> [W]e expect that the Decoder Interface connector will be flexible enough to accommodate most, if not all, of the attributes of cable digital service.

The FCC realized that there is an extensive installed base of more than 11,000 cable systems, more than 250 million television receivers, and more than 150 million VCRs. In addition, approximately 100 million TV households purchase around 25 million color TVs and about 15 million VCRs each year. No law could bring compatibility to this situation in a short time. The FCC divided its approach into three periods: short-term measures, intermediate-term measures, and long-term measures.

In the short term, cable operators were required to avoid scrambling the basic services, which consisted of mostly the broadcast channels carried on cable. In addition, the cable operators were required to provide certain equipment such as switches to bypass the entire spectrum around the set top terminal when the terminal is not required for descrambling, and timers in set top terminals to change channels at specific times. Cable operators were given a year to make available

multiple tuners and descramblers in set top terminals. They were allowed to charge for the costs of this equipment and its installation. Additionally, the consumer electronics industry was given a year to discontinue the use of the terms "cable ready" or "cable compatible" or any similar terminology in the sale of TVs or VCRs that did not fully comply with the FCC's promised technical definition of "cable ready." A rule was included that would have required a label on devices that tuned cable channels but did not fully comply with the FCC's definition of "cable ready," notifying consumers that this device may under some circumstances fail to provide satisfactory performance when connected to cable. The consumer electronics industry managed to get this rule reversed and gained a relaxation of other rules in a subsequent petition for reconsideration submitted to the FCC. The FCC imposed a bizarre rule prohibiting the cable industry from changing infrared codes in each set top terminal installed in each home in order to protect the home owner's "investment" in purchased remote controls. This rule was all but impossible to follow because of the large number of different codes. It was removed in response to a subsequent petition for reconsideration. Also, the FCC recognized the aggressive work of those who would steal signals:

> [W]e are clarifying that the availability of set-top devices from retailers applies only to basic converters without descrambling or other access control function.

> [W]e do not believe it is desirable to require cable operators to make their security technology available for ownership by the public.

The FCC then separately considered longer-term measures centered around the decoder interface (described in the previous chapter) being developed by the Joint Engineering Committee of the National Cable Television Association and the Electronic Industries Association: "We . . . emphasize . . . the Decoder Interface connector and associated component descrambler/decoders to be an important part of . . . equipment compatibility."

The FCC envisioned a digital world that would ultimately be based on standards and methods that avoided the problems of compatibility. The FCC intends to consider that environment separately in future actions.

19.8.3 The 1996 Telecommunications Act

The Communications Act of 1934 created the Federal Communications Commission and set the original public policy on communications. There have been multiple amendments to that act over the years, making it a quilt of patches. Additionally, technology had changed so dramatically that a complete rewrite became necessary. The result is the 1996 Telecommunications Act.[8] Although the act has a wide range of impact on cable television practice, only the portions pertaining to compatibility will be discussed here.

Probably the most important provision of the 1996 Telecommunications Act pertaining to compatibility is Section 304, "Competitive Availability of Navigation Devices," by Representative Thomas J. Bliley Jr., of Virginia:[9]

> (a) The Commission . . . in consultation with . . . industry standard-setting orga-
> nizations . . . [shall] assure the commercial availability, to consumers of . . . ser-
> vices offered over multichannel video programming systems, of converter boxes,
> interactive communications equipment, and other equipment . . . from manufac-
> turers, retailers, and other vendors not affiliated with any multichannel video
> programming distributor. Such regulations shall not prohibit any multichannel
> video programming distributor from also offering . . . equipment . . . if the system
> operator's charges . . . are separately stated and not subsidized.

The national headquarters of Circuit City stores is in Congressman Thomas
Bliley's district, and his efforts to further the interests of an important constituent
are clearly evident in this legislation.

The term *navigation devices* is a strange one. It has not been used in the cable
or other related business except in reference to electronic program guides and re-
lated means of helping consumers find their way—navigate—through the wide
variety of services available on systems with hundreds of channels. The law uses
this term to mean any converter or set top terminal intended to go between the
service provider's terminal and the TV or VCR.

Several important points should be noted. Although the 1992 Cable Act was
restricted to cable operators, the 1996 Telecommunications Act applies to any
multichannel video programming distributor. Additionally, the services covered
are not limited to video.

Limits were placed on what must be sold in order to protect the interests of
programmers and system operators.

> (b) The Commission shall not prescribe regulations . . . which would jeopardize
> security of multichannel video programming and other services . . . or impede the
> legal rights of a provider of such services to prevent theft.

In order to accommodate experimentation with new services and technologies,
the law provides for waivers allowing a period in which hardware can be un-
der trial.

> (c) The Commission shall waive a regulation adopted under subsection (a) for a
> limited time upon an appropriate showing . . . that such waiver is necessary to as-
> sist the development . . . of a new or improved . . . service.

A number of significant hazards are brought about by this legislation. The
most important hazard is that the other users of cable systems and of the over-
the-air spectrum could suffer interference and degradation of service when con-
sumers purchase their own navigation devices. Second, consumers are subject to
confusion with these complex devices. Money may be wasted on features that are
unusable or not needed. When technology changes, consumers will be stuck with
orphaned hardware. In some cases, the ownership of equipment will create a re-
luctance to try new services that require additional hardware that partially dupli-
cates what has already been purchased.

It is likely that many, if not most, subscribers will continue to lease equip-
ment because of the many benefits that come from such an arrangement. The

equipment is matched to the services provided. New equipment replaces old when technology evolves, service of equipment is included in the lease, and assistance on its use is readily available.

The other major portion of the 1996 Telecommunications Act pertaining to compatibility is Section 301(f), the "Eschoo Amendment" by California Congresswoman Anna Eschoo. This amendment modifies the compatibility rules of the 1992 Cable Act by putting restrictions on the FCC's actions. Specifically, it calls for mandating "a minimum degree of common design and operation, leaving all features, functions, protocols, and other product and service options for selection through open competition in the market." The goal of this amendment is to prevent the FCC's rules from picking winners and losers in a competitive environment where a variety of different approaches—some proprietary and others public—are vying for the same customer. Though that is a laudatory objective, it does make a difficult and complex task more challenging.

19.9 Standards

19.9.1 The Need for Standards

The cable industry has traditionally had little need for additional standards because it provides signals usable on the consumer's ordinary television or radio. The goal has always been to allow the consumer to take cable service without having to purchase additional equipment. A major attraction to cable is the ability to try a subscription with little investment and the assurance that it is easy to drop the subscription if it is disappointing. So the most important standard is NTSC. When the cable system provides services beyond the ability of the consumer's TV, an adapter is provided to allow access to the additional services. The earliest adapters were just tuners that gave access to channels beyond the range of the TV receiver. Since a cable system is a closed universe unto itself, there is no need for interoperability with adjacent cable systems.

Conditional access and scrambling are deliberately not standardized. Standardization of these functions would give the signal thief a larger market to sell "pirate boxes." By having multiple types of conditional access and scrambling, the garage manufacturer of pirate boxes is frustrated. Since the service provider has ownership of the conditional access equipment, he or she can decide when the economic losses due to a breach of the security system are of sufficient magnitude to require replacing the equipment. When the service provider replaces the equipment, the consumers are not adversely affected.

The 1996 Telecommunications Act includes a section that requires the FCC to make rules that will regulate how multichannel video service providers will make in-home equipment available at retail. Now, for the first time, there is a need for standards. Consumers who purchase equipment for connection to cable need to be assured that it will work when they move to another cable system. Fortunately,

the law and the FCC have recognized the problem of signal theft and have in-cluded a clarification that security systems do not have to be made available for subscriber ownership. There are three types of standards: mandatory, official voluntary, and de facto.

19.9.2 Mandatory Standards

Mandatory standards are those imposed by law or as a result of law. In the former case, the law itself contains the standards; in the latter, more common case, the law directs a government agency to do the difficult work and create rules. The FCC is the source of rules that govern the cable industry. There are two major sets of rules: (1) Part 78, the Community Antenna Relay Service (CARS) rules, which regulate the use of microwave links to convey signals across natural or man-made barriers or simply to transport the signals from one location to another where the geography in between does not have the cable service, and (2) Part 76, which governs the operation of the cable system itself. The items regulated include safety, signal leakage, and minimum performance requirements. These were first put into place in 1972 and then immediately stayed while further deliberations took place regarding the details. The performance rules cover frequency stabil-ity, carrier-to-noise ratio, allowable interference, in-band frequency response, hum content, adjacent channel response, cross modulation, must-carry and re-transmission consent, and signal leakage. The rules (with the exception of signal leakage) apply only to Class 1 signals, which are defined as signals originally broadcast. However, the rules cover only additional performance deficiencies. The cable operator is not responsible for problems in the signal as delivered to the cable headend. Class 2 signals are satellite-delivered programming, and Class 3 signals are locally originated. The rules were suspended by the 1984 Cable Act, which deregulated cable. At that time, they were converted into a recommended practice. Local governments were permitted to create their own rules, but those rules could not be more severe than the recommended practice. This was chal-lenged by the League of Cities. A consequence of this was that a panel of repre-sentatives from the cable industry, the FCC, and the League of Cities negotiated a set of rules that has become the current FCC Part 76 in 1991.

19.9.3 Official Voluntary Standards

Voluntary standards do not have the force of law. The main motivation for com-pliance with the standard is the ability to use the fact of compliance in marketing efforts. Official voluntary standards have the sanction of a standards-setting orga-nization. The primary standards-setting organization in the United States is the American National Standards Institute (ANSI). ANSI has very rigorous rules for the procedures involved in setting these standards. The rules are intended to en-sure that participation is available to all interested parties and that due process is followed. ANSI certifies organizations as complying with these rules. Currently,

the primary ANSI-certified cable organization is the Society Cable of Telecommunication Engineers (SCTE). Another organization that has worked to develop standards that govern both the cable industry and the consumer electronics industry is the Consumer Electronics Manufacturers Association (CEMA) sector of the Electronic Industries Association (EIA). Although CEMA is the official ANSI standards-setting organization, its efforts are under the auspices of the Joint Engineering Committee (JEC) of the EIA and the National Cable Television Association (NCTA) Engineering Committee.

The SCTE has about 100 standards under development at the time of this writing. Five of those standards are in a condition where they are suitable for submission to ANSI. One of those standards has been approved as ANSI/SCTE PS SP 400 1996 F-Port (Female Outdoor) Physical Dimensions. Next will be the 5/8-24 RF Outdoor Fitting Physical Dimensions standard. The Digital Video Transmission Standards for Cable Television is expected to be submitted before the end of 1998.

The JEC efforts began in the very early 1980s and have not been nearly as successful in completing standards. This lack of results is an indication of the difficulty involved in the interface between cable and consumer electronics. Its first standard, the ANSI/EIA-563 Decoder Interface Standard,[10,11] was in place for several years and then officially withdrawn. Another standard, EIA-542 Cable Television Channel Identification Plan,[12] lists channel numbers and frequencies for three different channel plans: the standard plan related to the broadcast frequency plan, the harmonically related carrier plan, and the incrementally related carrier plan. Though this standard has been slow to make its way through the official process, it has been used by both the cable and the consumer electronics industry as a de facto standard for more than a decade. Both industries have found it a useful means of cooperation.

One of the most important JEC standards is the EIA-IS-105[13] Decoder Interface Standard described in the previous chapter. Other standards in the works include the National Renewable Security Standard (NRSS)[14] and the RF Interface Standard IS-23.[15] An effort to create a digital television standard has been transferred to the SCTE. The IS-23 standard governs such parameters as the physical connection to cable, the frequency response, the signal levels, and direct pickup by inadequately shielded components.

19.9.4 De Facto Standards

De facto standards are the result of marketplace forces where the cable industry has settled on a particular approach because it works well. The types of connectors, cables, amplifiers, and other equipment used in cable have been sorted out by the marketplace, and a limited set of models have evolved on a de facto basis. Some of these de facto standards are being converted into official voluntary standards by the SCTE. This approach has served the industry well and will continue to be in use.

19.10 OpenCable

Cable Television Laboratories, or CableLabs, has undertaken a program intended to provide options for alleviating many of the interface problems. The goals of the program, called OpenCable, are to provide sufficient standardization of features and functions so that a multivendor interoperable environment can be created. CableLabs will certify compliance with OpenCable.

Initially, OpenCable will allow cable operators to utilize set top terminals from multiple vendors. Though OpenCable is primarily directed at digital set top terminals, it optionally includes analog descrambling. OpenCable set top terminals that CableLabs certifies as compliant will eventually be available at retail. Finally, it can be expected that OpenCable functions and features will be included in consumer electronics products. OpenCable is described in more detail in Chapter 3.

19.11 National Renewable Security System

The Joint Engineering Committee formed a subcommittee called the NRSS subcommittee to consider methods that could overcome a security breach. The motivation is to facilitate the sale of set top terminals or consumer electronics products while protecting the conditional access system by having it renewable in the event the system is defeated. The initial approach involved the Smart Card system, which was used for conditional access and for credit and debit cards. The Smart Card was intended to include a digital key that would decrypt the digital signal.

After some discussion, it was decided that the only safe approach is to have every transistor involved in the security function replaceable in a module that plugged into the set top terminal. A wide variety of proposals were considered. One of the proposals used the Smart Card physical format but put a lot more electronics on it. This NRSS Type A card complied with the International Standards Organization (ISO) 7816 standard. Several limitations flow from this. There are only eight contact pads, forcing the MPEG signal to be transported serially. There is no mechanism for handling cable's out-of-band signaling. Eventually, the committee settled on an optional Type B approach based on the Personal Computer Manufacturers Communications Interface Adapter (PCMCIA) Type II standard card. This card has 68 pins, which is adequate to allow the set top terminal to convey out-of-band signals received by a built-in modem.

The NRSS standard, designated the EIA-679, has been balloted and passed. It is under the EIA's auspices since the NCTA is not a standards-setting organization and the work began before the SCTE became accredited to set standards by ANSI. Work continues on optional extensions to the standard. Proposals for an optional PCMCIA card that itself could accept an ISO 7816 card have been made. Another

manufacturer suggested an extended PCMCIA card with an on-board single chip tuner. The 1394 "Fire Wire" communications protocols are being studied.

It is recognized that the NRSS system is a necessary but not sufficient condition for the implementation of digital set top terminals and consumer electronics products that can be sold at retail. At least two issues remain. The first is the requirement that upstream signals be possible for two-way interactive services. This can be handled by having the cable connection first go to the NRSS card, loop through the card, and then go to the set top terminal or consumer electronics product. The second requirement involves the graphical user interface (GUI) of the cable service. Since the NRSS does not process the signal as a baseband video signal, it is not possible to simply cut graphics into the video. Another access point is required for GUI processing.

19.12 Summary

The interface between cable service and consumer electronics products affects the quality of service and the ease of use of a service. In severe cases, the interface prevents access to a service. This same problem interferes with the ease of use of some of the functions and features of consumer electronics products. These problems are complex because of the wide variety of services, functions, and features possible with modern methods.

The JEC and its subcommittees have been working diligently for more than a decade on ways of alleviating the interface problems. Just the increased understanding between the technologists of the two industries has helped. Products and cable practices now more fully accommodate the interface issues.

Although a great deal of progress has been made, much is left to do. We are hopeful that the solutions found in digital implementation will be more satisfactory because of the lessons learned in the struggle with analog systems.

Endnotes

1. Walter Ciciora, Consumer Electronics Friendly Cable, *1992 NCTA Convention Technical Papers.* NCTA, Washington, D.C.

2. David Large, Options for Friendly Delivery of Cable Services. *CED,* April 1993, pp. 62–74, 83.

3. Walter Ciciora, The Realities of the Consumer Electronic Compatibility Interface Regulations, *1995 NCTA Convention Technical Papers.* NCTA, Washington, D.C.

4. David Large, *Connecting Cable Systems to Subscribers' TVs and VCRs—Guidelines for the Cable Television Industry.* NCTA Engineering Committee, NCTA, Washington, D.C., January 1987.

5. Walter Ciciora, Regulation and the Continuing Struggle for Consumer Electronic Compatibility, *1995 Montreux International Television Symposium Technical Record.* Montreux, Switzerland, June 1995, pp. 597–611.

6. Walter Ciciora, The 1992 Cable Act and the Consumer Electronics Interface, *1994 NCTA Convention Technical Papers.* NCTA, Washington, D.C.

7. Cable Television Consumer Protection and Competition Act of 1992, House of Representatives Report 102-862, 102nd Congress, 2nd Session.

8. Telecommunications Act of 1996, House of Representatives Report 104-458, 104th Congress, 2nd Session.

9. Walter Ciciora, The Realities of the Retail Sale of "Navigation Devices," *1996 NCTA Convention Technical Papers.* NCTA, Washington, D.C.

10. *EIA/IS-15-A, Standard Baseband (Audio/Video) Interface Between NTSC Television Receiving Devices and Peripheral Devices,* March 1989. Electronic Industries Association, Engineering Department, Arlington, VA.

11. *The EIA-653 Decoder Interface Standard,* 1990. Electronic Industries Association, Arlington, VA. Available from Global Engineering Documents, Englewood, CO, 800-854-7179.

12. *EIA-542 Cable Television Channel Identification Plan,* 1997. Electronic Industries Association, Engineering Department, Arlington, VA. Available from Global Engineering Documents, Englewood, CO, 800-854-7179.

13. *EIA/IS-105 Decoder Interface Standard,* 1997. Electronic Industries Association, Engineering Department, Arlington, VA. Available from Global Engineering Documents, Englewood, CO, 800-854-7179.

14. *EIA/IS-679 National Renewable Security Standard,* 1997. Electronic Industries Association, Engineering Department, Arlington, VA. Available from Global Engineering Documents, Englewood, CO, 800-854-7179.

15. *EIA/IS-23 RF Interface Specification for Television Receiving Devices and Cable Television Systems,* 1997. Electronic Industries Association, Engineering Department, Arlington, VA. Available from Global Engineering Documents, Englewood, CO, 800-854-7179.

Other References of Interest

David Large, Defining "Cable Ready" Consumer Equipment. *CED,* April 1993, pp. 84–90.

EIA-608-A Recommended Practice for Line 21 Data Service, 1994. Electronic Industries Association, Engineering Department, Arlington, VA. Available from Global Engineering Documents, Englewood, CO, 800-854-7179.

Several feature articles on the decoder interface. *Communications Technology,* August 1989.

The Consumer Interface Problem, A Status Report, CableLabs, Louisville, CO, February 1990.

Customer Premises Equipment Performance and Compatibility Testing, CableLabs, Louisville, CO, 1993.

Subjective Assessment of Cable Impairments on Television Picture Quality, CableLabs, Louisville, CO, 1992.

Chapter 20
Equipment Compatibility

20.1 Introduction

The previous chapter described the nature of the challenges regarding the interface between cable and consumer electronics equipment. That chapter concentrated on the way the equipment is connected and used, as well as on the impact on usability of features and services. The legal requirements and FCC rules pertaining to usage of the equipment were also covered.

This chapter will cover the minimum technical requirements for consumer-owned equipment that is to be connected to cable. There are two broad aspects to this issue: (1) the quality of the pictures and sound delivered to the subscriber, and (2) the potential for interference with other devices connected to cable and with other users of the radio spectrum. The first aspect is considered a marketplace issue to be determined by the purchaser of the equipment. But this aspect of compatibility unfortunately also affects the enjoyment of cable-delivered services and, to some extent, the likelihood that the consumer will subscribe to premium services. Often, the consumer purchases his or her equipment in an environment where its performance on cable is impossible to judge and factor into the purchase decision.

20.1.1 Broadcast Versus Broadband Receivers

The technical requirements of tuners for broadcast channels versus broadband cable channels are significantly different. It is possible to design an excellent broadcast tuner that performs poorly on cable channels, and vice versa. Designing a cost-effective tuner that performs well in both environments is a significant challenge.

The broadcast environment consists of channels with gaps. Local channels are almost never in adjacent frequencies. However, the viewer may wish to receive a weak signal with frequency between two strong local signals. The weak signal may be from a distant location. This is called *fringe reception* and was an important selling feature before cable became popular. The tuner must function well in that environment. Note that the lower channel numbers, the VHF channels, are designated by the FCC so that the frequency of no one channel is a second harmonic of any other. This reduces the requirements on linearity in the tuner circuits.

Broadcast channels are not located on an "image" channel of another broadcast channel. The image channel problem will be described later in this chapter. Image interference problems are avoided in the broadcast environment. An inspection of the frequencies assigned to the VHF channels reveals that no channels are assigned so that they are images of valid channel frequencies. The higher channel numbers, the UHF channels, have an elaborate set of "taboos" in the FCC's rules that are intended to relax the design requirements of practical consumer electronics tuners. Finally, in the broadcast environment, there is no particular need to prevent direct pickup of the broadcast channel. Direct pickup simply adds to the signal brought in from the antenna and causes no harm. Also,

since only broadcast channel frequencies are found in the tuner, there is no need to prevent their radiation from the tuner terminals or from the tuner itself.

The cable tuner is very different. The cable spectrum is fully loaded with no significant gaps up to the highest frequency. The signal levels are strong and relatively uniform across the frequency band. The challenge is in handling a lot of energy and in avoiding distortion and cross modulation of the signals.

A consequence of these differences is that a properly designed cable tuner will probably not perform well in the broadcast environment, and a properly designed broadcast tuner will probably not perform well in the cable environment.

Essentially, all cable converter tuners are double conversion tuners. That is, they apply the superheterodyne principle twice. They have two local oscillators and two intermediate frequency amplifiers and filters. This gives improved selectivity and blocks the image channel. Also, the first local oscillator frequency is usually above the highest channel tuned, and the second local oscillator signal is isolated from the cable by the first conversion stage. So backfeed of internally generated spurious signals is not a problem with cable converters. The use of a double conversion tuner in a "cable-ready" TV or VCR is extremely rare. Nearly all consumer electronics tuners are single conversion devices.

The deficiencies of "cable-ready" tuners will become more apparent as cable systems expand bandwidths beyond 450 MHz on to 750 MHz and even 1.0 GHz and beyond.

20.2 Levels of Interference

Four kinds of interference have been considered. In order of severity, they are (1) interference with the product's own performance, (2) interference with other products in the same home, (3) interference with another subscriber's reception, and (4) interference with other users of the electromagnetic spectrum.[1]

Interference with the product's own performance results in a dissatisfied customer for the product and probably for the services delivered. Unfortunately, in many cases, the poor performance of the equipment can be confused with a problem with the service. This results in unnecessary and unproductive service calls, missed work while waiting for the service call, and the potential for finger-pointing between the cable service provider and the retailer of the consumer product. Too often the consumer gives up. The problem is not resolved, and the consumer is suspicious of both the consumer electronics product and the cable operator.

The next level of severity occurs when the product backfeeds interfering signals into the cable, causing problems with another TV or VCR in the same home. This is often a transitory problem occurring only when certain combinations of channels are tuned by the two devices. This is likely to be blamed on the cable operator. Yet one step worse is when another subscriber suffers interference from a product not even in the same home. This will almost always be blamed on the cable operator.

The most severe situation is when interference is caused to other users of the radio frequency spectrum, particularly safety-related users such as aircraft communications and navigation, fire, police, and emergency communications. This level of interference goes beyond merely reducing the entertainment value of cable and can become dangerous to others' well-being.

20.3 Direct Pickup Interference

One of the most troublesome problems is direct pickup (DPU) interference.[2] DPU in equipment connected to cable is caused by inadequate shielding of tuner and related circuits and loose connections that allow signal ingress on channel frequencies. A common DPU occurrence is the location of a receiver in the strong field of a transmitter whose signal is also carried on cable at or near its broadcast frequency. Since the propagation time in cable is more than the propagation time in free space, the over-the-air signal gets to the receive site first. A short while later, the cable signal arrives. The two mix in the receiver, and a deteriorated picture results. The consequences can range from a usable but degraded picture to a loss of the channel. The degradation can take the form of a ghost of the channel and/or beats between offset visual carriers. If the interference is relatively weak, only the sync pulse will be visible as a vertical bar in the picture. If the broadcast signal is not carried on the same frequency in the cable, the cable video will be different from the ingress signal. The result can be a moving background pattern. When the pattern is weak, the sync bar of the interferer wipes through the picture like a windshield wiper. When the interfering signal is strong, both pictures may be visible moving past each other. Depending on the phase of the ghost's visual carrier, the image may appear as a distorted color negative.

DPU is not limited to the simultaneous reception of a broadcast television signal over the air and through cable. In other parts of the spectrum, signal such as from amateur radio or pagers can make its way into the receiver and cause problems. In these cases, the interference manifests itself as alternating dark and light bands that usually move through the picture.

It is important to appreciate that a receiver properly designed for broadcast reception may have no need for extensive shielding. It may perform well in the broadcast environment. When connected to cable, the result might be unacceptable. The consumer may then draw the wrong conclusion. He or she may deduce that since antenna operation yields good performance and the connection to cable has difficulties, there must be something wrong with the cable system. A technically sophisticated observer is not likely to fall prey to such a failure in logic. An intelligent, but nonexpert, consumer is likely to make such a mistake and complain to the cable operator.

20.3.1 Estimate of Exposure

CableLabs and the EIA funded research by consultants to study the problem. Stern Telecommunications Corporation of New York City did a study of the poten-

Table 20.1 Exposure to various levels of fields.

Frequency Band	80 dBu 10 mV/m	90 dBu 31.6 mV/m	100 dBu 100 mV/m	120 dBu 1 V/m
VHF	61.5%	41.0%	19.9%	0.9%
VHF and UHF	65.4%	54.8%	40.8%	6.0%

tial impact of DPU. The study included a literature search, measurement of representative recent "cable-ready" receivers, and a computer model of transmitter power levels at various distances from the transmitter.

The receiver testing was done at the Carl T. Jones Corporation in Springfield, Virginia, in July 1992. Five receivers were tested. The testing revealed that there is a range of susceptibility to DPU. This susceptibility is a complex phenomenon that depends on the orientation of the receiver relative to the signal's field, the way in which the coaxial cable and the power cord are oriented with respect to the signal's field, and the channel involved.

The computer modeling considered the 10 top broadcast-relevant geographic areas, called areas of dominant influence (ADIs). The ADI terminology is important in considering advertising reach and effectiveness. For the purposes of this study, the ADIs have well-documented data on the distribution and density of households. This geographic data is important to the estimates of exposure to DPU. The 10 ADIs include about 30% of U.S. households. The field strength contours were computed with a commercially available program by Communications Data Service, Inc., which is based on the FCC's propagation curves. The study considered only UHF channels up to channel 27 (visual carrier at 549.25 MHz) on the assumption that most cable systems are limited to 550 MHz. As such, the study is a little out of date, but the general trend is clear. Field strengths of 80 dBu (10 mV/m), 90 dBu (31.6 mV/m), 100 dBu (100 mV/m), and 120 dBu (1 V/m) were considered.

Data was presented for each of the 10 top ADIs and extrapolated to all U.S. households. Table 20.1 lists the percentage of all U.S. households that may be subject to at least the field strengths listed from stations operating up to 550 MHz.

This data shows that the previously used Canadian specification of 100 mV/m will leave a significant number of receivers vulnerable to DPU. Even at an immunity level set at 1 V/m, a modest number of subscribers will still either have to accept DPU or take a set top terminal with better shielding.

20.3.2 Estimate of Susceptibility

In December 1992, CableLabs solicited proposals to develop test methods for measuring DPU susceptibility and to measure a representative grouping of current products used in the home. The Carl T. Jones Corporation was hired for both of these tasks. Thirty-five television receivers, 8 VCRs, and 13 cable set top

Table 20.2 Performance of units tested.

Susceptibility	TV	VCR	Set Top Terminal
Best	193.4 mV/m	162.73 mV/m	732.33 mV/m
Median	51.57 mV/m	57.27 mV/m	292.73 mV/m
Worst	1.17 mV/m	29.9 mV/m	77.23 mV/m
Best/Worst	165	5.4	9.5
Sample Size	35	8	14

terminals were tested for DPU susceptibility. The products were subjected to fields of increasing strength until a DPU 55 dB below the desired channel's sync tip was achieved. The 55-dB figure was confirmed to be reasonable by perception of interference studies. The discussion in Chapter 2 on the W curve and DPU indicated the 1993 tests generally confirmed the results of the 1986 CBS experiments. Across the three channels, four frequencies, two carrier-to-noise ratios, and two desired signal levels, the average value of the sine wave strength attenuation relative to the carrier strength to achieve "just perceptible" impairment levels was 56 dB.

Table 20.2 summarizes the results for the sample set of products. The mV/m figures required to breach the −55-dB threshold for slightly worse than "just perceptible" interference are listed for the best, worst, and median products. As can be seen, for the 1993 model products tested, a wide range of results were obtained.

An important difference between set top terminals supplied by cable operators and consumer electronics products is that the cable technician has several units with him or her and will keep trying them until the DPU problem is solved. Since the consumer purchases a TV or VCR and takes it home, the ability to pick the best from a collection is not available. Also, when the consumer makes a purchase decision, that selection will usually be in a different neighborhood from his or her residence. The retailer may not be using a cable feed for display purposes. Often, a laser disc is used as the source in order to provide the best possible pictures to encourage a purchase.

20.4 Receiver Performance

The Joint Engineering Committee (JEC) of the National Cable Television Association (NCTA) Engineering Committee and Consumer Electronics Manufacturers Association (CEMA) of the Electronic Industries Association (EIA) have been working on an RF Interface Specification for Television Receiving Devices and Cable Television Systems. This specification has the designation EIA-IS-23.[3] Eight

parameters of customer premises equipment are important to this specification: (1) reradiation of cable signals, (2) local oscillator leakage and backfeed, (3) A/B switch isolation, (4) DPU backfeed, (5) VCR through-loss, (6) adjacent channel rejection, (7) image rejection, and (8) tuner overload distortions.

Carl T. Jones Corporation developed tests for these eight parameters. The test procedures were reviewed by the JEC in October 1993. Although no formal acceptance of the test procedures was solicited, comments were noted, and the tests adjusted where appropriate. There were two goals for this work: to develop test procedures for these parameters, and to determine the current state of the art.[4]

20.4.1 Reradiation of Cable Signals

Cable systems operate in an enclosed spectral environment. If this environment is sealed, the same frequencies that are used in the over-the-air spectrum for other purposes can be reused for cable channels. In fact, multiple enclosed cable systems can be simultaneously operated to further increase the number of signals delivered to subscribers. As long as the cable system remains sealed, no harm is done to other users of the over-the-air spectrum.

A potential source of signal leakage exists in any equipment connected to the cable system. This is because the entire cable spectrum is conveyed to the input terminals of the connected device. If leakage of these signals occurs because of inadequate shielding of the circuits, the potential for interference with other users of the spectrum exists. This is especially serious in the case of frequencies used for aircraft navigation and communications and for emergency messages. Cable systems occupy these frequencies in their closed systems and cause no harm as long as the integrity of the system remains intact.

Equipment connected to the cable system must also be designed to prevent signal leakage. If this equipment leaks signals, interference may result. Set top terminals purchased by cable operators are specifically designed to prevent such leakage. The output of these set top terminals is just one television channel that is on a frequency not used locally for over-the-air service, usually channel 3 or 4. If leakage of this channel occurs from the TV or VCR connected to it, no harm is done since this is an unused local frequency. Consumer electronics products designed for broadcast applications do not have this requirement. When used for broadcast reception, they do not tune frequencies that are used for other purposes; they just tune the broadcast channels. Because of these differences, it is important to impose additional requirements on products connected to cable and prevent the connection of equipment that is not specifically designed to avoid leakage.

The FCC rules, Part 76.605 (a)12, limit the leakage or egress of RF energy. Cable systems must not emit energy that causes more than 15 μV per meter at a distance of 30 meters for frequencies less than 54 MHz or greater than 216 MHz. Between these frequencies, the limit is 20 μV per meter at 3 meters distance.

Of the units tested, only two television receivers failed.

20.4.2 Local Oscillator Leakage and Backfeed

Television receivers generate internal signals that are used to process the received spectrum. In addition, other internal signals are generated to display information on the screen and to operate functions and features of the product. Switch mode power supplies are also a source of interference. If these signals backfeed into the cable system, they can cause interference and degradation of performance in other TVs and VCRs connected to the cable system. They can add to the background noise, which makes two-way cable operation difficult. The other TVs and VCRs on the same signal splitter in the same residence are most at risk since there is a minimum of attenuation of the interfering signals. However, neighbors' receivers on the same cable system tap are also subject to this interference. Consequently, signal backfeed is carefully specified in equipment used by the service provider. Consumer equipment requires the same degree of care if problems are to be avoided.

The local oscillator in the single conversion tuners commonly used in TVs and VCRs operates at 45.75 MHz above the desired channel's visual carrier frequency. This frequency is in the frequency range of the cable system for most of the channels tuned. If this signal leaks back into the cable system, interference will result to other receivers according to the W curve discussed in Chapter 2. Fortunately, the splitters used by cable operators are required to have a minimum of 18-dB isolation between ports and usually have 20 dB or more. Thus if the local oscillator or other signal backfeed is no worse than 35 dB below the visual carrier, the 20 dB of the splitter will increase the signal attenuation to 55 dB below the visual carrier when it reaches the other receiver connected to the same splitter. At this level, it will be "just perceptible," but not annoying. Any further losses in the connecting cable will also help. Cable set top terminals avoid this problem by using double conversion tuners with the first local oscillator frequency higher than the cable spectrum tuned.

20.4.3 A/B Switch Isolation

A/B switches are part of the input of many receivers and VCRs, and allow remote-controlled selection of input sources. These sources include external antennas; cable signals; and the outputs of set top terminals, video games, and other playback devices. If insufficient isolation exists between ports of the switch, cross feed of signals will occur. The FCC regulations dated 1987 in Part 15.606 require a minimum of 80-dB isolation for frequencies between 54 and 216 MHz; 60 dB is required above that to 550 MHz. Since 1987, cable systems have gone beyond 550 MHz. Switches that have inadequate isolation at these higher frequencies will be a source of difficulty.

When an antenna is used, two kinds of problems may result. Antenna input signals can exceed +25 dBmV, whereas cable signals can be as low as 0 dBmV— lower if the subscriber does his or her own wiring and has split the signal excessively without appropriate amplification. In those situations, the isolation must

be adequate to prevent interference. This explains why the cable industry uses switches that generally exceed 90 dB and often 100 dB at all used frequencies. The other problem arises when strong cable signals are not sufficiently isolated from a rooftop antenna. If cross feed of the cable signal occurred in many homes, the cumulative effect of efficient radiators all pointing in the same direction, that is, toward the broadcaster's transmitting tower, could yield unacceptable interference to aircraft traversing that path.

Although most of the devices tested had no problems with this specification, there were three instances that failed.

20.4.4 DPU Backfeed

The DPU backfeed issue is related to the DPU phenomenon but is not the same problem. A receiver that has a DPU problem may also backfeed such signals into the cable to cause difficulties with other devices. Again, there is a benefit from the 20-dB isolation between signal splitter ports. DPU backfeed of less than −35 dB will be at the perceptible but not annoying point.

Unfortunately, 37% of TVs had DPU backfeed above the perceptibility point on channel 6, 34% on channel 12, 26% on channel 78, and 17% on channel 59. Understandably, units with DPU problems had DPU backfeed problems as well.

20.4.5 VCR Through-Loss

Designing VCRs with adequate playback signal quality is a difficult technical challenge. A technique used in early VCRs to mitigate the problem was to unsymmetrically split the signal at the input terminals of the device, giving the larger share to the recording circuits and shortchanging the "through" side of the splitter. The through side of the splitter provides an attenuated version of the entire input spectrum to the RF output terminal of the VCR when neither playback nor the VCR tuner is used. The attenuation is due to the use of a splitter without first amplifying the signal. This trick helped the noise performance of the recording circuits while providing a less disadvantageous comparison for the viewer when watching unrecorded materials.

As the last chapter discussed, the most common configuration of set top terminal, VCR, and TV connects the cable system to the set top terminal, which is connected to the VCR, which is connected to the TV. The TV suffers the additional through-loss of the VCR. This through-loss should be no greater than 5 dB. All eight tested VCRs passed this hurdle.

20.4.6 Adjacent Channel Rejection

The broadcast environment avoids adjacent channels. Cable utilizes as much of the spectrum as possible with nearly all channels having adjacent neighbors. An important issue for cable tuners is the ability to separate the desired channel from the upper adjacent visual carrier and the lower adjacent aural and chroma

carriers. Cable practice is to operate the aural carriers some 17 dB below the visual carriers. This additional attenuation relaxes the difficulty somewhat. The criteria again is the 55 dB used to bring the signal to the not-annoying level.

The passband of the receiver is a combination of the IF response and the tuner response, which differs slightly from channel to channel and is under the influence of the automatic gain control (AGC) strategy. Consequently, devices may have isolated points of failure on some channels but not on others.

Of the devices tested, three television receivers and one VCR failed to sufficiently reject the lower adjacent color subcarrier. Two television receivers failed the aural rejection test on all channels, and two others had problems on just two channels. Of the eight VCRs, three had no adjacent aural problems, three others failed on the four channels tested, one failed on three channels, and one unit had difficulties with just one channel. All TVs and VCRs behaved adequately with the upper channel's visual carrier. Not surprisingly, cable set top terminals passed all tests since they are specifically designed for such service.

20.4.7 Image Rejection

Image channels are a result of the superheterodyne receiver design, which uses a balanced mixer with two inputs followed by a filter for the purpose of separating out the desired signal from the spectrum at its input terminals. The inputs to the balanced mixer are the received spectrum and a local oscillator. The output of the mixer consists of the sum and differences of the input frequencies.

TV local oscillators for VHF are 45.75 MHz above the desired visual carrier. The difference between the local oscillator frequency and the desired frequency is the intermediate frequency (IF), which passes through the filter characteristic of the IF amplifier. The signal is then detected and becomes the source for the picture and sound selected by the viewer. The sum of the local oscillator and the desired channel is blocked by the filtering of the IF amplifier.

There is another difference frequency equal to 45.75 MHz. Signals that are 45.75 MHz above the local oscillator frequency will also have a difference of 45.75 MHz relative to the local oscillator frequency. This signal, if present at the input of the mixer, will be heterodyned to a frequency that will pass through the filter characteristic of the IF amplifier and provide interference to the picture and/or sound of the desired program. The visual carrier of the desired channel is 45.75 MHz below the local oscillator frequency. The image signal will be 45.75 MHz above the local oscillator frequency or 91.5 MHz above the visual carrier of the desired channel, about 15 6-MHz channels away. The visual carrier is the strongest part of the interfering channel. Visual carriers are spaced 6 MHz apart in cable practice. The visual carrier 15 channels above the desired carrier will be 90 MHz higher in frequency. When local oscillator frequency is subtracted from it, the visual carrier in the image region will be 1.5 MHz above the desired visual carrier and produce an interfering beat in the picture. The aural carrier of the channel that is 14 channels above the desired channel will also fall into the video bandwidth of the desired channel.

Since the tuner determines the rejection at the image frequency, the level of rejection varies with the channel. All receivers did well on channel 3, and only four had problems with channel 12. Half the receivers had interference on channel 53, and 67% failed on channel 74.

20.4.8 Tuner Overload Performance

Tuner problems and other deficiencies may result in impaired performance on the TV or VCR itself. In addition to the DPU problem previously described, nonlinearity in the receiver's circuits can give rise to the production of new frequencies not present in the original signal. The signals thus generated produce either diagonal bars in the picture or even a discernable image from another channel.

Tuner overload is a serious problem that will become more serious as cable bandwidth expands. Of the TVs, 66% had problems on at least one channel, 50% of VCRs suffered as well. Surprisingly, even 36% of the set top terminals had troubles.

20.5 The Federal Communications Commission's Role

The Federal Communications Commission (FCC) was created by Congress through the 1934 Communications Act to regulate and promote communications used in interstate commerce. Over the years, Congress has given the FCC assignments by passing laws. The laws usually indicate the desired result and leave the details to the FCC to figure out. In some cases, Congress asks the FCC to create rules that attempt to accomplish something not physically possible. In other cases, the FCC is simply given an extremely difficult task. In these situations, the burden on the FCC is heavy indeed.

The FCC attempts to set most rules by a process of consensus among all affected parties, including the public. If there are objections from a major participant, these objections are studied carefully.

Usually, the FCC first issues a notice of inquiry (NOI). After studying the information harvested from the NOI, the FCC will issue a notice of proposed rule making (NPRM). The NPRM has a deadline for comments and recommended changes. After that deadline, the commenters are allowed to comment on each others' submissions. After the FCC has an opportunity to analyze the complete set of comments, it issues a Report & Order that makes specific changes and additions to the FCC's rules. In all its publications, the FCC is very careful to discuss its reasons for taking its actions. Only rarely are there surprises. Most of the time, it is possible to see where things are going by carefully reading the documents. Since there are so many documents, this requires a lot of time and a serious intent to be involved in the process. Though any citizen can respond to the FCC's requests for information or comments, the vast majority of response filings come from companies affected by the FCC's regulations and by lobbying organizations that represent those companies. Those who are not in agreement with the rules or

who feel that the situation giving rise to the rules has changed can file a petition for reconsideration. In individual cases where the rule causes a hardship, a request for a waiver can be made.

When equipment is connected to a cable system, the potential exists for several modes of interference. The seriousness of the interference depends on the intrusion caused and on the services disrupted. The FCC has rules that limit some of the potential interference, and good practice requires other interference to be controlled as well. This control is implemented in technical specifications placed on equipment purchased by service providers for use in their systems. When consumers purchase equipment for direct connection to the system, control is jeopardized and the hazard of interference exists. Section 17 of the 1992 Cable Act requires the FCC to create rules:[5,6] "to specify the technical requirement with which a television receiver or video cassette recorder must comply in order to be sold as 'cable compatible' or 'cable ready.'" Unfortunately, there is no provision to preclude the connection of devices that are not "cable ready" to the system and thereby cause interference.

20.6 Summary

There is a substantial difference in the technical requirements for good performance in the broadcast environment and in the cable environment. Receivers that perform well in one environment will likely have difficulties in the other. Designing a cost-effective product to do well with both types of signals is very difficult. We have described the difficulties and the consequence of failure. Testing reveals that much of the equipment available at the time of the tests did not make the grade in several aspects.

In summary, the relevant parameters for equipment connected to cable are at a minimum:

- Immunity to external fields of at least 1 V/m.
- No more reradiation of cable signals than is allowed by FCC leakage rules.
- At least 80-dB isolation between switch ports.
- DPU backfeed into the cable system at least 35 dB below the desired visual carrier.
- Through-switch attenuation of no more than 5 dB.
- Rejection of adjacent channel carriers to levels at least 55 dB below the desired visual carrier.
- Image signal rejection of at least 55 dB below the desired visual carrier.
- Tuner overload performance so that spurious signals are at least 55 dB below the desired visual carrier.

The challenge of making products that function well in both environments is a heavy one. There is no formal set of standards for awarding the title "cable

ready." There is no compliance testing service. The consumer electronics industry unilaterally named its products "cable ready" without the concurrence of the cable industry. Much of the consumer's difficulty stems from not appreciating that "cable ready" is merely a consumer electronics marketing term without cable industry approval.

Endnotes

1. *The Consumer Interface Problem, A Status Report,* CableLabs, Louisville, CO, February 1990.

2. *Customer Premises Equipment Performance and Compatibility Testing,* CableLabs, Louisville, CO, 1993.

3. *EIA/IS-23 RF Interface Specification for Television Receiving Devices and Cable Television Systems,* 1997. Electronic Industries Association, Engineering Department, Arlington, VA. Available from Global Engineering Documents, Englewood, CO, 800-854-7179.

4. David Large, Defining "Cable Ready" Consumer Equipment. *CED,* April 1993, pp. 84–90.

5. *Cable Television Consumer Protection and Competition Act of 1992,* House of Representatives Report 102-862, 102nd Congress, 2nd Session.

6. Walter Ciciora, The 1992 Cable Act and the Consumer Electronics Interface, *1994 NCTA Convention Technical Papers.* NCTA, Washington, D.C.

Appendix A
Channel Allocation

The channel allocation shown here conforms to the requirements of EIA/ANSI-542, which specifies the correspondence between channel number and frequency for North America. This standard was developed by the joint EIA/NCTA engineering committee and has since been adopted by the FCC as the standard for channel allocations in the United States. (As of this writing, the FCC Rules refer to the standard by its old working designation, IS-132.) Picture carrier frequencies are shown. The channel extends from 1.25 MHz below the picture carrier to 4.75 MHz above. Three channelization plans are shown. They are explained in Section 8.2.

The complete standard is available from the EIA's duplication contractor, Global Engineering, at 800-854-7179. This is reprinted with permission of the EIA.

Channel	STD P.C. (MHz)	HRC P.C. (MHz)	IRC P.C. (MHz)	Channel	STD P.C. (MHz)	HRC P.C. (MHz)	IRC P.C. (MHz)
1		72.0036	73.2625	21	163.2500	162.0081	163.2625
2	55.2500	54.0027	55.2625	22	169.2500	168.0084	169.2625
3	61.2500	60.0030	61.2625	23	217.2500	216.0108	217.2625
4	67.2500	66.0033	67.2625	24	223.2500	222.0111	223.2625
5	77.2500	78.0039	79.2625	25	229.2625	228.0114	229.2625
6	83.2500	84.0042	85.2625	26	235.2625	234.0117	235.2625
7	175.2500	174.0087	175.2625	27	241.2625	240.0120	241.2625
8	181.2500	180.0090	181.2625	28	247.2625	246.0123	247.2625
9	187.2500	186.0093	187.2625	29	253.2625	252.0126	253.2625
10	193.2500	192.0096	193.2625	30	259.2625	258.0129	259.2625
11	199.2500	198.0099	199.2625	31	265.2625	264.0132	265.2625
12	205.2500	204.0102	205.2625	32	271.2625	270.0135	271.2625
13	211.2500	210.0105	211.2625	33	277.2625	276.0138	277.2625
14	121.2625	120.0060	121.2625	34	283.2625	282.0141	283.2625
15	127.2625	126.0063	127.2625	35	289.2625	288.0144	289.2625
16	133.2625	132.0066	133.2625	36	295.2625	294.0147	295.2625
17	139.2500	138.0069	139.2625	37	301.2625	300.0150	301.2625
18	145.2500	144.0072	145.2625	38	307.2625	306.0153	307.2625
19	151.2500	150.0075	151.2625	39	313.2625	312.0156	313.2625
20	157.2500	156.0078	157.2625	40	319.2625	318.0159	319.2625

Channel	STD P.C. (MHz)	HRC P.C. (MHz)	IRC P.C. (MHz)	Channel	STD P.C. (MHz)	HRC P.C. (MHz)	IRC P.C. (MHz)
41	325.2625	324.0162	325.2625	81	565.2500	564.0282	565.2625
42	331.2750	330.0165	331.2750*	82	571.2500	570.0285	571.2625
43	337.2625	336.0168	337.2625	83	577.2500	576.0288	577.2625
44	343.2625	342.0171	343.2625	84	583.2500	582.0291	583.2625
45	349.2625	348.0174	349.2625	85	589.2500	588.0294	589.2625
46	355.2625	354.0177	355.2625	86	595.2500	594.0297	595.2625
47	361.2625	360.0180	361.2625	87	601.2500	600.0300	601.2625
48	367.2625	366.0183	367.2625	88	607.2500	606.0303	607.2625
49	373.2625	372.0186	373.2625	89	613.2500	612.0306	613.2625
50	379.2625	378.0189	379.2625	90	619.2500	618.0309	619.2625
51	385.2625	384.0192	385.2625	91	625.2500	624.0312	625.2625
52	391.2625	390.0195	391.2625	92	631.2500	630.0315	631.2625
53	397.2625	396.0198	397.2625	93	637.2500	636.0318	637.2625
54	403.2500	402.0201	403.2625	94	643.2500	642.0321	643.2625
55	409.2500	408.0204	409.2625	95	91.2500	90.0045	91.2625
56	415.2500	414.0207	415.2625	96	97.2500	96.0048	97.2625
57	421.2500	420.0210	421.2625	97	103.2500	102.0051	103.2625
58	427.2500	426.0213	427.2625	98	109.2750	108.0250	109.2750*
59	433.2500	432.0216	433.2625	99	115.2750	114.0250	115.2750*
60	439.2500	438.0219	439.2625	100	649.2500	648.0324	649.2625
61	445.2500	444.0222	445.2625	101	655.2500	654.0327	655.2625
62	451.2500	450.0225	451.2625	102	661.2500	660.0330	661.2625
63	457.2500	456.0228	457.2625	103	667.2500	666.0333	667.2625
64	463.2500	462.0231	463.2625	104	673.2500	672.0336	673.2625
65	469.2500	468.0234	469.2625	105	679.2500	678.0339	679.2625
66	475.2500	474.0237	475.2625	106	685.2500	684.0342	685.2625
67	481.2500	480.0240	481.2625	107	691.2500	690.0345	691.2625
68	487.2500	486.0243	487.2625	108	697.2500	696.0348	697.2625
69	493.2500	492.0246	493.2625	109	703.2500	702.0351	703.2625
70	499.2500	498.0249	499.2625	110	709.2500	708.0354	709.2625
71	505.2500	504.0252	505.2625	111	715.2500	714.0357	715.2625
72	511.2500	510.0255	511.2625	112	721.2500	720.0360	721.2625
73	517.2500	516.0258	517.2625	113	727.2500	726.0363	727.2625
74	523.2500	522.0261	523.2625	114	733.2500	732.0366	733.2625
75	529.2500	528.0264	529.2625	115	739.2500	738.0369	739.2625
76	535.2500	534.0267	535.2625	116	745.2500	744.0372	745.2625
77	541.2500	540.0270	541.2625	117	751.2500	750.0375	751.2625
78	547.2500	546.0273	547.2625	118	757.2500	756.0378	757.2625
79	553.2500	552.0276	553.2625	119	763.2500	762.0381	763.2625
80	559.2500	558.0279	559.2625	120	769.2500	768.0384	769.2625

*These channels must be omitted from the comb in order to comply with the required FCC frequency tolerance.

Channel	STD P.C. (MHz)	HRC P.C. (MHz)	IRC P.C. (MHz)	Channel	STD P.C. (MHz)	HRC P.C. (MHz)	IRC P.C. (MHz)
121	775.2500	774.0387	775.2625	141	895.2500	894.0447	895.2625
122	781.2500	780.0390	781.2625	142	901.2500	900.0450	901.2625
123	787.2500	786.0393	787.2625	143	907.2500	906.0453	907.2625
124	793.2500	792.0396	793.2625	144	913.2500	912.0456	913.2625
125	799.2500	798.0399	799.2625	145	919.2500	918.0459	919.2625
126	805.2500	804.0402	805.2625	146	925.2500	924.0462	925.2625
127	811.2500	810.0405	811.2625	147	931.2500	930.0465	931.2625
128	817.2500	816.0408	817.2625	148	937.2500	936.0468	937.2625
129	823.2500	822.0411	823.2625	149	943.2500	942.0471	943.2625
130	829.2500	828.0414	829.2625	150	949.2500	948.0474	949.2625
131	835.2500	834.0417	835.2625	151	955.2500	954.0477	955.2625
132	841.2500	840.0420	841.2625	152	961.2500	960.0480	961.2625
133	847.2500	846.0423	847.2625	153	967.2500	966.0483	967.2625
134	853.2500	852.0426	853.2625	154	973.2500	972.0486	973.2625
135	859.2500	858.0429	859.2625	155	979.2500	978.0489	979.2625
136	865.2500	864.0432	865.2625	156	985.2500	984.0492	985.2625
137	871.2500	870.0435	871.2625	157	991.2500	990.0495	991.2625
138	877.2500	876.0438	877.2625	158	997.2500	996.0498	997.2625
139	883.2500	882.0441	883.2625				
140	889.2500	888.0444	889.2625				

Appendix B
Video Waveforms

This appendix illustrates the NTSC video waveform and is taken from EIA/TIA 250C, an earlier version of which formed the basis for the video waveform as specified in the FCC rules for broadcast television. See Chapter 2 for a more detailed description of the NTSC video waveform. The figures as shown in this appendix are drawn to conform as closely as possible to the defined amplitude and times of the NTSC signal. Other references frequently distort the time relationships.

The basic frequency in an NTSC waveform is the color subcarrier. All other video frequencies are defined in terms of the color subcarrier and for broadcast purposes are required to be locked to the subcarrier. The color subcarrier frequency is defined in the FCC rules as being

$$f_{COLOR} = \left(\frac{63}{88}\right)5 \text{ MHz} = 3.5795454545 \ldots \text{ MHz}$$

The digits . . . 45 . . . repeat continuously. This is often called the 3.58 signal.

The horizontal line rate is

$$f_H = \frac{2}{455} f_{COLOR} = 15.734 \text{ kHz}$$

This frequency is sometimes referred to as 15.75 kHz, which was the old horizontal frequency prior to color transmission. At that time the horizontal frequency was exactly 15.75 kHz but was changed with the addition of color to reduce the effect of a beat between the color subcarrier and the sound carrier.

The time of one horizontal line is given by the reciprocal,

$$H = \frac{1}{15.734} = 63.5556 \text{ }\mu s$$

Since there are 525 horizontal lines in one frame and two fields in a frame, the vertical frequency is given by

$$f_V = \frac{2}{525} f_H = 59.94 \text{ Hz}$$

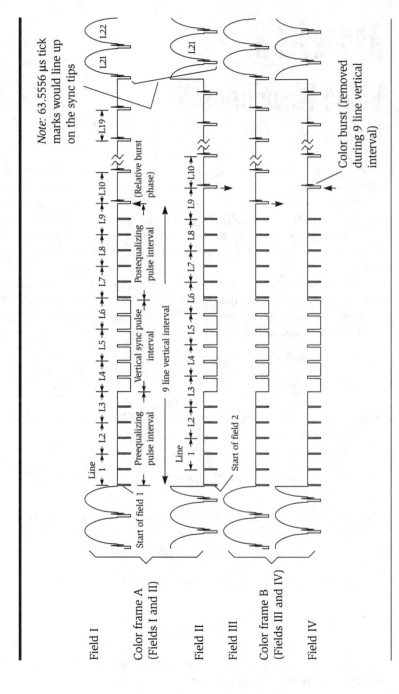

Figure B.1 Horizontal blanking interval.

The resultant period of one field is then

$$V = \frac{1}{f_V} = 16.6833 \text{ ms}$$

From the above relationships, you can show that there are 227.5 cycles of the color subcarrier in one horizontal line, so when you trigger an oscilloscope to the horizontal sync, you will see an apparent phase reversal in the color burst from one line to the next. The color subcarrier is not really changing phase, but the extra half cycle per line results in the apparent phase reversal.

Figure B.1 shows the vertical interval of four consecutive fields. Together the four fields make up two frames, and it takes the two frames to return to the initial relation between sync and color subcarrier. Shown to the left of each line are the last active video lines at the bottom of the picture for each of four consecutive fields. The vertical interval begins with sync pulses spaced every half horizontal line rather than every one line. Each pulse is one half the normal width of 4.7 µs. These half-width, twice-occurring pulses are called *equalizing pulses.* All sync pulses are defined as starting at the negative-going edge of the pulse, which is the zero time reference for the blanking interval. If you start counting horizontal time intervals of 63.5556 µs from the negative-going edge of line 1, then all interval edges will fall on negative-going edges of horizontal sync pulses.

The start of odd fields is identified by the beginning of the equalizing pulses following a full width horizontal line. Even fields start with the first equalizing pulse following a half line. This is the way interlace is achieved. Notice that line 1 in even fields doesn't start immediately after the half active video line that ends the odd field.

Vertical sync is defined by three lines of video (lines 4–6), in which the voltage waveform is at the lowest level for all except 4.7 µs in the center of the line and at the end of the line. Even in these three lines, the line begins at the negative-going transition of the waveform. Following common practice, the four vertical intervals are shown with the vertical sync pulses lined up. Vertical sync is followed by three more lines of equalizing pulses. During the nine line vertical interval, the color burst is not present. It resumes on line 10, occurring immediately after the sync pulse. Lines 10–20 are normal horizontal lines with sync and burst but no video. This is the *retrace interval,* which allows time for the electron beam to move from the bottom of the screen to the top. Video is never transmitted during these lines, but digital data and test signals may optionally be transmitted. Lines 17 and 18 are frequently used for vertical interval test signals (VITS), which may consist of various signals used to test the quality of the transmission path. Line 19 was at one time reserved for a vertical interval reference signal (VIRS), used to overcome some problems with processing signals in a network distribution system. Today, it may be used for a ghost-canceling reference signal.

Active video may start again on line 21, but most commonly today, line 21 is used for slow speed data transmission using a standard popularly known as the closed captioning standard. This is used for transmission of captioning data for

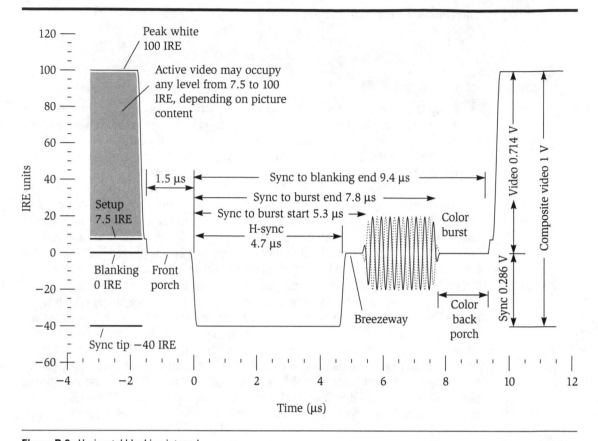

Figure B.2 Horizontal blanking interval.

viewers who are hard of hearing and also for transmission of a number of different types of information data for viewers.

Due to the way the color subcarrier is related to sync, the phase of the subcarrier relative to the start of sync changes between odd and even fields and from one odd (or even) field to the next odd (or even) field. This is indicated by the up and down arrows under the start of the sync pulse on line 10 in each field. This relationship is important when switching between two video sources. Fields I and III, the odd fields, are identical except that the phase of the color subcarrier is opposite in the two. Similarly, the even fields are identical except for the color phase.

Figure B.2 illustrates a typical horizontal blanking interval. All times are measured with respect to the 50% (−20 IRE) point of the negative-going edge of the horizontal sync pulse. The video at the right side of the picture tube ends 1.5 μs before the start of sync. The period from the end of active video to start of sync is known as the *front porch*. It is during the front porch that most sync suppression scrambling systems switch to suppressing the sync pulse, creating an artifact as shown in Chapter 18.

The actual sync pulse is 4.7-µs-wide measured at the 50% (−20 IRE) points. The transitions from one level to another follow a sine squared shape with a duration of about 140 ns, limited by the 4.18-MHz baseband bandwidth of the NTSC signal. The sync is followed by a short *breezeway* between the end of sync and beginning of color burst. The color burst consists of 9 cycles of the 3.579-MHz color subcarrier. As explained above, due to the manner in which the sync relates to the color subcarrier, the subcarrier phase shifts with respect to the sync from line to line and field to field. It is shown here in both phases, because this is typically the way you will see the burst when observing it on a waveform monitor or oscilloscope. The color channel is limited in bandwidth, which is why the envelope of the color burst increases over about one cycle at the beginning of the burst and decreases over one cycle at the end. If the burst started and ended abruptly, the bandwidth of the signal would exceed the color channel bandwidth. The color burst is needed by the TV in order to demodulate the color subcarrier properly.

Following the color burst is another short interval before active video begins, called the *color back porch*. (In the black-and-white days, the color burst didn't exist, and the entire interval from sync tip end to active video begin was called the back porch. When the color burst was added, the name breezeway was used to define the interval between the sync and burst.) The color back porch is the region in which most sync suppression systems stop suppressing and expanding the sync, leaving another artifact. Normally the front and back porches are off-screen in the TV receiver, so the artifacts are not seen.

NTSC video levels are measured in IRE units, named for one of the predecessor organizations to today's IEEE, the Institute of Radio Engineers, which sponsored much of the early standardization of the NTSC system. The scale is referenced to the blanking level, occupied by the video waveform during the front and back porches. Sync extends from 0 IRE to −40 IRE. Active video occupies the positive portion of the IRE scale. It rides on a 7.5 IRE *setup* level, which in the early days was thought to improve the operation of low-cost tube sync separator circuits. Black corresponds to 7.5 IRE and white to 100 IRE, the maximum level of the video signal (sometimes the color subcarrier will extend somewhat above 100 IRE but not much above when the signal is transmitted). Video can begin at the same time setup begins, so the 7.5 IRE setup level may not be visible in the waveform.

The voltage amplitude of the composite (video and sync) signal is 1 volt peak to peak when the signal is interfaced from one piece of equipment to another. Sync is always negative-going. This means that the active video (including setup) occupies 0.714 volts and the sync tip 0.286 volts.

Traditionally video interfaces between two pieces of equipment are ac coupled. This means that the sync tip level will move up and down as the signal's average value changes with video content. 0 volts will be at the average value. Some modern systems clamp their output video so that 0 volts occur at the blanking level. The video is said to be dc coupled. Because each piece of equipment is capable of clamping the video as it requires, equipment with ac or dc coupling can usually be intermixed without regard for the coupling.

Glossary

Note that the terms contained herein are not intended to be rigorous definitions but rather reflect common usage in the lexicon of the cable television industry.

absorption (optical) Attenuation of optical signals due to conversion of optical energy into heat energy.

actives Components that handle signals and that require power. Examples are line amplifiers and nodes.

addressability The ability to remotely control parameters of the subscriber's reception of signals by transmitting signals addressed to that subscriber's equipment.

AGC Automatic gain control. Circuitry in an amplifier that senses the level of a signal at the output and adjusts the gain such that the signal level is constant regardless of input level.

AGL Above ground level. Used in microwave systems to designate the elevation of various antennas relative to the ground level at the foot of the tower.

alpha In data transmission, the excess width of a modulated RF signal compared with the minimum theoretical width needed to transmit data at a particular symbol rate.

AM/FM conversion AM to FM conversion. A type of distortion to a modulated signal, in which amplitude modulation is translated into frequency modulation. Also known as AM/PM conversion.

AML Amplitude modulated link. A microwave technology in which the cable television spectrum is up-converted to a microwave band for transmission, then down-converted at the receiver for transmission through a cable distribution system.

AM/PM conversion AM to PM conversion. A type of distortion to a modulated signal, in which amplitude changes in the signal are translated into phase changes. Also known as AM/FM conversion.

AMSL Above mean sea level. The official designation of elevation used in microwave systems.

analog television The system of television transmission that has been practiced since the medium was invented. Picture levels are represented by a continuously varying voltage level.

ANI Automatic number identification. A capability of most telephone systems that allows the number of the calling party to be known to the receiving party. Used in some pay-per-view ordering systems.

ANSI American National Standards Institute. An umbrella standards organization for the United States.

APC Angle polished connector. A fiber-optic connector that exhibits reduced reflections due to an angled fiber end.

artifact As used commonly in television technology, any artificial element introduced into a picture or sound signal as a result of the processing the signal undergoes.

ARU Audio response unit. A telephone answering system that guides the caller through audio menus and accepts inputs via touch-tone key presses. Used in some pay-per-view ordering systems.

ASC Automatic slope control, also known as dual pilot AGC. A system in which one low-frequency carrier and one high-frequency carrier are both monitored for AGC purposes. By monitoring two carrier levels in different parts of the spectrum, the system can compensate for changes in gain over the entire spectrum. ASC systems tend not to be used in newer designs because of cost and the reduced need due to the relatively short cascades used with HFC architectures.

ASG Automatic slope and gain. A control system used in many modern pieces of distribution equipment, where an AGC function is combined with compensation for varying cable loss with frequency. As the gain at a pilot frequency is changed, the gain at higher frequencies is changed more, and at lower frequencies is changed less, matching an assumed profile for the coaxial cable.

aspect ratio The ratio of width to height of a picture. Standard definition television uses a 4:3 aspect ratio, and most HDTV systems use a 16:9 aspect ratio.

ATM Asynchronous transfer mode. A set of standards for transmitting many types of baseband digital signals over a network. Very efficient for sharing channels among bitstreams of varying bit rates.

ATTC Advanced Television Test Center. An organization established to provide impartial testing for proponent digital video systems, which more recently has coordinated much of the testing required to make digital television and, especially, HDTV a reality.

attenuation A measure of the RF or optical signal loss (generally in decibels) of a component, circuit, or cable.

audio A baseband signal conveying at least one channel of sound information.

aural An adjective implying an RF or IF carrier that is modulated with an audio signal.

availability A measure of the fraction of time that a circuit or system is available to carry signals.

azimuth The compass heading of an antenna, measured in degrees clockwise from north.

backbone Point-to-point or ring connections between headends or linking headends and major hubs in a cable system. Often signals traveling in the backbone are not in the same format used to deliver services to homes (for example, digital baseband).

balanced transmission A technique of coupling signals between two pieces of equipment using two conductors, having equal and opposite signal voltage, neither of which is grounded. Used extensively for audio, and infrequently for video and RF.

bandstop filter A filter that passes all frequencies except those within some band.

bandwidth efficiency The ratio of bits per second transmitted through a communications channel to the channel bandwidth in Hz. The units are b/s/Hz (bits per second per hertz).

barker channel A channel used to promote programming on other channels.

baseband A signal not modulated onto an RF carrier. In video, refers to the visual signal as it is handled before being modulated onto an RF carrier.

basic service The foundation service all subscribers are required to take. Usually consists of at least local off-air channels and public channels, and often is a reasonably complete set of signals.

baud In digital transmission, the number of symbols transmitted per second. Except for a few simple modulation formats, the baud rate is less than the bit rate.

Bellcore Bell Communications Research. The research and development arm of the Bell Telephone companies.

BER Bit error rate. A measure of the proportion of the bits in a datastream, which are corrupted in transmission.

BERT Bit error rate test set. A test instrument used in data communications to insert a known data pattern at a transmitter and to count the number of errors received per second at a receiver.

binary Representation of a number by a series of elements (usually called "bits"), which can take on only one of two values. Contrast binary representation with our normal base 10 representation of numbers, where each element representing the number can take on any of 10 values.

bit A single binary data element having a value of either 1 or 0, usually represented by two different voltage levels.

bit rate The number of bits of data transmitted per second.

block conversion A technique for putting more signals on a return optical path by converting some incoming signals to alternative frequency bands.

BPSK Binary phase shift keying. A digital modulation method in which the phase of a carrier is changed by 180° to represent transmission of a 1 or a 0.

bridger An amplifier (or module within a larger amplifier station) that is fed a sample of the trunk input signal and that is used to create a distribution leg.

broadband As used in the cable industry, refers to the ability to frequency division multiplex many signals in a wide bandwidth of RF frequencies, using an HFC network.

BTSC Broadcast Television System Committee. An entity that developed the analog TV stereo system used for transmission in North America and elsewhere.

byte A collection of eight bits.

CableLabs Cable Television Laboratories. The research consortium of the cable television operating companies.

caching A technique of retrieving information from Web sites and storing it on a local server for the purpose of sending Web pages to users from that local server. The technique speeds up access to popular sites by eliminating the Internet delays involved in going back to the source for information. (The term also has an analogous meaning in retrieving data from a computer's memory.)

carrier Or RF carrier. A signal on which another, lower-frequency signal is modulated in order to transport the lower-frequency signal to another location.

carrier-to-noise (C/N) ratio In analog visual transmission, the ratio, usually expressed in decibels, between the RMS power present in the visual carrier during sync tips and the RMS thermal noise power in the video bandwidth. In cable television work, the term implies a measurement made at radio frequencies. Contrast with S/N, measured at baseband. See noise.

CARS Cable antenna relay service. The official FCC designation for a set of microwave frequencies reserved for the use of cable television companies.

cascade A configuration in which a signal passes through more than one stage or system. In cable usage, it usually refers to the number of series-connected line amplifiers through which the RF signals pass. It can also refer to multiple stages of amplification or filtering through which a signal passes.

Cassegrain feed An antenna for microwave (including satellite) reception consisting of a parabolic main reflector that focuses power on a hyperbolic subreflector, which in turn focuses the power to the feed.

CBR Constant bit rate. An adjective describing a service or communications channel that carries a constant number of bits per second.

CCS 1. Custom calling services. A particular collection of enhanced telephone services, distinct from CLASS services. 2. Centum call seconds. Usage of a telephone resource (such as a phone line) for 100 seconds per hour.

CDMA Code division multiple access. A data transmission method in which a number of individual transmitters occupy the same spectrum at the same time, each using a different spreading sequence.

CEMA Consumer Electronics Manufacturers Association. A division of the EIA that represents manufacturers of consumer electronics.

channel A range of frequencies assigned to a signal in an FDM transmission system. In North America, analog video and most other downstream channels are 6 MHz wide and correspond to one of the allocation forms defined in the ANSI/EIA-542 standard.

channel mapping A feature of some set top terminals, TVs, and VCRs that allows programmable correspondence between channel number and received frequency. Useful where an off-air station must be channel shifted because of direct pickup problems. The station loses its channel identity to cable subscribers, but if channel mapping is available, it is possible to make the station appear to be on its off-air channel even though it is carried on another.

characteristic impedance The impedance of a transmission line such that, if terminated in a pure resistance of the same value, the return loss would be infinite. That is, the cable would "look" to a driving circuit, as a resistor of the characteristic impedance.

chroma Or chrominance. Referring to the color information in a television signal, as opposed to the luminance (black and white) and aural (sound) portions.

chroma/luma delay A form of distortion of analog video signals in which the times of arrival of the luma (black and white) and corresponding chroma (color) signals are offset, resulting in color information being shifted with respect to the black-and-white outline. Also called chrominance delay.

CIN Composite intermodulation noise. Noiselike distortion products resulting from even- or odd-order mixing among RF signals, at least one of which is digitally modulated.

clamping See dc restoration.

CLASS Custom local area signaling services. A particular collection of enhanced telephone services.

clipping A phenomenon where the voltage or current peaks of a signal are sharply limited at some value (for example, in directly modulated DFB laser diode transmitters).

closed captioning A system used to transmit captioning information to hard-of-hearing viewers, allowing text to overlay the picture. The system has been expanded to permit transmission of other types of data, most notably the V-chip rating data.

CMI Cable microcell integrator. See RAD.

coax Short name for coaxial cable, which is used extensively for transmission of RF and baseband signals. The cable consists of an inner conductor on which signal voltage is impressed with respect to the shield. The center conductor is surrounded by a dielectric, then a shield. Frequently, an insulation layer will surround the shield.

coaxial distribution The portion of the broadband cable television distribution network that starts at the node and terminates at the tap. Sometimes the drop system is included in the definition.

coding gain A measure of the efficiency with which a datastream is prepared for transmission. The coding gain is the ratio of the input to output bit rate of the encoder.

color difference signals The two signals in a TV picture that add color information. They are derived from the original red, green, and blue primary signals, and are called color difference signals because of the way in which they are defined. They are modulated onto a color subcarrier.

color subcarrier A signal in an analog TV signal that carries color information. Also known as the chrominance signal.

combiner A three or more port device used to combine signals without regard for the frequency of the signals. It is identical to a splitter, except for the direction of use.

community access Programming produced locally over which the cable operator has no editorial control. It may be produced by community groups, schools, and the like.

composite video A baseband signal that contains multiple components of the signal required to reproduce a picture. The term has multiple meanings depending on context. It may mean a video signal including synchronizing (sync) information. It is also used to mean the signal including chroma (color) information.

compression A number of techniques for reducing the bandwidth needed to transmit a signal by removing unnecessary or redundant information from the signal.

concentration In telephony, the sharing of a resource among many users, based on the presumption that not all users will demand the resource at the same time.

conditional access A generic term used to indicate that access to a program or channel is limited to subscribers who have paid additional money for it.

constellation diagram In data transmission, a diagram that plots the amplitude and phase of each state in a symbol.

CSO Composite second order. A form of distortion generated by amplifiers and other devices handling multiple RF signals. It is second-order distortion that combines signals at frequencies A and B, as A ± B. Most A + B products fall 1.25 MHz above picture carriers, while the A − B products fall 1.25 MHz below picture carriers in the standard channel allocation. The ratio of carrier to distortion product is often called C/CSO, carrier to composite second order.

CSR Customer service representative. An employee of a cable system who handles incoming telephone calls from customers.

CTB Composite triple beat. A form of distortion generated by amplifiers and other devices handling multiple RF signals. It is a third-order distortion product that combines signals A, B, and C as A + B − C. Most CTB products fall on the nominal frequency of carriers in the standard channel allocation. Also known as C/CTB.

cutoff frequency In filter terminology, a frequency at which the response of a filter is reduced by some specified amount, usually 3 dB.

data slicer In the recovery of a transmitted data signal, a circuit that compares the input signal voltage with a predetermined level. If the input signal is above the threshold, it is assigned one of the two logic levels. If it is below the threshold, it is assigned the other logic level.

dB Decibel. A measure of the relative strength of two signals. Defined as 10 times the logarithm (to the base 10) of the ratio of the two power levels in any consistent units, such as watts, or 20 times the logarithm of the ratio of the voltage or current levels.

dBm Decibels with respect to one milliwatt. A unit of RF signal strength used in satellite work and other communications applications.

dBmV Decibels with respect to 1 millivolt in a 75-ohm system. The unit of RF power used in cable television work in North America. In a 75-ohm system, 0 dBmV = −48.75 dBm.

dc restoration In analog television transmission, the process of recovering a true black level in the picture by forcing the sync tip or back porch (see Chapter 2) to be at a desired voltage. Also called clamping.

decibel See dB.

deemphasis In FM transmission, the complementary process to preemphasis. A frequency selective reduction of the amplitude of higher-modulating frequencies after reception to restore flat response. Used to improve the signal-to-noise ratio.

demodulation The complementary process to modulation. Recovery of information modulated onto a carrier.

demodulator A device used by cable television systems to convert an RF modulated video and/or audio signal to baseband.

DFB laser Distributed feedback. A laser technology in which the wavelength of the laser is set by the distance between reflecting surfaces at each end of the laser cavity, augmented by a grating running the length of the cavity.

differential gain A measure of the variance in gain at the chrominance frequency, as a function of luminance level.

differential phase A measure of the variance in phase at the chrominance frequency, as a function of luminance level.

digital repeating ring An architectural structure in which a baseband digital signal is received and retransmitted at each terminal point around a closed ring.

digital television Television signals transmitted using digital techniques, in which video levels are converted to digital states that approximate the analog level of the original signal. As used in the consumer world, it also implies compressing the signal to minimize the transmission bandwidth required.

digital video Representation of video signals in digital format. It does not necessarily imply a compressed video signal.

diplex filter A filter used to separate RF signals into one of two paths, based on the frequency of the signals. Can also be used to combine signals that exist in different frequency bands.

dipole A basic antenna, consisting of a straight conductor, usually broken in the middle to connect to a feedline.

direct pickup Or DPU. The characteristic of a TV or VCR that a signal is picked up off the air in the device, creating interference with the signal being received via cable.

directional coupler A three-port device for unequally dividing signal between two paths. Can also be used to combine signals with more loss in one input than in the other, with the signal from the high-loss port being routed to only one of the other two ports.

directivity In a tap or directional coupler, the difference in sensitivity to upstream and downstream signals, as measured at the side port.

discrete cosine transform DCT. A fundamental process used in video compression, in which two-dimensional blocks of pixels are converted to a frequency domain representation, with the option of dropping higher-frequency components if they are small.

dispersion The variation of transmission time due to some controlling parameter.

distribution The broadband portion of a cable television network extending from the point where the broadband spectrum is created in the same format as it will be received by subscribers (for example, in a simple star HFC network, the distribution system starts at the input to the linear optical transmitter).

DLC Digital loop carrier. A telephony system in which the analog interfaces of a switch are extended closer to the using telephones through use of a datastream to a remotely located interface.

DOCSIS Data-over-cable service interface specification. The formal name of the cable modem standard produced by a consortium led by CableLabs.

double sideband modulation Amplitude modulation in which both the upper and lower sideband are retained. Also called DSB.

downstream Signal flow from a headend toward subscribers. Also known as the forward direction.

drop The portion of the distribution plant between the tap and the individual subscriber's home. The term sometimes also includes the in-home wiring.

DSM Demand side management. A method used by utility companies to limit peak usage of electrical power by controlling customer equipment.

DWDM Dense wavelength division multiplexing. The technique of using fibers to carry many signals at slightly different wavelengths. See WDM.

EAS Emergency Alert System. A federally mandated system by which emergency information interrupts normal programming on program delivery systems, including cable television. See emergency override.

egress A measure of the degree to which signals from the nominally closed coaxial cable system are transmitted through the air. Also known as signal leakage.

EIA Electronic Industries Association. A trade organization in Washington, D.C., that represents companies that manufacture electronic equipment in the United States.

EIRP Effective isotropic radiated power. The total power supplied to an antenna, multiplied by the gain of the antenna in a given direction.

electronic program guide EPG. A display on the TV of a selection of programs available to the subscriber. The EPG may include other features such as information on programs or interactivity, including automatic tuning to selected programs or the ability to program a VCR.

elevation 1. The number of feet that a location is above mean sea level (AMSL). 2. The number of degrees by which an antenna is tilted upwards from the horizontal plane.

emergency override A technique whereby, in an emergency, normal audio and/or video programming on all or most channels may be replaced with emergency announcements generated by an authorized entity. See EAS.

encryption The process of rendering a digital signal unintelligible to any receiver that doesn't have some unique piece of information needed to recover that signal.

envelope In amplitude modulation, the shape that contains the amplitude modulated carrier. That is, the envelope is the outline of the carrier amplitude versus time.

envelope detection The process of recovering modulation from an amplitude modulated signal by recovering the envelope shape of the modulated signal. The most common envelope detector is a diode.

erlang In telephony, the maximum physical capacity of a line or trunk. One telephone line has a capacity of 1 erlang, or 36 CCS.

error correction In data transmission, the technique of adding extra bits to a transmitted signal, with the extra bits being used to detect and correct errors in the transmission.

expanded basic service A subscription service offering by a cable company that consists of basic service plus more advertiser-supported channels (as opposed to pay services).

express feeder An untapped coaxial line between two amplifiers in the coaxial portion of an HFC distribution system.

eye diagram In data transmission, a plot of amplitude versus time of either an in-phase or quadrature signal.

Fabry–Perot laser A laser technology in which the wavelength of the laser is set by the distance between reflecting surfaces at each end of the laser cavity.

failure rate The per-unit average rate at which a group of similar devices will randomly fail during their normal service life. Usually expressed in fraction or percentage per year.

F connector The RF connector commonly used to attach drop cables to taps and other devices.

FCC Federal Communications Commission. The governmental body in the United States that oversees telecommunications entities. In other countries, the corresponding body may be called the Department of Communications or may be part of a Ministry of Posts, Telegraph, and Telecommunications.

FDM Frequency division multiplexing. A data transmission method in which a number of transmitters share a transmission medium, each occupying a different frequency.

feeder The portion of the coaxial subnetworks that starts at an amplifier fed by an express feeder or trunk and supplies signals directly to subscriber taps.

field In interlaced television transmission, such as NTSC, one-half of a TV picture, consisting of alternate lines. Two fields make up a complete frame.

filtering In communications, usually refers to the transmission of power in some frequency range(s), combined with the nontransmission of power in other frequency range(s). In computer use, may also refer to the elimination of certain types of information, with passage of other types.

FITL Fiber in the loop. Used by the telephone industry to describe the deployment of fiber on the subscriber side of class 5 telephone switches. See loop.

flat loss The loss in a cable plant that is nominally not a function of frequency. Typically, taps and attenuators have flat loss, as contrasted to coaxial cable, whose loss varies approximately as the square root of frequency.

FM/AM conversion FM to AM conversion. A type of distortion to a modulated signal, in which frequency modulation is translated into amplitude modulation.

forward Transmission of cable signals toward subscribers. See downstream.

forward error correction FEC. In data transmission, a process by which additional data is added that is derived from the payload by an assigned algorithm. It allows the receiver to determine if certain classes of errors have occurred in transmission and, in some cases, allows other classes of errors to be corrected.

Fourier series A mathematical expression relating the time domain representation of a periodic waveform with the frequency domain representation. It shows that any periodic waveform may be expressed as a series of harmonically related sine and cosine waves. (Sine and cosine waves have the same shape but are shifted in phase by 90°.)

frame One complete TV picture consisting of two fields.

franchise An agreement with a local governing body that permits a cable operator to construct a cable system and spells out the conditions that must be met by that system.

frequency response The gain as a function of frequency through a system or device relative to the defined ideal. In the case of a cable distribution system, this is also known as peak-to-valley response or P/V.

Fresnel zone In a microwave path, this describes a hypothetical ring centered on the path such that the distance from transmitter to ring to receiver is one-half wavelength longer than the distance from transmitter to receiver.

FSK Frequency shift keying. A low-cost, spectrum-inefficient, very robust form of digital modulation in which the frequency of a carrier is changed to represent data.

FTF or **FTTC** Fiber to the feeder. A distribution architecture in which fiber nodes feed coaxial feeder lines without any intervening coaxial trunk links.

FTTC Fiber to the curb. A distribution architecture in which directly fiber-fed termination units are shared by several dwellings.

FTTH Fiber to the home. A distribution architecture in which fibers directly connect to termination equipment in each dwelling.

ground block (or bonding block) A passive device located on the subscriber's premises that allows a safety ground to be connected to the drop.

group delay The deviation from linear phase shift versus frequency through a circuit or network. This can occur because of reflections in a network. Also, it is a property of many practical filters. Signal power near the edge of the passband is delayed in traversing the network, with respect to signals away from the edge. This gives rise to waveform distortion, which can damage both analog and digital signals.

G/T "G over T." A figure of merit for a satellite earth station, expressing the ratio of the gain of the station to its effective noise temperature. G/T allows computation of the signal strength needed to receive a signal satisfactorily. The higher the station G/T, the lower the signal level required.

hard cable Sometimes called hard line. An informal term applied to the solid-jacketed coaxial cable used in the coaxial distribution portion of a cable system ahead of the subscriber tap. Also applied as an adjective to that portion of the distribution system.

HDT Host digital terminal. Digital headend equipment used to interface with a distribution plant.

HDTV High-definition television. Television with significantly more picture information (resolution) than that provided by a good NTSC or PAL television signal. Though the definition of HDTV is nonspecific, it usually implies about twice the resolution in both the horizontal and vertical direction, and wide aspect ratio.

headend The point at which all programming is collected and formatted for placement on the cable system.

HFC Hybrid fiber-coax. A network for transmitting signals modulated onto RF carriers that includes a linearly amplitude modulated optical link followed by a coaxial distribution network.

high-pass filter A filter that passes all frequencies above its cutoff frequency but not below. An example is a filter that goes in a drop and whose purpose is to prevent low-frequency interference from entering the cable return path.

horizontal blanking interval (HBI) That time in a television signal during which the electron beam in the receiver is moving from the right side of the

screen back to the left to start a new line. The HBI normally consists of a sync pulse and color burst.

HRC Harmonically related coherent (or carrier). A method of establishing picture carriers on a cable plant such that all carriers are harmonics of a single fundamental frequency close to 6 MHz(NTSC system).

Hub A point in a distribution system where signals are converted to the form used to transmit them to subscribers. If a hub system is used, typically each hub will supply signals to a number of nodes.

Huffman coding A digital compression technique in which more frequently occurring input sequences are represented by shorter codes, and less common sequences are replaced by longer codes.

IEEE Institute of Electrical and Electronics Engineers. A professional organization of technologists interested in all aspects of electrical engineering.

IIN Interferometric intensity noise. Noise in an optical link caused by mixing of direct and doubly reflected signals in the optical detector diode.

image response The response of a tuner to signals in a portion of the spectrum separated from the IF by the local oscillator frequency but on the wrong side of the local oscillator frequency.

index of refraction The ratio of the speed at which light is transmitted in a vacuum to its speed in a material. It is a property of any material that can transmit light.

ingress Over-the-air signals that are inadvertently coupled into the nominally closed coaxial cable distribution system.

intercarrier Refers to a process in television reception in which the picture and sound carriers are mixed together to form a sound IF at the difference frequency (called the intercarrier frequency).

interdiction A program denial technique in which a jamming signal is inserted very near the picture carrier of programs for which the subscriber has not paid. The interdiction equipment is usually integrated with the tap. Signals are sent in the clear from the headend.

interlaced scanning The process of scanning every other horizontal line in a picture during one-half of the picture transmission, followed by scanning the intervening lines. The purpose of doing so is to reduce the flicker perceived by the viewer.

interleaving In data transmission, a technique by which the order of bits transmitted is changed. The bits are returned to the original order before error correction. If a burst of errors occurs, then the errored bits will be "spread out" before error correction is applied. This makes error correction more effective.

IPPV Impulse pay per piew. Enhanced pay-per-view technology in which a subscriber can make a last-minute impulsive purchase of a program rather than having to order it in advance.

IR blaster A technique in which one device controls another through the controlled device's infrared (IR) input. For example, a set top terminal with electronic program guide may include an IR transmitter that can program a VCR.

IRC Incrementally related coherent (or carrier). A frequency plan in which all carriers are locked to a master oscillator and offset 1.25 MHz from harmonics of 6 MHz.

IRD Integrated receiver-decoder. A device used in a cable television headend to receive a scrambled signal from a satellite and convert it to baseband. It consists of a satellite signal receiver and a descrambler.

isolation The signal loss between nominally isolated ports of a coaxial or optical device (for instance, the output ports of a coaxial splitter). In a tap or directional coupler, it is the difference between the amount of signal coupled to the side port and the directivity.

ISP Internet service provider. An organization whose business it is to connect users to the Internet.

IXC Interexchange carrier. An acronym for a long-distance telephone carrier.

JEC Joint Engineering Committee. A joint committee of the EIA-CEMA and the NCTA that meets periodically to discuss technical issues of interest to both the cable television and the consumer electronics industries. It has been responsible for developing several standards in use by the two industries, such as the ANSI/EIA-542 band plan.

Kell factor The vertical resolution of an interlace-scanned picture is less than the number of active lines in the picture, due to interlace. The Kell factor estimates the practical reduction and is commonly taken as 0.7.

laser In the context of cable systems, the device used to convert electrical signals to optical signals.

lashing The wire used to attach cable to a strand or, as a verb, the process of installing it.

latency A net delay in the time information is delivered from when it is transmitted. A satellite circuit introduces latency due to the propagation time to the satellite and back. A more common use of the term is with respect to video compression, where practical delays in the compression system delay the reception of the signal with respect to the origination.

LE Line extender. A single-input, single-output amplifier used to boost signals in a coaxial distribution plant.

LEC Local exchange carrier. An acronym used to identify a local telephone company.

local origination Programming over which the cable operator has editorial control. It is usually produced in the system.

local oscillator An oscillator that generates a frequency that is mixed with a signal to produce a signal at a desired third frequency.

log periodic An antenna type consisting of multiple driven elements having a specified ratio of length and spacing, and used to direct energy along a single axis. Performance is similar to that of a Yagi, except the bandwidth can be much larger.

loop Local office outside plant. Used to describe that portion of a telephone network extending from the local (class 5) switch to the network termination points at individual subscribers.

low-pass filter A filter that passes all frequencies below its cutoff frequency but none above.

luminance Or luma. Refers to the black-and-white portion of the modulated television signal, as opposed to the chrominance (color) and aural (sound) portions.

Manchester encoding In data transmission, a method of embedding clock into the data such that recovery is simple.

MCNS Multimedia Cable Network System. An industry consortium that produced the DOCSIS procurement standard for modems to be used on cable television networks.

mesh An architecture in which terminal points are connected to several other terminal points directly rather than just to a central point (a star) or the adjacent terminals in a ring.

messenger A support member in an electronic or fiber cable.

microreflections Small reflections occurring within both the coaxial distribution and drop portions of a cable plant. They can lead to impaired analog reception, group delay, and errors in digital signals.

mixer A circuit that accepts two signals at different frequencies and produces a third frequency equal to the sum or the difference of the two input frequencies.

modal dispersion In a multimode fiber, the variation of transmission time among the various possible paths for the light.

modulation The process of imposing information on an RF carrier by varying some parameter of the carrier. The parameters that can be varied are the amplitude, frequency, and phase.

modulator A device used by cable television systems to convert a baseband video and/or audio signal from baseband to the RF channel on which it is to be transmitted.

MPEG Moving Picture Experts Group. A group that developed standards for digital compression of television pictures. The second generation of this standard, MPEG-2, is used in one form or another for transmission of digital television in all parts of the world that have introduced such transmission.

MSO Multiple system operator. A company that owns more than one separate cable system.

MTBF Mean time between failure. A measure of the average time interval between circuit restoration from a previous failure and the onset of the next failure in a network.

MTTR Mean time to restore. A measure of the average time required to restore proper network or circuit operation after a failure.

multimode An optical fiber that supports more than one transmission path and thus has a transmission delay whose uncertainty increases with length, thereby limiting bandwidth.

multipath A condition where an over-the-air signal may reach a receiver via more than one path from a transmitter, leading to degraded reception. Applies to all transmission systems and frequencies.

multiplex Use of a single data path for more than one set of data.

multiplexing The method by which a transmission system is shared by more than one signal. Cable systems use *frequency division multiplexing* in which many signals at different frequencies are combined in one signal path. Other forms of multiplexing, also used in cable systems, include *time division multiplexing,* where many signals share the same path but are segregated in terms of time, and *space division multiplexing,* where different portions of the network are used to simultaneously carry independent signals.

NCTA National Cable Television Association. A trade organization that represents companies in the cable television industry.

negative trap A technology for denying analog television programs to subscribers who have not purchased them. Negative trap systems work by transmitting normal signals on the cable plant. At the tap of subscribers who have not bought the program, a trap is inserted to remove the picture carrier.

NICAM Near instantaneous amplitude companding and modulation. A system of transmitting stereo audio digitally in a TV signal, developed in the United Kingdom and used there and in a number of other countries.

NID Network interface device. One of many names for a box on the side of the house that contains any required interfaces between a telephone line or cable plant and inside wiring.

node 1. The portion of a coaxial cable plant served from one optical receiver. 2. The equipment that receives the linearly modulated FDM spectrum, converts it back to RF form, and couples it into one or more coaxial distribution legs.

noise Generally defined as undesired signals in a transmission channel. In television, the term is commonly applied to broadband thermal noise *excluding* discrete carriers and distortion products.

noise figure The ratio, expressed in decibels, of the noise produced by a device, over and above the thermal noise produced in a matched resistor at the input to that device. Higher noise figure devices require a higher signal level to deliver a given signal-to-noise ratio.

NPR Noise power ratio. A method of specifying the quality of a transmission path by loading it with noise, eliminating the noise in a narrow band, and measuring the amount of noise filled in by the transmission path.

NRZ Nonreturn to zero. A method of transmitting digital data bit by bit, with each bit immediately following the preceding one with no time between.

NTSC National Television System Committee. An entity that developed the analog TV system used in North America and elsewhere. There were actually two NTSCs: the first developed the monochrome transmission system, and the second added color.

NVOD Near video on demand. A PPV service that offers sufficiently frequent start times for popular events (movies) such that it approaches on-demand availability. Half- to quarter-hour incremented start times are common.

Nyquist frequency In a sampled data system, the highest frequency that can be sampled and then recovered without distortion.

Nyquist slope filter A filter applied to a VSB modulated signal just prior to detection to restore flat frequency response.

OAM&P Operations, administration, maintenance, and provisioning. A generic term for software suites that allow centralized administering of a telephone network. It is a superset of what the cable industry knows as status monitoring.

offset feed An antenna for microwave reception consisting of a parabolic reflector and feed, shaped such that the axis of the focus passes in front of the edge of the reflector. Offset feeds are popular with very small aperture antennas because of reduced blockage of the transmission path.

optical modulation index (OMI) A measure of how much optical modulation is applied to a laser. It is usually stated in terms of how much of the range from the

laser's quiescent bias point to cutoff is taken by the signal. The OMI may be expressed in terms of per-channel or composite signal.

orthogonal Literally, at right angles. Often used by the engineering community to describe two functions that are independent of each other.

oscilloscope A piece of test equipment that allows you to display the time value of a waveform.

overhead In data transmission, the portion of the data that is included to enable transmission. The remaining data transmitted is the payload.

overload In a receiver, the condition of supplying too much signal to that receiver, resulting in distortion being generated in the receiver.

packetized elementary stream PES. In MPEG-2 transmission, a collection of bits containing data from one source only. PESs may contain video, audio, or data.

PAL Phase alternating line. The television system used in most of Europe and many other places, in which the color carrier phase definition changes in alternate scan lines.

passive A component that handles signals but that requires no power of its own. Examples are taps and power inserters.

pay per view PPV. The selling of programs to subscribers on a per-program basis as opposed to a subscription basis.

pay services Nonadvertiser-supported channels for which a subscriber pays on a subscription basis. Also called premium services.

payload In data transmission, the portion of the data that is desired to be transmitted. Other transmitted data, such as error correction and headers, is considered overhead.

PCS Personal communications services. An alternative service to cellular telephone that uses a different frequency range and generally smaller cell sizes.

peak-to-average ratio In data transmission, the ratio, usually expressed in decibels, between the peak power of a signal and the average power.

peak to valley P/V. The peak-to-peak variation in the frequency response of a cable distribution network relative to the ideal response curve.

PEG Public, educational, and government. A set of locally generated channels many systems are required to carry as a public service.

phase noise A distortion in which a carrier signal is subjected to random variations in phase.

picture-in-picture A feature of some television sets and set top terminals that allows display of a small second picture within the main picture.

picture-to-sound ratio The ratio between the amplitude of the picture carrier and the associated sound carrier. Usually set to 15 dB in cable television work.

PID Program identification. A header preceding a PES in an MPEG-2 transport stream that identifies the purpose of the PES. In a multiplex of several different TV programs, for example, each audio and video will contain its unique PID, which allows for the selection of the desired program out of the multiplex.

pilot A signal on the distribution plant that is used for AGC or ASG purposes. In most cases, the pilot is one of the normal analog picture carriers that has been so designated.

pixel Picture element. The smallest unit of a picture that can be represented as combinations of the three primary colors only, with no unique shape. Frequently, it is assumed that a pixel has either a square or rectangular shape.

PMD Polarization mode dispersion. In an optical fiber, a measure of the transmission time as a function of the orientation of the E field of the light relative to the fiber.

POE Point of entry. Equipment that is placed between the drop and inside wiring of a dwelling and that processes some or all of the signals for specific services. One application is cable telephony where the POE device interfaces standard analog in-home wiring to the RF distribution system. See NID.

polarization An expression of the orientation of the electric field in a radiated or guided RF or optical signal.

polarization diversity The use of the same frequency band to radiate two sets of signals by using orthogonal polarizations, usually either vertical and horizontal or left- and right-hand circular.

positive trap A technology for denying analog television programs to subscribers who have not purchased them. Positive trap systems work by adding a high-level interfering carrier between the picture and sound carriers. Subscribers who have paid for the program are supplied with a filter (trap) to remove the interfering signal.

POTS Plain old telephone service.

power inserter A passive device used to add ac power to the center conductor of a coaxial cable for the purpose of powering active devices.

power pack In a cable system, the module within an amplifier station (or other active equipment) that receives the power multiplexed with the RF signals on the distribution system and creates the voltages required by the station's internal circuits.

power supply In a cable system, the device that receives power from the utility, processes it for use in the cable system, and inserts it into the coaxial plant through a power inserter.

preemphasis In FM transmission, a technique for reducing noise by selectively increasing the deviation of higher frequencies before transmission. Deemphasis is used to restore flat response at the receiver.

primary colors The three colors that can be combined in differing amounts to yield any color that can be perceived by the human visual system. For the emission spectrum used in television, the primary colors are specific shades of red, green, and blue.

prime focus antenna An antenna for microwave reception consisting of a parabolic reflector and a feed assembly (antenna and, usually, preamplifier and converter) located at the focus of the parabola.

processor Or signal processor. In cable television terminology, a piece of equipment used in the headend to receive a television signal and reformat it for transmission on cable. Reformatting consists of amplification, bandpass filtering, shifting the relative levels of the picture and sound carriers, and frequently, changing the carrier frequency (channel).

Protocol In data transmission, a set of rules defining how various pieces of equipment or software will interact cooperatively.

QAM Quadrature amplitude modulation. A digital modulation method in which the value of a symbol consisting of multiple bits is represented by amplitude and phase states of a carrier. Typical types of QAM include 16-QAM (four bits per symbol), 32-QAM (five bits), 64-QAM (six bits), and 256-QAM (eight bits).

QPSK Quadrature phase shift keying. A digital modulation method in which the state of a two-bit symbol is represented by one of four possible phase states.

quadrature At right angles. In transmission, often used to express independence between two signals. See orthogonal.

quantization noise The noiselike uncertainty imparted to an analog signal when it is digitized. An alternative term is quantization error. The more bits used in an analog-to-digital conversion process, the smaller the quantization noise.

RAD Remote antenna driver. A technique by which signals from PCS antennas are frequency converted and transmitted through cable systems to remotely located base stations.

rain fade In a terrestrial or satellite microwave link, signal loss due to absorption and reflections caused by raindrops within the signal path.

reflectometer A device used to determine the location of a fault in a cable by propagating a signal down the cable and reading any signal reflected. May be used for RF or optical transmission. See TDR.

refraction The process by which optical signals are transmitted through the interface between two dissimilar materials. See index of refraction.

resolution A measure of the "sharpness" of a picture. Usually measured in pixels in the horizontal direction and lines or pixels in the vertical. It is frequently the case that horizontal and vertical resolution are different.

return loss A measure of the amount of incident power that is reflected toward the opposite end of a cable from a termination or other discontinuity in the characteristic impedance of the cable.

reverse See upstream.

RIN Relative intensity noise. A measure of the residual internal noise modulation of an optical transmitter. RIN limits the attainable C/N of RF signals transmitted through an optical link.

ring network An architecture in which terminal points are connected in a continuous ring.

router In data communications, a device used to switch packets from any number of input ports to any one of a number of output ports.

run-in A signal preceding some asynchronous data transmissions that permits the receiver to synchronize itself with the transmission.

SBS Stimulated Brillouin scattering. A mechanism that limits the amount of optical power at a single wavelength that can be transmitted through an optical fiber.

scanning The process of sequentially picking up or "painting" a picture, wherein each picture element is delivered one at a time in a set sequence over the entire picture. Commonly, the scanning sequence is top to bottom and left to right, as viewed from in front of the picture tube.

scattering In an optical network, scattering is any process by which light is briefly delayed, then retransmitted. Forward scattering causes an overall slowing of the signal (see index of refraction), whereas scattering in other directions leads to signal loss and, sometimes, distortion.

scrambling In cable television, the process of modifying an analog TV signal so that it cannot be received by a television receiver. In data transmission, scrambling refers to the process of randomizing the bit pattern of a transmitted signal to prevent peaks in the spectrum of the modulated signal. Note that the cable television and data transmission definitions are *not* equivalent. The cable television industry usually uses the data transmission term *encryption* for program denial of digital signals.

SCTE Society of Cable Telecommunications Engineers. A professional organization of technologists interested in cable telecommunications technology.

SDM Space division multiplexing. The simultaneous use of different physical portions of a network (for example, nodes) to carry different signals at the same frequencies.

SECAM A television system used in France and some former Soviet bloc countries.

second audio program In BTSC stereo signals, a second, monaural, audio channel that is available to carry a second audio channel, such as a foreign-language soundtrack.

set top terminal Also known as a set top converter. A device used with a subscriber's TV to allow reception of television programs. It may tune channels the TV does not tune and may include descrambling circuitry. It also may include a digital decoder and auxiliary functions such as an electronic program guide.

sheath ring network A ring architecture in which separate fibers are used in each direction from a common terminal to each subterminal. Sheath rings require more fiber than repeating rings but are more secure and offer higher potential bandwidth to a common location.

shielding An expression of the ability of a device either to prevent radiation of signals contained within that device or to prevent undesired pickup of signals outside the device.

shot noise In an optical link, noise at the optical receiver that is caused by the statistical variation in the arrival of photons.

signal level In analog television work, the RMS level of the picture carrier measured over the occupied bandwidth of the signal, during sync tips. In digital communications, as practiced by the cable industry, it is the average level of the modulated signal, or the level that would be read on a thermocouple type RF power meter.

signal level meter See SLM.

signal-to-noise (S/N) ratio In video work, the ratio between the video signal amplitude and the noise power measured in a reference bandwidth. Many variations in the definition of "signal" and "noise" exist.

single mode fiber Optical fibers that support only a single transmission path and thus offer a better-defined transmission delay.

single sideband modulation Amplitude modulation in which one or the other of the modulation sidebands is removed.

skin effect The tendency of RF signals to travel near the surface of conductors.

SLM Signal level meter. A tunable power meter commonly used to measure signal levels in an FDM network.

SONET Synchronous optical network, a set of standards describing an optical baseband network designed for carrying many types of data.

spectrum analyzer A piece of test equipment used to plot signal amplitude versus frequency.

splitter A device having three or more ports, used to divide an RF signal equally among multiple paths without regard for the frequency of the signal. Can also be used to combine two signals.

SPM Self-phase modulation. In an optical fiber, incidental phase modulation at high optical levels caused by variation of the index of refraction with the instantaneous power level of the optical signal.

spread spectrum Any of several techniques used to increase the transmitted spectrum of a signal with respect to the minimum spectrum that can be used for the transmission. The purposes are usually to reduce noise and interference pickup, to allow more than one signal to occupy the same spectrum, to mitigate transmission problems in one portion of the spectrum, and/or for security.

SRL Structural return loss. A measure of the degree to which signals are reflected from a length of cable as a function of frequency.

standard definition Refers to today's NTSC, PAL, and SECAM transmissions, with their normal resolution.

star network An architecture in which each terminal is directly connected through separate transmission paths to a common master terminal. In a simple HFC cable system, the master terminal is the headend, and the other terminals are nodes.

start bit(s) One or more bits that precede a digital word in asynchronous transmission and that are used to identify the beginning of the transmission.

statistical multiplexing In multiplexed video compression systems, the process of assigning more or less of the available channel bandwidth to each of the several signals being multiplexed, according to the instantaneous needs of each.

strand A heavy wire attached to utility poles to which coaxial and fiber-optic cables are attached.

supertrunk In a cable system, the signal transportation link used to connect the point where the FDM signal complex is created to local coaxial distribution networks. In an HFC network, linearly modulated optical links are used, but other systems use AML microwave or coaxial supertrunks.

switch In telephony, the device used to connect one telephone line to another for the duration of a call.

symbol A collection of some number of bits of data that are transmitted together.

sync The portion of the analog television signal used to synchronize scanning in the receiver to scanning in the camera. Corresponds to the most negative portion of the baseband waveform ("blacker than black") and to the highest power portion of the modulated signal. In digital transmission, a set of bits of known value used to locate a position in a datastream.

sync suppression scrambling A scrambling technique in which the synchronization portion of the picture is hidden usually by moving it to a level normally occupied by picture information.

synchronization In data transmission, the process of ensuring that data recovery at the receiver is timed properly with the data being transmitted. In video, the process of ensuring that the camera pickup and the receiver display are scanning the same part of the image at the same time.

synchronous detection The process of recovering modulation from an amplitude modulated signal by mixing the signal with a replica of the original unmodulated carrier. In modulation methods with asymmetrical sidebands (such as analog television transmission), synchronous detection can be shown to provide more accurate detection than can the simpler envelope detection.

taboo channels In FCC assignment of off-air UHF channels, certain channels that cannot be used in a particular location because of the use of other channels. The taboo channels exist because of certain limitations in UHF converter performance that are brought about by the need to keep cost low.

tap The component installed in the distribution cable that diverts a portion of the downstream distribution signal and splits it to feed two to eight individual subscribers. In the reverse direction, it combines the signals from subscribers and inserts them into the upstream direction of the distribution cable.

TASO Television Allocations Study Organization. A group that, in the late 1950s, studied the subjective effects of noise and distortion on the television-viewing experience. It used a six-point rating scale that is still referred to today.

TDM Time division multiplexing. A data transmission method in which a number of individual digital datastreams share a transmission channel, each occupying the channel for a portion of the total time.

TDMA Time division multiple access. A data transmission method in which a number of individual transmitters in different locations share a transmission channel, each occupying the channel for a portion of the total time.

TDR Time domain reflectometer. A device that measures the quality of a transmission line by transmitting a pulse, then measuring signals reflected as a function of time. The reflections indicate the distance to the point of reflection. See reflectometer.

TEM Transverse electric and magnetic. A description of the normal field configuration for signals transmitted through coaxial cables.

termination 1. The impedance at the end of a coaxial or fiber-optic cable that accepts power being transmitted on the cable. 2. A resistance connected to an unused device port used to reduce reflections.

transport stream In MPEG-2 systems, the data formatted for transport. Such data comprises one or (usually) more PESs, plus overhead.

trap A passive coaxial component inserted in series with a drop line that blocks transmission of a specific frequency or range of frequencies. Used to block or enable specific channels or to tailor the services available in homes.

trunk 1. The untapped portion of a coaxial distribution network. Trunk links are typically operated at lower levels than distribution links to minimize distortions. 2. An amplifier module used in the trunk portion of a coaxial distribution system. 3. An amplifier station that contains a trunk module. It may also contain a bridger module if it is desired to create a distribution leg at that location.

twisted pair A type of cable used extensively for transmission of balanced audio signals. The cable consists of two insulated conductors twisted on each other. It may or may not have a shield over the twisted pair.

unavailability The fraction of time that a network or channel is unavailable for use. See availability.

universal remote control A remote control transmitter that can control more than one brand of TV, VCR, set top terminal, and so on. Often used to control all of a consumer's entertainment equipment.

UPC Ultra-polished connector. An optical connector that minimizes loss and reflections by careful preparation and polishing of the fiber end and alignment with the mating fiber.

upstream Signal flow from subscribers toward a headend.

VBI Vertical blanking interval. That time in a television signal during which the electron beam in the receiver is moving from the bottom of the screen back to the top in order to start a new field. In NTSC, the VBI is 21 horizontal lines long, and these lines are frequently used to insert test signals and data.

VBR Variable bit rate. A signal or communications channel that generates or accommodates a digital signal whose bit rate varies with time.

video In television work, the word *video* is reserved to mean a baseband electrical signal that conveys information regarding picture content.

video compression The reduction of bandwidth needed to transmit or store a video signal by removing information in the signal that is not necessary for satisfactory perception of the signal or that can be inferred from other elements of the transmitted signal.

video inversion A scrambling technique in which analog video is inverted so that black corresponds to the minimum carrier amplitude, as opposed to the maximum amplitude (in NTSC and PAL transmission). Sync may or may not be inverted.

visual An adjective implying an RF or IF carrier that is modulated with a video signal.

VITS Vertical interval test signals. Signals that are inserted into the VBI to allow testing of the quality of the transmission link independently of the video programming.

VOD Video on demand. A video service that allows users to select the program and exact start time interactively. In some embodiments, it allows VCR-like control of the playback (for example, pause, rewind, fast forward).

VSB Vestigial sideband. The elimination of a large portion, but not all, of one sideband of an amplitude modulated signal. May be done with either analog or digital modulation.

W curve A description of the variation of subscriber perceptibility of interfering signals in an analog television channel as a function of frequency within the channel boundaries. So called because of the shape of the curve, which is caused by the extreme sensitivity to interfering signals occurring near the visual and chrominance carriers.

wavelength The distance traveled through a transmission medium by an RF or optical signal in a time equal to the inverse of the frequency (for example, for a 1-GHz signal, the distance traveled in 1 ns). It is equal to the speed of propagation in the medium, divided by the frequency.

wavelength dispersion In an optical fiber, the variation in transmission time as a function of the wavelength of the light.

WDM Wavelength division multiplexing. The use of an optical fiber to simultaneously carry signals at different wavelengths.

XMOD Cross modulation. A type of distortion in which the amplitude modulation of one signal causes incidental amplitude modulation to another signal. XMOD is an odd-order distortion.

Yagi Or Yagi-Uda. A gain antenna consisting of a single driven element and one or more parasitic elements used to direct energy along a single axis.

Index